Jim GEE

HANDBOOK
FOR
ELECTRONICS
ENGINEERING
TECHNICIANS

OTHER McGRAW-HILL HANDBOOKS OF INTEREST

BAUMEISTER AND MARKS Standard Handbook for Mechanical Engineers

BEEMAN Industrial Power Systems Handbook

COOMBS Basic Electronic Instrument Handbook

COOMBS Printed Circuits Handbook

CROFT, CARR, AND WATT American Electricians' Handbook

FINK AND BEATY Electronics Engineers' Handbook

FINK AND CARROLL Standard Handbook for Electrical Engineers

GIACOLETTO Electronic Designer's Handbook

HARPER Handbook of Electronic Packaging

HARPER Handbook for Materials and Processes for Electronics

HARPER Handbook of Thick Film Hybrid Microelectronics

HARPER Handbook of Wiring, Cabling, and Interconnecting for Electronics

HICKS Standard Handbook of Engineering Calculations

JOHNSON AND JASIK Antenna Engineering Handbook

JURAN Quality Control Handbook

KAUFMAN AND SEIDMAN Handbook of Electronics Calculations

KURTZ Handbook of Engineering Economics

MARKUS Electronics Dictionary

PERRY Engineering Manual

STOUT Handbook of Microprocessor Design and Applications

STOUT AND KAUFMAN Handbook of Microcircuit Design and Application

STOUT AND KAUFMAN Handbook of Operational Amplifier Design

TUMA Engineering Mathematics Handbook

WILLIAMS Designer's Handbook of Integrated Circuits

WILLIAMS Electronic Filter Design Handbook

HANDBOOK
FOR
ELECTRONICS
ENGINEERING
TECHNICIANS

SECOND EDITION

Editors in Chief
MILTON KAUFMAN
President
Electronic Writers and Editors, Inc.

ARTHUR H. SEIDMAN
Professor of Electrical Engineering
Acting Dean
Pratt Institute, School of Engineering

PERRY J. SHENEMAN
Associate Editor

McGRAW-HILL BOOK COMPANY
New York   St. Louis   San Francisco   Auckland   Bogotá
Hamburg   Johannesburg   London   Madrid   Mexico
Montreal   New Delhi   Panama   Paris   São Paulo
Singapore   Sydney   Tokyo   Toronto

Library of Congress Cataloging in Publication Data

Main entry under title:
Handbook for electronics engineering technicians.
Includes index.
1. Electronics—Handbooks, manuals, etc.
I. Kaufman, Milton. II. Seidman, Arthur H.
III. Sheneman, Perry J.
TK7825.J37 1984   621.38   83-23901
ISBN 0-07-033408-0

1 2 3 4 5 6 7 8 9 0   DOC/DOC   8 9 8 7 6 5 4

ISBN 0-07-033408-0

The editors for this book were Harold B. Crawford and Janet B. Davis, the designer was Mark E. Safran, and the production supervisor was Reiko F. Okamura. It was set in Caledonia by University Graphics, Inc.

Printed and bound by R. R. Donnelley & Sons Company.

To the fond memory of
David M. Kaufman
and
Rose and Isidore Sieser

Contents

2. CHARACTERISTICS OF CAPACITORS

3. COILS

4. MAGNETIC CIRCUITS

6. PRACTICAL CIRCUIT ANALYSIS

7. METERS AND MEASUREMENTS

8. SEMICONDUCTOR DEVICES AND TRANSISTORS

9. INTEGRATED CIRCUIT TECHNOLOGY

10. TUNED CIRCUITS

11. FILTERS

12. TRANSISTOR AMPLIFIERS AND OSCILLATORS

13. OPERATIONAL AMPLIFIERS

14. DIGITAL CIRCUIT FUNDAMENTALS

15. DIGITAL INTEGRATED CIRCUITS

16. POWER SUPPLIES

17. BATTERIES

18. MICROPROCESSORS AND MICROCOMPUTERS

19. LOGIC ANALYZERS, LOGIC PROBES, AND SIGNATURE ANALYSIS

20. MICROWAVES

Contents

Contents xxvii

22.4 State-Variable Filters 22-22
Basic Circuit 22-22
IC filters 22-24
Low- and high-pass functions 22-24
Bandpass function 22-26
Bandstop (notch) function 22-27
Higher-order filters 22-27
Low-frequency operation 22-27
Voltage-tuned filters 22-28
22.5 Other Filters 22-29
Elliptic Filters 22-29
All-Pass Networks 22-29
Digital Filters 22-30
22.6 References and Bibliography 22-31

23. DIGITAL MULTIMETERS

23.1 Basic Building Blocks 23-1
23.2 DC Input Amplifier/Attenuator 23-2
23.3 Internal Reference Voltage 23-3
23.4 AC Conversion 23-4
23.5 Measuring Resistance 23-6
23.6 A/D Conversion Techniques 23-7
23.7 Noise Rejection 23-10
23.8 How To Read DMM Specifications 23-14

24. OSCILLOSCOPE MEASUREMENTS

24.1 Introduction 24-1
24.2 Oscilloscope Controls 24-2
Display System 24-2
Beam finder 24-4
Intensity 24-4
Focus 24-4
Trace rotation 24-4
Vertical System 24-4
Vertical position 24-5
Input coupling 24-5
VOLTS/DIV switch 24-5
Variable volts per division 24-5
Channel 2 inversion 24-5
Vertical operating modes 24-5
Horizontal System 24-6
Horizontal position 24-6
Seconds per division 24-6
Variable seconds per division 24-7
Horizontal magnification 24-7
Horizontal operating mode 24-7

APPENDIX

Index follows the Appendix

List of Contributors

JOSEPH R. CLEVELAND
MCI
Richardson, Texas

CHET HEYBERGER
Tektronix, Inc.
Beaverton, Oregon

KENNETH JESSEN
Hewlett-Packard Co.
Loveland, Colorado

GARY C. MEYER
Philips Test and Measuring Instruments, Inc.
Minnetonka, Minnesota

RICHARD J. PANKOW
Norden Systems
Norwalk, Connecticut

ROGER C. THIELKING
Onondaga Community College
Syracuse, New York

PAT ZAGARIA
Philips Test and Measuring Instruments, Inc.
Mahwah, New Jersey

Preface

This is the first fully comprehensive handbook designed to meet the day-to-day needs of electronics technicians. While useful to engineers as well, it does not require an extensive background in high-level engineering principles and techniques. All that is required is a background no more extensive than an education in an electronics technical school or a two-year community college. And in many cases, because of the handbook's simplified approach, even a lesser background would be sufficient.

The handbook treats fundamental topics in discrete circuits, and also in analog and digital integrated circuits, from the point of view of practical applications. Each topic is illustrated with practical, numerical, worked-out examples that can be applied to the reader's own particular problems. Numerous practical examples illustrate, for example, how to choose the proper resistor, capacitor, transistor, integrated circuit, or operational amplifier and how to find current in, or voltage across, an element in dc and ac circuits.

Each one of the twenty-four in-depth sections follows the same practical, concise format that has been developed to help readers find the information they need quickly and easily. The general approach to each topic in a section is as follows:
1. Definition of terms and parameters
2. Breakdown of the types and characteristics of components
3. Analysis of the basic and special functions
4. Detailed practical problems and clearly worked-out solutions
5. Clarifying charts, tables, nomographs, and illustrations

With this easy-to-follow format, every effort has been made to ensure that this handbook will have the greatest usefulness to its readers.

All mathematics, both in theory sections and in worked-out problems, has been kept to the level of relatively simple algebra or arithmetic. No calculus is employed.

In this second edition, some important changes have been made. The chapter on vacuum tubes has been deleted, and Chapter 6 has been revised.

Seven new chapters have been added to assist the readers with state-of-the-art theoretical and practical information. Each of these new chapters has been written by a recognized authority in his particular field. (See separate list of contributors.)

The new chapters are as follows:
18. Microprocessors
19. Logic Analyzers and Signature Analysis
20. Microwaves
21. Fiber-Optic Systems
22. Active Filters
23. Digital Multimeters
24. Oscilloscopes

The following chapters were written by Professor Seidman: Chapters 1, 2, 6, 8, 9, and 12 to 15.

The following chapters have been written or edited by Mr. Kaufman: Chapters 3 to 5, 7, 10, 11, 16, and 17.

The editors wish to gratefully acknowledge the cooperation and assistance of various individuals and companies in providing manuscript material, photographs, tables, graphs, and other technical data. Due credit is given within the book as applicable. We wish particularly to thank the following persons for their invaluable contributions: Donald R. Phillips of RCL Electronics, Inc; Dr. Howard J. Strauss of Gould, Inc.; J. A. "Sam" Wilson of Bank Wilson Services; Joshua A. Hauser of Lambda Electronics, Inc.; Raymond C. Miles; Bill Stutz; and the late Louis Toth. In addition, we wish to thank Sharon Wilson for her considerable efforts in typing and editorial services.

Milton Kaufman
Arthur H. Seidman

Chapter 1

Characteristics of Resistors

1.1 INTRODUCTION

Every physical material impedes the flow of electric current to some degree. Materials such as copper offer hardly any resistance to current flow; copper, therefore, is called a *conductor*, or a material having negligible resistance. Other materials, such as ceramic, which offer extremely high resistance to current flow are referred to as *insulators*.

In electric and electronic circuits, there is a need for materials with specific values of resistance in the range between that of a conductor and an insulator. These materials are called *resistors* and their values of resistance are expressed in ohms (represented by the Greek letter omega, Ω). The large variety of types and forms of discrete resistors available are indicated in Fig. 1.1.

Resistors may be classified as being *fixed* or *variable* in their value. Variable resistors are commonly referred to as *potentiometers*, or *pots*. Electrical symbols for fixed resistors and pots are illustrated in Fig. 1.2.

Resistors may also be classified as *linear* and *nonlinear*. For a linear resistor, as the voltage across it varies, the current flowing varies by a proportionate amount. The behavior of a nonlinear resistor is such that as the voltage varies, the current change is not proportional to the voltage change. Resistors generally used in circuits are linear. For special applications, to be described later, nonlinear resistors are available.

1.2 GENERAL DESCRIPTION

The resistance of any material is given by the following expression:

$$R = \frac{\rho L}{A} \tag{1.1}$$

where R = resistance, Ω
ρ = resistivity of the material, Ω-cm
L = length of material, cm
A = cross-sectional area of material, cm^2

Resistivity ρ (Greek letter rho) is an inherent property of materials. Values of ρ for some commonly used materials are summarized in Table 1.1.

Equation (1.1) illustrates two important facts. For a material of a given resistivity, the resistance varies *directly* with length L and *inversely* with cross-sectional area A. For example, a wire of long length has greater resistance than a short-length wire. Also, a wire of large diameter (large cross-sectional area) has less resistance than a wire with a small diameter.

Fig. 1.1 A sampling of some discrete resistors available to the user. (*Courtesy Victoreen Instrument Division*)

The voltage and current in a resistor are related by Ohm's law:

$$I = \frac{E}{R} \tag{1.2a}$$

$$E = IR \tag{1.2b}$$

$$R = \frac{E}{I} \tag{1.2c}$$

where E = voltage across resistor, V
   I = current flowing in resistor, A
   Power P (in watts, W) dissipated in a resistor may be expressed by any of the following expressions:

$$P = EI \tag{1.3a}$$

$$= I^2 R \tag{1.3b}$$

$$= \frac{E^2}{R} \tag{1.3c}$$

The application of the preceding equations is illustrated in the following examples.

example 1.1   Calculate the resistance of a wire whose length is 10 m and cross-section area is 0.1 cm² if the material of the wire is (*a*) copper and (*b*) Nichrome.

solution   From Table 1.1, for copper $\rho = 1.7 \times 10^{-6}$ Ω-cm and for Nichrome, $\rho = 100 \times 10^{-6}$. Because 1 m = 100 cm (units in an equation must always be consistent), $L = 10 \times 100$ cm = 1000

Fig. 1.2 Electrical symbols for resistors: (*a*) Fixed. (*b*) Variable (pot).

TABLE 1.1   Resistivities of Some Commonly Used Materials

Material	ρ, Ω-cm
Silver	1.5×10^{-6}
Copper	1.7×10^{-6}
Aluminum	2.6×10^{-6}
Carbon (graphite)	30×10^{-6} to 190×10^{-6}
Nichrome	100×10^{-6}
Glass	$10^{10}-10^{14}$

cm. Substitution of the given values in Eq. (1.1), $R = \rho L/A$, yields

(a) $R_{copper} = 1.7 \times 10^{-6}\ \Omega$ cm $\times$ (1000 cm)/(0.1 cm^2)
    $= 1.7 \times 10^{-2}\ \Omega$
(b) $R_{Nichrome} = 100 \times 10^{-6} \times 1000/0.1 = 1\ \Omega$

example 1.2   If the current flowing in the wires of Example 1.1 is 3 A, calculate the dissipated power in each wire.

solution   Using Eq. (1.3b) $P = I^2R$. $P_{copper} = 3^2 \times 1.76 \times 10^{-2} = 9 \times 1.76 \times 10^{-2} = 0.158$ W. $P_{Nichrome} = 3^2 \times 1 = 9 \times 1 = 9$ W.

1.3   RESISTOR TERMS AND PARAMETERS

In this section, commonly used terms and parameters for characterizing fixed and variable resistors are defined. Where appropriate, typical curves showing the variation in resistance with temperature and other quantities of interest are supplied.

RESISTANCE   The unit of resistance is the ohm (Ω). Generally, resistance values in thousands of ohms is expressed in kilohms (kΩ) and millions of ohms in megohms (MΩ). Nominal values of resistors are generally based on 25°C (room temperature) operation.

TOLERANCE   Tolerance expresses the maximum deviation in resistance from its nominal value. For example, if the tolerance of a 1000-Ω resistor is ±10 percent, this denotes that the actual value of resistance is in the range of $1000 - 0.1 \times 1000 = 900\ \Omega$ to $1000 + 0.1 \times 1000 = 1100\ \Omega$.

TEMPERATURE COEFFICIENT OF RESISTANCE   The temperature coefficient of resistance (abbreviated TCR, or *Tempco*) indicates how resistance changes with temperature. It is expressed as a percentage change in the nominal value at 25°C for each degree Celsius or in parts per million per degree Celsius (ppm/°C) of the resistor. The TCR may be positive or negative. Typical curves showing the variation in resistance with temperature are illustrated in Fig. 1.3. In this figure, a *logarithmic* (log) *scale* is used for plotting nominal resistance (ohms). A log scale permits a large range of values (in this case, 10^1 to $10^8\ \Omega$) to be compressed in a convenient scale.

example 1.3   A 1000-Ω resistor has a TCR of +1000 ppm/°C. Calculate its resistance at 125°C.

solution   The quantity 1000 ppm is equal to $10^3/10^6 = 0.001$. The change in resistance from room temperature (25°C) is equal to the product of the difference in temperature (125 − 25°), the TCR, and the nominal value of resistance at 25°C (1000 Ω). Hence, the change in resistance $= (125 - 25) \times 0.001 \times 1000 = 100\ \Omega$. The resistance value at 125°C is, therefore, $1000 + 100 = 1100\ \Omega$.

POWER RATING   Power rating is the maximum continuous power, in watts, that a resistor can dissipate at a temperature as high as 70°C. At temperatures beyond 70°C, the power rating of the resistor is reduced, or *derated*. A typical derating curve is given in Fig. 1.4.

example 1.4   A particular resistor is rated at 60 W. Using the derating curve of Fig. 1.4, determine the power rating of the resistor when operating at 130°C.

Fig. 1.3 Typical curves showing the variation in nominal resistance with temperature. (*Courtesy Allen-Bradley*)

solution From Fig. 1.4, at 130°C, the rated load is 30 percent. Expressing 30 percent by the decimal 0.3, the rating of the resistor at 130°C is $0.3 \times 60 = 18$ W.

In specifying the power rating of a resistor, for conservative operation it is usual practice to increase the actual dissipated power by a factor of 2. For example, if the calculated dissipation is 0.5 W, specify a $0.5 \times 2 = 1$-W resistor.

Fig. 1.4 Typical derating curve for a resistor. Ambient temperature refers to the temperature of the environment in which the resistor is operating.

RATED CONTINUOUS WORKING VOLTAGE The rated continuous working voltage (RCWV) is the maximum voltage that can be safely applied to a resistor. Typical curves for RCWV versus resistance are shown in Fig. 1.5. For example, the maximum rated continuous working voltage for a 50-Ω 2-W resistor is 10 V.

CRITICAL RESISTANCE The critical resistance R_c is the value of resistance where the maximum voltage and power rating occur simultaneously. For example, for a resistor rated at 500 V and 1 W, from Eq. (1.3c), $R_c = E^2/P = 500^2/1 = 250\,000$ Ω.

NOISE Because of the molecular structure of matter, electrons in a material exhibit a random motion. As a result, random voltages are produced which are referred to as *noise*. The noise increases with increasing resistance values, operating temperature, and the bandwidth of the circuit to which the resistor is connected. Typical curves of noise voltage (microvolts) at 25°C versus bandwidth (hertz) are illustrated in Fig. 1.6. It is seen that as the resistance value or bandwidth is increased, the noise voltage becomes greater.

FREQUENCY EFFECTS A *model* (equivalent circuit) of a resistor operating at high frequencies is illustrated in Fig. 1.7. In series with the desired resistance R is induct-

Fig. 1.5 Typical curves of RCWV for different power ratings as a function of resistance. (*Courtesy Allen-Bradley*)

ance L; shunted across R and L is capacitance C. Inductance L arises from the type of construction and connecting leads of the resistor. Capacitance C is present also because of the resistor's construction and the capacitance of the connecting leads. These undesired elements, L and C, are referred to as *parasitics*.

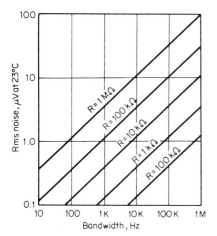

Fig. 1.6 Typical curves of noise for different values of resistance as a function of bandwidth. (*Courtesy MEPCO/ELECTRA, Inc.*)

Fig. 1.7 A model of a resistor operating at high frequencies.

Owing to the parasitics in a resistor, the useful frequency range is limited. Typical high-frequency characteristics of a resistor are illustrated in Fig. 1.8. Expression $R_{h\text{-}f}/R_{dc}$ is the ratio of the resistance at high frequencies, $R_{h\text{-}f}$ to the resistance at direct current and low frequencies, R_{dc}. It is seen that as frequency is increased, $R_{h\text{-}f}/R_{dc}$ is reduced.

DRIFT   Drift, or *time stability*, refers to the change in resistance value over a time interval of use, such as 1000 h. High-value resistors exhibit more drift than low-value units. Short-term drift is generally more pronounced than long-time drift.

TAPER   The taper of a pot refers to the variation of its resistance as a function of the rotation of the wiper, or slider, arm. Examples of taper curves are illustrated in Fig. 1.9.

Fig. 1.8 Typical high-frequency characteristics of different-type fixed resistors. (*Courtesy Allen-Bradley*)

A *linear taper* is exhibited by curve *a*. For this taper, the variation in resistance is directly proportional to the displacement of the wiper. *Nonlinear tapers* are indicated by curves *b* and *c*. Curve *b* represents a logarithmic taper, and curve *c* a special nonlinear taper.

RESOLUTION The resolution of a pot is the smallest change in resistance that can be realized as the wiper arm is rotated.

END RESISTANCE End resistance is the resistance between the wiper and end terminals with the wiper positioned at the corresponding end point.

CONTACT RESISTANCE Contact resistance is the resistance between the wiper terminal and the resistance element in contact with the wiper.

WIPER (SLIDER) CURRENT Wiper current is the maximum current that may flow in or out of the wiper terminal.

SETTING STABILITY This parameter indicates the *repeatability* of a pot setting to a given resistance value.

Fig. 1.9 Examples of taper curves for potentiometers. (*Courtesy MEPCO/ELECTRA, Inc.*)

TABLE 1.2 Summary of Fixed and Variable Resistor Types

Category	Type	Key property	Power	Temperature coefficient, °C	Resistance range
General purpose ≥ 5% tolerance ≥ 200 ppm	Carbon composition	Cost	$1/8$–2 W	> 500	1 Ω ≥ 100 MΩ
	Molded wire-wound	Tempco	$1/2$–2 W	≥ 200	0.1 Ω–2.4 kΩ
	Ceramic wire-wound	Low voltage W	2–50 W	≥ 200	0.1 Ω–30 kΩ
	Metal glaze	Flexibility	$1/8$–5 W	200	4.3 Ω–1.5 MΩ
	Tin oxide	Reliability	$1/8$–20 W	200	4.3 Ω–2.5 MΩ
	Carbon film (import)	Cost	$1/4$–2 W	> 200	10 Ω ≥ 1 MΩ
	Cermet film	Stability	$1/4$–3 W	150	10 Ω–10 MΩ
Semiprecision > 1 < 5% ≤ 200 ppm	Metal glaze	Flexibility	$1/8$–2 W	≤ 200	1 Ω–1.5 MΩ
	Tin oxide	Stability	$1/8$–2 W	≤ 200	4.3 Ω–1.5 MΩ
Power ≥ 2 W	Ceramic wire-wound	Cost	2–50 W	≥ 200	0.1 Ω–30 kΩ
	Axial lead coated WW	Auto insertion	$1/2$–15 W	≤ 50	0.1 Ω–175 kΩ
	Tubular and flat WW		4–250 W	≤ 100	0.1 Ω ≥ 1 mΩ
Precision ≤ 1% ≤ 100 ppm	Metal film	Tolerance	$1/10$–1 W	20	0.1 Ω–1 MΩ
	Metal glaze	Environment	$1/10$–1 W	≤ 100	1 Ω–1 MΩ
	Tin oxide	Power	$1/10$–$1/2$ W	≤ 100	10 Ω–1 MΩ
	Thin film	Size, networks	$1/20$–5 W	≤ 100	10 Ω–100 MΩ
	Encaps. wire-wound	Power, Tempco	$1/20$–1 W	≤ 20	0.1 Ω ≥ 1 MΩ
Ultraprecision ≤ 0.5% ≤ 25 ppm	Thin film	Flexibility	$1/20$–$1/2$ W	≤ 25	20 Ω–1 MΩ
	Encaps. wire-wound	Noise	$1/20$–1 W	≤ 20	0.1 Ω ≥ 1 MΩ
Variable devices (pots, trimmers)	Wire-wound	Tempco	5 at 70°C	± 20	10 Ω–100 kΩ
	Conductive plastic	Rotational life	2 at 70°C	± 250–500	1 kΩ–100 kΩ
	Cermet	Environmental	12 at 70°C	± 250–500	500 Ω–2 MΩ
	Carbon	Cost	5 at 70°C	± 300–2000	100 Ω–2 MΩ
Networks	Thick film	Cost	≤ 2 W/pkg.	≤ 200	10 Ω–10 MΩ
	Thin film	Performance	≤ 2 W/pkg.	≤ 100	10 Ω–1 MΩ

Reprinted from *Electronics*, May 30, 1974; copyright McGraw-Hill, Inc., 1974.

1.4   FIXED RESISTORS

Fixed resistors are manufactured in four basic types:
1. Carbon composition
2. Metal film
3. Carbon film
4. Wire-wound

Resistors can also be classified in terms of their tolerance:
1. General purpose: tolerance 5 percent or greater
2. Semiprecision: tolerance between 1 and 5 percent
3. Precision: tolerance between 0.5 and 1 percent
4. Ultraprecision: tolerance better than 0.5 percent

A summary of different resistor types and their key properties appears in Table 1.2.

CARBON COMPOSITION   The carbon-composition resistor is perhaps the most widely used fixed resistor in discrete circuits. Composition resistors are available in resistance values from 1 Ω to 100 MΩ and typical power ratings of $^1/_8$ to 2 W. Their temperature coefficient is high (greater than 500 ppm/°C), and their cost is low.

A cutaway view of a carbon resistor is shown in Fig. 1.10. The resistance material is a form of carbon, such as graphite, embedded in a binder. The resistance and in-

Fig. 1.10 Cutaway view of a carbon-composition resistor. (*Courtesy Allen-Bradley*)

sulation material, and the wire leads, are molded simultaneously under high temperature and pressure. The result is referred to as a slug-type structure. Different resistance values are obtained by varying the carbon and filler content.

METAL FILM   Metal-film resistors are available as thin- and thick-film-type components. In the thick-film category are the tin-oxide, metal-glaze, cermet, and bulk-film resistors. Each type will be considered in this section.

Thin film   The resistance element in a thin-film resistor is a film having a thickness in the order of one-millionth of an inch. (A thick film has a thickness greater than one-millionth of an inch.) Typically, the thin film is deposited on a ceramic substrate under a high vacuum; the technique is referred to as *vacuum deposition* (see Chap. 9). Metals used for deposition include nickel and chromium. Some characteristics of the thin-film resistor are: resistance range of 10 Ω to 1 MΩ; tolerance better than 0.5 percent, TCR less than 25 ppm/°C, power rating up to 5 W; low noise.

Tin oxide   Tin oxide, in vapor form, is usually deposited on a ceramic substrate under high temperature. The vapor reacting with the substrate, which is heated, results in a tightly formed resistance film. Some characteristics of the tin-oxide resistor are: resistance range of a few ohms to 2.5 MΩ, tolerance better than 1 percent, TCR less than 200 ppm/°C, power rating up to 2 W, good stability.

Metal glaze   A powdered glass and fine metal particle (palladium and silver) mixture is deposited on a ceramic substrate. The combination is then fired at a high temperature (typically 800°C). This results in a fusion of metal particles to the substrate.

Some characteristics of the metal-glaze resistor are resistance range of a few ohms to 1.5 MΩ, tolerance better than 1 percent, TCR as low as 20 ppm/°C, power rating up to 5 W, good stability.

Cermet   A cermet-film resistor is made by screening (see Chap. 9) a mixture of precious metals and binder material on a ceramic substrate. Similar to the manufacture of a metal-glaze resistor, the combination is then fired at a high temperature. An example of a cermet film resistor is illustrated in Fig. 1.11. Some characteristics of this type of resistor are: resistance range of 10 Ω to 10 MΩ, tolerance as low as 1 percent, TCR in the order of 100 ppm/°C, power rating up to 3 W, good stability.

Fig. 1.11   Cutaway view of a cermet-film resistor. (*Courtesy Allen-Bradley*)

Bulk film   In this resistor, the metal film is etched on a glass substrate (Fig. 1.12). Because of their unequal coefficients of expansion, the metal film is compressed slightly by the glass substrate. The compressed film has a negative temperature coefficient, which cancels out the inherent positive temperature coefficient of the film. As a result, the bulk-film resistor has a TCR close to zero. Some characteristics of the bulk-film resistor are resistance range of 30 Ω to 600 kΩ, tolerance as low as 0.005 percent, TCR in the order of 1 ppm/°C, power rating up to 1 W, very low noise, good stability.

Fig. 1.12   Cutaway view of a bulk-film resistor. (*Courtesy Vishay Resistor Products*)

CARBON FILM   This resistor is manufactured by depositing a carbon film on a ceramic substrate (Fig. 1.13). Some characteristics of carbon-film resistors are resistance range of 10 Ω to 10 MΩ, tolerance 5 percent or greater, TCR in the order of 150 ppm/°C, power rating up to 2 W, generally less noisy than the carbon-composition resistor, low cost.

Fig. 1.13   Cutaway view of a carbon-film resistor. (*Courtesy Piher International Corporation*)

Fig. 1.14 Cutaway view of a power-style molded vitreous enamel wire-wound resistor. (*Courtesy Ohmite Manufacturing Company*)

Fig. 1.15 Adjustable power-style wire-wound resistor. (*Courtesy Ohmite Manufacturing Company*)

WIRE-WOUND   The wire-wound resistor enjoys a broad spectrum of applications. Made in many shapes and sizes, it is used as an ultraprecision resistor in instrumentation and as a power resistor in industrial applications.   Examples of its construction are illustrated in Figs. 1.14 to 1.16.

The *power-style* wire-wound resistor is made by winding a single-layer length of special alloy wire, in the form of a coil, around an insulating core.   The unit is then covered with a coating, such as vitreous enamel (an inorganic glasslike mixture) or silicone.   The coating protects the winding against moisture and breakage.

The resistance wire used must have carefully controlled resistance per unit length of wire and low-temperature coefficient, and be able to operate at high temperatures. Alloys used include nickel-chromium-aluminum (800 alloy) and nickel-chromium-iron (Nichrome).   The cylindrical core is ceramic, steatite, or a vitreous material.   In the *precision-style* wire-wound resistor, typically a multilayer coil is wound on an epoxy form, or *bobbin*.

Because the wire-wound resistor acts like a coil, its inductance (as well as the capacitance between coil turns) is a problem at high-frequency operation.   A number of techniques are utilized to minimize inductance.   In one method, one-half the resistance wire is wound in one direction and the other half in the opposite direction.   This type of winding is referred to as *bifilar*.

In another method, used in precision-style resistors, a thick-film *serpentine pattern* is deposited on a ceramic core (Fig. 1.17). Similar in behavior to the bifilar winding,

Fig. 1.16 Examples of precision-style wire-wound resistors. (*Courtesy Shallcross*)

Fig. 1.17 A serpentine resistor pattern used to reduce inductance. (*Courtesy Caddock Electronics*)

opposite magnetic fields produced by current flowing in adjacent resistance paths cancel each other.   The result is a resistor with virtually no inductance.

Precision-style wire-wound resistors are available having the following characteristics: resistance range of a fraction of an ohm to 10 MΩ, tolerance better than 0.5 percent, TCR less than 20 ppm/°C, power rating in the order of 2 W.

Some characteristics of power-style wire-wound resistors are: resistance range of less than an ohm to greater than a megohm; tolerance in the range of ±5 to ±20 percent; TCR as low as 5 ppm/°C; power rating as high as 1500 W.   To increase its power rating, the resistor is sometimes placed in a metal housing, such as aluminum, illustrated in Fig. 1.18.   For some types of resistors, the power rating of a resistor in a metal housing is increased by a factor of 2.   For the realization of the increased power rating, however, the housing must be mounted on a metal chassis or suitable heat sink.

1.5 COLOR CODING OF RESISTORS

Figure 1.19 illustrates the standard color code for composition and some axial-type resistors.   Different color bands are used to designate the resistance value and toler-

First band – 1st digit
Second band – 2nd digit

Color	Digit	Multiplier	Tolerance
Black	0	1	–
Brown	1	10	±1 %
Red	2	100	±2 %
Orange	3	1000	–
Yellow	4	10 000	–
Green	5	100 000	–
Blue	6	1 000 000	–
Violet	7	10 000 000	–
Gray	8	–	–
White	9	–	–
Gold	–	0.1	± 5 %
Silver	–	–	± 10 %
No color	–	–	±20 %

Fig. 1.18 A power resistor mounted in an aluminum housing.   (*Courtesy RCL Electronics, Inc.*)

Fig. 1.19 Standard color code for composition and some axial-type resistors.

ance.   The first two bands denote the first and second digits of the resistance value and the third band indicates how many zeros follow the first two digits.   Tolerance is given by the fourth band.   For example, a resistor with the following colored bands, yellow-violet-orange-silver, denotes a 47 000-Ω ±10 percent tolerance resistor.

For film and wire-wound resistors, in general, the resistance value and tolerance are stamped on the body of the resistor.   Occasionally, a manufacturer may use his own code.   For this reason, it is good practice to consult the manufacturer's catalog or data sheet.

1.6 CONNECTING RESISTORS

In this section we examine how fixed resistors are connected in series and parallel.

RESISTORS IN SERIES   Figure 1.20 shows n resistors connected in series.   In a series circuit, the current I flowing in each resistor is the same.   The total equivalent resistance of the series circuit R_{TS} is equal to the sum of the individual resistors:

$$R_{TS} = R_1 + R_2 + \cdots + R_n \tag{1.4}$$

If n equal resistors are connected in series, the total equivalent resistance is equal to the product of the value of an individual resistor and n:

$$R_{TS} = nR \qquad (1.5)$$

The maximum voltage E_{max} that can be applied across a series circuit of resistors is equal to the sum of the rated continuous working voltage RCWV of each resistor:

$$E_{max} = (RCWV)_1 + (RCWV)_2 + \cdots + (RCWV)_n \qquad (1.6)$$

example 1.5 Three resistors of 10, 20, and 30 Ω are connected in series. The current flowing in the circuit is 2 A, and the RCWV for each resistor is 500 V. Determine (*a*) the total equivalent resistance of the circuit, (*b*) the power dissipated in each resistor, (*c*) the maximum voltage that can be impressed across the series circuit. If the three resistors are replaced by a single equivalent resistor, determine (*d*) its power and maximum voltage ratings.

solution   (*a*) By Eq. (1.4) $R_{TS} = 10 + 20 + 30 = 60$ Ω.
(*b*) By Eq. (1.3*b*), $P = I^2R$, we have

$$P_{10} = 2^2 \times 10 = 4 \times 10 = 40 \text{ W}$$
$$P_{20} = 2^2 \times 20 = 4 \times 20 = 80 \text{ W}$$
$$P_{30} = 2^2 \times 30 = 4 \times 30 = 120 \text{ W}$$

(*c*) By Eq. (1.6), $E_{max} = 500 + 500 + 500 = 1500$ V.
(*d*) In part *a* we found that the equivalent resistance is 60 Ω. The power rating, therefore, is $2^2 \times 60 = 4 \times 60 = 240$ W. (*Note:* This value must equal the sum of the dissipated power in each individual resistor found in part *b*.) From part *c*, the maximum voltage rating is 1500 V.

Fig. 1.20 Resistors connected in series.

Fig. 1.21 Resistors connected in parallel (shunt).

example 1.6 A 90- and a 10-Ω resistor are connected in series. The impressed voltage across the series combination is 1000 V. Determine the voltage across each resistor.

solution By voltage division, the voltage across each resistor is proportional to its resistance value. For the 10-Ω resistor, $E_{10} = 1000 \times 10/(10 + 90) = 100$ V; for the 90-Ω resistor, $E_{90} = 1000 \times 90/(10 + 90) = 900$ V.
What is important in this example is that in a series circuit the voltage across the larger resistor (90 Ω) is much greater than that across the smaller (10 Ω) resistor. The RCWV of the 90-Ω resistor, therefore, must be at least 900 V. For the 10-Ω resistor, an RCWV of 100 V is adequate.

RESISTORS IN PARALLEL   Figure 1.21 shows n resistors connected in parallel. In a parallel circuit, the voltage E across each resistor is the same. It is convenient to consider two resistors at a time when calculating the equivalent resistance R_{TP} of a parallel circuit. The equivalent resistance R of two resistors R_1 and R_2 in parallel (denoted as $R_1 \| R_2$) is equal to their product divided by their sum:

$$R = \frac{R_1 R_2}{R_1 + R_2} \qquad (1.7)$$

For n equal resistors in parallel, the equivalent resistance is equal to the value of an individual resistor R_1 divided by n:

$$R_{TP} = \frac{R_1}{n} \qquad (1.8)$$

The maximum voltage that can be applied across resistors in parallel is limited by the smallest value of RCWV for a resistor in the circuit.

example 1.7   Three resistors of 120, 120, and 12 Ω are connected in parallel. The voltage across the parallel circuit is 240 V, and the RCWV of the 120-Ω resistors is 1000 V and 750 V for the 12-Ω resistor. Determine (a) the total equivalent resistance of the circuit, (b) the power dissipated in each resistor, (c) the maximum voltage that can be impressed across the parallel circuit. If the three resistors are replaced by a single equivalent resistor, determine (d) its power rating.

solution   (a) Considering the two 120-Ω resistors, by Eq. (1.8), $R_{TP} = {}^{120}/_2 = 60$ Ω. By Eq. (1.7), the equivalent parallel resistance of the circuit is $60\|12 = (60 \times 12)/(60 + 12) = 10$ Ω.

(b) By Eq. (1.3c), $P = E^2/R$, we have

$$P_{120} = \frac{(240)^2}{120} = 480 \text{ W}$$

$$P_{12} = \frac{(240)^2}{12} = 4800 \text{ W}$$

(c) The maximum voltage, limited by the smaller value of RCWV, is 750 V.

(d) In a we found that the equivalent resistance is 10 Ω. The power rating, therefore, is $(240)^2/10 = 5760$ W.

Examples 1.5 and 1.7 illustrate that a high power rating can be realized by connecting lower power-rated resistors in series or in parallel.

1.7   SPECIAL RESISTORS

A number of special resistors and resistive networks are available. Some of these components are considered in this section.

HIGH-VOLTAGE RESISTORS   Resistors that operate at voltages as high as 40 000 V are available. In one type of construction, a carbon-film resistance element is vacuum-sealed in a glass envelope.

HIGH-MEGOHM RESISTORS   Using a semiconductor glass for the resistance element, resistance values as high as 1 000 000 MΩ (a million, million ohms) are obtainable. Measuring in length in the order of 0.25 in. and about 0.05 in. in diameter, the application for high-megohm resistors includes radiation detectors, electrometers, and use in FET circuits.

DIP NETWORKS   Resistors, in various configurations packaged in the *dual-in-line package* (DIP), used for integrated circuits, constitute a DIP network. The DIP contains typically 14 or 16 pins. An example of a DIP ladder network is illustrated in Fig. 1.22. Innumerable resistor configurations are available. The resistors are generally thin or thick film.

1.8   VARIABLE RESISTORS (POTS)

A variable resistor, commonly referred to as a potentiometer or pot, converts the rotation of a shaft to an output voltage. There are basically three types of pots:

1. Single turn
2. Multiple turn
3. Trimmer

Examples of the three types of pots are illustrated in Figs. 1.23 to 1.25.

The single-turn pot may be regarded as the "workhorse" of variable resistors. Available in resistance ranges from 50 Ω to 5 MΩ and higher, in tolerances of ± 10 and ± 20 percent, and in power ratings of 2 and 3 W, the single-turn pot enjoys a wide spectrum of applications. It is used, for example, as a gain, treble, or base control in an amplifier and as the brightness and contrast controls in a TV receiver. In some applications, two or more pots sharing a common rotating shaft are required. Such a combination, referred to as a *ganged pot*, is illustrated in Fig. 1.26.

Multiple-turn pots are used in applications that require the precise setting of a resistance value. An example is the setting of coefficient pots in an analog computer.

Resistors: R_1, R_3, R_5, R_7, R_9, R_{11}, R_{13}, R_{15}, R_{17}, R_{19}, R_{21}, R_{23}, R_{25}, R_{27}, R_{28}; each = 2R

Resistors: R_2, R_4, R_6, R_8, R_{10}, R_{12}, R_{14}, R_{16}, R_{18}, R_{20}, R_{22}, R_{24}, R_{26}; each = R

(a)

(b)

Fig. 1.22 An example of a DIP network: (a) Ladder network. (b) DIP with 14 pins. (*Courtesy RCL Electronics, Inc.*)

These pots generally have 10 turns and are available in typical resistance ranges of 50 Ω to 250 kΩ, in tolerances of ± 3 percent, and in power ratings up to 5 W.

The trimmer pot, which may be viewed as a "set and forget me" pot, is used generally for a one-time adjustment of resistance. Its rotational life, compared to that of other pots, is therefore limited. Trimmer pots are available as single- and multiple-turn units. Typically, their resistance range is from a few ohms to 5 MΩ, tolerance is ± 10 percent, and power rating is 1 W.

Fig. 1.23 An example of a single-turn pot that enjoys a wide variety of application. (*Courtesy Ohmite Manufacturing Company.*)

Fig. 1.24 An example of a multiple-turn pot. (*Courtesy Beckman Instruments, Inc., Helipot Division*)

Fig. 1.25 Examples of trimmer pots. (*Courtesy Beckman Instruments, Inc., Helipot Division*)

TYPES OF MATERIALS Four basic types of materials are used in the construction of pots:
1. Carbon
2. Cermet
3. Conductive plastic
4. Wire-wound

Pots other than wire-wound are referred to as non-wire-wound. A comparison of different types of pots is provided in Table 1.2.

Carbon The resistance element is made by spraying or brushing a carbon-resistance compound on an insulating material, such as laminated plastic. Its chief merit is its low cost. The carbon pot exhibits a high TCR and is susceptible to moisture, and its maximum operating temperature is generally limited to 85°C.

Variations of the carbon pot that exhibit somewhat improved characteristics are the *carbon-ceramic* and *molded-carbon* types. In the carbon-ceramic pot, the carbon compound is screened on a ceramic substrate. Molded-carbon pots are made by molding, simultaneously, the carbon-resistance element and other parts of the pot.

Cermet In the cermet pot, a mixture of precious metals and glass or ceramic powder is screened on a ceramic substrate. The TCR of a cermet pot may be as low as 50 ppm/°C and its operating temperature as high as 150°C. The cermet pot is very stable and can withstand overloads.

Conductive plastic One method of making a conductive plastic pot is by spraying, under pressure, a special liquid suspension on a plastic material, such as Mylar. The suspension contains carbon, a solvent, and a filler. The resistance value depends on the mix of carbon and filler. Conductive plastic pots exhibit good linearity and long rotational life. They are, however, susceptible to humidity.

Fig. 1.26 A dual-ganged pot. (*Courtesy Ohmite Manufacturing Company*)

Wire-wound   In the wire-wound pot, the resistance element is wound on a phenolic, plastic, or glass-filled laminate card.   Low-resistance wire-wound pots are generally wound with a copper alloy wire and high-resistance pots with a nickel-chromium wire. The wire-wound pot exhibits the lowest TCR (20 ppm/°C), operates at temperatures as high as 150°C, and is available in high-power ratings.   Because of its construction, the wire-wound pot has appreciable stray inductance and capacitance which may be a problem at high-frequency operation.

RHEOSTAT   A wire-wound pot that can dissipate 5 and more watts is often referred to as a *rheostat*.   The resistance wire is wound on an open ring of ceramic which is covered with vitreous enamel, except for the track of the wiper arm.   Rheostats are used to control motor speed, x-ray tube voltages, welding current, ovens, and in many other high-power applications.

1.9   CHIP RESISTORS

A variety of thin- and thick-film resistor chips are available for hybrid microelectronic circuits (see Chap. 9).   They measure as little as 30 by 30 mils (0.03 by 0.03 in.) and have resistance values from a few ohms to 1000 MΩ.   Typical power rating is 0.25 W, and TCR is in the order of 20 to 200 ppm/°C.

1.10   THERMISTORS

A thermistor is a *nonlinear resistor*, made of semiconductor material, that is extremely sensitive to changes in temperature.   For a small change in body temperature of a

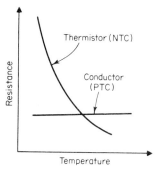

Fig. 1.27   The variation in resistance of a conductor (PTC) and a thermistor (NTC) with temperature.

Fig. 1.28   Electrical symbol for a thermistor.

thermistor, there is an appreciable change in its resistance.   Whereas most conductors have a positive temperature coefficient (PTC), the thermistor can exhibit a positive or negative temperature coefficient (NTC).   Of particular interest is thermistors having a negative temperature coefficient.

A comparison of a conductor and an NTC thermistor is illustrated in Fig. 1.27.   For the thermistor, the resistance decreases rapidly for elevated temperatures.   The conductor having a positive temperature coefficient, however, exhibits a small increase in resistance with rising temperatures.

Thermistors are used in a wide variety of applications.   These include the measurement and control of temperature, time delay, temperature compensation, and as liquid-level indicators.   Thermistors are available as disk, washer, bead, and bolted assembly packages.   The electrical symbol for a thermistor is illustrated in Fig. 1.28.

CHARACTERISTIC CURVES AND PARAMETERS   In this section, typical characteristic curves and parameters for the thermistor are considered.

Resistance-temperature characteristic   A typical resistance-temperature characteristic of a thermistor is illustrated in Fig. 1.29.   Plotted is resistance as a function of temperature. The resistance, for zero power dissipation, decreases rapidly with elevated temperature.

Three useful parameters for characterizing the thermistor are the time constant, dissipation constant, and resistance ratio. The time constant is the time for a thermistor to change its resistance by 63 percent of its initial value, for zero-power dissipation. Typical values of time constant range from 1 to 50 s.

Fig. 1.29 Typical resistance-temperature characteristic of a thermistor. (*Courtesy N L Industries, Inc., Electronics Department*)

The dissipation factor is the power necessary to increase the temperature of a thermistor by 1°C. Expressed in milliwatts per degree Celsius, typical values of dissipation factor are 1 to 10 mW/°C.

Resistance ratio is the ratio of the resistance at 25°C to that at 125°C. Its range is approximately 3 to 60.

Static voltampere characteristics A typical voltampere characteristic is shown in Fig. 1.30. Plotted is the voltage across the thermistor as a function of current flow. For

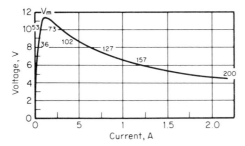

Fig. 1.30 Typical static voltampere characteristic of a thermistor. (*Courtesy N L Industries, Inc., Electronics Department*)

minute currents, the dissipated power is too small to heat the thermistor. In this region Ohm's law is obeyed. As the current is increased, the thermistor temperature rises and the resistance begins to decrease. At the *self-heating voltage* V_m, the characteristic begins to exhibit a negative slope. The temperature of the thermistor (in degrees Celsius) at various values of voltage and current is denoted by the numbers plotted along the curve.

Current-time characteristic Because of its mass, it takes a finite time for a current in a thermistor to reach its final value. This is illustrated in the current-time characteristic curves of Fig. 1.31a for three different thermistors.

The test circuit of Fig. 1.31b contains a 100-Ω resistor in series with a thermistor

Fig. 1.31 Current-time characteristics of a thermistor: (*a*) Typical family of curves. (*b*) Test circuit. (*Courtesy N L Industries, Inc., Electronics Department*)

whose resistance is 1000 Ω at 25°C. A voltage of 24 V dc is impressed across the circuit. For example, the thermistor having the characteristic of the top curve reaches its final value of approximately 200 mA in 20 s. The thermistor having the bottom characteristic, however, takes 120 s to approach its final value of 160 mA.

example 1.8 Show how a thermistor may be used as a liquid-level indicator.

solution The circuit for a simple liquid-level indicator employing a thermistor is given in Fig. 1.32. Suspended above the liquid is a thermistor in series with an indicator, such as a lamp, and battery E. In air, the resistance of the thermistor is relatively low, thereby allowing sufficient current I to flow and light the lamp. When the liquid level reaches the thermistor, it

Fig. 1.32 A simple liquid-level indicator using a thermistor.

cools the device and its resistance is increased. Current I is reduced and the lamp is extinguished.

1.11   VARISTORS

The varistor, another example of a nonlinear resistor, is a device in which the current varies as a power of the impressed voltage. The resistance, therefore, is also reduced. In an ordinary resistor, the current is directly proportional to the impressed voltage and Ohm's law is obeyed. In the varistor, however, the current is proportional to the power of the impressed voltage E^n, where n is in the range of 2 to 6. A typical voltage-current characteristic curve and electrical symbol for the varistor are shown in Fig. 1.33.

Fig. 1.33 The varistor: (*a*) Typical voltampere characteristic. (*b*) Electrical symbol. (*Courtesy N L Industries, Inc., Electronics Department*)

Fig. 1.34 A simple surge-protection circuit using a varistor.

Applications for the varistor include voltage surge and protective circuits and the generation of nonsinusoidal waveforms. The varistor is made of silicon carbide and is available in disk, rod, and washer forms. It can withstand dc voltages as high as 10 000 V.

example 1.9   Explain the operation of the surge protection circuit of Fig. 1.34.

solution   Without the varistor connected across coil L, when the switch is opened, the high energy in the magnetic field develops a dangerously high voltage across L. This can result in the puncturing of the coil winding insulation. With a varistor in the circuit, the high voltage across it causes appreciable current to flow. The resistance of the device is thereby reduced and the magnetic field energy is dissipated in the varistor, instead of in the coil.

Chapter 2

Characteristics of Capacitors

2.1 INTRODUCTION

Next to resistors, capacitors are the most widely used passive element in circuits. They are available as fixed and variable units having capacitance values from a few picofarads (pF) to thousands of microfarads (μF). For achieving unique characteristics, a large variety of materials are used in their construction. The abundance of different fixed capacitors available to the user is vividly portrayed in the photograph of Fig. 2.1. Because of such a plethora of units, the technician and engineer are often in a dilemma in selecting a capacitor for a given application.

APPLICATIONS The applications for capacitors may be broadly categorized as follows:
1. Blocking direct current. A capacitor cannot conduct direct current.
2. Coupling a signal from one circuit, or system, to another.
3. Bypassing a resistor to permit the easy flow of alternating current.
4. Filtering.
5. Tuning.
6. Generation of nonsinusoidal waveforms, such as a sawtooth wave.
7. Energy storage. Capacitors are used to build up sufficient electric charge to fire, for example, a flash tube or a laser.

2.2 GENERAL DESCRIPTION

The basic fixed capacitor consists of two metal plates, called *electrodes*, separated by an insulator, called a *dielectric*, as shown in Fig. 2.2a. In a variable capacitor, the relative position of one or more plates is changed with respect to a set of fixed plates, thereby varying the capacitance. Air is generally used as the dielectric for variable capacitors. Symbols for fixed and variable capacitors are shown in Fig. 2.2b and c, respectively.

The capacitance of parallel-plate capacitors is directly proportional to the relative dielectric constant of the insulator and to the area of the plates. Further, the capacitance is greater if the separation between plates is small, and vice versa. These facts are conveniently expressed by the following equation:

$$C = \frac{k\epsilon_0 A}{d} \tag{2.1}$$

where C = capacitance in farads, F
  ϵ_0 = a constant, called the *permittivity of free space* and is equal to 8.85×10^{-12} F/m
  k = relative dielectric constant of insulator

Fig. 2.1 A wide variety of types and styles of fixed capacitors are available to the technician and engineer. (*Courtesy Sprague Electric Company*)

Fig. 2.2 The basic parallel-plate capacitor: (*a*) Construction. (*b*) Symbol for fixed capacitor. (*c*) Symbol for variable capacitor.

A = area of plates, m²
d = separation between plates, m

From Eq. (2.1) it is seen that capacitance may be increased by increasing the area of the plates or the dielectric constant, and by decreasing the separation between plates.

RELATIVE DIELECTRIC CONSTANT   The relative dielectric constant k compares different dielectrics with that of a vacuum for which $k = 1$. The dielectric constants for some commonly used materials are given in Table 2.1. Note that the dielectric constant of air is approximately equal to that of a vacuum.

TABLE 2.1 Dielectric Constants of Some Commonly Used Materials

Dielectric	k
Vacuum	1
Air	1.0006
Teflon	2
Polystyrene	2.5
Mylar	3
Paper, paraffin	4
Mica	5
Aluminum oxide	7
Tantalum oxide	25
Ceramic (low k)	10
Ceramic (high k)	100–10 000

example 2.1   A parallel-plate capacitor has the following dimensions: A = 100 cm² and d = 1 mm. Determine the values of capacitance if the dielectric used is: (a) air; (b) paper, paraffin; and (c) ceramic ($k = 200$).

solution   The given dimensions are first converted to meters. One square centimeter equals 10^{-4} m²; hence, 100 cm² = 100×10^{-4} = 10^{-2} m². One millimeter equals 10^{-3} m. Substitution of these values in Eq. (2.1) yields

$$C = k \times 8.5 \times 10^{-12} \times \frac{10^{-2}}{10^{-3}} = k \times 85 \times 10^{-12} \text{ F}$$

(a) The dielectric constant for air is approximately one. Hence, $C_{air} = 85 \times 10^{-12} \times 1 = 85 \times 10^{-12}$ F = 85 pF, since 10^{-12} F is equal to 1 pF.

(b) For paper coated with paraffin, from Table 2.1, $k = 4$. Therefore, $C_{paper} = 85 \times 10^{-12} \times 4 = 350 \times 10^{-12}$ F = 350 pF.

(c) $C_{ceramic} = 85 \times 10^{-12} \times 200 = 17 \times 10^{-9}$ F = 0.017 μF, since 10^{-6} F is equal to 1 μF.

The particular dielectric used exerts a strong influence on the value of a capacitor. As we shall see, capacitors are designated by the dielectric used in their construction.

CHARGE AND ENERGY   The charge Q (coulombs) stored in a capacitor equals the product of the capacitance C (farads) and the voltage E (volts) across the capacitor:

$$Q = CE \qquad (2.2a)$$

Solving Eq. (2.2a) for E and C,

$$E = \frac{Q}{C} \qquad (2.2b)$$

$$C = \frac{Q}{E} \qquad (2.2c)$$

From the preceding equations it is seen that to store a given charge, more voltage is needed for a small capacitor and less voltage for a large capacitor.

The maximum voltage that can be impressed across a capacitor, without causing irreparable damage, depends on the dielectric used and the separation d of the plates. For a given dielectric, the allowable maximum voltage increases as the separation between plates increases. From Eq. (2.1), as d grows large for a given plate area and dielectric the capacitance is reduced. One may conclude that for a given dielectric and physical size, a capacitor with a high-voltage rating has less capacitance than one with a low-voltage rating.

A charged capacitor has *energy stored* in the electric field existing between its plates. The expression for stored energy is

$$w = \frac{Ce^2}{2} \qquad (2.3)$$

where w = stored energy in Joules, J (1 J = 1 W-s)
   C = capacitance, F
   e = voltage across capacitor terminals, V

example 2.2   Determine the energy stored in a 100-μF capacitor if the impressed voltage is 1 kV.

solution   From Eq. (2.3) $w = 100 \times 10^{-6} \times (1000)^2/2 = 50$ J.

2.3 CAPACITOR TERMS AND PARAMETERS

In this section commonly used terms and parameters for characterizing capacitors are defined. Where appropriate, typical curves showing the variation of parameters with temperature and frequency are included.

CAPACITANCE   The basic unit for capacitance is the farad. Because practical capacitors have a capacitance much less than a farad, typical units employed are the microfarad and picofarad.

AMBIENT TEMPERATURE   The actual value of capacitance depends on the temperature of the medium surrounding the capacitor. This is referred to as the *ambient temperature*. A typical curve showing the variation in capacitance as a function of temperature is given in Fig. 2.3. Note that *percent capacitance change* is plotted as a

Fig. 2.3   A typical curve showing how capacitance varies with temperature. (*Courtesy Electrocube*)

function of temperature. At 25°C, which corresponds to room temperature, the percentage change is zero. This indicates that the changes in capacitance values are taken with respect to room temperature.

example 2.3   A capacitor is rated at 100 μF at 25°C. Using the curve of Fig. 2.3, determine the value of capacitance at −55 and +125°C.

solution   From Fig. 2.3, at −55°C the change is approximately −6 percent. Hence, the value of C is $100 - 100 \times 0.06 = 94$ μF. At +125°C, the change is approximately +13 percent; the value of C at 125°C then is $100 + 100 \times 0.13 = 113$ μF.

TOLERANCE   Tolerance is the variation in capacitance expressed as a percentage of its specified value at 25°C. The specified value at 25°C is referred to as the *nominal value*. For example, 20 μF $\pm$ 10% means that the actual value is between $20 - 20 \times 0.1 = 18$ μF and $20 + 20 \times 0.1 = 22$ μF.

TEMPERATURE COEFFICIENT   The temperature coefficient TC is the change in capacitance per degree change in temperature. It is generally expressed in parts per million per degree Celsius (ppm/°C). The temperature coefficient may be positive, negative, or zero. If the TC is positive, the letter P precedes the coefficient; if negative, N precedes the coefficient. The designation NPO is used for a zero-temperature coefficient capacitor.

WORKING VOLTAGE   The working voltage is the maximum voltage that can be impressed across a capacitor for continuous operation. This rating must indicate whether the voltage is dc or ac; in general, the rating is not the same.

BREAKDOWN VOLTAGE   The breakdown voltage is the maximum capacitor voltage that causes the dielectric to become damaged. It is also referred to as the *surge* and *test* voltage.

DC LEAKAGE   Dc leakage refers to a minute direct current that flows in a capacitor at a specified direct voltage. Leakage is due to the presence of a few free carriers of charge in the dielectric. For this reason, a charge in a capacitor cannot be stored indefinitely, and it ultimately leaks off.

INSULATION RESISTANCE   The insulation resistance IR is the resistance of the dielectric. The greater is the resistance, the less is the leakage current. Insulation resistance decreases with increasing temperature, as illustrated in Fig. 2.4.

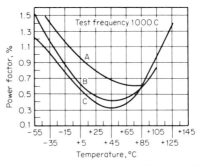

Fig. 2.4   The variation of insulation resistance with temperature. (*Courtesy Electrocube*)

Fig. 2.5   Power factor as a function of temperature for three types of fixed capacitors. (*a*) Paper-dipped type (Type PD). (*b*) Mylar-paper-dipped (Type MPD). (*c*) Mylar. (*Courtesy The Electro Motive Manufacturing Company, Inc.*)

CAPACITIVE REACTANCE   The capacitive reactance X_C (ohms) of a capacitor is equal to 0.159 divided by the product of the frequency (hertz) and capacitance (farads):

$$X_C = \frac{0.159}{fC} \qquad (2.4)$$

The capacitive reactance increases as the frequency or capacitance decreases, and vice versa.

POWER FACTOR   The power factor PF expresses the ratio of energy wasted to the energy stored in a capacitor. If, for example, PF = 2 percent, then 98 percent of the applied energy can serve a useful purpose and the remaining 2 percent is lost as heat in the capacitor. The variation of PF as a function of temperature is illustrated in Fig. 2.5. The power factor also increases with rising frequency.

EQUIVALENT SERIES RESISTANCE Owing to the resistance of the capacitor plates and leads, the equivalent series resistance, ESR, is the net series resistance of a capacitor. An example of variation in ESR for capacitors designed to operate at UHF and microwave frequencies is shown in Fig. 2.6. The ESR increases with increasing frequency.

IMPEDANCE The impedance Z (ohms) of a capacitor is the total opposition to the flow of alternating current. It is equal to the square root of the sum of the capacitive reactance squared and the equivalent series resistance squared:

$$Z = \sqrt{X_C^2 + (\text{ESR})^2} \tag{2.5}$$

Typical curves of impedance as a function of frequency and temperature are illustrated in Fig. 2.7. Impedance of a capacitor decreases with increasing frequency and temperature. For a high-quality capacitor, the impedance is approximately equal to the capacitive reactance.

Fig. 2-6 Variation in ESR with frequency for a porcelain capacitor designed to operate at ultrahigh and microwave frequencies. (*Courtesy American Technical Ceramics*)

Fig. 2.7 The impedance of a fixed capacitor decreases with increasing frequency and temperature. (*Courtesy Sprague Electric Company*)

DISSIPATION FACTOR The dissipation factor DF is another parameter used to describe the quality of a capacitor. The DF is expressed as a percentage and defined as the ratio of ESR to X_C:

$$\text{DF} = \frac{\text{ESR}}{X_C} \times 100\% \tag{2.6}$$

The lower the value of DF is, the better is the capacitor. A typical curve showing how DF varies with temperature is given in Fig. 2.8.

Fig. 2.8 The variation of dissipation factor DF with temperature. (*Courtesy Electrocube*)

QUALITY FACTOR (Q) The quality factor is one over the dissipation factor:

$$Q = \frac{1}{DF} = \frac{X_c}{ESR} \qquad (2.7)$$

The greater the value of Q is, the better is the capacitor. Parameters DF and Q are relatively simple to determine on an impedance bridge.

RIPPLE CURRENT AND VOLTAGE Ripple current (voltage) is the ac component of a unidirectional current (voltage). Ripple is present, for example, in the output of a dc power supply. If the ripple component in a capacitor is less than 5 percent, it may be considered negligible. For values greater than 5 percent, the ripple contributes to the heating of the capacitor.

2.4 FIXED CAPACITORS

Although the types, shapes, and sizes of capacitors are nearly countless (see Fig. 2.1), the basic structure of Fig. 2.2, namely two plates separated by a dielectric, is common to all fixed capacitors. The plates and dielectric may be rolled in a tube, placed in a

TABLE 2.2 Typical Characteristics of Commonly Used Fixed Capacitors

Type	Capacitance range	Maximum working voltage, V	Maximum operating temperature, °C	Tolerance, %	Insulation resistance, MΩ
Mica	1 pF – 0.1 μF	50 000	150	±0.25 to ±5	>100 000
Silvered mica	1 pF – 0.1 μF	75 000	125	±1 to ±20	1000
Paper	500 pF – 50 μF	100 000	125	±10 to ±20	100
Polystyrene	500 pF – 10 μF	1000	85	±0.5	10 000
Polycarbonate	0.001 – 1 μF	600	140	±1	10 000
Polyester	5000 pF – 10 μF	600	125	±10	10 000
Ceramic:					
Low k	1 pF – 0.001 μF	6000	125	±5 to ±20	1000
High k	100 pF – 2.2 μF	100	85	+100 to −20	100
Glass	10 pF – 0.15 μF	6000	125	±1 to ±20	>100 000
Vacuum	1 – 5000 pF	60 000	85	±5	>100 000
Energy storage	0.5 – 250 μF	50 000	100	±10 to ±20	100
Electrolytic:					
Aluminum	1 μF – 1 F	700	85	+100 to −20	<1
Tantalum	0.001 – 1000 μF	100	125	±5 to ±20	>1

disk or rectangular package, or otherwise to reduce the overall size. There are nearly 200 capacitor manufacturers in the United States. Their general thrust has been to develop larger capacitance values in smaller packages. The ratio of the capacitance value to the volume of the package is the *volumetric efficiency* of the capacitor.

Capacitors may be grouped into four broad categories:
1. Fixed
2. Variable
3. Chip
4. Voltage-variable (varactor)

Because the dielectric is a prime influence on capacitance size, a host of dielectrics are used in the construction of capacitors. Consequently, capacitors are classified according to their dielectrics. The characteristics of the most commonly used capacitors are summarized in Table 2.2.

MICA One of the earliest made, and still used, capacitors employs mica for the dielectric. Mica, a natural mineral, is a chemically inert and highly stable dielectric. The mica capacitor generally has a "sandwich" structure (Fig. 2.9), consisting of interleaving layers of tin-lead foil and mica. Mica capacitors are used over a wide temperature range (−55 to +150°C), and they have a high insulation resistance. Their capacitance values range from approximately 1 pF to 0.1 μF.

SILVERED MICA In the silvered mica a thin layer of silver, which is screened and fired onto the surface of the mica, is the conducting electrode. Among the advantages of a silvered mica over a mica capacitor are greater mechanical stability and more uniform characteristics.

PAPER The dielectric in paper capacitors is kraft paper impregnated with a wax or resin. It is generally packaged as a "rolled sandwich," shown in Fig. 2.10. Paper capacitors come in a large variety of values (500 pF to 50 μF), can operate in ambient temperatures as high as 125°C, are low cost, and can withstand high voltages. They are, however, temperature-sensitive. A 10°C rise in ambient temperature may reduce the life of the capacitor by as much as 50 percent.

Fig. 2.9 The "sandwich" structure mica capacitor.

Fig. 2.10 The "rolled-sandwich" structure is commonly used for paper and plastic film capacitors.

The black band, usually printed on the package, is near the lead that is connected to the outer metal foil of the capacitor. To minimize noise, this lead should be connected to the lowest potential point in a circuit.

PLASTIC FILM Various plastics are used as the dielectrics of capacitors. Plastic capacitors are available in typical ranges of 500 pF to 10 μF. The plastics used include *polystyrene*, *polycarbonate*, and *polyester* (Mylar).
Polystyrene capacitors exhibit a low dissipation factor, small capacitance change with temperature, and very good stability. They tend to be large in size, and their maximum operating temperature is 85°C.
Polycarbonate capacitors have a dissipation factor and stability which approach those of polystyrene capacitors. They can operate at temperatures as high as 140°C.
Polyester capacitors, also referred to as Mylar capacitors, are the most widely used plastic capacitors. They are highly stable and exhibit a greater resistance to moisture than kraft paper. The maximum operating temperature of polyester capacitors is 125°C.

CERAMIC There are two basic types of ceramic dielectrics: low k and high k. Low-k ceramic capacitors can be made to exhibit zero temperature coefficient, and they find wide use in temperature-compensation networks. The low-k capacitor operates at voltages as high as 6 kV and up to 125°C. Its capacitance value is generally limited to 0.001 μF.
The high-k ceramic capacitor has capacitance values up to 2.2 μF and maximum working voltage of 100 V. These capacitors, however, change their value appreciably with temperature, dc voltage, and frequency.

GLASS This capacitor is made by stacking alternate layers of aluminum foil and glass ribbon in a sandwich structure which is similar to the mica capacitor. These capacitors are extremely reliable and stable. It is claimed that any two glass capacitors, regardless of their value or size, exhibit a temperature coefficient within 10 ppm/°C of each other. Glass capacitors can operate at voltages as high as 6000 V and up to 125°C. The range of capacitor values is 10 pF to 0.15 μF.

VACUUM For very high voltages, a vacuum is often used as the dielectric. Vacuum capacitors range from 1 to 5000 pF and can operate at voltages as high as 60 kV.

ENERGY STORAGE Energy-storage capacitors store energy which can be discharged in a time interval from a fraction of a microsecond to several hundred microseconds. The dielectric used for energy storage capacitors often consists of a specially prepared and impregnated kraft paper. Capacitance values of 0.5 to 250 μF and voltage ratings up to 50 kV are available.

METALLIZED DIELECTRICS If the dielectric, such as paper or plastic, is deposited with a thin metallic film instead of separate metal foils, volume savings up to 75 percent may be realized. Because of their low insulation resistance, however, these capacitors are not recommended for coupling and logic circuits.

ELECTROLYTIC In applications where a large value of capacitance in a small volume (high volumetric efficiency) is required, such as a filter in a power supply, the electrolytic capacitor is the choice. There are two basic types, *aluminum* and *tantalum* electrolytics, and are either *polarized* or *nonpolarized*.

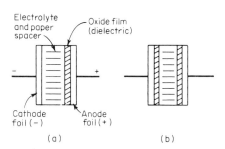

Fig. 2.11 An elementary electrolytic capacitor: (a) Polarized. (b) Nonpolarized.

Fig. 2.12 The rolled-sandwich structure of an electrolytic foil capacitor.

In the polarized electrolytic capacitor, a plus sign is printed on the package near one of the two leads. When the capacitor is put in a circuit, the lead near the plus sign *must* be connected to a higher dc potential than the unmarked lead. If this is not done, the capacitor may be short-circuited and permanently damaged. No such restraint exists for connecting nonpolarized electrolytics in a circuit.

The basic structure of an electrolytic foil capacitor is illustrated in Fig. 2.11. For the polarized type of Fig. 2.11a, to increase its surface area, the aluminum (or tantalum) foil is etched. Then, a very thin layer of aluminum (or tantalum) oxide is electrochemically formed on the anode foil. The oxide layer becomes the dielectric for the capacitor. Because the oxide film is exceedingly thin and the surface area of the etched foil is large, the volumetric efficiency of electrolytic capacitors is high.

Separating the cathode and oxide-coated anode is a paper spacer which is soaked in an electrolyte solution. The spacer is required to prevent short circuiting between the cathode and anode foils. For high-voltage ratings, the spacer is thicker than for low-voltage units. The most common construction for electrolytic capacitors is the rolled sandwich, shown in Fig. 2.12.

The nonpolarized electrolytic capacitor in Fig. 2.11b has two oxide-coated anodes. For an equal-size polarized capacitor, the nonpolarized type has one-half the capacitance for the same voltage rating. Nonpolarized electrolytics should be used for such applications as ac motor starting, crossover networks, and large-pulse signals.

Aluminum electrolytic capacitors have high dc leakage and low insulation resistance.

Their shelf life is limited and their capacitance deteriorates with time and use. They enjoy, however, a high volumetric efficiency and are low in cost.

In addition to being very stable, tantalum oxide has nearly twice the dielectric constant of aluminum oxide. Tantalum electrolytic capacitors have a long shelf life, stable operating characteristics, increased operating temperature range, and a greater volumetric efficiency. The chief disadvantages of the tantalum, in comparison to the aluminum electrolytic, are its greater cost and lower voltage rating (see Table 2.2).

There are three types of tantalum electrolytics: *foil, wet anode*, and *solid anode*. The tantalum foil type is similar in construction to the aluminum foil electrolytic. The wet-anode capacitor is made by first molding, usually into the shape of a pellet, a mixture of tantalum powder and a binder. Under high temperature and in a vacuum, the pellet mixture is welded together (*sintered*), and the binder and impurities are driven off. The result is a porous pellet on which a layer of tantalum oxide is electrochemically formed. The volumetric efficiency of a wet-anode tantalum is about three times as great as the foil type.

The solid-anode tantalum is made by sintering an anode pellet on which is a layer of tantalum oxide. The pellet is then covered with a layer of manganese dioxide which serves as a solid electrolyte and cathode of the capacitor. This is followed by a layer of carbon and of silver paint, completing the cathode connection. The solid-anode construction is most widely used. It has the longest life and lowest leakage current of the three types of tantalum capacitors.

2.5   CONNECTING CAPACITORS

In this section we examine how fixed capacitors are connected in parallel and in series.

CAPACITORS IN PARALLEL   One may think of connecting capacitors in parallel (Fig. 2.13) as effectively increasing the area of the plates. Because capacitance in-

Fig. 2.13   Capacitors in parallel.

Fig. 2.14   Capacitors in series.

creases with plate area (more charge is stored in the capacitor), the total capacitance of capacitors connected in parallel, C_{TP}, is equal to the sum of the individual capacitances:

$$C_{TP} = C_1 + C_2 + \cdots + C_n \tag{2.8}$$

The working voltage of the parallel combination is limited by the smallest working voltage of the individual capacitors.

   example 2.4   In Fig. 2.13 assume that three capacitors are connected in parallel ($n = 3$): $C_1 = 40\ \mu F$, 300 V dc; $C_2 = 25\ \mu F$, 450 V dc; $C_3 = 80\ \mu F$, 200 V dc. Determine the capacitance and dc working voltage of the parallel combination.

   solution   The total capacitance C_{TP} is equal to the sum of C_1, C_2, and C_3. Hence, $C_{TP} = 40 + 25 + 80 = 145\ \mu F$. The dc working voltage is equal to the smallest rating of the individual capacitors. In this example, it is equal to the voltage rating of the 80-μF capacitor, 200 V dc.

CAPACITORS IN SERIES   Connecting capacitors in series may be thought of as increasing the separation of the outer plates of the combination (Fig. 2.14). The result is that the total capacitance C_{TS} is less than the smallest capacitance of the individual capacitors. The relationship for C_{TS} is similar to that for resistors in parallel. For two capacitors in series, C_1 and C_2, we have

$$C_{TS} = \frac{C_1 C_2}{C_1 + C_2} \tag{2.9}$$

If equal-valued capacitors are connected in series, the net value of capacitance equals the value of an individual capacitor divided by the number connected in series. For example, if three 150-pF capacitors are in series, the net capacitance is $^{150}/_3 = 50$ pF.

The total dc working voltage rating of capacitors in series is equal to the sum of the dc ratings of the capacitors. This fact is sometimes used, for example, in the design of filters for high-voltage power supplies. For this purpose, electrolytics are generally chosen and connected in series to increase the dc working voltage. Because the insulation resistance of electrolytics is relatively low and variable with age and use, it is possible that one of the series capacitors will see a much greater voltage than it could withstand.

To prevent this from happening, *equalizing resistors*, of values between 20 and 50 kΩ, are placed across the capacitors (Fig. 2.15). The resistors form a voltage divider.

Fig. 2.15 Equalizing resistors R_1 and R_2 across electrolytic capacitors.

If they are equal, the voltage across each capacitor is $E/2$ V. If they are unequal, the voltage across a capacitor is proportional to the resistance value.

2.6 COLOR CODING OF CAPACITORS

For large-size capacitors, the values of capacitance, working voltage, and tolerance are usually stamped on the packages. Color coding, based on the resistor code, is used for small capacitors. Figure 2.16 illustrates the coding schemes for various shapes of ceramic and mica capacitors.

2.7 VARIABLE CAPACITORS

The need for a variable capacitor arises in tuning circuits found in receivers, transmitters, and oscillators. A commonly used type is the *air-variable capacitor*. The capacitor of Fig. 2.17a is a *single-gang* type; the *dual-gang capacitor* of Fig. 2.17b is used in tuning, for example, a superheterodyne receiver.

Capacitance values for air-variable capacitors range from a few picofarads up to 500 pF; their maximum voltage rating is approximately 9 kV. For higher operating voltages (up to 60 kV), a *vacuum-variable capacitor* is used. A cross-sectional view of such a capacitor is illustrated in Fig. 2.18.

The air-variable capacitor is composed of two sets of plates, usually made of aluminum. One set, referred to as the *rotor*, is mounted on a shaft and meshes with a set of fixed metal plates, referred to as the *stator*. The stator and rotor plates are never permitted to touch each other.

Upon rotation of the shaft, either more or less area exists between the rotor and stator plates. Because capacitance is directly proportional to the area of the plates, the capacitance is thereby varied. The shape of the plates determines the manner in

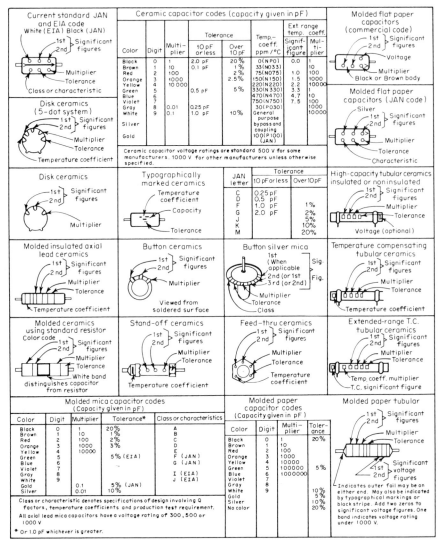

Fig. 2.16   Color-coding schemes used for fixed capacitors.

which capacitance varies with shaft rotation.   For example, a linear or a nonlinear variation in capacitance with shaft position can be obtained.

TRIMMERS   A variety of small variable capacitors using air, ceramic, mica, quartz, and other dielectrics are available for circuit applications.   Referred to as *trimmer capacitors*, they are utilized for fine tuning and in hybrid microelectronic circuits. Their capacitance values range from a few picofarads to about 100 pF.

The capacitor of Fig. 2.19 is designed for hybrid circuits.   It has a working voltage of 200 V dc, and in the 15- to 50-pF range, a Q of greater than 300 at 100 MHz.   The ceramic trimmer of Fig. 2.20 has a working voltage of 250 V dc, and for the 4.5- to 50-pF range, a minimum Q of 100 at 100 MHz.

At high voltages and frequencies, the piston trimmer, with quartz or glass as the

(b)

(a)

Fig. 2.17 Examples of air-variable capacitors: (a) Single-gang. (b) Dual-gang. (*Courtesy E. F. Johnson Company*)

dielectric, is suitable (Fig. 2.21). A 1.5- to 38-pF piston trimmer with a quartz dielectric has a Q greater than 750 at 100 MHz and a working voltage of 1000 V dc.

2.8 CHIP CAPACITORS

Often no larger than a match head, chip capacitors ranging in values from a few pico-farads to 100 μF are available for hybrid microcircuits. Because of their high volu-

Capacitance adjustment screw

Variable end

Turning head

Heavy-copper mounting surface

Sleeve-type bearing

Long-life bellows allows plate movement in vacuum

Shaft provides plate movement

Patented concentric-cylinder reentrant flange construction

Ceramic or glass envelope

Heavy-copper plates

Fixed end

High strength vacuum dielectric

Fig. 2.18 Construction details of a vacuum-variable capacitor. (*Courtesy ITT Jennings*)

Fig. 2.19 A high-Q ceramic tuning capacitor. It is used, for example, in electronic wristwatches. (*Courtesy Johanson Manufacturing Corp.*)

Fig. 2.20 Side-tuned ceramic disk trimmer capacitor. (*Courtesy Johanson Manufacturing Corp.*)

Fig. 2.21 Examples of piston tuning capacitors. (*Courtesy Johanson Manufacturing Corp.*)

Fig. 2.22 Ceramic chip capacitors. (*Courtesy Sprague Electric Company*)

Fig. 2.23 Tantalum chip capacitors. (*Courtesy Sprague Electric Company*)

Fig. 2.24 Construction features of a ceramic chip capacitor. (*Courtesy West-Cap Division, San Fernando Electric Manufacturing Company*)

metric efficiency, *ceramic* (Fig. 2.22) and *tantalum* (Fig. 2.23) chips are commonly used. Porcelain chip capacitors are employed in microwave circuits.

The ceramic chip capacitor is usually constructed as a multilayered sandwich of ceramic and screened-on conductor, illustrated in Fig. 2.24. In the porcelain chip capacitor, porcelain is used instead of ceramic for low loss at microwave frequencies. The tantalum chip capacitor is constructed in essentially the same manner as the solid-anode electrolytic.

A comparison of chip capacitors is provided in Table 2.3. They are all capable of operating at temperatures from −55 to +125°C. Using low-k ceramic, chip capacitors up to approximately 0.047 μF with zero temperature coefficient are available. Although the maximum voltage rating of tantalum and porcelain chip capacitors is in the order of 50 V, this is no serious disadvantage. Transistors in monolithic and hybrid circuits generally operate at voltages that are well below 50 V. In terms of cost, tantalum chip capacitors tend to be the least expensive and porcelain chips the most expensive.

TABLE 2.3 Typical Characteristics of Chip Capacitors

Type	Capacitance range	Tolerance range, %	Maximum working voltage, V	Insulation resistance, MΩ
Ceramic	10 pF − 3.5 μF	±1 to ±20	200	1000
Tantalum	100 pF − 100 μF	±5 to ±20	50	1500
Porcelain	1 − 330 pF	±1 to ±10	50	100 000

2.9 VOLTAGE-VARIABLE CAPACITORS (VARACTORS)

Consider the pn junction diode of Fig. 2.25a. Separating the p and n regions is the depletion layer. When the pn junction is reverse-biased (the p region is negative with respect to the n region), the depletion region becomes devoid of free electrons and holes. One may therefore view the depletion layer of a reverse-biased diode as the dielectric, and the p and n regions as the plates of a capacitor.

As the reverse bias is increased, the depletion layer widens and the capacitance decreases, because the separation of the p and n regions has increased. If the reverse bias is reduced, the depletion layer contracts and the capacitance is increased. We thus have a diode whose capacitance varies inversely with the impressed reverse bias. The greater the magnitude of the reverse bias, the less the capacitance, and vice versa. Such a device is called a *voltage-variable* capacitor, or *varactor*. The symbol for this device is given in Fig. 2.25b.

Curves showing the variation in capacitance with reverse voltage for the varactor are illustrated in Fig. 2.26. Variation in capacitance is nonlinear. Over a range of

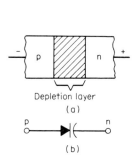

Fig. 2.25 Voltage-variable capacitor (varactor): (a) Reverse-biased p-n junction diode. (b) Symbol for voltage-variable capacitor.

Fig. 2.26 Variations of varactor capacitance with reverse voltage. (*Courtesy Codi Semiconductor*)

reverse voltage from 1 to 60 V, for example, the variation in capacitance for the 1N5139A is approximately 80 to 14 pF. Diode capacitances as high as 2000 pF are possible for low-frequency applications. Microwave varactors may have a maximum capacitance of only 0.4 pF.

Applications for varactors are numerous and include TV tuners, high-power frequency multipliers, and frequency control. They can handle from 100 mW to hundreds of watts of r-f power. Their breakdown voltage ranges from approximately −6 to −300 V. Examples of various case styles used for varactors are illustrated in Fig. 2.27.

(a)   (b)   (c)   (d)   (e)

Fig. 2.27 Examples of case styles for varactors: (*a*) General purpose. (*b*) High capacitance. (*c*) Power. (*d*) UHF power. (*e*) Microwave.

PARAMETERS   Some of the significant parameters for characterizing the varactor will now be defined.

Total diode capacitance (C_t)   The total diode capacitance C_t is the sum of the junction and case capacitances. It is generally specified at a reverse voltage E_R of −4 or −6 V. This parameter is important because it indicates whether the diode can be used at a particular frequency of interest. Variation in C_t with E_R is illustrated in Fig. 2.26.

Series resistance (R_S)   Resistance R_S is the resistance in series with the junction of the diode. It varies with the reverse voltage, decreasing in value as the reverse voltage is increased.

Figure of merit (Q)   The Q of a varactor is equal to

$$Q = \frac{0.159}{fR_S C_t} \tag{2.10}$$

where R_S is in ohms, f in hertz, and C_t in farads. The variation in Q with reverse voltage and frequency is illustrated in Fig. 2.28. As the reverse voltage increases, the values of R_S and C_t decrease, and Q rises. For rising frequencies, the losses in the diode increase and Q falls.

Cutoff frequency (f_{co})   Cutoff frequency f_{co} is the frequency where the Q of the device is unity.

Variation of capacitance with temperature   The variation of C_t with junction temperature T_j is often presented as a normalized plot shown in Fig. 2.29. At a junction temperature corresponding to room temperature (25°C), the normalized value of C_t is one. As the temperature decreases, the capacitance falls and increases with increasing temperature.

C_t ratio   The ratio of capacitance variation at a reverse voltage of −4 or −6 V to the capacitance at approximately 80 percent of the breakdown voltage is referred to as the C_t ratio.

Fig. 2.28 Variation in Q of a varactor with (a) reverse voltage and (b) frequency. (*Courtesy Codi Semiconductor*)

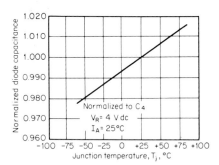

Fig. 2.29 A normalized plot of varactor capacitance as a function of junction temperature. (*Courtesy Codi Semiconductor*)

Conversion efficiency (η) The conversion efficiency defines the performance of a varactor used as a frequency multiplier. This parameter is defined as the ratio of the output power P_o to the input power P_i and is expressed as a percentage:

$$\eta = \frac{P_o}{P_i} \times 100\% \tag{2.11}$$

example 2.5 A tripler using a varactor multiplies an input frequency of 50 MHz to an output frequency of 150 MHz. If the r-f input power is 52.5 W and the output r-f power is 40 W, what is conversion efficiency of the tripler?

solution From Eq. (2.11) $\eta = (40/52.5) \times 100\% = 76\%$.

Chapter **3**

Coils

3.1 INTRODUCTION

Capacitors and resistors are available in a wide variety of sizes and component values. It is possible to purchase almost any value, tolerance, or other rating in these components without special orders.

Coils, on the other hand, may not be as readily available. Technicians may find it necessary to wind their own coils when they want a specific value of inductance. This chapter gives the equations and characteristics of the most frequently used coils. A nomogram is included for simplifying the design of coils for specific applications.

3.2 TYPES OF FERROUS CORES AND THEIR EFFECTS IN COILS

When a core of magnetic material is inserted into an air-core coil, there is a great increase in coil inductance. A magnetic material is any kind of substance that is attracted by a magnet, and may be classified as being either hard or soft. The hard magnetic materials retain their magnetism after the magnetizing source has been removed, but the soft magnetic materials immediately lose magnetism upon removal of the magnetizing source. Saying that a magnetic material is soft does not mean that it is physically soft. A magnetic material may be physically extremely hard and brittle and yet be magnetically very soft. In other words, it very quickly loses all traces of magnetism as soon as the magnetizing source is removed. Soft magnetic materials are used as core material in radio-frequency and in audio-frequency coils and transformers.

IRON-POWDER CORES Iron-powder cores are composed of finely powdered iron or iron alloys mixed with a plastic binder. The cores are formed and pressed into final shapes at pressures ranging from 10 to 50 tons/in², and then baked to set the binder.

Iron-powder cores are available in a large variety of different shapes and sizes and with different magnetic properties to meet different requirements.

FERRITE CORES Ferrites are also magnetic substances. They are unlike iron and steel and their various alloys in that ferrites are insulators. There are many different types of ferrites, and the exact chemical composition of each different type depends on its use; some ferrites that are useful at low frequencies cannot be employed at high frequencies and vice versa.

Table 3.1 gives the magnetic and physical properties of a number of different types of ferrites that are used as core material.

LAMINATED CORES The cores utilized in transformers and chokes that operate at power-line and audio frequencies are made of sheet steel laminations of various grades and thicknesses. These are treated in considerable detail in Chap. 5.

TABLE 3.1 Magnetic and Physical Properties of Ferrites Used as Core Material*

Material	Initial permeability, μ_o	Maximum permeability, μ_{max}	Flux density @ 14, B@ H	Residual magnetism, Br	Coercive force, Hc	Frequency range, mHz	Loss factor, $\frac{1}{\mu_o Q}$	Curie temperature, °C	Volume resistivity, Ω-cm	Specific gravity, g/cm³
AM04	40	160	2200	400	4.0	10–80	@50 mHz 1.8×10^{-4}	500	10^9	4.3
AM12	125	340	2300	1000	1.9	0.2–20	@10 mHz 1.7×10^{-4}	300	10^8	4.4
AM20	200	640	2700	1800	1.5	0.05–8	@2.5 mHz 5×10^{-5}	280	10^8	4.4
AK04	400	1300	3500	1500	0.75	0.01–2	@100 kHz 2.5×10^{-5}	230	10^3	4.6
AK08	800	1500	3400	1500	0.45	0.01–1	@100 kHz 2.0×10^{-5}	200	10^2	4.6
AK16	1600	2800	3900	1000	0.20	1–400 kHz	@100 kHz 8.0×10^{-6}	140	10^2	4.5
AK20	2000	3400	3900	1800	0.16	1–250 kHz	@100 kHz 7.0×10^{-6}	135	10^1	4.6
AK30	3000	4500	4200	1500	0.18	1–250 kHz	@100 kHz 5.0×10^{-6}	200	10^1	4.6

* Typical values as measured on toroidal cores.
Courtesy The Arnold Engineering Company.

Fig. 3.1   Ferrite tuning cores, toroids, special forms, and slugs.
(*Courtesy The Arnold Engineering Company*)

TUNING CORES, SLUGS, AND TOROIDS   When the core of a coil is adjustable within the coil, it is called a tuning core, or tuning slug.   The nonadjustable cores are simply slugs or toroids.   Ferrite tuning cores, toroids, special forms, and slugs of various sizes are shown in Fig. 3.1.

CORE PERMEABILITY   Some idea of the increase in inductance achieved by the use of a core of magnetic material may be learned from a consideration of the permeability of the magnetic material — iron-powder, ferrite, or other.   The permeability of air is always one, but the permeability of an iron-powder or ferrite core may be hundreds and even thousands of times greater than this.   If the permeability of a magnetic core material is 100, its use increases the magnetic flux 100 times; this means that the inductance of the coil will also be increased 100 times.

Computations involving iron-powder and ferrite-core coils usually require technical data from core manufacturers.   This will be discussed in a later paragraph.

3.3   TYPES OF NONFERROUS CORES

Besides cores of magnetic materials, cores of nonmagnetic metals (brass, copper, silver) are also sometimes employed at frequencies of about 50 mHz and above.   Nonmagnetic cores have an effect on coil inductance just the opposite of that produced by a magnetic core; that is, they reduce, instead of increase, the inductance of a coil.   When nonmagnetic cores are used, measures must be taken to reduce r-f core losses as much as possible.   Since high-frequency currents flow mainly near the surface of a conductor, a thin metal sleeve or a silver-plated plastic core is sometimes utilized in the wire center.   Solid nonmagnetic cores may be used when high coil efficiency is not a circuit requirement.

The effect of nonferrous cores on coil inductance is much less than that produced by cores of magnetic material.   Nonferrous tuning cores are useful in equipment that requires a final adjustment of coil inductance before being put into service.

3.4   COMPUTING THE INDUCTANCE OF AIR-CORE COILS

A number of different formulas may be used for computing the inductance of air-core coils.   Although these formulas take different forms, since they do give the same results, they may all be reduced to certain basic forms.   The ones for the coil forms of Fig. 3.2a will give results sufficiently accurate for all practical purposes.   The equations are listed here for convenience.   Typical transmitter-type air-core coils are shown in Fig. 3.2b.

$$L = \frac{r^2 N^2}{9r + 10l} \tag{3.1}$$

$$L = \frac{0.8 r^2 N^2}{6r + 9l + 10d} \tag{3.2}$$

$$L = \frac{r^2 N^2}{8r + 11d} \tag{3.3}$$

If coil dimensions are given in centimeters, instead of in inches, use the following:

$$L = \frac{0.394r^2N^2}{9r + 10l} \tag{3.4}$$

$$L = \frac{0.315r^2N^2}{6r + 9l + 10d} \tag{3.5}$$

$$L = \frac{0.394r^2N^2}{8r + 11d} \tag{3.6}$$

The inductance of single-layer coils may also be determined by means of nomograms and by the use of special slide rules designed for this purpose.

The formulas of Fig. 3.2a all show that inductance is directly proportional to the square of the number of turns. Thus, if the number of turns in a coil is doubled, and

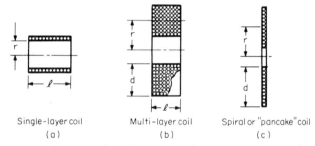

Single-layer coil   Multi-layer coil   Spiral or "pancake" coil
(a)   (b)   (c)

Fig. 3.2   Computing the self-inductance of some typical air-core coils:

$(a)\ L = \dfrac{r^2N^2}{9r + 10l}$ $(b)\ L = \dfrac{0.8r^2N^2}{6r + 9l + 10d}$ $(c)\ L = \dfrac{r^2N^2}{8r + 11d}$

where L = coil inductance, μH
        N = number of turns
        l = length of coil, inches
        d = depth of coil, inches
        r = mean radius of coil, inches
If coil dimensions are given in centimeters instead of inches, use the following:

$(a)\ L = \dfrac{0.394r^2N^2}{9r + 10l}$ $(b)\ L = \dfrac{0.315r^2N^2}{6r + 9l + 10d}$ $(c)\ L = \dfrac{0.394r^2N^2}{8r + 11d}$

coil dimensions are kept unchanged, inductance will be increased four times. This is strictly true only if there is very close coupling between turns, as there is between the turns and between the windings of iron-core transformers and chokes. However, for air-core coils, it is only approximately true.

example 3.1   A single-layer coil like the one shown in Fig. 3.2a is to be wound in a space 2 in. long. How many turns must be wound in the 2-in. space if coil diameter is 1 in., and the required coil inductance is 250 μH?

solution

$$L = \frac{r^2N^2}{9r + 10l} = 250$$

$$= \frac{(0.5)^2N^2}{9(0.5) + 10(2)}$$

Solving for the number of turns,

$$N = \frac{250[9(0.5) + 10(2)]}{(0.5)^2}$$

$$= 156.5 \text{ turns}$$

example 3.2 A multilayer coil, such as that shown in Fig. 3.2*b*, is wound in a slot 1 in. square. Mean radius *r* of the coil is 2.25 in., and $d = 1$. If the coil contains 900 turns, what is the inductance of the coil?

solution

$$L = \frac{0.8r^2N^2}{6r + 9l + 10d}$$

$$= \frac{(0.8)(2.25)^2(900)^2}{6(2.25) + 9(1) + 10(1)}$$

$$= 100\,938 \ \mu\text{H}$$

$$= 100.938 \ \text{mH}$$

3.5 MUTUAL INDUCTANCE

Mutual induction, or the mutual inductive effect between two magnetically coupled coils, is due to the magnetic flux of the first coil acting on the turns of the second coil, as at the same time, the magnetic flux of the second coil acts on the turns of the first coil. To put it another way, there is a mutual interaction between the magnetic fields of the two coils.

Mutual inductance is a measure of the amount of mutual induction that exists between two magnetically coupled coils. Figure 3.3 illustrates the meaning of mutual induction. If the two coils are wound on separate cores and placed some distance from each other, as shown in Fig. 3.3*a*, there will be no magnetic coupling between the mag-

(a) Mutual induction zero

(b) Low mutual induction; low mutual inductance

(c) Maximum mutual induction between L_1, and L_2

Fig. 3.3 Mutual induction is the mutual interaction between the magnetic fluxes of L_1 and L_2. The amount of mutual inductance depends on the coupling between L_1 and L_2, which, in turn, depends on the reluctance of the magnetic circuit between L_1 and L_2 and on the self-inductances of the coils.

netic fields. Hence, the two coils will not have any mutually inductive effect on each other, and their mutual inductance will be zero.

If the two coils are brought nearer to each other, as in Fig. 3.3b, the magnetic field of coil L_1 will start inducing a voltage in coil L_2, and a small current will flow in coil L_2 which will then generate a magnetic field of its own because of the current. However, since the reluctance of the air path between the two coils will still be very high, the magnetic coupling between them will be very low, and they will have only a very low mutual inductance.

If the coils are next brought in close contact with each other, as in Fig. 3.3c, the magnetic coupling between the two coils will be a maximum; thus, mutual induction and mutual inductance between the two coils will also be a maximum.

It may be seen, therefore, that mutual inductance depends on the distance between two magnetically coupled coils. It also depends on the self-inductances of the coils and on the reluctance of the magnetic path between the coils. All these factors have an effect on the amount of coupling between the coils.

If we again consider the very closely coupled coils of Fig. 3.3c, and denote the inductance of coil L_1 by L_1 and that of coil L_2 by L_2, then the combined inductance of the two coils connected in series aiding equals $L_1 + L_2$ plus the mutual inductance due to the magnetic lines of force of L_1 acting on the turns of L_2, plus the mutual inductance due to the magnetic lines of force of L_2 acting on the turns of L_1. Since L_1 and L_2 are identical, their mutual induction effects are equal. Therefore, the total inductance is determined by the equation

$$L = L_1 + L_2 + 2M \tag{3.7}$$

where M = mutual inductance. Since $L_1 = L_2$,

$$L = 2L_1 + 2M = 4L_1 = 4 \text{ H}$$

as determined by an actual measurement of the combined inductance of the two coils. Then, since $L_1 = 1$ H,

$$M = \tfrac{1}{2}(L - 2L_1)$$
$$M = \tfrac{1}{2}(4 - 2 \times 1) = 1 \text{ H}$$

In any other case in which two identical coils are as closely coupled as the two coils of Fig. 3.3c, mutual inductance will always be approximately the same as the inductance of either coil.

It should be noted that even if L_1 and L_2 are not equal to each other, it cannot be said that one of the coils has a mutual inductance that differs from the mutual inductance of the other. In fact, neither coil by itself has any mutual inductance. It is only when the two coils are magnetically coupled to each other that a mutual inductive effect exists between them, and it is never correct to say that one of the coils has a mutual inductance.

COUPLED COILS IN SERIES AND IN PARALLEL   Not all coils will be as closely coupled magnetically as the two coils of Fig. 3.3c. For some applications this would not even be a desirable condition. In any case, it would be impossible to achieve with air-core coils or with any coils in which the coupling medium has a low permeability or high reluctance. Moreover, two coupled coils may have entirely different values of self-inductance. Hence in these cases the combined inductance of two coils connected in series aiding will be less than four times the inductance of one coil, even if the coupled coils are identical. The combined inductance of any two coils connected in series aiding and having self-inductances of L_1 and L_2, respectively, will be given by

$$L = L_1 + L_2 + 2M \tag{3.8}$$

where L = combined inductance, and M = mutual inductance.

But if the coils are connected in series opposing, their combined inductance will be

$$L = L_1 + L_2 - 2M \tag{3.9}$$

And if two coils are connected in parallel with fields aiding,

$$L = \frac{1}{1/(L_1 + M) + 1/(L_2 + M)} \tag{3.10}$$

And for two coils in parallel with fields opposing,

$$L = \frac{1}{1/(L_1 - M) + 1/(L_2 - M)}$$ (3.11)

COMPUTING THE MUTUAL INDUCTANCE OF SINGLE-LAYER AIR-CORE COILS

An important case for consideration is the mutual inductance between coils having air cores. Figure 3.4 shows two important examples, one with no spacing between the coils and the other with a space between them.

CASE 1: If there is no separation between the windings of two coupled coils wound on the same form, as shown in Fig. 3.4a, their combined inductance with the coils connected in series aiding is given by Eq. (3.8), and also by the equation for computing

Fig. 3.4 Typical single-layer air-core coils: (a) L_1 and L_2 on the same form and touching; mutual inductance, $M = \frac{1}{2}(L - L_1 - L_2)$. (b) With L_1 and L_2 on the same form but separated by an air space, computing mutual inductance is more difficult.

Fig. 3.5 L_1 and L_2 are single-layer coils wound on the same form. L_c is an assumed winding in the space between L_1 and L_2. Mutual inductance between L_1 and L_2 is computed from the following formula: $M = \frac{1}{2}(L + L_c - L_{1c} - L_{2c})$.

the inductance of a single-layer coil [Eq. (3.1)]. Since the separate inductances of the two coils may also each be computed by using Eq. (3.1), the only unknown in Eq. (3.8) will be M. This can be determined from the equation

$$M = \frac{1}{2}(L - L_1 - L_2)$$ (3.12)

CASE 2: When the windings of two coupled coils are separated by an air space, as in Fig. 3.4b, imagine that the space between the coils is also wound with wire that has the same size as the windings of L_1 and L_2. The fictitious center winding is called L_c. Also, assume that L_1, L_c, and L_2 are connected in series aiding, as shown in Fig. 3.5, and their combined inductance is found by using Eq. (3.1). Using the same equation, the inductance of L_1 in series with the imaginary center winding L_c is found, then of L_2 in series with L_c, and of L_c separately. L, the combined inductance of L_1, L_c, and L_2, is then given by

$$L = L_{1c} + L_{2c} - L_c + 2M$$

Then

$$2M = L + L_c - L_{1c} - L_{2c}$$

and

$$M = \frac{1}{2}(L + L_c - L_{1c} - L_{2c})$$ (3.13)

example 3.3 Two single-layer coils, L_1 and L_2, are wound side by side on the same form, as in Fig. 3.4a. Coil L_1 has 40 turns and L_2 has 60. The total winding space is 2 in long, and coil radius r is 0.5 in. Compute:
(a) Total inductance with the coils connected in series aiding
(b) Mutual inductance
(c) Total inductance with the coils connected series opposing
(d) Total inductance with L_1 and L_2 connected in parallel and fields *aiding*
(e) Total inductance of L_1 and L_2 in parallel with fields opposing

solution (a) Using Eq. (3.1),

$$L = \frac{(0.5)^2 100^2}{9(0.5) + 10(2)}$$

$$= \frac{0.25(10\ 000)}{4.5 + 20} = 102.04\ \mu H$$

(b) Since L_1 has 40 turns, and the total number of turns is 100 in a winding space 2 in long, the length of L_1 is 0.4 of 2, or 0.8 in. Therefore,

$$L_1 = \frac{0.5^2(40)^2}{9(0.5) + 10(0.8)} = 32\ \mu H$$

and

$$L_2 = \frac{0.5^2(60)^2}{9(0.5) + 10(1.2)} = 54.54\ \mu H$$

Since the series-aiding inductance,

$$L = 102.08 = 32 + 54.54 + 2M$$

it follows that

$$M = \tfrac{1}{2}(102.08 - 32 - 54.54) = 7.77\ \mu H$$

(c) With the coils connected in series opposing,

$$L = 32 + 54.54 - 2M$$
$$= 86.54 - 2(7.77) = 71\ \mu H$$

(d) With the coils connected in parallel and fields *aiding*,

$$L = \frac{1}{1/(L_1 + M) + 1/(L_2 + M)}$$

$$= \frac{1}{1/(32 + 7.77) + 1/(54.54 + 7.77)} \qquad (3.10)$$

$$= 24.28\ \mu H$$

(e) When the coils are connected in parallel with fields *opposing*,

$$L = \frac{1}{1/(L_1 - M) + 1/(L_2 - M)}$$

$$= \frac{1}{1/(32 - 7.77) + 1/(54.54 - 7.77)} \qquad (3.11)$$

$$= 15.96\ \mu H$$

example 3.4 Two single-layer coils, L_1 and L_2, are wound on the same form, as in Fig. 3.4b. The coils are separated by an air space of $\frac{1}{4}$ in, and coil radius, r is 1 in. Coil L_1 contains 24 turns and is 1 in long; L_2 contains 48 turns and is 2 in long. Compute:
(a) Mutual inductance
(b) Total inductance of L_1 and L_2 connected in series aiding
(c) Total inductance of L_1 and L_2 connected in parallel with fields *aiding*.

solution To determine the mutual inductance, the four inductances, L, L_{1c}, L_{2c}, and L_c must first be computed. Since L_1 contains 24 turns and is 1 in long, the $\frac{1}{4}$-in air space would contain six turns; total number of turns in the $3\frac{1}{4}$-in winding space would then be $24 + 6 + 48 = 78$ turns, and for 78 turns,

$$L = \frac{1(78)}{9(1)10(3\frac{1}{4})} = 146.6\ \mu H$$

$$L_{1c} = \frac{1^2(30)^2}{9(1) + 10(1\frac{1}{4})} = 41.86\ \mu H$$

$$L_{2c} = \frac{1^2(54)^2}{9(1) + 10(2\frac{1}{4})} = 92.57\ \mu H$$

$$L_c = \frac{1^2(6)^2}{9(1) + 10(\frac{1}{4})} = 3.13\ \mu H$$

(a) The mutual inductance,

$$M = \tfrac{1}{2}(L + L_c - L_{1c} - L_{2c}) = \tfrac{1}{2}(146.6 + 3.13 - 41.86 - 92.57) = 7.65\ \mu H$$

(b)
$$L_1 = \frac{1^2(24)^2}{9(1) + 10(1)} = 30.32 \ \mu\text{H}$$

$$L_2 = \frac{1^2(48)^2}{9(1) + 10(2)} = 79.45 \ \mu\text{H}$$

$$L = L_1 + L_2 + 2M$$
$$= 30.32 + 79.45 + 2(7.65) = 125.07 \ \mu\text{H}$$

(c)
$$L = \frac{1}{1/(L_1 + M) + 1/(L_2 + M)}$$

$$= \frac{1}{1/(30.32 + 7.65) + 1/(79.45 + 7.65)}$$

$$= 26.44 \ \mu\text{H}$$

3.6 COMPUTING THE VALUE OF INDUCTANCE FROM MEASUREMENTS

In some cases it is very difficult to determine mutual inductance by calculation. Two examples are shown in Fig. 3.6.

Fig. 3.6 (a) Two multilayer coils on the same form. (b) Ferrite core r-f chokes separated by an air gap. Mutual inductance is computed from $M = \frac{1}{4}(L_A - L_o)$, where M = mutual inductance, L_A = total inductance of L_1 and L_2 in series aiding, L_o = total inductance of L_1 and L_2 in series opposing.

The two r-f coils of Fig. 3.6a are wound on the same iron-powder core but separated from each other. Figure 3.6b shows two r-f chokes wound on separate ferrite cores and separated from each other by a small air gap. In both of these, and in many similar instances, it would be extremely difficult to calculate mutual inductance from such known factors as the self-inductance of the coils. It is possible, however, to measure the combined inductance of the coils connected in series aiding and then in series opposing, denoting the series-aiding inductance by L_A and the series-opposing inductance by L_o. Two simultaneous equations are then formed.

$$L_A = L_1 + L_2 + 2M$$
$$L_o = L_1 + L_2 - 2M$$

Subtracting the second equation from the first,

$$L_A - L_o = 4M$$

from which

$$M = \frac{1}{4}(L_A - L_o) \qquad (3.14)$$

where M = mutual inductance
   L_A = combined inductance of L_1 and L_2 connected in series aiding
   L_o = combined inductance of L_1 and L_2 connected in series opposing

example 3.5 Two radio-frequency coils are separated by a small air gap, as shown in Fig. 3.6b. When the coils are connected in series aiding, their combined inductance, as measured on an inductance bridge, is 17 mH, and when the coils are connected in series opposing, combined inductance is 3 mH. What is the mutual inductance?

solution

$$M = \frac{17 - 3}{4} = 3.5 \text{ mH}$$

3.7 THE COEFFICIENT OF COUPLING

The coefficient of coupling, or coupling coefficient, is a measure of the amount of coupling between two coils; it shows whether the coupling is "tight" or "loose." The greatest possible value of the coefficient of coupling is one, in which case the *coupling* may be said to be 100 percent. This is very tight coupling. Coupling coefficient may be computed from the equation

$$k = \frac{M}{\sqrt{L_1 L_2}} \tag{3.15}$$

where k = coefficient of coupling (There are no units for k)
    M = mutual inductance in the same units as L_1 and L_2
    L_1, L_2 = self-inductances of the separate coils
A coupling coefficient of 1, or 100 percent coupling, is approached only in well-designed iron-core transformers and chokes. In air-core transformers and coils it is very much less, but increases in transformers with iron-powder and ferrite cores.

> **example 3.6** When the primary of an r-f antenna coil, such as that shown in Fig. 3.6a, is connected to the secondary in series aiding, the measured inductance is 3085 μH; when the primary and secondary are in series opposing, the measured inductance is 1405 μH. If primary and secondary self-inductances are 2000 and 245 μH, respectively, what is the coefficient of coupling between the primary and secondary?
>
> **solution**
>
> $$M = \frac{1}{4}(L_A - L_o)$$
> $$= \frac{1}{4}(3085 - 1405) = 420 \ \mu\text{H}$$
>
> and the coefficient of coupling,
>
> $$k = \frac{M}{\sqrt{L_1 L_2}} = \frac{420}{\sqrt{2000(245)}} = 0.6$$

3.8 COIL AND CONDUCTOR RESISTANCE AT RADIO FREQUENCIES

The resistance that a conductor offers to the flow of high-frequency alternating current through it is greater than the resistance of the same conductor to direct current. This is due to a characteristic of coils and conductors that is known as the skin effect. It is a characteristic of conductors that carry alternating current, but since it is a magnetic effect, it is much more pronounced in coils than in straight conductors.

Skin effect causes the current in a conductor to be much denser near the surface of the conductor than at its center, and is a result of the magnetic lines of force associated with any current-carrying conductor. These magnetic lines of force, or flux lines, not only surround the conductor but also exist in the interior of the conductor and are densest at its center. Consequently, as the instantaneous value of the alternating current changes, the inductance of a conductor in its interior is much greater than the inductance near its surface. The result is that more current flows near the surface, or "skin," of the conductor than in its interior. Since inductive effects increase with increasing frequency, skin effect also increases as frequency is increased.

EDDY CURRENTS IN CONDUCTORS Besides the skin effect, every conductor that carries alternating current also has eddy currents induced in it by its own changing magnetic field—just as a changing magnetic field induces eddy currents in the core material of a transformer or choke. These eddy currents constitute an additional loss which causes the effective ac or r-f resistance of a conductor (or coil) to be increased.

At very high frequencies, because of the skin effect and induced eddy currents in a conductor, a hollow, thin-walled tube is really a very much better conductor than a solid conductor of the same diameter. Some short-wave (high-frequency) coils are actually made with copper tubing, instead of with solid wire.

Since the resistance of a coil at ac frequencies depends on the frequency, in all ac problems this resistance must be considered at the operating frequency of the coil.

DIELECTRIC LOSSES Another source of loss in coils at high frequencies is that which occurs in the forms on which the coils are wound and in the insulating covering of the wire. Since both of these are dielectric materials, any loss of this sort is called a dielectric loss. Even though a coil form may be made of the highest quality dielectric material, such as molded mica, steatite, resinite, or polystyrene, there is no perfect insulator, and dielectric losses cannot be entirely avoided at very high frequencies.

Dielectric loss, like skin effect and the eddy current loss in a conductor, also tends to increase the effective resistance of a coil at high frequencies.

DISTRIBUTED CAPACITANCE   Since every turn of a coil is separated from its adjacent turns by an insulating material—which may be air in certain cases—and a difference of potential always exists between turns when the coil is conducting current, there is always a capacitive effect between the turns. This is shown in Fig. 3.7. This

Fig. 3.7   Capacitive effects between the turns of a coil (distributed capacitance).

capacitive effect is known as the distributed capacitance of the coil, because it is distributed throughout the entire length of the coil.

Distributed capacitance causes a decrease in the effective inductance of a coil and an increase in its effective resistance. Hence, it is another undesirable effect, and measures are usually taken to decrease it as much as possible.

3.9   METHODS USED TO DECREASE SKIN EFFECT, EDDY CURRENT LOSS, DIELECTRIC LOSS, AND DISTRIBUTED CAPACITANCE IN COILS

LITZENDRAHT, OR LITZ, WIRE   One of the earliest methods for reducing skin effect in conductors and coils at high frequencies was the use of litzendraht, commonly known as litz wire. Litz is a braided conductor made up of a large number of very thin strands of enamel insulated wire. The individual strands are all connected in parallel at the start and finish of a conductor or coil, but because of the enamel insulation, there is no connection between the individual strands of the conductor at any other points along the conductor. The braided conductor is also twisted at regular intervals so that a particular insulated strand will at some points be at the exact center of the conductor, and at the surface of the conductor at other points.

One of the reasons why skin effect is reduced with litz wire is that skin effect is not as pronounced in thin wires as it is in thick ones. Another reason is the twisting of the conductor, which tends to keep the inductance of a litz conductor at its center the same as it is at its surface.

A very large number of individual insulated strands in a litz conductor are most advantageous at frequencies below 600 kHz; but for frequencies above this, more than 25 strands are unfavorable. According to some authorities, there is a certain critical frequency above which a solid conductor of the same diameter again has less high-frequency resistance than litz wire.

FLAT COPPER STRIP   A flat copper strip conductor, instead of a round one, is also sometimes used to reduce skin-effect losses. Although skin effect still exists in a flat strip conductor, it is much less than in a round conductor of the same cross-sectional area.

Inductors wound with flat copper strip are used in many high-power radio transmitters; the strip is wound edgewise, and the turns are widely spaced so that the distributed capacitance between turns is not excessive.

TUBULAR CONDUCTORS   A thin-walled hollow tube actually offers less resistance to high-frequency currents than a solid conductor of the same diameter. Tubular conductors may be used most advantageously in coils designed for high-frequency operation. An added advantage is that coils wound with tubular conductors generally require no supporting forms, because the tubing is rigid enough to make such coils self-supporting.

SPECIAL WINDING METHODS USED TO REDUCE DISTRIBUTED CAPACITANCE
Figure 3.8 shows a number of different ways of winding coils so as to reduce the distributed capacitance. In any ordinary multilayer coil, such as that shown in Fig. 3.8*a*, since the first and last turns are adjacent, and the second and next to the last turn, etc.,

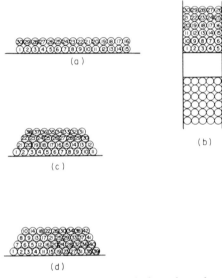

Fig. 3.8   Special winding methods used to reduce distributed capacitance: (*a*) Ordinary multilayer coil has high distributed capacitance. (*b*) Distributed capacitance may be reduced by winding the coil in a narrow slot. (*c*) and (*d*) Bank windings used to reduce distributed capacitance. Note difference between the windings of (*c*) and (*d*).

Fig. 3.9   The "universal" or "honeycomb" winding, also known as a pie winding.

are all adjacent, the distributed capacitance tends to be high. If the same number of turns is wound in a narrow slot, as in Fig. 3.8*b*, the first and last turns will not be adjacent, and the coil will have a lower distributed capacitance. Other methods of winding coils to reduce capacitance are the bank windings shown in Fig. 3.8*c* and *d*, and the universal or honeycomb winding of Fig. 3.9.

THE USE OF POWDERED IRON AND FERRITE CORES   Since all the undesirable effects, skin effect, eddy currents, dielectric losses, and distributed capacitance in a coil are proportional to the amount, or length, of wire in a coil, all these unwanted effects may be greatly reduced by the use of powdered iron and ferrite cores. The permeability of powdered iron and ferrite-core materials is very much greater than the permeability of air. Thus, coils that are wound over such cores have a much higher inductance than air-core coils of the same diameter and the same number of turns; accordingly, much less wire may be used in a powdered iron or ferrite-core coil than in an air-core coil which has the same inductance, and all the aforementioned undesirable effects and losses in a coil may be reduced to a minimum.

At frequencies above those where core losses with powdered iron outweigh its advantages, ferrite cores may be most advantageously substituted for the powdered iron ones.

3.10   THE Q OR FIGURE OF MERIT OF A COIL

If a curve is plotted showing how the current through a series-tuned coil varies as the coil is tuned through resonance while the impressed voltage is kept constant, it will take a form similar to the curves of Fig. 3.10.   If the ratio of coil reactance to effective coil resistance is high, the curve will also be high and steep, like curve a.   However, if the coil has more losses, such as core, copper, dielectric, skin effect, its effective re-

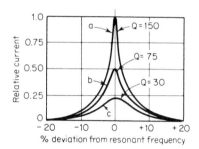

Fig. 3.10   Selectivity curves of tuned circuits which show how selectivity depends on coil Q.

sistance will be higher, and the ratio of coil reactance will be lower.   The current at resonance will also be lower because of the higher effective resistance, and the curve will then take a form similar to curve b, which is less steep than curve a.   If a coil has still more losses, and consequently an even lower reactance to resistance ratio, the resonance curve will take form c.

Since increased losses in a coil will always yield a lower reactance to resistance ratio, and the efficiency of a coil—just as the efficiency of any device—decreases as losses increase, a coil with a high reactance to resistance ratio will always be more efficient than one with a low reactance to resistance ratio.   Hence, the reactance to resistance ratio of a coil may be used as a figure of the merit, or Q of a coil.

In the form of an equation;

$$Q = \frac{X}{R} \qquad (3.16)$$

where X = reactance of the coil or capacitor at resonance

R = effective resistance of the coil

Figure 3.10 shows how the selectivity of a tuned circuit, or the sharpness of resonance, depends on Q.

example 3.7   In a series-tuned circuit the current at resonance with an impressed voltage of one volt is 50 mA.   Circuit capacitance is 160 pF, and inductance is 250 μH.   What is the Q of this circuit?

solution   Effective resistance,

$$R = \frac{E}{I} = \frac{1}{0.05} = 20 \ \Omega$$

Resonant frequency,

$$f_r = \frac{1}{2\pi \sqrt{LC}}$$

$$= \frac{1}{2\pi \sqrt{(250 \times 10^{-6} \times 160 \times 10^{-12})}}$$

$$= \frac{10^8}{2\pi \sqrt{16 \times 25}} = \frac{10^8}{2\pi(20)} = \frac{10^7}{2\pi(2)}$$

At resonance,

$$X_L = 2\pi f_r L$$

$$= 2\pi \left(\frac{10^7}{4\pi}\right)(250 \times 10^{-6}) = 1250 \ \Omega$$

$$Q = \frac{X}{R} = \frac{1250}{20} = 62.5$$

The letter Q is sometimes the Quality factor, which implies that the quality of a tuned circuit is dependent only upon this value. However, there are many applications where a high-Q coil or circuit is very undesirable. For example, if a circuit is to pass a band of frequencies, then the coil with curve c of Fig. 3.10 would be better than the coil with curve a. Even though the current is smaller at resonance, the falloff in response at the skirts of curve c is not as rapid.

In tuned circuits where a lower Q (that is, a broader frequency response) is desired, the coil of the tuned circuit may be placed in parallel with a resistor. This resistor, which is sometimes called a *swamping resistor*, is used to deliberately introduce resistive losses into the tuned circuit and lower its Q.

3.11 TRUE INDUCTANCE AND APPARENT INDUCTANCE

When inductance measurements are made on a coil, and the test frequency is at or near the self-resonant frequency of the coil, the results may be misleading by very wide margins. The self-resonant frequency of a coil is the frequency at which the inductance of the coil and its distributed capacitance are in resonance. Since coil impedance rapidly increases as the self-resonant frequency of a coil is approached, and since impedance also increases as inductance increases, the result, when measurements are made at or near the self-resonant frequency of a coil will provide an inaccurate inductance reading.

Any measured inductance at a frequency at or near the self-resonant frequency of a coil is called the *apparent inductance*, to distinguish it from *true inductance*, which should always be measured at a frequency far below the self-resonant frequency of a coil.

A formula for determining true inductance from an apparent measured inductance is the following:

$$L_{\text{appar}} = \frac{L_{\text{true}}}{1 - (f_{\text{test}}/f_o)^2} \tag{3.17}$$

where L_{appar} = apparent inductance of the coil, μH
   L_{true} = true inductance of the coil, μH
   f_{test} = frequency at which the apparent inductance measurement is made, MHz
   f_o = self-resonant frequency of the coil, MHz

Equation (3.17) shows that at test frequencies above the self-resonant frequency of a coil the computed (not measured) apparent inductance would be negative. Since there is no such thing as a negative inductance, a negative result merely shows that the reactance of the coil has become negative; in other words, above its self-resonant frequency, the coil acts as a capacitor, and the current leads the impressed voltage.

example 3.8   The self-resonant frequency of a 100-μH choke is 5 MHz. What would be the apparent inductance of this choke if tested on an inductance bridge at a test frequency of 4.5 MHz?

solution   Using (3.17),

$$L_{\text{appar}} = \frac{100}{1 - (4.5/5)^2}$$

$$= \frac{100}{1 - 0.81} = \frac{100}{0.19} = 526 \ \mu\text{H}$$

3.12 DETERMINING THE SELF-RESONANT FREQUENCY OF AN INDUCTOR

Figure 3.11 shows the test setup for measuring the self-resonant frequency of an inductor. A resistor is placed in series with the coil and its distributed capacitance C_d.

The combination is placed across a variable-frequency signal generator. At resonance, the parallel-tuned circuit will have maximum impedance. Thus, the current through R will be minimum, and the voltage across R will also be minimum. This is indicated by a dip in the VTVM reading. The response curve in Fig. 3.11 shows the voltage across R as the signal generator is tuned through resonance.

For low-Q coils the exact resonance point may be hard to determine, because the impedance curve of a low-Q coil is rather broad and not very steep.

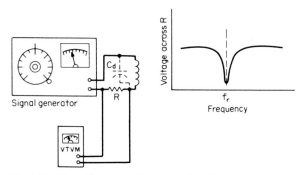

Fig. 3.11   The self-resonant frequency of a coil. Any coil, in conjunction with its distributed capacitance, forms a parallel-resonant circuit.

It is also necessary to watch for resonance points at frequencies that are harmonics of the fundamental. All these will cause some dip in the voltmeter reading across the resistor, but the dip caused by the fundamental frequency will be very much greater than any dip resulting from a harmonic of the fundamental. Harmonics are always either an even or an odd number times the frequency of the fundamental.

If the coil being measured has a core that changes permeability considerably with changes in applied voltage, the signal generator output voltage should be adjusted for a voltage across the coil that will be the same as the voltage across the coil under actual operating conditions.

3.13 DETERMINING THE DISTRIBUTED CAPACITANCE OF A COIL BY MEASUREMENT

For this measurement a grid-dip oscillator and two precise, calibrated capacitors are required—or a standard variable calibrated capacitor may be used. The capacitors should be high-quality low-loss units such as silvered mica or polystyrene types. For coils in the standard broadcast range, capacitors having values of 100 and 150 pF will be suitable.

One of the capacitors, C_1, is connected across the coil, which is loosely coupled to the pickup loop of the grid-dip oscillator, as shown in Fig. 3.12. The oscillator frequency is adjusted for resonance, which will be indicated by the grid-dip oscillator meter, and the oscillator frequency is recorded as f_1. The second capacitor, C_2, is then substituted

Fig. 3.12   Equipment and circuit used to determine the distributed capacitance of a coil.

for C_1, and the coil is again resonated at the new frequency, f_2. The distributed capacitance of the coil is then computed from the equation

$$C_d = \frac{(f_1/f_2)^2 \, C_1 - C_2}{1 - (f_1/f_2)^2}$$
(3.18)

where C_d = distributed capacitance of coil, pF
f_1, f_2 = two resonant frequencies, kHz
C_1, C_2 = two capacitor values, pF

If, instead of the fixed capacitors, a calibrated variable capacitor is used, it is only necessary to resonate the coil at two different settings of the variable capacitor; substituting the observed values of capacitance and oscillator frequencies in Eq. (3.18) will then give the distributed capacitance.

example 3.9   The resonant frequency of a coil when shunted with a 100-pF capacitor is 1200 kHz, but if the capacitor is changed to 155 pF, the resonant frequency becomes 996 kHz. What is the distributed capacitance of the coil?

solution   Using Eq. (3.18),

$$C_d = \frac{(1200/996)^2 \, 100 - 155}{1 - (1200/996)^2} = 21.79 \, \text{pF}$$

3.14 MEASURING COIL INDUCTANCE

Figure 3.13 shows how a grid-dip meter can be utilized for measuring inductance. A high-quality (high-Q) capacitor, silvered mica or polystyrene, is connected across

Fig. 3.13 Method used to measure coil inductance.

the coil whose inductance is to be measured. For coils that are to be used at standard broadcast frequencies—500 to 1600 kHz—any capacitance value from about 100 to 200 pF will be suitable. For coils to be used at higher frequencies capacitance values should be reduced in proportion, and should be increased for coils to be used at frequencies below 500 kHz. For approximate inductance measurements, capacitors having a tolerance rating of ±5 percent will do, but for more accurate closer tolerance, calibrated capacitors should be used.

The coil is loosely coupled to the pickup loop of the grid-dip oscillator, as shown in Fig. 3.13. Only the minimum amount of coupling that will give an indication on the grid-dip meter should be used. The oscillator frequency is adjusted for resonance which will be indicated on the grid-dip meter, and the coil inductance is computed from

$$L = \frac{25 \ 330}{C f^2}$$
(3.19)

where L = inductance of coil, μH
C = tuning capacitance, pF
f = frequency, MHz

Since the distributed capacitance of the coil is ignored in the above method, the inductance, as given by Eq. (3.19), will be slightly higher than the true inductance of the coil.

For more precise results, the capacitance of the capacitor used to resonate the coil for the inductance test should also be more precisely known; a calibrated capacitor should be used, and the distributed capacitance of the coil should be determined.

The distributed capacitance of the coil is added to the capacitance of the calibrated capacitor, and their sum is then substituted for C in Eq. (3.19).

MEASURING HIGH INDUCTANCE R-F CHOKES   For high-inductance r-f chokes out of the frequency range of grid-dip oscillators, the circuit of Fig. 3.14 may be employed.   For some tests it may be necessary to use an audio frequency instead of an r-f

Fig. 3.14 Method used to measure the inductance of chokes having high inductive values—beyond the measuring range of grid-dip oscillators.

generator.   Shunting capacitor values should have reactance values in proportion to the inductive reactance values.

Resonance will be indicated by a very pronounced dip in the voltage across the series resistor R.   The inductance may then be computed by using Eq. (3.19) and solving for L.

example 3.10   When a capacitor that has a capacitance of 198 pF is shunted across a coil, the resonant frequency of the combination is 800 kHz.   The distributed capacitance of the coil is 14 pF.   Compute:

(a) The approximate inductance of the coil by neglecting the distributed capacitance of the coil

(b) The inductance more precisely by taking the distributed capacitance into consideration

solution   (a) Using Eq. (3.19), and without taking distributed capacitance into consideration:

$$L = \frac{25\ 330}{198 \times 0.8^2} = 199.89\ \mu\text{H}$$

(b) Taking the distributed capacitance into consideration:

$$L = \frac{25\ 330}{(198 + 14) \times 0.8^2} = 186.68\ \mu\text{H}$$

3.15   DETERMINING THE Q OF A COIL

A circuit for determining coil Q is shown in Fig. 3.15.   The capacitance of capacitor C need not be precisely known, but it should have a value sufficient to resonate the coil, and it should be of the best quality low-loss type—silvered mica or polystyrene.   The voltmeters should have high input impedance.   If low-impedance meters are employed, they should be used with high impedance probes.   The input capacitance of V_2 should be very much lower than the capacitance C.

When resonance occurs, the voltage indicated by V_1 will dip to a minimum, and V_2 will peak.   The Q of the coil at resonance is given by

At resonance

$$Q = \frac{V_2}{V_1} \qquad\qquad (3.20)$$

where V_1 = voltage across capacitor and coil in series
$\quad\quad V_2$ = voltage across capacitor

example 3.11   If the voltmeter across the capacitor of Fig. 3.15 indicates 144 V while V_1 indicates 0.6 V at the resonant frequency of the coil and capacitor, what is the Q of the coil?

solution   Using Eq. (3.20), the Q at resonance is equal to

$$\frac{144}{0.6} = 240$$

Fig. 3.15   A very convenient method used to measure the Q of a coil. At the resonant frequency, Q equals V_2/V_1.

example 3.12   If the inductance of the coil of Example 3.11 is 300 μH, what is the effective resistance of the coil if the resonant frequency is 750 kHz?

solution   The inductive reactance of the coil is obtained by the equation $X_L = 2\pi f L$. At 750 kHz,

$$2\pi f L = 6.2832 \times 750 \times 10^3 \times 300 \times 10^{-6}$$
$$= 1413.72 \ \Omega$$

Then

$$Q = 240 = \frac{X}{R} = \frac{1413.72}{R}$$

Solving for R, the result is

$$R = \frac{1413.72}{240} = 5.89 \ \Omega \qquad\qquad Answer$$

3.16   COILS WITH POWDERED IRON OR FERRITE CORES

Unlike air, which always has a permeability of 1, the permeability of iron-powder and ferrite-core materials may have values ranging from just a few times to many thousands of times the permeability of air. Even though manufacturers do produce iron-powder and ferrite cores in certain standard permeabilities and sizes, there are so many of these that it would be extremely difficult to develop formulas for computing coil characteristics that would be applicable to all the different types of iron-powder and ferrite cores. Even if such were available, they would be extremely voluminous – and difficult to apply, except to single-layer coils, such as those used in the tuning circuits of radio receivers.

In any case, when computations involving iron-powder or ferrite-core coils are to be made, it is best to obtain the required computational data directly from the manufacturer of the particular core material that is to be used.

Iron-powder and ferrite-core materials are constantly being improved, and some of the core material produced only a few years ago is already obsolete. Newest types of core materials are more stable against environmental changes – changes of coil inductance with changes in temperature, induction, direct current and frequency.

3.17   COMPUTING IRON-POWDER AND FERRITE-CORE COILS

Typical of the technical data available from manufacturers are those given on pages 3–19 to 3–22 taken from a manufacturer's° technical bulletin which contains more than

° The Arnold Engineering Company.

PHYSICAL SPECIFICATIONS

CORE DIMENSIONS	INCHES			MILLIMETERS		
	O.D.	I.D.	HT.	O.D.	I.D.	HT.
NOMINAL	2.000	1.250	0.530	50.8	32.0	13.46
AFTER FINISH	2.035 MAX.	1.218 MIN.	0.565 MAX.	51.7 MAX.	30.9 MIN.	14.35 MAX.

MEAN LENGTH OF MAGNETIC PATH		CROSS SECTIONAL AREA OF MAGNETIC PATH A		APPROXIMATE WEIGHT OF FINISHED CORE		MINIMUM WINDOW AREA		
in.	cm.	in.^{2}	cm.^{2}	LBS.	GRAMS	CIRCULAR MILS	in.^{2}	cm.^{2}
5.13	13.03	0.1922	1.24	0.298	135	1,480,000	1.165	7.52

	CORE HEIGHT	in.	cm.
MEAN LENGTH OF TURN FOR FULL WINDING	0.550	2.01	5.11

ELECTRICAL SPECIFICATIONS

	CORE SPECIFICATIONS			WINDING DATA FOR FULL-WOUND CORE		
PART NUMBER	NOM. PERM. μ	INDUCTANCE mh FOR 1000 TURNS ±8%	NOMINAL DC RESISTANCE OHMS PER mh	AWG WIRE SIZE	TURNS	DC RESISTANCE OHMS
A-424426-2	350	426	0.0060	12	113	0.0324
A-404365-2	300	365	0.0070	13	141	0.0505
A-382304-2	250	304	0.0084	14	177	0.0788
A-217249-2	205	249	0.0103	15	221	0.1227
D-217249-4	205	249	0.0103	16	276	0.1922
W-217249-4	205	249	0.0103	17	344	0.298
A-181210-2	173	210	0.0122	18	430	0.467
D-181210-4	173	210	0.0122	19	537	0.728
W-181210-4	173	210	0.0122	20	668	1.133
A-327195-2	160	195	0.0131	21	832	1.768
D-327195-4	160	195	0.0131	22	1042	2.79
W-327195-4	160	195	0.0131	23	1289	4.31
A-154179-2	147	179	0.0142	24	1608	6.76
D-154179-4	147	179	0.0142	25	1999	10.55
W-154179-4	147	179	0.0142	26	2495	16.62
A-715152-2	125	152	0.0169	27	3084	25.7
D-163152-4	125	152	0.0169	28	3855	40.6
W-715152-4	125	152	0.0169	29	4730	61.8
A-106073-2	60	73	0.035	30	5941	98.8
D-164073-4	60	73	0.035	31	7391	154.8
W-106073-4	60	73	0.035	32	9048	234.
A-348032-2	26	32	0.080	33	11330	371.
A-349017-2	14	17	0.151			

(Reproduced by permission from the catalogue of The Arnold Engineering Company)

Frequency–kiloHertz

(Reproduced by permission of The Arnold Engineering Company)

Formulae and Mathematics of MAGNETICS

PERMEABILITY

Permeability of a test core can be calculated by either formula below (inches or millimeters). Inductance is measured on a sensitive bridge with a uniformly distributed winding on the core.

H = core height (in. or mm.)
r = corner radius (in. or mm.)
H_e = effective height before finish, corrected for corner radii (in. or mm.)
L = measured inductance (henries)
OD = mean outside diameter before finish (in. or mm.)
ID = mean inside diameter before finish (in. or mm.)
N = number of turns in test winding.

INCHES

$$\mu = \frac{L \times 10^9}{11.7 \left[\log_{10} \frac{OD}{ID} \right] H_e N^2}$$

$$H_e = H - \frac{1.717\ r^2}{OD - ID}$$

MILLIMETERS

$$\mu = \frac{L \times 10^9}{.4606 \left[\log_{10} \frac{OD}{ID} \right] H_e N^2}$$

$$H_e = H - \frac{.0675\ r^2}{OD - ID}$$

TURNS VS INDUCTANCE

The number of turns (N) necessary to obtain a specific inductance (L) can be calculated by:

$$N = 1000 \sqrt{\frac{L}{L_{1000}}}$$

L = desired inductance (millihenries)
L_{1000} = core inductance (millihenries for 1,000 turns)

"Q" FORMULA

The Q formula calculates the ratio of reactance to effective resistance for an inductor and thus indicates its quality. For electrical wave filters, an increase in Q provides sharper cut-off, higher attenuation ratios, and better defined resonance. Q is affected by the distributed capacitance of an inductor's winding.

Neglecting the effects of self-resonance caused by the distributed capacitance, Q can be calculated, when designing inductors, by this formula:*

$$Q = \frac{\omega L}{R_{dc} + R_{ac} + R_{cd}}$$

Q = quality factor
L = inductance (henries)
ω = $2\pi \times$ frequency (frequency in hertz)
R_{dc} = dc winding resistance (ohms)
R_{ac} = resistance due to core losses (ohms)
R_{cd} = resistance due to dielectric losses in winding (ohms)

DC WINDING RESISTANCE

DC winding resistance for an average winding can be calculated by:

$$R_{dc} = \frac{\ell_w N r}{12000}$$

ℓ_w = mean length of turn (in.)
N = number of turns
r = resistance of wire in ohms per 1000 feet

AC CORE LOSS RESISTANCE

AC core loss resistance is calculated from:

$$R_{ac} = \mu Lf \, (a \, B_{max} + c + ef),$$

which is Legg's equation.

DIELECTRIC LOSS RESISTANCE

Dielectric loss resistance is significant at higher frequencies and can be calculated from the equation found in Terman's Handbook:[**]

$$R_{cd} = d\omega^3 \, L^2 C_d$$

d = power factor of distributed capacitance

$\omega = 2\pi \times$ frequency (frequency in hertz)

L = inductance (henries)

C_d = distributed capacitance.

NOTES ON "Q" CURVES

The Q curves published in this manual have been developed through the use of digital computer techniques. They have been checked with various core sizes and inductances to assure reasonable correspondence to the real world of wire, insulation, and winding. The user's ability to get equivalent results depends in part upon his ability to duplicate the assumed conditions. These are:

1. A "full-wound core" is defined to be one in which the minimum winding ID or residual hole left after winding is one-half of the inside diameter of the core.

2. This leaves a useful winding area which is three-quarters of the available window area. It was assumed that 70% of this space would be filled with copper wire including heavy synthetic film insulation.

3. The dc resistance of a full wound core varies as the square of the number of turns in the same manner that the resultant inductance varies as the square of the turns. Therefore, each core size has a table of calculated ohms per millihenry based on the "full-wound core" definition above. This resistance determines the positive slope of the low frequency portion of the Q curve and is assumed to be independent of inductance.

4. Three factors affect the high frequency performance of an inductor.

  a. The most fundamental is the eddy current loss of the core material which is mostly responsible for the negative slope of the low inductance curves at frequencies above the frequency of maximum Q. This is calculated from Legg's equation.

  b. The second factor is caused by dielectric loss and is calculated as R_{cd} in the formula above.

  c. The most dramatic factor is the ef-

fect of self resonance of the distributed capacitance and the inductance. For small inductances, such as the 0.001 henry or the 0.01 henry curve for each core, the self resonant frequency f_o is well above the normal useful frequency range of the component. Therefore, these curves tend to indicate the component performance with a negligible effect of self resonance. The distributed capacitance and the self inductance determine a self resonant frequency according to:

$$f_o = \frac{1}{2\pi \sqrt{LC_d}} \text{ hertz.}$$

At some lower frequency, f, the value Q_f can be calculated from:

$$Q_f = Q \left[1 - \left(\frac{f}{f_o}\right)^2 \right]$$

where Q is calculated from determined values of loss resistances as indicated above, and Q_f is the apparent Q taking into account the effect of the distributed capacitance. It should be noted that when f is 20% of f_o, Q_f is 96% of its original value. However, when f is 70% of f_o, Q_f drops to 51% of its original value. The apparent value of the inductance, L_a, is also affected as follows:

$$L_a = \frac{L}{1 - \left(\frac{f}{f_o}\right)^2}$$

5. Because the distributed capacitance is determined by the winding method, the user can obtain different results from those plotted, depending on this value of the capacitance.

[*] This analysis follows Herman Blinchikoff, "Toroidal Inductor Design," *Electro-Technology*, November, 1964

[**] *Radio Engineer's Handbook*, F. E. Terman, McGraw-Hill, Inc. New York (1943), p. 84

70 pages of data pertaining to the toroidal, MPP iron-power cores manufactured by this company.

MPP is a special magnetic alloy composed of approximately 2 percent molybdenum, 81 percent nickel, and 17 percent iron. After being pulverized, the resulting powder is pressed into cores of various sizes. The cores are available in standard permeabilities ranging from 14 to 350, and in toroidal core sizes having an outer diameter of less than 0.2 to greater than 5 in. The data on page 3–19 are for a core having an outer diameter of 2 in. These data show at once how difficult it might be to compute the characteristics of iron-powder and ferrite-core coils without such data, and how easy the computations are when such data are available.

Similar data are available for other types of iron-powder and ferrite cores.

The Q curves show that beyond a certain frequency, which is different for cores having different permeabilities, the Q rapidly decreases, owing to the increased core and other losses in a coil. Below the frequency at which Q is a maximum, although core losses are not excessive, the ratio of coil reactance to resistance is lower, and the result is a lower Q.

example 3.13 Use the manufacturer's data on page 3–19 and determine the number of turns required to produce an inductance of 100 mH in an MPP OD 2,000 toroidal core that has a nominal permeability of 147.

solution From the turns versus inductance formula on page 3–21 and the data for a 147-μ core on page 3–19,

$$N = 1000 \sqrt{\frac{100}{179}} = 748 \text{ turns}$$

example 3.14 What size wire should be used to wind the coil of Example 3.13 for highest Q?

solution The dc resistance of the winding should be kept as low as possible. From the winding data for a full-wound core, we see that if we used number 20 wire, a full-wound core would contain only 668 turns. Therefore, we must use the next smaller size—higher gage number—wire, number 21, to get the 748 turns into the available winding space.

3.18 CHANGE OF COIL INDUCTANCE WITH DIRECT CURRENT

Anyone familiar with TV receivers and circuits may have noted that most of the coils used in TV receivers have adjustable tuning slugs. One reason for this is that if a coil's inductance is above or below its rated or computed value, a slight adjustment of the tuning slug quickly brings the inductance to the required value; but another reason is that the permeability, and hence the inductance, of an iron-powder or ferrite-core coil may change appreciably with changes of direct current in the coil. Since many of these coils are used in the plate circuits of vacuum tubes, and plate currents for the same type of tubes may differ widely from tube to tube, a slug-tuned coil allows immediate readjustment of a coil's inductance to correspond to the dc plate current in the coil winding.

This same sort of inductance change with current is also one of the reasons why a coil in a TV circuit may have to be readjusted when a tube is changed.

3.19 TESTING AND MEASURING IRON-POWDER AND FERRITE-CORE COILS

Since the inductance of some iron-powder and ferrite-core coils is likely to change considerably with changes in coil current, either ac or dc, it is best to make inductance measurements under actual operating conditions for the coil. Special measuring equipment is now available which will measure iron-core and ferrite-core inductors under almost any combination of ac and dc voltages. In the absence of such equipment, a combination of other equipment may be set up which will do the job just as well—but not as conveniently.

When coils are merely to be tested, rather than measured, it is usually not necessary to test them under actual operating conditions. For instance, a test instrument may be designed merely to indicate whether or not the inductance of a coil falls within certain limits. In this case, it is not necessary to know the actual inductance of the coil being tested.

3.20   MEASURING METHOD FOR COILS CARRYING BOTH R-F and DC

A convenient method that may be used to measure both the Q and the inductance of a coil that carries direct current combined with a certain amount of r-f is shown in Fig. 3.16.   The inductance of the choke coil L_2 should be very much greater than the inductance of L_1, which is to be measured.   A convenient value for L_2 when L_1 is about 100 mH would be 10 H.   This would be 100 times the inductance of L_1, but the greater

Test oscillator

V_1 L_1 L_2

C V_2 A

Fig. 3.16   Measuring the inductance of a coil that carries both r-f and direct currents.   This method should be used if coil inductance is likely to change considerably when carrying current.

the inductance of L_2, the less will it load the ac impedance of L_1.   The current through the inductors is adjusted to the level at which the inductance of L_1 is to be measured. The rheostat R, in series with the battery and coils, is used to adjust the current.

At resonance $Q = E_2/E_1$, in which E_1 is the voltage across the coil and capacitor in series (the generator output voltage), and E_2 is the voltage across the capacitor.

The inductance,

$$L = \frac{1}{(2\pi f_r)^2 C} \tag{3.21}$$

where L = coil inductance, H
f_r = frequency at resonance, Hz
C = capacitance, F

example 3.15   What will be the inductance of the coil L_1 in the test circuit of Fig. 3.16 if a 0.05-μf capacitor is needed to resonate the coil at a frequency of 2 kHz?

solution   Using Eq. (3.21),

$$L = \frac{1}{(6.28 \times 2000)^2 \times 0.05 \times 10^{-6}}$$
$$= 0.126\ 78\ \text{H} = 126.78\ \text{mH}$$

3.21   NOMOGRAMS

Nomograms provide a very convenient method of quickly designing a coil to meet some specified design requirement, for solving coil design problems, or for checking the results of pencil and paper computations.

The use of most types of nomograms may be mastered in just a few minutes; nevertheless, the results obtained are remarkably accurate — accurate enough for most practical purposes.

A very useful coil design nomogram is shown in Fig. 3.17.   This is a dual-purpose nomogram which makes it doubly useful.   To use the nomogram in coil design problems, three values must be known.   A straightedge is placed across the nomogram to connect two of the known values on the scale according to the key.   A light pencil line is drawn at the point where the straightedge intersects the index line.   The straightedge is then run from the third unknown value across the intersecting point to the fourth scale which will give the quantity to be determined at its point of intersection with the straightedge.

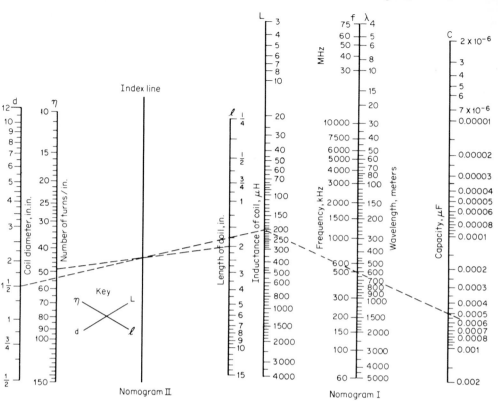

Fig. 3.17 Coil design nomogram.

Nomogram I is used to determine the inductance required to tune to a certain frequency with a known value of capacitance or vice versa or to find the frequency to which a given inductance will tune with a given capacitance. Here it is merely necessary to run a straightedge from the capacitance scale across the frequency, or wavelength, scale at the given frequency value. The required inductance value will then be found on the inductance scale at the point where the straightedge intersects it.

> **example 3.16**   (a) Determine the required value of inductance to tune to a frequency of 500 kHz with a 500-pF capacitor.
> (b) How many turns must a coil contain for the required inductance if the coil is to be $1\frac{1}{2}$ in in diameter and 2 in long?

> **solution**   (a) A straightedge from the given capacitor value (500 pF) on scale C through the frequency scale (f) at 500 kHz and over to the inductance scale (L) shows L to be 203 μH.
> (b) The straightedge is placed on the nomogram to intersect the known values of coil inductance (203 μH) and coil diameter ($1\frac{1}{2}$ in) on scales L and d, and a thin penciled line is drawn at the point where the straightedge crosses the index line; the straightedge is then run from 2 in (the coil length) on scale (l) through the marked line of intersection on the index line and over to the scale n, which shows turns per inch to be 49. Total number of coil turns will therefore be

$$2 \times 49 = 98 \text{ turns}$$

Chapter **4**

Magnetic Circuits

4.1 INTRODUCTION

Today it is hardly possible to be concerned with electronics without at the same time being concerned with magnetism or electromagnetism in one form or another. Consider, for instance, the numerous applications of electromagnetics in the modern television receiver with its loudspeaker, centering magnet, purity and focus magnets, deflection yoke, and various types of transformers and chokes. In other branches of electronics such as computers or navigational instruments — to mention only two — the importance of magnetism and electromagnetism is well established. The electronics industry is dependent (at least in part) on the supporting roles played by companies that specialize in the manufacture of magnetic and electromagnetic materials and components. Figure 4.1 shows a comprehensive selection of various magnetic materials including magnets, cores, bobbins, and iron-powder cores.

4.2 MAGNETS, NATURAL AND ARTIFICIAL

Natural magnets are found in the form of ore deposits known as *magnetite*, or more commonly, *lodestone*. This form of magnetic substance was known as early as 600 B.C. All other magnets are artificial. The first artificial magnets were made by contacting pieces of iron with magnetite. Even today, for some very minor applications, we may occasionally make a magnet by stroking a piece of iron or steel with any conveniently available magnet.

If a piece of soft iron is magnetized by induction from any magnetic source, the soft iron very quickly loses most of its induced magnetism after the magnetizing source is removed. However, if a piece of hardened steel or cast iron is magnetized, the induced magnetism will be retained for extremely long periods of time. Magnetized materials that very quickly lose most of their induced magnetism are called *temporary* magnets. Those that retain their magnetism are *permanent* magnets. Permanent magnets are made of wrought iron, hardened steel, or the various kinds of alloyed steels used for permanent magnets in loudspeakers and electric meters.

Figure 4.2 shows the difference between temporary and permanent magnets. Figure 4.3 illustrates a variety of permanent magnets which are used for making electric motors and other electric and electronic components.

4.3 MAGNETIC MATERIALS

When we speak of magnetic materials, we usually mean substances that are very strongly attracted by magnets or magnetic fields. Substances of this type are said to be either *ferromagnetic* or *ferrimagnetic*. Ferromagnetic substances are the various kinds

Fig. 4.1 A selection of various types of magnets, cores, bobbins, and iron-powder cores. (*Courtesy The Arnold Engineering Company*)

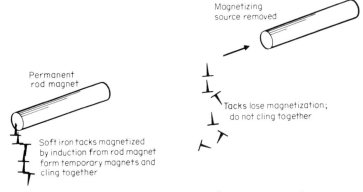

Fig. 4.2 Comparison of permanent and temporary magnetism.

Fig. 4.3 Permanent-magnet devices, used for motors and for other applications.

of iron and steel, including the powdered iron core materials that are used in some radio-frequency coils. Ferrimagnetic substances are the ferrites and other magnetic oxides that are used as core materials in coils operating at microwave frequencies, in high-frequency pulsing transformers such as the TV flyback transformer, and in the memory and switching units of high-speed computers.

Magnetic materials may also be classified as being either magnetically *soft* or magnetically *hard*. The magnetically soft materials are the kinds used in the cores of power and audio-frequency transformers and chokes. Magnetically hard materials are used for making permanent magnets.

Substances, such as aluminum, chromium, manganese, and air on which even very intense magnetic fields have only a very mild, barely detectable, attractive effect are called *paramagnetic*. Some substances, such as bismuth, antimony, copper, silver, and a few others on which very intense magnetic fields have a scarcely perceptible repellent effect, are said to be *diamagnetic*. Most substances, with the exception of those that are either ferromagnetic or ferrimagnetic, are commonly said to be *nonmagnetic*. Nonmagnetic substances do allow magnetism to pass through them, but they never become magnetized to any noticeable degree. This is illustrated in Fig. 4.4.

Fig. 4.4 Magnetic lines of force from the bar magnet pass through the nonmagnetic material (sheet aluminum) and attract the magnetic material (iron tack).

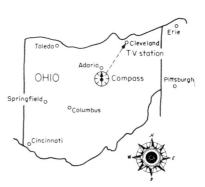

Fig. 4.5 A compass and a map can be used for orienting a TV antenna.

4.4 THE MAGNETIC COMPASS

If a piece of magnetite is suspended by a thread, it will align itself so that one part of it will always point in the general direction of the Earth's north magnetic pole. This was the very earliest form of compass, to be followed later by an artificial magnet suspended by a string. The modern magnetic compass takes the form of a magnetized steel needle balanced on a jeweled pivot and free to rotate horizontally. The end of the compass needle which points toward the north is called the *north seeking pole*. The other end, of course, points toward the south pole.

The compass, in conjunction with a map, is sometimes used to orient a TV antenna for reception from some distant station. The technique is illustrated in Fig. 4.5. The antenna being oriented is located at the point where the compass is lying on the map, and the desired station is in Cleveland. The dotted line shows the direction in which the antenna must be pointed to receive the signal directly from the station.

A compass may also be used in this manner to beam a radio signal toward some specific location.

4.5 MAGNETIC POLES

When the north pole of one magnet is placed near to the north pole of another magnet, there is a repelling force between them. Likewise, there is a repelling force between two south magnetic poles, but a north pole and a south pole are strongly attracted. This characteristic of magnetism is expressed in the law of magnetic attraction and repulsion which states:

RULE: *Magnetic poles that are alike will repel each other, while those that are unlike will attract each other.*

COULOMB'S LAW   Coulomb's law, which defines the *amount* of attraction or repulsion, may be stated as follows:

RULE: *The force of attraction or repulsion between two magnetic poles is inversely proportional to the square of the distance between the poles and directly proportional to the product of the pole strengths.*

THE UNIT MAGNETIC POLE   To use Coulomb's law in computations, we must have some method of precisely designating the strength of a particular magnetic pole in terms of some unit.   This unit is the unit pole, derived from Coulomb's law and based on the concept that two magnetic poles of equal strength could be so selected that they would repel each other with a force of one dyne when the distance between the poles is exactly one centimeter.   It is from this concept that the following definition of the unit magnetic pole is obtained:

The unit magnetic pole is of such strength that it will repel an exactly similar pole with a force of one dyne when the distance between the poles is one centimeter.

MAGNETIC POLE COMPUTATIONS   From Coulomb's law we get the following equation for computing the force of attraction or repulsion between two magnetic poles:

$$F = \frac{M_1 M_2}{d^2} \tag{4.1}$$

where F = force between poles, dyn
M_1 = strength of first pole, unit poles
M_2 = strength of second pole, unit poles
d = distance between poles, cm

example 4.1   The force of repulsion between the north poles of two rod magnets is 100 dyn when the distance between them is 2 cm.   What is the force of repulsion between the rod magnets if the distance between them is decreased to 1 cm?

solution   In order to solve this type of problem, it is necessary to assume that the south magnetic poles of the rods are at a sufficient distance from the north poles so that their effects are negligible.   In other words, the south pole of one of the rods is not noticeably attracted by the north pole of the other.   Denoting the quantity to be determined by X, and using Coulomb's law, an inverse proportion can be written as

$$100 : X = 1^2 : 2^2$$

or

$$\frac{100}{X} = \frac{1^2}{2^2}$$

Solving for X, the result is

$$\frac{(100)(2^2)}{1^2} = \frac{(100)(4)}{1} = 400 \text{ dyn}$$

example 4.2   The force of repulsion between the north poles of two bar magnets is 75 dyn when the distance between them is 10 cm.   If the north pole of one of the bar magnets has a strength of 100 unit poles, what is the strength of the other bar magnet?

solution   Again, it must be assumed that only the north poles of the magnets are exerting a measurable force, and the attraction or repulsion of the south poles can be ignored.   Substituting into Eq. (4.1),

$$75 = \frac{100 \times M_2}{10^2}$$

and solving this for M_2, the result is

$$\frac{10^2 \times 75}{100} = \frac{100 \times 75}{100} = 75 \text{ unit poles}$$

4.6 THE EARTH'S MAGNETIC FIELD

From what is known about magnetic poles, it is evident that the only reason why a magnetic compass functions as it does is that the Earth itself is a huge magnet, with one of its poles in the north and the other just opposite, in the south. The Earth's magnetic pole that is located in the *geographical* north is actually a south *magnetic* pole. This is evident because it attracts the north pole of the magnetic compass.

Because the magnetic poles of the Earth are some distance from its geographic poles, the magnetic compass does not point precisely north from all points of the Earth's surface. In the eastern parts of the United States the compass will point west of geographic or true north; and in the western parts of the United States, it will point east of true north. The angle by which a compass points away from geographic or true north is known as *magnetic declination*. Tables and charts showing the magnetic declination in degrees at various points of the Earth's surface may be found in handbooks on surveying and navigation.

Some knowledge of the Earth's magnetic field and the compass is very often of importance to those working with or on electronic navigational equipment, radar or other locating equipment, missile tracking, and radio astronomy.

example 4.3   In making tests on a gyrocompass, it is necessary to orient it so that it points to geographic or true north. A precise magnetic compass is available. What procedure should be followed?

solution   The gyrocompass should first be oriented to point to magnetic north, using the magnetic compass. The necessary correction should then be made by consulting a table of magnetic declinations.

4.7 ELECTROMAGNETISM

If a compass is brought near a conductor carrying current, as shown in Fig. 4.6, the compass needle will point in the direction shown. If the current through the conductor

Fig. 4.6 The magnetic field encircling a current-carrying conductor. (Conventional current shown by arrows in circuit.)

is reversed, the compass needle will turn and point in the opposite direction. From this it may be seen that a conductor carrying electric current is surrounded by a magnetic field, and this field exhibits bipolar effects. In other words, the field has direction, just as the magnetic field of a permanent magnet.

The magnetic field is said to come out of the magnet at its north pole and enter at its south pole. Actually the field lines do not move. Instead, their direction is defined as the direction that a unit north pole would move if placed in the field. There is no such thing in real life as a unit north pole, but it is a useful imaginary concept for defining the direction of magnetic fields.

To summarize, the magnetic field around a current-carrying wire has a north-to-south direction just as the field around a permanent magnet has direction. This direction is defined as the direction that a unit north pole will move, and is indicated by the direction a compass points when placed in the field.

THE RIGHT-HAND RULE AND THE LEFT-HAND RULE FOR MAGNETIC FIELDS AROUND WIRES   To determine the direction of the magnetic field surrounding any current-carrying conductor, use the following rule which is illustrated in Fig. 4.7.

RULE: *Grasp the wire in the right hand, and let the thumb point in the direction in which conventional current (that is, current from positive to negative) is flowing through the wire. The fingers will then point in the same direction as the direction of the magnetic field, or lines of force, that encircle the wire.*

If electron current is assumed, the left-hand rule is used.

RULE: *Grasp the wire with the left hand so that the thumb points in the direction of electron current flow and the fingers will encircle the wire in the direction of the magnetic field.*

4.8   THE RIGHT-HAND RULE AND THE LEFT-HAND RULE FOR COILS

Any coil or solenoid that carries current exhibits magnetic effects, and magnetic effects are always of a bipolar nature. Any current-carrying coil must always have both a north and a south pole. To determine the magnetic polarity of a current-carrying coil, use the right-hand rule for coils (see Fig. 4.8):

Fig. 4.7   The right-hand rule for determining the direction of the magnetic field around a current-carrying conductor. (Conventional current shown.)

Fig. 4.8   The right-hand rule for determining the magnetic polarity of a coil.

RULE: *Grasp the coil in the right hand, and let the fingers point in the direction of the conventional current around the coil; the thumb will then point in the direction of the north pole of the coil.*

If the coil is grasped with the left hand so that the fingers encircle the coil in the same direction that *electron* current is flowing, then the thumb of the left hand will point in the direction of the north pole.

> **example 4.4**   Two identical coils, A and B, are wound on an iron core of rectangular cross section, as shown in Fig. 4.9. The coils are wound in opposite directions, and the start of coil A is connected to the positive terminal of a battery. How should the other terminals of the two coils be connected together in series and to the negative terminal of the battery so that their magnetic fields will aid each other?

Fig. 4.9   Connect the coils to the battery so that their magnetic fields combine.

Fig. 4.10   Coils A and B are connected so that their fields are aiding.

solution Terminal F of coil A should go to terminal S_1 of coil B, and F_1 of coil B should be connected to the negative terminal of the battery, as shown in Fig. 4.10. Check this illustration with the right-hand (or left-hand) rule to make sure that the fields of the coils are aiding.

example 4.5 What would be the effect if the two coils, A and B, were to be connected in parallel and to a battery, as in Fig. 4.11?

solution In this case also, the magnetic fields of the two coils would aid each other. Application of either the right-hand rule or the left-hand rule for coils will prove this.

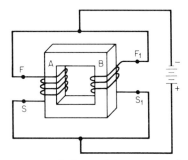

Fig. 4.11 Coils A and B are connected in parallel. Do the magnetic fields of the coils aid or oppose?

4.9 UNITS OF MEASUREMENT IN MAGNETICS

The basic units of measurement in electricity are amperes, volts, and ohms. It would be convenient if there were also only three basic units of magnetism, but unfortunately such is not the case. Instead, there are several different systems of measurement, and there is no universally used system. (The mks system has been suggested as a logical universal system of measurement in magnetism, but at the time of this writing the literature from American manufacturers does not define their products in terms of the mks system exclusively.)

One way to understand the myriad of units for magnetic measurements would be to describe all the units in a particular system. For example, the method of measuring flux, flux density, magnetomotive force, etc., in the cgs system could be described carefully, and then a conversion table could be included at the end of the discussion for changing from cgs to any other system.

Another method — the one used in this book — is to describe each term and its unit of measurement in the system that describes that particular term most simply. Thus, we will define the unit of *flux* in terms of the cgs system, and the unit of magnetomotive force in terms of the mks system. A table will then be included for converting from one system to another.

MAGNETIC FLUX Magnetic fields exist in lines of force, called flux. In the cgs system, each individual line of flux is called a *maxwell*. If there are three lines of flux, the amount of flux is said to be three maxwells in the cgs system, whereas in the British system it is simply called three *lines*. The greater the number of lines of flux, the stronger the magnetic field. The lines are imaginary, and the concept of lines of flux probably comes from the popular experiment in which iron filings are sprinkled on a paper that is placed over a magnet. The iron filings tend to arrange themselves in a group of lines, as shown in Fig. 4.12. These are actually lines of equal magnetic intensity.

UNIT POLES All magnets contain a north and a south pole. Some magnets can be made to have more than one north and south pole, but there is always a south pole for each north pole. For the purpose of discussion, however, let us suppose that it is possible to make an extremely small north pole with no associated south pole. This very small imaginary north pole is called a unit north pole in the cgs system. This unit north pole is helpful in describing some of the terms used in magnetism, and some of the units of measurement. However, it must always be remembered that it does not actually exist anywhere in the world physically.

How small should this unit north pole be? It is not enough to just say small, or very small; it is necessary to say exactly how small if it is going to be useful for making measurements, and for describing units of measurement. Figure 4.13 shows how the unit north pole is defined. If two unit north poles are placed in a vacuum one centimeter apart, they will *repel* each other with a force of one dyne. This defines, then, how small a unit north pole is.

If a unit north pole was placed on the piece of paper in Fig. 4.12, it would move away from the north pole of the magnet and toward the south pole. Its path will be along one of the flux lines shown in the illustration. For this reason, you will often see arrows along the lines of flux. These arrows are supposed to show the direction of a unit north

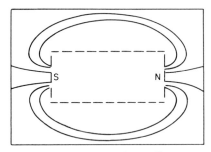

Fig. 4.12 Arrangement of iron filings on a piece of paper covering a bar magnet.

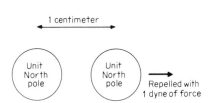

Fig. 4.13 Definition of unit north pole.

pole along the flux line, but often give the mistaken idea that the flux line itself is supposed to be moving. The direction of a line of flux is defined as the direction that a unit north pole will move if placed on the flux line. One line of flux, or one maxwell, will act upon a unit north pole with a force of one dyne. Thus, the unit north pole is helpful in defining the magnetic strength of a line of flux.

FLUX DENSITY If one line of flux passes perpendicularly through a square centimeter of this page, then there is a certain amount of magnetic field strength present. If *two* lines of flux pass through that same square centimeter, then the field strength will be twice as great as when one flux was present. If three lines of flux pass through that square centimeter, the field strength will be three times as great, and so on. It should be easy to visualize that three lines of flux passing through one square centimeter will produce a greater concentration of magnetic field strength than would be produced if the two lines of flux pass through one square centimeter. An important way of defining the strength of a magnetic field, then, is in terms of the number of flux lines that pass through a square centimeter of area. Obviously, a very strong magnet will cause more flux lines to pass through a square centimeter of area than a weak magnet will. This is illustrated in Fig. 4.14.

The number of lines of flux that pass through one square centimeter of area is called the flux density. In the cgs system, the unit of flux density is the *gauss*. When one maxwell (that is, one line of flux) passes through one square centimeter of cross-sectional area, the flux density is said to be one gauss. Figure 4.15 illustrates one gauss of flux density. The gauss is a unit of measurement in the cgs system.

MAGNETOMOTIVE FORCE Flux lines can be established in a number of ways. One of the easiest ways is to use an electric current. For every electric current there is an associated magnetic field. The force that causes a magnetic flux to be established is called the magnetomotive force. It can be compared to the voltage in electricity which is sometimes considered to be the "force" that produces an electric current. (Actually, of course, voltage is a measure of work, *not* a unit of force.)

The easiest unit of magnetomotive force to understand is the *ampere-turn*. If one ampere of current flows through one turn of wire, then the magnetomotive force is said to be one ampere-turn.

Figure 4.16 illustrates the ampere-turn. The magnetomotive force is always equal to the number of amperes of current multiplied by the number of turns of wire. The ampere-turn is a mks unit of measurement. (In the cgs system, a gilbert is used as the unit of mmf. One gilbert = $1.257 \times$ ampere-turns.)

RELUCTANCE   When a magnetomotive force is used to establish flux in a material, there is always some opposition to the flux. This opposition is called *reluctance*. There

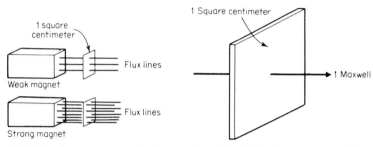

Fig. 4.14 The strong magnet will cause more lines of flux per square centimeter than the weak magnet.

Fig. 4.15 Illustrating one gauss of flux density.

Fig. 4.16 The ampere-turn is a unit of magneto-motive force.

are no *English* units for measuring reluctance. The term *rel* has been proposed but is not in common usage. In the cgs system reluctance is measured in *gilberts per maxwell*, but there is no common name for the unit of reluctance, and there is no name for it in the mks system.

In magnetic circuits, it is sometimes more convenient to know the ease with which flux lines can be established in a material, rather than the opposition that the material offers to establishment of flux. The ease with which flux lines can be established in a material is called the *permeance* of that material, and it is the reciprocal of reluctance.

COMPARISON OF ELECTRIC AND MAGNETIC CIRCUITS   There is a similarity between electric and magnetic units. Table 4.1 lists the more important electrical terms and the comparable magnetic terms.

TABLE 4.1   Comparison of Electric and Magnetic Terms

Electric units	Magnetic units
Electric current I	Magnetic flux ϕ
Electromotive force E	Magnetomotive force, mmf
Resistance R	Reluctance R
Conductance G	Permeance P

There are certain things about electric circuits that are quite different from magnetic circuits, and should always be kept in mind. The more important differences will now be discussed.

In an electric current, all the current is normally confined to the conductors carrying it. In a magnetic circuit, some of the flux lines actually leave the circuit along the route, so that it is somewhat more difficult to calculate the amount of flux at a certain point in comparison to calculating current in a circuit.

The resistance of a wire or resistor is a fixed value for a particular circuit over the range of currents for which it is intended to be used. In a magnetic circuit, the amount of reluctance is dependent upon the amount of flux already present. The greater the amount of flux already present in a piece of iron, the greater the reluctance becomes — that is, the greater the opposition to a further increase in flux.

Fig. 4.17 Comparison of an electric and a magnetic circuit.

In electric circuits, there are insulating materials that offer great opposition to current flow. In magnetic circuits there are no materials that can prevent flux from being established. In order to prevent magnetic flux lines from entering a particular area, the only course that can be taken is to employ a very low reluctance path around that area. Nearly all the flux lines will follow the low-reluctance path in such cases, and therefore they will not enter the region being shielded.

Figure 4.17 compares an electric and a magnetic circuit. A battery supplies the power to the electric circuit. In the magnetic circuit the coil of current-carrying wire provides the magnetomotive force that establishes flux in the circuit.

Electric current flows through the conductor in the electric circuit. In the magnetic circuit a low-reluctance iron provides a path for the flux. As mentioned before, some of the flux may leave this low-reluctance path and return to the source of mmf by another path. If the circuit is properly designed, the flux leakage can be kept to a minimum.

A resistor in the electric circuit offers opposition to the flow of current. In the magnetic circuit, the air gap offers a greater opposition to establishment of flux than the iron path. Therefore, the air gap may be thought of as being the resistor in the magnetic circuit.

Table 4.2 lists some of the units of magnetism, and conversion factors between units.

MAGNETIZING FORCE   One ampere of current flowing through one turn of wire will give a magnetomotive force of one ampere-turn. (Mmf in ampere-turns = number of amperes × number of turns of wire.) Likewise, one ampere of current flowing through ten turns of wire will give a magnetomotive force of ten ampere-turns. Now, suppose you wind the ten turns very closely together so that each turn is tight against the adjacent turns. The magnetomotive force will then be concentrated along a short length of the coil form. On the other hand, if you spread the turns out so that they are evenly distributed along a length of one foot, then the magnetomotive force will be

TABLE 4.2 Magnetic Units

Unit	cgs	mks	British
Flux	Maxwells	Webers	Lines
Magnetomotive force	Gilberts	Ampere-turns	Ampere-turns
Magnetizing force	Oersteds	Ampere-turns per meter	Ampere-turns per foot
Reluctance	Gilberts per maxwell	Ampere-turns per weber	No unit (rel suggested)
Flux density (also called magnetic induction)	Maxwells per square centimeter°	Webers per square meter	Lines per square inch

CONVERSION FACTORS

Inches × 2.54 = centimeters   Gilberts × $10/4\pi$ = ampere-turns
Maxwells × 10^{-8} = webers   Oersteds × $10^3/\pi$ = ampere-turns per meter
Gauss (maxwells per square centimeter) × 10^{-4} = webers per square meter
1 gamma = 10^{-5} gauss

° One maxwell per square centimeter is called a gauss; so flux density in the cgs system can be given in either the number of maxwells per square centimeter, or gauss.

distributed along a path one foot long. In both cases *the magnetomotive force* is ten ampere-turns, but in one the mmf is concentrated in a short length, whereas in the other it is spread out along a length of one foot.

It is important to know the amount of magnetomotive force, but it is also essential to know how it is distributed. The distribution of mmf along a length is called the *magnetic gradient*. It is also known as the *magnetizing force*. In the above example, ten turns per inch will produce a greater magnetizing force than ten turns per foot. In the mks system, magnetizing force is measured in ampere-turns per meter, while in the British system either ampere-turns per foot or ampere-turns per inch may be used. In the cgs system, the oersted is used. One gilbert per centimeter gives a magnetizing force of one oersted.

RELATIVE PERMEABILITY Relative permeability, which is often simply called permeability, is the ratio of the flux density in a material for a given magnetizing force to the flux density that would be produced in a vacuum for the same amount of magnetizing force. Mathematically, the relative permeability in the cgs system is expressed by the equation

$$\mu = \frac{B}{H} \tag{4.2}$$

where μ = relative permeability of the material
B = flux density
H = magnetizing force

There are no units for relative permeability. Relative permeability depends on the type of material. In other words, every different type of material will have a different value of relative permeability. For example, the relative permeability of air, like that of a vacuum, is one. For all paramagnetic materials, the relative permeability is slightly greater than one. The relative permeability of diamagnetic materials is less than one, while the relative permeability of ferromagnetic materials is much greater than one (and also much greater than the relative permeability of paramagnetic materials).

A relative permeability greater than one simply means that it is easier to establish flux lines in that material than it is to establish them in a vacuum. A material having a relative permeability less than one indicates that it is more difficult to establish flux lines in that material than it is to establish them in a vacuum.

4.10 ROWLAND'S LAW

Rowland's law expresses the relationship among magnetomotive force, flux, and reluctance. It is sometimes called *Ohm's law for magnetic circuits*.

The three fundamental quantities, the maxwell (ϕ), the gilbert (F), and the unit of reluctance (R), are used in Rowland's law as follows:

$$F = \phi R \qquad (4.3)$$

where F = magnetomotive force, Gb
ϕ = flux, Mx
R = reluctance, Gb/Mx
If any two factors are known, the third can be obtained by solving Eq. (4.3). Thus,

$$\phi = \frac{F}{R} \qquad (4.4)$$

$$R = \frac{F}{\phi} \qquad (4.5)$$

example 4.6   What current flows through the coils of an electromagnet that has 400 turns, develops a flux of 250 000 lines and has a core reluctance of 0.04 unit?

solution   The magnetomotive force developed by the coils is

$$1.257 \times I \times N = 1.257 \times I \times 400$$

And since the flux $\phi = F/R$,

$$250\ 000 = \frac{1.257 \times I \times 400}{0.04}$$

Solving for I, the result is

$$\frac{0.04 \times 250\ 000}{400 \times 1.257}$$

or 19.89 A.

example 4.7   The magnetizing coil of a magnetic circuit has 300 turns and carries a current of 0.5 A. The flux in the magnetic circuit is 100 000 Mx. What is the reluctance of the circuit?

solution   The magnetomotive force

$$F = 1.257 \times 0.5 \times 300 = 188.55 \text{ Gb}$$

The reluctance is

$$R = \frac{F}{\phi} = \frac{188.55}{100\ 000}$$
$$= 0.001\ 885\ 5 \text{ unit or Gb/Mx}$$

Owing to the fact that the reluctance in a magnetic circuit is not linear, Rowland's law is not as versatile as Ohm's law for electric circuits. The reluctance in the circuit is only a given value when a certain amount of flux is present. If you change the flux, you change the reluctance. The situation is very similar to that of an electric circuit with a thermistor. The resistance of the thermistor depends on its temperature which, in turn, depends on the amount of current through it. You can use Ohm's law only when you know the resistance of the thermistor for some given amount of current. If you change the current (or temperature), the resistance changes, and a new set of conditions exist.

To summarize, magnetic circuits (like thermistor circuits) are nonlinear. You cannot apply a linear equation, such as $I = E/R$ or $\phi = \text{mmf}/R$, to nonlinear circuits for all possible conditions in the circuit.

4.11   RELUCTANCES IN SERIES

To find the total reluctance of any number of reluctances in series, we merely add them together, just as we would add resistances to find the total resistance of a number of resistors in series.

example 4.8   Figure 4.18 shows a magnetic structure that is designed for a special-purpose application. The total flux in the core material, as determined by means of a flux-meter, is 50 000 lines with an applied magnetomotive force of 1600 Gb. The reluctance of section A is 0.001 25 unit, and the reluctance of section B is 0.000 75 unit. One of the air gaps has a length of 0.01 in, while the other is 0.02 in long. Compute (a) the combined reluctance of the air gaps, and (b) the reluctance of each air gap taken separately.

solution (*a*) We denote the unknown reluctance of the air gaps by Z. Since this is a series magnetic circuit, the combined reluctance of the magnetic circuit is

$$0.001\ 25 + 0.000\ 75 + Z$$

Total flux is

$$\phi = \frac{F}{R} = 50\ 000$$
$$= \frac{1600}{0.001\ 25 + 0.000\ 75 + Z}$$
$$= \frac{1600}{0.002 + Z}$$

Solving for Z, the combined reluctance of the two air gaps is

$$\frac{1500}{50\ 000} = 0.03\ \text{unit}$$

(*b*) Since one air gap is twice as long as the other, its reluctance will also be twice as great as the reluctance of the other air gap, just as the resistance of a wire that is 2 ft long is twice as

Air gap
0.01 in.

Air gap
0.02 in.

Fig. 4.18 Special-purpose magnetic structure.

great as the resistance of a wire that is only 1 ft long. Denoting the reluctance of one air gap by Y, the reluctance of the other will be 2Y; then $Y + 2Y = 0.03$ unit. Solving this for Y, the reluctance of one air gap is $0.03/3 = 0.01$ unit; and the reluctance of the other air gap is 2Y, or 0.02 unit.

4.12 RELUCTANCES IN PARALLEL

The permeance of any magnetic circuit is equal to 1 divided by the total reluctance of the circuit, or

$$\text{Permeance } P = \frac{1}{\text{reluctance}} = \frac{1}{R} \tag{4.6}$$

Permeance in magnetic circuits may be likened to conductance in electric circuits and is used to compute the combined reluctance of a number of reluctances in parallel, just as conductance is used in parallel resistance calculations. To find the combined reluctance of a number of reluctances in parallel, we take the reciprocal of the sum of the permeances of the separate paths. Thus,

$$R = \frac{1}{1/R_1 + 1/R_2 + 1/R_3 + \cdots + 1/R_N} \tag{4.7}$$

where R = combined reluctance
$R_1, R_2, R_3, \cdots, R_N$ = reluctances of the separate paths
For two reluctances in parallel, Eq. (4.7) becomes

$$R = \frac{1}{1/R + 1/R} = \frac{R_1 R_2}{R_1 + R_2} \tag{4.8}$$

and for three reluctances in parallel,

$$R = \frac{R_1 R_2 R_3}{R_1 R_2 + R_1 R_3 + R_2 R_3}$$

example 4.9   What current flows through the coils of an electromagnet that has 400 turns, develops a flux of 250 000 lines, and has a core reluctance of 0.04 unit?

solution   From Eq. (4.4) and the equation for magnetomotive force

$$\phi = \frac{F}{R} = 250\ 000 = \frac{I \times 400 \times 1.257}{0.04}$$

Solving for I, the result is

$$\frac{0.04 \times 250\ 000}{400 \times 1.257}$$

or 19.89 A.

example 4.10   The core of a hastily improvised choke coil is made up of three different grades of steel laminations having reluctance values of 0.05, 0.08, and 0.12, respectively. What is the reluctance of the core?

solution   Using Eq. (4.9) for computing the reluctance of three reluctances in parallel, we get

$$R = \frac{0.05 \times 0.08 \times 0.12}{(0.05 \times 0.08) + (0.05 \times 0.12) + (0.08 \times 0.12)}$$
$$= 0.003 \text{ unit}$$

4.13   VARIATION OF PERMEABILITY WITH MAGNETIZING FORCE

Since the reluctance of a magnetic substance is not constant but changes with changes in flux density, the permeability of a magnetic substance will also change with every change in flux density.   The reason for this is that the reluctance of a magnetic material changes with every change in the flux.   The reluctance of only a 1-cm cube of that material will, of course, also change at the same time.   The reluctance of a 1-cm cube of any material is its *reluctivity*, and permeability is equal to one divided by reluctivity. It follows, then, that the permeability of a magnetic material will also change with every change in flux.   Since the change in reluctance will *not* be *directly proportional* to the change in flux, the change in permeability will also not be directly proportional to changes in flux.   These facts should always be taken into consideration when magnetic circuit calculations are to be made.

MAGNETIZATION CURVES   Magnetization curves and tables showing the relations between B and H or among B, H, and μ at various stages of magnetization for

Fig. 4.19   Typical dc magnetization curve of 12-mil Silectron uncut core—not impregnated.   (*Courtesy The Arnold Engineering Company*)

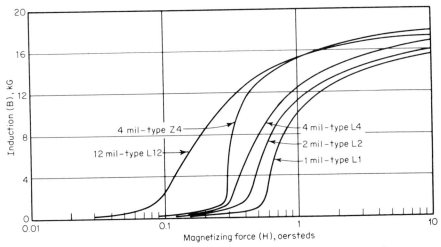

Fig. 4.20 Typical dc magnetization curves for 1-, 2-, 4-, and 12-mil Silectron toroids—not impregnated. (*Courtesy The Arnold Engineering Company*)

different grades of iron and steel and various other kinds of magnetic materials are available from manufacturers of transformer steel and other core materials.

Magnetization curves may take various forms. Some may be graphed on ordinary, linear graph paper, while others may be on logarithmic or on semilogarithmic paper. The use of magnetization curves greatly simplifies the solution of magnetic circuit problems. Typical magnetization curves are shown in Figs. 4.19 to 4.21. These

Fig. 4.21 Magnetization curves for 0.014-in 4750 induction and permeability versus magnetizing force: type AL 4750, gage 0.014 in, sample from rings 1.9 by 2.5 in. (*Courtesy The Arnold Engineering Company*)

magnetization curves are something like the characteristic curves of electron tubes, and they are just as easy to interpret. To determine permeability from a *B–H* curve, such as is shown in Figs. 4.19 and 4.20, it is necessary to divide the induction *B* on the vertical scale by the corresponding value of the magnetizing force *H* on the horizontal scale. Figure 4.21 gives the value of permeability for any corresponding values of *B* and *H* without any calculations. Figure 4.21 also gives the values of *B*, *H*, and μ at different frequencies up to 1600 Hz.

example 4.11 Using the *B–H* curve of Fig. 4.20, determine (a) the magnetizing force needed for an induction (flux density) of 12 kG (12 000 lines/cm²) in a 12-mil-type L12 toroid, and (b) the permeability of the core material for this value of induction.

solution (a) Find the value of the induction, 12 kG, on the vertical scale, and from this point follow the horizontal line to its point of intersection with the curve for 12-mil-type 12 material; then find the corresponding value of the magnetizing force, 0.25 Oe, on the horizontal scale.

(b)
$$\text{Permeability } \mu = \frac{B}{H}$$
$$= \frac{12\ 000}{0.25} = 48\ 000$$

example 4.12 Using the *B–H–μ* curve of Fig. 4.21, determine (a) the dc induction in the core material when the magnetizing force is 0.008 Oe, and (b) the permeability of the core for the given value of magnetizing force.

solution (a) Find 0.008 on the horizontal scale, and follow the vertical line up from this point to its intersection with the dc curve; the corresponding value of the induction on the vertical scale is 200 G. (b) From the point on the dc curve corresponding to the determined values of *B* and *H*, place a straightedge parallel to and between the diagonal lines leading to the upper scale; and the corresponding permeability value of 25 000 will be found on the upper scale.

4.14 MAGNETIC CIRCUIT COMPUTATIONS

Just as the resistance of an electric conductor is equal to the specific resistance, or resistivity, of the conductor multiplied by its length and divided by its cross-sectional area, the reluctance of a magnetic material (or conductor of flux or lines of force) is similarly equal to the specific reluctance, or reluctivity, of the magnetic material multiplied by the length of the material and divided by its cross-sectional area. In other words, as the length of a magnetic circuit increases, its reluctance also increases in proportion; but as the cross-sectional area increases, the reluctance decreases in proportion. Then the reluctance of a magnetic material is directly proportional to its length and inversely proportional to its cross-sectional area. And since permeability is equal to 1 divided by reluctivity, the reluctivity of a magnetic material is equal to

$$\frac{1}{\text{Permeability}} = \frac{1}{\mu}$$

and the equation for finding reluctance may be written in the form

$$R = \frac{(1/\mu) \times L}{A} = \frac{L}{\mu A} \tag{4.9}$$

where *R* = reluctance in units
  μ = permeability
  L = length of magnetic circuit
  A = cross-sectional area of magnetic circuit

If *L* is given in inches, *A* is in square inches; but if *L* is in centimeters, *A* will be in square centimeters.

Dividing 1 by the permeability of a magnetic substance to obtain its reluctivity is a procedure similar to that used to obtain the resistivity, or specific resistance, of an electric conductor by taking the reciprocal of its conductivity.

Our basic fundamental equation for finding total flux may now be restated as

$$\text{Total flux in maxwells } (\phi) = \frac{F}{R} = \frac{1.257 \times IN}{L/\mu A}$$

or

$$\phi = \frac{1.257 \times IN \times \mu A}{L} \tag{4.10}$$

where IN = ampere-turns
μ = permeability
A = cross-sectional area, cm^2
L = length, cm

4.15 ELECTROMAGNETS

The pulling power or tractive force of an electromagnet may be computed from an equation derived from the fundamental definition of a unit magnetic pole. A unit magnetic pole is a pole of such strength that it will repel an exactly similar pole with a force of one dyne when the distance between the poles is one centimeter.

The fundamental equation of electromagnet computations is

$$T = \frac{AB^2}{72\ 134\ 000} \tag{4.11}$$

where T = pull, lb
A = cross-sectional area of pole faces, in^2
B = flux density, lines/in^2

To find A, when T and B are known, the equation is

$$A = \frac{72\ 134\ 000T}{B^2} \tag{4.12}$$

and to find B, when T and A are known, we use

$$B = \sqrt{\frac{72\ 134\ 000T}{A}} \tag{4.13}$$

example 4.13   The current in the windings of an electromagnet induces a flux density of 110 000 lines/in^2 in the core. If the core and pole face have a cross-sectional area of 20 in^2, what weight will the electromagnet support?

solution

$$T = \frac{AB^2}{72\ 134\ 000} = \frac{20 \times (110\ 000)^2}{72\ 134\ 000}$$

$$T = \frac{242\ 000\ 000}{72\ 134} = 3355\ \text{lb}$$

4.16 SATURATION

Any magnetic material is said to be saturated when any further increase in the magnetizing force, or field intensity, no longer causes any appreciable increase in the flux density in the material.

The value of the flux density in the core material of any electromagnetic circuit is of considerable importance; this value will depend upon the use of the magnetic material. An iron-core choke in which the core material is highly saturated loses most of its inductive properties and becomes almost wholly resistive; for this reason, inductive chokes that carry direct current are usually designed with an air gap in the magnetic structure to keep the core from becoming magnetically saturated. If the core of a transformer in the plate circuit of a power output tube becomes magnetically saturated, the result will be severe distortion and poor transfer of energy from the primary to the secondary. Or if the biasing current in the recording head of a tape recorder is either too high or too low, any recorded material will be badly distorted. On the other hand, some electromagnetic devices will operate only very poorly, or not at all, if the initial flux density in the core material is not at its maximum value. The saturation point of any magnetic iron or steel may easily be determined from its B–H magnetization curve.

4.17 RETENTIVITY

A measure of retentivity is the amount of magnetism, or value of B, that remains in a magnetic material after the magnetizing force has been raised to a value more than sufficient to saturate the material and then reduced to zero; the value of B remaining in the material is called *residual magnetism* or *remanence*. If two different kinds of permanent magnet materials of the same size are both magnetized to saturation, the one that makes the stronger permanent will, of course, have the higher retentivity. However, this does not necessarily mean that the stronger magnet will be harder to demagnetize, since this depends upon other factors besides retentivity.

4.18 COERCIVITY OR COERCIVE FORCE

Coercivity is the value of the opposing magnetizing force that is needed to reduce the residual magnetism, or remanence, in a magnetic substance down to zero.

A clear concept of coercivity may be obtained by referring to Fig. 4.22. Assume that the iron core of the coil is not magnetized, and that the potentiometer setting is at

Fig. 4.22 Illustrating coercivity, or coercive force H_c, and residual magnetism, or remanence B_r.

its midpoint, or zero. As the potentiometer setting is shifted toward point a, the core material will become magnetized to saturation along the line $A-B$ of the magnetization curve. If the potentiometer setting is then shifted back toward the center, or zero position, the flux density B in the core will not fall back to zero along the line $A-B$ on the curve, but instead along the line $B-C-D$; and when the potentiometer setting is again at zero, the flux density will still be at point c. This will be the residual magnetism, or remanence, denoted by the symbol B_r. To reduce the remanence to zero, it is necessary to shift the potentiometer setting toward point b to produce a magnetizing force opposing the residual field in the core. This will cause the residual field to fall back to zero along the line from C to D. The negative value of H needed to reduce the residual magnetism in the core to zero is the coercivity of the core material. Coercivity is denoted by the symbol H_c.

If the potentiometer setting is shifted still further to the left, the core will again become magnetized, but the magnetization will now be in a negative direction. (This is based on the assumption that the original direction was positive.) Saturation will next occur at point E of the magnetization curve. If the negative value of H is now reduced to zero and then increased again to its highest positive value, by shifting the potentiometer setting toward the right, the flux density in the core will follow a new line from E to F, or zero, and then back up to B.

4.19 HYSTERESIS

The loop formed by the curves of Fig. 4.22 is known as a hysteresis loop, and the area enclosed by the loop is a measure of the energy used in reversing the magnetic molecules of the core. The energy thus lost is called the *hysteresis loss*. Hysteresis loss is of importance only when the core material is being subjected to pulsating dc or ac

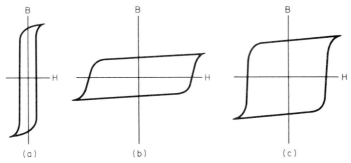

Fig. 4.23 Typical hysteresis loops: (*a*) Typical magnetic core material; small coercive force H_c required to eliminate large residual magnetism B_r. (*b*) Typical of a class of permanent-magnet material; large coercive force H_c required to eliminate small residual magnetism. (*c*) Typical of another class of permanent-magnet material; large coercive force H_c required to eliminate large residual magnetism B_r. (*Courtesy The Arnold Engineering Company*)

fields. Since the hysteresis loop is repeated over and over again for every cycle, core losses become greater and greater as the frequency is increased. For this reason, the core materials used for transformers should have very narrow hysteresis loops. Some typical hysteresis loops are shown in Fig. 4.23.

Core-loss data may be obtained in both tabular and graphical form from manufacturers of transformer core materials. An example of core-loss data in graphical form is shown in Fig. 4.24, which gives the total core loss (hysteresis loss plus eddy current loss) at various frequencies up to 10 kHz and inductions up to 18 kG. The form in which the data are presented may depend upon the engineering practices of the companies supplying the data. Some older publications give equations for computing hysteresis loss, but such equations would hardly be applicable to the large variety of core materials now available.

4.20 EDDY CURRENTS

Currents that are set up in magnetic core materials by induced voltages because of changing magnetic fields are known as eddy currents. The energy lost in the form of heat as a result of eddy current circulating in the core material is called the eddy-current loss. Since eddy-current loss is proportional to the square of the inducing frequency, such losses become very appreciable as the frequency is increased.

Eddy currents may be reduced by stacking the core laminations so that they are parallel to the flux. Natural oxidation of the laminate surfaces also helps greatly in reducing eddy currents; or the laminations may be coated with a thin layer of insulating varnish. Figure 4.25 shows how eddy currents are reduced by laminated cores. But even with laminated cores there is still some circulation of eddy currents in the in-

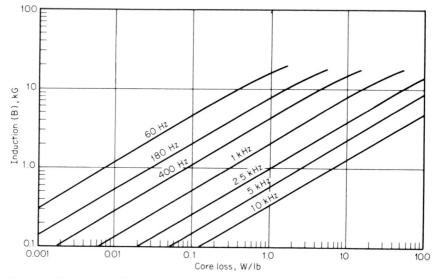

Fig. 4.24 Design curves showing maximum core loss for 4-mil Silectron C cores, type AH. Core loss guarantee 10.0 W/lb maximum at 400 Hz, 15 kg. (*Courtesy The Arnold Engineering Company*)

Fig. 4.25 Eddy currents circulating in solid core are stopped by laminated core.

dividual laminations. Therefore, at radio frequencies, the powdered iron-core materials are used. As the frequency is further increased, even the powdered iron cores are no longer effective in reducing eddy-current loss to a sufficient degree. Then the ferrite-core materials must be used. Ferrites, unlike iron or steel, are not electric conductors.

Some equations are available for computing eddy-current loss in ordinary transformer steel, but such equations could hardly be used for computing the eddy-current loss in the large variety of other core materials that have become available in recent years.

4.21 IRON-POWDER CORES

Iron-powder cores for radio coils first came into existence in the early 1930s, when they were used in i-f transformers. Since that time there has been an ever-increasing demand for iron-powder cores and great improvement in production techniques.

Fig. 4.26 Iron powder (ferrite) cores for TV horizontal output transformers, and some ring cores for coils.

Iron-powder cores are now available for a wide variety of applications and for use at frequencies as high as 250 mHz. Every TV receiver now contains a number of coils that have iron-powder cores. (See Fig. 4.26.) Some of these may be fixed-tuned, but others may be of the so-called slug-tuned variety, such as the horizontal width and the horizontal frequency-control coils.

The powdered iron used for iron-powder cores is first subjected to certain chemical processes which cause the individual tiny particles of iron to take on an insulating layer. Then the binder, or glue, used to bind the particles of iron together adds an additional insulating layer over the insulating layer formed by the chemical processing. Thus, in the finished core, the individual tiny particles of iron are effectively insulated from one another; and, as a result, eddy current cannot circulate in the core. Still there will be some eddy-current loss, because the individual particles of iron are electric conductors and not insulators; eddy current can still circulate within each individual particle.

The fineness of the individual iron particles, the amount of binder used in proportion to the amount of iron powder, will be different for different applications and frequency ranges. The iron powder may also be made up of two or three different grades of iron mixed in various proportions.

Some users of iron-powder cores mistakenly assume that all iron-powder cores have the same characteristics; for example, in some electronic construction article, it may be stated that a particular coil was wound on a one-half-inch-diameter form that had an iron-powder core. But then nothing is said about the type of core material. The result is that many of those following the data presented in the article are often disappointed by the results. They wonder what has gone wrong, since they, too, are unaware of the fact that iron-powder cores are not all alike.

Chapter 5

Transformers

5.1 INTRODUCTION

The function of every type of transformer is to transform or change electricity in some manner. This is true whether the transformer is only a tiny sub-ouncer unit or is one of the huge devices used at the substations and terminal points of high-voltage electric-power distribution systems. For example, the tiny output transformer of a transistor pocket-portable radio transforms the current in the output stage to a form that is suitable for actuating the loudspeaker. As another example, the large transformer at the power distribution substation transforms (steps down) the high voltage of the power lines so that it may safely be used in homes and industry.

Since the iron-cored transformer and choke came into use much earlier than other kinds of transformers and chokes, it seems only fitting to take up the study of such transformers and chokes (see Fig. 5.1) before taking up the study of other types. Figure 5.1*a* shows a variety of ferrite-core coils. An iron-core transformer is shown in Fig. 5.1*b*.

5.2 THE BASIC TRANSFORMER

The basic transformer is shown in Fig. 5.2. It consists of two coils wound upon the same iron core which forms a closed magnetic circuit. The flux path is shown by dotted lines around the core. The transformer will not work on pure direct current, but instead must be supplied with power from a source of alternating current or from a source of pulsating direct current.

The winding to which power is supplied is called the *primary*, and the winding from which power is taken is the *secondary*. Any change in the primary current results in a change in flux (ϕ). In accordance with Faraday's law, a change in the flux linking the secondary winding will induce a voltage across the secondary. If a path for current flow exists across the secondary winding, then in accordance with Lenz's law the direction of current flow will be such that its magnetic field opposes the changing flux that produced it.

If the secondary voltage is the same as the primary voltage, the transformer is called a *one-to-one transformer*; if the secondary voltage is greater than the primary, it is a *step-up transformer*, and when the secondary voltage is less than the primary, it is a *step-down transformer*. Power supplied to the primary at a specific frequency produces power at the same frequency in the secondary.

5.3 MUTUAL INDUCTION

If the transformer primary of Fig. 5.2 is connected to a source of alternating current, as in Fig. 5.3, the instantaneous values of the primary voltage and current and the resulting

Fig. 5.1a A variety of ferrite-core coils and chokes. (*Courtesy James Millen Manufacturing Company, Inc.*)

Fig. 5.1b Iron-core transformer with end bell construction. (*Courtesy Acme Electric Corporation.*)

Fig. 5.2 Illustrating the basic transformer. Power is transferred from the primary to the secondary winding by means of the continuously changing flux (ϕ) in the core.

Fig. 5.3 When the transformer is connected to a source of alternating current, the continuously changing magnetic flux in the core induces a voltage in the secondary winding and at the same time also induces a back emf (or counter-emf) in the primary winding.

magnetic field will all be continuously changing and inducing a back voltage in the primary. The instantaneous value of the back voltage will be very nearly equal to the instantaneous value of the applied voltage, but it will always be just a bit lower than the applied voltage. If the applied and back voltages were exactly equal to each other, it would be impossible to get the current flowing in the circuit.

Since the core of the transformer forms a closed magnetic circuit, the primary and secondary windings are interlinked through the mutual magnetic flux in the core. The changing flux in the core induces a voltage in the secondary at the same time that it is inducing a back voltage in the primary, and the combined induction effect is called mutual induction.

The secondary induced voltage may be the same as the primary voltage, or it may be higher or lower than the primary voltage, depending on the ratio of the number of turns of wire between primary and secondary.

When the secondary is connected to a load, current flows from the secondary into the load just as it would from any other voltage source. According to Lenz's law, the current flow in the secondary must always be in such a direction that it will set up an opposition to any *change* in the flux of primary current. For example, if the secondary has a load connected to it, the load current in the secondary winding creates a magnetic field, and as this field changes, it induces a voltage in the primary in opposition to the applied voltage.

example 5.1   Figure 5.4 represents a transformer supplying power to a load connected to its secondary. A waveform of the current is also shown. Draw a diagram showing terminal polarities of both primary and secondary windings and the direction of current flow in each

Fig. 5.4   Transformer with sine-wave current input supplying a load.

Fig. 5.5   Terminal polarities and the directions of the magnetic fields produced by currents in the primary and secondary windings at the instant when the primary current is rising and is at point x on the current curve.

winding at the instant when the primary current is rising from zero to its maximum positive value and has reached the point x on the current curve. Also show the directions of the magnetic fields developed by each winding at this instant. Use the right-hand rule for coils and Lenz's law.

solution   Figure 5-5 shows that as the current is rising from zero to its maximum positive value, the upper terminal of the primary will be positive. Therefore, current flow through the primary winding will be in the direction shown by the arrows, and by the right-hand rule for coils, the direction of the resulting magnetic field will be clockwise in the core. Since the field is building up, the direction of the magnetic field developed in the secondary must, according to Lenz's law, oppose the building up of the field developed by the primary; hence, the magnetic field developed by the secondary must be in a counterclockwise direction in the core. The right-hand rule for coils will show that the magnetic field developed by the secondary will be in a counterclockwise direction in the core when current flows through the coil windings in the direction indicated by the arrows. The upper terminal of the secondary will then be positive.

example 5.2   Draw a diagram showing conditions in the transformer of Fig. 5.4 at the instant when the primary current is falling from its maximum positive value toward zero and has reached the point y on the current curve.

solution   The upper terminal of the primary winding is still positive and the primary current will be in the direction indicated. Since the current is now decreasing, its associated magnetic field will be collapsing. According to Lenz's law, the magnetic field developed in the secondary must oppose any *decrease* of the primary field. Therefore, the magnetic field of the secondary winding at this instant will be in the same direction in the core as the primary field, or clockwise, as shown in Fig. 5.6. Application of the right-hand rule for coils will then show that current flow in the secondary will be in the direction indicated by the arrows, and the lower terminal of the secondary winding will be positive.

Fig. 5.6 Showing terminal polarities and the directions of the magnetic fields produced by currents in the primary and secondary windings of a transformer at the instant when the current is falling and has reached point y on the curve of current.

5.4 TRANSFORMER VOLTAGE RELATIONSHIPS

Since the mutual flux in the core of any transformer interlinks the turns of the primary and secondary windings, any change in this flux that induces a certain voltage in *each* turn of the primary winding must also induce the same voltage in *each* turn of the secondary winding. Then, if the secondary winding has only half as many turns as the primary, total secondary voltage will be only half the induced or impressed primary voltage. If the secondary has twice as many turns as the primary, the total induced secondary voltage will be twice as great as the impressed primary voltage.

It may be concluded, then, that the total voltage induced in each winding is always proportional to the number of turns in the winding:

$$\frac{E_p}{E_s} = \frac{N_p}{N_s} \tag{5.1}$$

where E_p = impressed primary voltage
E_s = secondary voltage
N_p = number of primary turns
N_s = number of secondary turns

When there is no load connected to the secondary, the induced primary voltage has practically the same value as the impressed voltage; and even when the transformer is supplying a load, if the transformer is properly designed, there will be only a very slight difference between the impressed and induced primary voltages. In an efficient, well-designed transformer the voltage drops due to resistance and reactance will be very low. Therefore, the ratio E_p/E_s or N_p/N_s may be taken as the voltage ratio of the transformer.

example 5.3   What is the secondary voltage of the transformer shown in Fig. 5.7?

Fig. 5.7 What is the voltage across the secondary?

solution   Substituting in Eq. (5.1),

$$\frac{E_p}{E_s} = \frac{N_p}{N_s}$$

$$\frac{120}{E_s} = \frac{200}{700}$$

Solving for the unknown,

$$E_s = \frac{120 \times 700}{200} = 420 \text{ V}$$ *Answer*

example 5.4   The primary winding of a transformer contains 100 turns. How many turns must the secondary contain to step down a voltage of 120 to 12.6 V?

solution   By substitution in Eq. (5.1),

$$\frac{120}{12.6} = \frac{100}{N_s}$$

Solving for N_s,

$$N_s = \frac{100 \times 12.6}{120} = 10.5 \text{ turns}$$

5.5   TRANSFORMER CURRENT RELATIONSHIPS

The modern transformer is a highly efficient device. In fact, some large iron-core transformers have an efficiency exceeding 99 percent. This means that almost as much power may be taken out of the transformer as is put into it. Therefore, for practical purposes we may assume that the transformer has an efficiency of 100 percent. If the efficiency is 100 percent, primary and secondary power factors are equal. Power input to the transformer is given by $E_p I_p \times$ (power factor), and the output power is $E_s I_s \times$ (power factor). Assuming that the input and powers are equal,

$$E_p I_p \times \text{(power factor)} = E_s I_s \times \text{(power factor)}$$

Since power factor in the first member has the same value in the second member of the equation, it cancels out, leaving $E_p I_p = E_s I_s$ or

$$\frac{E_p}{E_s} = \frac{I_s}{I_p} \tag{5.2}$$

where E_p = primary impressed voltage
I_s = secondary current
E_s = secondary voltage
I_p = primary current

In a step-up transformer, if the secondary voltage is twice the primary voltage, secondary current is only one half the primary current.
Since

$$\frac{E_p}{E_s} = \frac{N_p}{N_s}$$

it follows that

$$\frac{N_p}{N_s} = \frac{I_s}{I_p} \tag{5.3}$$

Equation (5.3) shows that the currents in the primary and secondary windings are inversely proportional to the corresponding turns.

example 5.5   A transformer supplies 2.4 A at 6.3 V to a secondary load. What is the primary current if the primary is connected to a 120-V ac source?

solution   By substitution in Eq. (5.2),

$$\frac{E_p}{E_s} = \frac{I_s}{I_p}$$

$$\frac{120}{6.3} = \frac{2.4}{I_p}$$

Solving for I_p.

$$I_p = \frac{6.3 \times 2.4}{120}$$
$$= 0.126 \text{ or } 126 \text{ mA} \qquad\qquad Answer$$

example 5.6 A 120-V generator supplies a current of 20 A to a load some distance from the source. The total resistance of the transmission lines from the generator to the load is 2 Ω. The line drop, in this case, is therefore 2×20, or 40 V, and the voltage across the load is only $120 - 40$, or 80 V. The power loss in the lines is $20^2 \times 2$, or 800 W. These conditions are shown in Fig. 5.8.

Line drop = 2 X 20 = 40 V
Power loss in lines = 20^2 X 2 = 800 W

Fig. 5.8 Resistance of line wires between generator and load causes large power loss.

Fig. 5.9 Stepping up the line voltage at the generator results in reduction of line loss to insignificant value. Line current = 2 A; line drop = $2 \times 2 = 4$ V; power loss in lines = $2^2 \times 2 = 8$ W.

Using the same transmission lines, what will be the line drop and the line loss if a transformer is used to step up the voltage at the generator to 1200 V, and another transformer at the load is then used to step down the line voltage from 1200 to 120 V, as shown in Fig. 5.9.

solution Since the load takes 20 A at 120 V, and the primary of the load transformer operates at 1200 V,

$$\frac{E_p}{E_s} = \frac{I_s}{I_p} \qquad\qquad (5.2)$$

$$\frac{120}{1200} = \frac{I_s}{20}$$

$$I_s = \frac{20 \times 120}{1200} = 2 \text{ A} \qquad\qquad Answer$$

The line drop is found by Ohm's law:

$$E = IR = 2 \times 2 = 4 \text{ V} \qquad\qquad Answer$$

The power loss is

$$P = I^2R = (2)^2 \times 2 = 8 \text{ W} \qquad\qquad Answer$$

5.6 THE FUNDAMENTAL EQUATION OF THE TRANSFORMER

The average of any varying quantity is the average of all the instantaneous values that the variable will assume in changing from one value to another.

If the voltage impressed on the primary of a transformer has a sine waveform, as shown in Fig. 5.10, this will produce similar sine waves of both current and flux. (This assumes that the flux in the magnetic circuit is linearly related to the current in the winding. Although this is usually not true, it is a reasonable assumption when the flux does not approach the saturation point.) The average value of the induced voltage in either the primary or secondary of a transformer is given by

$$E_A = \frac{N\phi_m}{10^8 t} \qquad (5.4)$$

where E_A = average induced voltage
$\quad N$ = number of turns
$\quad \phi_m$ = maximum flux
$\quad t$ = time in seconds for the flux to rise from zero to its maximum value

Figure 5.10 shows that a sine wave rises from zero to maximum in one-fourth the time it takes to complete one cycle. Therefore, for a sine wave of frequency f

$$t = \frac{1}{4f}$$

Equation (5.4) may be written as

$$E_A = \frac{N\phi_m}{10^8(1/4f)} = \frac{4fN\phi_m}{10^8} \qquad (5.5)$$

Fig. 5.10 A voltage or current having a sine-wave form rises from zero to maximum in one-fourth of the time it takes to complete 1 Hz.

To change Eq. (5.5) to a form that gives effective or root-mean-square voltage E, instead of average values, we multiply by

$$\frac{0.707}{0.636}$$

Thus,

$$E = \frac{0.707}{0.636} \times \frac{4fN\phi_m}{10^8}$$

$$= \frac{1.11 \times 4fN\phi_m}{10^8}$$

$$= \frac{4.44fN\phi_m}{10^8} \qquad (5.6)$$

These equations show that the induced voltage is directly proportional to the maximum flux in the core, the frequency of the applied voltage, and the number of turns in the winding.

The permissible flux in a magnetic circuit is usually expressed in terms of the flux density B_m.

$$B_m = \frac{\phi_m}{A}$$

where $\quad B$ = lines per square inch if the area A is in square inches
$\quad B_m$ = gausses if the area is in square centimeters

Then Eq. (5.6) may be written as

$$E = \frac{4.44fNB_mA}{10^8} \tag{5.7}$$

Equation (5.7) is the fundamental equation of the transformer. In the transformer primary, the effective value of the induced voltage,

$$E_p = \frac{4.44fN_pB_mA}{10^8} \tag{5.8}$$

In the transformer secondary, the effective value of induced voltage,

$$E_s = \frac{4.44fN_sB_mA}{10^8} \tag{5.9}$$

Effective values are of interest since transformer ratings are in these values, and all meters, unless marked otherwise, indicate effective values.

example 5.7 The primary winding of a transformer contains 300 turns. Determine (a) at what maximum flux (ϕ) must the core operate if the impressed primary voltage is 120 V at 60 Hz? (b) How many turns must the secondary winding have to develop 12.6 V?

solution (a) The maximum flux $\phi = B \times A$; by substitution in Eq. (5.8),

$$E_p = \frac{4.44 \times 60 \times 300 \times \phi_m}{10^8} = 120$$

Solving for ϕ_m, we get

$$\phi_m = \frac{120 \times 10^8}{4.44 \times 60 \times 300} = 150\ 150 \text{ Mx} \qquad \qquad Answer$$

(b) By substitution in Eq. (5.9),

$$E_s = 12.6 \frac{4.44 \times 60 \times N_s \times 150\ 150}{10^8}$$

Solving for N_s, the result is

$$\frac{12.6 \times 10^8}{4.44 \times 60 \times 150\ 150} = 31.5 \text{ turns} \qquad \qquad Answer$$

example 5.8 A transformer primary winding has 450 turns, and its secondary has 2700. The cross-sectional area of the core is 1 in², and maximum flux density B_m in the core is 100 000 lines in². Line frequency is 60 Hz. Compute: (a) rated primary voltage E_p, (b) rated secondary voltage E_s, (c) effective voltage induced in each primary turn, (d) effective voltage induced in each secondary turn.

solution (a) Substituting in Eq. (5.8),

$$E_p = \frac{4.44 \times 60 \times 450 \times 100\ 000 \times 1}{10^8} = 119.88$$

(b) Substituting in Eq. (5.9),

$$E_s = \frac{4.44 \times 60 \times 1200 \times 100\ 000 \times 1}{10^8} = 719.28$$

(c) $\dfrac{119.88}{450} = 0.266$ V

(d) $\dfrac{719.28}{2700} = 0.266$ V

5.7 DISCUSSION OF NO-LOAD CONDITIONS

If the secondary winding of a transformer is left open-circuited, the primary current is very small, since the primary current is then opposed not only by the resistance of the winding but also by its inductive reactance which has a *choking* effect on the primary current. The no-load current produces the magnetic flux, and at the same time supplies the eddy current and hysteresis losses in the core. The no-load current, therefore, consists of two components: the magnetizing current component and a power-loss component. The magnetizing current component lags the applied primary voltage by 90°, but is always in phase with the flux. The core loss, or power component, is always in phase with the applied voltage.

Since the magnetizing current component lags the applied primary voltage by 90°, it must lead the induced primary voltage by 90°, because the primary applied and induced voltages are 180° out of phase with each other.

According to Lenz's law, the magnetizing current must lead the induced secondary voltage by 90°, since as the magnetizing current and flux decrease from their maximum positive values and pass through zero in a negative direction, both current and flux

Fig. 5.11   Use of dot notation to indicate phase of input and output signals: (*a*) The transformer symbol does not indicate the phase of input and output signals. (*b*) The dots indicate leads having the same phase. (*c*) Connection of a transformer for phase reversal.

will be changing at their greatest rates at the exact instant when they are passing through their zero values. Therefore, the greatest rate of change of flux in the core occurs at this very instant, and the induced voltages in primary and secondary windings will both reach their maximum values at this instant. This should not be misconstrued to mean that the voltage across the secondary winding must always be 180° out of phase with the voltage across the primary.

Figure 5.11 shows how the phase of the secondary voltage is related to the phase of the primary voltage. The symbol for a transformer in Fig. 5.11*a* does not give any indication of the phase of voltage across the secondary, since the phase of that voltage actually depends on the direction of the winding around the core. To solve this problem, the *dot notation* is used when the phase of the secondary is important. This notation is shown in Fig. 5.11*b* where the primary and secondary currents are in phase, and Fig. 5.11*c* where the currents are 180° out of phase.

5.8   VECTOR REPRESENTATION OF NO-LOAD CONDITIONS

Figure 5.12 is a vector diagram of conditions in a typical, unloaded step-down transformer. The flux lags the applied primary voltage by 90° and leads the induced primary and secondary voltages by 90°. The total no-load current is represented by I_E, and resolved into its core loss and magnetizing components by I_H and I_M, respectively. Since the core-loss current I_H is small in comparison with I_M, the magnetizing current is very nearly equal to the total primary current. Thus, the total no-load current is often called the *magnetizing current* or the *exciting current*. The cosine of the angle θ by which the no-load current lags the applied voltage is the power factor of the transformer at no load. Power factor is given by

$$\text{Power factor} = \frac{I_H}{I_E} \qquad (5.10a)$$

$$= \frac{I_H}{\sqrt{I_H^2 + I_M^2}} \qquad (5.10b)$$

where I_H = core-loss current
I_E = total no-load current
I_M = magnetizing current

The total power expended in the transformer at no load is equal to $EI \times \cos \theta$. The copper, or I^2R, loss at no load is very small and need not be taken into account in most practical problems.

For the following problem draw a vector diagram approximating the conditions of the problem. Fill in the known values and then the unknown values as they are determined.

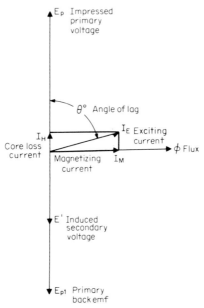

Fig. 5.12   Vector representation of no-load conditions.

example 5.9   When the secondary of a 120- to 240-V transformer is open, the primary current from a 120-V source is 0.24 A at a power factor of 0.16. The transformer is rated at 2 kVA. Calculate: (a) the primary no-load exciting current in percentage of full-load current; (b) the core-loss current in amperes and in percentage of full-load current; (c) the magnetizing current in amperes and in percentage of full-load current.

solution   (a) The transformer being rated at 2000 kVA, full-load primary current is

$$\frac{2000}{120} = 16.666 \text{ A}$$

$$\frac{\text{No-load primary current} \times 100}{\text{Full-load primary current}} = \% \text{ of full-load current}$$

Then,

$$\frac{0.24 \times 100}{16.666} = 1.44\%$$

(b) Since the power factor is the cosine of the angle of lag between the exciting current and the core-loss current, core-loss current equals

$$I_E \cos \theta = I_E \times \text{PF}$$
$$= 0.24 \times 0.16 = 0.0384 \text{ A}$$

and

$$\frac{0.0384 \times 100}{16.666} = 0.23\%$$

(c) Reference to Fig. 5.12 will show that the magnetizing current I_m may be determined by use of the Pythagorean theorem. (The square of the hypotenuse of a right triangle equals the sum of the squares of the other two sides.) In this case,

$$I_E^2 = I_M^2 + I_H^2$$

or

$$0.24^2 = I_M^2 + 0.0384^2$$

Solving for I_M results in

$$I_M = \sqrt{0.24^2 - 0.0384^2} = 0.2369 \text{ A}$$

and

$$\frac{0.2369 \times 100}{16.666} = 1.4214\% \text{ of full-load current}$$

In these results, it should be noted that the magnetizing current has nearly the same value as the no-load primary current. A vector diagram showing the conditions of this problem is given in Fig. 5.13.

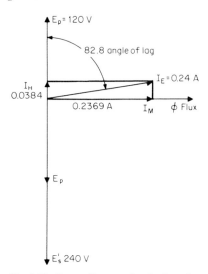

Fig. 5.13 Vector diagram of no-load conditions in the transformer of Example 5.9.

5.9 CONDITIONS UNDER LOAD

Figure 5.14 shows the transformer with a load. This illustration indicates the conditions existing at the moment when the terminal P_1 of the transformer is positive, and is increasing. Application of the right-hand rule for coils shows that the magnetic field produced by the primary is then in the direction shown in the illustration. By Lenz's law, the current in the secondary must be in such a direction that its associated magnetic field must oppose the building up of a magnetic field by the primary current. In other words, the magnetic field of the secondary will oppose the field of the primary when the field due to the primary current is building up.

Also, by Lenz's law, the induced secondary voltage must be in a direction that opposes the change that produced it. Hence, when a load is connected to the secondary

Fig. 5.14 As secondary load current increases, opposing magnetic field of secondary winding reduces inductance of primary, and more current flows into primary, thus maintaining core flux at its normal operating value.

winding, the current flowing in the secondary produces a magnetomotive force which opposes, or tends to oppose, the flux ϕ produced by the primary. Since an opposing flux, according to Eq. (5.6), results in a decrease in the back voltage of the primary, more current will flow into the primary as the secondary current increases! Therefore, any decrease of magnetization in the transformer core caused by current flowing in the secondary winding will immediately be counterbalanced by an increase in primary current. In this manner, the transformer automatically adjusts itself to load conditions, and the core flux is practically the same with full load as with no load.

Since the transformer is a highly efficient device, almost as much power is used up in the secondary as is fed to the primary. The power factor of the secondary, therefore, differs only slightly from that of the primary, and in most practical computations primary and secondary power factors may be considered equal to each other.

Because the core-loss and magnetizing current components of primary current are very much less than the load-current component, the core-loss and magnetizing current components may be added to the load-current component directly, instead of vectorially, except in those cases requiring the greatest accuracy. Such cases will be mostly of theoretical interest only.

example 5.10   The no-load current taken by a 120- to 240-V transformer is 0.6 A. The transformer is rated at 2.4 kVA. Assuming that the primary and secondary power factors are equal to each other, determine the primary current when the secondary is supplying its rated kilovoltamperes to a load that has a power factor of 0.9.

solution   Full-load secondary current:

$$I_x = \frac{2400}{240} = 10 \text{ A}$$

Since it is assumed that full-load primary and secondary power factors are equal, power factor may be disregarded in the calculations.

The transformer voltage ratio being 1 to 2, the primary component of the load current will be

$$2 \times 10 = 20 \text{ A}$$

Adding this directly to the 0.6-A no-load, or core-loss and magnetizing current component, the result is 20.6 A.

example 5.11   Determine the ratio of full-load secondary power to full-load primary power for the transformer of Example 5.10.

solution

$$\text{Secondary power} = 240 \times 10 \times 0.9$$
$$= 2160 \text{ W}$$
$$\text{Primary power} = 120 \times 20.6 \times 0.9$$
$$= 2224.8 \text{ W}$$

Ratio of full-load secondary power to full-load primary:

$$\frac{2400}{2472} = 0.9708$$

5.10 OPERATING A TRANSFORMER ON LOWER OR HIGHER THAN RATED FREQUENCY

Solving Eq. (5.8) for B_m (the maximum flux density in the core), we obtain

$$B_m = \frac{E_p \times 10^8}{4.44 \times f \times N_p \times A}$$

From this it may be seen that if we increase the voltage applied to the transformer primary, the flux will increase in proportion. The increased flux is, of course, due to the increased magnetizing ampere-turns, or magnetizing current.

For efficient operation, it is necessary to operate a transformer with a *maximum* flux density below the knee of the core magnetization curve, as shown in Fig. 5.15. Below the knee of the curve, an increase in magnetizing current will be out of proportion to the resulting increase in B_m. For example, with a magnetizing current of 0.5 A, the flux density will be 70 kG. If the applied primary voltage is increased to raise the flux density to 105 kG, the magnetizing current will rise to 2.0 A. Thus, as the flux density increased by 50 percent, the magnetizing current increased by 300 percent.

Equation (5.8) also shows that if the frequency f is decreased, B_m will again increase — and so will the primary current. Operating a transformer at considerably lower than rated frequency *could* cause the primary current to rise excessively, even to a value far in excess of the full-load primary current. The increased current would then burn out the winding.

From Eq. (5.8) it is evident that if the number of primary turns is increased, B_m decreases; but if the number of turns is decreased, B_m increases. Therefore, since B_m is directly proportional to the applied voltage and inversely proportional to the number of turns, changing the applied voltage and the number of turns in direct proportion would leave B_m and the primary current unchanged.

If we wanted to operate a transformer on a lower or higher than rated frequency, it would be impractical to change the number of its turns. However, since B_m is also directly proportional to the applied voltage and inversely proportional to the frequency,

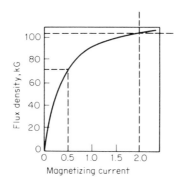

Fig. 5.15 Above the knee of the curve a large increase in magnetizing current results in a comparatively small increase in induction.

we could operate a transformer on lower than rated frequency, with normal flux, by decreasing the applied voltage in direct proportion to the decrease in frequency. For example, if we wished to operate a 120-V 60-Hz transformer at a frequency of only 25 Hz, it would be necessary to reduce the applied voltage to $^{25}\!/_{60}$ of 120, or to 50 V. The transformer would then still operate at its normal flux and safe current levels, but its power output — and input — capacity and its core losses would all be lower in almost direct proportion to the decrease in operating frequency. It follows that if a transformer is to be operated at normal flux and current levels but at a higher than rated frequency, its primary applied voltage must be increased. Its power output capacity and core losses would, consequently, also increase.

From the above relations between operating frequencies and voltages, we get the proportion

$$F_1 : F_2 = E_1 : E_2 \tag{5.11}$$

where F_1 = rated operating frequency of transformer
 F_2 = changed operating frequency
 E_1 = rated operating voltage of transformer
 E_2 = changed operating voltage

A transformer may be safely operated at normal flux and current levels at any frequency if the ratio of applied voltage to frequency is kept at approximately its same, normal operating value.

example 5.12 A 120-V 60-Hz transformer is to be operated at a frequency of 400 Hz. What voltage should be applied to the primary if the transformer is to operate at normal flux and current levels?

solution   By substitution in Eq. (5.11),

$$60{:}400 = 120{:}E_2$$

and solving for E_2;

$$E_2 = \frac{120 \times 400}{60} = 800 \text{ V} \qquad\qquad Answer$$

5.11 CONNECTING TRANSFORMERS IN SERIES AND PARALLEL

Occasions may often arise when it is necessary to connect two or more transformers in series or parallel, usually to obtain more current or more voltage than is available from only one transformer. It is essential that the transformers be properly phased when they are connected in series or parallel. Transformers may be damaged or ruined if they are connected together improperly.

SERIES OPERATION   Connecting two transformers together to obtain a higher voltage is a very simple matter. It is only necessary to connect the primaries together in parallel, and connect their secondaries in series and to a voltmeter, as shown in Fig. 5.16. If the secondaries are properly phased, the meter will indicate the sum of

Fig. 5.16 Phasing two transformers for series or parallel operation.

the secondary voltages. If the meter indicates the difference of the secondary voltages, the connection to one of the secondaries may be reversed, or one of the primary windings may be reversed.

PARALLEL OPERATION   If two transformers are to be connected in parallel, the transformers should be alike in every respect. They should have the same number of primary and secondary turns and the same secondary voltages, and the secondary voltages should remain the same under load. If the secondaries of two transformers having unequal voltages were to be connected in parallel, the transformer with the higher secondary voltage would, in addition to supplying its load, also send a current through the secondary winding of the other transformer. The result would be a large copper (I^2R) loss in the windings, which might produce enough heat to damage the winding insulation.

Even though two transformers may have the same primary and secondary voltages, they may differ in other respects. One winding may have a greater internal drop than the other when the transformers are supplying a load, and this would cause a difference in their terminal voltages under load.

The usual reason for connecting two transformers for parallel operation is that one does not have the necessary current-delivering capacity.

PHASING TRANSFORMERS FOR PARALLEL OPERATION   If the secondaries of two transformers are not properly phased for parallel operation, the result is a short circuit that burns out the windings. Even if two transformers of the same make have the same numbered or lettered terminals, it should not be assumed that they will be properly phased for parallel operation.

To phase two transformers for parallel operation, we connect the primaries in parallel and the secondaries in series — just as we would in phasing the transformers for series operation — as shown in Fig. 5.16. The secondary terminals of one of the transformers are arbitrarily tagged S_1 and F_1 and those of the other transformer S_2 and F_2. If the voltmeter across the terminals, S_1 and S_2, reads twice the voltage of either transformer, we connect terminal S_1 to F_2 and terminal F_1 to S_2. However, if the voltmeter reading is zero, S_1 and F_2 have opposite polarities, and we connect S_1 to S_2 and F_1 to F_2.

The phasing of transformers with high-voltage secondaries or primaries should be done with lowered excitation voltages in order to reduce high-voltage shock hazard.

Fig. 5.17 What will be the result if connections to one of the primaries are reversed?

example 5.13 Two transformers are connected together for parallel operation, as in Fig. 5.17. If the connections to the primary of transformer A are accidentally reversed, what will be the result?

solution The polarity of the secondary winding of transformer A will also be reversed, and the result will be a short circuit of the secondary windings.

5.12 TRANSFORMER LOSSES

Power losses in a transformer occur in the transformer core material and in the transformer primary and secondary windings.

Eddy-current losses, as explained in Chap. 4, are the result of currents circulating in the core and producing heat. The circulating current in the core is set up by induced voltages in the core material. These voltages are the result of the changing magnetic flux.

Hysteresis loss is the energy lost by reversing the magnetic field in the core as the magnetizing alternating current rises and falls and reverses direction. Hysteresis loss was also described in Chap. 4.

COPPER LOSS Copper loss is the power lost in the primary and secondary windings of the transformer. It is due to the ohmic resistance of the windings. Copper loss is obtained by the equation

$$\text{Copper loss} = I_p^2 R_p + I_s^2 R_s$$

where R_p = resistance of primary
           R_s = resistance of secondary
      I_p, I_s = respective primary and secondary currents

DETERMINATION OF COPPER LOSS Copper loss may be determined by measuring the ohmic resistance of each winding and then computing the I^2R loss in both primary and secondary windings for any given secondary load current.

The copper loss in both windings may also be determined by means of a wattmeter, and a circuit diagram showing the proper connections for this method is given in Fig. 5.18. If a transformer has multiple primaries, secondaries, or both, all the coils must be connected into the circuit for this test.

At the start of the test, the variable-voltage autotransformer is set at zero; its setting is then gradually raised until the ammeter connected across the secondary indicates rated full-load secondary current (assuming that we wish to determine copper loss at full load). At this point the wattmeter will indicate the power loss in both windings and also a very small core loss. Usually the core loss is too low to be indicated on the wattmeter.

Fig. 5.18 Short-circuit test for copper loss.

Although rated primary and secondary currents flow through the windings when this method is used, core loss is negligible because the voltage impressed on the primary has only a very low value—perhaps less than 5 percent of normal. This low primary voltage is still enough to cause rated secondary current to flow in the secondary winding, which is practically short-circuited by the ammeter connected across its terminals. Since the transformer is operating at only about five percent of rated voltage, the maximum flux density B_m in the core is also only about five percent of normal. Since core loss is almost directly proportional to B_m^2, the core loss with five percent of normal impressed primary voltage will be 0.05×0.05 or 0.0025 ($^{25}/_{100}$ of 1 percent) of normal. As an example, if normal core loss in a transformer is 50 W, the loss on short circuit with five percent of rated primary voltage will be $0.05^2 \times 50$, or 0.125 W, a value too small to be noticeable on the wattmeter.

The above test for copper loss is known as the *short-circuit test* method. Copper loss by this method may be determined from either the primary or secondary side of the transformer; the high side will usually be more convenient because on short circuit it would take only about five percent of full-rated voltage to cause full-load current to flow through the windings, but it might not always be convenient to adjust the applied voltage to the proper low value needed for a test from the low side.

CORE-LOSS TEST Core loss in a transformer may be determined by applying the rated voltage to the transformer primary at its rated frequency and measuring the power input with the secondary open-circuited. Power input is measured by means of a wattmeter. Figure 5.19 shows a diagram of the necessary connections. The variable voltage autotransformer is adjusted to apply rated voltage to the transformer primary. This will be indicated by the VTVM (or other high-resistance meter) connected across the primary. (If a low-resistance voltmeter were to be used to measure the applied voltage, its power consumption would be added to the core-loss power, and the wattmeter would then give a false core-loss reading. Therefore, if a low-resistance voltmeter is used to measure the applied voltage, the wattmeter reading should be taken with the circuit to the voltmeter open. The input power measured by the wattmeter will include a small copper loss, but at no load this is of no consequence and may be ignored.

If the rated primary voltage of the transformer has a very high value, or whenever it is higher than the secondary voltage, core loss may be determined by applying rated secondary voltage to the transformer. This is called the *open-circuit test*, and it is

Ac amperes

Watts

To ac

Variac

VTVM

Volts

Low side   High side

Fig. 5.19 Open-circuit test for core loss.

usually more convenient and safer for any step-down transformer. It gives the same results as a core-loss measurement taken on the high or primary side of the transformer.

example 5.14 A 240- to 24-V step-down transformer has a full-load secondary current rating of 50 A. A short-circuit test for copper loss at full load gives a wattmeter reading of 100 W. If the resistance of the primary is 0.8 Ω, what is the resistance of the secondary, and what is the secondary power loss? (Ignore the small core loss.)

solution Since full-load secondary current is 50 A, and the voltage ratio of the transformer is 240 to 24, or 10 to 1, the full-load primary current is $^{50}/_{10}$ or 5 A. Total copper loss equals

$$100 \text{ W} = I_p^2 R_p + I_s^2 R_s = 5^2$$

Then,

$$R_s = \frac{100 - 5^2(0.8)}{50^2} = \frac{100 - 20}{2500}$$

$$= \frac{80}{2500} = 0.032 \ \Omega$$

Secondary power loss is

$$50^2 \times 0.032 = 80 \text{ W}$$

example 5.15 On an open-circuit test for core loss, the transformer of Example 5.14 takes 1.2 A from a 240-V ac source. If the wattmeter reading is 80 W, determine (a) the copper loss at zero load; (b) the core loss.

solution (a) Since the primary resistance is 0.8 Ω, and the secondary is an open circuit, copper loss occurs only in the primary and is equal to

$$I_p^2 R_p = 1.2^2 \times 0.8 = 1.15 \text{ W}$$

(b) The wattmeter reading minus the copper loss equals the core loss, or

$$80 - 1.15 = 78.8 \text{ W}$$

5.13 EFFICIENCY

The efficiency of a transformer, like the efficiency of any machine, is the ratio of power output to power input, or in the form of an equation:

$$\text{Efficiency} = \frac{\text{power output}}{\text{power input}}$$

$$= \frac{\text{power output}}{\text{power output} + \text{copper loss} + \text{core loss}}$$

$$= \frac{E_s I_s \times \text{PF}}{(E_s I_s \times \text{PF}) + \text{copper loss} + \text{core loss}} \tag{5.12}$$

where PF = power factor of the load.

It might be assumed that transformer efficiency may be determined by connecting wattmeters in the primary and secondary of a transformer—that is, supplying a load, and then dividing the secondary wattmeter reading by that of the primary. However, this would be an incorrect assumption. Transformer efficiency is very high, and therefore the copper and core losses are very low. A wattmeter connected in the secondary circuit of a transformer would read very nearly the same as a wattmeter connected in the primary, and because of meter inaccuracies and errors introduced in reading the meters, it would be impossible to determine power input and output with the degree of precision necessary for computing transformer efficiency.

To compute efficiency, core loss must be determined. This can be done by the core-loss test method described earlier in this chapter. Copper loss may be determined by the measured resistance method, or by the short-circuit test. These core and copper loss factors are substituted in Eq. (5.12), which will then give the efficiency.

example 5.16 An open-circuit test for core loss in a 240- to 720-V 5-kVA transformer gives a value of 60 W. The measured resistance of the low side winding (primary) is 0.025 Ω, and that of the high side (secondary) is 1.25 Ω. If the power factor of the load is 0.84, what is the efficiency of the transformer at full load?

solution Full-load secondary current 5000/720; and full-load secondary copper loss,

$$\left(\frac{5000}{720}\right)^2 1.25 = 60.28 \text{ W}$$

Full-load primary current 5000/240; and full-load primary copper loss.

$$\left(\frac{5000}{240}\right)^2 \times 0.025 = 10.85 \text{ W}$$

Total copper loss,

$$60.28 + 10.85 = 71.1 \text{ W}$$

Efficiency,

$$\frac{5000 \times 0.84}{(5000 \times 0.84) + 60 + 71.1} = 0.971 \text{ or } 97.1\%$$

5.14 EQUIVALENT RESISTANCES

In certain computations it is sometimes useful to assume that the copper loss in both primary and secondary windings of a transformer actually occurs in only the primary; this is equivalent to assuming that the wattmeter in a short-circuit test for copper loss actually indicates power being expended in a purely resistive load. In the form of an equation this may be written as

$$W = I_p^2 R_{ep}$$

or

$$R_{ep} = \frac{W}{I_p^2} \tag{5.13}$$

where W = wattmeter reading on a short-circuit test
$\quad\quad I_p^2$ = primary current squared
$\quad\quad R_{ep}$ = equivalent resistance of transformer

It is important to note that R_{ep} in this case is *not* the dc resistance of the primary, but merely an assumed resistance that would give the same *IR* drop as the *combined IR* drops of both primary and secondary windings.

It is also possible to express the equivalent primary resistance in terms of the resistances of both primary and secondary windings in the form

$$R_{ep} = R_p + R_s \left(\frac{N_p}{N_s}\right)^2 \tag{5.14}$$

And it is possible to express the *equivalent secondary resistance* R_{es} also in terms of both primary and secondary resistances, as follows:

$$R_{es} = R_s + R_p \left(\frac{N_s}{N_p}\right)^2 \tag{5.15}$$

From Eqs. (5.14) and (5.15) we also get

$$\frac{R_{ep}}{R_{es}} = \frac{N_p^2}{N_s^2} \tag{5.16}$$

example 5.17 A 1200- to 120-V transformer is rated at 5 kVA. The primary resistance is 1 Ω, and secondary resistance is 0.05 Ω. Determine (a) the equivalent primary resistance R_{ep}; (b) the equivalent secondary resistance R_{es}; (c) the total copper loss I^2R, using the determined equivalent secondary resistance; (e) the total copper loss, using the dc resistances of the primary and secondary windings.

solution (a) The equivalent primary resistance,

$$R_{ep} = 1 + 0.05 \left(\frac{N_p}{N_s}\right)^2$$

Since the voltage ratio of the transformer is 1200/120, or $^{10}/_1$, N_p/N_s also equals $^{10}/_1$, or 10. Therefore

$$R_{ep} = 1 + 0.05(10)^2$$
$$= 6 \ \Omega$$

(b) The equivalent secondary resistance,

$$R_{es} = 0.05 + 1 \left(\frac{N_s}{N_p}\right)^2$$

$$= 0.05 + (^1/_{10})^2 = 0.06 \ \Omega$$

(c) Primary current,

$$I_p = \frac{5000}{1200} = 4.175 \ A$$

Copper loss computed from R_{ep} and I_p values equals

$$4.175^2 \times 6 = 104.58 \ W$$

(d) Secondary current,

$$I_s = \frac{5000}{120} = 41.75 \ A$$

Copper loss computed from R_{es} and I_s values equals

$$41.75^2 \times 0.06 = 104.58 \ W$$

(e) Primary copper loss equals

$$(I_p^2 R_p) = 4.175^2 \times 1 = 17.43 \ W$$

Secondary copper loss equals

$$(I_s^2 R_s) = 41.75^2 \times 0.05 = 87.15 \ W$$

Total copper loss,

$$I_p^2 R_p + I_s^2 R_s = 17.43 + 87.15 = 104.58 \ W$$

5.15 VOLTAGE REGULATION

The voltage regulation of a transformer in percent of the rated full-load secondary voltage is given by the equation

$$\text{Percent regulation} = \frac{\text{no-load secondary volts} - \text{full-load secondary volts}}{\text{full-load secondary volts}} \times 100 \quad (5.17)$$

The voltage regulation of small transformers may easily be determined by adjusting the primary voltage so that the transformer delivers its rated kilovoltamperes at its rated secondary voltage to a load at some specified power factor. Then, disconnecting the load but keeping the primary voltage constant will cause the secondary voltage to rise to its no-load value. Substituting the observed values of full-load and no-load voltages in Eq. (5.17) will give the percent regulation.

The above method, however, would not be very convenient for large heavy-duty transformers, especially when such transformers may also have high-voltage primary or secondary windings. For such transformers it is better to compute regulation from the short-circuit test data. The computation for a load having either a lagging or leading power factor is somewhat more complicated than for a load having unity power factor.

Regulation may be computed in terms of either the primary or secondary winding, and the first step, in either case, is to compute the equivalent IR and IX drops for the winding from the short-circuit test data. Determining regulation in terms of the secondary winding is equivalent to assuming that the secondary is really the primary; and regardless of whether we compute regulation in terms of either the primary or secondary winding, the no-load voltage will always be greater than the rated full-load voltage of the transformer.

Vector diagrams for computing regulation from known values of $\mathbf{E}_R$, I_p, and $\mathbf{IX}$ are shown in Fig. 5.20. A vector diagram for a transformer supplying a unity power factor load is shown in Fig. 5.20a and for a load with a lagging power-factor in Fig. 5.20b. The vector $\mathbf{OE}_R$ represents the rated full-load voltage (primary or secondary) of the transformer. Vectors $\mathbf{IR}$, $\mathbf{IX}$, and $\mathbf{IZ}$ are the equivalent (primary or secondary) resistance, reactance, and impedance drops. Vector $\mathbf{OE}_I$ is the vector sum of $\mathbf{E}_R$ and $\mathbf{IZ}$, and it represents the no-load voltage of the transformer. In (a) the current I is in phase with $\mathbf{E}_R$, since the load has a power factor of unity. In (b) the current I lags $\mathbf{E}_R$ by the angle θ, corresponding to the power factor (cosine of the angle of lag) of the load. In

(a)

(a)

(b)

(b)

Fig. 5.20 Vector diagrams for computing regulation from known values of $\mathbf{E}_R$, $\mathbf{I}_p$, and $\mathbf{IX}$. $\mathbf{E}_R$ equals the rated voltage of the transformer. $\mathbf{IR}$, $\mathbf{IX}$, and $\mathbf{IZ}$ are the equivalent values of resistance, reactance, and impedance. (a) Unity power-factor load. (b) Lagging power-factor load.

Fig. 5.21 Vector diagrams for Example 5.18. Computing $\mathbf{OE}_I$, the no-load voltage, is a problem in simple trigonometry. (a) Partially solved problem. (b) Completed problem.

either case, the problem is to find the value of the vector $\mathbf{OE}_I$, the no-load voltage of the transformer, from the known values of $\mathbf{E}_R$, I_p, and $\mathbf{IX}$.

example 5.18 On a short-circuit test, a 2400- to 240-V, 10 kVA transformer takes 4.17 A at an applied primary voltage of 110 V. The wattmeter reading is 221 W. Assuming a unity power factor load, compute the regulation in terms of the primary. Draw a vector diagram illustrating the conditions of the problem, and fill in the unknown values as they are calculated.

solution Since the power factor of the load is unity, the applied voltage and the current will be in phase. A vector diagram of circuit conditions will, therefore, be similar to the one shown in Fig. 5.20a. The voltmeter reading is the equivalent primary impedance times the primary current:

$$I_p Z_{ep} = 110 \text{ V}$$

The equivalent primary resistance in volts is

$$I_p R_{ep} = \frac{W}{I_p} = \frac{221}{4.17} = 53 \text{ V}$$

Figure 5.21 shows the vector diagrams for this problem. Figure 5.21a illustrates the conditions of the problem at this stage. By simple trigonometry.

$$(I_pX_p)^2 = 110^2 - 53^2$$

Hence the equivalent primary reactance in volts is

$$I_pX_p = \sqrt{110^2 - 53^2} = 96.39 \text{ V}$$

Since $\mathbf{E}_R$ is in phase with I_pR_p, we may add $\mathbf{E}_R$ to I_pR_p directly, or

$$\mathbf{E}_R + I_pR_p = 2400 + 53 = 2453$$

and again using simple trigonometry,

$$\begin{aligned}(\mathbf{OE}_l)^2 &= 2453^2 + (I_pX_p)^2 \\ &= 2453^2 + 96.39^2\end{aligned}$$

Solving for the no-load voltage

$$\begin{aligned}\mathbf{OE}_l &= \sqrt{2453^2 + 96.39^2} \\ &= 2451.53 \text{ V}\end{aligned}$$

$$\text{Regulation} = \frac{2451.53 - 2400}{2400} \times 100 = 2.147\%$$

Figure 5.21b is a vector diagram of the completed problem.

5.16 THE ISOLATION TRANSFORMER

An isolation transformer is used whenever it is necessary to avoid a direct electric connection between a piece of electric equipment and the power lines or other source of power. This is useful and very often even necessary where equipment must be grounded for increased efficiency, to reduce noise, or for other reasons. Since one side of the ac power line is grounded, connecting a piece of grounded equipment to the lines could result in a short circuit unless precautions are taken to assure that the grounded side of the equipment is connected to the grounded side of the power lines. There are also certain types of equipment — such as those with ac-dc power supplies — in which a lethal shock *could* be received when the chassis is exposed. (It would be necessary to touch a ground point *and* the exposed chassis to receive such a shock.) By using an isolation transformer, the chassis becomes a floating ground, and it is nearly impossible to receive a deadly shock under such conditions.

Since equipment may frequently be used in different locations and because power outlets in the same building or even in the same room are not usually polarized, a transformer between the equipment and the power lines is a very convenient and practical method of avoiding the possibility of damaging a piece of electric equipment that must be grounded. Since the transformer serves to isolate the equipment from the power lines, it is called an isolation transformer; and it is evident that any transformer may be used as an isolation transformer if it has the necessary voltampere rating and the required primary to secondary turns or voltage ratio.

5.17 THE AUTOTRANSFORMER

In an ordinary transformer the primary and secondary windings are magnetically coupled, but there is no direct electric connection between the primary and secondary windings. In an autotransformer, the primary and secondary windings are not only magnetically coupled, but there is also a direct electric connection between the windings. In fact, the secondary winding may be merely a continuation of the primary.

Any ordinary transformer may be used as an autotransformer. For example, consider the 1-to-1 isolation transformer of Fig. 5.22 in which there is a direct series connection between the primary and secondary windings. With the transformer phased as shown, the impressed primary voltage is added to the induced secondary voltage, and the total voltage across the load, connected to terminals A and Y, is

$$120 + 120 = 240 \text{ V}$$

Thus we have a step-up ratio of 1 to 2 in a transformer with a primary to secondary turns ratio of only 1 to 1. Figure 5.23 shows some additional connections of autotransformers in which a transformer with a 1-to-1 ratio is used to step up the voltage (Fig. 5.23a), and to step down the voltage (Fig. 5.23b). In Fig. 5.23a the windings of the transformer are wound on a straight, open core, and this more clearly illustrates the conditions of the transformer of Fig. 5.22. This same transformer may also be used as a step-down autotransformer by connecting the load across terminals A and B and connecting terminals A and Y to a power source, as shown in Fig. 5.23b which merely interchanges the line and load connections of Fig. 5.23a. In other words, the transformer of Fig. 5.23a is operated in reverse—just as we might operate any transformer in reverse by using the secondary, with rated applied voltage, as the primary and then connecting the load to the normal primary.

Fig. 5.22   A 1-to-1 isolation transformer, connected for use as an autotransformer, gives a stepped-up voltage ratio of 1 to 2.

Fig. 5.23   Some additional connections of autotransformers: (a) Autotransformer connection steps up impressed primary voltage from 120 to 240 V. (b) Same 1:1 ratio transformer used in reverse steps down impressed line voltage of 240 to 120 V.

Fig. 5.24   Autotransformer supplying a load. Current *transformed* is 5 A, and in this case the current *transferred* is also 5 A.

A schematic diagram of an autotransformer supplying a load is shown in Fig. 5.24. In this case we have a 2-to-1 step-down transformer. The number of turns between A and B is the primary, and the turns between B and C the secondary. If we assume that the transformer is 100 percent efficient and supplies a load having unity power factor, power input to the transformer will be

$$120 \times 5 = 600 \text{ W}$$

Power output will be

$$60 \times 10$$

which also equals 600 W, but the transformer primary power is only 60 × 5, or 300 W. This is also the transformer secondary power. Hence, the power *transformed*, by transformer action, is only 300 W; and the power *transferred* directly to the secondary load, by means of the connection between the primary and secondary windings, is the remaining 300 W, making the total of 600 W supplied to the load.

Since, in this case, each half of the autotransformer supplies five amperes to the load, copper loss is only one-half the copper loss that would occur in a transformer with separate primary and secondary windings and supplying the same load with the same current.

As the primary to secondary voltage ratio of an autotransformer decreases, the copper loss decreases in proportion, and the amount of power *transferred* directly – instead of by transformer action – increases in proportion.

Autotransformers are seldom used when it is necessary to reduce a very high voltage to a much lower value. For example, a voltage ratio of 1200 to 120, or 10 to 1, would result in a small reduction of copper loss, and the power *transferred* directly would also be small. Furthermore, a dangerous shock hazard would exist in the secondary, since there would be no isolation between the high-voltage primary and the low-voltage secondary.

If an ordinary transformer having several windings is to be used as an autotransformer by connecting all the windings together, the current carried by any winding must not be greater than its normal rated current.

example 5.19 The windings of an ordinary 1200- to 120-V 5-kVA transformer are connected in series to form a step-up autotransformer, as shown in Fig. 5.25. Assuming that the

Fig. 5.25 Ordinary transformer connected as autotransformer; in this case power capacity is increased for 5000 to 55 000 W.

transformer is 100 percent efficient and operates at a power factor of unity with rated secondary current, determine (a) the power delivered to the load; (b) the total current supplied by the source; (c) the increase in the power rating of the transformer when it is used as an autotransformer; (d) the power transferred.

solution (a) Since rated secondary current is

$$\frac{5000}{120} = 41.66 \text{ A}$$

the current in the load may also be 41.66 without overloading the secondary winding, XY. Also, since the voltage across the load is 1200 + 120 or 1320 V, the power supplied to the load is

$$1320 \times 41.66 = 55\ 000 \text{ W}$$

(b) Since we have assumed 100 percent efficiency, power supplied by the source is equal to the power delivered; current taken from the source is, therefore,

$$\frac{55\ 000}{1200} = 45.83 \text{ A}$$

(c) Normal power rating of the transformer is 5000 W. Power rating of the transformer when used as an autotransformer is 55 000 W. Percentage increase in power rating,

$$\frac{55\ 000}{5000} \times 100 = 1100\%$$

(d) Power transferred is

$$55\ 000 - 5000 = 50\ 000 \text{ W}$$

5.18   THE VARIABLE AUTOTRANSFORMER

The variable autotransformer is usually better known under one of its commercial trade names, such as Variac, Powerstat, Adjust-A-Volt. But regardless of what we call it, its basic principle of operation remains the same. Figure 5.26 is a photograph of a commonly used Variac. The exploded view clearly shows all the basic parts. Figure 5.27 shows several autotransformer circuits.

Figure 5.27a illustrates the most basic type of variable autotransformer. The coil is a single-layer winding over a toroidal core of high permeability. The sliding brush contact, which is controlled by its attached knob or dial, may be set to tap the winding at any point about its periphery. Thus, the output voltage across the load, connected to terminals 1 and 2, may be varied from zero to full line voltage. A slight modification of the design, as shown in Fig. 5.27b, permits the voltage to be varied from zero to slightly (approximately 17 percent) above line voltage. This feature is useful when-

Fig. 5.26 An exploded view of a Variac. The brush, shown at lower left, would normally contact the front outer perimeter of the winding. (*Courtesy General Radio Company.*)

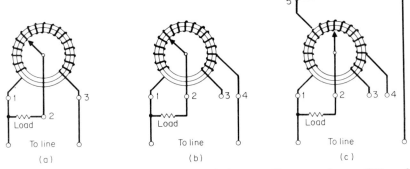

Fig. 5.27 Connections for autotransformers: (a) The basic variable autotransformer. (b) Tapped winding permits adjustment from zero to slightly above line voltage. (c) Tapping the winding at a lower point permits adjustment from zero to more than twice the line voltage.

ever it is necessary to compensate for the drop in line voltage that occurs when the power lines are heavily loaded. But if necessary, this design may also be connected for a range of zero to line voltage by joining terminals 1 and 3, instead of 1 and 4, to the lines.

An additional modification of the basic design, as shown in Fig. 5.27c, gives a voltage range adjustable from zero to slightly more than twice the line voltage (from 0 to 280 V for one model). Terminal 5 is connected to a tap on the winding at a point slightly less than midway between the total number of turns.

The rated current of a variable autotransformer should not be exceeded at any voltage setting of the transformer, and it should not be assumed that greater than rated current may safely be drawn at voltage settings below line voltage — as we might use an ordinary step-down transformer to supply a much greater current to a secondary load than the current supplied to the primary.

When the variable autotransformer is used to supply a fixed, known load, without using the overvoltage connection, maximum current may be supplied to the load at full line voltage. Then as the voltage setting is reduced, the decrease in current with decreasing output voltage will tend to keep the load current within safe limits for any voltage setting. Accidental overload damage may easily be avoided by fusing the variable autotransformer for rated current.

When it is necessary to supply a load at some reduced voltage but with a current considerably in excess of the normal current rating of a particular variable autotransformer, an auxiliary transformer, connected to the output terminals of the variable autotransformer, may be used, as shown in Fig. 5.28. For the transformer and circuit of Fig. 5.28, the primary-to-secondary voltage ratio of the transformer is 12 to 1, and normal rated current of the variable autotransformer is two amperes. Current taken by the load may, therefore, be as great as 24 A at voltages between 0 and 10 V. In this case, the use of an auxiliary transformer has increased current capacity 12 times. In any application of this

Fig. 5.28 The use of supplementary transformer in conjunction with low-current variable autotransformer increases current capacity 12 times.

sort, the voltampere rating of the transformer should be equal to the voltampere rating of the variable autotransformer.

USING VARIABLE AUTOTRANSFORMERS AT HIGHER THAN RATED VOLTAGES

A single variable autotransformer should not, of course, be operated at higher than normal rated voltage. However, two variable autotransformers may be ganged for high-voltage operation, as shown in Fig. 5.29. Thus, if the normal voltage rating of each

Fig. 5.29 Two 120-volt variable autotransformers ganged for operation on 240 V.

transformer is 120 V, the ganged transformers may be used at 240 V, or two 240-V models may be used at 480 V.

SWITCHING Before a variable autotransformer is switched into or out of a circuit, its setting should be reduced to zero to prevent excessive surge currents from damaging the windings. This is especially important when a variable autotransformer is used to supply a load that has a very high ratio of hot-to-cold resistance. For example, the hot-to-cold resistance ratio of the heaters or filaments of vacuum tubes or of incandescent lamps may be as great as 15 to 1.

example 5.20 A 120-V 1-to-1 isolation transformer is available. Draw a diagram showing how this transformer should be connected to a 120-V Variac and to a load that is to be operated at various voltages ranging from 0 to 240 V.

solution The primary winding of the transformer is connected to the Variac, and then the transformer itself is connected to form an autotransformer, as shown in Fig. 5.30. This gives a

Fig. 5.30 Increasing the voltage range of a Variac.

step-up ratio of 1 to 2. If the voltmeter indicates zero as the Variac setting is increased, the connections to either terminals 1 and 2 or 3 and 4 should be reversed — for proper phasing.

5.19 INSTRUMENT TRANSFORMERS

Instrument transformers are designed for use with measuring instruments, relays, or control devices. They are especially designed to maintain a specific relationship in magnitude and phase between the primary and secondary currents or voltages.

THE POTENTIAL TRANSFORMER Since it would hardly be feasible to construct voltmeters to directly measure the extremely high voltages encountered in power distribution systems, such high voltages cannot be measured by connecting a voltmeter directly across the lines. Even if voltmeters having the necessary range and insulation resistance were available, it would be very dangerous to bring leads from the high-voltage lines directly into the instrument switchboards and to the meters. A dangerous high-voltage shock hazard would then be a constant threat to personnel at the switchboard. Therefore, instead of bringing the high-voltage leads directly to a meter, these leads are connected to one side of a small transformer, called a potential transformer, as shown in Fig. 5.31. In Fig. 5.31*a* the potential transformer steps down the high volt-

Fig. 5.31 Instrument transformers connected to measure high voltage and current: (*a*) Potential transformer. (*b*) Current transformer.

age to a very much lower value that may be safely brought to a low-voltage meter. Usually, the meter has its scale calibrated to indicate the high voltage across the transformer's primary terminals. Since the potential transformer has very high primary-to-secondary insulation resistance, there is no high-voltage shock hazard at the meter terminals.

THE CURRENT TRANSFORMER A current transformer, instead of being used to step down a high voltage, is used to step down a high current to a much smaller and more conveniently measurable value.

The method of connecting a current transformer into the line is shown in Fig. 5.31*b*. Since a current transformer is always connected in series with the line, it is also known as a *series* transformer. The primary, or line side, of a current transformer usually has only a few turns—sometimes only one—since it must carry a very large current. The secondary, which is connected to a low-range ammeter, has a much larger number of turns. The meter scale is calibrated to indicate the line current in the primary of the current transformer. This line current is, of course, very much greater than the current in the secondary of the current transformer.

Although the use of a current transformer completely insulates the secondary and the ammeter from the high-voltage lines, the secondary of a current transformer should never be left open-circuited. To do so may result in a dangerous high voltage being induced in the secondary. Therefore, if on any occasion it should be necessary to disconnect the meter used with a current transformer, the transformer secondary terminals should first be short-circuited, and this can do no harm because the primary current will still be limited to the amount that the load will permit to flow.

example 5.21 The voltage across the secondary of a potential transformer is 180 V. The primary, or line side, of the transformer has 24 000 turns, and the secondary has 600 turns. What is the line voltage?

solution Dividing 24 000 by 600 shows that the turns or voltage ratio of the transformer is 40 to 1. Then, since the secondary voltage is 180, the primary or line voltage is

$$40 \times 180 = 7200 \text{ V}$$

example 5.22 The current ratio of a current transformer is 100 to 1. What is the line current when an ammeter connected to the secondary indicates 3.5 A?

solution The current ratio being 100 to 1, the line current is

$$100 \times 3.5 = 350 \text{ A}$$

5.20 TRANSFORMERS IN TWO- AND THREE-PHASE SYSTEMS

Power transformation in a three-phase system may be accomplished by using either three separate but exactly similar single-phase transformers or a single three-phase transformer. In either case, the transformers may be of the closed core type or of the shell type.

A three-phase system may also be changed to two phase, or a two phase to three phase. Secondary voltage of the changed system may be the same as its primary voltage, or it may be stepped up or stepped down as the phase is being changed.

CONNECTING THREE SEPARATE TRANSFORMERS FOR THREE-PHASE OPERATION Figure 5.32 shows separate transformers of both the closed core (Fig. 5.32*a*) and shell types stacked for three-phase operation (Fig. 5.32*b*). The three primary windings

(*a*)   (*b*)

Fig. 5.32 Three separate single-phase transformers stacked for three-phase operation: (*a*) Core types. (*b*) Shell types.

and the three secondaries may be connected together in two different ways. Thus, there is a total of four different ways in which a bank of three transformers may be used for transforming three-phase power. These are the four ways: (*a*) wye (Y) primary to wye (Y) secondary; (*b*) delta (Δ) primary to delta (Δ) secondary; (*c*) wye (Y) primary to delta (Δ) secondary; (*d*) delta (Δ) primary to wye (Y) secondary.

In using the three separate transformers, the primaries may be connected together in either wye or delta in any arbitrary manner; but the secondaries will then have to be phased, and terminal polarities must be taken into consideration in connecting them together.

WYE-TO-WYE TRANSFORMATION   Figure 5.33 illustrates the two different methods that may be used to show a Y-to-Y connected transformer bank (Fig. 5.33*a*). This

(a)   (b)

Fig. 5.33 A wye-to-wye transformer connection: (*a*) The wye-to-wye connection of three separate but identical transformers for three-phase to three-phase transformation. (*b*) Alternate method of representing a wye-to-wye-connected three-phase bank.

connection is called a Y *connection* because of its resemblance to the letter Y when the connections are drawn as in Fig. 5.33*b*. A vector diagram of conditions in a Y-connected primary or secondary also takes the form of a Y.

If we assume that the transformer bank of Fig. 5.33 is connected to a three-phase 2200 V source, as shown, the actual voltage across each transformer primary winding will be only

$$\frac{2200}{\sqrt{3}}$$

or $0.5773 \times 2200 = 1270$ V

For any Y-connected primary the voltages across the transformer windings will differ in phase by 120°, and the vector sum of any two of these voltages, or the line voltage, will be $\sqrt{3}$ times the voltage across any primary. The current in each primary winding will be equal to the line current.

These same relations will also hold true for the secondary windings. In other words, the voltage between any two secondary output leads, L_1–L_2, L_2–L_3, or L_1–L_3, will be $\sqrt{3}$ times the voltage across any secondary winding, and the current in each secondary winding will be the same as the secondary line current.

DELTA-TO-DELTA TRANSFORMATION   Two methods of representing the transformer windings for a delta (Δ)-to-delta (Δ) transformation are presented in Fig. 5.34. Figure 5.34*a* shows the connections of the primary and secondary windings. Drawing the connected windings as in Fig. 5.34*b* shows that this connection resembles the

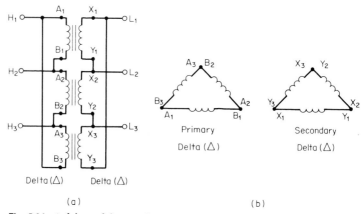

Fig. 5.34 A delta-to-delta transformer connection: (*a*) The delta-to-delta connection for three-phase transformation. The transformers must be identical. (*b*) Alternate method of representing a delta-to-delta three-phase bank.

Greek letter Δ. Hence, it is called a *delta connection*. Here, it is evident that the total voltage across any winding equals the line voltage because each winding is directly across the line. The secondary line voltages will be equal to the transformer secondary voltages. But the current in each winding, with a delta connection, will be

$$\frac{\text{Line current}}{\sqrt{3}}$$

The primaries in a delta-connected bank may be connected together in any random order, but the secondaries must be phased before they are connected together.

WYE-TO-DELTA TRANSFORMATION When the primaries are connected in wye and the secondaries in delta, as in Fig. 5.35, the primary-to-secondary voltage ratio of each transformer will not be the same as the ratio of primary-to-secondary line voltages. One method of illustrating this is shown in Fig. 5.35*a* and an alternative method is illustrated in Fig. 5.35*b*. In this case, the ratio of line voltages will be $\sqrt{3}$ times the voltage ratio of each transformer. For example, in transforming from a three-phase line

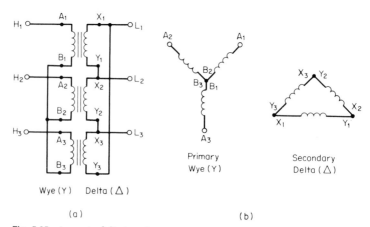

Fig. 5.35 A wye-to-delta transformer connection: (*a*) A wye-to-delta three-phase transformation. (*b*) Alternate method of representing a wye-to-delta transformation.

voltage of 210 V to a secondary line voltage of 105 V, the primary-to-secondary turns ratio of each transformer in a wye-to-delta transformer bank will be

$$\frac{210}{105\sqrt{3}} = \frac{2}{1\sqrt{3}} = \frac{1.156}{1} = 1.156{:}1$$

DELTA-TO-WYE TRANSFORMATION   The delta-to-wye connection, shown in Fig. 5.36, is merely a wye-to-delta connection in reverse. Therefore, in this case, the ratio of line voltages will be equal to

$$\frac{\text{Voltage ratio of each transformer}}{\sqrt{3}}$$

One way to illustrate a delta-to-wye connection is shown in Fig. 5.36a, and the alternative method is shown in Fig. 5.36b.

(a)   (b)

Fig. 5.36   A delta-to-wye transformer connection: (a) Delta-to-wye-connected transformer bank for three-phase operation. (b) Alternate method of representing the delta-to-wye connection.

Tables 5.1 and 5.2 give the voltage and current relations for both wye- and delta-connected transformers in a convenient form.

example 5.23   A bank of three transformers is connected for wye-to-delta transformation. Each transformer has a voltage ratio of 3 to 1, and the primary line voltage is 650 V. De-

TABLE 5.1   Current and Voltage Relationships in Δ and Y Transformers

	Wye (Y)	Delta (Δ)
Line voltage	Coil voltage × $\sqrt{3}$	Same as coil voltage
Line current	Same as coil current	Coil current/ $\sqrt{3}$

TABLE 5.2   Voltage Ratios for Three ϕ Transformers

	Transformer voltage ratios
Y to Y	Same as ratio of line voltages
Δ to Δ	Same as ratio of line voltages
Y to Δ	Ratio of line voltages/ $\sqrt{3}$
Δ to Y	Ratio of line voltages × $\sqrt{3}$

termine (*a*) the secondary line voltage; (*b*) the voltage across each primary winding; (*c*) the voltage across each secondary winding.

solution   (*a*) Transformer voltage ratio is

$$\frac{3}{1} = \frac{\text{ratio of line voltages}}{\sqrt{3}}$$

The ratio of line voltage is

$$\frac{3\sqrt{3}}{1} = \frac{5.1963}{1}$$

$$= \frac{650}{\text{secondary line voltage}}$$

Solving for secondary line voltage, we obtain

$$\frac{650}{5.1963} = 125.08 \text{ V}$$

(*b*) Since the primary line voltage is equal to

$$650 = \text{coil voltage} \times \sqrt{3}$$

primary coil voltage equals

$$\frac{650}{\sqrt{2}}$$

or

$$650 \times 0.5773 = 375.25 \text{ V}$$

(*c*) Secondary line voltage equals 125.08 V, and since the secondaries are connected in delta, the voltage across each secondary winding is also 125.08 V.

example 5.24   The secondary line voltage of a delta-to-wye transformer bank is 416 V. (*a*) If the transformers have a primary-to-secondary ratio of 5 to 1, what is the primary line voltage? (*b*) If the current in each secondary line is 40 A, what is the current in each secondary coil or winding? (*c*) Assuming that power delivered to the load equals the power taken from the lines, what is the primary line current?

solution   (*a*) Transformer ratio is 5 to 1:

$$\text{Ratio of line voltages} \times \sqrt{3} = \frac{\text{primary line voltage}}{\text{secondary line voltage}} \times \sqrt{3}$$

$$= \frac{\text{primary line voltage}}{416} \times \sqrt{3}$$

or

$$\text{Primary line voltage} = \frac{5 \times 416}{\sqrt{3}} = 1200.78 \text{ V}$$

(*b*) When the secondaries are connected in Y, the current in each secondary coil equals the line current, or 40 A.

(*c*) Since the transformer ratio is 5 to 1, and secondary current is 40 A, primary coil current is

$$\frac{40}{5} = 8 \text{ A}$$

The primary line current is

$$\frac{\text{primary coil current}}{\sqrt{3}} = \frac{8}{\sqrt{3}} = 4.618 \text{ A}$$

5.21   THE SCOTT OR T CONNECTION

For some industrial control or servosystem applications, it is at times necessary to convert from three-phase power to two phase. Use is then made of the Scott or T connection, which may not only transform from three phase to three phase but also from three phase to two phase or from two phase to three phase.

Transformer ratio 10 : 1

(a)

Primary   Secondary

(b)

Fig. 5.37 Characteristics of the Scott connection: (*a*) The Scott or T connection for three-phase to three-phase transformation. (*b*) Voltage relations in the 10:1 ratio transformer. Drawing the connection here shows why this connection is also called the T connection.

TWO-PHASE TO THREE-PHASE TRANSFORMATION Two exactly similar transformers, having a primary-to-secondary voltage ratio of 10 to 1, and connected for three-phase to three-phase transformation, are shown in Fig. 5.37. Figure 5.37*a* shows a circuit configuration and Fig. 5.37*b* shows the voltage relationships. Transformer T_1 is called the main transformer, and the center taps of its primary and secondary windings are connected to terminals C and c of transformer T_2, called the *teaser*.

The connections for transforming from two-phase power to three phase are shown in Fig. 5.38. In this connection (Fig. 5.38*a*) also, the two transformers T_1 and T_2 should have the same number of primary and the same number of secondary turns—and also the same regulation. The secondary of T_1 is tapped at its midpoint and connects to terminal c of the other transformer, T_2, which is tapped at a point that gives only 86.6 percent of its rated secondary voltage; thus, only 86.6 percent of the secondary turns of transformer T_2 are used. (See Fig. 5.38*b*.)

THREE-PHASE TO TWO-PHASE TRANSFORMATION The connections for changing from three phase to two phase are merely the reverse of the connections for changing from two phase to three phase. For example, applying a three-phase voltage of 100

Transformer ratio 10:1

(a)

Not used

Secondary

(b)

Fig. 5.38 The connections for transforming from two-phase power to three-phase: (*a*) Two Scott-connected transformers being used for two-phase to three-phase conversion. (*b*) Voltage relations in the secondary windings.

Fig. 5.39 Using two Scott-connected transformers to transform from three-phase power to two-phase reduces current capacity of each transformer to 86.6 percent of normal current rating—from 16 to 13.856 A in this case.

V to the secondaries of the transformers of Fig. 5.38 would result in a 1000-V two-phase voltage across the primary terminals.

For any Scott-connected transformation, the power capacity of the two transformers is only 86.6 percent of their normal single-phase rating.

   example 5.25 Two similar transformers that have a normal 1-kVA rating are connected to change from three-phase power to two phase, as shown in Fig. 5.39. Assuming a unity power factor load, what current may be supplied by each of the two transformers, if the primary three-phase line voltage is 250 V, and the primary-to-secondary turns ratio of each transformer is 4 to 1?

   solution Normal single-phase rated primary current of either transformer,

$$\frac{1000}{250} = 4 \text{ A}$$

Since the primary-to-secondary voltage ratio of the transformers is 4 to 1, normal secondary current of each transformer is

$$4 \times 4 = 16 \text{ A}$$

The reduced current rating for each of the Scott-connected secondaries is

$$0.866 \times 16 = 13.856 \text{ A}$$

5.22 FILTER CHOKES

An important application of chokes is in the filter sections of power supplies, where their function is to help smooth out the pulsating dc wave from the rectifier and change it to approximately pure direct current that is needed for the operation of radio, TV, and other electronic equipment.

Typical power-supply filter circuits are shown in Fig. 5.40. One of the functions of a choke in these circuits is to store energy as the filter input voltage and current are rising. Then, as the filter input decreases from maximum to zero, the stored energy is released to the load by inductive action. Thus, the output from the filter remains fairly constant. The inductance of the choke coil tends to keep the current flowing through it constant even though the dc input to the filter is a constantly varying quantity.

A filter choke is made with a tiny air gap in the core to keep the core from becoming magnetically saturated at high currents and thereby losing inductance to some extent

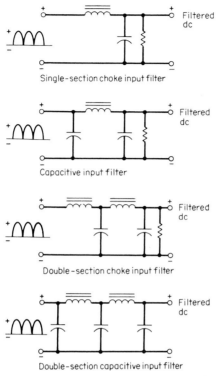

Single-section choke input filter

Capacitive input filter

Double-section choke input filter

Double-section capacitive input filter

Fig. 5.40 Use of filter chokes in power-supply filter circuits.

or even entirely with excessively high currents. But, even with an air gap in the core, there will still be some change in inductance as current changes, and it is customary to specify choke inductance at the current which the choke will normally carry.

The dc output voltage of a choke input filter is about twenty percent less than the output of a capacitor input filter, and its hum level is greater. However, the choke input filter does have better regulation.

A small ac component always gets through to the output of any of the filters shown in Fig. 5.40. This ac component is called the *ripple.* For a full-wave rectifier ripple frequency is twice the line frequency, and for a half-wave rectifier it is the same as line frequency. When the ripple frequency is 60 Hz, ripple frequency with full-wave rectification will be 120 Hz.

A formula for determining the approximate ripple percentage for a single-section choke-input filter is for a choke-input single-section filter at a ripple frequency of 120 Hz.

$$\text{Percent ripple} = \frac{100}{LC} \qquad (5.18)$$

where L = inductance, H
  C = capacitance, μF

The lower the value of percent ripple, the more effective the filter. For a 60-Hz ripple, twice as much inductance and capacitance will be needed to reduce ripple percentage to the same value.

Adding more filter sections greatly improves filtering and reduces ripple to very low values. The amount of tolerable ripple will depend on the equipment to be supplied from the power supply. For a high-gain high-fidelity amplifier, it may be necessary to reduce ripple below 0.05 percent.

example 5.26 In a single-section choke-input filter, the capacitor has a capacitance of 32 μF. What should be the inductance of the input choke for an output ripple of 0.05 percent if the ripple frequency is 120 Hz?

solution

$$0.05 = \frac{100}{LC}$$

or

$$L = \frac{100}{0.05 \times 32} = \frac{2000}{32} = 62.5 \text{ H}$$

5.23 SWINGING CHOKES

A swinging choke, instead of being designed to maintain constant inductance with changes in current, is devised to change inductance with changes in current. Thus, its inductance will decrease with increases in current, and increase as the current decreases. For example, a choke may have an inductance of 5 H when choke current is 250 mA, but will increase to 25 H when choke current is only 25 mA.

Another function of chokes, besides filtering, is to keep the dc output voltage of a power supply below the average value of the *rectified* ac input to the filter. The critical value of inductance required for this purpose increases as load current decreases, and decreases as load current increases. Hence, if a choke is so designed that it will have its critical inductance with only the filter bleeder load, it will not be necessary to use a choke that will have this critical inductance at higher load currents. Therefore, a smaller, less expensive choke may be used.

An approximate formula for determining critical inductance is given by

$$L_{crit} = \frac{\text{load resistance, ohms} \times 120}{1000 \times f} \qquad (5.19)$$

where f = ripple frequency

example 5.27 If the bleeder resistance in a power-supply filter is 25 000 Ω, and full-load resistance which includes the bleeder is 5000 Ω, between what values of inductance must a choke "swing" to maintain critical inductance at a ripple frequency of 120 Hz?

solution Critical inductance at 25 000 Ω is

$$\frac{25\ 000 \times 120}{1000 \times 120} = 25 \text{ H}$$

Critical inductance at full load is

$$\frac{5000 \times 120}{1000 \times 120} = 5 \text{ H}$$

Therefore, a swinging choke that "swings" from 25 to 5 H and has the required current rating will meet requirements.

5.24 THE TUNED FILTER CHOKE

In a power supply that is designed to deliver a large but well-filtered current to a connected load, the filter chokes would have to be very large and bulky. For instance, assuming that a power supply is capable of supplying six amperes of smooth, low-hum current to a load, this same six amperes must also flow through the filter chokes. Hence, the filter chokes must be wound with wire large enough to carry the six amperes without overheating. Furthermore, in order to obtain sufficient inductance for smooth filtering, the chokes must have a large number of turns. To keep the cores from becoming magnetically saturated, they must have a large cross-sectional area. Thus, a large wire size, combined with a large number of turns and a bulky core, results in a large and costly choke.

By using a tuned filter choke, instead of an ordinary choke, the same amount of filtering may be obtained with much smaller and less expensive units. Figure 5.41 shows a pair of tuned filter chokes connected into the filter section of a high-current

Fig. 5.41 By using tuned filter chokes in a high-current power supply, very effective filtering may be attained without the necessity of bulky, heavy-duty, expensive chokes. The tuned primaries reflect a high impedance back into the secondaries at the 120-Hz ripple frequency.

power supply. The principle of operation is very simple. The transformers that are used as tuned chokes may have a primary-to-secondary turns ratio as high as 20 to 1. Even ordinary filament transformers with a primary-to-secondary turns ratio of 117 to 6.3 will be suitable, if secondary current capacity is large enough to carry the load current. The primaries of the transformers are shunted with capacitors having the necessary capacitance values to obtain resonance at 120 Hz (the ripple frequency of the full-wave rectifier). It may be necessary to try several different combinations of smaller and larger capacitors before resonance occurs. Shunted capacitance will range from about 0.1 to 0.3 μF. Exact values, in any case, will have to be determined experimentally and will depend on transformer primary inductance. Hum output will be lowest at resonance.

When the transformer primaries are tuned to resonate at 120 Hz, a high impedance is reflected back into the secondaries, and the secondary windings then function as high-impedance chokes at the ripple frequency. Since the secondary windings do not have a large number of turns, they are well able to carry the high load current without saturating the core or overheating.

The condensers across the primaries should have a voltage rating great enough to safely withstand the high voltage developed across them at the 120-Hz ripple frequency.

5.25 TRANSFORMER CORES

The three basic types of transformer cores are the *open core*, the *closed core*, and the *shell core*. These three types are illustrated in Fig. 5.42. The open-core type is far

Open core

Closed core or core type core   Shell–type core

Fig. 5.42 Core types.

less efficient than the closed-core or shell-core types, but it is still used in some special applications in which efficiency is not an important factor. The closed-core type is commonly called the core type; it may have its windings on only one leg of the core or on both legs. In the shell type the windings are on the center leg.

LAMINATIONS   Some of the various types of laminations used for stacking transformer cores are shown in Fig. 5.43. Laminations are available in various thicknesses

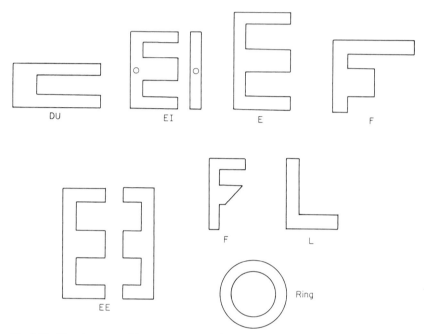

Fig. 5.43   Showing some of the various types of laminations used for the cores of transformers and chokes.

Fig. 5.44   This photo shows a variety of the laminations used to construct various types of transformers. (*Courtesy The Arnold Engineering Company.*)

Window Area = 5.25 sq in
.265″ dia holes
⟶ *Indicates Grain Direction*

PROPERTIES OF SQUARE STACK

V = 646.5 cu cm = 39.43 cu in
A = 19.74 sq cm = 3.063 sq in
l = 33.0 cm = 13.0   in

$B_{max} = \dfrac{19.0 \times 10^3}{K_1 N}$ gausses per volt at 60 cycles (N is number of turns)

H_o = (.038 × 10⁻³) N oersteds per milliampere of direct current in winding

L = (.750 × 10⁻⁸) $K_1 N^2 \mu_{ac}$ henries

Solid Core Weight

High Silicon = 4935 g = 10.86 lb
50% Nickel = 5325 g = 11.72 lb
80% Nickel = 5638 g = 12.40 lb

SETS PER INCH	HIGH SILICON WEIGHT			50% NICKEL WEIGHT			80% NICKEL WEIGHT		
	Ga	Lb/M Prs	Prs/Lb	Ga	Lb/M Prs	Prs/Lb	Ga	Lb/M Prs	Prs/Lb
71	.014″	86.8	11.5	.014″	93.6	10.7	.014″	99.2	10.1
54	.0185″	114.7	8.7						
40	.025″	155.0	6.4						

Fig. 5.45 Typical dimensions of EI laminations. (*Courtesy The Arnold Engineering Company.*)

ranging from about 0.004 to 0.025 in. The L and DU laminations are used in the core-type transformer; all others are used in stacking the cores of the shell type. Figure 5.44 shows some of the actual laminations used in transformer construction.

Figures 5.45 and 5.46 give actual dimensions of some typical laminations. The accompanying tables list the properties of the square stacks, and also give design data. The center leg of any square-stacked core is just as thick as it is wide. The core weights for the square stacks are given in both pounds and grams. This is useful in determining core loss from core-loss data curves. Core volume is in cubic centimeters and in cubic inches. Core cross-sectional area is given in square centimeters and in square inches. The length of the magnetic circuit l is in centimeters and in inches. The equation for finding B_{max} is similar to

$$B_m = \frac{E_p \times 10^8}{4.44 \times f \times N_p \times A}$$

given in Sec. 5.10, where B_m is the maximum flux density for the total applied voltage, instead of for only one volt. The *stacking factor* K may be taken equal to one for most

Window Area = 2.0 sq in
.014″ Laminations have ¼″ dia holes
.006″ and .004″ Laminations have 3/16″ dia holes
◄──► *Indicates Grain Direction*

PROPERTIES OF SQUARE STACK

V = 40.15 cu cm = 2.45 cu in
A = 1.61 sq cm = .25 sq in
l = 22.9 cm = 9.0 in

$B_{max} = \dfrac{233 \times 10^3}{K_1\,N}$ gausses per volt at 60 cycles (N is number of turns)

H_0 = (.055 x 10^{-3}) N oersteds per milliampere of direct current in winding
L = (.089 x 10^{-8}) K_1 N^2 μ_{ac} henries

Solid Core Weight

High Silicon = 307.0 g = .675 lb
50% Nickel = 331.0 g = .728 lb
80% Nickel = 351.0 g = .773 lb

SETS PER INCH	HIGH SILICON WEIGHT			50% NICKEL WEIGHT			80% NICKEL WEIGHT		
	Ga	Lb/M Pcs	Pcs/Lb	Ga	Lb/M Pcs	Pcs/Lb	Ga	Lb/M Pcs	Pcs/Lb
71	.014″	18.95	52.8	.014″	20.45	48.8	.014″	21.65	46.1
167	.006″	8.13	123.0	.006″	8.76	114.0	.006″	9.28	107.8
250	.004″	5.42	184.5	.004″	5.84	171.0	.004″	6.20	161.2

Fig. 5.46 Typical dimensions of DU laminations. (*Courtesy The Arnold Engineering Company.*)

practical applications, but for greater accuracy the following factors may be used for the corresponding lamination gages:

$$0.025 - 97\%$$
$$0.0185 - 96\%$$
$$0.014 - 95\%$$
$$0.006 - 93\%$$
$$0.004 - 88\%$$

In the formula for L, the inductance of the core in henrys μ_{ac} is the ac permeability of the core.

GRAIN-ORIENTED LAMINATIONS The arrows on the laminations shown in Figs. 5.45 and 5.46 show the direction of grain orientation in the steel. Grain orientation is brought about by the hot- and cold-rolling processes used in the manufacture of sheet steel for transformers. These processes cause the individual crystals of silicon steel to become aligned in the direction of rolling. Grain-oriented steel is much easier to magnetize in the direction of grain orientation. Hence, grain-oriented laminations will have their maximum permeabilities when the magnetic fields induced in them are parallel to grain orientation.

A recent development has been the production of a magnetic steel that has two directions of easy magnetization. One is in the direction of rolling, and the other is perpendicular to it.

TAPE-WOUND CORES   Tape-wound cores are wound from long strips of transformer sheet steel having the necessary width and thickness for a particular application.
Since grain orientation is always in the direction in which the tape is wound, tape-
wound cores may be made to have extremely high permeabilities. Insulating varnish
between layers not only serves to insulate the layers from each other—to prevent the
formation of transverse eddy currents—but also binds the layers together and adds
rigidity to the core.

Uncut tape-wound cores are shown in Fig. 5.47. To use these cores in the construction of transformers, the cores are cut and then reassembled about the transformer

Core type   Shell type

Fig. 5.47   Uncut tape-wound cores.

Cut tape-wound core   Single-coil core-   Double-coil core-   Single-coil shell-
                        type assembly   type assembly   type assembly

Fig. 5.48   Cores cut and reassembled about the transformer windings to form either a core or a shell-type
core.

windings to form either a core or a shell-type core, as shown in Fig. 5.48. Besides
improving performance and efficiency, the use of tape-wound cores eliminates much
of the tedious and time-consuming labor involved in stacking ordinary laminated cores.

TAPE-WOUND BOBBIN CORES   Not all tape-wound cores are cut; some small tape-
wound cores are made of such extremely thin and narrow magnetic steel tape that cutting would ruin them. In fact, to maintain their mechanical rigidity, it is even necessary
to wind them on bobbins. Magnetic steel tape for tape-wound bobbin cores is manufactured in standard gage sizes from $\frac{1}{8}$ to 1 mil (0.000 125 to 0.001 in), and tape widths
may be as narrow as $\frac{1}{32}$ in. Core magnetization in tape-wound bobbin cores is usually
produced by only a few turns of wire—about 1 to 10. Tape-wound bobbin cores operate
at low power levels and are used in many computer circuits, shift registers, ring counters, high-frequency magnetic amplifiers, etc.

example 5.28   A 60-Hz transformer operates with an applied primary voltage of 120 V.
The primary winding contains 152 turns. The core is square-stacked and is made up of the
EI-19 laminations shown in Fig. 5.45. Use the appended design data and determine (a) the
maximum flux density in the core in gauss and in lines per square inch; (b) the total maximum
flux in the core. (Assume that stacking factor $K = 1$.)

solution   Maximum flux density B_{max} in gauss per volt equals

$$\frac{19 \times 10^3}{1 \times 152} = 125$$

Then maximum flux density B_m with an applied primary voltage of

$$120 \text{ V} = 120 \times 125 = 15\,000 \text{ G}$$

In lines per square inch

$$15\,000 \text{ G} = 6.45 \times 15\,000 = 96\,750$$

(b) Since the center leg of a lamination is $1^3/_4$ in. wide, the cross-sectional area of a square-stacked core is

$$(1^3/_4)^2 = {}^{49}/_{16} \text{ in}^2$$

Therefore, maximum core flux ϕ_m equals

$${}^{49}/_{16} \times 96\,750 = 296\,297 \text{ Mx}$$

example 5.29 Use the data supplied for the DU-50 laminations of Fig. 5.46 and determine the magnetizing force in oersteds developed in a square-stacked core wound with a 200-turn coil that carries a direct current of 100 mA.

solution From the data, the magnetizing force in oersteds for each milliampere of direct current in the coil is

$$(0.055 \times 10^{-3})200 = 0.011$$

Then, with a direct current of 100 mA, the magnetizing force will be

$$100 \times 0.011 = 1.1 \text{ Oe}$$

Chapter 6

Practical Circuit Analysis

6.1 INTRODUCTION

Electronic technicians are often required to analyze simple circuits energized by dc or ac sources. Generally, only elementary algebra and trigonometry are required in their analysis. The object of this chapter is to review some basic circuit laws and their application in the analysis of dc and ac circuits. Topics discussed are illustrated by pertinent examples and their complete solution.

6.2 NOTATION

By convention, for time-varying quantities, and often in general discussions, *lowercase* letters are used. Voltage is designated by e, current by i, energy by w, and power by p. *For specific dc and ac quantities, uppercase letters are used:* voltage E, current I, energy W, and power P.

6.3 DEFINITIONS

We first define some basic terms used in the analysis of circuits.

PASSIVE AND ACTIVE ELEMENTS An element is considered *passive* if it is capable of only *absorbing* electric energy. A resistor, an inductor, and a capacitor are passive elements. Elements that *provide* electric energy are *active elements*. Examples of active elements include the battery (used to energize circuits) and the transistor (which may be used to energize a loudspeaker).

IDEAL ELEMENTS An ideal element *exhibits a specific property*, regardless of the frequency of operation or the impressed voltage and current flow through the element. For example, an ideal resistor of 50 Ω remains 50 Ω whether it operates at direct current or at hundreds of gigahertz (GHz) or has 1 or 10 000 V impressed across its terminals. Ideal elements, in terms of terminal voltage and current, are defined as follows:
a. Resistor R Current i (amperes) in an ideal resistor of R (ohms) is equal to the voltage e (volts) across R divided by R:

$$i = \frac{e}{R} \qquad (6.1a)$$

Equation (6.1a) is the basic expression of Ohm's law. By algebraic manipulation, one also may write

$$e = iR \tag{6.1b}$$

$$R = \frac{e}{i} \tag{6.1c}$$

b. Inductor L   The voltage e_L (volts) across an ideal inductor equals the value of the inductance L (henrys) multiplied by the rate of change of current with respect to time (amperes per second). A symbol used for the rate of change of current is $\Delta i/\Delta t$, where Δ (*delta*) denotes a *small change*. Hence,

$$e_L = L\frac{\Delta i}{\Delta t} \tag{6.2}$$

Assume, for example, that at time $t_1 = 6$ s, current i_1 in the inductor is 1 A. At $t_2 = 8$ s, the current is $i_2 = 2$ A. Hence, $\Delta i = 2 - 1 = 1$ A, and $\Delta t = 8 - 6 = 2$ s. The rate of change of current is, therefore, $\Delta i/\Delta t = \frac{1}{2}$ A/s.

c. Capacitor C   The current i_C (amperes) in an ideal capacitor equals the value of capacitance C (farads) multiplied by the rate of change of the impressed voltage with respect to time (volts per second). The symbol for the rate of change of the impressed voltage is $\Delta e/\Delta t$. Hence,

$$i_C = C\frac{\Delta e}{\Delta t} \tag{6.3}$$

d. Voltage source e   An ideal voltage source is a source whose output voltage is constant, regardless of the current it supplies to a circuit. A practical voltage source, with a low output resistance, approximates an ideal voltage source.

e. Current source i   An ideal current source is a source whose output current is constant, regardless of the resultant voltage appearing across the circuit to which the source is connected. Symbols used for ideal voltage and current sources are given in Fig. 6.1.

The symbol of Fig. 6.1*a* is general and can be used to denote a dc or time-varying voltage source. Figure 6.1*b* shows an ideal dc voltage source. The symbol for the ideal current source of Fig. 6.1*c* is used for dc and time-varying current sources.

f. Dependent (controlled) sources   The sources in Fig. 6.1 are *independent* because their value does not depend on the circuit to which they are connected. In the analysis of transistor amplifiers, for example, the model of a transistor may contain a current or voltage source whose value *depends* on a specific current or a voltage elsewhere in the circuit. These are called *dependent*, or *controlled*, sources. Electrical symbols for controlled sources are provided in Fig. 6.2. Either the diamond or the circle symbol may be used.

In Fig. 6.2*a*, the value of the controlled voltage source e_d depends on current i_x flowing in another part of the circuit; this is an example of a *current-controlled (dependent) voltage source*. If e_d depended on a voltage $e_d = ke_x$, it would be called a *voltage-controlled voltage source*.

Refer to Fig. 6.2*b*. The value of the controlled current source i_d depends on voltage e_x; this is an example of a *voltage-controlled current source*. If i_d depended on a current $i_d = ki_x$, it would be called a *current-controlled current source*.

MODEL   A model is a representation of a *physical* device or circuit by *ideal elements*. Consider, for example, a resistor of R ohms. At very high frequencies, such as in the giga-

Fig. 6.1 Symbols for ideal sources. (*a*) Dc or time-varying voltage source. (*b*) Dc voltage source. (*c*) Dc or time-varying current source.

Fig. 6.2 Examples of controlled (dependent) sources. (*a*) Current-controlled voltage source. (*b*) Voltage-controlled current source.

Fig. 6.3 A possible model of a resistor operating at very high frequencies.

Fig. 6.4 Series *RC* circuit excited by a dc source.

hertz range, the lead inductance and shunt capacitance enter into the picture. Hence, a possible model of a resistor operating at high frequencies may appear as in Fig. 6.3. Note that the model is composed of ideal elements *R* and *L* in series, shunted by an ideal capacitor *C*.

LINEAR ELEMENT   A linear element is an element in which a change of its input (such as voltage) results in a proportionate change in its response, or output (such as current). If, for example, the voltage across an 8-Ω resistor is 16 V, then $I_1 = {}^{16}\!/\!_8 = 2$ A. If the voltage is *doubled* to 32 V, then $I_2 = {}^{32}\!/\!_8 = 4$ A, and the current is also *doubled*.

NONLINEAR ELEMENT   An element that is not linear is said to be nonlinear. In a nonlinear element, its response is not proportional to a change in its input. An example of a nonlinear element is a *thermistor*, whose resistance depends on its current flow.

LINEAR CIRCUIT   A circuit that contains only linear elements is called linear. (The term *network* is also used for circuit.)

NONLINEAR CIRCUIT   A circuit that contains one or more nonlinear elements is nonlinear.

SERIES CIRCUIT   A circuit in which the current in each element is the same is a series circuit.

PARALLEL CIRCUIT   A circuit in which the voltage across each element is the same is a parallel circuit.

6.4   *RC* and *RL* CIRCUITS

Consider the series *RC* circuit of Fig. 6.4. Initially, that is, at time equals zero $(t = 0)$, capacitor *C* acts as a *short circuit*. Hence, the current *I* at $t = 0$, denoted by $I(0)$, is

$$I(0) = \frac{E}{R} \qquad (6.4)$$

The product *RC* is the *time constant T* (seconds) of the circuit,

$$T = RC \qquad (6.5)$$

where *R* is in ohms and *C* in farads. After an elapsed time of approximately 5 time constants $(5T)$, the circuit is said to be in *steady state*, denoted by $t = \infty$. In steady state, the voltages and currents in the circuit are not changing and have settled down to a fixed value. For dc excitation in steady state, $\Delta e/\Delta t = 0$. From Eq. (6.3), if $\Delta e/\Delta t = 0$, the current in the capacitor is also zero $(i_C = 0)$. Therefore, a capacitor acts as an *open circuit* to direct current in the steady state.

   example 6.1   In Fig. 6.4 assume that $E = 100$ V, $R = 1$ kΩ, and $C = 1$ μF. Determine (*a*) time constant *T*, (*b*) initial and steady-state values of current, and (*c*) initial and steady-state values of voltage across the capacitor.

   solution   (*a*) From Eq. (6.5), $T = RC = 10^3 \times 10^{-6} = 10^{-3}$ s (1 ms).
   (*b*) Because at $t = 0$ the capacitor behaves as a short circuit to direct current, $I(0) = E/R = 100/1000 = 0.1$ A. In steady state the capacitor acts as an open circuit to direct current, and $I(\infty) = 0$.

(c) Because C acts as a short circuit at $t = 0$ (and has no charge), $E_C(0) = 0$ V. In steady state, no current flows in C, and the capacitor is fully charged by the source voltage; therefore, $E(\infty) = 100$ V.

Consider now the series RL circuit of Fig. 6.5. To direct current, L acts as an open circuit at $t = 0$ and as a short circuit in steady state ($t = \infty$). The current in the circuit is, therefore, maximum in steady state for dc excitation. The time constant T (seconds) is

$$T = \frac{L}{R} \tag{6.6}$$

Fig. 6.5 Series RL circuit excited by a dc source.

where L is in henrys and R in ohms. At $t = 0$, the voltage across the inductor is $E_L(0) = E$ volts. In steady state, $\Delta i/\Delta t = 0$; hence, from Eq. (6.2), $E_L(\infty) = 0$ V.

example 6.2 In Fig. 6.5, let $E = 25$ V, $R = 100$ Ω, and $L = 1$ H. Determine (a) T, (b) $I(0)$ and $I(\infty)$, (c) $E_L(0)$ and $E_L(\infty)$.

solution (a) From Eq. (6.6), $T = L/R = \frac{1}{100} = 0.01$ s.
(b) At $t = 0$, L acts as an open circuit to direct current; hence, $I(0) = 0$. In steady state, L acts as a short circuit to direct current, and $I(\infty) = E/R = \frac{25}{100} = 0.25$ A.
(c) $E_L(0) = 25$ V and $E_L(\infty) = 0$ V.

For times between approximately 0 and $5T$, the current i_C and voltage e_C as a function of time for a series RC circuit excited by a dc source (Fig. 6.4) are

$$i_C = \frac{E}{R} \epsilon^{-t/(RC)} \tag{6.7}$$

and

$$e_C = E(1 - \epsilon^{-t/(RC)}) \tag{6.8}$$

For the series RL circuit excited by a dc source (Fig. 6.5),

$$i_L = \frac{E}{R}(1 - \epsilon^{-tR/L}) \tag{6.9}$$

and

$$e_L = E\epsilon^{-tR/L} \tag{6.10}$$

for times between 0 and $5T$.

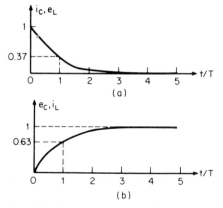

Fig. 6.6 Performance of series RC and RL circuits energized by a dc voltage source: (a) i_c and e_L as a function of time. (b) e_c and i_L as a function of time.

Equations (6.7), i_C, and (6.10), e_L, are of the same form and so are plotted in Fig. 6.6a. For convenience, the maximum value of either i_C or e_L is denoted by 1. (This is called *normalizing* the maximum value to 1.) The unit for the x axis is t/T. Note that when t is equal to five time constants ($5T$), the current i_C in a capacitor and the voltage e_L across an inductor are approximately equal to zero. In a single time constant ($t = T$), i_C (or e_L) falls to approximately 0.37, or 37 percent of its maximum value.

Figure 6.6b is a plot of e_C and i_L [Eqs. (6.8) and (6.9)]. The maximum value is again denoted by 1, and in $t = 5T$ steady state has been reached. For $t = T$, e_C (or i_L) has reached 63 percent of its maximum value.

6.5 ENERGY AND POWER

Energy is the ability to do work. In SI units, the unit of energy is the *joule* (J); 1 joule equals 1 *watt-second* (W·s). Power is the *rate of expending energy;* its unit is *joule per second* which equals 1 *watt.* Energy w equals the product of power p and time t:

$$w = pt \tag{6.11}$$

For electric circuits, power p may be expressed by

$$p = ei \tag{6.12a}$$

$$= \frac{e^2}{R} \tag{6.12b}$$

$$= i^2R \tag{6.12c}$$

where R is the resistance, e the impressed voltage across the resistance, and i the current flowing in the resistance.

Only resistors are capable of *dissipating energy;* capacitors and inductors only *store energy.* The energy stored in a capacitor w_C is

$$w_C = \frac{Ce^2}{2} \tag{6.13}$$

where e is the voltage across C. For an inductor the stored energy w_L is

$$w_L = \frac{Li^2}{2} \tag{6.14}$$

where i is the current flowing in the inductor.

example 6.3   Referring to Example 6.1, determine the energy stored in the capacitor in steady state.

solution   In Example 6.1, $C = 10^{-6}$ F, and the voltage across C equals 100 V. From Eq. (6.13), $w_C = Ce^2/2 = 10^{-6} \times 100^2/2 = 5 \times 10^{-3}$ J.

example 6.4   Referring to Example 6.2, determine (a) the power dissipated in R and (b) the energy stored in L, in steady state.

solution   (a) In Example 6.2, $I = 0.25$ A in steady state; $L = 1$ H; and $R = 100\ \Omega$. From Eq. (6.12c), by letting $i = I$, $p = I^2R = (0.25)^2 \times 100 = 6.25$ W.
(b) From Eq. (6.14), $w_L = LI^2/2 = 1 \times (0.25)^2/2 = 31.75 \times 10^{-3}$ J.

6.6 CIRCUIT LAWS AND THEOREMS

A number of simple, yet powerful, laws and theorems for the analysis of circuits are available to the technician. Although the application of these laws and theorems is demonstrated first in the analysis of dc circuits, they are also applicable in the analysis of ac circuits. For dc circuits the quantities of concern, such as voltage and current, are always expressed by real numbers. For ac circuits, however, these quantities are generally represented by complex numbers. Complex numbers and the application of the circuit laws and theorems to ac circuits are reviewed later in the chapter.

KIRCHHOFF'S LAWS   Kirchhoff's voltage law (KVL) and current law (KCL) provide the technician with two useful tools for the analysis of circuits, regardless of their complexity. The voltage law states that the algebraic sum of voltages in a closed-circuit path is always equal to zero. The current law states that the algebraic sum of currents leaving a circuit node is always equal to zero. A circuit node may be thought of as a connecting point that is common to two or more circuit elements.

In the application of KVL and KCL, it is essential that we designate the polarities of voltages across elements and the direction of electron flow at a circuit node. In this chapter we adopt the following conventions:

1. In traversing a closed-circuit path, a *plus sign* precedes a voltage quantity, such as *e* or *iR*, in going *with* the direction of electron flow or from a *low* to a *high* potential. In going *against* the electron flow or from a *high* to a *low* potential, a *minus sign* precedes the voltage quantity.

2. Current flowing *away* from a circuit node will have a *negative sign;* current flowing *toward* a circuit node will have a *positive sign.* A current source pointing to a node will have a *negative sign.*

The application of KVL and KCL is illustrated in the next group of examples.

Fig. 6.7   Finding current *I* in a series circuit by KVL (Example 6.5).

example 6.5   Find current *I* in Fig. 6.7.

solution   Starting at point 2 and traversing the circuit in the direction (counterclockwise) of current *I*, we have, according to KVL and our conventions,

$$-100 + 30I + 20I + 50I = 0$$

From point 2 to point 1, we go from a *high* to a *low* potential, hence the minus sign before 100. In traversing the remainder of the circuit, we go in the direction of electron flow. The voltage across each resistor, therefore, has a positive sign.

Bringing the -100 term to the right side of the equation and simplifying, we have

$$(30 + 20 + 50)I = 100$$

or

$$100I = 100$$

Solving yields $I = {}^{100}\!/_{100} = 1$ A.

example 6.6   Figure 6.8 shows a circuit having two closed-circuit paths. This configuration is referred to as a *two-mesh,* or *two-loop,* circuit. Circuits containing two or more meshes are fairly common.

In Fig. 6.8, currents I_1 and I_2 are assumed to be flowing counterclockwise. (If it turns out that the actual current flows in a direction opposite to that assumed, the solved current will have a minus sign.) Applying KVL, we write the necessary equations and solve for the currents.

solution   Starting at point 1 for the first mesh, we obtain

$$-100 + (2 + 6)I_1 - 6I_2 = 0$$

or

$$100 = 8I_1 - 6I_2$$

Dividing by 2 yields

$$50 = 4I_1 - 3I_2 \qquad (a)$$

Starting at point 2 in the second mesh, we have

$$-6I_1 + (12 + 6)I_2 + 90 = 0$$

or

$$-90 = -6I_1 + 18I_2$$

Fig. 6.8   Finding currents I_1 and I_2 in a two-mesh circuit by KVL (Example 6.6).

Thus

$$-45 = -3I_1 + 9I_2 \tag{b}$$

Equations (a) and (b) constitute a pair of linear simultaneous equations. One may use determinants or substitution to find the currents. With the latter procedure and by solving Eq. (b) for I_2, we have

$$I_2 = \frac{-45 + 3I_1}{9} = -5 + \frac{I_1}{3} \tag{c}$$

Substitution of Eq. (c) in Eq. (a) yields

$$50 = 4I_1 - 3\left(-5 + \frac{I_1}{3}\right)$$
$$= 4I_1 + 15 - I_1 = 15 + 3I_1$$

Solving gives $I_1 = (50 - 15)/3 = {}^{35}\!/_3 = 11.67$ A. Substituting $I_1 = 11.67$ A in Eq. (b) gives

$$-45 = -3(11.67) + 9I_2 = -35 + 9I_2$$

Solving yields $I_2 = -{}^{10}\!/_9 = -1.11$ A. The minus sign denotes that the actual direction of current flow in the second mesh is *clockwise*.

In the literature, the sum of resistances in a mesh is referred to as the *self-resistance* of the mesh. The resistance *common* to two meshes is called the *mutual resistance*. To return to Fig. 6.8, the sum $R_1 + R_2$ is the self-resistance of mesh I and is denoted by $R_{11} = R_1 + R_2$. Resistors $R_2 + R_3$ constitute the self-resistance of mesh II ($R_{22} = R_2 + R_3$). Resistor R_2 is the mutual resistance linking meshes I and II ($R_{12} = R_{21} = R_2$).

example 6.7   Apply KCL and solve for E_1 and E_2 in Fig. 6.9.

solution   There are two independent nodes: 1 and 2. Node N is a *common*, or *reference*, node for E_1 and E_2. Taking the direction of currents as flowing from right to left and from bottom to top, according to our conventions, at node 1 we have

$$-\frac{10 - E_1}{1} + \frac{E_1}{2} + \frac{E_1 - E_2}{1} = 0 \tag{a}$$

The current flowing in a resistor is equal to the voltage difference across the resistor divided by its value. Thus, the current in R_1 is $(10 - E_1)/1$; in R_2, it is $E_1/2$; etc.

Simplifying Eq. (a), we get

$$10 = E_1(1 + \tfrac{1}{2} + 1) - \frac{E_2}{1} \tag{b}$$
$$= 2.5E_1 - E_2$$

At node 2,

$$-\frac{E_1 - E_2}{1} + \frac{E_2}{2} - 10 = 0$$

or

$$10 = -E_1 + 1.5E_2 \tag{c}$$

Solving Eq. (c) for E_1 gives

$$E_1 = -10 + 1.5E_2 \tag{d}$$

Substituting Eq. (d) in Eq. (b) yields

$$10 = 2.5(10 + 1.5E_2) - E_2$$
$$= 25 + 2.75E_2$$

Fig. 6.9   Finding voltages E_1 and E_2 by KCL (Example 6.7).

Hence,
$$E_2 = \frac{10 + 25}{2.75} = 12.7 \text{ V}$$

and
$$E_1 = -10 + 1.5(12.7) = 9.05 \text{ V}$$

CONDUCTANCE   Conductance G is equal to 1 divided by the resistance R:

$$G = \frac{1}{R} \tag{6.15}$$

The unit for conductance is the *siemen* (SI).

By letting $G_1 = 1/R_1$, $G_2 = 1/R_2$, $G_3 = 1/R_3$, and $G_4 = 1/R_4$, Eqs. (b) and (c) in Example 6.7 may be expressed by

$$10 = (G_1 + G_2 + G_3)E_1 - G_3 E_2$$
$$10 = -G_3 E_1 + (G_3 + G_4)E_2$$

where $G_1 + G_2 + G_3 = G_{11} = 1 + 0.5 + 1 = 2.5$ S is the *self-conductance* of node 1, $G_3 = G_{12} = G_{21} = 1$ S is the *mutual conductance* between nodes 1 and 2, and $G_3 + G_4 = G_{22} = 1 + 0.5 = 1.5$ S is the *self-conductance* of node 2.

6.7 TOPOLOGY OF NETWORKS

In complex networks containing many nodes and meshes, it can be exceedingly difficult to choose mesh or nodal analysis for fewest equations. Furthermore, one may not always be certain whether all the nodes or meshes have been considered. By borrowing some results from a branch of mathematics called *topology*, it becomes relatively simple to remove these doubts. The procedure is as follows:

1. All sources, independent as well as dependent, in the circuit are set to zero. In setting a voltage source to zero, it is *short-circuited out* of the circuit. In setting a current source to zero, it is *open-circuited*, or *removed*, from the circuit.

2. All circuit elements, such as a resistor, are replaced by line segments, called *branches*. The result is referred to as a *graph*.

The minimum number of nodal equations, referred to as *independent equations N*, is

$$N = n - 1 \tag{6.16}$$

where n is the number of nodes contained in the graph. The number of independent mesh (loop) equations M is

$$M = b - n + 1 \tag{6.17}$$

where b is the number of branches contained in the graph.

example 6.8   Draw the graphs and determine the number of independent nodal and mesh equations required for the circuits of (a) Fig. 6.8 and (b) Fig. 6.9.

solution   (a) The graph of Fig. 6.8 is drawn in Fig. 6.10a. Note that in setting E_1 and E_2 to zero, resistors R_1 and R_2 become connected between nodes A and 1, 2. Also, the resistors are represented by line segments (branches). The number of branches is $b = 3$; the number of nodes is $n = 2$. Hence, from Eqs. (6.17) and (6.16), $M = 3 - 2 + 1 = 2$ mesh equations, and $N = 2 - 1 = 1$ nodal equation.

In Example 6.6 we saw that two mesh equations employing KVL were required in the solution of Fig. 6.8. If KCL were used, only one nodal equation would have been required.

(b) The graph of Fig. 6.9 is given in Fig. 6.10b. Voltage source E is short-circuited, thereby placing resistor R_1 between nodes 1 and N. Current source I is removed from the circuit. From the graph, $b = 4$ and $n = 3$. Hence, $N = 3 - 1 = 2$ nodal equations, and $M = 4 - 3 + 1 = 2$ mesh equations.

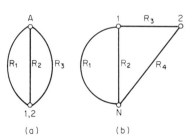

Fig. 6.10   Graphs for the circuits of (a) Fig. 6.8 and (b) Fig. 6.9 (Example 6.8).

6.8 EQUIVALENT RESISTANCE

RESISTORS IN SERIES   For n resistors in series (Fig. 6.11), current I flowing in each resistor is the same. Applying KVL to the circuit, we have

$$E = I(R_1 + R_2 + \cdots + R_n) = IR_{TS}$$

where the equivalent resistance for n resistors in series R_{TS} is

$$R_{TS} = R_1 + R_2 + \cdots + R_n \tag{6.18}$$

This result shows that the equivalent resistance of a number of resistors in series is equal to the *sum of the individual resistors.*

If the resistors are *equal*, the equivalent resistance is the product of the number of equal resistors n and the value of the resistor R:

$$R_{TS} = nR \tag{6.19}$$

example 6.9   In a particular prototype it is found that a 1.5-, a 2.7-, and a 5.8-kΩ resistor are connected in series. What single equivalent resistor can be used in their place?

solution   From Eq. (6.18), $R_{TS} = 1.5 + 2.7 + 5.8 = 10$ kΩ.

RESISTORS IN PARALLEL   For n resistors in parallel, or in *shunt* (Fig. 6.12), the voltage across each resistor is the same. Applying KCL yields

$$I = E \left(\frac{1}{R_1} + \frac{1}{R_2} + \cdots + \frac{1}{R_n} \right)$$

Letting $G_1 = 1/R_1$, $G_2 = 1/R_2, \ldots, G_n = 1/R_n$, we obtain

$$I = E(G_1 + G_2 + \cdots + G_n) = EG_{TP}$$

where the equivalent conductance G_{TP} is

$$G_{TP} = G_1 + G_2 + \cdots + G_n \tag{6.20}$$

The equivalent conductance of a number of resistors in parallel is thus equal to the sum of their conductances. It is usual, however, to think in terms of resistances. Letting $G_{TP} = 1/R_{TP}$, we see that Eq. (6.20) may be expressed as

$$\frac{1}{R_{TP}} = \frac{1}{R_1} + \frac{1}{R_2} + \cdots + \frac{1}{R_n}$$

Solving for R_{TP} gives

$$R_{TP} = \frac{1}{1/R_1 + 1/R_2 + \cdots + 1/R_n} \tag{6.21}$$

For two resistors in parallel, $n = 2$, Eq. (6.21) simplifies to

$$R_{TP} = \frac{R_1 R_2}{R_1 + R_2} \tag{6.22}$$

Hence, for two resistors R_1 and R_2 in parallel (often denoted by $R_1 \| R_2$), their equivalent resistance is equal to their product divided by their sum. For more than two resistors in

Fig. 6.11   A series circuit containing n resistors.

Fig. 6.12   A parallel circuit containing n resistors.

parallel, it is practical to determine their equivalent resistance by considering two resistors at a time.

For *n equal* resistors in parallel, the equivalent resistance is equal to one of the resistors *R* divided by *n*:

$$R_{TP} = \frac{R}{n} \tag{6.23}$$

example 6.10   A 60-, a 12-, and a 10-Ω resistor are in parallel. Determine their equivalent resistance.

solution   Taking first 60‖12, from Eq. (6.22), $R_{TP1} = (60 \times 12)/(60 + 12) = 10$ Ω. By combining R_{TP1} (10 Ω) with the remaining 10-Ω resistor, from Eq. (6.23), $R_{TP} = {}^{10}\!/_2 = 5$ Ω.

Fig. 6.13   A series-parallel circuit (Example 6.11).

RESISTORS   IN   SERIES-PARALLEL
Consider a circuit containing some resistors in parallel connected to other resistors in series. To find the equivalent resistance, first reduce the parallel combinations to their equivalent values and then add in series with the remaining resistors in the circuit.

example 6.11   Replace the resistors in the circuit of Fig. 6.13 with a single equivalent resistor.

solution   We reduce the parallel combinations to their equivalent resistance values. From Eq. (6.23), 20‖20 = ${}^{20}\!/_2$ = 10 kΩ. From Eq. (6.22), 60‖12 = 60 × 12/(60 + 12) = 10 kΩ. Combining this with the remaining 30-kΩ resistor yields 10‖30 = 10 × 30(10 + 30) = 7.5 kΩ. Hence, R = 5 + 10 + 7.5 + 2 = 24.5 kΩ.

6.9   VOLTAGE DIVIDER

Voltage dividers are used to tap off a portion of an impressed voltage across a circuit. An example of a simple series-resistance voltage divider is illustrated in Fig. 6.14. Because the resistors are connected in series, the same current flows in each resistor. So voltage E_n across resistor R_n is

$$E_n = \frac{R_n E}{R_1 + R_2 + \cdots + R_n} \tag{6.24}$$

Voltage E_n is said to be *proportional* to R_n. Remember that Eq. (6.24) is valid only if the *same current* flows in each resistor.

example 6.12   Determine E_2 across $R_2 = 15$ Ω in Fig. 6.15.

solution   Since the same current flows in each resistor, we can use Eq. (6.24) with $n = 2$. Hence, $E_2 = 15 \times 10/(10 + 15) = 6$ V.

6.10   CURRENT DIVIDER

Current division arises when a current is forced to divide among a number of resistors connected in parallel (Fig. 6.16). Because the voltage across each resistor is the same for *n* resistors in parallel, the current I_n in resistor R_n is

Fig. 6.15   A simple voltage divider (Example 6.12).

Fig. 6.14   A generalized series voltage divider.

Fig. 6.16 A generalized parallel current divider.

Fig. 6.17 A simple current divider (Example 6.13).

$$I_n = \frac{G_n I}{G_1 + G_2 + \cdots + G_n} \tag{6.25}$$

where $G_1 = 1/R_1$, etc. For $n = 2$, in terms of resistance values,

$$I_1 = \frac{R_2 I}{R_1 + R_2} \tag{6.26a}$$

$$I_2 = \frac{R_1 I}{R_1 + R_2} \tag{6.26b}$$

The current is said to divide *inversely* with the resistances in a parallel circuit. Most current flows in the smallest resistor, and least current flows in the largest resistor.

example 6.13 Using current division, find I_1 and I_2 in Fig. 6.17.

solution From Eqs. (6.26a) and (6.26b), $I_1 = 3 \times 5/(2 + 3) = {}^{15}\!/_5 = 3$ A and $I_2 = 2 \times 5/(2 + 3) = {}^{10}\!/_5 = 2$ A. Note that the greater current, $I_1 = 3$ A, flows through the smaller resistor (2 Ω), and $I_2 = 2$ A flows through the larger resistor (3 Ω). Also, from KCL, $I = I_1 + I_2$ (5 A = 2 A + 3 A). In any problem, Kirchhoff's laws must always be satisfied.

6.11 SUPERPOSITION THEOREM

The superposition theorem, *which applies only to linear circuits*, provides a powerful tool for finding the response of a circuit that is energized by more than one source. In addition, the effect on a circuit to a particular source can be easily obtained.

The superposition theorem states that the response of a circuit energized by a number of independent sources is equal to the sum of the responses of each source with all remaining sources set to zero. Dependent sources are not set to zero—only independent sources. As mentioned earlier, an independent voltage source is set to zero by short-circuiting it; an independent current source is set to zero by removing it from the circuit.

example 6.14 Referring to Fig. 6.18, determine the current in the 60-Ω resistor I_{60}, using superposition.

solution Setting the 72-V source to zero yields Fig. 6.19a. By current division, I_1, due to the 6-A source, is $-6 \times 12/(60 + 12) = -1$ A.
Setting the 6-A source to zero yields Fig. 6.19b. Current I_2, due to the 72-V source, is $72/(12 + 60) = 1$ A. Hence, $I_{60} = I_1 + I_2 = -1 + 1 = 0$ A.

Fig. 6.18 Superposition used to find the current in the 60-Ω resistor I_{60} (Example 6.14).

Fig. 6.19 Resultant circuits after sources in Fig. 6.18 are set to zero: (a) Voltage source set to zero. (b) Current source set to zero.

(a)

(b)

Fig. 6.21 Resultant circuits after independent sources in Fig. 6.20 are set to zero: (a) Current source set to zero. (b) Voltage source set to zero.

Fig. 6.20 Superposition used to find current I_5 in the 5-Ω resistor (Example 6.15).

example 6.15   Using superposition, determine current I_5 flowing in the 5-Ω resistor in Fig. 6.20.

solution   The circuit of Fig. 6.20 contains two independent sources, 20 V and 10 A, and one current-dependent source, $2I$. Current I is the current in the 10-Ω resistor. In the application of the superposition theorem, a dependent source is never set to zero.

Setting first the 10-A source to zero results in Fig. 6.21a. Current $I = {}^{20}\!/_{10} = 2$ A; hence, $2I = I_1 = 2 \times 2 = 4$ A.

Setting the 20-V source to zero yields Fig. 6.21b. Because the 10-Ω resistor is now short-circuited, $I = 0$, and so $2I$ is also equal to zero. Current $I_2 = -10$ A (the minus sign signifies that the 10 A is flowing in a direction opposite to that assumed for I_2). Hence, $I_5 = I_1 + I_2 = 4 - 10 = -6$ A.

6.12   THEVENIN'S THEOREM

Consider the linear *two-terminal* (also referred to as a *one-port*) network of Fig. 6.22a. The network may contain dependent as well as independent sources. According to Thévenin's

theorem, network N, regardless of its complexity, may be replaced by a voltage source E_{TH} in series with a resistor R_{TH} (assuming that the sources in N are direct current) as far as the external terminals a–b are concerned. The resulting circuit, shown in Fig. 6.22b, is the Thévenin equivalent circuit of N.

Voltage E_{TH} is called the Thévenin, or *open-circuit*, voltage across terminals a–b. One may perceive E_{TH} as the voltage a voltmeter reads when connected across a–b. Resistance R_{TH} is called the Thévenin, or *equivalent*, resistance of the network. If the

Fig. 6.22 Thévenin's theorem. For the linear two-terminal network N of (a), the Thévenin equivalent circuit is given in (b).

network contains only independent sources, R_{TH} may be calculated as the resistance seen across a–b with all independent sources in N set to zero. If, however, the network also con-

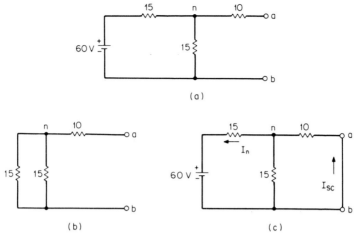

Fig. 6.23 Determining the Thévenin equivalent circuit of a network: (a) Given network. (b) Finding R_{TH}. (c) Finding I_{sc} (Example 6.16).

tains dependent sources, the procedure is different. Terminals a–b are connected (short-circuited), and the *short-circuit current* I_{sc} flowing in a–b is determined. Then

$$R_{TH} = \frac{E_{TH}}{I_{sc}} \tag{6.27}$$

Equation (6.27) may be used also for networks containing only independent sources.

example 6.16   Determine the Thévenin equivalent circuit for the network of Fig. 6.23a.

solution   Because no current flows in the 10-Ω resistor, the open-circuit (Thévenin) voltage at a with respect to b is equal to the voltage at n with respect to b. Hence, by voltage division, E_{TH} = 60 × 15/(15 + 15) = 30 V.

Setting the 60-V source to zero yields Fig. 6.23b. The equivalent resistance is R_{TH} = 15‖15 + 10 = ¹⁵⁄₂ + 10 = 17.5 Ω. Figure 6.24 shows the complete Thévenin equivalent circuit of Fig. 6.23a.

We may also find R_{TH} by obtaining I_{sc}. In Fig. 6.23c, terminals a–b are shown connected. The current I_n flowing away from n is equal to 60/(15 + 10‖15) = ⁶⁰⁄₂₁ A. By current division, I_{sc} = ⁶⁰⁄₂₁ × 15/(15 + 10) = ¹²⁄₇ = 1.7 A. By Eq. (6.27), $R_{TH} = E_{TH}/I_{sc}$ = 30/1.7 = 17.5 Ω. This is identical with the value found by setting the 60-V source to zero.

Although it requires additional calculations to find R_{TH} in terms of the open-circuit

Fig. 6.24   Thévenin equivalent circuit of the network of Fig. 6.23a.

voltage and short-circuit current, this is the only method that is permitted when dependent sources are present in a network.

example 6.17   Determine the Thévenin equivalent circuit for the network of Fig. 6.25a with respect to terminals S–N.

solution   Figure 6.25a is a simplified model of a FET amplifier. The dependent source is $50E_{GS}$, where E_{GS} is the voltage between G and S. Applying KVL to the mesh, we obtain $50E_{GS}$ = (1 + 9)I_D = 10I_D. From the figure, E_{GS} = 1 − 1(I_D) = 1 − I_D. Hence, substituting this expression for E_{GS} in the previous equation, we obtain 50(1 − I_D) = 10I_D. Solving yields I_D = ⅚ mA (the answer is in milliamperes because the resistance values are in kilohms). The open-circuit voltage is E_{TH} = 1000 × ⅚ × 10⁻³ = ⅚ V.

Connecting terminal S to N results in Fig. 6.25b. For this condition, E_{GS} = 1 V and I_{sc} = ⁵⁰⁄₉

(a)

(b)

Fig. 6.25 Determining the Thévenin equivalent circuit of a network containing a dependent source: (a) Network. (b) Finding I_{sc} (Example 6.17).

Fig. 6.26 Thévenin equivalent circuit for the network of Fig. 6.25a.

mA. Hence, $R_{TH} = E_{TH}/I_{sc} = \frac{5}{6}/(\frac{5}{9} \times 10^{-3}) = 150\ \Omega$. The Thévenin equivalent circuit for the FET amplifier is given in Fig. 6.26.

6.13 NORTON'S THEOREM

According to this theorem, the linear two-terminal network of Fig. 6.22a can be represented by the Norton equivalent circuit of Fig. 6.27, where current source I_{sc} is in parallel with the equivalent resistance R_{TH}. Regardless of the complexity of network N, the Norton equivalent circuit of Fig. 6.27 is identical to N with respect to terminals a–b. The procedures for determining I_{sc} and R_{TH} are the same as given for the Thévenin equivalent circuit.

Fig. 6.27 Norton equivalent circuit of a network.

example 6.18 Determine the Norton equivalent circuits for the networks of (a) Example 6.16 and (b) Example 6.17.

solution (a) From Fig. 6.24, $I_{sc} = 30/17.5 \times 1.7$ A, and $R_{TH} = 17.5\ \Omega$. The Norton equivalent circuit is shown in Fig. 6.28a.

(b) From Fig. 6.26, $I_{sc} = \frac{5}{6}/150 = 5.56$ mA, and $R_{TH} = 150\ \Omega$. The Norton equivalent circuit is given in Fig. 6.28b.

Note that the direction of the arrow in the current source I_{sc} always points to the negative terminal of the open-circuit voltage source E_{TH} for electron current flow.

SOURCE CONVERSION In the analysis of dc and ac circuits, it is often advantageous to convert a physical voltage source to a corresponding current source, or vice versa. A physical voltage source, for dc, is an ideal voltage source in series with a resistance, as indicated in

Fig. 6.28 Norton equivalent circuits for the networks of (a) Fig. 6.24 and (b) Fig. 6.26 (Example 6.18).

Fig. 6.29a. The equivalent current source comprises an ideal current source in parallel with a resistance (Fig. 6.29b). For ac circuits, in general, the resistance is an impedance (see Sec. 6.20).

Based on the preceding discussion of the Thévenin and Norton theorems, the current source I_s is obtained by dividing the source voltage E_s by the source resistance R_s:

$$I_s = \frac{E_s}{R_s}$$

In converting from a current to a voltage source, the voltage source is obtained by multiplying the current source by the source resistance:

$$E_s = I_s R_s$$

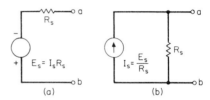

Fig. 6.29 Conversion of (a) a voltage source to (b) a current source, or vice versa.

The value of the resistance is the same in both cases.

6.14 PERIODIC WAVEFORMS

In addition to direct current, there are many waveforms generated and used for voltage and current sources in electronic circuits. Examples of some common waveforms are illustrated in Fig. 6.30. One feature in common with these examples is that their waveshape repeats itself with time. Waveforms having this property are called *periodic waves*. The *sine wave* of Fig. 6.30a, also referred to as a *sinusoid*, or *sinusoidal function*, is the most common periodic wave found in circuits. Alternating current (ac), for example, is a sine wave.

A periodic wave that is not a sine wave is said to be *nonsinusoidal*. The square and triangular waves of Fig. 6.30b and c, respectively, are examples of nonsinusoidal waves. It is possible to express a periodic nonsinusoidal wave as a sum of sine waves. Such an expression is called a Fourier series.

6.15 SINE WAVE

Because of its importance in the analysis of ac circuits, terms used in characterizing the sine wave are defined here.

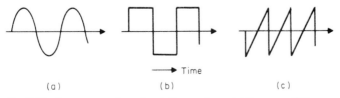

Fig. 6.30 Some commonly used periodic waveforms: (a) Sine wave. (b) Square wave. (c) Triangular wave.

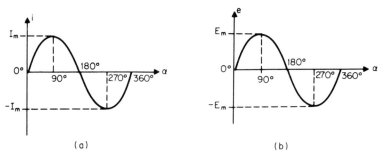

Fig. 6.31   One cycle of a sine wave: (*a*) Current sine wave. (*b*) Voltage sine wave.

BASIC DEFINITION   *One cycle* of a current and voltage sine wave is illustrated in Fig. 6.31. The current sine wave is expressed by

$$i = I_m \sin \alpha \tag{6.28a}$$

and the voltage sine wave by

$$e = E_m \sin \alpha \tag{6.28b}$$

Coefficient I_m is the *peak*, or *maximum amplitude*, of the current; similarly, E_m is the peak, or maximum amplitude, of the voltage. Angle α *(alpha)* is the *displacement angle* of the waveform. Values of sin α for 45° increments are given in Table 6.1. Note, for example, that when sin $\alpha = 1$, the sine wave reaches its maximum *positive* amplitude—$i = I_m$ and $e = E_m$. When sin $\alpha = -1$, the sine wave reaches its maximum *negative* amplitude—$i = -I_m$ and $e = -E_m$. These points are indicated in Fig. 6.31.

FREQUENCY   The frequency f of a sine wave is the number of times a cycle of the wave repeats itself in 1 s. The unit of frequency is the *hertz* (HZ) which equals *one cycle per second* (cps). In scientific notation, $1000 = 10^3$, $10\,000 = 10^4$, etc.; 10^3 Hz = 1 kilohertz (1 kHz); 10^6 Hz = 1 megahertz (1 MHz); and 10^9 Hz = 1 gigahertz (1 GHz). For example, household alternating current has a frequency of 60 Hz in the United States.

PERIOD   The period T is the time required for a periodic waveform to complete one cycle. The period is equal to 1 over the frequency:

$$T = \frac{1}{f} \tag{6.29}$$

If f is expressed in hertz, T is in seconds. For example, the period of 60 Hz is ⅟₆₀ = 0.0167 s = 16.7 milliseconds (ms).

TABLE 6.1   Some Values of sin α

α, degrees	sin α
0	0
30	0.500
45	0.707
90	1
135	0.707
180	0
225	−0.707
270	−1
315	−0.707
360	0

ANGULAR FREQUENCY   It is common practice to express the displacement angle of a sine wave by ωt, where ω *(omega)* is the *angular frequency* and is expressed in *radians per second* (rad/s). One radian = 57.3° and 2π rad = 360°. Omega is related to frequency by

$$\omega = 2\pi f \qquad (6.30a)$$

and

$$f = \frac{\omega}{2\pi} = 0.159\omega \qquad (6.30b)$$

Equations for current and voltage, written in terms of ω, appear as

$$i = I_m \sin \omega t \qquad (6.31a)$$
$$e = E_m \sin \omega t \qquad (6.31b)$$

example 6.19   Given $e = 141 \sin 377t$ volts, find (*a*) its frequency f and (*b*) peak amplitude E_m.

solution   (*a*) From the given expression, $\omega = 377$ rad/s. Applying Eq. (6.30b) gives $f = 0.159\omega = 0.159 \times 377 = 60$ Hz.
(*b*) The coefficient of the sine wave is the peak amplitude E_m. In this example, $E_m = 141$ V.

PHASE ANGLE   Consider the sine wave of Fig. 6.32a. Comparison of this figure with Fig. 6.31a at $\alpha = 0$ ($\omega t = 0$) reveals that $i = -0.707I_m$ instead of zero. The reason is that the sine wave of Fig. 6.32a has been *shifted to the right* by 45°. To express this mathematically, we write

$$i = I_m \sin (\omega t - 45°) \qquad (6.32a)$$

Similarly, for a voltage, we have

$$e = E_m \sin (\omega t - 45°) \qquad (6.32b)$$

We can verify Eqs. (6.32a) and (6.32b) with the aid of Table 6.1. If $\omega t = 0$, then $i = I_m \sin (-45°) = -0.707I_m$, and $e = -0.707E_m$. For $\omega t = 45°$, $i = I_m \sin (45° - 45°) = I_m \sin 0° = 0$. Likewise, $e = 0$.

The sine wave of Fig. 6.31b has been *shifted to the left* by 45°. This condition is represented by

$$i = I_m \sin (\omega t + 45°) \qquad (6.33a)$$

For a voltage,

$$e = E_m \sin (\omega t + 45°) \qquad (6.33b)$$

Refer again to Table 6.1. If $\omega t = -45°$, then $i = I_m \sin (-45° + 45°) = I_m \sin 0° = 0$. If $\omega t = 0°$, then $i = I_m \sin (0 + 45°) = 0.707I_m$. Similar results are found for voltage sine waves.

The angle by which a sine wave has been shifted to the right or to the left is called the

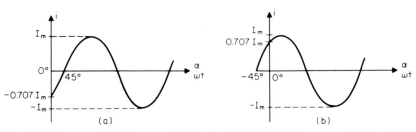

Fig. 6.32   Phase angle: (*a*) Sine wave shifted to the right by 45°. (*b*) Sine wave shifted to the left by 45°.

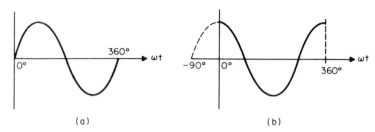

Fig. 6.33 A cosine wave viewed as a sine wave shifted 90° to the left: (*a*) Sine wave. (*b*) Cosine wave.

phase angle. Letting the phase angle be designated by θ *(theta)* in general terms, for current, we can write

$$i = I_m \sin (\omega t - \theta) \tag{6.34a}$$
$$i = I_m \sin (\omega t + \theta) \tag{6.34b}$$

Similar expressions can be written for voltage.

The relationship between sine and *cosine* waves is illustrated in Fig. 6.33. Figure 6.33*a* displays a sine wave. In Fig. 6.33*b* we have a cosine wave. The cosine wave can be viewed as a sine wave that has been shifted by 90° to the left. Mathematically, we may write

$$\cos \omega t = \sin (\omega t + 90°) \tag{6.35}$$

It is common practice to refer to *both* sine and cosine waves as sinusoidal waves.

PHASE RELATIONSHIPS   Consider the two sine waves of Fig. 6.34*a*. In this example, *e* is said to *lead i* by θ degrees, because the peak value of *e* is reached before *i*. An alternative statement is *i lags e* by θ degrees.

Because *i* reaches its peak value before *e* in Fig. 6.34*b*, *i leads e* by θ degrees (or *e lags i*

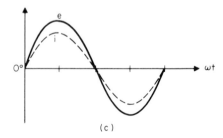

Fig. 6.34 Phase relationships between the sine waves of the same frequency: (*a*) *e* leads *i* by $\theta°$. (*b*) *i* leads *e* by $\theta°$. (*c*) *e* and *i* are in phase.

by θ degrees). In Fig. 6.34c the phase angle between e and i is $0°$. In this case, e and i are said to be *in phase*.

In characterizing the phase angle between sinusoids, it is necessary that the waveforms in question *have the same frequency*. If the frequencies of the waveforms differ, one cannot speak of phase angle between waves.

6.16 AVERAGE (DC) VALUE OF WAVEFORMS

The average, or dc, value of a periodic waveform is what one would read on a dc ammeter or voltmeter. The average value of a periodic waveform is equal to its *net area* over one cycle divided by its period. For example, because the positive half-cycle is equal to the negative half-cycle, the average (dc) value of the sine, square, and triangular waveforms of Fig. 6.30 is equal to zero.

6.17 EFFECTIVE (RMS) VALUE OF WAVEFORMS

The effective, or root-mean-squared (rms), value of a periodic waveform is equal to the direct current (or voltage) that dissipates the same energy in a given resistor. To read the effective values of any current or voltage waveform, a rms meter is required. A multimeter, such as a VOM, is generally calibrated to read the rms value of only a sine wave. For a nonsinusoidal wave, a true rms meter must be used.

The effective value of a sinusoidal voltage E is equal to 0.707 times its peak value E_m:

$$E = 0.707E_m \tag{6.36a}$$

Also,

$$E_m = \frac{E}{0.707} = \sqrt{2}E \simeq 1.4E \tag{6.36b}$$

Similarly, for current

$$I = 0.707I_m \tag{6.37a}$$
$$I_m = \sqrt{2}I \tag{6.37b}$$

In calculating the effective value of any periodic waveform, its peak value is squared. The squared area is determined and then divided by the period of the waveform. Finally, the square root is taken of the preceding result.

example 6.20   Determine for the periodic waveform of Fig. 6.35a the (a) dc and (b) rms values.

solution   (a) The area of the positive half-cycle is $4 \times 0.5 = 2$; the area of the negative half-cycle is $-3 \times 0.5 = -1.5$. Hence, the net area is $2 - 1.5 = 0.5$. Dividing by the period 1 s, we find the dc value is $0.5/1 = 0.5$ V.

(b) Squaring 4 yields $4^2 = 16$; also $(-3)^2 = 9$. The squared waveform is illustrated in Fig. 6.35b. Its area is $16 \times 0.5 + 9 \times (1 - 0.5) = 8 + 4.5 = 12.5$. Dividing by the period 1 s gives $12.5/1 = 12.5$. Taking the square root of 12.5, we find the rms voltage is $\sqrt{12.5} = 3.54$ V.

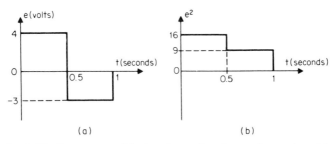

Fig. 6.35 Determination of the dc and rms values of a periodic waveform: (a) One cycle of a periodic waveform. (b) Squared area of waveform (Example 6.20).

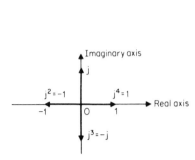

Fig. 6.36   Operator j and its raised powers.

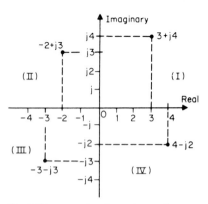

Fig. 6.37   Examples of complex numbers.

6.18   COMPLEX NUMBERS

Upon excitation of a circuit containing capacitors and/or inductors, in addition to resistors, changing waveforms of current and voltage, called *transients*, are produced. After the decay of these transients (as mentioned earlier in the chapter, after approximately five time constants), the circuit is said to be in *steady state*. What is of significance is that in steady state, networks energized by sinusoidal sources may be analyzed by using nothing more complicated than simple algebra. In fact, this is exactly what was required in the analysis of dc circuits. In general, however, currents and voltages in ac circuits are represented by *complex numbers*, instead of real numbers as in dc circuits.

A complex number has two parts: a *real part* and an *imaginary part*. Mathematically, a complex number A is represented as $A = a + jb$, where a is the real part and b is the imaginary part of the number. (This representation is called the *rectangular form* of a complex number.) Coefficient j is an *operator* and is equal to the square root of -1: $j = \sqrt{-1}$. Hence, $j^2 = -1$, $j^3 = -j$, $j^4 = 1$, etc. Operator j and its raised powers are illustrated in Fig. 6.36. In plotting complex numbers, the j axis is referred to as the *imaginary axis*, and the x axis as the *real axis*. Note that in multiplying a number by j, the number is located on the imaginary (or j) axis.

A complex number may be thought of as being a *general* number. If the imaginary part is equal to zero, the complex number becomes *real;* if the real part is zero, the complex number becomes *imaginary*. Examples of complex numbers are plotted in Fig. 6.37. The Roman numeral refers to the quadrant occupied by the complex number.

COMPLEX CONJUGATE   The complex conjugate of a complex number is the complex number with *only* the sign of its imaginary part changed: plus to minus or minus to plus. Examples of complex numbers and their conjugates are listed in Table 6.2.

In each example, only the sign of the imaginary term was changed; the real part was untouched. In general terms, if $A = a + jb$, then its complex conjugate, denoted by $A°$, is $A° = a - jb$. If a complex number is multiplied by its conjugate, a *real* number is obtained:

$$AA° = (a + jb)(a - jb) = a^2 + b^2 \tag{6.38}$$

TABLE 6.2   Examples of Complex Numbers and Their Conjugates

Complex number	Complex conjugate
$3 - j4$	$3 + j4$
$3 + j$	$3 - j$
$-2 + j6$	$-2 - j6$
$-4 - j5$	$-4 + j5$
5	5
$j2$	$-j2$

ADDING AND SUBTRACTING COMPLEX NUMBERS Adding or subtracting complex numbers is easily accomplished. The real and imaginary parts of the numbers are added (or subtracted) *separately.*

> **example 6.21** Perform the following additions and subtractions: (a) $(2 + j6) + (3 - j2)$; (b) $(9 + j10) - (-3 + j4)$.
>
> **solution** (a) The real and imaginary parts are added separately: $(2 + j6) + (3 - j2) = (2 + 3) + j(6 - 2) = 5 + j4$.
> (b) The real and imaginary parts are subtracted separately: $(9 + j10) - (-3 + j4) = [9 - (-3)] + j(10 - 4) = 12 + j6$.

MULTIPLYING AND DIVIDING COMPLEX NUMBERS Although one can multiply or divide, as well as raise to a power or take a root of, a complex number in rectangular form, it is easier to perform these operations when the complex number is expressed in *polar form.* In polar form, a complex number A is expressed as

$$A = |A| \underline{/\theta}$$

where $|A|$ is the *magnitude* of the complex number. The magnitude is equal to the square root of the sum of the real part squared and the imaginary part squared:

$$|A| = \sqrt{a^2 + b^2} \qquad (6.39a)$$

Fig. 6.38 A complex number represented in polar form.

and θ is the *angle measured from the positive real axis.* The angle is equal to the arc tangent of the imaginary part divided by the real part:

$$\theta = \tan^{-1} \frac{b}{a} \qquad (6.39b)$$

A complex number in polar form is depicted graphically in Fig. 6.38. From the figure, we see that the real part is equal to the magnitude multiplied by the cosine of the angle ($\cos \theta$):

$$a = |A| \cos \theta \qquad (6.40a)$$

The imaginary part is equal to the magnitude multiplied by the sine of the angle:

$$b = |A| \sin \theta \qquad (6.40b)$$

> **example 6.22** Express the complex numbers of Fig. 6.37 in polar form.
>
> **solution** (a) $3 + j4 = \sqrt{3^2 + 4^2} \underline{/\tan^{-1} \frac{4}{3}} = 5\underline{/53.1°}$.
> (b) $-2 + j3 = \sqrt{2^2 + 3^2} \underline{/\tan^{-1} (3/-2)} = 3.61\underline{/123.7°}$.
> (c) $-3 - j3 = \sqrt{3^2 + 3^2} \underline{/\tan^{-1} (-3/-3)} = 4.25\underline{/225°}$.
> The sign of the angle is always positive when it is taken in a *counterclockwise* direction with respect to the positive real axis. If the angle is taken in a *clockwise* direction, the sign of the angle is *negative.* In (c), therefore, the positive angle 225° may be expressed as a negative angle: $-(360° - 225°) = -135°$. Hence, one can also write $-3 - j3 = 4.25\underline{/-135°}$.
> (d) $4 - j2 = \sqrt{4^2 + 2^2} \underline{/\tan^{-1} (-\frac{2}{4})} = 4.5\underline{/331.3°} = 4.5\underline{/-28.7°}$.

With a number expressed in polar form, it is a simple task to multiply complex numbers. The rule is: Multiply the magnitudes, and add algebraically the angles. For example, $5\underline{/30°} + 12\underline{/-12°} = (5 \times 12)\underline{/30° - 12°} = 60\underline{/18°}$.

For division, the rule is: Divide the magnitudes, and subtract algebraically the angle of the divisor from the angle of the dividend. For example, $24\underline{/-30°} \div 4\underline{/-16°} = \frac{24}{4}\underline{/-30° - (-16°)} = 6\underline{/-14°}$.

For raising a complex number to the nth power, the rule is: Raise the magnitude to the nth power, and multiply the angle by n. For example, $(3\underline{/20°})^3 = (3^3)\underline{/20° \times 3} = 27\underline{/60°}$.

For finding the nth root of a complex number, the rule is: Take the nth root of the magnitude, and divide the angle by n. For example, $(16\underline{/36°})^{1/2} = (16^{1/2})\underline{/36°/2} = 4\underline{/18°}$.

Occasionally, a complex number is given in polar form, and one is required to convert it to rectangle form. This is accomplished with the aid of Eqs. (6.40a) and (6.40b).

example 6.23 Convert the following complex numbers from polar to rectangular form: (a) $18/\underline{30°}$, (b) $10/\underline{-45°}$.

solution (a) From Eq. (6.40a) the real part is $a = |A| \cos \theta$. Hence, $a = 18 \cos 30° = 18 \times 0.866 = 15.6$. From Eq. (6.40b) the imaginary part is $b = |A| \sin \theta$; therefore, $b = 18 \sin 30° = 18 \times 0.5 = 9$. Finally, $18/\underline{30°} = 15.6 + j9$.

(b) $10/\underline{-45°} = 10 \cos(-45°) + j \sin(-45°)$. Because $\cos(-\theta) = \cos \theta$ and $\sin(-\theta) = -\sin \theta$, we can express $10/\underline{-45°}$ by $10 \cos 45° - j10 \sin 45° = 10 \times 0.707 - j10 \times 0.707 = 7.07 - j7.07$.

6.19 REACTANCE

The resistance of an ideal resistor to alternating current is *constant* and *independent of frequency*. The resistance of an ideal inductor or capacitor to alternating current, referred to as *inductive reactance* and *capacitive reactance*, respectively, is *not constant* and *is dependent on frequency*.

Let L = henrys, C = farads, ω = radians per second, and f = hertz. The resistance of an ideal inductor to alternating current is $j\omega L$, where ωL is the inductive reactance X_L,

$$X_L = \omega L = 2\pi f L \qquad (6.41)$$

The resistance of an ideal capacitor to alternating current is $-j/(\omega C)$, where $1/(\omega C)$ is the capacitive reactance X_C:

$$X_C = \frac{1}{\omega C} = \frac{0.159}{fC} \qquad (6.42)$$

As for resistance, the unit for inductive and capacitive reactance is the *ohm*.

Note from Eq. (6.41) that the inductive reactance *increases* with frequency. At direct current, which can be taken as $f = 0$, the inductive reactance is zero. However, from Eq. (6.42), the capacitive reactance *decreases* with increasing frequency. At direct current, the capacitive reactance is infinite; a capacitor, therefore, *blocks* direct current.

EQUIVALENT REACTANCE The equivalent reactance of a number of inductors and capacitors in series, parallel, or series-parallel is calculated in the same manner as for resistors. If magnetic coupling exists between inductors, however, it must be considered in the calculation of equivalent inductive reactance. In this chapter we assume that no magnetic coupling is present.

example 6.24 Calculate the equivalent reactance for the circuit of Fig. 6.39a.

solution The first step is to reduce the parallel combinations to single reactances. For the inductors, $(j60)(j12)/j(60 + 12) = j10 \, \Omega$. For the capacitors, because they are equal, the equivalent capacitive reactance is $-j10/2 = -j5 \, \Omega$. The net reactance for the network, therefore, is $j(10 - 5 + 10 - 5) = j10 \, \Omega$, as illustrated in Fig. 6.39b.

6.20 IMPEDANCE AND ADMITTANCE

In general, ac circuits contain resistors, inductors, and capacitors. By letting the net resistance equal R and the net reactance $X_L - X_C = X$, the impedance Z in ohms is

$$Z = R + jX = |Z|/\underline{\theta}$$

(a)   (b)

Fig. 6.39 Calculating the equivalent reactance of a circuit: (a) Circuit. (b) Equivalent reactance (Example 6.24).

Fig. 6.40   Impedance triangle.

Fig. 6.41   Finding the impedance of a circuit (Example 6.25).

where

$$|Z| = \sqrt{R^2 + X^2} \qquad (6.43a)$$

$$\theta = \tan^{-1} \frac{X}{R} \qquad (6.43b)$$

The relationship of R, X, and $|Z|$ is illustrated by the *impedance triangle* of Fig. 6.40. It is seen from the triangle that when $X = 0$, the angle is zero and the impedance is equal to R. For this condition, the circuit is purely resistive. As the value of X increases, the angle also increases. When $R = 0$, $\theta = 90°$, $|Z| = X$, and the network is purely reactive.

example 6.25   Find the impedance of Fig. 6.41.

solution   The net resistance is $R = 5\ \Omega$, and the net reactance is $X = 15 - 3 = 12\ \Omega$. Hence, $Z = 5 + j12\ \Omega$.
Expressed in polar form, $|Z| = \sqrt{5^2 + 12^2} = \sqrt{169} = 13$; and $\theta = \tan^{-1} 12/5 = 67.4°$. In polar form, $Z = 13/\underline{67.4°}$.

ADMITTANCE   Admittance Y is defined as 1 over Z:

$$Y = \frac{1}{Z} \qquad (6.44)$$

As for conductance, the unit of admittance is the siemen (S). In general, Y may be expressed as

$$Y = G + jB = |Y|/\underline{\phi}$$

where G is the net conductance and B the net *susceptance* of the circuit. The conductance and susceptance, expressed in terms of R and X, are

$$G = \frac{R}{R^2 + X^2} \qquad (6.45a)$$

$$B = -\frac{X}{R^2 + X^2} \qquad (6.45b)$$

The magnitude of Y, $|Y|$, and angle ϕ are expressed by

$$|Y| = \sqrt{G^2 + B^2} \qquad (6.46a)$$

$$\phi = \tan^{-1} \frac{B}{G} \qquad (6.46b)$$

Fig. 6.42   Admittance triangle.

As for impedance, the *admittance triangle* of Fig. 6.42 can be drawn. If $B = 0$, then $Y = G = 1/R$, and the circuit is purely resistive. As B increases, ϕ increases; when $G = 0$, the angle is 90°, $Y = jB$, and the circuit is purely reactive.

example 6.26   Verify Eqs. (6.45a) and (6.45b).

solution   By definition, $Y = 1/Z$. Substituting $Z = R + jX$, we obtain $Y = 1/(R + jX)$. The complex conjugate of the denominator is $R - jX$. Multiplying the numerator and denominator by the complex conjugate of the denominator yields $Y = (R - jX)/(R^2 + X^2) = R/(R^2 + X^2) - jX/(R^2 + X^2)$. Hence, $G = R/(R^2 + X^2)$ and $B = -X/(R^2 + X^2)$.

OHM'S LAW FOR AC CIRCUITS   If Z is substituted for R, then Ohm's law is expressed by

$$I = \frac{E}{Z} \qquad (6.47a)$$

If G is replaced by Y, we have

$$I = EY \qquad (6.47b)$$

In general, current I and voltage E will be complex numbers.

6.21 PHASOR DIAGRAMS

A *phasor diagram* is a graphical description of the relationships of currents and voltages present in an ac circuit. Phasor diagrams provide insight into circuit performance and are useful in the graphical solution of some ac problems.

Figure 6.43a shows a voltage source of $E/0°$ volts impressed across an ideal resistor of R ohms. Note that the impressed voltage is expressed in polar form. Often, a sine wave symbol is drawn within the circle to indicate a sine wave source. The plus sign indicates that at the time we are examining the circuit, point b is *positive* with respect to point a.

By Ohm's law, current I is

$$I = \frac{E/0°}{R} = \frac{E}{R}\,/0°$$

Fig. 6.43 Phasor diagrams for an ideal resistor: (*a*) Circuit. (*b*) Voltage used as the reference phasor. (*c*) Current used as the reference phasor.

Because the angle of I is identical with that of E (0°) the voltage and current are in phase.

The relationship between E and I is illustrated in the phasor diagram of Fig. 6.43b. A directed line, to a convenient scale, is drawn to represent $E/0°$; this is called the *reference phasor*. Because the current and voltage are in phase, another directed line, to a convenient scale to represent the current phasor I, is superimposed on the voltage phasor. To distinguish between the voltage and current phasors, the current phasor is denoted by an open arrow.

Current may also be selected as the reference phasor. This is shown in Fig. 6.43c, where I and E are drawn to scales different from those of Fig. 6.43b.

In Fig. 6.44a a voltage source is impressed across an ideal inductor. By Ohm's law,

$$I = \frac{E}{jX_L}$$

Expressing jX_L in polar form, from Eqs. (6.43a) and (6.43b) we have $|Z| = X_L$, and $\theta = \tan^{-1}(X_L/0) = 90°$. Hence, $jX_L = X_L/90°$. Substitution of this quantity in the expression for I yields

$$I = \frac{E}{X_L/90°} = \frac{E}{X_L}\,/-90°$$

From the final equation it is seen that the current in an ideal inductor *lags* the impressed voltage by 90° (or the impressed voltage *leads* the current by 90°).

Phasor diagrams using voltage and current as the reference phasors are drawn in Fig. 6.44b, and c, respectively. One may think of a phasor diagram as a *rigid body*. By convention, positive rotation of a phasor is taken as being counterclockwise. It is seen, for example, that if the phasor diagram of Fig. 6.43b is rotated 90° counterclockwise, the resultant dia-

Fig. 6.44 Phasor diagrams for an ideal inductor: (*a*) Circuit. (*b*) Voltage used as the reference phasor. (*c*) Current used as the reference phasor.

Fig. 6.45 Phasor diagrams for an ideal capacitor: (a) Circuit. (b) Voltage used as the reference phasor. (c) Current used as the reference phasor.

gram appears as in Fig. 6.44c. The relation of I lagging E (or E leading I) by 90° is always maintained.

An ideal capacitor excited by a voltage source is illustrated in Fig. 6.45a. By Ohm's law,

$$I = \frac{E}{-jX_C} = \frac{E}{X_C}\underline{/90°}$$

since, in polar form, $-jX_C = X_C\underline{/-90°}$. The current flowing in an ideal capacitor *leads* the impressed voltage by 90° (or the voltage *lags* the current by 90°). This is opposite to that of the ideal inductor. Phasor diagrams, with voltage and current used as reference phasors, are given in Fig. 6.45b and c, respectively.

6.22 AC CIRCUIT ANALYSIS

The techniques for analyzing ac circuits in steady state are identical with those used for dc circuits. As mentioned earlier, the difference is that complex numbers are generally employed in ac analysis whereas real numbers are used in dc analysis. The following group of examples illustrates the methods of analyzing ac circuits.

example 6.27 Referring to Fig. 6.46: (a) find the rms current I; (b) draw the phasor diagrams with the impressed voltage and current as reference phasors; (c) write an expression for the current as a function of time.

solution (a) The impedance of the circuit is $Z = (2 + 2) + j(5 - 1) = 4 + j4\ \Omega$. In polar form, $|Z| = \sqrt{4^2 + 4^2} = \sqrt{32} \simeq 5.7;\ \theta = \tan^{-1}\frac{4}{4} = 45°$. Hence, $Z = 5.7\underline{/45°}$.

Current I is $100\underline{/0°}/5.7\underline{/45°} = 17.5\underline{/-45°}$ A. Expressed in rectangular form, this is $I = 17.5\cos(-45°) + j17.5\sin(-45°) = 17.5\cos 45° - j17.5\sin 45° = 17.5(0.707 - j0.707) = 12.4 - j12.4$ A.

(b) By using the voltage as the reference phasor, the phasor diagram appears as in Fig. 6.47a. With current as the reference phasor, the phasor diagram is shown in Fig. 6.47b. In either representation, the current lags the voltage by 45°.

(c) From Eq. (6.34b), current as a function of time for a lagging phase angle is $i = I_m \sin(\omega t - \theta)$. In (a) the magnitude of the rms current is $I = 17.5$ A. From Eq. (6.37b), $I_m = \sqrt{2}I = 1.4 \times 17.5 = 24$ A. The angular frequency is $\omega = 377$ rad/s, and $\theta = 45°$. Therefore, $i = 24(377t - 45°)$.

example 6.28 Write the required loop equations for the network of Fig. 6.48. Assume that no magnetic coupling exists between L_1 and L_2.

solution Based on the topology of the network, three independent equations are required. Using KVL, we obtain

For mesh I: $E_1\underline{/0°} = (R_1 + j\omega L_1)I_1 - j\omega L_1 I_2 - R_1 I_3$

For mesh II: $0 = -j\omega L_1 I_1 + [R_2 + j\omega(L_1 + L_2)]I_2 - R_2 I_3$

For mesh III: $0 = -R_1 I_1 - R_2 I_2 + (R_1 + R_2 - j/\omega C)I_3$

Fig. 6.46 Analyzing an ac circuit (Example 6.27).

Fig. 6.47 Phasor diagrams for Fig. 6.46: (a) Voltage used as the reference phasor. (b) Current used as the reference phasor.

Fig. 6.48 Writing loop equations for a three-mesh circuit (Example 6.28).

Fig. 6.49 Finding current i by superposition (Example 6.29).

example 6.29   Using superposition, find the steady-state current as a function of time for the circuit of Fig. 6.49.

solution   In steady state, the inductor acts as a short circuit to direct current. Setting the ac source to zero results in Fig. 6.50a. Current I_1, owing to the dc source, is $I_1 = {}^{100}\!/_{100} = 1$ A.

Setting the dc source to zero yields Fig. 6.50b. The angular frequency (from 5000 sin 1000t) is 1000 rad/s; hence, $X_L = \omega L = 1000 \times 0.1 = 100\ \Omega$. Note that the peak value of the ac source, $5000\underline{/0°}$ V, is shown instead of its rms value. The reason is that we are solving directly for the peak current.

The impedance of the circuit of Fig. 6.50b is $Z = 100 + j100 = 141\underline{/45°}$ in polar form. The peak value of i_2, I_{m2}, is $5000\underline{/0°}/141\underline{/45°} = 35.4\underline{/-45°}$. Therefore, $i_2 = 35.4$ (sin 1000t − 45°). Finally, by superposition, $i = I_1 + i_2 = 1 + 35.4$ sin $(1000t − 45°)$ A.

example 6.30   With respect to terminals a–b, determine the Thévenin and Norton equivalent circuits of Fig. 6.51.

solution   Because no current flows in the 5-Ω resistor, the open-circuit (Thévenin) voltage across terminals a–b is equal to the voltage across the 10-Ω resistor. By voltage division, $E_{TH} = 10 \times 10\underline{/0°}/(10 + j10) = 100\underline{/0°}/14.1\underline{/45°} = 7.07\underline{/-45°}$ V.

By setting the voltage source to zero, the equivalent (Thévenin) *impedance*, Z_{TH}, is $10 \| j10 + 5 = j100/(10 + j10) + 5$. Multiplying the numerator and denominator by the complex conjugate of the denominator of the fractional term, we have

$$\frac{j100(10 - j10)}{(10 + j10)(10 - j10)} = \frac{j1000 + 1000}{100 + 100} = \frac{1000}{200} + \frac{j1000}{200} = 5 + j5$$

Hence, $Z_{TH} = 5 + j5 + 5 = 10 + j5\ \Omega$. The Thévenin equivalent circuit is shown in Fig. 6.52a.

To arrive at the Norton equivalent circuit, the short-circuit current I_{sc} is determined from the expression $I_{sc} = E_{TH}/Z_{TH}$. By expressing Z_{TH} in polar form, $|Z_{TH}| = \sqrt{10^2 + 5^2} = \sqrt{125} = 11.2;\ \theta = \tan^{-1} {}^{5}\!/_{10} = 28.7°$. Hence, $Z_{TH} = 11.2\underline{/28.7°}$. The short-circuit current is $I_{sc} = 7.07\underline{/-45°}/11.2\underline{/28.7°} = 0.63\underline{/-45°} - 28.7° = 0.63\underline{/-73.7°}$ A. The Norton equivalent circuit is shown in Fig. 6.52b.

6.23 POWER IN AC CIRCUITS

The average *(real)* power P dissipated in an ac circuit is

$$P = EI \cos \theta \tag{6.48}$$

Fig. 6.50 Resultant circuit obtained after sources in Fig. 6.49 are set to zero: (a) Ac source set to zero. (b) Dc source set to zero.

Fig. 6.51 Finding the Thévenin and Norton equivalent circuits of the network with respect to terminals a–b (Example 6.30).

where E is the rms voltage across the circuit, I is the rms current flowing in the circuit, and θ is the phase angle between E and I. The term $\cos \theta$ is defined as the *power factor* (PF) of the circuit:

$$PF = \cos \theta \qquad (6.49)$$

Maximum power occurs when the power factor is equal to *unity* ($\cos \theta = 1$). Referring to Table 6.3, we see that $\cos \theta = 1$ when θ is equal to 0° or 360° (or a multiple of 360°). This occurs only in a *purely resistive circuit* where the voltage and current are in phase. In a purely inductive circuit, the current lags the voltage by 90° ($\theta = -90°$), and $\cos (-90°) = 0$. The average power dissipated in a purely inductive circuit is, therefore, equal to zero. Similarly, the current in a purely capacitive circuit leads the voltage by 90° ($\cos 90° = 0$), and the average power dissipated is also zero.

Fig. 6.52 Equivalent circuits for the network of Fig. 6.51: (*a*) Thévenin. (*b*) Norton.

In summary, only resistors and resistive circuits dissipate power, and it is equal to EI (the power factor is unity). For inductors, capacitors, or circuits containing only capacitors and inductors, no power is dissipated (the power factor is zero). For a circuit containing resistance *and* reactance, the phase angle is between 0 and 90°. If resistance predominates, the angle will be close to 0°; if reactance predominates, the angle will be close to 90°.

Average power dissipated in an ac circuit also equals I^2R and E^2/R, where I is the rms current flowing in the resistor and E is the rms voltage across the resistor.

REACTIVE POWER A useful concept in ac circuit analysis is *reactive power P_q*. The unit of reactive power is the *var*. Reactive power is equal to

$$P_q = EI \sin \theta \qquad (6.50)$$

where E, I, and θ have been previously defined. Reactive power is also equal to I^2X and E^2/X, where I is the rms current in the reactance, E is the voltage across the reactance, and X is the net reactance.

APPARENT POWER Apparent power P_a is the product of the voltage E impressed across the circuit and the current I flowing in the circuit:

$$P_a = EI \qquad (6.51)$$

The unit of apparent power is the voltampere (VA).

The relationships among real, reactive, and apparent power may be displayed graphically by a *power triangle*. The power triangle of Fig. 6.53*a* is for a circuit with net inductive reactance. For a circuit with net capacitive reactance, the power triangle appears as in Fig. 6.53*b*.

By the pythagorean theorem, the relationship of P, P_q, and P_a is expressed by

$$P_a^2 = P^2 + P_q^2 \qquad (6.52)$$

example 6.31 For an ac circuit, $E = 100$ V rms, and $I = 10$ A rms with a lagging phase angle of 30°. Determine (*a*) power factor, (*b*) average power, (*c*) reactive power, and (*d*) apparent power.

TABLE 6.3 Some Values of cos θ

θ, degrees	$\cos \theta$
0	1
±30	0.866
±45	0.707
±60	0.5
±90	0
±360	1

(a)

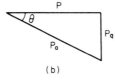

(b)

Fig. 6.53 Power triangles: (a) For a circuit with net inductive reactance. (b) For a circuit with net capacitive reactance.

(a)   (b)

Fig. 6.54 Application of the maximum power transfer theorem in (a) a dc and (b) an ac circuit.

solution   (a) The power factor is PF = $\cos \theta$. From Table 6.3, $\cos 30° = 0.866$. Hence, PF = 0.866 lagging.

(b) $P = EI \cos \theta = 100 \times 10 \times 0.866 = 866$ W.

(c) $P_q = EI \sin \theta$; from Table 6.1, $\sin 30° = 0.5$. Hence, $P_q = 100 \times 10 \times 0.5 = 500$ var.

(d) $P_a = EI = 100 \times 10 = 1000$ W = 1 kW.

The results may be verified by Eq. (6.52): $1000^2 = 866^2 + 500^2$.

6.24 MAXIMUM POWER TRANSFER THEOREM

The maximum power transfer theorem states that for a dc source having a source resistance of R_s ohms, the maximum power is delivered to load R_L if $R_L = R_s$ ohms. For an ac source having an impedance of Z_s ohms, the maximum power is delivered to load Z_L if Z_L is the complex conjugate of Z_s ($Z_L = Z_s$ ohms).

example 6.32   (a) Determine the values of the load for maximum power transfer in the circuits of Fig. 6.54a and b. (b) Calculate the maximum power absorbed by each load.

solution   (a) $R_L = 10$ Ω, and $Z_L = 10 - j10$ Ω. Note that the effective impedance for the ac circuit is now equal to $10 + j10 + 10 - j10 = 20$ Ω, which is purely resistive. The power factor, therefore, is unity, and maximum power is delivered to the load.

(b) In each circuit, the voltage across the load is $100 \times 10/(10 + 10) = 50$ V. Power absorbed, therefore, is $50^2/10 = 250$ W.

Chapter 7

Meters and Measurements

7.1 INTRODUCTION

It is important for a technician to have an understanding of the difficulties involved in making exact measurements if he is to accurately evaluate the readings on his meter. The first part of this chapter is devoted to the problems of obtaining and interpreting accurate measurements. This information is also valuable for someone who is planning to purchase meters.

Measurements of current are made indirectly—either by the effect of their magnetic field or by the heat produced when the current flows through a resistance. A *meter movement* is the electromechanical part of a meter that operates in accordance with either the magnetic or heat effect of the current. A number of different meter movements are discussed in this chapter.

By using a sensitive meter movement for the indicator, it is possible to make ammeters, voltmeters, ohmmeters, and electronic multimeters for a wide variety of electrical measurements. The required circuitry is discussed in this chapter, including the procedure for calculating the shunt and series resistance values for the meters.

When accurate measurements of resistance, inductance, or capacitance are to be made, a bridge circuit is generally used. The theory of bridge circuits and the circuits for various bridges are included in this chapter.

7.2 PARALLAX AND OTHER PROBLEMS IN USING INSTRUMENTS

Consider the problem of reading an ammeter. If two different people sight along the needle of the meter movement, the slight difference in the position of their eyes while they are making the reading will make a difference in their interpretation of the current value. You may have encountered the same problem when reading a speedometer in an automobile. The driver, who sits *behind* the speedometer, will see the speed as one value, whereas the passenger looking at the speedometer from his position will interpret its reading as being entirely different. This problem is known as *parallax*. It can be reduced somewhat by placing a mirror behind the meter.

Figure 7.1 shows a meter with a mirrored scale. The proper way to use this meter is to sight along the needle with one eye in such a way that you cannot see the reflection of the needle in the mirror. When you do this, your eye is looking exactly perpendicular to the needle, and your measurements will be most accurate.

Even though you eliminate the problem of parallax, you still have the problem of interpreting the position of the meter when it falls between marks on the scale. As with the yardstick, one person will see a reading as six-tenths of the way between the marks whereas another person will see it as seven-tenths of the way between the marks. The interpretation of the reading is very important. To improve accuracy in

Fig. 7.1 A meter with a mirrored scale. (*Courtesy Triplett Electrical Instrument Company*)

this regard some practice will be required. This is one of the advantages of a *digital voltmeter* like the one shown in Fig. 7.2. Instead of using a meter movement with a needle, this type of meter actually shows the digits of the reading accurately to four or five decimal places. This eliminates the problem of parallax.

The most difficult part of making a reading with a meter is calibration. Here we encounter the same problem that we had with the yardstick. Two different voltmeters will take the same voltage measurements and give two different readings. Which one is correct? The answer is that probably neither one is 100 percent correct. Again, if we wanted a very, very, precise measurement, we would need to use a more expensive meter, and it would have to be accurately calibrated at the Bureau of Standards.

Another problem in making accurate instrument measurements is that when the instrument is inserted into the circuit, changes in circuit current, frequency, voltage, and resistance may occur.

When all these factors are taken into consideration, we are faced with a difficulty in making highly accurate measurements. If a vacuum tube is supposed to have 100 V on its plate, it will operate satisfactorily if the voltage is 101 or 99 V. In fact, if the voltmeter reading at the plate indicates that the voltage is within 5 or even 10 percent of the manufacturer's stated value, we can usually presume that the stage is operating satis-

Fig. 7.2 A digital, solid-state volt-ohm-milliammeter. (*Courtesy Triplett Electrical Instrument Company*)

factorily. The same is true of transistor circuits. When we make a voltage, current, or resistor check during a troubleshooting procedure, we must learn to interpret the reading we get in terms of the circuit complexity and decide whether the inaccuracy is sufficient to cause the trouble we are looking for. Learning to read the instruments is the easiest part. Learning to interpret the readings is more difficult.

7.3 PERCENT ACCURACY

When the percent accuracy of an instrument is stated by a manufacturer, the value given usually refers to the accuracy at the *full-scale reading*. As an example, suppose that the accuracy of a 0- to 10-mA meter movement is given as ± 10 percent. This means that a full-scale reading of 10 mA may actually be due to a current that is

$$10 \text{ mA} + 10\% \text{ of } 10 \text{ mA} = 11 \text{ mA}$$

or

$$10 \text{ mA} - 10\% \text{ of } 10 \text{ mA} = 9 \text{ mA}$$

The manufacturer is saying, in so many words, that his instrument may be in error by as much as 1 mA for the full-scale reading. To show why this is misleading (unless we fully understand the rating), suppose that the meter in the above example is used to measure a current of 5 mA, and the reading is in error by 1 mA—that is, the meter indicates 6 mA when it is actually measuring a current of 5 mA. Now the percent error is much greater than the ± 10 percent stated for the meter at a full-scale reading. The percent error is calculated as follows:

$$\% \text{ error} = \frac{\left(\begin{array}{c}\text{current indicated} \\ \text{by meter}\end{array}\right) - \left(\begin{array}{c}\text{actual current} \\ \text{being measured}\end{array}\right)}{\text{current indicated by meter}} \qquad (7.1)$$

By using the *indicated* value of current of 6 mA for an *actual* current flow of 5 mA, the percent error is calculated as follows:

$$\% \text{ error} = \frac{\left(\begin{array}{c}\text{current indicated} \\ \text{by meter}\end{array}\right) - \left(\begin{array}{c}\text{actual current} \\ \text{being measured}\end{array}\right)}{\text{current indicated by meter}} \qquad (7.1)$$

$$\% \text{ error} = \frac{6 - 5 \text{ mA}}{6 \text{ mA}} = 16.1\%$$

Therefore, *an error of 1 mA* will cause an error of 10 percent at full scale (which is the manufacturer's rating) but will cause an error of 16.6 percent at half scale. The manufacturer's rating, then, is for the *maximum possible accuracy* when the meter is deflected full scale.

It might be argued that an error of 1 mA at full scale is the maximum possible error of the instrument, and that the error when reading at half scale will be correspondingly less. However, this is not necessarily true. In fact, the reading error at half scale *may* actually be *greater* than at full scale. For this reason the calibration of an instrument should be checked (whenever possible) in the part of the scale where most of the readings are going to take place.

7.4 BASIC INSTRUMENTS AND THEIR USE FOR SERVICING ELECTRONICS EQUIPMENT

Ammeters, voltmeters, and ohmmeters are the three basic instruments used for measurements in electric and electronic circuits.

AMMETERS   The instrument used for measuring electric current is an *ammeter*. See Fig. 7.3. (When designed for measuring very small currents, it is called a milliammeter or a microammeter.) In order to measure a current flow with an ordinary ammeter, it is necessary to have the current flow *through* the instrument, as shown in Fig. 7.3a. A special instrument, called a clamp-on ammeter, operates by measuring the

Fig. 7.3a To measure current with an ordinary ammeter it is necessary to open the circuit and insert the ammeter so that the current being measured flows *through* it.

Fig. 7.3b The clamp-on ammeter can measure current without the need for opening the circuit to insert the instrument.

Fig. 7.3 Two instruments used for measuring current.

strength of the magnetic field around the current-carrying wire, as shown in Fig. 7.3b. The clamp-on ammeter is more convenient for general servicing because it does not require that the circuit be opened in order to insert the ammeter in series with the circuit components. Clamp-on ammeters are a relatively recent introduction in the instrument field, and they are becoming more and more popular with electronic servicemen.

It is seldom necessary to measure the current in vacuum-tube receivers. It is usually easier to measure the voltage across a resistor, and if it is desired to find the current through that resistor, it is a simple matter to apply Ohm's law ($I = E/R$). Vacuum tubes are voltage-operated devices, and therefore we are interested in the plate and grid voltages when troubleshooting this type of receiver. In transistor receivers, however, current measurements are sometimes important. Specifically, we may be concerned about the total amount of current drawn by a transistor radio because this is an indication of whether the transistor amplifiers are operating properly.

VOLTMETERS Manufacturer's literature often gives values of voltages that should be measured at various points in a receiver. In order to make these measurements, a voltmeter is used. To measure a voltage, the voltmeter must be placed *across* the circuit or component (not in series with it). Figure 7.4 shows how a voltmeter is employed to measure the voltage across a resistor.

Generally the voltage measurements in a receiver are taken with respect to a *common point*, often called the *ground point*. This means that one of the meter probes is connected to the ground point, and the other probe is connected to the point where the voltage is being measured. If the meter reads +15 V, it means that the point in question (that is, the point where the voltage is being measured) is 15 V positive with respect to ground. There are exceptions, however. If we want to measure the actual grid voltage of a vacuum tube, we must connect the voltmeter between the grid and cathode because, by definition, the grid voltage is the voltage on the grid *with respect to the cathode* (not with respect to ground).

Fig. 7.4 To measure a voltage the meter probes are placed across the difference of potential.

Fig. 7.5 To measure a resistance the ohmmeter is placed across the resistor. The circuit must be opened so there will be no voltage across the resistor.

Many voltmeters are not sensitive enough to measure the difference in voltage between the emitter and the base of a properly operating transistor amplifier. However, the base or emitter voltage is sometimes measured with respect to ground for the purpose of troubleshooting.

OHMMETERS   An ohmmeter is an instrument used to measure resistance. Most ohmmeters operate by sending a current through the unknown resistance and measuring the resulting voltage drop across that resistor. A voltmeter, having a scale that is actually marked in ohms, is used for the indicator. In using an ohmmeter it is very important that it be in a circuit that has *no voltage applied*—that is, no voltage other than the voltage source inside the meter. To make a resistance measurement, the ohmmeter is placed *across* the component to be measured, as shown in Fig. 7.5.

In using an ohmmeter it is always important to be sure that the component being measured is not in parallel with another component. If the components are in parallel, as shown in Fig. 7.6, the ohmmeter will measure the equivalent resistance of the parallel combination. If both resistors in this circuit are 1000 Ω in value, the equivalent resistance of the parallel combination is 500 Ω; and this is the resistance that will be indicated by the ohmmeter. It will appear, then, that the 1000-Ω resistor is defective because it measures only half of its rated value. Parallel branches of resistance are not always so obvious as the one shown in Fig. 7.6. It may be necessary to check the circuit schematic diagram to make sure that there is not a parallel path. If there is a parallel path, then it is necessary to disconnect the resistor before measuring its ohmic value.

Ohmmeters are also convenient instruments for use in measuring *continuity*. Continuity simply means that there is a continuous path for current to flow from one point to another. As an example of a continuity measurement, consider the multiwire cable of Fig. 7.7. There are five wires coming out of the cable at each end. It is desired to trace one wire through the cable. To do this, one probe of the meter is connected to one of the wires at one side of the cable. Then, the other probe is touched to the wires (one at a time) at the other side until the ohmmeter reads zero resistance. The wire that indicates zero resistance is obviously connected to the other ohmmeter probe.

Fig. 7.6 When making a resistance measurement, the component being measured must not be in parallel with another resistance path.

Fig. 7.7 An ohmmeter can be used for continuity checks.

Fig. 7.8 Using an ohmmeter to check a printed-circuit board for cracks in the metal foils.

Continuity measurements are a valuable aid in servicing printed-circuit boards. If there is a tiny crack in a conductor on the board, it may not be visible to the naked eye. However, by placing an ohmmeter across the conductor and flexing the board, the ohmmeter will show that the path is being intermittently opened and closed. This type of test is shown in Fig. 7.8.

MULTIMETERS   Instead of having separate instruments for measuring current, voltage, and resistance, all three may be included in a single instrument, as shown in Fig. 7.9a. Figure 7.9a presents an example of a volt-ohm-milliammeter (VOM). It can measure current, voltage, and resistance and has high input impedance.

The transistorized volt-ohmmeter of Fig. 7.9b has a very high input impedance (usually about 20 MΩ). The solid-state equivalent of a VTVM (vacuum-tube voltmeter), is the FET volt-meter shown in Fig. 7.9c. This instrument is battery-operated, and therefore it has the advantage of portability over the VTVM. Like the VTVM, it has the advantage of a very high input impedance. Both the VTVM and FET voltmeters have a higher input impedance than the VOM. They can measure voltages without noticeably changing the circuit voltage. Figure 7.10 shows why this is important. The 5-kΩ resistor of Fig. 7.10a has two volts across it. When the VTVM is placed across it, we have 5 kΩ and 20 MΩ in parallel. For all practical purposes, the voltmeter is an open circuit, and only 5 kΩ is in the circuit during measurement.

Figure 7.10b shows what happens when the voltage across the 5-kΩ resistor is measured with a voltmeter that has only 5000 Ω of resistance. The two resistances in parallel have a combined resistance of 2.5 kΩ. Since the circuit resistance is decreased by the presence of the voltmeter, both the voltage across the resistor and the circuit current are changed during measurement. In general, the higher the input resistance of the measuring device, the less its effect on the circuit being measured, and the more accurate the measurement.

(a)

(b)

(c)

Fig. 7.9 Three examples of multimeters: (a) Volt-ohm-milliammeter. (*Courtesy Triplett Corporation*) (b) Transistorized volt-ohmmeter. (*Courtesy Triplett Corporation*) (c) FET meter. (*Courtesy Sencore, Inc.*)

Fig. 7.10 This shows the importance of a high-impedance voltmeter: (*a*) The VTVM does not disturb the circuit when making measurements. (*b*) The low-impedance voltmeter reduces the circuit voltage.

7.5 METER MOVEMENTS IN MEASURING INSTRUMENTS

An ampere may be defined as a flow of electric current equal to one coulomb per second. In other words, an ampere of current is flowing whenever 6 250 000 000 000 000-000.0 electrons (that is: 6.25×10^{18} electrons) flow past a point on a conductor every second. Obviously, it would not be possible to count this number of electrons directly even if they could be seen. Therefore, some form of indirect measurement must be used to determine the amount of current flow. By "indirect method of measurement," we mean that the current is measured according to some effect that it produces in a circuit.

When an electric current flows through a circuit, there are several effects that are always produced. For example, heat is produced when current flows through a resistance. The amount of heat is directly related to: (*a*) the square of the amount of current flow, and (*b*) the amount of resistance in the circuit. This is evident from the equation for power dissipated by a resistor given in equation form as

$$P = I^2R \qquad (7.2)$$

where P = power dissipated in the form of heat, W
       I = current flow, A
      R = resistance, Ω

If the amount of heat produced in a resistor can be accurately measured, and if, at the same time, the value of resistance is known, then the current can be determined by transposing the above equation:

$$I = \sqrt{\frac{P}{R}} \qquad (7.2a)$$

We cannot directly substitute the temperature for P in this equation. Temperature and power dissipation are two different things, and they must be treated as such. Thus, if we measure the amount of temperature rise due to a current flowing in a resistor, this temperature rise must be converted to the power dissipated in watts. The method of measuring power dissipation by the amount of heat generated is called the *calorimetric method.* Once the power is known, the current can be obtained from Eq. (7.2*a*).

The calorimetric method may seem like a rather roundabout method of measuring current, but it is very useful in certain applications. For example, a high-powered transmitter may be radiating 50 000 or even 100 000 W of power. This amount of power is difficult to measure accurately with ordinary measuring instruments, but the calorimetric method can be used. The output of the transmitter is fed through a *dummy load* (an artificial load) which is immersed in water. The temperature rise of the wire is accurately measured, and from this information the output power can be calculated quite accurately. The manufacturer of equipment for making caloric-type measurements usually supplies charts for interpreting temperature rise into power.

The calorimetric method of measuring current can be used in almost any application where it is necessary to measure a current, specifically a large current, by precise measurements. However, it is much too inconvenient for such applications as servicing electronic equipment. (Later we will discuss another method of measuring current by utilizing the amount of heat produced by the current.)

In addition to the heat generated by a current, there is also a magnetic field produced around the current-carrying conductor. The strength of this magnetic field is related to the amount of current flow. One way to measure the strength of a magnetic field around a current-carrying conductor is to measure the effects of a force produced when that magnetic field is inserted into another magnetic field with a known intensity. The force produced on a current-carrying conductor when placed in a magnetic field is given by Ampere's law. Stated in words, Ampere's law tells us that when a conductor carries an electric current in a magnetic field, there is a force on the conductor that is directly proportional to the current, directly proportional to the length of the conductor in the magnetic field, and directly proportional to the magnetic flux density of the field. Mathematically this law is stated as

$$F = \frac{BIL \sin \Theta}{10} \qquad (7.3)$$

where F = force on conductor, dyn
$\quad I$ = current flowing through conductor, A
$\quad L$ = length of conductor, cm
$\quad B$ = flux density of magnetic field into which conductor is inserted, G
$\quad \Theta$ = angle between direction of current flow and direction of flux

Ampere's law is the principle upon which a number of measuring instruments are designed. In fact, most of the meter movements used in voltmeters, ammeters, ohmmeters, and multimeters are based on the principle stated by Ampere's law. The needle of the meter movement is moved upscale as a result of the magnetic field of a current flowing through a coil which reacts with either a permanent magnetic field or the field of another coil.

METERS THAT UTILIZE THE MAGNETIC FIELD OF A CURRENT FOR MEASUREMENT   To summarize, instruments for measuring current employ indirect measurements based on either of two principles: the effect of heat produced by the current, or the effect of the magnetic field produced by the current. With the knowledge of these basic principles we can now study meter movements and see how they operate.

First, we will discuss measurements based on the use of a magnetic field accompanying a current; then we will discuss measurements based on the heat generated by a current.

MOVING-COIL METERS   Instead of holding the coil stationary in a meter and having a piece of soft iron move in response to a magnetic field produced by a coil, some instruments are made in which a current-carrying coil itself is moved in a permanent-magnet field. Figure 7.11 shows an example of a moving-coil meter. In this case a movable coil is inserted into a permanent magnetic field. As current flows through the coil, the magnetic field surrounding the coil reacts with the permanent magnetic field and causes the coil to turn. The coil is connected in such a way that its magnetic field will always cause the pointer to move upscale. This is an important characteristic of this type of meter: in order to get an upscale movement of the pointer, it is necessary that the current flow through the coil in a specified direction! If the direction of current is reversed, the coil will try to move in the reverse direction. Very often this will destroy the meter movement. In order to measure an alternating current with this type of movement it is necessary to employ some type of a rectifying device to convert the alternating current to direct current flowing in the proper direction.

The name D'Arsonval meter movement is often used to describe the moving-coil meter movement of Fig. 7.11. In the strictest sense of the word, this is not a D'Arsonval type of movement. However, the terminology appears now to be firmly established. This type of meter movement is also referred to as a *jeweled-type* instrument because the pivot of the coil rests in tiny near-frictionless jewels. Instead of a jeweled pivot, many of the newer meters use what is known as a *taut-band movement*. Physically, the taut band is a flat metal band that twists as the pointer moves upscale.

Figure 7.12 shows the construction of a taut-band meter. The force that returns the

Fig. 7.11   A moving-coil meter and the parts that combine to make it.

Fig. 7.12   Construction details of a taut-band moving-coil meter.

pointer to zero is due to the twist of the taut band. Taut-band meters are usually more accurate than jeweled types because of the reduction in the number of moving parts.

METERS THAT DEPEND ON THE HEATING EFFECT OF A CURRENT   The meter movements that have been discussed so far depend for their operation on the interaction of magnetic fields. In general, the magnetic field surrounding a wire or a coil carrying an unknown current reacts with a magnetic material or a magnetic field in such a way that the attraction or repulsion moves a meter pointer. We will now discuss a type of meter that depends for its operation on the heat produced by a current flow.

Fig. 7.13   A thermocouple ammeter.

The instrument of Fig. 7.13 is called a *thermocouple meter.* (A thermocouple is simply a junction of two dissimilar metals. When the junction is heated, a voltage is generated across the junction.) In the meter movement of Fig. 7.13, the junction is heated by current flowing in wire A, and the voltage appears across points C and D. The amount of voltage is proportional to the amount of heat at the junction of the thermocouple. This voltage generated by the thermocouple causes a current to flow through a moving-coil meter. A moving-coil meter is used here because it is one of the most sensitive available.

Wire A in the thermocouple meter movement will be heated regardless of whether a direct or an alternating current flows through it. An especially valuable application of this instrument is for measuring r-f current. R-f currents flowing through the wire cause it to heat, and therefore the current can be measured by the action of the thermocouple. However, since the r-f current flows through a straight piece of wire, there are no reactive effects that would change the magnitude of the current. A simple moving-coil meter cannot be used to measure radio frequency because the inductance in the moving coil would cause an inductive reactance in the circuit that would change the amount of r-f current flow.

7.6   METER CIRCUITS

The meter movements described in the previous section are designed for measuring very small currents. In order to measure larger currents, or to measure voltage and resistance, the meter movement must be placed in a circuit that will protect it from an *overload*—that is, from an excessive current which would destroy the meter movement. The most frequently encountered meter movement is the moving-coil permanent-magnet type. A typical movement of this type will deflect to its maximum full-scale reading when there are only 50 μA flowing through it. Obviously, it cannot be used directly for most of the measurements required for servicing electronic equipment.

CALCULATION OF METER SHUNTS   Figure 7.14 shows the circuit for converting a sensitive microammeter to read larger current values, such as amperes and milliamperes. The meter M and its internal resistance R_M are paralleled with a shunt resistor R_{sh}. The purpose of the shunt resistor is to bypass excessive current around the meter movement. The arrows on the illustration show how the current divides. In the operation of the circuit, the portion of the current that flows through the meter does not exceed the value required for full-scale deflection. If the value of the shunt resistance was equal to the value of the meter resistance, then the current would divide evenly so that half the current being measured would flow through each branch. However, for measuring large currents, such as an ampere, it is necessary to divide the currents in such a way that the amount flowing through the shunt resistance is, by far, *greater* than the amount flowing through the meter. Thus, the shunt resistor has a very low resistance value—often as low as a fraction of an ohm. At first appearance, the shunt may often seem to be a short circuit across the meter movements.

Fig. 7.14 A circuit for reading milliamperes or amperes.

The resistance value of R_{sh} (Fig. 7.14) can be calculated by application of Ohm's law and basic reasoning. The current being measured, I_T, will divide so that part of it flows through the meter, I_M, and the rest flows through the shunt I_{sh}. Mathematically, according to Kirchhoff's current law, it can be stated that

$$I_T = I_M + I_{sh} \qquad (7.4)$$

where I_M = maximum current that can flow through the meter, A
I_{sh} = current that must flow through the shunt, A
I_T = sum of I_M and I_{sh}

The maximum value of I_M is fixed by the maximum reading of the meter M. Except for the small amount of current flowing through the meter, all the current being measured must flow through the shunt. By subtracting I_M from both sides of Eq. (7.4), we obtain

$$I_{sh} = I_T - I_M \qquad (7.4a)$$

(The value of I_{sh} will be needed later.)

Our basic knowledge of electricity tells us that the voltage across all branches of a parallel circuit is the same. In the circuit of Fig. 7.14 the voltage across the meter branch is, by Ohm's law, the meter current times the meter resistance $I_M R_M$. This must be equal to the voltage across the shunt $I_{sh} R_{sh}$; so we can set them equal in equation form:

$$I_{sh} R_{sh} = I_M R_M \qquad (7.5)$$

where I_M and I_{sh} = the same meaning as for Eq. (7.4)
R_{sh} = current that must flow through the shunt, A
R_M = sum of I_M and I_{sh}

Since we are interested in calculating the value of R_{sh}, we will isolate it by dividing both sides of the equation by I_{sh}:

$$R_{sh} = \frac{I_M R_M}{I_{sh}} \qquad (7.5a)$$

When calculating the value of a meter shunt, we normally have the following information: the *meter resistance* R_M, the *meter current* I_M, and the *total (maximum value of) current to be measured* I_T. We usually do *not* know the value of shunt current I_{sh}; so Eq (7.5a) cannot be used directly. However, Eq. (7.4a) tells us that $I_T - I_M$ can be substituted for I_{sh}. Making this substitution into (7.5a) gives

$$R_{sh} = \frac{I_M R_M}{I_T - I_M} \text{ ohms} \tag{7.5b}$$

The currents must be in amperes, and the resistances in ohms. Equation (7.5b) can be used directly for calculating the required value of shunt resistance to convert a meter movement for measuring higher current values. However, instead of memorizing the equation, it is a better practice to make the calculations by the same reasoning that was used for the derivation.

example 7.1 A certain 50-μA meter movement has a resistance of 90 Ω. What value of shunt resistance is needed to convert the meter so that it can measure a (maximum) current of 0.5 mA?

solution Given: $I_M = 50 \ \mu A = 0.000\ 05$ A, $I_T = 0.5$ mA $= 0.0005$ A, and $R_M = 90 \ \Omega$. The current through the shunt is

$$I_{sh} = I_T - I_M \tag{7.4a}$$
$$= 0.0005 - 0.000\ 05 = 0.000\ 45 \text{ A}$$

$$R_{sh} = \frac{I_M I_M}{I_{sh}}$$

$$= \frac{0.000\ 05 \times 90}{0.000\ 45} = \frac{0.0045}{0.000\ 45} = 10 \ \Omega \ \textit{Answer} \tag{7.5a}$$

It is important to remember that this type of meter must be placed *in series* with the circuit current being measured. The very low value of shunt resistance—less than a fraction of an ohm in many instances—makes the meter circuit resistance negligible. Therefore, when the meter is placed in series with the current being measured, this does not noticeably affect its value.

CALIBRATION OF CURRENT-READING METERS Occasionally it becomes necessary to check the accuracy of an ammeter, a milliammeter, or a microammeter. The circuit of Fig. 7.15 shows how this can be done. It is based on the principle that the same current flows through all parts of a series circuit. Since meter M_1 and M_2 are in series, it follows that they should indicate the same amount of current flow. Meter M_1 is known to be accurate, whereas M_2 is being checked for accuracy. If there is a discrepancy between the two readings, then M_2 can (in many cases) be adjusted to read correctly, and this process is known as calibration. Resistor R_2 is a series-limiting resistor, to prevent the current through the meters from becoming excessive, as would happen if resistor R_1 were accidentally adjusted to zero ohms in the circuit. Resistor R_1 can be adjusted for an exact reading for meter M_1, and the reading of M_2 can be compared with this value. It is not necessary to know precise values of resistors R_1 and R_2 since we are only interested in the current value produced by the circuit.

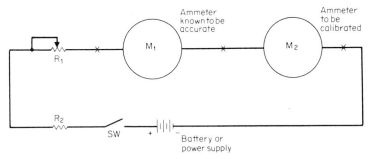

Fig. 7.15 Circuit used for calibrating ammeters.

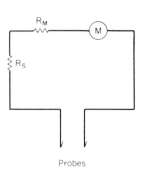

Probes

Fig. 7.16 A variable transformer that is useful for calibrating ac meters. (*Courtesy General Radio Company*)

Fig. 7.17 A simple voltmeter circuit.

If the meters in the circuit of Fig. 7.15 are designed for measuring alternating current, then the battery should be replaced with an ac power supply. A variable power transformer, such as a *Variac* (see Fig. 7.16), is useful for this purpose. If a Variac is not available, a step-down filament transformer can be used.

CALCULATION OF METER MULTIPLIERS Though it is desirable for an ammeter to have an extremely low resistance so that it will not alter the circuit current value, a *voltmeter* must have a very high resistance. The voltmeter is placed across the voltage source, and ideally it will draw very little current. A typical moving-coil permanent-magnet meter movement may have a coil resistance less than 100 Ω. If such a meter movement were placed across a 20-V source, the amount of meter current flow would be so large that it would damage the meter movement. Therefore, a series resistor R_S is placed in the circuit to limit the current flow through the meter. The series resistor is sometimes called a multiplier. Figure 7.17 shows the basic voltmeter circuit. The resistance value of R_S may be quite high, and it is often a precision resistance value.

In the circuit of Fig. 7.17, a sensitive meter movement M with an internal resistance value of R_M is placed in series with the multiplier R_S. When the probes are placed across the voltage to be measured, V, current flows in the meter circuit. This current causes a voltage drop V_S across R_S and a voltage drop V_M across the meter movement resistance R_M.

According to Kirchhoff's voltage law, the sum of the voltage drops around a circuit, $V_M + V_S$, must be equal to the applied voltage V. Applying this law to the circuit of Fig. 7.17 results in

$$V = V_S + V_M \tag{7.6}$$

where V = voltage being measured, V
V_S = resistance of multiplier, Ω
V_M = resistance of meter movement, Ω

We are interested in the value of V_S at the moment. If we know the voltage across R_S (which is V_S), and we know the current through R_S (which is the meter current I_M), then we can calculate the resistance value of R_S.

Solving Eq. (7.6) for V_S gives

$$V_S = V - V_M \tag{7.6a}$$

Since $V_S = I_M R_S$, and $V_M = I_M R_M$, we can substitute these values into the equation to get

$$I_M R_S = V - I_M R_M \tag{7.6b}$$

where V = voltage being measured, V
   I_M = maximum allowable meter current, A
   R_S = multiplier resistance, Ω
   R_M = resistance of meter movement, Ω
Dividing both sides of the equation by I_M to get an equation for R_S,

$$R_S = \frac{V - I_M R_M}{I_M} \text{ ohms} \tag{7.7}$$

All the values on the right side of Eq. (7.7) are normally known for determining the value of a meter multiplier.

example 7.2   A certain 50-μA meter movement has a resistance of 90 Ω. What value of series multiplier is needed to make an instrument that will read 5 V (maximum)?

solution   Given: $I_M = 50$ μA = 0.000 05 A, $R_M = 90$ Ω, and $V = 5$ V. Therefore,

$$R_S = \frac{V - I_M R_M}{I_M} \tag{7.7}$$

$$= \frac{5 - (0.000\ 05 \times 90)}{0.000\ 05} = 99\ 910\ \Omega \qquad Answer$$

If a moving-coil permanent-magnet meter movement is used in an ac voltmeter circuit, provision must be made for converting ac to dc voltage. You will remember that the pointer of this type of meter deflects in a direction that depends upon the direction of current flow through it. The circuit of Fig. 7.18 shows an ac voltmeter arrangement

Fig. 7.18   A meter circuit for making ac voltage measurements.

for three different ranges of voltage values. In this arrangement, the meter range is determined by the point where the probe is inserted into the circuit. For example, when the probe is plugged into the 150-V terminal, then resistors R_1, R_2, and R_3 serve the same purpose as R_S in Fig. 7.17. When the probe is plugged into the 1.5-V terminal, R_3 alone serves as the series resistor.

On one half-cycle of input alternating current, diode X_1 of Fig. 7.25 conducts and places an effective short circuit across the meter. On the next half-cycle diode X_1 is cut off, and current flows through X_2, R_4, and the meter movement. Resistor R_4 serves as a meter shunt.

It is desirable to make the series resistance of the voltmeter as large as possible to prevent loading the circuit that is being measured. Consider, for example, the circuit of Fig. 7.17. The value of R_S cannot be so large that it prevents the meter from deflecting on full scale when the maximum voltage is being measured. The greater the current required for full-scale deflection of M, the smaller the required value of R_S, and the greater the amount of loading of the circuit by the voltmeter.

A commonly used method of rating a voltmeter according to its sensitivity is the ohms-per-volt rating. The higher the ohms-per-volt rating, the greater the sensitivity of the meter movement, and hence the lower the loading effect of the meter. The reciprocal of the ohms-per-volt rating gives the amount of current required for full-scale deflection of the meter movement.

$$\left(\begin{array}{c}\text{Amperes of current required}\\\text{for full-scale deflection}\end{array}\right) = \frac{1}{\text{ohms-per-volt rating}} \tag{7.8}$$

example 7.3   A certain meter is advertised as having a sensitivity of 50 000 Ω/V. How much current is required to deflect the meter movement to full scale?

solution

$$\left(\begin{array}{c}\text{Amperes of current required}\\\text{for full-scale deflection}\end{array}\right) = \frac{1}{\text{ohms-per-volt rating}} \tag{7.8}$$

$$= \frac{1}{50\ 000\ \Omega/\text{V}} = 0.000\ 02\ \text{A} = 20\ \mu\text{A} \qquad Answer$$

7.7   VACUUM-TUBE VOLTMETERS

The vacuum-tube voltmeter arrangement of Fig. 7.19 causes very little loading of the circuit in which a voltage is being measured. The probe is connected directly into the grid circuit of a vacuum tube which has an extremely high input resistance. The voltage of E_1 is sufficient to hold the vacuum tube at cutoff when the probes are connected together. This means that the current flow through M will be zero when the probes are short-circuited together. Capacitor C_1 provides an ac bypass around the grid power supply, while capacitor C_2 provides a bypass around plate power supply E_2. When a positive voltage is applied to probe a, it partially cancels the negative voltage due to E_1, and the tube begins to conduct. The amount of plate current is proportional to the amount of positive voltage at a, and therefore the meter deflection is proportional to that voltage. Resistor R_L limits the plate-current flow through the meter movement.

The positive voltage applied to probe a must never exceed E_1 or the tube will conduct to saturation. Under the condition of saturation, the meter deflection will no longer be proportional to the input voltage. Therefore, when changing scales of the voltmeter, it will also be necessary to change the value of the grid voltage E.

The circuit of Fig. 7.19 can be used to measure either a dc or an ac voltage. When an ac voltage is applied, point a becomes alternately positive and negative. On one half-cycle, probe a will be negative with respect to probe b, and this voltage will be in series with that of E_1. Under this condition the tube is driven *beyond cutoff*. On the next half-cycle, probe a is positive with respect to probe b, producing a voltage that partially counteracts grid voltage E_1. This unblocks the tube and allows current to flow through the meter movement. Although the current that flows through the meter movement is flowing only on alternate half-cycles, the inertia of the needle prevents it from dropping back to zero during the half-cycles when no plate current flows. The

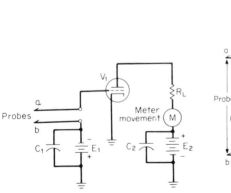

Fig. 7.19   Circuit for a simple VTVM.

Fig. 7.20   A better VTVM circuit.

needle deflection is read on a scale that is related to the value of input ac voltage.

An alternative method of measuring ac voltage is to rectify the input voltage before it is applied to the probes. Full-wave rectification—such as obtained by a bridge rectifier—is preferable because plate current will then flow through the meter during both half-cycles of input voltage. Since the plate current does not drop to zero for a full half-cycle, it follows that, for a given ac input, the meter deflection of the meter movement will be greater for a given ac voltage than would be obtained without full-wave rectification.

In vacuum-tube voltmeters the circuit arrangement shown in Fig. 7.20 is often used. As in the case of the simple VTVM shown in Fig. 7.19, field-effect transistors may be used instead of tubes for this circuit. Tubes and FETs are interchangeable because their input impedances are quite high. Transistors, on the other hand, require current flow in their base circuit, and therefore cannot be directly interchanged with tubes or FETs.

The tube circuit of Fig. 7.20 is sometimes called a *difference amplifier;* it is also known by the names *bridge amplifier, differential amplifier,* and *dc amplifier.* Assuming that there is no input voltage to the probes, we see that there are two identical tube circuits, V_1 and V_2. The resistance R_1 and R_4 are equal, R_2 equals R_3, and V_1 and V_2 are presumed to be identically matched. If it should turn out that there are slight differences in the two tube circuits, they can be compensated for by adjusting variable resistor R_6 until conduction through the two tubes is identical. Under this condition, the two cathode voltages will be identical, and therefore no current can flow through R_5 and M. Assume that during measurement of a dc voltage, point a is made positive with respect to point b. This will cause V_1 to conduct harder than V_2, and a cathode current through R_2 will be greater than that through R_3. As a result, point c becomes more positive than point d, and current will flow through the meter. The amount of deflection of the meter needle is dependent on the amount of positive voltage applied through the probe.

If an ac voltage is applied between points a and b during the measurement, V_1 will conduct harder than V_2 on alternate half-cycles. Assuming that V_1 is biased near the cutoff point, the negative half-cycles will quickly drive it to cutoff. Thus, for all practical purposes there will be current flow only during positive half-cycles of input ac voltages. (A better arrangement is to full-wave-rectify the alternating current before it is applied to the input probe so that V_1 will conduct during both half-cycles of input ac voltage. This will increase the sensitivity of the meter.)

The circuit of Fig. 7.20 is often used in modern-day electronic systems. One very important advantage of the circuit is that it is not *tube-sensitive.* In other words, as the tubes age—assuming that the effects of their aging are identical—the circuit will still be balanced. If both tubes are placed in the same envelope, they must both be replaced when one fails, thus assuring that the tubes are matched. In the circuit of Fig. 7.19, as the tube ages, the cutoff voltage changes, and furthermore the total plate current tends to decrease with aging of the cathode. Operation of the simpler circuit, then, depends somewhat upon the age of the tubes.

Let's assume that the cathode emission of the two tubes in the difference amplifier decreases by one-third. This will still mean that point c will be identical with point d when no input voltage is applied to the probes. There will, however, be some effect on the meter reading with a voltage applied, because the amount by which V_1 conducts (compared to V_2 conduction) will decrease somewhat with tube aging. However, it is not nearly so critical as a single tube circuit.

Differential amplifiers—employing transistors instead of tubes—are used in many integrated-circuit applications. Transistors, like tubes, tend to change their characteristics with age. By using differential amplifiers, the problem of aging is reduced. Since both transistors are mounted on the same chip, they will age equally, and their characteristics change simultaneously.

The FET voltmeter circuit shown in Fig. 7.21 is similar to the simple tube circuit shown in Fig. 7.19. (It should be reemphasized that an FET meter may also employ a difference amplifier like the one shown in Fig. 7.27.) In the circuit of Fig. 7.21 an N-channel FET is used. When there is no input voltage to the circuit, conduction in the FET is held at approximately cutoff value. The positive voltage on the drain makes the transistor ready for conduction. The negative voltage, E_2, produces a small loop

Fig. 7.21 Circuit for a FET voltmeter.

current, as shown by the arrows. This loop current exactly equals and cancels the current through the FET with no input voltage applied to the probes.

When a dc input voltage is applied to the probes in such a way that the positive terminal is at the gate, the FET will conduct. This conduction overrides the bucking current produced by E_2, and the resulting current flowing through meter M will cause the meter to deflect. The amount of deflection is proportional to the input voltage delivered to the gate. Resistor R_1 is part of a voltage divider; resistor R_2 forms the other part of the divider. Any input signal will appear as a voltage drop across the two resistors. The part of the voltage that appears across R_2 is delivered to the gate electrode. Capacitor C_1 serves as a noise signal bypass to ground.

When an ac voltage is applied to the probes, it will cause the FET to conduct on alternate half-cycles. In a better circuit arrangement, the ac voltage is full-wave-rectified with a bridge before being applied to the input probes. This would cause the FET to conduct during both half-cycles, resulting in a higher meter needle deflection for a given input ac voltage. As with other types of meters, full-wave rectification of the input ac signal being measured results in a more sensitive measuring instrument. The *sensitivity* of a measuring instrument refers to the ability of that instrument to measure very small values.

The ganged switch (SW) is used in the circuit to simultaneously remove the bucking voltage and drain voltage from the circuit when the meter is turned off, thus prolonging battery life.

An important advantage of the FET circuit over the simple circuit of Fig. 7.19 is that it does not require any filament voltage for its operation. This makes a completely portable high-impedance meter more feasible.

7.8 DIGITAL MULTIMETERS

All the meters discussed previously in this chapter may be referred to as "analog"-type meters. That is, they all require scales and a pointer to read a particular point on the appropriate scale. The scale readings are all continuous and change smoothly.

The analog meters all have several disadvantages in common, namely, difficulty in actual reading of the scale, the need to interpolate between printed numbers, the possibility of parallax errors, the need for multiplying factors on some ranges, limited accuracy, need for zero pointer setting, and the possibility of damage to the pointer or meter movement because of overload.

Digital multimeters overcome all the above-stated disadvantages. (A typical solid-state digital multimeter is shown in Fig. 7.2.) Such multimeters have large, easily read numbers and automatic placing of the decimal point. In some models a plus or minus sign appears at the left of the numbers, indicating the polarity of the voltage being measured. This feature is automatic and it is not necessary to switch leads or change a switch position when measuring either polarity or voltage. All measurements appear as numbers, and obviously there is no need for interpolation, no parallax error, no need for multiplying factors, and no such thing as zero pointer setting.

The accuracy of a digital multimeter on dc is commonly 0.1 to 1.0 percent, as compared with 2.0 percent for the usual good-quality vacuum-tube (or FET) voltmeter. Generally no damage can result to a digital meter from overload. Also, most of these meters have some sort of overload indication, ranging from a separate indicator to a scheme which causes all the digits to blink. This scheme operates in the case where the applied potential exceeds the selected scale. Some of the more expensive models also feature an "autoranger," which automatically selects the correct range required without the need to operate a range switch. Although the unit shown in Fig. 7.2 contains a read-out of four digits, other units are available with six or more digits. Obviously these latter units would be somewhat more expensive.

DIGITAL DISPLAYS An early type of digital display unit, which is still in fairly wide use, is the so-called "Nixie" tube. This is a small vacuum tube having a common anode and ten individual cathodes. The cathodes take the form of numbers from 0 through 9. These are stacked front to rear in the tube, which has a transparent front. The tube is a sort of glorified neon bulb. The separate cathodes are connected to the control circuitry and when a cathode is energized, its number is illuminated with an orange glow which is clearly visible. Obviously there will be as many tubes used in a particular display as the desired number of digits.

The Nixie tubes generally provide a display which is brighter and has larger numbers than the LED types (discussed below). However, they require a relatively high exciting voltage and draw too much current for practical battery operation. Thus these units are generally line-operated. The Nixie-tube display is not as visible under high ambient light conditions as the reflective liquid crystal type (discussed below).

A very popular digital display is the one which is illuminated by LEDs (light emitting diodes). This is generally found in one of two forms: the 7-segment or the 35-segment (5 × 7) display. These configurations are shown in Fig. 7.22. Although the number 8 is shown in both parts of this figure, any number between 0 and 9 could be shown. It is also possible to configure the displays so that a plus or minus sign may be produced. A plus sign is shown in Fig. 7.2.

The LED displays draw much less power than the Nixie types. The LEDs operate from a 1.5-V source and draw approximately 25 mA per segment. Many LED-type displays are "strobed" (intermittent, rapid lighting). This further reduces the power consumption, so that battery operation becomes very practical. The types of batteries in common use are the throwaway types, rechargeable types, and external battery packs. The readability of LED displays in high ambient light is not as good as that of the Nixie or the reflective liquid crystal types.

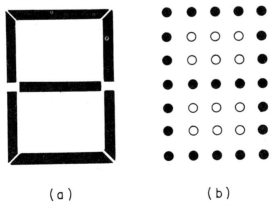

(a)　　　　　　(b)

Fig. 7.22 Two common methods of producing digital segmented number displays. (a) Seven-segment display showing number 8. (b) A 5 × 7 (35-segment) display also showing number 8.

The liquid crystal displays draw by far the least current of the three types described here. Each liquid crystal segment draws only a few microwatts. Consequently this type is the most suitable for battery operation. This device is used in the seven-segment digit display shown in Fig. 7.22a. The liquid crystal displays come in two forms: the transmissive (back-lighted) type and the reflective (ambient-lighted) type. The transmissive type results in a greater power drain because of the back lighting and thus produces a shorter battery life. The reflective type has a very low power drain and is also used in devices such as watches and clocks, where maximum battery life is required.

Each segment of a liquid crystal display consists of two clear glass plates, with a thin layer of liquid crystal material between them. The glass plates each have conductive coatings which connect to the energizing source. When energized, the character of the liquid crystal material in the appropriate segments changes with regard to its light-transmission properties. In this manner the desired number appears, either from back lighting or from reflective lighting.

THEORY OF OPERATION   The operation of a typical digital multimeter is explained here only in general terms. The reader should refer to Chap. 14, "Digital Circuit Fundamentals," and Chap. 15, "Digital Integrated Circuits," for the various details involved in such circuits.

The operation of a typical multimeter will be described with the aid of the simplified block diagram of Fig. 7.23. The quantity to be measured, ac or dc voltage, ac or dc current, or resistance, is fed into the input selector. The output of this selector is always a dc voltage, which is proportional to the quantity being measured. (The input selector also contains the range switching.) This proportional dc voltage is then fed to the ramp generator-comparator, which is the analog to digital converter. The output of this block will consist of digital pulses which are related to the proportional dc voltage, but only in terms of the time between them.

The ramp generator produces a linear, negative-going ramp, which begins at a positive value, goes through zero, and continues to an equally negative value. This is shown in Fig. 7.24. Note that the ramp (an upside-down sawtooth wave) voltage is proportional to time. Now assume that the voltage to be measured is 5 V. When the negative-going ramp reaches the same value as the measured voltage (5 V), the comparator puts out a pulse, often called a "start" pulse, which permits the LSI (large-scale integrated circuit) to begin counting pulses generated by the clock oscillator. The counting continues until the ramp decreases to 0 V. At this time the comparator puts out a "stop" pulse, which tells the LSI (logic circuits) to stop counting the clock oscillator pulses. Now note that the original voltage to be measured (5 V) has been translated from an anolog quantity to a time interval and then to a number of clock pulses, which is exactly proportional to the original 5 V. Obviously both the ramp repetition rate and the clock oscillator frequency are quite high, to minimize the time required to display readings (about 1 ms) and to permit "strobing" (or multiplexing) of the displays. This feature will now be explained.

Fig. 7.23 Simplified block diagram of a digital multimeter. Corresponding segments of each digit are connected in parallel.

Fig. 7.24 When the ramp reaches the measured voltage (5 V), a "start" pulse is generated. When the ramp reaches zero volts, a "stop" pulse is generated (see text).

In the block diagram of Fig. 7.23 (and Fig. 7.2), the display is seen to consist of four digits. This means the maximum display number is 9999. However, an automatic decimal-point placement is included in the logic circuits, so that exact value is immediately evident. Note the placement of the decimal point in Fig. 7.2, where the reading is +17.76 V. When the strobing or multiplexing feature is present, as shown in Fig. 7.23, each of the four digits is lighted separately, in turn. However, this is done at such a high rate of speed that they all appear to be lighted simultaneously. Strobing saves an appreciable amount of battery power, as previously mentioned.

Now to tie it all together. The multiplexing (strobing) signal is fed to the LSI and affects both the operation of the digit drivers and the segment drivers. Thus, as one particular digit is caused to be energized, at that same exact time the segment drivers cause the proper combination of segments for that digit to appear. Since all corresponding segments of each digit are tied in parallel and connected to the segment drivers in this way, it can be seen that in order to have only one digit at a time light up, showing the correct number, the strobing or multiplexing scheme just described is necessary.

The ½ digit In Figs. 7.2 and 7.23, four-digit displays are shown, which means that the maximum number which can be displayed is 9999. However, for relatively little additional expense and circuitry it is possible to provide an additional digit in front of the above-mentioned four. This added digit can only be programmed (as in this case) to be either unlighted (no reading) or to show a 1. Thus, if energized, this so-called ½ digit can increase the above reading from 9999 to 19 999, or practically double the original reading. Thus a 1-V range could now read 1.9999, or a 1000-V range could now read 1 999. When the ½ digit is added to a four-digit display, it is called a 4½-digit display. When the ½ digit is switched on, the meter is said to be "overranging."

Digital meter specifications There are certain types of specifications which are unique to digital meters. Some of the more common ones are: sensitivity, resolution, and accuracy.

Sensitivity defines the capability of a digital voltmeter to respond to the smallest changes of measured voltage. This, of course, varies with different types. However, a rule of thumb is that the sensitivity is approximately equal to the least significant digit (extreme right-hand digit) of the reading. For example, if we are reading (as in Fig. 7.2), 17.76 V, the sensitivity would be 0.06 V.

Resolution is the inherent ability of a display to resolve one digit out of the total number capable of being displayed. For example, in a four-digit display the largest number which can be displayed is 9999 (for practical reasons, 10 000). Since the display can resolve 1 digit (least significant) out of 10 000, the resolution is 1 part in 10 000, or 0.01 percent.

The accuracy of the conventional D'Arsonal-type multimeter is expressed as a percentage of the full-scale reading. For example, a conventional meter with an accuracy rating of 2 percent, when used on the 10-V scale, would have an accuracy within plus or minus 0.2 V at full scale. However, the accuracy at lower readings is not usually known, although it must be not worse than 2 percent.

The accuracy of a digital multimeter is expressed in a somewhat more complex manner. It is given in two parts. The first part is a percentage of the actual reading; the second part (or modifier) is given as a percentage of the full scale reading, or a number of digits (least significant digit). This is best illustrated by the following three examples taken from commercial units.

Unit A:
  DCV: $\pm 1\%$ of reading, ± 1 digit
  DCI and ACI and ACV (20 Hz–1 kHz): $\pm 1.5\%$, ± 1 digit
  Ohms: $\pm 2\%$, ± 1 digit
Unit B:
  DCV: $\pm 0.1\%$ of reading, $+.05\%$ of full scale
  ACV (50–500 Hz): 0.5% of reading, $+0.1\%$ of full scale
  Ohms: $\pm 0.15\%$ of reading, $+0.1\%$ of full scale
Unit C:
  DCV: $\pm 0.1\%$ of reading, $+1$ digit
  DCI: $\pm 0.3\%$ of reading, $+1$ digit
  ACV (45 Hz–10 kHz): 0.5% of reading, plus 2 digits
  ACI: $\pm 1\%$ of reading, plus 2 digits

7.9 CALIBRATION OF VOLTMETERS

The circuit of Fig. 7.25 shows how the accuracy of a voltmeter can be checked by comparing its reading with the reading of a meter known to be accurate. Meter M_1 is the one known to be accurate, and M_2 is to be calibrated. When the switch SW is closed, current flows through resistors R_1 and R_2 due to the applied voltage E. This may be either a dc voltage supplied by a battery or other power supply; or it may be an ac voltage if the voltmeters are being compared for ac voltage readings. Variable resistor R_1 is adjusted to vary the circuit current until the voltage drop across R_2 is some convenient value; then the meter readings are compared. In most VTVM's and FET voltmeters, provisions are made for adjusting the reading of M_2 until it is identical with that of M_1 during calibration.

Another method of calibrating a voltmeter (not illustrated) is to connect it across a voltage source of known value. An example of such a source is a dry cell that is manufactured for this purpose. Such batteries should never be employed as power sources in ordinary circuitry. They are to be utilized only for calibrating the voltage readings of voltmeters—a situation in which they are not required to supply any appreciable current. When properly used, calibration batteries may last a very long time.

OHMMETER CIRCUITS   Most ohmmeters are actually voltmeters with a power supply included. One example is the circuit of Fig. 7.26. Note that the meter and its internal resistance combine with R_A and R_B to form a simple voltmeter circuit. When a voltage E is added to one leg of this circuit, it becomes a series ohmmeter.

When the ohmmeter probes are touched together—producing *zero* resistance between them—the battery will produce a full-scale deflection on meter M. Therefore zero ohms will be marked at full scale in this type of ohmmeter. Variable resistor R_B is used to adjust the full-scale deflection so that it reads exactly zero when the probes

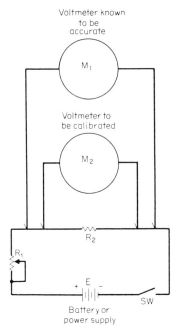

Fig. 7.25 Circuit for calibrating voltmeters.

Fig. 7.26 Circuit for a series ohmmeter. The resistance being measured is R_X.

are touched together. This variable resistor is sometimes called "zero ohms." When the probes are apart, no current can flow through the circuit, and the meter needle does not deflect. Since there is nothing between the probes, we can assume that the resistance is infinitely large, and that is the marking on the left-hand side of the scale for ohmmeters of this type.

Figure 7.27 shows an ohmmeter scale with the zero value on the right side. Some beginning students have trouble with this type of scale because it is necessary to read it backward. When an unknown resistance (R_X in Fig. 7.26) is placed across the probes, the meter will deflect to some value between zero and infinity (that is, to some value between full-scale deflection and zero deflection). The ohmmeter scale is calibrated to read the resistance value of R_X.

Figure 7.28 shows a different kind of ohmmeter. It is called a *shunt ohmmeter* because the unknown resistance value R_X is placed in parallel with the meter movement. When the meter probes of this ohmmeter are touched together — producing *zero* resistance — the current of the circuit is shunted around the ohmmeter movement, and no deflection occurs. This means that at zero ohms, the needle of the scale is at the left side (zero deflection). When the probes are not touching, all the current from battery E will flow through the meter movement. With the probes open, we can assume that there is infinitely large resistance between them, and the right side of the scale will be marked infinity. Variable resistor R_B is adjusted to read full scale when the ohmmeter probes are not touching.

Of the instruments used by radio technicians, the ohmmeter is usually the most inaccurate. An accuracy of 10 to 20 percent of the reading is acceptable for most ohmmeter designs. The accuracy of the ohmmeter tends to deteriorate somewhat as the battery ages. This is because the internal resistance of the battery increases with an increase in battery age, and therefore the unknown resistance being measured represents a smaller percentage of the total resistance in the circuit. When the battery has deteriorated to the point where it can no longer produce full-scale deflection on the

Fig. 7.27 Meter with ohmmeter scale that reads backward.

Fig. 7.28 Circuit for a shunt ohmmeter. The resistance being measured is R_X.

ohmmeter scale, it must be replaced. A VTVM, which operates on house current, uses batteries for the ohmmeter scale.

The best way to make a quick check of ohmmeter accuracy is to measure a precision resistance value with it.

7.10 DETERMINATION OF METER RESISTANCE

In order to calculate the value of the multiplier or shunt resistance needed to convert a sensitive meter movement into an ammeter or voltmeter, it is necessary to know the resistance of the meter movement. Before explaining how this can be done, it is very important to emphasize that an ohmmeter must *never* be used in an attempt to measure meter resistance. It has been shown that an ohmmeter contains a source of voltage. This voltage can send an excessive amount of current through a meter movement, thus causing damage to the coil or moving parts.

The circuit of Fig. 7.29 can be used for determining the resistance of meter M by the *half-deflection method*. With switch SW open, variable resistor R_1 is adjusted until meter M deflects to its full-scale reading. The reading of meter I is carefully noted. (Meter I should have a higher full-scale reading than meter M.)

Next, switch SW is closed. This will, of course, reduce the total resistance and increase the circuit current as measured by meter I. At the same time, it will reduce the amount of current flowing through meter M. Variable resistor R_X is adjusted until meter M reads its half-scale value. At the same time, variable resistor R_1 is adjusted to keep the total circuit current I equal to the value that it was with switch SW open. When the reading of M is equal to one-half its original value, then the resistance of R_X is equal

Fig. 7.29 Circuit for measuring the resistance of meter M.

Fig. 7.30 A circuit used for explaining the proportional method of finding current.

to the meter resistance R_M. Without disturbing the setting of R_X when it has been adjusted for a half-scale reading of M, it is removed from the circuit and its resistance value is measured.

Another means of determining the value of meter resistance is based on the proportional method for finding current in a parallel branch. Figure 7.30 shows a parallel branch comprised of R_M and R_X, with a current I flowing into the junction of the branches. By the proportional method, the following equations are used for determining I_M and I_X:

$$I_M = I_T \frac{R_X}{R_M + R_X} \tag{7.9}$$

$$I_X = I_T \frac{R_M}{R_M + R_X} \tag{7.10}$$

Figure 7.31 shows a circuit that can be used for determining the meter resistance by using the proportional current equation. The current through meter M is

$$I_M = I_T \frac{R_X}{R_M + R_X} \tag{7.9}$$

Solving this equation for R_M gives

$$R_M = \frac{(I_T - I_M)R_X}{I_M} \tag{7.11}$$

In the circuit of Fig. 7.31 any precision resistance value that is comparable to the meter resistance can be used. Variable resistor R_Y is adjusted to provide a convenient reading on I_T and M.

Fig. 7.31 An alternate method of finding the value of meter resistance.

7.11 THE WHEATSTONE BRIDGE

The ohmmeter is one of the least accurate of the instruments used by electronics technicians. If a more accurate method of making measurements is needed, a Wheatstone bridge may be used. These instruments may be made commercially, or they may be constructed with a few parts. In either case a highly accurate measurement is possible with a bridge circuit.

Before discussing the theory and use of the Wheatstone bridge, we will review the proportional method of calculating voltage drops. This will be useful for understanding the bridge circuit.

Figure 7.32 shows two resistors, R_1 and R_2, connected across a voltage source E. It does not matter whether this is a dc or an ac source; the relationships to be discussed hold true in either case. If the values of R_1, R_2, and E are known, then the circuit current I can be found by Ohm's law. By knowing the current through the resistors, the voltage drop across each resistor, V_1 and V_2, can be found.

Using the proportional method, the values of V_1 and V_2 can be found directly without the need for determining the current first. The derivation of the proportional voltage equations is quite simple, and will be given here.

Fig. 7.32 Circuit used for obtaining a formula for V_1 and V_2.

Fig. 7.33 Circuit for demonstrating the proportional method of calculating voltage drops.

Circuit total resistance:

$$R_T = R_1 + R_2$$

Circuit current:

$$I = \frac{E}{R_T} = \frac{E}{R_1 + R_2}$$

Voltage across R_1:

$$V_1 = IR_1 = \frac{E}{R_1 + R_2} \times R_1 = \frac{ER_1}{R_1 + R_2} \tag{7.12}$$

Voltage across R_2:

$$V_2 = IR_2 = \frac{E}{R_1 \pm R_2} \times R_2 = \frac{ER_2}{R_1 \pm R_2} \tag{7.13}$$

example 7.4   Calculate the values of V_1 and V_2 for the circuit of Fig. 7.33.
solution

$$V_1 = \frac{ER_1}{R_1 + R_2} = \frac{12.6 \times 33 \text{ k}\Omega}{33 + 47 \text{ k}\Omega} \cong 5.2 \text{ V}$$

$$V_2 = \frac{ER_2}{R_1 + R_2} = \frac{12.6 \times 47 \text{ k}\Omega}{33 + 47 \text{ k}\Omega} \cong 7.4 \text{ V}$$

Check: $E = V_1 + V_2$
$12.6 = 5.2 + 7.4$ V

It is obvious from the proportional method of calculating voltage that the amount of voltage across a resistor (such as R_1) in a branch (such as R_1 and R_2) depends upon the *proportion* of total resistance represented by the resistor R_1. In other words, $R_1/(R_1 + R_2)$ of the total applied voltage appears across R_1, and $R_2/(R_1 + R_2)$ of the applied voltage appears across R_2.

In the circuit of Fig. 7.34, the voltage at point a with respect to point c is

$$V_{a-c} = \frac{ER_2}{R_1 + R_2}$$

and the voltage at point b with respect to point c is

$$V_{b-c} = \frac{ER_4}{R_3 + R_4}$$

If $V_a = V_b$, then it follows that

$$\frac{\not{E}R_2}{R_1 + R_2} = \frac{\not{E}R_4}{R_3 + R_4}$$

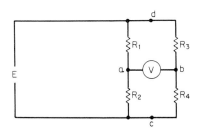

Fig. 7.34 The basic Wheatstone bridge circuit.

Fig. 7.35 A circuit for a commercial bridge.

This equation may be rewritten by a basic rule of algebra:

$$\frac{R_2}{R_4} = \frac{R_1 + R_2}{R_3 + R_4}$$

The voltage at point a with respect to point d is

$$V_{a-d} = \frac{R_1}{R_1 + R_2}$$

The voltage at point b with respect to point d is

$$V_{b-d} = \frac{R_3}{R_3 + R_4}$$

Under the condition that the voltage at point a is equal to the voltage at point b,

$$V_{a-d} = V_{b-d} \quad \text{and} \quad \frac{R_1}{R_1 + R_2} = \frac{R_3}{R_3 + R_4}$$

By using the same rule of algebra, this equation becomes

$$\frac{R_1}{R_3} = \frac{R_1 + R_2}{R_3 + R_4}$$

Since R_2/R_4 and R_1/R_3 are both equal to $(R_1 + R_2)/(R_3 + R_4)$, it follows that they are equal to each other.

$$\frac{R_1}{R_3} = \frac{R_2}{R_4} \tag{7.14}$$

This important equation holds true only when the voltage at point a is equal to the voltage at point b! This condition means that the bridge is *balanced*.

Figure 7.35 shows the Wheatstone bridge circuit used for measuring a resistance. The values of R_1 and R_2 are accurately known. Variable resistor R_4 has an accurately calibrated dial that makes it possible to know the exact resistance at any setting. The unknown resistance value—that is, the resistance being measured—is R_X. An ac or a dc voltage E is applied, and variable resistor R_A regulates the amount of current flowing to the bridge.

During a resistance measurement, the unknown resistance value is connected to terminals X and Y of Fig. 7.35. Then variable resistor R_4 is adjusted until no current flows through the galvanometer G—thus indicating that there is no voltage difference between points a and b. In other words, R_4 is adjusted *until the bridge is balanced*.

When the unknown resistance is first placed in the bridge circuit, there may be a considerable amount of unbalance, which would result in a large current flow through the galvanometer. To prevent damage to the meter movement, a shunt R_{SH} may be switched into the circuit. The switch is usually marked *sensitivity*. It is spring-loaded so that the shunt is always across the meter except when SW is depressed. This prevents the meter from being accidentally damaged during times when the bridge is greatly unbalanced.

When the bridge is balanced, the following equation holds true:

$$\frac{R_1}{R_X} = \frac{R_2}{R_4}$$

Solving this equation for R_X gives

$$R_X = \frac{R_1 R_4}{R_2} \tag{7.15}$$

example 7.5   The values of R_1 and R_2 in the circuit of Fig. 7.35 are 33 and 47 kΩ, respectively. The bridge is balanced when R_4 is set to 94 kΩ, with an unknown resistance R_X inserted into the bridge. What is the exact value of R_X?

solution

$$R_X = \frac{R_1 R_4}{R_2} = \frac{33 \text{ k}\Omega \times 94 \text{ k}\Omega}{47 \text{ k}\Omega} = 66 \text{ k}\Omega$$

7.12 AC BRIDGES

Variations of the Wheatstone bridge circuit can be used for measuring inductance, capacitance, impedance, and resonant frequency. In order to make these measurements, the input voltage to the bridge must be ac, and some minor changes must be made to the ratio arms. A few of the more important but commonly encountered ac bridge circuits will now be discussed.

RESONANCE AND FREQUENCY BRIDGE   Figure 7.36 shows a bridge circuit that can be used for measuring resonance or frequency. The series-resonant circuit comprised of C_X and L_X is placed across one arm of the bridge. Resistance R_X may be the resistance of this series-resonance circuit, or it may be a resistance added to the bridge in order to make it possible to balance the bridge. An audio or r-f generator is applied across the bridge input terminals. The indicator in this case may be a pair of earphones, an ac voltmeter, or an oscilloscope. In some commercially manufactured bridges, tuning-eye tubes are employed as indicators.

For this discussion we will assume that an audio generator is used across the bridge terminals, and the indicator is a pair of earphones. As the generator frequency is varied so that its output frequency passes through the resonant frequency of $C_X - L_X$, the volume of sound heard will vary. The volume will be minimum when the audio generator is tuned to the resonant frequency. In other words, the signal generator is tuned for a null in the earphones. The resonant frequency of L_X and C_X is then read from the frequency dial of the audio generator. In order to get a sharper null, it may be necessary to adjust the resistance value of R_C so that the bridge is balanced.

The procedure just described is especially useful for finding the resonant frequency of an audio filter. The filter is connected into the bridge circuit in place of R_X, L_X, and C_X. The bridge will be balanced when the following conditions are met:

$$\frac{R_A}{R_B} = \frac{R_C}{R_X} \tag{7.16}$$

$$f_r = \frac{1}{2\pi \sqrt{L_X C_X}} \tag{7.17}$$

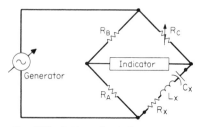

Fig. 7.36 Bridge circuit for measuring resonance and impedance.

where R_A, R_B, R_C, R_X = resistance values, Ω
$\qquad\quad f_r$ = frequency of audio generator, Hz
$\qquad\quad L_X$ = inductance of series-tuned circuit (or effective inductance of filter), H
$\qquad\quad C_X$ = capacitance of series-tuned circuit (or effective capacitance of filter), F

It is important to note that *both* conditions described by Eqs. (7.16) and (7.17) must be met simultaneously in order for a true null to be obtained.

If R_C is a calibrated resistor, the effective series resistance of the resonant circuit will be known by its setting when a true null is obtained. If R_X is in actuality a combination of the series resistance of the resonant circuit and a resistance value added to it to effect a balance, then the resistance of R_X must be subtracted from the calculated value to find the actual value of the resistance of the series-resonant circuit. In most cases R_X will be simply the resistance of the coil. If the resistance value of R_X is very small, then it might not be possible to obtain its value on this type of bridge. The reason is that the junction connections will affect the balance more than the small resistance.

If an r-f generator is employed in this circuit, the r-f signals must be modulated and a detector circuit used in conjunction with earphones (or other type of indicator).

Equation 7.17 shows that the frequency of the ac generator can be accurately determined if the values of C_X and L_X (as well as the resistance values) are accurately known. In this application the usual procedure is to vary C_X in series with an accurately known inductance L_X. The frequency is read directly from the dial of C_X.

INDUCTANCE BRIDGES The inductance of a coil can be accurately obtained by means of an ac bridge. Figure 7.37 shows a *resistance ratio bridge* that can be used

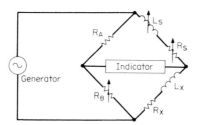

Fig. 7.37 Measurement of inductance can be accomplished with this resistance ratio bridge.

for this purpose. As in all the ac bridges discussed in this section, the indicator may be a set of earphones, an oscilloscope, or an ac meter, provided frequency response of the indicator is not exceeded by the frequency of the ac generator. In this simple circuit, R_X and L_X comprise the unknown inductor and its equivalent dc resistance value. These are represented schematically as a coil and a resistor in series, although it must be understood that the resistance is actually the resistance of the wire in the coil. Resistor R_S and inductor L_S represent the *Standard resistance* and *Standard inductance*, respectively. They must be accurately calibrated so that when they are adjusted for a balance, the value of R_X and L_X can be read directly from their dials. In order to effect a balance with this bridge, the following conditions must be met simultaneously:

$$L_X = \frac{R_B}{R_A} L_S \qquad\qquad (7.18)$$

$$R_X = \frac{R_B}{R_A} R_S \qquad\qquad (7.19)$$

where R_A, R_B, R_S, R_X = resistance values, Ω
$\qquad\quad L_S$, L_X = inductance values, H

Fig. 7.38 Commonly used inductance bridges: (a) Maxwell bridge. (b) Hay bridge.

In the circuit of Fig. 7.37 resistor R_b is made variable to obtain the larger number of resistance ratios, and therefore increase the range of inductance values that can be measured with the bridge. Normally R_b is comprised of four or five precision resistors, any one of which can be connected into the circuit with a selector switch.

The most important disadvantage of the resistance ratio bridge of Fig. 7.34 is that it is difficult to obtain precision variable inductors. Such inductors are considerably more expensive than precision capacitance values. For this reason, the bridge circuits of Fig. 7.38 are more often encountered. In these circuits precision variable capacitors C_S are used for balancing the bridge (instead of precision inductors).

Although the two circuits of Fig. 7.38 are quite similar in construction, they actually have different applications. The Maxwell bridge (Fig. 7.38a) is used whenever the Q of the inductance being measured is relatively low—that is, when the dc resistance value of R_X is relatively high. When the Q of the inductor is high (indicating a low value of R_X), the Hay bridge (Fig. 7.38b) is employed for inductance measurement. The equations for L_X and R_X under the condition of a balanced bridge must be *simultaneously* met.

For the Maxwell bridge:

$$L_X = R_A R_B C_S \tag{7.20}$$

and

$$R_X = \frac{R_A R_B}{R_S} \tag{7.21}$$

The Q of the coil can be obtained from the equation

$$Q = \frac{\omega L_X}{R_X} = \omega C_S R_S \tag{7.22}$$

For the Hay bridge,

$$R_X = \frac{R_A R_B R_S(\omega C_S)}{1 + (R_S \omega C_S)^2} \tag{7.23}$$

and

$$L_X = \frac{R_A R_B C_S}{1 + (R_S \omega C_S)^2} \tag{7.24}$$

The Q of the coil can be obtained from the equation

$$Q = \frac{\omega L_X}{R_X} = \frac{1}{\omega C_S R_S} \tag{7.25}$$

where L_X = unknown inductance value, H
R_A, R_B, R_S, R_X = resistance values, Ω
$\omega = 2\pi f$
f = frequency of ac generator, Hz
C_S = capacitance value of precision variable capacitor, F

In the case of the Maxwell bridge, if the value of C_S is accurately known, then the value of Q will be related to the setting of R_S. In most commercially manufactured

Fig. 7.39 Combination of Maxwell and Hay bridges. (*a*) Circuit for switching between series and parallel *RC* connections. (*b*) Circuit for combination bridge.

Maxwell bridge circuits, R_S is calibrated directly to read the value of Q. For the Hay bridge, the setting of R_S is proportional to the *reciprocal* of the Q of the coil, and it is normally marked D on the bridge dial. (The letter D stands for *dissipation factor*, which is defined as the reciprocal of the Q.) Once the value of D is obtained, the Q is easily determined by taking the reciprocal.

In a variation of the Hay bridge, resistor R_4 is replaced with a variable capacitor in the ratio arm. The operation of the circuit, however, is the same. This version is called an *Owen bridge*.

It would be inconvenient to have two different conductance bridges — one for measuring the inductance of coils with high Q and one for measuring the inductance of coils with low Q. The only difference between Maxwell and Hay circuits is that the variable resistor is in parallel with the capacitor in one case, and in series with the capacitor in the other case. It is a simple matter to arrange a switching circuit for converting the same bridge circuit to either a Maxwell bridge or a Hay bridge. Figure 7.39*a* shows the switching arrangement. The dotted line between R_h and R_m (which stands for *Hay resistance* or *Maxwell resistance*) indicates that these two variable resistors are actually connected onto the same shaft. This simplifies the operation, but it requires that both the Q values and the D values be marked on the variable-resistor indicator.

Figure 7.39*b* shows the actual combination bridge circuit. It will be noted that the resistances in the ratio arm are actually marked on the dial according to the inductance range that can be measured for each setting of range switch S_5. The switching circuit that converts the bridge from a Hay to a Maxwell configuration is comprised of S_3 and S_4 in conjunction with resistors R_h and R_m and capacitor C_S. The unknown inductance is placed in the circuit, and variable resistor R_A is adjusted for a null in the detector. The actual inductance value of the unknown inductance is read from the calibrated dial of R_A. The Q or dissipation factor is read from the dial of R_m or R_h—depending on the type of bridge circuit in use at the time.

CAPACITANCE BRIDGES   Figure 7.40 shows a simple *resistance ratio bridge* for measuring capacitance values. The unknown capacitance C_X and its resistance R_X

Fig. 7.40 Resistance ratio bridge for measuring capacitance.

are balanced by calibrated capacitors C_S and R_S. When the bridge is balanced, as indicated by a null, the following equations must simultaneously hold true:

$$C_X = \frac{R_A}{R_B} C_S \tag{7.26}$$

and

$$R_X = \frac{R_B}{R_A} R_S \tag{7.27}$$

where R_A, R_B, R_S, R_X = resistance values, Ω
$\qquad$ C_S, C_X = capacitance values, F

The capacitance bridges shown in Fig. 7.41 are variations of the resistance ratio bridge. The Wien bridge of Fig. 7.41*a* accurately measures the capacitance in terms of resistance and frequency values. This circuit can be used for precision determination of the capacitance value if the frequency of the generator and the resistance values are accurately known. The equations for the bridge under the condition of balance are as follows:

$$C_X = \frac{1}{\omega^2 R_C R_X C_C} \tag{7.28}$$

$\qquad$ (a) $\qquad\qquad\qquad\qquad\qquad\qquad\qquad\qquad$ (b)

Fig. 7.41 Two popular bridge circuits: (*a*) Wien bridge. (*b*) Schering bridge.

The following alternate equation is sometimes more useful:

$$C_X = C_C \left(\frac{R_B}{R_A} - \frac{R_C}{R_X} \right) \qquad (7.28a)$$

where R_A, R_B, R_C, and R_X are in ohms
$\quad$ C_X and C_C are in farads
$\quad$ $\omega = 2\pi f$ and f are in hertz

The Schering bridge of Fig. 7.41b is useful when both the capacitance and the power factor (or dissipation factor) of the capacitor are to be determined. When the bridge is balanced, the following equations must hold true for this circuit:

$$R_X = \frac{C_B}{C_A} R_C \qquad (7.29)$$

$$C_X = \frac{R_B}{R_A} C_A \qquad (7.30)$$

The Q of the capacitor C_X can be determined by the equation

$$Q = \frac{1}{\omega C_X R_X} \qquad (7.31)$$

$$Q = \frac{1}{\omega C_B R_B} \qquad (7.31a)$$

Fig. 7.42 Circuit for a manufactured Wien bridge.

Since a Schering bridge is easily calibrated to read the dissipation factor of the dielectric directly (by a dial connected to R_B), this circuit is sometimes used for comparing the properties of dielectrics. Two parallel plates are connected in the position of C_X, and various dielectrics are inserted between the plates. The bridge is balanced in each case. The dissipation factors are then tabulated.

Figure 7.42 shows a schematic diagram of a commercially manufactured capacitance bridge. In this case the Wien bridge circuit is used. Except for the fact that switching arrangements are provided for changing the range of capacitance values, the circuit is identical with the one shown in Fig. 7.41a.

Chapter **8**

Semiconductor Devices and Transistors

8.1 INTRODUCTION

The variety of semiconductor devices available to the technician and engineer is pro-
digious. Included are rectifying, regulating, tunnel, light-emitting, light-sensing,
switching, r-f, and microwave diodes. Diodes are used for converting alternating current to a unidirectional current (rectification), regulating voltage levels, detecting and
mixing r-f signals, switching, clamping, in displays, and in the generation of microwave power. Some devices based partially on diode action are the silicon-controlled
rectifier (SCR) and the unijunction transistor (UJT). The SCR and UJT find wide use
in controlling efficiently power delivered to a load.

Transistors fall into two broad categories: the *bipolar junction transistor* (BJT) and
the *field-effect transistor* (FET). Bipolar junction transistors, usually referred to simply as *transistors*, are available as npn and pnp types. Field-effect devices can be divided into two basic types: the *junction* FET (JFET) and the *metal-oxide semiconductor* FET (MOSFET). The MOSFET is also referred to as the *insulated-gate* FET
(IGFET). Figure 8.1 provides a sampling of the available shapes and packages of discrete diodes and transistors.

In this chapter we shall examine the operation of semiconductor devices and transistors, and their characteristics, and define the significant parameters used for their char-
acterization. A discussion of understanding diode and transistor data sheets concludes
the chapter. To comprehend the operation of semiconductor devices, however, an
elementary knowledge of semiconductor theory is vital.

8.2 ELEMENTARY SEMICONDUCTOR THEORY

Electrons appear to exhibit a dual nature, behaving as *particles* at one time and as *waves*
at other times. The particle picture is useful in describing the operation of transistors
and most diodes. For a few devices, such as the tunnel diode, the concept of electrons
being wavelike in nature seems best suited for describing their operation.

ATOMIC STRUCTURE According to the particle picture, the structure of an atom is
analogous to a miniature solar system. In the center of the atom is the *nucleus* which
contains positive-charge particles called *protons* and, for many atoms, electrically neutral particles referred to as *neutrons*. Rotating around the nucleus are *electrons*, which
have a negative charge. Because the number of electrons equals the number of protons,
the atom is normally electrically neutral. The mass of the proton and neutron is approximately the same; the mass of the electron is about 1/2000th that of a proton.

Fig. 8.1 Discrete diodes and transistors are available in many sizes and shapes. (*Courtesy General Electric Company, Semiconductor Department, Syracuse, N.Y.*)

The simplest atomic structure is hydrogen (H), depicted in the two-dimensional sketch of Fig. 8.2a. Its *atomic number* Z is 1 because a single electron orbits around a nucleus containing a single proton. The orbit, or *shell*, nearest to the nucleus can hold no more than two electrons. Helium (He), for example, has an atomic number Z = 2. This denotes that exactly two electrons are rotating around the nucleus (which contains two protons), and the shell is completely filled. For this reason helium is *chemically inert* and does not react with other elements. An element whose shell is unfilled, such as hydrogen, reacts with other elements. The number of electrons in an unfilled shell is called the *valence* electrons; for example, hydrogen has one valence electron.

For atoms of greater structural complexity, the electrons are contained in shells located farther away from the nucleus. The second shell can hold up to 8 electrons, the third up to 18, and so on. The atomic structure of silicon (Si), the basic material used for discrete semiconductor devices and integrated circuits, is illustrated in Fig. 8.2b. Its atomic number $Z = 14$; consequently, the third shell is unfilled and contains *four* valence electrons. Germanium (Ge), another semiconductor material, also has four valence electrons. Elements containing four valence electrons are said to be *tetravalent*.

Fig. 8.2 Examples of atomic structures: (a) Hydrogen, H. (b) Silicon, Si.

Fig. 8.3 Representation of covalent bonding.

Silicon for semiconductor devices is grown to have a *single-crystal* structure. In simple terms, a single crystal has an ordered array of atoms. In absolutely pure, or *intrinsic*, silicon, each atom shares an electron with its four neighboring atoms, as illustrated in Fig. 8.3. This type of electron sharing results in *covalent bonds*. Because only the valence electrons are involved in covalent bonds, the silicon atom is represented simply by four valence electrons around a nucleus of +4 positive charge for electrical neutrality.

ENERGY BANDS   The resistance of a semiconductor is greater than that of a conductor, but considerably less than that of an insulator. With the aid of an *energy-band diagram*, shown in Fig. 8.4a, one can readily understand the differences in conductors, semiconductors, and insulators. According to modern atomic theory, the energies of electrons in a material, such as silicon, are grouped in *energy bands*. The *valence band* of energies corresponds to electrons that are attracted strongly by the nucleus. As a

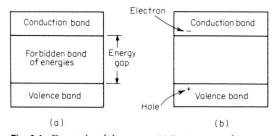

Fig. 8.4 Energy-band diagrams: (a) Basic energy diagram. (b) An electron reaching the conduction band of energies leaves a hole in the valence band of energies.

result, these electrons cannot serve as carriers of current. To become carriers of current, the electrons must occupy a higher band of energies, referred to as the *conduction band.* No electron, however, can ever occupy a region, referred to as the *forbidden band of energies,* which separates the valence and conduction bands. An electron in the valence band, therefore, must have a minimum energy before it occupies the conduction band. This minimum energy is the *energy gap* of the material.

For conductors such as copper the valence and conduction bands overlap. Consequently, there is a plentiful supply of electrons to serve as carriers of current. In an insulator, such as diamond, the energy gap is so great that there are virtually no electrons in the conduction band. The energy gap for semiconductors, however, is such that some electrons are found in the conduction band.

An electron that has reached the conduction band of energies in a semiconductor leaves a vacancy, or *hole,* in the valence band of energies, as illustrated in Fig. 8.4b. What is interesting is that the hole behaves like a positively charged particle whose charge is equal and opposite to the electron. It is influenced by an electric field, like an electron, but travels in an opposite direction and more slowly than an electron.

Because of the effects of energy, such as heat or light, equal numbers of electrons and holes are produced in a semiconductor. This action is referred to as the *intrinsic generation of electron-hole pairs.* As the temperature of a semiconductor rises, the electrons in the valence band gain energy and the generation of electron-hole pairs increases rapidly. In fact, this is what limits the operating temperature of silicon devices to 150°C. Because the energy gap of germanium is nearly one-half that of silicon, the maximum operating temperature of germanium is approximately 85°C.

DOPING   Intrinsic silicon, which contains equal numbers of electrons and holes, is useless for semiconductor devices. One needs to alter intrinsic silicon so that it has either an excess of electrons or of holes. This is accomplished by introducing suitable impurities in the silicon. The process is called *doping,* and the impurity is referred to as a *dopant.*

n-TYPE SILICON   Assume that an impurity, such as phosphorus (P) which has five valence electrons, is introduced in silicon. (Elements with five valence electrons are called *pentavalent* elements.) The impurity concentration required for semiconductor devices is extremely minute, in the order of 1 impurity atom for 10^8 silicon (*host*) atoms. The impurity atom replaces a silicon atom in the single-crystal lattice. Because for every 10^8 silicon atoms there is only one impurity atom, the probability is high that the impurity atom will be surrounded by silicon neighbors in the lattice. Silicon doped with a pentavalent impurity is called *n-type silicon.*

A two-dimensional view of n-type silicon is illustrated in Fig. 8.5a. Because phosphorus is a pentavalent dopant, it has five valence electrons. Four of the valence electrons form covalent bonds with four neighboring silicon atoms. The fifth valence electron is free and serves as a carrier of current. When a voltage is impressed across an

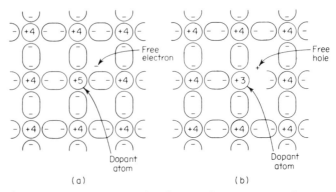

Fig. 8.5   Covalent bonding in a doped semiconductor: (*a*) n-type. (*b*) p-type.

n-type specimen, the free electrons are attracted toward the positive terminal, and current flows. If the doping is increased, more free electrons exist, and greater current flows.

Electrons in an n-type semiconductor are referred to as *majority carriers.* Regardless of the fact that the semiconductor is doped, the intrinsic generation of electron-hole pairs still continues, unabated. Hence, in addition to electrons, there are a few holes. Holes in an n-type material are referred to as *minority carriers.* At normal operating temperatures, the minority carriers are negligible. As the temperature rises, however, the minority carriers increase.

p-TYPE SILICON   Assume that an element having three valence electrons (*trivalent* element), such as boron (B), is introduced in silicon. The result is pictured in Fig. 8.5*b*. The trivalent impurity now shares its three valence electrons with three neighboring silicon atoms. The deficiency of a fourth electron results in a hole. As mentioned earlier, the hole acts like a particle having a positive charge, equal and opposite to that of an electron, and serves as a carrier of current.

When a voltage is impressed across a p-type semiconductor, the holes are attracted to the negative terminal of the source. As for an n-type semiconductor, the greater the doping is, the more holes are present and greater current flows. Holes in a p-type semiconductor are now the majority carriers, and electrons the minority carriers.

Having this basic knowledge of semiconductor theory, we can proceed now to consider the operation of diodes and transistors.

8.3   JUNCTION DIODE

The simplest semiconductor device is the junction diode. Its basic physical construction and electrical symbol are shown in Fig. 8.6. The junction diode is formed by doping one-half of intrinsic silicon (or germanium) with a p-type dopant and the other half with an n-type dopant. The boundary at the p and n regions, where all the action occurs, is called the *pn junction.* Connections made to the p and n regions are referred to as the *anode* and *cathode*, respectively.

When the anode is made positive with respect to the cathode, as in Fig. 8.7*a*, the diode is said to be *forward-biased.* Electrons in the n region are attracted toward the anode end, and holes in the p region toward the cathode end of the diode. The electrons and holes, therefore, cross the junction, and current flows in the circuit.

Fig. 8.6   Junction diode: (*a*) Basic physical construction. (*b*) Electrical symbol.

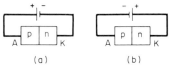

Fig. 8.7   Biasing a junction diode: (*a*) Forward-biased. (*b*) Reverse-biased.

In Fig. 8.7*b* the anode is made negative with respect to the cathode. Under this condition, the diode is *reverse*, or *back, biased.* Minority holes and electrons near the junction are initially attracted to the negative and positive terminals of the battery, respectively. As a result, the junction becomes depleted of carriers, and no further current flows. This region of depleted carriers is referred to as the *depletion region*, or *layer.* (The *voltage-variable capacitor*, or *varactor*, whose basis of operation is the depletion layer, is described in Chap. 2.)

DIODE CHARACTERISTICS   In the previous section we saw that the junction diode behaves like a "valve." When forward-biased, the diode permits current to flow; in the reverse-biased state, it prevents the flow of current. If it were possible to make the *perfect*, or *ideal, diode*, its characteristics might appear as shown in Fig. 8.8*a*.

When the diode is forward-biased, current I_F (dc forward current) flows and voltage V_F (dc forward voltage) is zero, regardless of the forward current. (*Note:* It is common practice to use V instead of E in denoting voltage for diode and transistor characteristics.

This will be adhered to in this and other chapters for describing the characteristics of semiconductor devices.) When the diode is reverse-biased, I_R (dc reverse current) is zero, regardless of the value of V_R (dc reverse voltage).

A typical characteristic curve for a *physical diode* appears in Fig. 8.8*b*. In forward bias, for voltages less than the threshold voltage V_t, virtually no current flows. (At room temperature, $V_t = 0.6$ V for silicon and 0.25 V for germanium. The threshold voltage decreases at the rate of 2 mV for each degree Celsius rise in temperature.) For forward voltages greater than V_t, the diode conducts, and the voltage across the diode is somewhat greater than the threshold value.

In reverse bias, a minute reverse current, in the order of microamperes for silicon,

(a)   (b)

Fig. 8.8 Voltampere characteristics: (*a*) For an ideal diode. (*b*) For a physical diode.

Fig. 8.9 A diode used for rectification and demodulation: (*a*) Basic circuit. (*b*) Input and output waveforms for rectification. (*c*) Input and output waveforms for demodulation. (*d*) Modulating signal obtained from demodulator. (See Example 8.1.)

flows.   The reverse current increases as the reverse bias voltage is increased.   At a voltage V_{br} (the *breakdown voltage*), the reverse current increases rapidly for a very slight increase in reverse voltage.   This characteristic is used in voltage-reference and regulating diodes, to be considered in the next section.

RECTIFIER, R-F, AND SWITCHING DIODES   A diode used for rectification, or in r-f and switching circuits, is basically a pn junction diode.   What distinguishes it is its physical size, geometry, and encapsulation.   Because it handles many milliamperes or amperes of current, a rectifier diode is generally larger in size than a r-f or switching diode.   A larger physical size implies a large junction area and capacitance.   As a result, rectifier diodes cannot be used effectively in high-frequency or switching applications.

R-f diodes are physically small and designed to minimize capacitance and inductance. Switching diodes also exhibit small physical size and are designed to switch rapidly from the reverse- to the forward-biased state.   Recently switching diodes have been developed that are capable of switching a few amperes in the order of a tenth of a microsecond.

example 8.1   Show how a junction diode may be used for (a) rectification of alternating current, and (b) demodulation (detection) of an amplitude-modulated (AM) signal.

solution   A basic circuit for both applications is illustrated in Fig. 8.9a.

(a) Referring to Fig. 8.9b for rectification, the input to the circuit is typically a 60-Hz sine wave.   During the positive half-cycle, the diode is forward-biased (the anode is positive with respect to the cathode), and current flows.   Neglecting the small forward voltage across the diode, the output is equal to the input.   On the negative half-cycle, the diode is reverse-biased (the anode is negative with respect to the cathode), and no current flows.   The output is *unidirectional*, that is, always positive.   With suitable filtering (see Chap. 16), the unidirectional waveform is converted to a dc voltage.

(b) An example of an amplitude-modulated input signal is shown in Fig. 8.9c.   As in rectification, on the positive half-cycle the diode conducts, and on the negative half-cycle the diode does not conduct.   The output, therefore, is a group of unidirectional waves of different amplitudes.   If a suitable capacitor is placed in parallel with resistor R in Fig. 8.9a, the modulating signal is obtained, as shown in Fig. 8.9d.

DIODE PARAMETERS   Commonly used parameters for characterizing junction diodes are summarized in Table 8.1.   In specifying their values, the manufacturer must

TABLE 8.1   Commonly Used Parameters for Junction Diodes

Parameter	Symbol	Meaning
Dc forward voltage	V_F	Voltage across a forward-biased diode (anode positive with respect to cathode)
Dc forward current	I_F	Direct current flowing in a forward-biased diode
Dc reverse voltage	V_R	Voltage across a reverse-biased diode (anode negative with respect to cathode)
Dc reverse current	I_R	Leakage current flowing in reverse-biased diode
Reverse breakdown voltage	V_{br}, B_V, PRV, PIV	Maximum reverse voltage across diode before it breaks down
Power dissipation	P	Maximum power that may be dissipated in a diode
Operating junction temperature	T_j	Temperature of the pn junction
Capacitance	C	Capacitance across diode in its forward- or reverse-biased state
Reverse recovery time	t_{rr}	Time required for reverse current or voltage to reach a specified value after switching diode from forward- to reverse-biased state
Forward recovery time	t_{fr}	Time required for forward voltage or current to reach a specified value after switching diode from its reverse- to forward-biased state
Noise figure	NF_O	Ratio of rms output noise power of receiver in which diode is used to that of an ideal receiver of same gain and bandwidth
Conversion loss	L_C	Power lost in mixer diode when converting an r-f signal to an i-f (intermediate frequency) signal
Video resistance	R_v	Low-level impedance of a detector diode

state the test conditions and ambient (operating) temperature at which measurements were made.

8.4 ZENER DIODES

The reverse characteristics of a physical-junction diode illustrated in Fig. 8.8b is redrawn in Fig. 8.10a. It was noted that as the reverse voltage increases, at a certain level, V_{br}, the diode breaks down. In the operating region of reverse current from I_{ZK} to I_{ZM}, the voltage across the diode is essentially constant.

Current I_{ZK}, the knee current, is the minimum current, and I_{ZM} is the maximum current limited by device dissipation. The voltage below the knee V_Z is the zener voltage of the device. Over the operating range of reverse current, the voltage across the diode is nearly constant and equal to the zener voltage.

Diodes made to exhibit specific zener voltages are referred to as *zener diodes;* their symbol is shown in Fig. 8.10b. Zener diodes are available having zener voltages of approximately 2 to 200 volts.

A major application for zener diodes is their use as a *voltage regulator,* shown in Fig. 8.11a. The zener diode is connected across the load, which is represented by

(a)   (b)   (a)   (b)

Fig. 8.10 The zener diode: (a) Voltampere characteristics. (b) Electrical symbol.

Fig. 8.11 A simple voltage regulator: (a) Circuit. (b) Model of a zener diode.

variable resistance R_L. Connected across R_L is the zener diode, which is reverse-biased (its cathode is positive with respect to the anode). The zener diode may be represented by a battery of V_Z volts in series with zener resistance R_Z, as shown in Fig. 8.11b. If the zener diode were ideal, its resistance would be zero, and the voltage across R_L would always equal the zener voltage. Because of the resistance, however, the voltage across the diode increases slightly with increasing zener current.

In Fig. 8.11a, current I flowing in resistance R always equals the sum of the zener and load currents, I_Z and I_L, respectively: $I = I_Z + I_L$. Assume that with E fixed, R_L is reduced. As a result, load current I_L increases, and the zener current decreases so that their sum is equal to I. If R_L is increased, the load current decreases and the zener current increases. During the changes in zener current, the zener voltage is hardly affected, and the voltage across the load resistance is nearly constant.

If the voltage source E should increase or decrease, the zener current will change accordingly. As before, the load voltage will hardly change its value.

example 8.2   For the voltage regulator of Fig. 8.11a, the zener diode regulates at 20 V over a current range of 5 to 30 mA. (a) If voltage source $E = 50$ V, determine the value of R if I_L varies from 0 to its maximum value. (b) What is the maximum load current?

solution   (a) For zero load current, current I is equal to the maximum zener current I_{ZM}; hence, $I = I_{ZM} = 30$ mA. The value of R equals the voltage across it $(50 - 20 = 30$ V$)$ divided by its current (30 mA); hence, $R = 30/(30 \times 10^{-3}) = 1$ k Ω.

(b) The maximum load current equals the value of I less the minimum zener current. Therefore, $I_{L,\max} = I - I_{ZK} = 30 - 5 = 25$ mA.

ZENER AND AVALANCHE BREAKDOWN   A reverse-biased zener diode appears to break down by two different mechanisms. For zener voltages below approximately 6 V, the high electric field across the pn junction results in the rupture of covalent bonds and a rapid increase of electron-hole pairs. This process is referred to as zener breakdown.

In avalanche breakdown a few carriers, having acquired sufficient energy, collide with silicon atoms in the crystal lattice. As a result of the collision, new carriers are released which in turn collide with other atoms, releasing more carriers. Thus, a rapid increase of free electrons and holes occurs.

For either mechanism, the diode may be referred to as a zener or avalanche device. True zener diode action exhibits a negative temperature coefficient (breakdown voltage decreases with increasing temperature). True avalanche diode action displays a positive temperature coefficient (breakdown voltage increases with increasing temperature).

REGULATOR AND REFERENCE DIODES   Zener diodes designed to maintain a relatively constant voltage over a wide range of load current, as in the simple regulator of Fig. 8.11a, are sometimes referred to as regulator diodes. A diode designed to exhibit a relatively fixed voltage and fixed current, regardless of temperature changes, is referred to as a reference diode.

DOUBLE-ANODE REGULATOR   Physically, the double-anode regulator contains a single cathode which is common to two anodes (Fig. 8.12). This device is used for

Anode   Anode
Cathode

Fig. 8.12 Electrical symbol for double-anode regulating diode.

symmetrical clipping of an ac waveform and in protective circuits where negative and positive overloads may occur.

ZENER DIODE PARAMETERS   Commonly used parameters for characterizing zener diodes are summarized in Table 8.2.

TABLE 8.2   Commonly Used Parameters for Zener Diodes

Parameter	Symbol	Meaning
Zener voltage	V_Z	Nominal voltage at which zener diode regulates
Knee current	I_{ZK}	Minimum current necessary for operation of zener diode
Maximum zener current	I_{ZM}	Maximum current that can flow in zener diode
Zener impedance	Z_Z	Indicates change in zener voltage for small changes in zener current with respect to a specified test current I_{ZT}.

8.5   BIPOLAR JUNCTION TRANSISTOR

The bipolar junction transistor (BJT), usually referred to simply as a transistor, is a device with three terminals. It comes in two types: npn and pnp. In the npn device, a "sandwich" of a thin p region is surrounded by two thicker n regions as shown in Fig. 8.13a. In the pnp device, illustrated in Fig. 8.13b, a thin n region is surrounded by two thicker p regions. Electrical symbols for both types of transistors are given in Fig. 8.14.

Connected to the inner (p or n) region is the *base* (B) lead. To the outer n (or p) regions are connected the *collector* (C) and *emitter* (E) leads. For either the npn or the pnp transistor, two junctions exist: the *base-emitter* and *collector-base junctions*.

In normal operation, as for an amplifier, the BJT is connected so the base-emitter

Fig. 8.13 Basic physical construction of a bipolar junction transistor (BJT): (a) npn type. (b) pnp type.

Fig. 8.14 Electrical symbols for a transistor: (a) npn type. (b) pnp type.

junction is *forward*-biased and the collector-base junction is *reverse*-biased. Referring to Fig. 8.15a, for an npn transistor the negative terminal of battery E_{BB} is connected to the emitter, and the positive terminal to the base. The base-emitter junction is therefore forward-biased like a pn junction diode. To the positive terminal of battery E_{CC} is connected the collector, and to the negative terminal the base. Similar to a reverse-biased *pn* junction diode, the collector-base junction is thereby reverse-biased.

To bias properly a pnp transistor, batteries E_{BB} and E_{CC} are reversed, as illustrated in Fig. 8.15b. In this case, the positive terminal of E_{BB} is connected to the emitter,

Fig. 8.15 For amplifier operation, the base-emitter junction is forward-biased and the collector-base junction is reverse-biased: (a) For an npn transistor. (b) For a pnp transistor.

and the negative terminal to the base. The collector is connected to the negative terminal of E_{CC}, and the positive terminal is returned to the base.

For either the npn or pnp transistor, the emitter current I_E always equals the sum of the base I_B and collector I_C currents:

$$I_E = I_B + I_C \qquad (8.1)$$

OPERATION OF THE BJT   To explain the operation of a bipolar junction transistor, the npn device will be considered. The operation of the pnp transistor is identical with the npn when the word "hole" is substituted for the word "electron" in the following description.

Consider the elementary transistor amplifier of Fig. 8.16. Load resistance R_C which, for example, can represent a loudspeaker, is connected to the collector in series with E_{CC}. The output voltage is designated by E_o. Signal source E_s, which may be the output of a phonograph cartridge, is connected in series with the base and E_{BB}. From a

Fig. 8.16 Illustrating the operation of a common-base amplifier using an npn transistor.

signal standpoint, the batteries act like short circuits. The base, therefore, is common to the input E_s and output E_o signals. Such a configuration is called a *common-base* (CB) amplifier.

Because the base-emitter junction is forward-biased, electrons from signal E_s are readily injected into the emitter. The electrons then cross the base-emitter junction into the base region. The base, which is quite thin, has a much greater resistivity than either the emitter or the collector region. Hence, very few electrons are lost in the base, and most of them cross the base-collector junction into the collector region. The electrons then flow through resistance R_C and are attracted by the positive terminal of battery E_{CC}.

The voltage gain of an amplifier A_v is defined as the signal output voltage E_o divided by the signal input voltage E_s:

$$A_v = \frac{E_o}{E_s} \tag{8.2}$$

Output voltage E_o may be expressed as the product of output signal current I_c and load resistance R_C:

$$E_o = I_c R_C \tag{8.3}$$

Voltage E_s equals the product of input signal current I_e and the resistance of a forward-biased base-emitter junction. Letting the value of the diode resistance be designated by R_d, we have

$$E_s = I_e R_d \tag{8.4}$$

Substitution of Eqs. (8.3) and (8.4) in Eq. (8.2) yields

$$A_v = \left(\frac{I_c}{I_o}\right)\left(\frac{R_C}{R_d}\right) \tag{8.5}$$

The ratio I_c/I_e is generally designated by the Greek letter *alpha* (α) and is called the *short-circuit current gain* of a transistor in the common-base configuration. (Another symbol commonly used for α is h_{fb}). Substituting α for I_c/I_e, Eq. (8.5) appears as

$$A_v = \frac{\alpha R_C}{R_d} \tag{8.6}$$

Because I_c is always slightly less than I_e, α can never exceed unity; a typical value of α is 0.95. The value of R_C, however, is generally much greater than that of R_d. Therefore, even though alpha can never exceed unity, the voltage gain can be much greater than one.

example 8.3 In the amplifier of Fig. 8.16, assume that $\alpha = 0.95$, $R_C = 1000\ \Omega$, and $R_d = 20\ \Omega$. Find the voltage gain.

solution Substituting the given values in Eq. (8.6), we obtain

$$A_v = 0.95 \times \frac{1000}{20} = 47.5$$

The most useful connection for the BJT is the *common-emitter* (CE) configuration of Fig. 8.17. In this circuit, emitter terminal E is common to the input and output

Fig. 8.17 An elementary common-emitter amplifier using an npn transistor.

signals. Because the base current I_b is minute, the current gain I_c/I_b is much greater than unity. The ratio, defined as the short-circuit current gain of a transistor in the common-emitter configuration, is designated by the Greek letter *beta* (β) or by the commonly used symbol, h_{fe}.

BJT PARAMETERS AND CURVES   The most commonly used parameters for characterizing the BJT are summarized in Table 8.3. Bipolar devices are classified as

TABLE 8.3   Commonly Used Parameters for Bipolar Junction Transistors

Parameter	Symbol	Meaning
Collector-base voltage (emitter open)	V_{CBO}	Maximum voltage that can be impressed across collector and base of a transistor with emitter open
Collector-emitter voltage (base short-circuited to emitter)	V_{CES}	Maximum voltage that can be impressed across collector and emitter of a transistor with base short-circuited to emitter. Its value is in the order of one-half of V_{CBO}
Collector-emitter voltage (with specified resistor between base and emitter)	V_{CER}	For this condition, maximum collector-emitter voltage is greater than V_{CES} but less than V_{CBO}
Emitter-base voltage (collector open)	V_{EBO}	Maximum voltage that can be impressed across emitter and base of a transistor with collector open
Collector saturation voltage	$V_{CE,\text{sat}}$	Collector-emitter voltage of a transistor that is fully conducting, as in a transistor switch
Small-signal input resistance	h_{ib}, h_{ie}	Input resistance of a transistor with output short-circuited for the signal. This and other h parameters are called *hybrid parameters*. The second subscript refers to the transistor configuration: b for common base and e for common emitter.
Small-signal output admittance	h_{ob}, h_{oe}	Output admittance of a transistor with input open-circuited for the signal
Small-signal reverse-voltage transfer ratio	h_{rb}, h_{re}	Ratio of voltage developed across input of a transistor to voltage present at output, with input open-circuited for the signal
Small-signal forward-current gain	h_{fb}, h_{fe}	Ratio of output to input signal currents with output short-circuited
Dc forward-current gain	h_{FE}	Ratio of dc collector current to dc base current for transistor in common-emitter configuration
Collector dissipation	P_C	Power dissipated in collector of a transistor. It is equal to the product of the dc collector current and dc collector-emitter voltage.
Gain-bandwidth product	f_T	Frequency at which common-emitter forward current gain is unity
Cutoff frequency	f_{hfb}, f_{hfe}	Frequency at which h_{fb} (or h_{fe}) is 0.707 times its value at 1 kHz
Collector-cutoff current (emitter open)	I_{CBO}, I_{CO}	Reverse saturation (leakage) current flowing between collector and base with emitter open
Collector-cutoff current (base open)	I_{CEO}	Reverse saturation current flowing between collector and emitter with base open
Collector-base capacitance	C_{ob}, C_{cb}	Transistor capacitance across collector and base. This capacitance influences, to a great degree, the high-frequency performance of a transistor amplifier

small-signal, medium-power, power, r-f, and switching transistors. Several available transistors can operate at collector voltages and currents as high as 500 V and 100 A, respectively.

A useful set of curves is the common-emitter *collector characteristics* of Fig. 8.18. On the y axis is plotted the dc collector current i_C, and along the x axis the dc collector-emitter voltage, v_{CE}. A family of curves, for different values of dc base current I_B, is drawn in the figure. The reason for plotting base current is that the BJT is a *current-operated* device. As mentioned earlier, the base-emitter junction is forward-biased for normal transistor operation. Base current therefore flows and is a key quantity in establishing the operating point of a transistor.

Fig. 8.18 Typical common-emitter collector characteristics.

8.6 FIELD-EFFECT TRANSISTOR

The field-effect transistor (FET) is basically a three-terminal semiconductor device having characteristics similar to that of a pentode vacuum tube. Unlike the bipolar transistor, the FET is a *voltage-operated* device. Instead of being biased by a current, the FET is biased by a voltage, and no input current flows. Its input resistance, therefore, is virtually infinite.

There are two kinds of field-effect devices: the *junction* FET (JFET) and the *metal-oxide semiconductor* FET (MOSFET). The MOSFET is also referred to as an *insulated-gate* FET (IGFET) in the literature.

JUNCTION FET A cross-sectional view of an elementary *n-channel* JFET is shown in Fig. 8.19*a*. It contains two p-type regions in an n-type silicon bar. The two p regions are connected and called the *gate, G.* Ohmic connections are made to each end of the silicon bar for the *source* (S) and *drain* (D) leads.

A cross-sectional view of a *p-channel* JFET is shown in Fig. 8.19*b*. In this structure, the two n-type regions for the gate are formed in a p-type silicon bar. The biasing voltages of a p-channel JFET are opposite to those of the n-channel type (similar to npn and pnp transistors). Electrical symbols for n- and p-channel JFET's are given in Fig. 8.20.

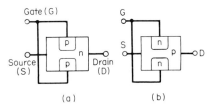

Fig. 8.19 Basic physical construction of an elementary JFET: (*a*) n-channel. (*b*) p-channel.

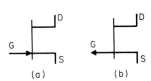

Fig. 8.20 Electrical symbols for a JFET: (*a*) n-channel. (*b*) p-channel.

Fig. 8.21 Biasing an n-channel JFET in the common-source configuration.

Fig. 8.22 Depletion region developed along the channel of a JFET.

Operation Consider the n-channel JFET of Fig. 8.21 biased for normal operation. Because source S is common to the output (drain D) and input (gate G) voltages, the connection is referred to as the *common-source* configuration. The drain is held positive with respect to the source, and the gate negative with respect to the source. Assume that the gate-source voltage is initially set to zero ($E_{GG} = 0$).

As the drain-source voltage E_{DD} is increased, more electrons are attracted from the source to the drain, and the drain current I_D increases. Because of the voltage drop along the n channel, a depletion region is developed along the pn junction formed by the p-type gate and n-type channel regions (Fig. 8.22). This is the same kind of depletion region that is formed in a pn junction diode. As you recall, a depletion region is devoid of current carriers. Because the channel becomes more depleted of carriers as E_{DD} is increased, the drain current levels off at a maximum value I_{DSS} (the drain current with $E_{GG} = 0$), as illustrated in Fig. 8.23. Voltage V_{PO} may be defined as the minimum value of drain-source voltage, V_{DS}, for which maximum drain current flows.

As the gate-source voltage is made increasingly negative, the depletion region is

Fig. 8.23 Drain current I_D as a function of drain-source voltage V_{DS}.

Fig. 8.24 Typical curves for a JFET: (a) Drain characteristics. (Operation must be restricted to the area below dissipation curve.) (b) Transfer characteristics. (*Courtesy Siliconix Incorporated.*)

formed more rapidly than with zero bias. As a result, V_{PO} occurs at a lower value of V_{DS}. A typical family of *drain characteristics* for different values of bias voltage V_{GS} is given in Fig. 8.24a. Plotted on the y and x axes are the drain current and voltage, respectively. Because the FET is a voltage-operated device, curves for different values of gate-source bias, V_{GS}, are plotted in the figure.

Another useful curve for field-effect transistors is the *transfer characteristics* of Fig. 8.24b. This is a plot of drain current I_D as a function of gate-source bias voltage V_{GS} for a given drain-source voltage, V_{DS}.

MOSFET   In the MOSFET, the gate is a metallic electrode separated from the channel by a thin insulating material, such as silicon dioxide (SiO_2). There are two types of MOSFET's: *depletion* and *enhancement*. A cross-sectional view of an n-channel depletion-type MOSFET is shown in Fig. 8.25a. The silicon bar, referred to as the substrate, is p-type material. To one of the n+ regions (the + sign denotes a heavily doped region) is connected the source lead, and to the other the drain lead. Between the n+ regions is an n region which forms the channel.

Fig. 8.25   Cross-sectional views of a depletion-type MOSFET: (a) n-channel. (b) p-channel.

Fig. 8.26   Electrical symbols for a depletion-type MOSFET: (a) n-channel. (b) p-channel.

A cross-sectional view of a p-channel MOSFET is shown in Fig. 8.25b. The substrate in this case is n-type material, and the drain and source connections are made to p+-type material. The channel is p-type. Electrical symbols for both types of depletion MOSFET's are shown in Fig. 8.26. The substrate (SUB) is generally connected to the source terminal.

Operation of depletion-type MOSFET   The gate, insulator, and channel of a MOSFET act like a capacitor. When, for example, a positive charge is on the gate, an opposite (negative) and equal charge is induced in the channel. If the gate has a negative charge, an equal positive charge is induced in the channel.

Consider the n-channel depletion-type MOSFET in Fig. 8.27a biased for normal operation. (A p-channel depletion-type MOSFET is biased with opposite polarities.) The drain is positive with respect to the source. If the gate is biased negatively with respect to S, a positive charge is induced in the n channel, as illustrated in Fig. 8.27b. This results in a reduction of free electrons in the n channel, and drain current I_D is reduced. As the gate is made more negative with respect to the source, the drain current is reduced further. This is illustrated in the typical drain and transfer characteristics of Fig. 8.28.

Fig. 8.27   Operation of an n-channel depletion-type MOSFET: (a) Circuit. (b) Gate biased negatively with respect to source. (c) Gate biased positively with respect to source.

Fig. 8.28 Typical curves for an n-channel depletion-type MOSFET: (*a*) Drain characteristics. (*b*) Transfer characteristics. (*Courtesy RCA.*)

Fig. 8.29 Cross-sectional views of an enhancement-type MOSFET: (*a*) p-channel. (*b*) n-channel.

Fig. 8.30 Electrical symbols for an enhancement-type MOSFET: (*a*) p-channel. (*b*) n-channel.

If the gate is now made positive with respect to the source, as in Fig. 8.27c, electrons are induced in the n channel. Conduction is *enhanced*, and greater drain current flows (see Fig. 8.28). Because of the insulated gate electrode, the input resistance of a MOSFET, whether the gate is positive or negative with respect to the source, approaches infinity.

Operation of enhancement-type MOSFET The depletion-type MOSFET may be regarded as a *normally ON* device. The reason for this designation is that for $V_{GS} = 0$, appreciable drain current flows (see Fig. 8.28). For many applications a *normally OFF* MOSFET is desirable, so that, for $V_{GS} = 0$, no drain current flows. This kind of device is realized in the *enhancement-type* MOSFET.

Cross-sectional views of p- and n-channel enhancement-type MOSFET's are illustrated in Fig. 8.29; their electrical symbols are given in Fig. 8.30. In either type, no physical channel, as in the depletion MOSFET, exists. The device, therefore, is

(a)   (b)

Fig. 8.31 Operation of a p-channel enhancement-type MOSFET: (*a*) Circuit. (*b*) Gate biased negatively with respect to the source.

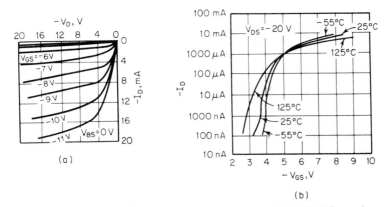

(a)   (b)

Fig. 8.32 Typical curves of a p-channel enhancement-type MOSFET: (*a*) Drain characteristics. (*b*) Transfer characteristics. (*Courtesy General Instrument Corporation.*)

normally OFF. Before a current can flow between the drain and source, a suitable bias is required across the gate and source terminals.

A p-channel enhancement MOSFET biased for normal operation is illustrated in Fig. 8.31. (For an n-channel device, the polarities of sources E_{DD} and E_{GG} are reversed.) The gate is held negative with respect to the source. Positive charges (holes) are induced in the region below the gate electrode, and drain current flows. As V_{GS} is increased, the drain current also increases. Typical drain and transfer characteristics for the enhancement MOSFET are illustrated in Fig. 8.32.

FET PARAMETERS Commonly used parameters for characterizing FET devices are summarized in Table 8.4.

TABLE 8.4 Commonly Used Parameters for the JFET and MOSFET

Parameter	Symbol	Meaning
Drain current for zero bias	I_{DSS}	Drain current flowing when gate is short-circuited to source ($V_{GS} = 0$)
Gate reverse current	I_{GSS}	Leakage current flowing between gate and source for a specified reverse bias across gate and source terminals
Drain cutoff current	$I_{D,off}$	Drain current flowing when device is biased in its OFF state
Gate-source breakdown voltage	BV_{GSS}	Maximum reverse voltage that may be impressed across gate and source terminals without damaging the device
Gate-source pinchoff voltage	V_P	Gate-to-source voltage which reduces I_{DSS} to 1% or less of maximum value at a specified drain-to-source voltage
Small-signal forward transconductance	g_{fs}, g_m	Ratio of a small change in signal drain current to a small change in signal gate-to-source voltage in common-source configuration. Parameter g_{fs} is an indication of the gain for the device.
Dc drain-source ON resistance	r_{DS}	Ratio of dc drain-source voltage to the dc drain current, generally measured at $V_{GS} = 0$
Input capacitance	C_{iss}	Small-signal input capacitance for device in the common-source configuration with $V_{DS} = 0$
Reverse transfer capacitance	C_{rss}	Capacitance between drain and gate in common-source configuration with $V_{DS} = 0$

8.7 SILICON-CONTROLLED RECTIFIER

The silicon-controlled rectifier (SCR), also referred to as a *thyristor*, is a four-layer pnpn "sandwich" device having three terminals, as illustrated in Fig. 8.33a. Connected to the outer p region is the *anode*, and to the outer n region the *cathode*. The *gate* terminal is connected to the inner p region. The electrical symbol for the SCR is given in Fig. 8.33b.

What makes the SCR unique and useful is its switching characteristics. With the anode held positive with respect to the cathode, a small pulse of current (*trigger*) is applied to the gate terminal. The SCR turns on (*conducts*). In its conducting state, it behaves like a rectifier diode. Furthermore, it remains conducting even after the trigger is reduced to zero.

Fig. 8.33 Silicon-controlled rectifier (SCR): (*a*) pnpn "sandwich" construction. (*b*) Electrical symbol.

The SCR turns off (becomes *nonconducting*) only when the current flowing between the anode and the cathode is reduced to a level that is less than a minute critical value of current, called the *holding current*. Because of these characteristics, the SCR is extremely efficient in controlling, for example, the power to a motor or in the dimming of theater lights.

example 8.4 Show how a SCR may be used to rectify alternating current and control the power delivered to a dc motor.

solution A simplified circuit for accomplishing this is illustrated in Fig. 8.34a, and the waveforms are shown in Fig. 8.34b and c. The point at which the trigger is applied during the positive half-cycle of the ac source is controlled by suitable circuitry (to be described later) in the box labeled *phase control*. Once the SCR turns on, it conducts like a rectifier diode. It stops conducting when the anode current falls below the holding current, which is close to zero. During the negative half-cycle, the SCR is reverse-biased, and no current flows.

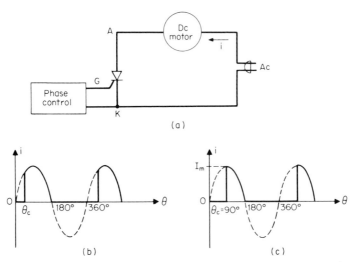

Fig. 8.34 A SCR controlling power delivered to a dc motor: (a) Circuit. (b) Typical current waveform. (c) Current waveform for a phase angle = 90°. (See Example 8.4.)

The angle at which conduction occurs, *phase angle* θ_c (Greek letter *theta*), may be anywhere between approximately 0 and 180° (Fig. 8.34b). In the example of Fig. 8.34c, where $\theta_c = 90°$, it is extremely simple to calculate the average (dc) current flowing in the motor. In a half-wave rectifier, current flows for 180° and the average current is $0.318I_m$ amperes, where I_m is the peak current. If current flows only during half of a positive cycle (90°), the average current is therefore equal to $0.318I_m/2 = 0.159I_m$ A.

OPERATION OF THE SCR For an understanding of SCR operation, it is convenient to visualize the pnpn sandwich structure of Fig. 8.33a as shown in Fig. 8.35a. The SCR is depicted here as a pnp transistor connected to an npn transistor and shown schematically in Fig. 8.35b.

Assume that anode A is positive with respect to cathode K, as indicated in Fig. 8.35b. With no positive trigger, both transistors are nonconducting, and anode current I_A is equal to a minute leakage current. Upon the application of a small positive trigger to the base of Q_2 (gate terminal G), the npn transistor begins to conduct. Collector current in the npn transistor is equal to the current flowing toward the base of pnp transistor Q_1. Hence, $I_{B1} = I_{C2}$, and Q_1 begins to conduct. Both transistors assume proper bias levels to ensure that the device conducts after the termination of the trigger. As mentioned earlier, the SCR turns off when the anode current is reduced to a value less than the holding current.

(a)   (b)

Fig. 8.35 Viewing the SCR as being comprised of an npn and a pnp transistor: (a) Connections of the transistors. (b) Schematic representation.

Fig. 8.36 Electrical symbol for a LASCR.

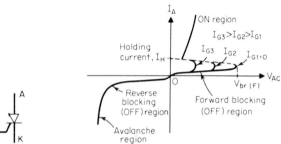

Fig. 8.37 Voltampere characteristics of a typical SCR.

LIGHT-ACTIVATED SCR A variation in the SCR is the light-activated SCR, referred to as the LASCR. Light passing through a translucent window in the package containing the device acts as a positive trigger. The LASCR, in other respects, behaves as a SCR. The electrical symbol for a LASCR is illustrated in Fig. 8.36.

CHARACTERISTIC CURVES Typical voltampere characteristics of a SCR are given in Fig. 8.37. Plotted along the vertical axis is anode current I_A and along the horizontal

Fig. 8.38 Gate-characteristic curves for a typical SCR. (*Courtesy RCA.*)

axis the anode-cathode voltage V_{AC}. When V_{AC} is positive, the SCR is forward-biased. Depending on the value of gate current I_G, the device turns on, or *fires*, at a specific *forward breakdown voltage*, $V_{(br)F}$. Note that for $I_{G1} = 0$, the breakdown voltage is greater than for I_{G2} or I_{G3}, which are greater than zero. When the anode current is less than the holding current, the SCR reverts to its forward blocking (OFF) region. In the forward conducting and reverse regions of operation, the SCR resembles a junction diode.

Because the gate trigger current and voltage required vary with temperature and with units of the same type, manufacturers often supply *gate-characteristic curves* for their devices. An example of gate-characteristic curves is illustrated in Fig. 8.38. Gate voltage versus gate current is plotted for different temperatures. Note that smaller trigger voltages and currents are required at higher temperatures. For example, the maximum voltage that does not trigger the SCR at 100°C is approximately 0.2 V. Curve A represents the characteristics of a typical SCR.

SCR PARAMETERS   Commonly used parameters for characterizing the SCR are summarized in Table 8.5.

TABLE 8.5 Commonly Used Parameters for the SCR and TRIAC

Parameter	Symbol	Meaning
Forward breakdown voltage	$V_{(br)F}$ V_{BO}	Forward voltage at which device fires
Reverse breakdown voltage	$V_{(br)R}$	Maximum reverse voltage that causes device to go into avalanche
On-state voltage	V_T, V_F	Voltage across device when it is conducting (ON state)
On-state current	I_T, I_F	Current flowing between anode and cathode in ON state
Holding current	I_H	Minimum current for device to be in ON state
Latching current	I_L	Minimum current to maintain device in ON state, after switching from OFF to ON state with trigger removed
Gate trigger current	I_{GT}	Minimum gate current for switching device from OFF to ON state
Gate trigger voltage	V_{GT}	Gate voltage needed to produce required gate current
Gate turn-on time	t_{on}	Time for device to turn on
Commutated turn-off time	t_{off}, t_q	Time for device to turn off

8.8 BIDIRECTIONAL TRIODE THYRISTOR

The electrical symbol for the bidirectional triode thyristor, or TRIAC, is shown in Fig. 8.39a, and its equivalent circuit in Fig. 8.39b. Referring to Fig. 8.39b, the TRIAC is viewed as two SCR's in parallel with the anode of one connected to the cathode of the other. The voltampere characteristics of a TRIAC (Fig. 8.40) exhibit the same forward blocking characteristics of the SCR when main terminal one is positive with respect to main terminal two, or vice versa. The parameters for characterizing the TRIAC are essentially the same as for the SCR, summarized in Table 8.5.

example 8.5   Show how a TRIAC may be used as a lamp dimmer.

Fig. 8.39 The TRIAC: (a) Electrical symbol. (b) Equivalent circuit.

Fig. 8.40 Voltampere characteristics of a typical TRIAC.

Fig. 8.41 The TRIAC used as a lamp dimmer: (*a*) Basic circuit. (*b*) Current waveform. (See Example 8.5.)

   solution   A basic circuit of a lamp dimmer is given in Fig. 8.41*a*. Depending on the setting of the phase control, the firing angle, as in the SCR, varies from approximately 0 to 180° in each half-cycle. For example, in Fig. 8.41*b* current flows only for 45° (180–135) during the positive half-cycle and another 45° (360–315) during the negative half-cycle.

8.9   UNIJUNCTION TRANSISTOR

The unijunction transistor (UJT) is a three-terminal device that differs from the BJT in two important respects:
   1. It has only one junction.
   2. It exhibits a region of *negative resistance*. A device is said to have negative resistance when, for increasing current, the voltage across the device decreases. Because of its negative resistance, the UJT finds application in oscillator, timing, and SCR trigger circuits.
   The basic physical construction of a UJT is illustrated in Fig. 8.42*a*, and its electrical symbol is given in Fig. 8.42*b*. Two ohmic (nonrectifying) contacts, B_1 and B_2, are made to the ends of an n-type silicon bar. The resistance of the bar, referred to as the *interbase resistance* R_{BB}, is typically of the order of 4000 to 10 000 Ω. A rectifying pn junction is formed close to terminal B_2 to which is connected emitter lead E.

Fig. 8.42   The unijunction transistor (UJT): (*a*) Basic physical construction. (*b*) Electrical symbol.

Fig. 8.43   Equivalent circuit of a UJT.

OPERATION OF THE UJT   An equivalent circuit of a UJT is shown in Fig. 8.43. Between B_2 and B_1 are two resistances, R_{B1} and R_{B2}. Their sum is equal to the interbase resistance: $R_{B1} + R_{B2} = R_{BB}$. Because the value of R_{B1} varies as a function of emitter current I_E, it is represented as a variable resistance. The pn junction is denoted by diode D and the voltage across it by V_D. The voltage V_A at point A with respect to B_1, by voltage division, is

$$V_A = \frac{R_{B1}V_{BB}}{R_{B1} + R_{B2}}$$
$$= \eta V_{BB}$$

(8.7)

where η (Greek letter *eta*) is referred to as the *intrinsic standoff ratio* of the device

$$\eta = \frac{R_{B1}}{R_{B1} + R_{B2}} = \frac{R_{B1}}{R_{BB}}$$

(8.8)

Fig. 8.44 Voltampere characteristics of a typical UJT.

Fig. 8.45 A circuit using a UJT that generates sawtooth and trigger waveforms. (See Example 8.6.)

A plot of emitter voltage versus emitter current is illustrated in Fig. 8.44. For $V_E = 0$, diode D is reverse-biased, and a small reverse saturation current, $-I_{EO}$, flows. As V_E is increased, the diode becomes less reverse-biased and I_E less negative. At a sufficiently high emitter voltage the diode conducts and holes are injected into the bar region denoted by R_{B1}. Because of the excess holes, the value of R_{B1} decreases and I_E increases. For example, when $I_E = 0$, $R_{B1} = 4000\ \Omega$, and when $I_E = 50$ mA, $R_{B1} = 40\ \Omega$. Owing to the reduced resistance (increased conductance) of R_{B1}, V_E decreases as I_E increases. This behavior is referred to as negative resistance.

The voltage and current at the peak point of the characteristic curve are referred to as the *peak voltage* V_P and the *peak current*, I_P, respectively. Voltage V_P is the emitter voltage at which the UJT makes a transition from the cutoff region to the negative resistance region. Peak current I_P is the minimum current required to turn on the UJT.

At the valley point ($V_V = $ *valley voltage*, $I_V = $ *valley current*), the UJT makes a transition from the negative resistance region to the saturation region. In the saturation region R_{B1} acts as a positive resistance; that is, as I_E increases, V_E increases, too. Voltage $V_{BE1,sat}$ is the voltage across the emitter and base B_1 corresponding to a specified emitter current. An important application of the UJT and the significance of the peak and valley points are illustrated in Example 8.6.

example 8.6   Analyze the operation of the UJT circuit of Fig. 8.45, and determine the waveforms across capacitor C and resistor R_1.

solution   Assume that the capacitor is initially uncharged. When the switch is closed, the capacitor begins to charge up, through R, toward E_{BB} volts. For V_E less than V_P, the UJT is in the cutoff region. As soon as the capacitor voltage reaches V_P, the UJT is turned on and R_{B1} is reduced to a small value. The capacitor discharges rapidly through R_{B1} and R_1 toward ground. But, when $V_E = V_V$, the UJT is turned off. The capacitor begins to charge up again and the cycle is repeated, as illustrated in Fig. 8.46a. The resulting waveform is referred to as a *sawtooth*, and its peak-to-peak amplitude is equal to the difference between the peak and valley voltages ($V_P = V_V$).

The waveform across R_1 is shown in Fig. 8.46b. During the time that C discharges, a current

Fig. 8.46 Waveforms obtained from the UJT circuit of Fig. 8.45: (*a*) Across capacitor C. (*b*) Across resistor R_1.

Fig. 8.47 Programmable uni-junction transistor (PUT): (*a*) Electrical symbol. (*b*) Connection of resistors R_1 and R_2 to alter device parameters.

Fig. 8.48 Static emitter characteristics of a typical UJT. (*Courtesy General Electric Company.*)

flows in R_1, and a trigger voltage is obtained which is suitable for firing a SCR. The amplitude of the trigger is determined primarily by the value of R_1. By changing the values of C and R, the time between triggers is made variable. The UJT circuit of Fig. 8.45 is commonly used as the phase control in the circuits of Figs. 8.34a and 8.41 for the SCR and TRIAC.

PROGRAMMABLE UNIJUNCTION TRANSISTOR The electrical symbol for the programmable unijunction transistor (PUT) is given in Fig. 8.47a. A device that is similar to the SCR, the gate of the PUT is brought out near the anode instead of near the cathode. By selecting suitable values for resistors R_1 and R_2 shown connected to the PUT in Fig. 8.47b, parameters such as η, R_{BB}, I_P, and I_V can be tailored to meet the user's needs. The PUT is used often in long-duration timer circuits.

UJT CURVES AND PARAMETERS A useful family of curves for the UJT is the *static emitter characteristics* of Fig. 8.48. Emitter voltage is plotted along the *y* axis and emitter current along the *x* axis. A family of curves for different values of interbase voltage V_{BB} is drawn in the figure. The greater the value of V_{BB}, the greater are the values of V_P, I_P, V_V, and I_V. Table 8.6 summarizes the important parameters for characterizing the UJT.

TABLE 8.6 Commonly Used Parameters for the UJT

Parameter	Symbol	Meaning
Peak emitter voltage	V_P	Maximum emitter voltage before UJT enters negative-resistance region
Peak emitter current	I_P	Maximum emitter current before UJT enters negative-resistance region. It may also be thought of as minimum emitter current to turn on a UJT.
Valley emitter voltage	V_V	Emitter voltage at valley point
Valley emitter current	I_V	Emitter current at valley point
Interbase voltage	V_{BB}	Voltage between base one and base two
Emitter saturation voltage	$V_{BE1,\text{sat}}$	Voltage across emitter and base one at a specified emitter current and interbase voltage
Interbase resistance	R_{BB}	Dc resistance between base one and base two with emitter open-circuited
Intrinsic standoff ratio	η	May be defined as ratio of emitter–base-one resistance R_{B1} to interbase resistance R_{BB}: $\eta = R_{B1}/R_{BB}$.

8.10 LIGHT-EMITTING DIODE

In a forward-biased pn junction, electrons traveling from the n region are injected into the p region. Some of the injected electrons recombine with the holes in the p region. As a result of the recombination, there is a release of radiant energy. For good efficiency and greater released energy, materials such as gallium (Ga), arsenic (As), and phosphorus (P) are used instead of silicon or germanium. A semiconductor diode designed to emit light is called a *light-emitting diode* (LED); its electrical symbol is shown in Fig. 8.49.

The wavelength of emitted light depends on the energy gap of the material. For example, gallium arsenide phosphide (GaAsP) emits red light, and gallium phosphide (GaP) emits green light. Light-emitting diodes are available in numeral and alphabetic characters measuring in height from $\frac{1}{8}$ to approximately $\frac{3}{4}$ in. Their greatest application is as indicators in, for example, electronic calculators.

CHARACTERISTIC CURVES A number of typical characteristic curves for a LED are illustrated in Fig. 8.50. In operation, the LED is forward-biased, like the junction

Fig. 8.49 Electrical symbol for a light-emitting diode (LED).

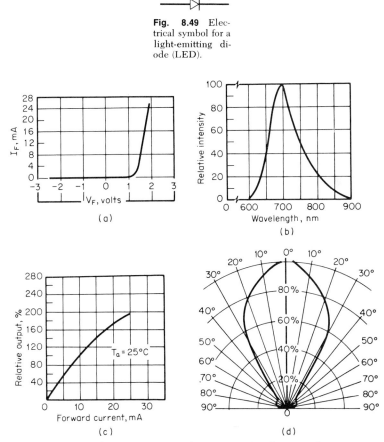

Fig. 8.50 Typical characteristic curves for a LED: (*a*) Voltampere characteristics. (*b*) Spectral distribution curve. (*c*) Visible output as a function of forward diode current. (*d*) Relative candlepower distribution curve. (*Courtesy General Electric Company.*)

diode. An example of the voltampere characteristics of a LED is shown in Fig. 8.50a. The threshold voltage is approximately 1.5 V.

Another curve of interest is the spectral distribution of light emanating from a LED. Such a curve is given in Fig. 8.50b. Relative intensity of the emitted light is plotted as a function of wavelength in nanometers (10^{-9} m). In this example, maximum intensity is obtained at a wavelength of 700 nm.

The dependency of the brightness of the emitted light and the forward current is illustrated by the curve of Fig. 8.50c. The curve of Fig. 8.50d shows the distribution of light for different viewing angles. *Candlepower* refers to the luminous intensity, that is, the brightness of the emitted light.

LED PARAMETERS Commonly used parameters for characterizing the LED are summarized in Table 8.7.

TABLE 8.7 Commonly Used Parameters for the LED

Parameter	Symbol	Meaning
Forward voltage	V_F	Dc forward voltage across an LED
Candlepower	CP	Measure of luminous intensity (brightness) of emitted light
Radian power output	P_o	Light power, or brightness, of an LED
Peak spectral emission	λ_{peak}	Wavelength of brightest emitted color. (Symbol λ is the Greek letter *lambda*.)
Spectral bandwidth	BW_λ	Indication of how concentrated is the emitted color

8.11 OTHER SEMICONDUCTOR DEVICES

In addition to the devices described, there is a host of other semiconductor devices that are of interest. In this section, many of these devices are described and some of their important characteristics examined.

TUNNEL DIODE The tunnel diode, also referred to as the *Esaki diode* after its inventor, is a two-terminal device that exhibits a region of negative resistance similar to that of a UJT. Materials used in its manufacture include silicon, germanium, gallium arsenide, and gallium antimony. Typical characteristics and the electrical symbol for the tunnel diode are illustrated in Fig. 8.51. The tunnel diode may be used as an amplifier, an oscillator, or a switch, and operates at very high frequencies (in the gigahertz region).

Referring to Fig. 8.51a, I_P and V_P are the peak current and voltage, respectively; I_V and V_V are the valley current and voltage, respectively. Two important parameters for the tunnel diode are the peak-to-valley current ratio, I_P/I_V, and the frequency where the

(a)

(b)

Fig. 8.51 Tunnel diode: (*a*) Typical voltampere characteristics. (*b*) Electrical symbol.

(a)   (b)

Fig. 8.52 Schottky diode: (*a*) Cross-sectional view of its basic physical construction. (*b*) Electrical symbol.

negative resistance is reduced to zero, f_{ro}. For switching, the device should exhibit a high I_P/I_V ratio; for high-frequency amplifier or oscillator operation, f_{ro} should be high.

SCHOTTKY BARRIER DIODE The Schottky barrier diode, also referred to as a *hot carrier diode*, achieves rectification by a metal semiconductor junction. A cross-sectional view of a basic Schottky diode is illustrated in Fig. 8.52a, and its electrical symbol in Fig. 8.52b. Contacts to the substrate are aluminum, a p-type impurity. The anode A is formed by the aluminum in contact with the n-type material. A pn junction is formed, which is referred to as a Schottky barrier. The cathode is a nonrectifying contact.

The Schottky diode is an example of a *majority carrier device* because only electrons are involved in its operation. (Both electrons and holes flow in a pn junction diode.) The Schottky diodes exhibit extremely fast switching capabilities and are also useful as detectors and mixers at microwave frequencies.

Typical forward and reverse characteristics of a Schottky diode are illustrated in Fig. 8.53. Currents I_F and I_R are the forward and reverse diode currents, respectively. Voltages V_F and V_R are the forward and reverse voltages across the diode, respectively. Note that at room temperature (25°C), the threshold voltage of a Schottky diode is approximately 0.1 V; for a silicon junction diode it is 0.6 V.

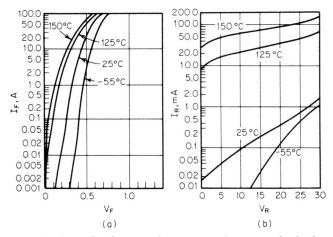

Fig. 8.53 Forward and reverse characteristics of a power Schottky diode: (a) Forward characteristics. (b) Reverse characteristics. (*Courtesy TRW Semiconductor Division.*)

PHOTODIODES There are two basic kinds of photodiodes: the *photovoltaic* and the *photoconductive* types. For the photovoltaic type, no external bias source is required; an external bias, however, is needed for the photoconductive types. The electrical symbol for the photodiode is given in Fig. 8.54.

From our earlier discussion of the operation of the pn junction diode, it was seen that its reverse saturation current (electron-hole pairs) increased with rising temperature (heat). Heat is a form of energy in the infrared region of the electromagnetic spectrum. Light, in the visible region of the spectrum, striking an exposed pn junction also results in the generation of electron-hole pairs. The greater the intensity and higher the frequency of the light, the greater is the diode current.

The diode is sensitive to certain colors, and its sensitivity is displayed by a spectral distribution curve similar to Fig. 8.50b. Sensitivity depends on the energy gap of the material used for the photodiode. In addition to germanium and silicon, cadmium selenide (CdSe), cadmium sulfide (CdS), and cadmium telluride (CdTe) are commonly used materials.

In connecting the conductive-type photodiode, the pn junction is reverse-biased, as

Fig. 8.54 Electrical symbol for a photodiode.

Fig. 8.55 Biasing a conductive-type photodiode.

illustrated by the simple circuit of Fig. 8.55. Variation in diode current with reverse voltage for different levels of illumination (expressed in footcandles, fc, a unit of illumination) is illustrated in Fig. 8.56. The *dark current* is a minute leakage current that flows for no incident radiation.

PHOTOTRANSISTOR The phototransistor operates like a photodiode, and, at the same time, the current generated by the light is amplified. The electrical symbol of a phototransistor is shown in Fig. 8.57a. One may view the device as an npn transistor with the incident radiation replacing the base of a BJT. Typical collector characteristics of a phototransistor are given in Fig. 8.57b for different levels of illumination. In this set of curves, illumination is expressed in milliwatts per square centimeter.

(a)

Fig. 8.56 Voltampere characteristics of a typical conductive-type photodiode.

Fig. 8.57 Phototransistor: (a) Electrical symbol. (b) Typical collector characteristics. (*Courtesy Motorola Semiconductor Products, Inc.*)

OPTO-ISOLATOR The opto-isolator, also referred to as an *optoelectronic coupler*, generally consists of an infrared LED and a silicon phototransistor combined in a single package. A significant advantage of this component is its high isolation resistance (in the order of 10^{11} Ω) between its input and output. Applications for the opto-isolator include the interfacing of different types of logic circuits and their use in level- and position-sensing circuits.

CHARGE-COUPLED DEVICE A cross-sectional view of a charge-coupled device (CCD) is illustrated in Fig. 8.58a. Structurally, it consists of a p- (or an n-) type silicon

Fig. 8.58 Charge-coupled device (CCD): (*a*) Cross-sectional view of its basic physical construction. (*b*) Potential well and electron charge under electrode 1. (*c*) Electron charge being transferred to under electrode 2.

substrate over which is a layer of silicon dioxide (SiO_2). On top of the silicon dioxide is an array of metallized electrodes which are connected to signal voltages. In Fig. 8.58, V_1, V_2, and V_3 constitute a three-phase system of voltages.

Assume that V_1 is greater than either V_2 or V_3. A *potential level*, or *well*, is formed and an electron charge exists below electrode 1, as shown in Fig. 8.58*b*. If V_2 is made greater than V_1, the charge moves and resides below electrode 2, as illustrated in Fig. 8.58*c*. In this manner, the charge in the substrate is transferred under one electrode to the next, and so on.

The CCD finds application as a dynamic shift register in computers and in solid-state imaging, such as video cameras. The packing density (the number of devices occupying the substrate) is in the order of 100 times as great as for other semiconductor devices. The CCD consumes very little power and is capable of operating at high frequencies.

MICROWAVE POWER DIODES   There are two basic families of diodes used in the generation of microwave frequencies: the *avalanche-type* and the *transferred-electron bulk* device. An example of a diode that operates in the avalanche mode is the IMPATT (IMPact Avalanche and Transit Time). The LSA (Limited Space-charge Accumulation) is representative of the transferred-electron bulk device.

A cross-sectional view of an IMPATT diode is given in Fig. 8.59. An n region is sandwiched between a thin n+ and p+ region. Generally the device is fabricated from gallium arsenide. In operation, the diode is reverse-biased, and, at a sufficiently high dc level, the n material is divided into two regions: the drift and avalanche regions. Because of the high electric field, electron-hole pairs are generated essentially in the avalanche region. The flow of carriers is proportional to the voltage in the drift region.

At the oscillating frequency, the phase shift between the two regions is 90°, and the device exhibits a negative resistance. When it is connected to a tuned circuit, the negative resistance may be thought of as canceling out the positive resistance of the tuned circuit, and the combination oscillates.

A cross-sectional view of an LSA diode is illustrated in Fig. 8.60. An n-type region is sandwiched between two alloyed metal contacts; the n-type material is gallium arsenide. Above a given threshold voltage, a charge of electrons (space charge) in the material is dissipated before it can build up appreciably. The current, therefore, tends to decrease as the impressed voltage is increased. This results in the device displaying negative resistance and serves as an oscillator, as for the IMPATT diode.

Fig. 8.59 Cross-sectional view of a basic IMPATT diode.

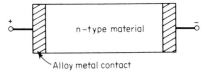

Fig. 8.60 Cross-sectional view of a basic LSA diode.

8.12 SEMICONDUCTOR CHIPS

With the growth of hybrid technology (see Chap. 9), *unencapsulated* diodes, transistors, and even monolithic integrated circuits are available for use in hybrid microelectronic circuits. The unencapsulated devices are referred to as *chips*. There are three basic types of chips: *flip chip, beam lead,* and *leadless inverted devices*. Miniature encapsulated devices, which may measure less than $0.1 \times 0.1 \times 0.05$ in, are also used occasionally.

Fig. 8.61 Connecting flip chip to conductor pads on a substrate.

Fig. 8.62 Connecting beam-lead device to conductor pads on a substrate.

FLIP CHIP Contacts on the flip chip are on one side of the device and may take the form of solder-coated metal pillars or spheres about 0.05 in. in diameter. Figure 8.61 shows how a flip chip is connected to conductor pads on a substrate. By heating the pad in contact with the chip, solder flows between the pad and sphere and an electric connection is made.

BEAM-LEAD DEVICE In a beam-lead device, relatively long leads, or beams, are attached to the semiconductor chip (see Fig. 8.62). The beam-lead device is connected to the pads on a substrate in a manner similar to that for a flip chip.

LEADLESS INVERTED DEVICE The leadless inverted device (LID), also referred to as a *ceramic flip chip,* is a ceramic block in which is mounted the semiconductor chip (Fig. 8.63). Wires from the chip are bonded to the metallized lands. The LID is positioned over the pads on the substrate, and a connection is made with the application of heat.

Fig. 8.63 Constructional features of an LID.

8.13 UNDERSTANDING DATA SHEETS

To develop an understanding of data sheets for semiconductor devices, three typical examples will be examined. The data sheets cover a medium-current silicon rectifier (Fig. 8.64), a germanium pnp transistor (Fig. 8.65), and an n-channel JFET (Fig. 8.66). The parameters specified on the sheets have been defined in appropriate tables contained in the chapter.

A study of the sheets reveals a commonality in the specifications of data and types of curves. Each data sheet has these common features:

1. A brief statement of the intended application and outstanding features of the device. For example, the rectifier is "... desirable in industrial applications that require high surge capabilities." The transistor is "... intended for industrial general-

Silicon

Rectifier

MEDIUM CURRENT

| IN3569-74 |

The 1N3569-74 series stud mounted device in the popular DO-4 package is designed especially for reliability and long life. This series is especially desirable in industrial applications that require high surge capabilities. In addition to this feature the 1N3569-74 units have:

- **Hermetically Sealed Housing**
- **Transient PRV Rating**
- **Low Forward Drop**
- **Operating Temperature up to 165°C**

OUTLINE DRAWING

RATINGS AND SPECIFICATIONS
60 cps Resistive or Inductive Load

	1N3569	1N3570	1N3571	1N3572	1N3573	1N3574
**Maximum Allowable Transient Peak Reverse Voltage (non-recurrent, 5 millisecond duration)	150	275	400	525	650	775 V
Maximum Allowable Repetitive Peak Reverse Voltage (PRV)	100	200	300	400	500	600 V
Maximum Allowable RMS Voltage	70	140	210	280	350	420 V
Maximum Allowable DC Blocking Voltage	100	200	300	400	500	600 V

*Maximum Allowable DC Output Current
    at 85°C Stud Temperature ⟵————— 3.50 —————⟶ Amps
    at 150°C Stud Temperature ⟵————— 1.25 —————⟶ Amps

**Maximum Allowable Peak One Cycle Surge Current
    (60 cps single phase, non-recurrent)—See Fig. 5 ⟵————— 35 —————⟶ Amps

**Minimum I²t Rating (non-recurrent)—See Fig. 6 ⟵————— 3 —————⟶ Amp² sec

Maximum Full Load Voltage Drop (full cycle average
    at rated load) 150°C Ambient ⟵————— 0.5 —————⟶ V

Maximum Reverse Current (full cycle average at
    rated voltage) 150°C Ambient ⟵————— 0.4 —————⟶ ma

Maximum Operating Temperature ⟵——— −65°C to +165°C ———⟶

Maximum Storage Temperature ⟵——— −65°C to +175°C ———⟶

Thermal Characteristics
    Typical Thermal Impedance: Junction to Stud ⟵————— 10°C/W —————⟶

Notes: *For current ratings on fins and at other stud temperatures, see Figures 1 and 2. Stud temperature refers to temperature on outside flat of hex.

    **Rectifier must return to normal thermal equilibrium before reapplication of these overloads.

Fig. 8.64 Typical data sheet for a rectifying diode. (*Courtesy General Electric Company.*)

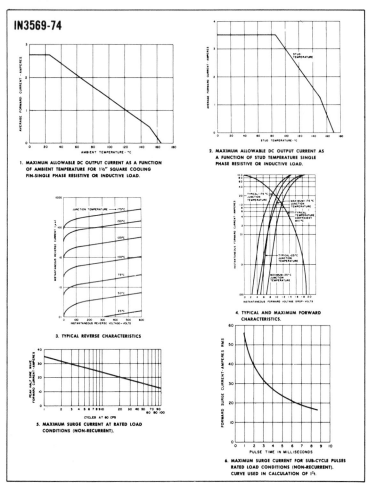

IN3569-74

1. MAXIMUM ALLOWABLE DC OUTPUT CURRENT AS A FUNCTION OF AMBIENT TEMPERATURE FOR 1½" SQUARE COOLING FIN-SINGLE PHASE RESISTIVE OR INDUCTIVE LOAD.

2. MAXIMUM ALLOWABLE DC OUTPUT CURRENT AS A FUNCTION OF STUD TEMPERATURE SINGLE PHASE RESISTIVE OR INDUCTIVE LOAD.

3. TYPICAL REVERSE CHARACTERISTICS

4. TYPICAL AND MAXIMUM FORWARD CHARACTERISTICS.

5. MAXIMUM SURGE CURRENT AT RATED LOAD CONDITIONS (NON-RECURRENT).

6. MAXIMUM SURGE CURRENT FOR SUB-CYCLE PULSES RATED LOAD CONDITIONS (NON-RECURRENT). CURVE USED IN CALCULATION OF I²t.

Fig. 8.64 *(Continued.)*

Germanium
Transcriptors

2N1175,A

The General Electric type 2N1175 is a PNP germanium alloy transistor intended for industrial general purpose applications where cost is of prime importance.
For extremely low noise applications, type 2N1175A is recommended. The 2N1175A is a 2N1175 selected for a maximum noise figure of 6 db.
Both types are similar to the 2N1415 but have a higher forward current transfer ratio.
All units are hermetically sealed and are subjected to 100 hours of high temperature bake as well as a detergent pressure test, thus assuring reliable performance under adverse environmental conditions. Efficient thermal characteristics are assured by welding the transistor base to the case.

absolute maximum ratings: (25°C)

Voltage

Collector to Base	V_{CBO}	-35 volts
Collector to Emitter	V_{CER} ($R_{BE} \leq 10K$)	-25 volts
Emitter to Base	V_{EBO}	-10 volts

Current

Collector	I_{CH}	-200 ma

Power

Collector Dissipation*	P_C	200 mw

Temperature

Storage	T_{STG}	-65 to $+85$ °C
Operating	T_J	$+85$ °C

*Derate 3.33 mw/°C increase in ambient temperature above 25°C.

electrical characteristics: (25°C)

SMALL SIGNAL CHARACTERISTICS

(Unless otherwise specified $V_c = -5V$ common base; $I_c = -1ma$; f = 1000 cps)

		Min.	Typ.	Max.	
Output Admittance (Input AC Open Circuited)	h_{ob}	.1	.45	.9	μmhos
Input Admittance (Output AC Short Circuited)	h_{ib}	26	28	31	ohms
Reverse Voltage Transfer Ratio (Input AC Open Circuited)	h_{rb}	1	5.9	14	x10⁻⁴
Forward Current Transfer Ratio (Common Emitter; Output AC Short Circuited)	h_{fe}	60	80	120	
Frequency Cutoff (f = 1mc; Input AC Open Circuited)	f_{hfb}	1.5	4.2		mc
Output Capacity (f = 1mc; Input AC Open Circuited)	C_{ob}		26	40	μμf
Noise Figure (f = 1kc; BW = 1 cycle)	NF			6	G°	db

D-C CHARACTERISTICS

		Min.	Typ.	Max.	
Forward Current Gain (Common Emitter, I_C/I_B)					
($V_{CE} = -1V$; $I_C = -20ma$)	h_{FE}	70	90	140	
($V_{CE} = -1V$; $I_C = -100ma$)	h_{FE}	62			
Collector Saturation Voltage ($I_C = -20ma$; I_B as indicated)	$V_{CE(SAT)}$ @ $I_B =$		-90		mv
			$-.67$		ma
Base Input Voltage, Common Emitter ($V_{CE} = -1V$; $I_C = -20ma$)	V_{BE}		$-.220$	$-.260$	volts
Collector Cutoff Current ($V_{CBO} = -30V$)	I_{CO}		-6	-12	μa
Emitter Cutoff Current ($V_{EBO} = -10V$)	I_{EO}		-4	-10	μa
Collector to Emitter Voltage ($R_{BE} = 10K$ ohms; $I_C = -.6ma$)	V_{CER}	-25			volts

°Type 2N1175A only.

Fig. 8.65 Typical data sheets for a bipolar junction transistor. (*Courtesy General Electric Company.*)

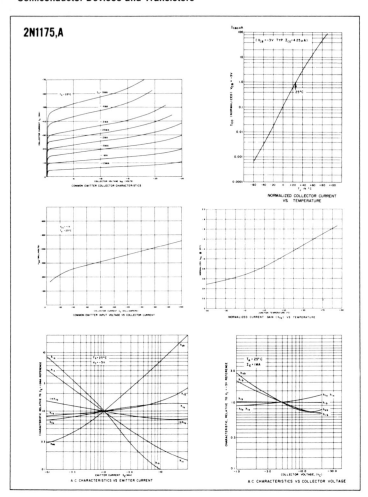

Fig. 8.65   *(Continued.)*

U244

N-CHANNEL SILICON JUNCTION FIELD-EFFECT TRANSISTOR

FOR POWER AMPLIFIERS AND SWITCHES

- High Power Capability — 10 W

- High g_{fs} — 80,000 μmhos Minimum

- High I_{DSS} — 300 mA Minimum

- No Thermal Runaway

- No Secondary Breakdown

All leads isolated from case.
TO-60

PRODUCT CONDITIONING

Units receive the following treatment before final electrical tests:

High Temp Storage: 24 Hours at 150°C   25,000g Acceleration/Impact in the Y_1 Plane
Thermal Shock: +100 to 0°C for 5 Cycles   Helium and/or Gross Leak Tests for Hermeticity

ABSOLUTE MAXIMUM RATINGS (25°C)

Gate-Drain Voltage and Gate-Source Voltage (Note 1) . . . . -25 V
Drain Current . . . . .900 mA
Total Device Dissipation at 25°C Case Temperature (Note 2) . . . . 10 W
Storage Temperature Range . . . . -65 to +150°C
Operating Junction Temperature Range . . . . -55 to +150°C

ELECTRICAL CHARACTERISTICS (25 C unless otherwise noted)

	Characteristic	Test Conditions		Min	Max	Unit
I_{GSS}	Gate Reverse Current	V_{GS} = -15 V, V_{DS} = 0	25°C		-1	nA
			150°C		-1	μA
$I_{D(OFF)}$	Drain Cutoff Current	V_{DS} = 5 V, V_{GS} = -10 V	25°C		1	nA
			150°C		1	μA
BV_{GSS}	Gate-Source Breakdown Voltage	I_G = -1 μA, V_{DS} = 0		-25		V
V_P	Gate-Source Pinch-Off Voltage	V_{DS} = 10 V, I_D = 1 mA		-3.5	-8	V
I_{DSS}	Drain Current at Zero Gate Voltage (Note 3)	V_{DS} = 10 V, V_{GS} = 0		300	900	mA
g_{fs}	Common-Source Forward Transconductance (Note 3)	V_{DS} = 10 V, V_{GS} = 0, f = 100 kHz		80,000	200,000	μmho
r_{DS}	Static Drain-Source On Resistance	V_{GS} = 0, I_D = 10 mA			10	ohms
C_{iss}	Common-Source Input Capacitance	V_{DS} = 0, V_{GS} = -10 V, f = 1 MHz			35	pF
C_{rss}	Common-Source Reverse Transfer Capacitance	V_{DS} = 0, V_{GS} = -10 V, f = 1 MHz			15	pF

NOTES: NI
1. Due to symmetrical geometry, these units may be operated with source and drain leads interchanged.
2. Derate linearly to 150°C case temperature at rate of 80 mW/°C.
3. Pulse test: duration 2 ms.
These devices are manufactured to meet or exceed the requirements of MIL-S-19500.

Fig. 8.66 Typical data sheets for a field-effect transistor. (*Courtesy Siliconix Incorporated.*)

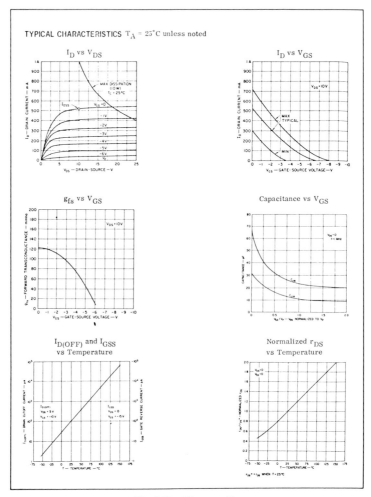

Fig. 8.66   (*Continued.*)

purpose applications where cost is of prime importance." The JFET is designed for power amplifier and switching applications.

2. A sketch with dimensions of the device: identification of connecting leads.

3. Maximum ratings, such as voltage, current, and temperature, that should never be exceeded.

4. Characteristics curves, such as the forward and reverse characteristics for a diode, collector characteristics for a BJT, and drain characteristics for a JFET.

5. Various curves of interest as a function of temperature, voltage, current, or frequency. For example, power derating and surge current curves are included for the diode. Interesting curves included for the BJT are the variation of the small-signal hybrid parameters with emitter current and collector voltage. The transfer characteristic (I_D versus V_{GS}) is of interest for the JFET.

6. The spread of parameters by denoting their minimum, typical, and maximum values.

In making a selection, the data sheets for the device should be carefully studied and understood. Test condition for each parameter should be stated on the sheets. If data that are important for the intended application are missing, the data should be obtained.

Chapter **9**

Integrated Circuit Technology

9.1 INTRODUCTION

Since the appearance of the integrated circuit (IC) in the early 1960s, the electronics industry has experienced monumental changes in design philosophy and manufacturing methods. This has inspired a host of new products, especially in the consumer sector, such as hand-held electronic calculators and electronic watches. In the future, besides innovations in the industrial market, many other products incorporating IC's will be developed for consumer use.

The IC evolved from attempts by semiconductor manufacturers to improve transistor performance and reduce costs. These efforts culminated in the late 1950s in the development of the planar diffused transistor. The techniques used in processing this device are essentially the same as those employed in making the most widely produced integrated circuit, the *monolithic* IC. Monolithic is derived from a Greek word meaning "single stone." In the monolithic IC, all components of the integrated circuit are formed in a single chip of silicon.

A family tree of integrated circuits manufactured is shown in Fig. 9.1. In addition to the monolithic IC, there is the *hybrid* which is available in two forms: *thin* and *thick film*. In either one, resistors, capacitors, and conductors are deposited on a substrate, such as ceramic. In the thin-film process, resistors, capacitors, and conductors are deposited on the substrate by evaporation of a suitable material in a vacuum. In the thick-film process, the components may be thought of as being "painted on" the substrate. Thin- and thick-film, as well as monolithic technology, will be described in this chapter.

Because of the initial high "tooling up" costs, monolithic technology is suited for mass production of IC devices. For small to moderate requirements, hybrid processing is profitable. If, for technical reasons, it is not feasible to develop a monolithic IC for a given circuit, the hybrid is made, regardless of quantity.

9.2 DIMENSIONING INTEGRATED CIRCUITS

A number of different dimensional units are utilized in defining the geometry of integrated circuits and their components. A comparison of these units is given in Table 9.1. Generally, *microns* are used for IC device dimensions and *mils* for the overall chip.

9.3 MONOLITHIC TECHNOLOGY

In the manufacture of IC's, hundreds to thousands of circuits are *processed simultaneously.* For this reason, the cost per circuit is low, making it possible for many IC am-

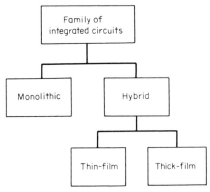

Fig. 9.1 A family tree of integrated circuits.

TABLE 9.1 Comparison of Dimensional Units for Integrated Circuits

	Mil	Centimeter	Micron	Angstrom
Mil	1	2.54×10^{-3}	25.4	2.54×10^{5}
Centimeter, cm	3.94×10^{2}	1	10^{4}	10^{8}
Micron, μ	3.94×10^{-2}	10^{-4}	1	10^{-4}
Angstrom, Å	3.94×10^{-6}	10^{-8}	10^{4}	1

For example, 1 mil $= 2.54 \times 10^{-3}$ cm $= 25.4 \ \mu$; $1 \ \mu = 10^{-4}$ cm; and 1 Å $= 10^{-4} \ \mu$.

plifiers to sell for less than a single transistor sold for a few years ago. The basic silicon material used in monolithic processing comes in a wafer, 1.5 to 3 in diameter and 6 mils thick, as shown in Fig. 9.2. The flat is used for reference, ensuring that the proper orientation of the wafer is maintained during the different processing steps.

Each of the squares in Fig. 9.2 is a chip that contains a complete integrated circuit. A typical size of a chip is $\frac{1}{16}$ by $\frac{1}{16}$ of an inch. Upon completion of processing, the wafer is *scribed* and then *fractured* along the vertical and horizontal lines. The chip is tested and packaged. Because of defects and loss of material along the periphery of the wafer, the number of good chips is less than the number theoretically possible from the wafer. The ratio of good chips to the maximum number of chips available from a wafer is referred to as the *yield* of the process.

BASIC PROCESSING STEPS To learn how a circuit is integrated, we shall first describe the processing steps required in integrating a single transistor (Fig. 9.3). Once

Fig. 9.2 A silicon wafer. Each chip contains a complete integrated circuit. The flat is used as a reference in the processing sequence of the wafer.

Fig. 9.3 A npn transistor to be integrated.

Fig. 9.4 Processing steps for integrating a transistor. (See text.)

the processing steps are understood, we shall see how an amplifier is integrated. Reference will be made to Fig. 9.4 in the following discussion.

P-type substrate The starting material is generally p-type silicon having a resistivity of about 10 Ω-cm. The material has a *single-crystal* structure. Single crystal denotes a material in which the atoms are arranged in an ordered pattern. Material in which the atoms are not in an ordered pattern is referred to as *polycrystalline*.

Epitaxial layer Upon the p-type silicon a well-defined n-type silicon region is grown in an epitaxial furnace. *Epitaxial growth* means that the n region formed has the same single-crystal structure as the p-type silicon (Fig. 9.4a).

Oxide layer The silicon wafer, seated in a *quartz boat*, is placed in an *oxidation furnace* of Fig. 9.5 and exposed to steam or dry oxygen at a temperature between 1000 and 1200°C. The surface of the silicon is oxidized, and a thin layer of silicon dioxide (SiO_2), which is glass, is formed (Fig. 9.4b).

An oxide layer has two important functions:

1. It serves as the surface for a light-sensitive film.

2. It protects the circuit against contamination, a serious problem in processing integrated circuits. This property is referred to as *passivation*.

Fig. 9.5   Cross-sectional view of a basic oxidation furnace.

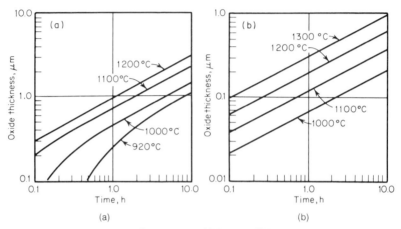

Fig. 9.6   Oxidation curves: (a) Steam. (b) Dry oxygen.

Curves of SiO_2 growth in steam and dry oxygen as a function of time for different temperatures are illustrated in Fig. 9.6. The application of the curves is illustrated in our first example.

example 9.1   Using the oxidation curves of Fig. 9.6, estimate the time needed to grow a 0.2-μ thickness of SiO_2 on silicon, using steam and dry oxygen. The temperature of the oxidation furnace is $1100°C$.

solution   It is noted that the vertical and horizontal scales of the oxidation curves are not linear. To cover a large range of values ($10/0.1 = 100:1$) in a compact manner, *logarithmic scales* are used. Logarithmic scales in electronics are nothing new. For example, the frequency-response curve of an amplifier is also logarithmic. Referring to the $1100°C$ steam curve of Fig. 9.6a, for an oxide thickness of 0.2 μ, 0.1 h (6 min) is required. If dry oxygen is used, from the $1100°C$ curve of Fig. 9.6b, 3 h (180 min) is necessary. Because it takes less time to grow an oxide layer, steam is generally used.

Photoresist   An organic substance that is sensitive to ultraviolet (UV) light, called *photoresist*, is applied over the oxide layer (Fig. 9.4c). A commonly used resist is *Kodak Photo Resist* (KPR). The wafer is held by vacuum on a *spinner*. Some drops of KPR are applied to the surface, and the wafer is spun at a few thousand rpm for 10 to 30 s. This ensures a uniform deposition of KPR over the wafer's surface.

Upon exposure to UV, the photoresist becomes *polymerized*. When this occurs, the molecular structure of the exposed resist is altered. The change results in the exposed material becoming *resistant* to chemicals used in subsequent processing such as hydrofluoric acid (HF) for etching and phosphorus (P) and boron (B) for doping. The unexposed areas, however, are affected by these materials.

Artwork   Each chip on the wafer of Fig. 9.2, which may measure $^1/_{16}$ by $^1/_{16}$ in or less, contains perhaps a dozen or more transistors and an equal number of resistors. It becomes apparent that the regions on a chip must be precisely defined. For an npn transistor, three regions are needed: collector, base, and emitter. The three regions are cut as masks, illustrated in Fig. 9.7. A Mylar base, which is dimensionally stable with

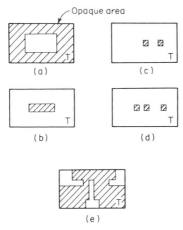

Fig. 9.7 Set of five masks required for integrating the transistor of Fig. 9.3. (See text.)

temperature and humidity, on which is a peelable red tinted layer, is used for the mask material. Commercially the material is referred to as *Rubylith*.

The pattern is formed by cutting through the tinted layer (and not the Mylar base) and peeling away the unwanted layer. To achieve the required precision, the pattern is cut on a mask of the size in the order of 20 by 30 in. This is accomplished on a *co-ordinatograph* (Fig. 9.8), a machine that provides the required cutting accuracy. The tee in each mask is used in mask alignment (to be described later).

The mask is ultimately reduced to the actual size of the chip (for example, $^1/_{16}$ by $^1/_{16}$-in chip size). In addition, the pattern must be repeated to cover the whole wafer. This is done in a *step-and-repeat camera*, which reduces the mask to its final size and repeats the pattern on a high-resolution emulsion glass plate. A plate is made for each of the masks.

Fig. 9.8 A coordinatograph is used for the precision cutting of masks. (*Courtesy Owen Jones Equipment Sales.*)

Exposure of collector mask   The cross-hatching in Fig. 9.7 denotes the opaque areas of the mask. When exposed, the UV is prevented by the opaque areas on the mask from reaching the resist; the UV, however, strikes the resist through the clear mask areas. The collector mask superimposed on the wafer is shown in Fig. 9.4d.

Diffusion   The key to monolithic IC technology is the diffusion of impurities, called *dopants*, into silicon. In diffusion, p-type such as boron (B), and n-type, such as phosphorus (P), dopants in high concentration are brought in contact with the silicon surface. The process occurs in a diffusion furnace, which is similar in basic construction to the oxidation furnace of Fig. 9.5. Diffusion occurs in the temperature range of 1100 to 1200°C. The dopant is typically in vapor or gaseous form. The penetration depth and concentration of impurity in the silicon depend on the type of dopant, time, and temperature of diffusion, and on how the dopant is introduced.

Silicon, like other materials, can absorb only a finite number of impurity atoms, called the *solid solubility* of the material. Solid solubility is the maximum number of atoms of one material that can be dissolved in another material.

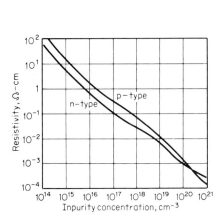

Fig. 9.9   Curves of silicon resistivity for p- and n-type impurity concentrations at room temperature (25°C).

Fig. 9.10   Erfc and gaussian distribution curves.

Curves of silicon resistivity (ohm-centimeters) for p- and n-type *impurity concentrations* (impurity atoms per cubic centimeter of silicon) are given in Fig. 9.9. It is seen that, as the impurity concentration is increased, the resistivity of the silicon is decreased.

There are two methods for applying dopants to the wafer in the diffusion furnace. In the first method, the wafer *always* "sees" a *constant* concentration, N_o atoms/cm^3, of dopant. The resulting distribution of the impurity in silicon is described by what is called the complementary error function (erfc) curve of Fig. 9.10. In the second method, the wafer sees a *fixed* amount of dopant atoms. The resulting distribution is gaussian and is also shown in Fig. 9.10.

For determining the location of a pn junction in silicon, Fig. 9.10 is extremely useful. Both curves are plotted as N_x/N_o versus $z = x/2\sqrt{Dt}$. Term N_x is the actual concentration of impurity atoms per cubic centimeter at a distance x from the surface of the silicon; N_o has already been defined. Term D is the diffusion coefficient in square centimeters per second; it indicates how fast the dopant spreads out in the silicon. Plots of D for different dopants as a function of temperature are given in Fig. 9.11.

example 9.2   A uniformly doped p-type silicon wafer is used in the manufacture of an integrated circuit. The resistivity of the wafer measures 10 Ω-cm. It is placed in a diffusion furnace at 1200°C and is exposed to a *constant source* of phosphorus, $N_o = 10^{19}$ atoms/cm³. How long does it take for a pn junction to form 3 μ away from the surface?

solution   Referring to the resistivity curves of Fig. 9.9, for p-type silicon of 10 Ω-cm resistivity, the impurity concentration is about 1.4×10^{15} atoms/cm³. This figure is called the *background concentration, N_B,* of impurity. In this example, then, $N_B = 1.4 \times 10^{15}$ atoms/cm³.

A constant source of dopant (phosphorus) implies a complementary error function (erfc) distribution. A pn junction is formed when the concentration of the n-type dopant (phosphorus) equals the p-type dopant (background concentration). Hence, $N_x = N_B = 1.4 \times 10^{15}$.

Fig. 9.11   Diffusion coefficients for various impurities in silicon.

The ratio $N_x/N_o = 1.4 \times 10^{15}/10^{19} = 1.4 \times 10^{-4} = 0.00014$. From the erfc curve of Fig. 9.10, it is estimated that $z = 2.6$.

At 1200°C, the diffusion coefficient for phosphorus, from Fig. 9.11, is $D_p \approx 3 \times 10^{-12}$ cm²/s. The junction depth x is 3 $\mu = 3 \times 10^{-4}$ cm (see Table 9.1). Substitution of these values in z yields

$$2.6 = \frac{3 \times 10^{-4}}{2\sqrt{3 \times 10^{-12}t}}$$

Squaring both sides of the equation, we obtain

$$6.75 = \frac{9 \times 10^{-8}}{4 \times 3 \times 10^{-12}t}$$

Solving for t,

$$t = \frac{9 \times 10^{-8}}{4 \times 3 \times 10^{-12} \times 6.75}$$
$$\approx 1100 \text{ s} \approx 0.3 \text{ h}$$

Etching   The unexposed SiO₂ is etched away by hydrofluoric acid (HF) (Fig. 9.4e).

p-type diffusion   The wafer is placed in a diffusion furnace, and sufficient p-type impurity, such as boron, is introduced. The impurity penetrates the silicon which was etched away by HF. After a length of time, an n region (or *island, tub*) for the collector is formed (Fig. 9.4f). Because the impurity diffuses *horizontally* (laterally), as well as vertically, the profile of Fig. 9.4f is the result. Note that some SiO₂ overlaps the n region. This is desirable because it helps to minimize contamination. The photoresist is removed, or *stripped,* after each diffusion.

Oxidation and resist   The wafer is placed again in the oxidation furnace, and a new layer of SiO_2 is grown. Photoresist is also applied (Fig. 9.4g). The wafer is now ready for exposure to the base mask.

Exposure to base mask   The base mask of Fig. 9.7b is now exposed to UV. The unexposed regions are etched away, and the result is an opening of a "window" for base diffusion (Fig. 9.4h).

To ensure perfect alignment between the base mask and the wafer, a *mask aligner* is used. With this equipment, the operator looks through a microscope and adjusts the position of the mask so that the tee of the base mask aligns with the tee on the wafer from its exposure to the collector mask.

Base diffusion   A p-type impurity is diffused through the window to form the base of the transistor (Fig. 9.4i).

Oxidation and resist   A new layer of oxide is grown and photoresist applied (Fig. 9.4j).

Forming the emitter   The emitter is formed by using a very high concentration of an n-type dopant, denoted by n+. (If p+ was indicated, this would mean a very high concentration of a p-type dopant.) Further, to ensure a *nonrectifying* or, *ohmic, contact* to the n-collector region, n+ doping is also used.

The connections between the diffused regions in a monolithic IC are generally made by a thin film of conducting material, such as aluminum. Aluminum is a p-type material. If a p material is in contact with an n material, a pn junction diode is formed. An n+ region in contact with aluminum, however, provides a nonrectifying contact.

Exposure to n+ mask   The mask of Fig. 9.7c has cutouts for the emitter region and for a contact to the collector region. The mask is exposed to UV, and, as in previous processing steps, the unexposed regions are etched away (Fig. 9.4k).

n+ diffusion   A high concentration of n-type dopant, such as phosphorus, is diffused through the windows to form the n+ regions (Fig. 9.4l).

Oxidation and resist   A new layer of oxide is grown and photoresist applied (Fig. 9.4m).

Exposure to contact mask   Windows must now be opened to permit electric connections to be made to the emitter, base, and collector regions of the transistor. The contact mask of Fig. 9.7d is exposed to UV. The unexposed regions are etched away, and the resist is stripped from the wafer (Fig. 9.4n).

Metallization   To provide connections between the diffused regions, a metallic

Fig. 9.12   An example of a prober. (*Courtesy Electroglas, Inc.*)

material is deposited over the entire wafer. This process is called *metallization*. Typically, aluminum is used and is deposited as a thin film on the substrate, which is accomplished in a vacuum. (Vacuum deposition of materials is considered later in the chapter.)

Exposure to interconnection mask   The deposited aluminum is covered with photoresist and exposed to the interconnection mask of Fig. 9.7e. The unexposed regions are etched away, and the resist is removed from the wafer. The result is a diffused (integrated) transistor (Fig. 9.4o).

Testing and packaging   The integrated circuit is tested and packaged. In testing, a *prober* (Fig. 9.12) is used. The wafer is vacuum-held and, under a microscope, fine points are brought into contact with the diffused regions. Through external electric connections to the points, voltages are impressed across the regions and measurements of device performance are made.

In summary, five different masks and three diffusions were required in the fabrication of a monolithic transistor. As we shall see, the same number of masks and diffusions are all that is generally required in the monolithic integration of a complete circuit. This feature is what makes monolithic technology so attractive for large-volume production of integrated circuits.

9.4 ELECTRICAL ISOLATION

In the monolithic IC all circuit elements, such as transistors and resistors, share the same silicon chip. It is necessary, therefore, that the elements be *electrically isolated* from one another. At present, three methods are used for isolation:

1. Junction
2. Dielectric
3. Beam lead

JUNCTION ISOLATION   We have seen how a single transistor is integrated. To illustrate junction isolation, consider the two transistors in a silicon chip shown in Fig. 9.13. Note that the p substrate is returned to a negative potential and the n regions to a positive potential. Because the n and p regions form pn junctions, they are *reverse-biased*, similar to a back-biased semiconductor diode. For this condition, no current flows between the n-collector regions through the p substrate, and electrical isolation is obtained.

DIELECTRIC ISOLATION   An example of two transistors dielectrically isolated is illustrated in Fig. 9.14. Surrounding each n-collector region and, separating it from the substrate, is a layer of silicon dioxide. Because SiO_2 is an insulator, electrical isolation is realized. The substrate does not have to be single-crystal silicon; instead, polycrystalline silicon is used.

BEAM-LEAD ISOLATION   In this method of isolation, the circuit elements are diffused as in the junction-isolated IC. The metallization deposited on the chip, however, is very thick (in the order of 1 mil). The superfluous silicon of the p substrate is re-

Fig. 9.13 Junction isolation for two diffused transistors.

Fig. 9.14 Dielectric isolation for two diffused transistors.

moved. The resulting structure is a semirigid connected circuit with all elements physically separated from one another.

Owing to lower production costs, most monolithic circuits use junction isolation.

PARASITICS A problem with junction isolation is the existence of parasitics. A parasitic may be defined as an *undesired element*. Referring to Fig. 9.13, we saw that

Fig. 9.15 Parasitic elements existing across a reverse-biased pn junction.

isolation was obtained by reverse-biasing the n-collector and p-substrate regions. In reverse-biasing a pn junction, a capacitance C_p, in parallel with a high-valued resistance R_p, exist as illustrated in Fig. 9.15. These two elements are examples of parasitics and, at very high frequencies, may create problems. Parasitics also exist for diffused resistors and capacitors in the junction-isolated IC.

9.5 PROPERTIES AND CHARACTERISTICS OF DIFFUSED ELEMENTS

BIPOLAR JUNCTION TRANSISTOR (BJT) A diffused BJT has typical values of short-circuit current gain h_{fe} as high as 300 and gain-bandwidth product f_T as high as 500 MHz. Because the gain of a junction transistor depends on the width of its base, the short-circuit current gain can be increased by reducing the base thickness. Transistors with very narrow base widths, exhibiting short-circuit current gains as high as 5000, are called *super-beta* and *punch-through* transistors. Owing to the narrow base width, the collector-emitter voltage for this type of transistor is limited to about 0.5 V. For a conventional BJT, the collector-emitter voltage is in the order of 20 to 30 V.

The gain-bandwidth product may be increased by decreasing the resistance of the collector region. A method for realizing this, without affecting appreciably the collector-emitter voltage rating of the transistor, is shown in Fig. 9.16. A heavily doped region, called an *n+ buried layer*, is diffused between the bottom of the n-collector region and the p substrate. Because the n+ layer is in parallel with the n-collector region, the effective collector resistance is reduced.

LATERAL pnp TRANSISTOR The diffusion of npn transistors is very compatible with the monolithic process. There are occasions, however, when a pnp transistor is needed in addition to an npn device. A structure that requires no additional diffusion steps is the *lateral pnp transistor* of Fig. 9.17.

Fig. 9.16 An n+ buried layer for reducing collector resistance.

Fig. 9.17 A view of a lateral pnp transistor.

Two p-type regions are diffused in the n island (tub). One of the p regions serves as the emitter and the other as the collector. Because the effective base width is greater than that of an npn transistor, the short-circuit current gain and gain-bandwidth product of a lateral pnp transistor are appreciably less. Typical maximum values are $h_{fe} = 20$ and $f_T = 2$ MHz.

FIELD-EFFECT TRANSISTOR (FET) In the integration of field-effect transistors, the enhancement-mode MOSFET (IGFET) is invariably chosen because
    1. It is self-isolating.
    2. It can be formed by a single diffusion.
    3. Because it is self-isolating, it has a much greater packing density than the BJT.
    A cross-sectional view of a p-channel enhancement-mode MOSFET is shown in Fig. 9.18. Two p-type regions are diffused simultaneously in the n substrate. The gate is formed by depositing aluminum over the oxide layer between the p regions. In operation, the substrate is returned to the most positive potential in the circuit. The source, gate, and drain are so biased that they are negative with respect to the substrate. Thus, reverse-biased pn junctions are formed and electrical isolation is achieved.

Fig. 9.18 A cross-sectional view of a p-channel enhancement MOSFET.

Fig. 9.19 A transistor may be connected in five ways to act like a diode (see text).

DIODE By utilizing only two terminals of a diffused npn transistor, five different diode configurations may be realized. Referring to Fig. 9.19, these are:
    a. Base-emitter junction diode; collector is open. Series diode resistance and reverse breakdown voltage are low.
    b. Base-emitter junction diode; collector short-circuited to base. Series diode resistance and reverse breakdown voltage are low.
    c. Base-collector junction diode; emitter open. Series diode resistance and reverse breakdown voltage are high.
    d. Base-collector junction diode; emitter short-circuited to base. Series diode resistance and reverse breakdown voltage are high.
    e. Base-collector junction diode; emitter short-circuited to collector. Series diode resistance is high, and reverse breakdown voltage is low.
    A series diode resistance which is low is less than 50 Ω; a high series diode resistance is greater than 50 Ω. Low reverse breakdown voltage is in the order of 7 V; high reverse breakdown voltage is about 40 V. Because of its good properties, the diode of Fig. 9.19*b* is commonly used.

RESISTOR Integrated resistors are generally formed by the diffusion of p-type material in an n tub, as illustrated in Fig. 9.20. This diffusion is done *simultaneously* with the diffusion of the p-base region of an npn transistor.
    Resistance, R (ohms), of a material is expressed by

$$R = \rho\left(\frac{L}{A}\right) \tag{9.1}$$

where ρ = resistivity of material, Ω-cm; L = length of material, cm; and A = cross-

Fig. 9.20 A cross-sectional view of a p-diffused resistor.

Fig. 9.21 The resistance measured across any two parallel edges is the same for either square.

sectional area of material, cm². The area may be expressed by the product of its width W and its thickness, T: $A = WT$. Substitution of this value for A in Eq. (9.1) yields

$$R = \frac{\rho}{T}\frac{L}{W} \tag{9.2}$$

If $L = W$, we have a *square* of material, and Eq. (9.2) reduces to

$$R = \frac{\rho}{T} \tag{9.3}$$

Assuming that thickness T is constant, which is reasonable, then the ratio ρ/T is also constant. Letting this quantity be denoted by R_S, the *sheet resistance*, we obtain

$$R = R_S \text{ ohms/square} \tag{9.4}$$

For any length and width of a diffused resistor,

$$R = R_S\left(\frac{L}{W}\right) \tag{9.5}$$

where L/W is the *aspect ratio*.

Consider the two square areas of Fig. 9.21. Both squares have the same thickness T. Although the surface area of square B is four times as large as that of square A, the resistance measures across their edges are the same and equal to the sheet resistance of the material R_S.

The last statement can be verified by noting that for square A, the cross-sectional area is $T(1) = T$. For square B, the area is $T(2) = 2T$. The length of A is 1 unit, and of B, 2 units. Hence, the length to cross-sectional areas are equal: $1/T = 2/2T = 1/T$. The resistance for each square is identical.

    example 9.3 For a diffused resistor of 2000 Ω, $R_S = 200$ Ω/square. Determine the aspect ratio and, if $L = 20$ μ, the width of the resistor.

    solution From Eq. (9.5), $R/R_S = L/W = 2000/200 = 10{:}1$. If $L = 20$ μ, $W = 20/10 = 2$ μ.

The practical range of p-diffused resistors is from 20 Ω to 30 kΩ. Their absolute tolerance, however, is poor, being in the order of ± 25 percent. The tolerance of the *ratio* of two diffused resistors can be as low as ± 2 percent. Resistors made with n+ diffusion are seldom used.

FOUR-POINT PROBE The four-point probe of Fig. 9.22 permits a simple measurement of the sheet resistance. A dc voltage source in series with a resistor, which simulates a *current source*, is connected to the outer points in series with a dc ammeter, I. The points are spaced about 0.1 cm apart and make physical contact with a silicon specimen whose sheet resistance is to be determined. The inner points are connected to a dc voltmeter which reads the voltage V developed across the points. Sheet resistance, in ohms per square, is given by

$$R_S = 4.5\frac{V}{I} \tag{9.6}$$

CAPACITOR Capacitors found in a monolithic IC are basically of two kinds: *junction* and *thin film*. A junction capacitor may be realized by using an npn transistor

Fig. 9.22 A four-point probe used for measuring sheet resistance.

Fig. 9.23 A cross-sectional view of a thin-film monolithic capacitor.

connected as a diode (Fig. 9.19). Application of a reverse voltage across any two terminals results in a variation in capacitance: the greater the reverse voltage, the less is the capacitance.

A cross-section view of a thin-film monolithic capacitor is shown in Fig. 9.23. An $n+$ region, diffused in the n tub, acts as one plate of the capacitor. A growth of silicon dioxide serves as the dielectric, and the deposition of aluminum over the oxide acts as the other plate of the capacitor.

Capacitors occupy a relatively large area on a chip and are limited to small values of capacitance, typically 200 pF. For these reasons, then, they are avoided wherever possible.

INDUCTOR   Currently there are no practical methods for diffusing inductors in a monolithic IC. Inductors, therefore, are avoided in the design of monolithic integrated circuits.

9.6 MAKING A MONOLITHIC IC

In this section we shall see how a circuit is transformed into a monolithic IC. Because it is most common, junction isolation will be assumed. In determining how many isolation tubs are required, the following guidelines are offered:

1. Transistor collectors at different potentials must be in separate isolation tubs.
2. Resistors connected to circuit points at different potentials can share the same isolation tub. The n tub should be connected to the most positive voltage available in the circuit.
3. Diffused capacitors generally require individual tubs.
4. Base-emitter diodes can generally share the same isolation tub.

Consider the two-stage direct-coupled amplifier of Fig. 9.24, which is to be integrated. Because the collector potentials of Q_1 and Q_2 are different, each transistor requires a separate isolation tub. Resistors R_1 and R_2 can share one tub. Therefore, three isolation regions are required.

A layout of the amplifier is given in Fig. 9.25a. Other layouts are also possible. A cross-sectional view of the integrated circuit is given in Fig. 9.25b. As in the integra-

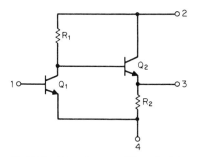

Fig. 9.24 A two-stage amplifier to be integrated.

Fig. 9.25 Monolithic integration of the amplifier of Fig. 9.24:
(a) Layout. (b) Cross-sectional view.

tion of a transistor, a p-type substrate with a grown n-epitaxial layer is the starting material. Also, five masks are required; referring to Fig. 9.26, these are:

Isolation mask Sufficient p-type impurity is diffused in the n-epitaxial region to ensure that the three n-isolation tubs are surrounded by p regions (Fig. 9.25b). The opaque areas of the isolation mask prevent UV from reaching the photoresist on the silicon wafer; consequently, windows are only opened for the diffusion of the p-type impurity.

p-diffusion mask This mask has windows for the transistor bases and resistors. The diffusion of the base and resistors occurs simultaneously.

n+ diffusion mask Windows are cut out for the diffusion of emitters and regions in n-type material to ensure nonrectifying contacts.

Contact and interconnection masks The contact mask has cutouts for connections, and the interconnection mask has the connection pattern for the diffused elements.

As for the integration of a transistor, described earlier in the chapter, oxidation, application of photoresist, etc., are used in similar sequence for integration of a circuit. An example of a commercial monolithic IC is illustrated in Fig. 9.27.

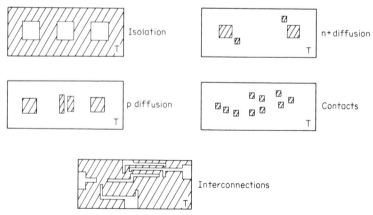

Fig. 9.26 Set of five masks required for the integration of the amplifier of Fig. 9.24 (see text).

(a)

(b)

(c)

Fig. 9.27 An example of a commercial IC: (a) Constructional details. (b) Cross-sectional view of transistor. (c) Approximate size of package.

BURIED CROSSOVER In the layout of an integrated circuit it is found occasionally that two conducting paths, which cannot be eliminated, intersect each other. To correct this situation, a buried crossover, illustrated in Fig. 9.28, is required.

An n+ region is diffused in a separate n-isolation tub. Conductor 1 completes its path through the n+ buried region. Conductor 2, shown perpendicular to conductor 1, is insulated from the buried region by the SiO₂ layer.

example 9.4 The circuit of Fig. 9.29 is to be integrated. Draw a layout of the circuit, showing the required isolation tubs; the devices in each tub may be shown schematically. The numbering of terminals must be maintained in sequence on the periphery of the layout.

solution The layout, shown in Fig. 9.30, requires four isolation tubs. Because they share a common connection, the three diodes are placed in one tub. Owing to different collector potentials, each transistor requires its own isolation tub. The three resistors can share a single tub.

example 9.5 A magnified drawing of an integrated-circuit layout is given in Fig. 9.31. Draw a schematic diagram of the circuit, and identify all components and the values of resistors. The sheet resistance $R_S = 200 \; \Omega/\text{square}$.

Assume that the transistor is saturated: that is, $E_{CE} = 0$ V. If the voltage across a forward-biased base-emitter junction or diode is 0.7 V, determine the base, emitter, and collector currents in the transistor.

solution The complete circuit is illustrated in Fig. 9.32. Resistance values were obtained by counting the number of squares in each p-diffused resistor in the isolation tub. Diode D is always reverse-biased. The base current, therefore, is $I_B = (6 - 0.7)/5.6 = 0.95$ mA. Col-

Fig. 9.28 A buried crossover.

Fig. 9.29 A circuit to be integrated. (See Example 9.4.)

Fig. 9.30 Schematic layout of the circuit of Fig. 9.29.

Fig. 9.31 A given layout of an integrated circuit. (See Example 9.5.)

Fig. 9.32 Schematic diagram of circuit determined from the IC layout of Fig. 9.31.

lector current, $I_C = (6 - 0)/1 = 6$ mA. The emitter current I_E is equal to the sum of base and collector currents: $I_E = I_B + I_C = 0.95 + 6 = 6.95$ mA.

9.7 LARGE-SCALE INTEGRATION (LSI)

Large-scale integration (LSI) is an extension of monolithic technology where 100 or more interconnected circuits are in a single chip of silicon. (Integrated circuits containing fewer than 100 interconnected circuits are often referred to as *medium-scale integration*, MSI.) Because of such a large number, invariably some diffused components are defective. What is necessary, therefore, is the elimination of the defective units before the interconnections are made. One commonly used technique for accomplishing this is called *discretionary wiring*.

In discretionary wiring, a number of circuits, called *unit cells*, in excess of the actual number required, are integrated. Generally, LSI is best suited where identical circuits are used, such as logic gates and computer memories. Each unit cell is checked on an automatic tester. Coupled to the tester is a computer which remembers the good cells. The computer generates an interconnection pattern that includes only good cells. Because the location of defective cells will be different for other chips of the same type, the testing and computer routine is repeated for each chip.

9.8 THIN-FILM TECHNOLOGY

In thin- and thick-film technology, resistors, small-valued capacitors, and inductors are deposited and interconnected on a substrate, such as ceramic. Transistors, large-value capacitors, and even monolithic IC's may be connected to the substrate to form a *hybrid* integrated circuit. Thin films generally refer to a layer thickness of 5 angstroms (1 Å = 10^{-8} cm) to 10 μ (1 μ = 10^{-4} cm). The thickness of thick films is 10 μ or greater.

PRODUCING THIN FILMS In the thin-film process, the deposition is generally accomplished by *vacuum evaporation* or *cathode sputtering*. A basic vacuum-evaporation system is illustrated in Fig. 9.33. The bell jar is evacuated to a pressure in the order of 10^{-5} *torr* (1 torr = 1 mm of mercury). A tungsten element, in which is placed the material to be evaporated, and the substrate are mounted inside the bell jar. Approximately 20 to 25 A of current is passed through the element, and the material, such as aluminum, is melted and vaporized instantaneously and deposited on the substrate. To ensure good adhesion of the evaporated film, the substrate is sometimes heated in the bell jar.

Cathode sputtering is used when materials, such as tantalum, are not easily melted. An elementary sputtering system is illustrated in Fig. 9.34. The stand holding the substrate acts as the *anode*, and the material to be deposited as the *cathode* of a gas

Fig. 9.33 A basic vacuum evaporation system for producing thin films.

Fig. 9.34 An elementary cathode sputtering system.

discharge tube. The bell jar is initially evacuated, as in vacuum evaporation. Argon, or another inert gas, is introduced in the jar until the pressure increases to approximately 0.05 torr.

Upon the application of 4 to 6 kV across the anode and cathode terminals, a glow discharge is initiated. The argon atoms become positively ionized and are attracted toward the cathode (which is negative). When the ions strike the cathode, atoms of the material to be deposited are liberated and attracted by the substrate.

The thickness of a deposited film can be measured with a quartz crystal oscillator. A quartz crystal, which is connected to an oscillator circuit, is placed near the substrate in the bell jar. The material deposited on the substrate is also deposited on the crystal. Because of the deposition, the mass of the crystal is increased and its frequency of oscillation reduced. The actual frequency is calibrated to read directly film thickness.

THIN-FILM COMPONENTS The electrical characteristics of thin-film components depend primarily on the film material and the geometry of the deposited component.

Resistor Commonly used materials for resistors include Nichrome ($R_S = 10$ to 400 Ω/square) and tantalum nitride ($R_S = 50$ to 500 Ω/square). The higher resistance values are obtained with thin deposits of material, and the low values with thick deposits.

Nichrome, which is deposited by vacuum evaporation, exhibits a good temperature coefficient and good adhesion properties. Sputtering is used for the deposition of tantalum nitride. This material has a good temperature coefficient, is highly stable, and adheres well to the substrate.

By means of the folded geometry of Fig. 9.35 (which is also used for diffused resistors), resistance values in megohms may be realized that occupy less than 0.5 by

Fig. 9.35 A folded geometry resistor allows a large-valued resistor to occupy a relatively small substrate area.

Fig. 9.36 A "sandwich" thin-film capacitor.

0.5 in on a substrate. The tolerance of thin-film resistors is in the order of 5 to 10 percent. With suitable trimming, such as laser trimming, tolerances better than one percent may be achieved. Thin-film resistors deposited on a glazed substrate can dissipate approximately one watt per square inch of substrate area.

Capacitor A thin-film capacitor is a "sandwich" of a bottom electrode, dielectric, and top electrode (Fig. 9.36). Generally, the top and bottom electrodes are evaporated aluminum. The dielectric used determines the value of capacitance. For example, if the deposited dielectric is a 1000-A layer of silicon dioxide, approximately 350 pF/mm^2 of surface area is realized.

The tolerance of thin-film capacitors before trimming is about 10 to 20 percent; after trimming, it is less than 5 percent. The quality factor Q seldom exceeds 200.

Inductor Thin-film inductors are generally deposited on the substrate in the form of a spiral. Because they occupy a large surface area, they are practical only for circuits operating in the gigahertz frequency range.

Fig. 9.37 An example of thin-film circuits used in a 1-GHz transceiver. (*Courtesy RCA.*)

THIN-FILM PROCESSING Many of the steps used in thin-film processing are similar to those used in monolithic technology. A layout of the circuit to be integrated is made. Four masks are generally required: resistor, bottom capacitor electrode, top capacitor electrode, and interconnection pattern masks.

The resistor material is deposited over the entire substrate. This is followed by a coating of photoresist which is exposed through the resistor mask to UV. The unexposed resist is dissolved, and the result is a resistor pattern on the substrate. Metal, such as aluminum, is now deposited over the entire substrate, and resist is applied. The resist is exposed to a mask containing the pattern for the bottom capacitor electrode. The same procedure is repeated for the top capacitor electrode, dielectric, and interconnection pattern.

Upon completion of the preceding steps, transistors, large-value capacitors and inductors, and occasionally monolithic IC chips are attached to the circuit. Transistors and capacitors for this purpose come in the form of a *flip chip*. A flip chip is a component having an upside-down mounting, making it relatively simple for attaching to a substrate. Finally, the completed hybrid circuit is tested and encapsulated. An example of a commercial thin-film circuit is shown in Fig. 9.37.

9.9 THICK-FILM TECHNOLOGY

Thick-film technology is a method for depositing resistors, capacitors, and interconnections on a substrate which is based on *screen-process printing*. One may think of the process as "painting on" components over a substrate. The advantages of thick- over thin-film technology are:

1. Lower cost of capital equipment.
2. Easier to train manufacturing personnel.
3. High production rates are realized.
4. Thick-film sheet resistance of 10 MΩ/square are possible.
5. Greater dissipation of thick-film resistors.

On the debit side, thick-film resistors are noisier, have poorer tolerance, and are wider than thin-film resistors.

SCREEN-PROCESS PRINTING The starting point for the process is a fine-mesh screen. Generally, a stainless steel screen mounted on an aluminum frame is used (Fig. 9.38). The *mesh count*, which equals the number of wires per inch of screen, comes in various sizes (80 to 400). Typically, a 200-mesh screen is used for resistors and a 300-mesh screen for conductors.

Fig. 9.38 An example of a round stainless steel screen being inserted in a screen printer. (*Courtesy The Sel-Rex Company.*)

Separate masks are needed for conductors, resistors (one for each different sheet resistance used), and dielectric for thick-film capacitors. The masks are reduced, step-and-repeated, and a positive is made of the pattern. The dark areas on a positive correspond to what is printed on the screen.

A number of methods are available for transferring the mask pattern to the screen (referred to as a *screen stencil*). The *direct-emulsion* method is most commonly used today. It has good life expectancy and is inexpensive and easy to prepare.

A direct-emulsion stencil is made by first applying a light-sensitive emulsion to the screen. After drying, a positive of the mask pattern is placed in contact with the emulsion and exposed to UV. The exposed regions are polymerized, and the unexposed regions are washed away. Areas on the screen which were polymerized are therefore sealed, and the remaining areas are opened.

To print the pattern, the substrate is positioned and secured on a printing platform (Fig. 9.39). The screen is then located parallel to and about 20 mils above the sub-

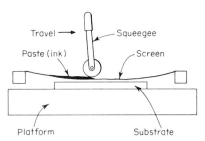

Fig. 9.39 Printing a screen pattern on a substance.

strate. Paste (ink) is applied and a squeegee is moved at a constant rate across the screen. The paste is forced through the openings in the screen onto the substrate.

In practice, a *screen printer* is used to ensure uniform patterns on the substrate. An example of a commercial screen printer is shown in Fig. 9.40.

After printing, the substrate remains at room temperature for a few minutes to allow the paste to level and even out impressions left by the screen. It is then dried for approximately 10 min at 100 to 150°C to drive volatile solvents out of the paste. Finally, the substrate is placed in a furnace between 700 and 1000°C for 30 min for *firing*. In addition to burning off the organic binder (filler) in the paste, firing forces the particles in the deposited conductors and resistors into a solid mass.

Upon the completion of firing, trimming may be used to tighten resistance values. Transistors and other chips are then connected to the circuit, as in the thin-film process, and the completed circuit is tested and encapsulated.

THICK-FILM PASTES (INKS) Pastes, or inks, for thick-film circuits are a blend of three materials:
1. Base material for conductor, resistor, or dielectric
2. Organic binder to carry the ink through the mesh openings
3. Volatile solvent to control the flow (viscosity) of the paste
Because they have low resistivity and do not oxidize, noble metals are used in pastes for conductors. Examples of conductive pastes are silver, palladium-silver, gold, and palladium-gold.

Basically, a mixture of finely divided glass and metal forms a resistor after firing. Four families of resistor pastes and their firing temperatures commonly used are:
1. Platinum: 980°C
2. Ruthenium oxide: 760°C
3. Palladium-silver: 760°C
4. Thallium oxide: 580°C
Sheet resistances of available pastes range from 1 Ω/square to 10 MΩ/square. By proper blending of material, nearly any sheet resistance can be obtained. Manufacturers of resistive pastes provide *blending curves* for this purpose.

Fig. 9.40 An example of a commercial screen printer. Also shown are various pastes and rectangular screens. (*Courtesy The Sel-Rex Company.*)

The geometry of a thick-film capacitor is similar to that of the thin-film capacitor of Fig. 9.36. The lower electrode is screened on the substrate. A dielectric paste, having a high glass content, is then screened on top of the lower electrode. Finally, the screening of the upper electrode over the dielectric completes the capacitor.

THICK-FILM COMPONENTS Resistance values range from a few ohms to tens of megohms. Power dissipation may be as high as 25 W/in^2 of surface. Tolerance before trimming is in the order of ±20 percent; after trimming, better than 1 percent.

The useful range for thick-film capacitors is 10 to 300 pF; tolerance is ±20 percent. As in thin-film circuits, thick-film inductors are not too practical.

Chapter 10

Tuned Circuits

10.1 INTRODUCTION

In previous chapters, the subject of *reactance* was discussed. It was noted that inductors *(coils)* have a quality that is called *inductive reactance*, the magnitude of which varies with the applied frequency. It was also noted that capacitors have a quality called *capacitive reactance*, the magnitude of which also varies as a function of frequency.

Of particular importance is the fact that inductive reactance, which is generally considered as being *positive* reactance, is vectorially opposite to capacitive reactance, generally considered as being *negative* reactance. That is, when they are combined in a circuit, inductive reactance tends to *cancel* capacitive reactance. Also, inductive and capacitive reactance behave oppositely with respect to changes in applied frequency. For a fixed value of inductance, inductive reactance *increases* as the applied frequency is increased, but the capacitive reactance of a fixed value of capacitance *decreases* as the applied frequency is increased.

The reference to inductive reactance as being positive and capacitive reactance as being negative should not be misinterpreted. In an inductive circuit, the voltage leads the current and, in a capacitive circuit the voltage lags behind the current. Therefore, the effect of an inductor is opposite to the effect of a capacitor. As a convenient way of distinguishing between the effects of the two reactances, we call one positive and the other negative.

Another way of distinguishing between the effects of inductive reactance and capacitive reactance is to represent them graphically as shown in Fig. 10.1. This illustration shows that the effects of X_L and X_C are in opposite directions. If the values of X_L and X_C are equal, their effects cancel, leaving only resistance in the circuit. If the three quantities shown in Fig. 10.1 (X_L, X_C, and R) are added vectorially, the resultant will be the circuit *impedance*. By definition, impedance is the combined opposition to the flow of alternating current that is offered by the inductance, capacitance, and resistance in that circuit.

The qualities of inductive and capacitive reactance give rise to an important area of applications in what are referred to as tuned circuits. The name *tuned circuits* generally refers to a combination of inductive and capacitive reactances arranged so that the impedance of the combination changes abruptly at some specific frequency, or narrow range of frequencies. Tuned circuits are made with inductors and capacitors. As with any circuit, they also have resistance. Resistance is frequently useful in establishing the desired performance characteristics of a tuned circuit, and it is present in all circuits. The frequency at which the tuned circuit exhibits its sharp change in impedance is referred to as the *resonant frequency*. There is also a term *antiresonant frequency* which, in some applications, should be used instead of the term resonant frequency

Fig. 10.1 Graphical representation of X_L, X_C, and R.

Fig. 10.2 A miniaturized tuned circuit, used for printed-circuit boards. In the upper view the microminiature, variable ceramic capacitor is at the top. Beneath this is the inductor, wound on a powdered iron toroidal core. (*Courtesy JFD Electronics Corp.*)

when referring to certain types of tuned circuits. There is a fine distinction between the terms which will be covered later in this chapter.

ANALOGIES There are many analogies between electronic tuned circuits and other physical phenomena. The relationship between the length and period of swing for a pendulum is one example. A tuning fork is another good example of a tuned mechanical device. A tuning fork will vibrate at only one frequency, regardless of how hard it is struck, or how it is set into vibration. In a like manner, an electrical tuned circuit will operate at only one frequency, or at only one narrow band of frequencies.

SOME USES OF TUNED CIRCUITS Tuned circuits are an extremely important factor in the overall art and science of electronics, as they make it possible to perform many functions that would otherwise be either impossible or extremely difficult. These functions include *selective amplification of certain frequencies* (or bands of frequencies) without amplifying other unwanted frequencies, and *rejection of unwanted frequencies* or bands of frequencies. Tuned circuits are utilized in radio broadcast receivers as the means of selecting the one desired station to be received from among the large number of stations that are simultaneously delivering signals to the receiver. Tuned circuits are employed in radio transmitters to establish the transmitter carrier frequency. In television receivers they are used to select the desired channels. In electronic circuits there are many other applications.

The importance of tuned circuits in the overall field of electronics can hardly be overestimated. It can safely be said that, were it not for tuned circuits, the art and science of electronics as we know it today would not exist. Figure 10.2 shows an example of a tuned circuit so small that it can be packaged into a transistor case. Much larger tuned circuits are used in transmitters where much larger powers are involved.

10.2 TYPES OF TUNED CIRCUITS

There are many different methods of classifying tuned circuits. For example, they may be classified as *tunable* or *fixed*—depending on whether the frequency of the tuned circuit can be varied, or whether it is always set at the same frequency. They may be called *narrow-band* or *wideband*—depending on the shape of their characteristic curves. They are referred to as being either *series-tuned* or *parallel-tuned* circuits — according to whether their components are series- or parallel-connected. Further, a given tuned circuit can be classified according to more than one of the several different methods of distinction. That is, a tuned circuit may be a tunable, narrow-band, parallel-tuned circuit.

A few examples of commonly used tuned circuits are illustrated in Fig. 10.3. The most often used circuit consists of a variable capacitor either in series or in parallel with an inductor (Fig. 10.3a and b). The minimum and maximum values of capacitance for the variable capacitor are chosen to ensure complete coverage of a band of frequencies. Where a number of different frequency bands must be tuned, as in a short-wave receiver, a switchable inductor may be used (Fig. 10.3c). If, for example, the switch is in position one, the capacitor and resulting inductance covers the lower-band (broadcast) frequencies. In position two or three, less inductance is present across the capacitor; for these positions the circuit is tuned to the shortwave frequencies.

The distinction between *fixed* (sometimes called *fixed-tuned*) and *tunable* (sometimes called *variable*) circuits refers to whether the circuit is tuned permanently to a single (*fixed*) frequency, or whether the frequency at which the circuit exhibits its unique impedance-response characteristic can be altered, or varied, according to the needs and desires of the user.

The many different methods of classifying tuned circuits, and the fact that a given tuned circuit can fit simultaneously into more than one classification, may seem confusing at first. However, there is a reason for each of the different methods of classification, and each has its own importance. This will become further apparent after the circuits have been discussed in greater detail.

FIXED-TUNED CIRCUITS   Fixed-tuned circuits are set permanently to a single frequency. They find many applications in electronic equipment where it is desired to respond to a known, predetermined frequency, either by enhancing the equipment response to that particular frequency as compared to the response to other frequencies, or by producing an especially *low* response to (rejecting) the frequency to which the circuit is resonant.

Some nominally fixed-tuned circuits may, however, be slightly *adjustable*. The term *adjustable* generally refers to the fact that one or more of the reactive elements of the tuned circuit can be varied (adjusted) over a narrow range. This adjustment is most often provided for the purpose of compensating for minor variations or imperfections

(a)   (b)   (c)

Fig. 10.3 Examples of tuned circuits: (*a*) A series-tuned circuit. (*b*) A parallel-tuned circuit. (*c*) Tuned circuit used in radio receiver circuit. Switch permits the receiver to be tuned to more than one band of frequency.

Fig. 10.4 A tuned circuit in which the resonant frequency is adjusted with a variable inductor.

of the circuit elements (or in some cases, minor variations in frequency of the input signal that is desired to be tuned). The frequency of the tuned circuit in Fig. 10.4 is adjusted with a slug-tuned inductor.

TUNABLE CIRCUITS   Tuned circuits which are slightly adjustable circuits should not be confused with tunable circuits, as the latter have a very different kind of application. In general, tunable, or variable, circuits are used in applications in which it is desirable to be able to change the frequency for which the circuit exhibits its characteristic *selective* response. Such changing may be done frequently and with relative ease. Perhaps the most common example of the use of tunable circuits is in radio broadcast receivers, in which tunable input circuits can be varied by the user to select the particular one station to be received out of the many that are broadcasting simultaneously on different frequencies.

CIRCUIT BANDWIDTH   Another method of classifying tuned circuits is as "narrowband" or "wideband"—whether the circuit responds only to frequencies within a very narrow range or *band* about the resonant frequency, or whether the response is to frequencies in a relatively wide band about the resonant frequency. (The terms *narrow band* and *wide band* are, of course, very general, and more sophisticated methods for describing the bandwidth of tuned circuits are needed in most practical situations.)

The fact that a tuned circuit is narrow-band or wideband may sometimes be a matter of necessity brought on by the practical limitations of the components with which the equipment designer must work. More often, however, it is a matter of deliberate design to fit the requirements of the application. In a radio broadcast receiver, for example, the tuned circuits that determine which of the many available stations is actually received are narrow-band circuits designed so that the one desired station can be received clearly, without interference such as might be produced by other stations broadcasting on frequencies near that of the desired station.

In many applications, on the other hand, including television broadcast receivers, radar receivers, and other equipment that must handle a wide band of signals, the tuned circuits are deliberately made so as to respond effectively to signals within a relatively wide band on both sides of the center frequency to which the tuned circuit is resonant.

SINGLE-TUNED AND DOUBLE-TUNED CIRCUITS   The distinction between single-tuned and double-tuned (or more generally, multiple-tuned) circuits refers to certain details of tuned-circuit configuration and/or adjustment. Generally speaking, the distinction concerns whether the tuned circuit is "resonant" to only one frequency (single-tuned), or whether it is resonant *simultaneously* to two or more frequencies (double- or multiple-tuned).

Multiple tuning of circuits may be done for a variety of reasons, but most often it is a means of obtaining wideband response without sacrificing other important circuit performance characteristics. One of these other performance characteristics is *selectivity*, which is a measure of the ability of a tuned circuit to respond to a desired frequency or band of frequencies, while *rejecting* frequencies outside this band, even though close to it.

SERIES-TUNED AND PARALLEL-TUNED CIRCUITS Perhaps the most important single method of classifying tuned circuits is based on the circuit configuration. As we shall see in the next section, a tuned circuit with the components connected in series may have a completely different function from one with the components connected in parallel.

10.3 COMPARISON OF SERIES-TUNED AND PARALLEL-TUNED CIRCUITS

The terms "series circuit" and "parallel circuit" refer to the arrangement of the capacitive and inductive elements that make up the tuned circuit. The distinction can readily be understood with the aid of Fig. 10.3. Figure 10.3*a* shows a simple series-tuned circuit, and Fig. 10.3*b* shows a parallel-tuned circuit.

In Fig. 10.3*a* the capacitor C_s and the inductor L_s that make up the tuned circuit are in *series* across the source of input voltage (shown schematically as a generator in Fig. 10.3*a*). Also in series with the other circuit elements are the resistances of the tuned circuit, shown schematically as a single lumped resistance R_s and the internal impedance of the generator, or *source generator*, Z_g.

In Fig. 10.3*b* the capacitor C_p and the inductor L_p are in parallel across the "source" generator with its internal impedance Z_g. In this parallel-tuned circuit the symbol R_p is used to denote the *total* effective resistance of the circuit. This lumping of resistance into a single equivalent-circuit element is common practice, and reflects the practical fact that, in the great majority of instances, the resistance of the inductor is so much greater than that of the capacitor that the latter is, in fact, negligible.

Figure 10.3, then, illustrates the basic configurations of both series- and parallel-tuned circuits. The origin of the terminologies for describing these two types of tuned circuits is apparent from the illustration. Note that the distinguishing feature of the series circuit is that all the generator current *must* flow through R, L, and C. In the parallel-tuned circuit, the generator current divides so that each branch receives only part.

10.4 SERIES-TUNED CIRCUITS

APPLICATIONS Series-tuned circuits have applications wherever it is desired to pass a signal of one frequency with minimum attenuation while rejecting signals of all other frequencies.

As one example, consider a radio transmitter that is to be coupled to its antenna in such a way that the desired transmitter output frequency is transmitted without attenuation, but the harmonics — that is, multiples — of the output frequency that may also be present do not reach the antenna. This can be accomplished by using a series-tuned circuit, resonant at the desired output frequency, between the transmitter output and the antenna, as shown in Fig. 10.5. The series-tuned circuit, being resonant at the desired carrier frequency, will have minimum effect on the transmitter output at this frequency. At higher frequencies, however, such as at the second harmonic (second multiple) of the desired output frequency, the series circuit will present an impedance approximately equal to the reactance of the inductor at the frequency in question, thus effectively preventing harmonics from reaching the antenna.

Another example of the use of series-tuned circuits may be found in what are frequently referred to as "trap" circuits. These are circuit applications in which a series-

Fig. 10.5 Use of a series-tuned circuit in transmitter output.

Fig. 10.6 An example of a series-tuned circuit used as a trap across the output of an amplifier stage.

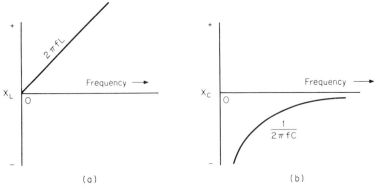

Fig. 10.7 Comparison of inductive and capacitive reactance curves: (a) Inductive reactance versus frequency. (b) Capacitive reactance versus frequency.

tuned circuit is used in parallel, or shunt, across a source of voltage, for the purpose of short-circuiting, or "trapping," signals of one frequency, while having negligible effect on signals of other frequencies. A specific example of the use of a series-tuned circuit as a trap circuit is shown in Fig. 10.6. This type of trap can be found in many television receivers.

The characteristic curves for inductors and capacitors are shown in Fig. 10.7.

FUNDAMENTALS   The *reactance* of an *inductor* varies with frequency according to the equation

$$X_L = 2\pi fL \tag{10.1}$$

where X_L = inductive reactance, Ω
$\qquad f$ = applied frequency, Hz
$\qquad L$ = value of inductance, H
$\qquad \pi$ = 3.14 (approximately)

Equation (10.1) states that for a given value of inductance L the inductive reactance X_L varies directly with frequency. This is shown graphically in Fig. 10.7a. At zero frequency (dc) the reactance is zero and the inductor acts as a short circuit. As the frequency increases, the inductive reactance increases in a direct manner.

The *reactance* of a *capacitor* varies with frequency according to the equation

$$X_C = -\frac{1}{2\pi fC} \tag{10.2}$$

where X_C = capacitive reactance, Ω
$\qquad C$ = value of capacitance, F

Equation (10.2) states that the capacitive reactance varies *inversely* with frequency; that is, as the frequency increases the capacitive reactance decreases — opposite to that for an inductor. Referring to Fig. 10.7b, at frequencies close to zero it can be seen that the capacitive reactance has a very large value, approaching infinity. The capacitor at zero frequency (dc), therefore, acts as an open circuit. At higher frequencies X_C approaches zero, and the capacitor is said to act as a short circuit.

The right side of Eq. (10.2) is preceded by a minus sign to denote the fact that capacitive reactance is opposite in effect to inductive reactance, and is considered to be *negative* reactance, whereas inductive reactance is considered to be *positive*. Equations (10.1) and (10.2) and also Fig. 10.7a and b both show that inductive reactance and capacitive reactance are oppositely affected by frequency. That is, *inductive reactance increases* as frequency increases, whereas the *value of capacitive reactance decreases* as frequency increases.

For a given combination of values of inductance L and capacitance C in a series circuit there will be some one frequency for which the inductive reactance X_L has the same magnitude as the capacitive reactance X_C. If we remember that inductive reactance and capacitive reactance are opposites, that is, the former is positive while the latter is negative, it follows that they tend to cancel each other when they are equal in value.

The frequency for which the value of inductive reactance X_L equals the value of capacitive reactance X_C is the *resonant* frequency of a tuned circuit consisting of an inductance of L henrys and a capacitance of C farads. In other words, the resonant frequency of an LC series circuit is the frequency for which

$$X_L = X_C \tag{10.3}$$

The negative sign is omitted because, by definition, the resonant frequency is that frequency for which the inductive reactance is *numerically* equal to the capacitive reactance.

Substituting $2\pi fL$ for X_L, and $\dfrac{1}{2\pi fC}$ for X_C, we get

$$2\pi f_r L = \frac{1}{2\pi f_r C} \tag{10.4}$$

where f_r = resonant frequency, Hz.

Solving Eq. (10.4) for f_r, we obtain

$$f_r = \frac{1}{2\pi\sqrt{LC}} \tag{10.5}$$

Equation (10.5) states that the resonant frequency f_r is inversely proportional to the square root of the LC product. Thus, for a high resonant frequency the LC product is small, and for a low resonant frequency it is large.

The behavior of the series-resonant circuit is summarized in Fig. 10.8. Figure 10.8a

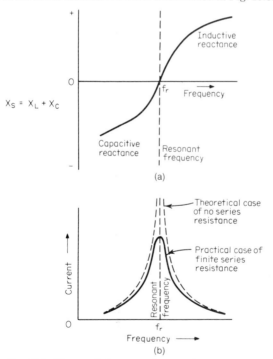

Fig. 10.8 Characteristic curves for a series-resonant circuit: (a) Reactance of a series-tuned circuit versus frequency. (b) Current through a series-tuned circuit at frequencies in the vicinity of resonance.

shows that at the resonant frequency the net reactance is zero, that is, $X_L = X_C$. Below the resonant frequency, capacitive reactance predominates; at frequencies above resonance, inductive reactance is predominant.

From Fig. 10.8b it is seen that the current in a series-tuned circuit is maximum at the resonant frequency. For frequencies less than or greater than the resonant frequency, the current falls and has a lower value. The magnitude of the current at resonance depends on the series resistance present. If there is zero resistance, the current approaches infinity; as the series resistance increases, the current at resonance decreases.

example 10-1 A series-tuned circuit, like the one shown in Fig. 10.3a, has an inductance of 3 mH, a capacitance of 30 μF, and a resistance of 300 Ω. What is the resonant frequency of this tuned circuit?

solution In a *series*-tuned circuit the amount of resistance in the circuit can be disregarded. (This is not always true in a parallel-tuned circuit.) To solve the problem, the inductance and capacitance values are expressed in henrys and farads.

$$3 \text{ mH} = 3 \times 10^{-3} \text{ H}$$
$$30 \ \mu\text{F} = 3 \times 10^{-5} \text{ F}$$

Substituting these values into Eq. (10.5), we get

$$f_r = \frac{1}{2\pi \sqrt{LC}} \tag{10.5}$$

$$= \frac{1}{(2 \times 3.14) \ \sqrt{3 \times 10^{-3} \ 3 \times 10^{-5}}}$$

$$= \frac{1}{6.28 \ \sqrt{9 \times 10^{-8}}}$$

$$= \frac{1}{6.28 \times 3 \times 10^{-4}} = 530 \text{ Hz}$$

example 10.2 It is desired to design a series-tuned circuit to have a resonant frequency of 5000 Hz. If the capacitor to be used has a value of 0.1 μF, what value of inductance is required?

solution Substitution of the given values in Eq. (10.4) yields

$$2\pi f_r L = \frac{1}{2\pi f_r C} \tag{10.4}$$

or
$$2\pi(5000)L = \frac{1}{2\pi(5000) \times 0.1 \times 10^{-6}}$$

Solving for L,

$$L = \frac{1}{4\pi^2 \times 25 \times 10^6 \times 0.1 \times 10^{-6}}$$

$$= \frac{1}{10\pi^2} = 0.010 \ 14 \text{ H} = 10.14 \text{ mH}$$

The simple tuned circuit in Example 10.1 will be resonant at a frequency of 530 Hz. At this frequency, the inductive and capacitive reactances *cancel*. This means that the only opposition to current flow will be what is offered by Z_g and R_s (see Fig. 10.3a). If these values are small, then there will be very little opposition to current flow, and the circuit current will be very high.

Equation (10.5) applies to series-tuned circuits with or without a series resistance R_s. However, the equation will only apply to parallel circuits if the resistances in the L_p and C_p branches are small enough to be neglected. If, for example, the resistance value of R_p in Fig. 10.3b is more than about 10 percent of the value of X_L, then Eq. (10.5) *cannot be used* to determine the value of resonant frequency for the circuit. The same is true if a resistance is placed in series with C_p in the circuit.

SERIES-TUNED VOLTAGES AT RESONANCE Another property of series-tuned circuits that deserves attention concerns the voltages that exist across the circuit elements at frequencies at or near the resonant frequency of the circuit.

If we remember that, at the resonant frequency, the capacitive reactances X_C of the inductor are equal and opposite to each other; it follows then that the net reactance of the series circuit will be zero at the resonant frequency. The current flowing through the circuit at the resonant frequency is thus given by the simple Ohm's law relationship,

$$I = \frac{E}{R_s}$$ (10.6)

where E = voltage applied across the entire series-tuned circuit, that is, the voltage developed by the voltage source less the voltage drop across the internal impedance of the source (Z_g in Fig. 10.3a)

I = current through the series-tuned circuit at resonance

R_s = total series resistance of the series-tuned circuit

If the total series resistance R_s is reasonably low, as it will be in a well-designed series-tuned circuit for most applications, the current I through the circuit will be quite large at resonance.

The voltage across the capacitor at resonance is given by the Ohm's law relationship,

$$E_C = I \times X_C$$ (10.7)

where E_C = voltage across the capacitor at resonance.

Similarly, the voltage across the inductor at resonance will be

$$E_L = I \times X_L$$ (10.8)

where E_L = voltage across the inductor at resonance.

The last two equations state that for a series-tuned circuit at resonance, the voltage across the capacitor (or inductor) is equal to the product of current I at resonance [defined by Eq. (10.6)] and the capacitive reactance (or inductive reactance). The practical significance of these effects will be considered shortly. Equations (10.7) and (10.8) both assume that whatever resistance is present in the series-tuned circuit is considered as a separate element. However, this assumption does not in any way affect the validity of the above equations.

Bearing in mind that, in a series-tuned circuit with a typically low value of series resistance R_s, the current through the circuit at resonance is quite large [see Eq. (10.6)], it follows that the voltages across the capacitive and inductive elements of the circuit are also quite large at resonance. These voltages, which will be equal in magnitude, will be greater than the voltage applied across the series-tuned circuit by a factor of X_C/R_s, or X_L/R_s. That is, if the voltage applied across the total circuit at resonance is E, the voltage across the capacitor and inductor may be also expressed as

$$E_C = E\left(\frac{X_C}{R_s}\right)$$ (10.9)

and

$$E_L = E\left(\frac{X_L}{R_s}\right)$$ (10.10)

The high voltages across the reactive components affect the voltage ratings. They must be considered if the circuit is to operate without arcing or short circuiting. The ratio of capacitive or inductive reactance at resonance to the equivalent series resistance may have values in the range of from 10 to perhaps 100; so it follows that the capacitor and inductor must be constructed to withstand voltages from 10 to 100 times as great as the voltage applied across the series-tuned circuit. (This refers, of course, to the r-f or ac voltage applied to the circuit, and not to any dc voltage that may be present across the series-tuned circuit.)

In low-voltage applications, such as a series-tuned circuit in a radio receiver, this is not a matter of great consequence. In other applications, such as in transmitters where the applied voltage itself may be reasonably large, the requirement that the individual elements of the series-tuned circuit withstand voltages that are several times the applied voltage can represent an important design requirement. This is illustrated in the next example.

example 10.3 In a series-tuned circuit $R_s = 10\ \Omega$, and the magnitude of the capacitive reactance is 1000 Ω at the resonant frequency. If the voltage source $E_g = E = 10$ V, $Z_g \approx 0$, what are the voltages across the inductor, capacitor, and resistor at the resonant frequency?

solution   From Eq. (10.6),

$$I = \frac{E}{R_s} = \frac{10}{10} = 1 \text{ A}$$

At resonance,

$$X_C = X_L = 1000 \ \Omega$$

Using Eqs. (10.7) and (10.8), we find

$$
\begin{aligned}
E_C &= I \times X_C \\
&= 1 \times 1000 \\
&= 1000 \text{ V}
\end{aligned}
\qquad \text{and} \qquad
\begin{aligned}
E_L &= I \times X_L \\
&= 1 \times 1000 \\
&= 1000 \text{ V}
\end{aligned}
$$

Also,
$$E = I \times R_s = 1 \times 10 = 10 \text{ V}$$

The significance of this example is that, with practical values of circuit elements, the individual voltages across the capacitor and inductor at resonance are *100 times* the voltage applied across the series-tuned circuit.

10.5   PARALLEL-TUNED CIRCUITS

In the preceding section, the characteristics and behavior of series-tuned circuits were described and discussed. We now consider the second major category of tuned circuits, the parallel-tuned circuit, which is also sometimes referred to as an *antiresonant circuit*.

In Fig. 10.3*b*, the parallel-tuned circuit consists essentially of a capacitor C_p and an inductor L_p connected in parallel (or *shunt*) across a source of voltage, a *generator*. The circuit in Fig. 10.3*b* shows resistance in the inductive branch, but not in the capacitive branch. This is a practical consideration because the resistance of the wires in the inductor is usually large compared to the ESR (Equivalent Series Resistance) of the capacitor. However, this is not to say that resistance will never be found in the capacitive branch. The equation for parallel resonance will be developed later in this chapter for all possible cases.

APPLICATIONS   Applications of parallel-tuned circuits are based on their impedance versus frequency characteristics, as illustrated in Fig. 10.9. At the resonant frequency the impedance is maximum, and at frequencies below and above resonance the impedance falls off. In many applications parallel-tuned circuits are used to obtain selective filtering of specific frequencies.

Figure 10.10 illustrates the use of a parallel-tuned circuit to obtain selective amplification of a desired frequency in a radio or television receiver. The antenna is connected to the parallel-tuned circuit, to which the input terminal of the first amplifier stage is also connected.

When signals arrive from the antenna at the resonant frequency of the parallel-tuned circuit, this presents a high impedance to ground. Thus, the signal current supplied

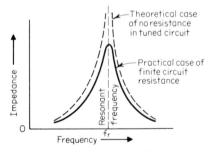

Fig. 10.9   Impedance of a parallel-tuned circuit at frequencies in the vicinity of resonance.

Fig. 10.10   Use of a parallel-tuned circuit for selective amplification at the receiver input.

by the antenna develops a relatively high voltage across the tuned circuit, this voltage being amplified by the first amplifier stage.

For signals of other frequencies, that is, frequencies relatively far from the resonant frequency, the impedance is much lower. The voltage developed in response to the current from the antenna, therefore, will also be lower, and the signals will not be amplified to any appreciable degree. (In fact, the circuit of Fig. 10.10 may be viewed as one that has the effect of *shunting* all signals (other than those at the resonant frequency) to ground. Since they are shunted to ground, they are not presented to the amplifier.

A similar application of parallel-tuned circuits is illustrated in Fig. 10.11. A parallel-tuned circuit is used at the output of an amplifier, instead of at the input as in Fig. 10.10.

Fig. 10.11 Use of a parallel-tuned circuit for selective amplification at the amplifier output.

Fig. 10.12 Use of a parallel-tuned circuit in series with amplifier output.

The amplifier in Fig. 10.11 may be a circuit in a transmitter, or in a receiver. In either case, the purpose of the circuit is to cause the amplifier to respond primarily to the resonant frequency of the tuned circuit, and to minimize the amplification of signals at other frequencies.

The voltage gain of an amplifier is directly proportional to the impedance of the load. In the circuit of Fig. 10.11, the parallel-tuned circuit is the amplifier load. At the resonant frequency this impedance will be quite high, and therefore the gain will be high. At all other frequencies the impedance of the parallel-tuned circuit will be low, and as a result the gain will be low. To summarize, the parallel-tuned circuit permits the amplifier to produce a large voltage gain at a specific frequency.

Another type of application of parallel-tuned circuits is illustrated in Fig. 10.12. In this circuit the parallel-tuned circuit is in series with an amplifier load (which in this case is an antenna). Its purpose is to prevent signals of an undesired frequency from reaching the load or the next stage. This is accomplished because the parallel-tuned circuit presents a high impedance, the signals of the (undesired) frequency to which it is tuned, while presenting a relatively low impedance to signals of the desired frequency. The circuit of Fig. 10.12 is used frequently in radio transmitters to prevent the second harmonic (second multiple) of the desired output frequency from reaching the antenna, from which it could be radiated and cause interference.

FUNDAMENTALS The equation for parallel resonance (which is sometimes called *antiresonance*) will be obtained for the circuit of Fig. 10.13. A few mathematical derivations are given in this section, but the overall results are summarized for those who want a less mathematical approach to the subject of parallel-tuned circuits. Note that the resistances in both the inductive and capacitive branches will be taken into consideration. Then the equation will be simplified for use in certain specialized cases such as the one shown in Fig. 10.3b.

When a circuit is in resonance, the current and the voltage in the circuit must be in phase. This is true for both series and parallel resonant circuits. Another way of stating the condition for resonance is that the circuit must have *unity power factor*.

The only way that unity power factor can occur in the circuit of Fig. 10.13 is if the phasor sum of the currents in the two branches produces a total current that is in phase with the applied voltage V. The mathematics required to establish the values of R_C, C, R_L, and L that are needed to produce resonance would be very tedious. Fortunately, there is a shortcut method.

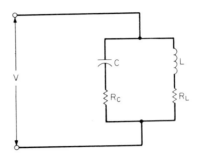

Fig. 10.13 A general circuit for determining parallel resonance.

The two series branches of Fig. 10.13 can be converted to the four parallel branches shown in Fig. 10.14. The equations required for finding the conductances G_C and G_L and the susceptances B_C and B_L are shown in the illustration. It is important to understand that the conversion from the circuit of Fig. 10.13 to the one in Fig. 10.14 is a standard procedure used for simplifying and solving parallel circuits. In other words, this is not a special case, but rather it is a standard procedure.

In the equivalent circuit of Fig. 10.14 there is no resistance in the capacitive and inductive branches. Resonance will occur when $B_C = B_L$, and this is the basis for finding the equation for parallel resonance. (Compare this with the solution of the series-resonant equation which is based on $X_C = X_L$.)

$$B_L = B_C$$

$$\frac{X_L}{R_L^2 + X_L^2} = \frac{X_C}{R_C^2 + X_C^2}$$

$$\frac{2\pi fL}{R_L^2 + (2\pi fL)^2} = \frac{1/2\pi fC}{R_C^2 + (1/2\pi fC)^2}$$

When this equation is solved for f_r, which is the resonant frequency of the circuit in Fig. 10.13, the result is

$$f_r = \frac{1}{2\pi \sqrt{LC}} \sqrt{\frac{L - CR_L^2}{L - CR_C^2}} \qquad (10.11)$$

where f_r = resonant frequency, Hz
L = circuit inductance, H
C = circuit capacitance, F
R_C = resistance in capacitive branch, Ω
R_L = resistance in inductive branch, Ω

This is the equation for the resonant frequency of the circuit in Fig. 10.13, even though it is derived from the equivalent circuit of Fig. 10.14.

Fig. 10.14 This circuit has the same impedance and will produce the same phase angle as the circuit in Fig. 10.13.

Several very important facts about parallel resonance can be determined from Eq. (10.11).

1. When the resistance in the two branches is negligible, the value zero can be substituted for both R_C and R_L in the equation. The result is

$$f_r = \frac{1}{2\pi \sqrt{LC}} \tag{10.12}$$

where f_r = resonant frequency, Hz
    L = inductance, H
    C = capacitance, F

In other words, the same equation can be used for both series and parallel resonance if there is negligible resistance in the parallel branches. In actual practice, if X_L is at least ten times the value of R_L, and X_C is at least ten times the value of R_C, the resistances are considered to be negligible.

2. Equation (10.11) shows that a parallel-resonant circuit can actually be tuned by varying the resistance in one of the branches – that is, by varying either R_C or R_L. This follows from the fact that changing the value of either R_C or R_L in the equation will change the value of f_r. (Of course, f_r can also be varied by changing either L or C.)
If we solve Eq. (10.11) for the value of R_L, the result is

$$R_L = \sqrt{\frac{X_L}{X_C} R_C^2 - X_L^2 + \frac{L}{C}} \tag{10.13}$$

where the letters have the same meanings as in Eq. (10.11).
If we solve Eq. (10.11) for the value of R_C, the result is

$$R_C = \sqrt{\frac{R_L^2}{(2\pi f)^2 LC} - \frac{1}{\omega^2 C^2} + \frac{L}{C}} \tag{10.14}$$

where the letters have the same meanings as in Eq. (10.11).

Equations (10.13) and (10.14) give the value of R_L or the value of R_C needed to bring the circuit of Fig. 10.13 into resonance.

3. It was mentioned earlier that the circuit of Fig. 10.3b is often given as a practical circuit because the dc resistance of the inductive branch cannot always be ignored. The equation for resonance in this circuit can be obtained by setting $R_C = 0$ in Eq. (10.11). The result is

$$f_r = \frac{1}{2\pi \sqrt{LC}} \sqrt{\frac{L - CR_L^2}{L}} \tag{10.15}$$

4. In a series LC circuit there is always a resonant frequency because there is always some frequency at which $X_L = X_C$. However, in the parallel-resonant circuit, it is possible to have a combination of components that are not resonant at any frequency. To understand how this is possible, look at the numerator and denominator under the radical in Eq. (10.11):

Numerator: $L - CR_L^2$
Denominator: $L - CR_C^2$

A very important point to consider is that the value under the radical will be negative if

                                CR_L^2 is greater than L
or CR_C^2 is greater than L

If the value under the radical is negative, no value of f_r exists because the result is an *imaginary* value. To summarize, then, it is possible to have a parallel circuit that has no resonant frequency!

5. Another important feature of parallel-resonant circuits is that they may be designed to be resonant at all frequencies. This can be demonstrated with a short mathematical derivation.
In the derivation of Eq. (10.11) this equation was given:

$$\frac{2\pi f L}{R_L^2 + (2\pi f L)^2} = \frac{1/2\pi f C}{R_C^2 + (1/2\pi f C)^2}$$

To simplify equations of this type, the symbol ω (omega) is substituted for $2\pi f$

$$\frac{\omega L}{R_L^2 + \omega^2 L^2} = \frac{1/\omega C}{R_C^2 + 1/\omega^2 C^2}$$

Setting the product of the means equal to the product of the extremes (sometimes called "cross-multiplying") gives

$$\omega L R_C^2 + \frac{\omega L}{\omega^2 C^2} = \frac{R_L^2}{\omega C} + \frac{\omega L^2}{C}$$

The lowest common denominator is ωC^2. Multiplying every term by this value gives

$$\omega^2 L C^2 R_C^2 + L = R_L^2 C + \omega^2 L^2 C$$

Now, if either the numerator or the denominator under the radical in Eq. (10.11) is equal to zero, then the resonant frequency cannot be given a numerical value.

For the numerator:

$$L - C R_L^2 = 0$$

$$R_L = \sqrt{\frac{L}{C}}$$

For the denominator:

$$L - C R_C^2 = 0$$

$$R_C = \sqrt{\frac{L}{C}}$$

Substituting the radical equivalent values for R_L and R_C in the above equations gives

$$\frac{\omega L^2}{C} + \frac{L}{\omega C^2} = \frac{\omega L^2}{C} + \frac{L}{\omega C^2}$$

Note that this equation is valid for *every value* of ω (and therefore every value of frequency f)—*except the value of $\omega = 0$.* The importance of this derivation is that the circuit is resonant at *all frequencies* whenever $R_L = R_C = \sqrt{L/C}$.

In a series-tuned circuit, if the applied frequency is above resonance, there is a greater voltage drop across the inductor than across the capacitor. This makes the circuit behave like an inductor. If the applied frequency is below the resonant frequency, on the other hand, the circuit behaves like a capacitor because the larger voltage drop is across the capacitor.

TABLE 10.1   Comparison of Series- and Parallel-Tuned Circuits

	Series-tuned circuit	Parallel-tuned circuit
Impedance	Minimum at the resonant frequency	Maximum at the resonant frequency
Current	Maximum at resonant frequency	Minimum at resonant frequency
Voltage	Minimum voltage across the circuit at resonant frequency	Maximum voltage across the circuit at resonant frequency
Resonance	There is one resonant frequency for each LC combination. Resistance does not affect resonance for all practical purposes.	a. Resistance affects resonant frequency. b. Circuit may resonate at all frequencies. c. Circuit may not have a resonant frequency.
When applied frequency is below resonance	Circuit is capacitive.	Circuit is inductive.
When applied frequency is above resonance	Circuit is inductive.	Circuit is capacitive.
Power factor at resonance	1.0	1.0
Special feature	Voltage across a component may be many times applied voltage.	Current through a component may be many times main current.

When the frequency of the applied signal across a parallel-tuned circuit is above the resonant frequency, more current flows through the capacitor because its reactance is lower than that of the inductor. This means that the circuit will be capacitive. Conversely, if the applied frequency is below resonance, the greater current flows through the inductor, and the circuit is inductive.

The characteristics of series- and parallel-tuned circuits are summarized in Table 10.1.

Fig. 10.15 Circuit for Example 10.5.

example 10.4 In the parallel-tuned circuit of Fig. 10.13 assume that $L_p = 100$ mH, $C_p = 0.1\ \mu$F, and $R_C = R_L = 0$. What is the resonant frequency of the circuit?

solution From Eq. (10.12),

$$f_r = \frac{1}{2\pi\ \sqrt{LC}}$$

$$= \frac{1}{6.28\ \sqrt{0.1 \times 0.1 \times 10^{-6}}} = 1590\ \text{Hz} \qquad (10.12)$$

example 10.5 In the circuit of Fig. 10.15, a variable resistor is placed in the capacitive branch. This resistor is used for varying the resonant frequency. What is the range of frequencies through which R_C can tune the circuit?

solution When $R_C = 0$,

$$f_r = \frac{1}{2\pi\ \sqrt{LC}} = 1590\ \text{Hz} \qquad (\text{See Example 10.4.})$$

When $R_C = 0.5$ kΩ,

$$f_r = \frac{1}{2\pi\ \sqrt{LC}}\ \frac{L}{L - CR_C^2}$$

$$= 1590\ \sqrt{\frac{100 \times 10^{-3}}{100 \times 10^{-3} - (0.1 \times 10^{-6})\ (0.5 \times 10^3)^2}} = 1828\ \text{Hz}$$

Thus a change of 500 Ω results in a frequency change of 238 Hz.

10.6 SELECTIVITY, BANDWIDTH, AND Q FACTOR

The ability of a tuned circuit to respond differently to signals of different frequencies is referred to as its *selectivity*. For quantitative purposes, the selectivity of a tuned circuit — that is, its ability to discriminate between signals of different frequencies — is measured in terms of a parameter called *bandwidth*.

BANDWIDTH The bandwidth of a tuned circuit is a range of frequencies between two points on its selectivity curve — one on each side of the resonant frequency — at which the circuit response is a stated fraction of the response at the resonant frequency. The curve of Fig. 10.16 is an example of a selectivity curve. Although the fractional response point on the selectivity curve at which the tuned circuit bandwidth is determined can be stated arbitrarily, it is almost universal practice to measure bandwidth at points on the selectivity curve at which the circuit power response is one-half the

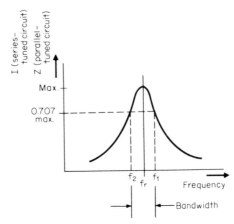

Fig. 10.16 The bandwidth is taken between the points where the current, voltage, or impedance is 70.7 percent of maximum, depending on whether it is a series- or parallel-tuned circuit. It may also be taken between points where the power is 50 percent of maximum.

power that is produced at resonance. The bandwidth so measured is called the *half-power* bandwidth, or the $-3dB$ bandwidth. The latter term is derived from the fact that one-half power corresponds to a power level three decibels below maximum power.

Although bandwidth is defined in terms of power, it is generally more convenient to measure it in terms of other circuit performance characteristics. The following relationships will be found especially useful in practice:

1. In a series-tuned circuit, the half-power bandwidth is the width (in frequency) of the portion of the selectivity curve between points at which the current through the circuit is equal to $1/\sqrt{2} = 0.707$ times the current at resonance. (This relationship should be self-evident when it is remembered that power P is proportional to the square of current, $P = I^2R$.)

2. In a parallel-tuned circuit, the half-power bandwidth is the width of the portion of the selectivity curve between points at which the effective parallel impedance of the circuit is equal to $1/\sqrt{2} = 0.707$ times the impedance at resonance.

It is evident that the bandwidth relationships for series-tuned and parallel-tuned circuits correspond, with the difference that bandwidth of a series-tuned circuit is expressed in terms of current flow while bandwidth of a parallel-tuned circuit is expressed in terms of effective parallel impedance. Figure 10-16 summarizes the definitions of bandwidth for series- and parallel-tuned circuits.

Q FACTOR. The bandwidth of a tuned circuit is determined by a circuit parameter called Q or Q factor. Circuit Q is a very important parameter used for describing the performance characteristics of a tuned circuit.

The Q of a tuned circuit is equal to the reactance of either the capacitor or the inductor at the resonant frequency of the circuit; the reactance of the capacitor and inductor are, as explained previously, equal at the resonant frequency, divided by the effective series resistance of the circuit. This relationship applies to both the series- and parallel-tuned circuits of Fig. 10.3. For the series-tuned circuit,

$$Q = \frac{2\pi f_r L_s}{R_s} \tag{10.16}$$

Similarly, for the parallel-tuned circuit with negligible resistance,

$$Q = \frac{2\pi f_r L_p}{R_p} \tag{10.17}$$

Expressions similar to Eqs. (10.16) and (10.17) can be written in terms of the capacitive reactances at resonance in place of the inductive reactances. However, expressing Q in terms of the inductive reactance is usually more convenient.

Another way of expressing Q is in terms of the half-power bandwidth of the tuned circuit as related to its resonant frequency:

$$Q = \frac{f_r}{f_1 - f_2} = \frac{f_r}{\Delta f} \qquad (10.18)$$

where f_1 = frequency above the resonant frequency at which the half-power relationship exists (see Fig. 10.16)

f_2 = frequency below the resonant frequency at which the half-power relationship exists

f_r = resonant frequency of a tuned circuit (either series or parallel)

$\Delta f = f_1 - f_2$ = bandwidth of the circuit

Equation (10.18) is especially convenient for determining the Q of a tuned circuit experimentally. It is necessary only to measure the frequencies at which the circuit exhibits a half-power response from which information, together with knowledge of the resonant frequency, the Q is readily calculated. (The required frequency measurement must, however, be made with some care for circuits with high Q's because errors in measuring either f_1 or f_2 can lead to serious errors in determining bandwidth Δf, and therefore in calculating the Q of the circuit.)

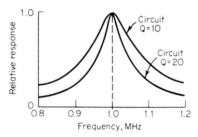

Fig. 10.17   Effect of circuit Q on bandwidth.

Measurements based on Eq. (10.18) are frequently superior to a method of determining Q according to Eq. (10.16) or (10.17). This is true because of the practical difficulty in determining the true value of R_s or R_p, inasmuch as their values are influenced by high-frequency losses in a manner that is not always easy to determine.

For use at many frequencies, however, there are instruments (usually called Q *meters*) that can measure the Q of an inductor, a capacitor, or a tuned circuit directly. This avoids the difficulty of determining the effective resistance by other means, and also the problems that sometimes arise in attempting to determine Q by frequency measurements.

It is evident from Eq. (10.16) or (10.17) that the resistance in a tuned circuit in relation to the reactance at resonance is the parameter that determines the circuit Q, and therefore the bandwidth. The effect of circuit Q on bandwidth is illustrated in Fig. 10.17. The higher the Q, the smaller the bandwidth.

The current in an inductor or a capacitor in a parallel-tuned circuit at resonance is Q times the external circuit current. Likewise, for a series-tuned circuit, the voltage across the inductor or capacitor at resonance is Q times the applied circuit voltage.

example 10.6   In a parallel-tuned circuit, $L = 100\ \mu\text{H}$, $R_L = 10\ \Omega$, $f_r = 1\ \text{mHz}$. Determine the values of (a) Q, and (b) bandwidth.

solution   (a) From Eq. (10.17),

$$Q = \frac{2\pi f_r L_p}{R_p}$$

$$= \frac{6.28 \times 10^6 \times 100 \times 10^{-6}}{10} = 62.8 \qquad (10.17)$$

(There are no units for measuring Q.)

(b) Solving Eq. (10.18) for the bandwidth Δf,

$$\Delta f = \frac{f_r}{Q} \qquad (10.19)$$

Therefore

$$\Delta f = \frac{10^6}{62.8} \approx 16 \text{ kHz}$$

example 10.7   Design a parallel-tuned circuit to cover the a-m broadcast band from 500 to 1600 kHz. At 500 kHz the minimum bandwidth should be 25 kHz. The minimum capacitance of the variable tuning capacitor is 30 pF, and stray wiring capacitance in parallel with the capacitor is on the order of 20 pF. Determine the value of L_p, Q, and the capacitance range of the variable capacitor.

solution   The value of L_p is determined at the upper frequency of 1600 kHz and at minimum capacitance of the capacitor, including stray capacitance (30 + 20 = 50 pF). From Eq. (10.12),

$$L = \frac{1}{4\pi^2 f_r^2 C}$$

Substituting known values,

$$\frac{1}{4\pi^2 (1600 \times 10^3)^2 \times 50 \times 10^{-12}} \approx 200 \ \mu\text{H}$$

From Eq. (10.18),

$$Q = \frac{f_r}{\Delta f} = \frac{500}{25} = 20$$

By using the value of $L = 200 \ \mu\text{H}$, the maximum value of C is determined at the lower frequency of 500 kHz. Solving Eq. (10.12) for C gives

$$C = \frac{1}{4\pi^2 f_r^2 L}$$

and substituting known values,

$$= \frac{1}{4\pi^2 (500 \times 10^3)^2 \times 200 \times 10^{-6}} \approx 500 \text{ pF}$$

But 20 pF of stray capacitance is already present; therefore, the maximum value of capacitance is 500 − 20 = 480 pF, and the capacitor range is 30 to 480 pF.

APPLICATION REQUIREMENTS   In some applications for tuned circuits the highest possible degree of selectivity—that is, the lowest bandwidth and the highest Q—is desired. This permits the circuit to do the best job of rejecting signals of unwanted frequencies, even though they may be very close to the desired frequency to which the circuit is tuned. In such cases, the tuned circuit is designed and constructed to preserve the highest possible Q by keeping the effective circuit resistance as low as possible.

There are also cases, namely in so-called wideband applications (of which television receiver circuits are an example), in which care must be taken lest the bandwidth of the tuned circuit be too narrow. This would result in some signals of desired frequencies being rejected, along with signals of undesired frequencies that may be farther from the resonant, or center, frequency of the circuit.

One method of keeping a tuned circuit from having a bandwidth that is too narrow is to deliberately introduce resistance into the circuit, thereby lowering the Q and producing a selectivity curve similar to the wider of the two curves in Fig. 10.17.

10.7 DOUBLE-TUNED (MULTIPLE-TUNED) CIRCUITS (ALSO CALLED STAGGER-TUNED)

As indicated in the preceding section, there are applications of tuned circuits where the circuit must be capable of responding to a wide band of frequencies about the center frequency (the term *center frequency* corresponds to *resonant frequency*, the latter term being used primarily in connection with circuits for narrow-band applications). In such applications, one solution is to resort to the use of double-tuned circuits, or more generally, to multiple-tuned circuits.

Fig. 10.18   Double-tuned (stagger-tuned) circuits.

The terms double-tuned and multiple-tuned circuits are used to refer to combinations of two (double-tuned) or more (multiple-tuned) circuits in which the individual circuits, instead of all being tuned to the same frequency, are tuned to slightly different frequencies for the purpose of producing the desired wideband response.   Figure 10.18 illustrates such a circuit configuration.   It is noted that the tuned circuits are isolated from each other by amplifier stages (other means of isolation are possible).   Although such isolation is not absolutely necessary, if the circuits are allowed to interact with each other directly (that is, if they are *coupled*), the response characteristics are somewhat different.   Coupled tuned circuits are discussed in the next section.

Figure 10.19 illustrates the response of the double-tuned circuit of Fig. 10.18.   Both response curves (selectivity curves) of the individual tuned circuits and their combined response are shown in Fig. 10.19.   This illustration clearly exhibits the desired wideband characteristic.

It is important to note that obtaining wideband response by using multiple-tuned circuits produces an overall selectivity curve which, while being broad at the top (to obtain uniform response), also has steep sides.   The response is said to have good *skirt selectivity*, giving the circuit the ability to reject unwanted frequencies, even though they may be very close to the desired frequency band.   The name skirt selectivity comes from the fact that the sides of a response curve are called "skirts."

The combination of wide bandwidth and good skirt selectivity is the primary reason why multiple-tuned circuits are often preferred over single-tuned circuits with low Q when a wideband response is required.   The advantage of multiple-tuned circuits, in terms of skirt selectivity, is apparent from a comparison of the overall response curve of Fig. 10.19 with the wider of the two curves of Fig. 10.17, which is the selectivity curve of a single-tuned circuit where wide bandwidth is obtained by reducing circuit Q.

There are applications for which the required bandwidth is so wide that two circuits used in a double-tuned configuration are not adequate.   This is because as the spacing between the separate frequencies to which the two circuits are tuned is increased to enlarge the overall bandwidth, a gap, or *valley*, begins to develop in the region between the frequencies to which the two circuits are individually tuned.   That this can occur is

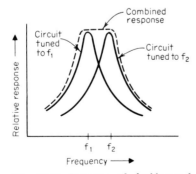

Fig. 10.19 Response of double-tuned (stagger-tuned) circuits.

evident from studying Fig. 10.19 and visualizing what happens to the combined response curve if the individual curves are separated further from each other.

To avoid this valley in the response curve, a three-circuit configuration can be used (triple-tuned circuit), in which a third circuit, tuned to the center frequency, is utilized to "fill in" the valley that otherwise exists when the original two circuits are separated further in frequency to increase overall bandwidth. Figure 10.20 illustrates the response of a triple-tuned circuit.

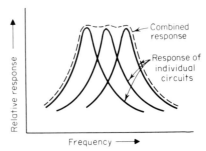

Fig. 10.20 Response of a triple-tuned (stagger-tuned) circuit.

Theoretically, of course, the concept of obtaining increased bandwidth through the use of multiple-tuned circuits can be extended to almost any number of circuits. In fact, arrangements of more than three such circuits are common. However, five circuits are about the usual practical limit. This is because, when bandwidths greater than those obtained with five circuits (quintuple-tuned circuits) are required, other methods exist which are superior from a practical standpoint.

10.8 COUPLED TUNED CIRCUITS

There are numerous applications of tuned circuits in which, instead of the individual circuits being isolated from each other by amplifiers or otherwise, they are deliberately coupled together in such a way that they interact directly on each other. Although the complete mathematical theory of coupled circuits is complex, there are certain phenomena that will be described as being of importance from a practical standpoint.

Figure 10.21 shows the general example of two parallel-tuned circuits coupled together. The symbol M denotes the *mutual inductance* that exists by virtue of the fact that the two circuits are inductively coupled. Although this mutual inductance does not correspond directly to any physical element in the circuit, as do the other circuit parameters L_1, L_2, C_1, and C_2, it is nonetheless important to the performance of the coupled circuit, and cannot be overlooked. Mathematically, mutual inductance is given by the expression

$$M = k \sqrt{L_1 L_2} \tag{10.20}$$

where M = mutual inductance, expressed in the same inductance units (e.g., henrys, millihenrys, etc.) as L_1 and L_2
  L_1 = inductance of the *primary* of the coupled circuit
  L_2 = inductance of the *secondary* of the coupled circuit
  k = coefficient of coupling. The coefficient of coupling is a measure of how many of the primary flux lines actually link with the secondary

It is important to note that the coefficient of coupling, k, has a value between 0 and 1. Thus, the mutual inductance can never be greater than the square root of the product of the two individual inductances, and is ordinarily less, inasmuch as k almost always has a value substantially less than 1.0. (In fact, a coefficient of coupling closely approaching the maximum value of 1.0 is usually found only in low-frequency transformers, in which magnetic cores are used. Coefficients of coupling approaching 1.0 are not found in high-frequency coupled circuits.)

Fig. 10.21 Coupled tuned circuits.

Fig. 10.22 Effect of coupling on selectivity of coupled tuned circuits.

CRITICAL COEFFICIENT OF COUPLING A related concept of importance is the critical coefficient of coupling k_c.

$$k_c = \frac{1}{\sqrt{Q_1 Q_2}} \qquad (10.21)$$

where k_c = critical coefficient of coupling
Q_1 = Q factor of the primary circuit
Q_2 = Q factor of the secondary circuit

The critical coefficient of coupling is that value of k which will produce a maximum current in the secondary winding of the transformer in Fig. 10.21. With the usual values of Q, the critical coefficient of coupling will be quite small.

example 10.8 Assume that the two identical circuits of Example 10.6, each with a Q of 62.8, are coupled together. What is the value of the critical coefficient of coupling k_c?

solution From Eq. (10.21),

$$k_c = \frac{1}{\sqrt{Q_1 Q_2}} = \frac{1}{\sqrt{62.8^2}} = 0.016$$

Practically, critical coupling is the condition for maximum energy transfer between the primary and secondary coupled circuits. This is of importance in many (but not all) applications. (For example, in transmitters or other applications where significant amounts of power are being handled, it is desired to transfer the maximum power from one stage to the next. Furthermore, critical coupling is the largest value of the coupling coefficient for which two circuits tuned to the same frequency will exhibit a single-peaked response. The significance of this point will become more apparent from the discussion that follows.

In some applications of coupled tuned circuits, the application is essentially one involving a transformer consisting of a primary winding L_1 and a secondary winding L_2. The primary objective is to transfer signals from one stage to the next. The transformer is chosen as the most convenient means of accomplishing this signal transfer, and the fact that the two transformer windings are tuned is hardly more than incidental.

In such a case, the coefficient of coupling is usually kept somewhere below the critical value, and the two tuned circuits behave more or less independently of each other, almost as though they are isolated by an intervening amplifier stage. Tuned interstage transformers, such as used in standard broadcast receivers, are an example of coupled tuned circuits with low values of coefficient of coupling.

As the coupling coefficient increases, the effective overall bandwidth also increases. When the coefficient of coupling is at the critical value, the overall bandwidth is the widest possible without departing from the single-peaked response that is typical of a single-tuned circuit. If the coefficient of coupling is increased to a value greater than the critical value, the overall bandwidth continues to increase, but it now has a two-peaked, or *double-hump*, shape. The effect of coefficient of coupling on the selectivity response (bandwidth) of coupled tuned circuits is illustrated in Fig. 10.22.

It is seen from the foregoing discussion and from Fig. 10.22 that coupled tuned circuits, with coefficients of coupling at or above the critical value, represent another means (in addition to the use of multiple-tuned circuits as discussed in the preceding section) of obtaining wideband circuit response while preserving good skirt selectivity. Even with coefficients of coupling so high as to produce a distinct double-humped response, the lack of flatness may not pose a serious overall performance problem. The increased bandwidth thus obtained may be well worth whatever slight performance sacrifice is involved by virtue of the lack of an absolutely flat response.

In other cases, where wide bandwidth is needed but the overall response must be kept flat, the double-humped response of the *overcoupled* (that is, coefficient of coupling in excess of the critical value) circuit can be combined with an isolated single-tuned circuit to produce an overall flat response. The result resembles the triple-tuned circuit response of Fig. 10.20.

10.9 TEMPERATURE COMPENSATION OF TUNED CIRCUITS

Inductors exhibit a *positive temperature coefficient*. This means that as temperature increases, the value of inductance also increases. Temperature coefficient, abbreviated TC, is generally expressed as a change in the value of a component in parts per million per degree Celsius change in temperature (ppm/°C). Capacitors can have a positive, negative, or zero temperature coefficient. The value of a component with a negative temperature coefficient *decreases* with increasing temperature and increases with decreasing temperature. A zero temperature coefficient means that there is no significant change in component value with reasonable changes in temperature.

Fig.   10.23 Temperature-compensating capacitor C_c across a parallel-tuned circuit.

With the proper application of temperature-coefficient capacitors in a tuned circuit, the resonant frequency of the circuit does not change with temperature. This is called *temperature compensation*. If the tuned circuit is not compensated, the values of L and C change with temperature, and this results in a change in resonant frequency. This undesired circuit behavior is referred to as *temperature drift*. Temperature drift is very detrimental to the performance of precisely tuned circuits, such as those used in oscillators.

If the temperature coefficient is positive, the letter P precedes the value of the coefficient. If the temperature coefficient is negative, the letter N precedes the coefficient. For zero temperature coefficient, designation NPO is used.

Consider the parallel-tuned circuit of Fig. 10.23 where it is assumed that there is negligible resistance. Capacitor C_c is a temperature-compensating capacitor. Let the total parallel capacitance be $C_T = C_p + C_c$. Including the effect of temperature coefficient, the value of the parallel capacitor combination is expressed by

$$C_T(1 + TC_T)$$

and the inductor by

$$L_p(1 + TC_L)$$

where TC_T = temperature coefficient for the parallel capacitors
    TC_L = temperature coefficient for the inductor

If these values are substituted in Eq. (10.12),

$$f_r = \frac{1}{2\pi \sqrt{LC}}$$

we obtain

$$f_r = \frac{1}{2\pi \sqrt{L_p(1 + TC_L)C_T(1 + TC_T)}}$$ (10.22)

Expanding the term under the radical sign of Eq. (10.22), we obtain

$$L_p C_T (1 + TC_L + TC_T + TC_L TC_T)$$ (10.23)

Because TC_L and TC_T are extremely small (ppm), their product $TC_L TC_T$ is much less than either TC_L or TC_T and, therefore, may be neglected in Eq. (10.23). This results in

$$L_p C_T (1 + TC_L + TC_T)$$ (10.24)

If $TC_L = -TC_T$, Eq. (10.24) is equal to $L_p C_T$, and the resonant frequency is *independent* of temperature.

SELECTING A TEMPERATURE-COMPENSATING CAPACITOR   Capacitors may be combined in series or parallel, as shown in Fig. 10.24 to obtain a desired temperature

(a)

(b)

Fig. 10.24 Capacitors in parallel and in series: (a) Capacitors in parallel. (b) Capacitors in series.

compensation. For a number of capacitors in parallel (Fig. 10.24a), the equivalent temperature coefficient TC_P for the resulting combination is

$$TC_P = \frac{C_1(TC_1) + C_2(TC_2) + \cdots + C_n(TC_n)}{C_1 + C_2 + \cdots + C_n}$$ (10.25)

For the capacitors in series of Fig. 10.24b, the equivalent temperature coefficient TC_S is

$$TC_S = C_T \left(\frac{TC_1}{C_1} + \frac{TC_2}{C_2} + \cdots + \frac{TC_n}{C_n} \right)$$ (10.26)

where TC_1 = temperature coefficient of capacitor C_1
  TC_2 = temperature coefficient of capacitor C_2, etc.

$$C_T = 1/ \left(\frac{1}{C_1} + \frac{1}{C_2} + \cdots + \frac{1}{C_n} \right)$$

example 10.9   Two capacitors connected in parallel have the following values: $C_1 = 50\,\text{pF}$, $TC_1 = P200$, $C_2 = 30\,\text{pF}$, and $TC_2 = N500$. Find the temperature coefficient for the combination and the total capacitance.

solution   Remember that P refers to a positive (+) temperature coefficient, and N refers to a negative (−) temperature coefficient.   For two capacitors in parallel, Eq. (10.25) reduces to

$$TC_p = \frac{C_1(TC_1) + C_2(TC_2)}{C_1 + C_2} \tag{10.27}$$

Substitution of the given problem values in Eq. (10.27) yields

$$TC_p = \frac{50 \times 200 + 30(-500)}{50 + 30}$$

$$= \frac{-5000}{80} = -62.5 \text{ ppm} = N62.5$$

The total capacitance $C_T = C_1 + C_2 = 50 + 30 = 80$ pF.

example 10.10   For the parallel-tuned circuit of Fig. 10.23, assume that inductor L_p has a positive temperature coefficient of +30 ppm, $C_p = 200$ pF, $TC_p = +50$ ppm, and $C_c = 30$ pF. Determine the temperature coefficient TC_c for the temperature-compensating capacitor.

solution   From Eq. (10.24) it was found that $TC_T = -TC_L$ for temperature compensation of a tuned circuit. Substitution of this and the problem values in Eq. (10.27) yields

$$-30 = \frac{20(TC_c) + 200(50)}{220}$$

Solving for TC_c, we obtain

$$-30 \times 220 = 20(TC_c) + 10\,000$$

or

$$TC_c = \frac{16\,600}{20} = -830 \text{ ppm} = N830$$

A 20 pF, $N830$ capacitor is therefore required for temperature compensation.

PRACTICAL CONSIDERATIONS   Ceramic capacitors are available in a wide range of temperature coefficient values, and they are, therefore, generally used for the temperature compensation of tuned circuits. Other types of capacitors have tolerances on their values, and their temperature coefficient may not be constant with changes in temperature value.   It is usually required to make further calculations or take temperature runs to determine the tracking of a compensated circuit with temperature.

Chapter 11

Filters

11.1 INTRODUCTION

The term *filter* is used to describe a wide variety of circuits that are frequency-selective. Certain frequencies will pass through a given filter while others will be attenuated. Electric filters are analogous to lint filters on automatic dryers that allow air to pass through but trap the lint. Another analogy can be made with the fuel-line filter of an automobile. It traps dirt, but allows gasoline to go to the carburetor. By way of comparison, electric filters will pass some frequencies and trap (or attenuate) others.

Knowing which frequencies are to pass and which are to be rejected, the circuit designer must decide which filter *configuration* is to be employed, and he must calculate the values of the components used. (The configuration is the method of electrically interconnecting the parts.) In this chapter we will discuss the advantages and disadvantages of the different configurations, and the methods used for determining their component values.

11.2 CLASSIFICATION OF FILTERS

There are many different ways of classifying or identifying filters. For example, they are sometimes identified by the layout of the components. Thus, a *pi filter* looks like the Greek letter π, and a *ladder circuit* looks like a ladder. Filters may be identified by the method used to design them. A *constant-k* filter is designed on the basis of a constant k which is related to the impedances of the circuit. An *m-derived filter* is based on a constant m which is derived from the constant k. Another way of identifying filters is by the shape of their characteristic curves. Thus, a *sharp cutoff filter* makes a very sharp distinction between which frequencies will pass through it and which will not. In some cases, the name of the person who originated the design or method of calculation is used for identifying the filter circuit. *Chebyshev filters* and *Butterworth filters* are examples of these.

One widely used scheme for classifying filters depends on the range of frequencies that the filter passes and rejects. According to this view, there are only four basic filter types: *low-pass, high-pass, bandpass,* and *band rejection*. All filter configurations can be placed in one of these categories.

The characteristic curves of Fig. 11.1 show the differences among the four basic categories. The low-pass filter of Fig. 11.1a permits all frequencies below f_c to pass, but rejects all frequencies above f_c. The high-pass filter (Fig. 11.1b) rejects all frequencies below f_c but passes all others. The bandpass filter (Fig. 11.1c) allows only a limited range of frequencies to pass. The band reject filter (Fig. 11.1d) permits all frequencies, except those in a limited range, to pass.

Fig. 11.1   Characteristic curves for four basic types of filters: (a) Low-pass filter. (b) High-pass filter. (c) Bandpass filter. (d) Band-reject filter; also called bandstop filter.

Filters composed of R, L, and C elements may be referred to as *passive filters*. Another type that utilizes a high-gain operational amplifier is the *active filter*. Besides an amplifier, an active filter also requires an external power source. The most commonly used active filter contains R and C elements.

The principal advantage of the active filter over the passive type is that inductors are not required in the former. This is especially important at the lower frequencies where inductors become excessively large and lossy.

APPLICATIONS   The applications for filters are so numerous in electronics that only a few examples will be cited here. In terms of frequency range, filters are used at frequencies from less than a hertz in seismology to many gigahertz in microwaves.

In the rectification of alternating current to a unidirectional current it is required to filter the ripple in order to obtain an output approximating pure direct current. The simplest filter for this purpose is a capacitor across the supply output terminals, as shown in Fig. 11.2. In this application, a large-value capacitor is placed in parallel with the load resistance (Fig. 11.2a). The capacitor appears as an "open" circuit to direct current, while to alternating current it looks almost like a "short" circuit. The capacitor filter is most effective for small load currents (large values of load resistance). An example of such an application is the high-voltage power supply for a cathode-ray tube in which the anode current is typically on the order of a few hundred microamperes. As load current increases, the ripple increases and the regulation becomes poor.

Fig. 11.2 Filters used with power supplies: (a) Pure C. (b) L section. (c) Pi.

Where appreciable load current is drawn from the supply, an L-section or pi filter (Fig. 11.2b and c) is required. In the L-section filter a large value of inductance, typically 10–30 H, referred to as a *choke*, is placed in series with the load resistance.

Placing a capacitor before the L-section configuration yields the pi filter of Fig. 11.2c. Greater ripple attenuation and dc output voltage are realized with this circuit than with the L-section filter.

Figure 11.3 illustrates some examples of filters found in communications circuits.

The RC-coupled network of Fig. 11.3a, employed in cascading ac amplifier stages, behaves as a filter. Low frequencies and direct current are attenuated, while the higher frequencies pass with minimum attenuation. Another example of a filter is the capacitor bypassing the emitter resistance in Fig. 11.3b. This circuit also tends to attenuate the lower frequencies. In fact, the low-frequency response of an ac amplifier is largely determined by the coupling and bypass networks which behave as filters.

At higher frequencies, the shunt capacitance of the active device, such as a transistor, and shunt stray and wiring capacitances, limit the high-frequency response of the amplifier. The shunting capacitance is lumped as capacitor C_s in Fig. 11.3c. Capacitor C_s and its associated resistors behave as a filter that attenuates the higher frequencies. This accounts for the falling off of high frequencies in an amplifier.

Fig. 11.3 Examples of filters found in communications circuits: (a) RC coupling. (b) Bypassing emitter resistor. (c) Shunt capacitance across amplifier output.

11.3   FILTER TERMINOLOGY

Before making a detailed study of filter theory, we will first review some of the more important terms that are applicable to this area.

The *configuration* of a circuit is simply the layout of its parts in relation to one another. An example is the pi filter shown in Fig. 11.2c. The parts are arranged on the schematic diagram in the form of the Greek letter π, and we can say that the filter has that configuration.

All filter circuits can be arranged into one of two possible configurations: the *lattice* and the *ladder*. These are illustrated in Fig. 11.4. The filter *elements* (that is, the inductors, capacitors, and resistors) are shown as separate impedances labeled Z. The most important difference between the ladder and the lattice is symmetry. The lattice,

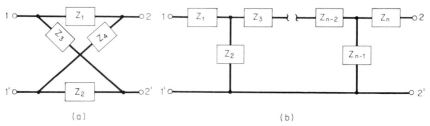

(a)   (b)

Fig. 11.4   Typical filter configurations: (*a*) The bridge, or lattice, circuit. (*b*) The ladder circuit.

or bridge, is *symmetrical.* This means that each terminal has a mirror-image counterpart on the opposite side of the circuit. An example is the equivalence of Z_1 and Z_2 or Z_3 and Z_4 in Fig. 11.4a. The ladder circuit does not have this property of symmetry. Because the lattice is symmetrical, it may drive (or be driven by) balanced loads, such as center-tapped transformers. The ladder can only drive unbalanced loads, such as a grounded resistor, tube grid, or transistor base.

The term *balance* always means with respect to ground. Figure 11.4b shows a ladder-circuit configuration. As an additional way of distinguishing between the circuits, any of the terminals of the lattice may be grounded without affecting the circuit performance while only terminals 1' and 2' of the ladder can be grounded.

Although there may be variations of these configurations, the ladder and the lattice are the basis of all filter structures. The type of element may change, i.e., capacitive, inductive, or resistive, but the configuration will not.

Insertion loss is the loss in signal strength that the frequencies in the passband experience in passing through the filter. (The *passband* of a filter is the range of frequencies that are supposed to be able to pass through the circuit.) If the filter were absent, and the source and load were connected directly to each other, the output signal would increase by the amount of the insertion loss.

The bandwidth of a filter is the frequency difference between two points on the filter-response curve that have a specified insertion loss. The points on the curve generally used to define bandwidth are the half-power, or -3-dB, points. These are the points where the filter response is down 3 dB from the maximum point on the curve. On the curve of Fig. 11.5, the bandwidth is between f_1 and f_2, where f_1 is the lower frequency at a point 3 dB down from maximum, and f_2 is the upper frequency at a point 3 dB down from maximum. The formula for bandwidth is

$$\text{Bandwidth, Hz} = f_2 - f_1 \qquad (11.1)$$

where f_1 and f_2 are at the -3-dB points.

example 11.1   Two points on a response curve that are at the -3-dB points are at 10 and 1000 Hz for f_1 and f_2 respectively. What is the bandwidth?

solution

$$\begin{aligned}
\text{Bandwidth, Hz} &= f_2 - f_1 \\
&= 1000 - 10 \text{ Hz} = 990 \text{ Hz}
\end{aligned}$$

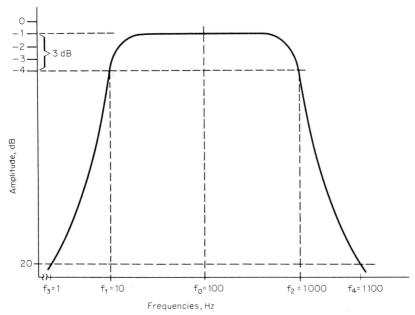

Fig. 11.5 A typical bandpass filter response.

Instead of expressing the bandwidth as the difference between two frequencies, it is sometimes expressed as a percent of frequency (*percent bandwidth*). For example, if a 100-kHz bandpass filter has a bandwidth of 10 percent, its bandwidth is 10 percent of 100 kHz, or 10 kHz. This can be expressed as ±5 percent, or ±5 kHz with respect to 100 kHz.

The *center frequency* F_0 may be defined as either the *arithmetic* or the *geometric center*. For narrow bandwidths these are practically equal. The formulas are:

$$F_0 = \frac{f_1 + f_2}{2} = \left(\begin{array}{c}\text{Arithmetic center} \\ \text{frequency, Hz}\end{array}\right) \tag{11.2}$$

$$F_0 = \sqrt{f_1 \times f_2} = \left(\begin{array}{c}\text{Geometric center} \\ \text{frequency, Hz}\end{array}\right) \tag{11.3}$$

Using our previous example,

$$F_0 = \frac{1000 + 10}{2} = 505 \text{ Hz} \qquad \text{(arithmetic center)}$$

$$F_0 = \sqrt{1000 \times 10} = 100 \text{ Hz} \qquad \text{(geometric center)}$$

If a filter-response curve is plotted on graph paper having the abscissa (horizontal axis) marked in evenly spaced frequencies, then the arithmetic center frequency is exactly halfway between the −3-dB points. As an example, for the response curve of Fig. 11.1c the arithmetic center is 455 kHz.

If a filter-response curve is plotted on graph paper having a logarithmic abscissa, then, the geometric center frequency will be exactly halfway between the −3-dB points. For the response curve of Fig. 11.5 the abscissa is logarithmically marked. Note that the same distances are used for 10 to 100 Hz as for 100 to 1000 Hz. The geometric center on this curve is 100 Hz, which is seen to be halfway between the −3-dB points.

The arithmetic center is used only for very narrow bandwidths; then the arithmetic and geometric centers are equal. In any case, the geometric center is always correct, and when we speak of the center frequency, we shall mean the geometric center.

Selectivity is a measure of the effectiveness of a filter in performing its function. If the filter were perfect, it would have no insertion loss in the passband and infinite attenuation at all other frequencies (the *stop band*). In practice, it takes a finite band of frequencies to change from very little attenuation to a greater amount.

One way to measure selectivity is by the *shape factor*. Shape factor is often defined as the ratio of the −3-dB bandwidth to the bandwidth where the insertion loss is 20 dB. (The −6- and −60-dB bandwidths are also used.) The closer this ratio is to a value of 1.0, the greater the selectivity of the filter. The shape factor for the response of Fig. 11.5 is

$$\text{Shape factor} = S = \frac{f_4 - f_3}{f_2 - f_1} \tag{11.4}$$

where f_4, f_3 = points on the response curve that are down 20 dB from maximum
$\quad\;\, f_2, f_1$ = points on the response curve that are down 3 dB from maximum

example 11.2 Using the values of $f_1, f_2, f_3,$ and f_4 in Fig. 11.5, what is the shape factor of the selectivity curve?

solution

$$S = \frac{f_4 - f_3}{f_2 - f_1} = \frac{1100 - 1}{1000 - 10} = 1.11 \qquad\qquad Answer$$

When lumped components are used to make a filter, there are a number of undesirable distributed elements — called *parasitic elements* — that cause attenuation of the *desired* frequencies. Parasitic elements may also change the shape of the response curve from the desired characteristic. A few examples of parasitic elements are: losses in iron cores of chokes, distributed wiring capacitance, capacitance between windings of a choke, resistance of connecting wires, and leakage inductance. Parasitic elements cannot be eliminated completely. In some cases, they may have to be considered during the design procedure.

This is not a complete list of all terms used to describe filters, but it includes the most important ones for evaluating filters.

11.4 FILTER COMPONENTS

Passive filters may be classed as *resistive-reactive* type or *totally reactive*. The resistive-reactive type is called *RC* or *RL*, depending whether an inductor and resistor (*RL*) or capacitor and resistor (*RC*) are used. The totally reactive filter uses capacitors and inductors, with resistance present only as an undesired distributed component. These filters are also called *LC* types because of the components utilized in their construction.

The *RL* and *RC* types are sometimes employed to generate high-pass or low-pass responses, but special modifications must be made to these in order to generate the bandpass or band-rejection response. Even then they are not as efficient as the *LC* type, owing to their inherent resistance dissipation.

11.5 METHODS OF DESIGNING LC FILTERS

Two different approaches are used for designing *LC* filters. One method is to treat the filter as a transmission line. This approach is called the *image-parameter design* method. For this method, the lumped components (which in *LC* filters are inductors and capacitors) are described in terms of the distributed components in a transmission line. The other approach to filter design is called the *network method of design*. (It is also called the *insertion loss method*, or *polynomial method*, or the *exact method*.) It deals directly with the filter as a circuit having parameters instead of dealing with the transmissionline.

IMAGE-PARAMETER DESIGN The image-parameter method of design was developed during the conduction of early experiments on telephone transmission lines. It was found that the attenuation of telephone lines increased with frequency, owing to the large amount of distributed capacitance, especially in the early types of lines. Inserting inductors reduced the attenuation and made it fairly constant over a range of frequencies. It was also noted that the attenuation increased very quickly at a particular frequency called the *cutoff frequency*.

As the requirements for bandwidth in the transmission line increased, various schemes were used to simultaneously increase bandwidth and keep the attenuation low. As a result of experiments, image-parameter filters were developed. When the equivalent circuits of loaded transmission lines were constructed with lumped components, it was found that they had different characteristics from those of the loaded line. The bandwidth and attenuation were not the same as in the transmission line because the filter is only an approximation of a transmission-line circuit. This is one of the difficulties in designing such filters.

The image-parameter filter has evolved into two basic types: *the constant-k type*, and *the m-derived type*. The constant-*k* type is the result of the first attempt to build an artificial transmission line. When component values of the constant-*k* type are modified in a certain way so as to vary the impedance and the shape factor, the result is an *m*-derived filter.

NETWORK METHOD OF DESIGN   Modern mathematics and the advent of the computer have done much to change filter design. Tables and nomographs obtained through the use of computers make it possible to obtain a desired filter characteristic with relative ease. Using only simple mathematics, it is possible to design a filter for almost any application and shape factor.

The image-parameter method of designing a filter is based on transmission lines. It attempts to simulate a transmission line with inductors and capacitors, that approximates the distributed constants in a coaxial cable, and obtain a certain impedance and frequency response simultaneously. For a long time it has been known that filters could be designed from response-curve formulas. The difficulty was that the equations were so long they were nearly impossible to solve. A computer was required for their solution, but at that time computers were not available for such applications.

When more and more computers became available, the required equations were solved, and tables of elements were generated that could be used in many different filter designs. It was desirable to make the tables useful for *any* frequency or impedance level; so they were developed in such a way that this information could be added later. Such tables are said to be *normalized*. The values are referenced to 1 Ω and 2π Hz (1 rad/s), meaning that the impedance and the cutoff frequency are 1 Ω and 2π Hz, respectively, for the filter elements listed. This is equivalent to the mechanical operation of *scaling* an object. To change the size of a scaled object, you simply divide or multiply by the number of times bigger or smaller that the object is to be. For example, if a scale model of a boat is $\frac{1}{25}$ of the actual size, then each and every dimension on the model must be multiplied by 25 to obtain its actual size. In a similar manner, the tables used for the network method of design can be considered to be scaled down. In order to get the desired values, it is necessary to multiply by the desired impedance in ohms and the desired frequency in hertz. By making the tables this way, they apply to *any* impedance and frequency.

We will give examples of each method of design, starting with the image-parameter approach. The constant-*k* and *m*-derived filters are designed by this method.

11.6 THE SIGNIFICANCE OF FILTER IMPEDANCE

Constant-*k* filters can be made in any of the three configurations shown in Fig. 11.6. Each of these is a form of ladder network, and is identified according to the alphabet letter that it forms. Thus we have an L, a T, and a Π section. A number of these sections can be added to obtain a sharper cutoff (or other desired characteristic).

An important property of constant-*k* filters—and for that matter, *all* types of filters—is their impedance. This is because of the *maximum power-transfer theorem*, which states that the maximum possible amount of power that can be delivered to a load occurs when the load impedance "matches" the internal impedance of the generator. For ac generators, the match must be achieved by using a *conjugate impedance*. Thus, if the internal impedance of the generator is $7 + j11$ Ω, then maximum power will be delivered to the load impedance when its value is $7 - j11$ Ω.

Figure 11.7 shows the effect of the filter on the impedance match. Suppose we have

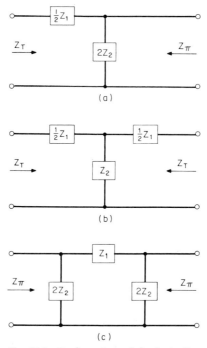

Fig. 11.6 Configurations of the basic filter sections: (*a*) An L section. (*b*) A T section. (*c*) A pi section.

a circuit like that of Fig. 11.7*a*, in which the load impedance Z_L is matched to the internal impedance Z_i of the generator G. In accordance with the maximum power-transfer theorem, the greatest amount of power (for this generator) is being delivered to the load. Now if we insert a filter circuit between the generator and load, as shown in Fig. 11.7*b*, then the presence of the filter can result in an improper match of impedances. (It is important to note that inserting a transmission line between a generator and a load can also produce a mismatch, and therefore transmission-line impedance and filter impedance are both important, and have a similar effect on the circuit into which they are inserted.)

Suppose that the internal impedance and the load impedance of the circuit in Fig.

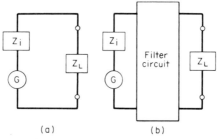

Fig. 11.7 Illustrating the importance of filter circuit impedance: (*a*) A circuit designed for maximum power transfer. (*b*) Unless properly designed, the filter circuit could unbalance the impedance match.

11.7a are conjugates, and they both have the same impedance — say 500 Ω. If the input impedance to the filter circuit is also 500 Ω — looking from either end — then, no impedance mismatch will occur between the generator and the load.

Now suppose the internal impedance of the generator is different from the load impedance. If the filter is designed in such a way that the generator operates into the correct impedance, *and* the load operates into the correct impedance, then the filter is actually serving two functions: elimination of unwanted frequencies and matching of impedances. Whenever a filter is designed in such a way that the generator "sees" the conjugate of its own internal impedance when looking into the filter output terminals, then the filter is said to be operating on an *image impedance basis.*

The *characteristic impedance* of a filter, and also of a transmission line, is the value of load impedance that must be connected across the output terminals in order to match the load to the filter (or line). This is also known as the *surge impedance.* If a filter, or transmission line, is terminated in a resistance that is equal to its impedance, then *all* the energy moving through the filter (or through the line) will be dissipated by that resistance.

The characteristic impedance Z_o of a filter or transmission line is given by the equation

$$Z_o = \sqrt{Z_{sc}Z_{oc}} \qquad (11.5)$$

where Z_{sc} = input impedance with the output terminals short-circuited together
Z_{oc} = input impedance with the output terminals open

11.7 DESIGN OF CONSTANT-k FILTERS

The application of the constant-k, as well as other LC filters, is generally practical over a range of frequency up to 100 mHz. Their applications in communication circuits include the limiting of the signal spectrum to a receiver (preselector), separating or combining the spectrum so that parts of it may be individually processed (diplexers and diversity receiver combiners), separating r-f signals, and reducing interference.

Although constant-k filters are easy to design and make, they have a number of shortcomings. Their attenuation is low near its cutoff frequency, and their image impedance is not constant with frequency. Also, the phase shift in the passband is not proportional to frequency.

It is apparent when we examine the circuits of Fig. 11.6 that the impedance of a filter circuit will normally change with frequency. As an example, assume that Z_1 is an inductor and that both impedances marked $2Z_2$ are capacitors in the circuit of Fig. 11.6c. If the input frequency increases, then the inductive reactance increases, and at the same time the capacitive reactance decreases. The result is a change in filter impedance. reactance decreases. The result is a change in filter impedance.

If the inductance and capacitance are chosen so that the increase in inductive reactance is exactly offset by the decrease in capacitive reactance, then their product will be a constant, which is called the constant k. Expressed mathematically,

$$k = \sqrt{Z_1Z_2} \qquad (11.6)$$

where k = an impedance independent of frequency
Z_1, Z_2 = series and shunt impedances of Fig. 11.6

The attenuation of all frequencies will be a constant up to the cutoff frequency. If the Q of the reactive components in the filter is high, then the cutoff will be quite sharp. As the Q is made lower and lower, the reduction in attenuation beyond cutoff becomes more and more gradual.

LOW-PASS FILTER DESIGN   Figure 11.8 shows three possible configurations for a constant-k low-pass filter. To design such a filter, it is necessary to know the cutoff frequency f_c—that is, the frequency the filter is to pass. This is designated f_c in Fig. 11.9. It is also necessary to know the impedance of the load to be connected across the output terminals of the filter. This load, which is designated as Z_o, must be the conjugate of the characteristic impedance of the filter in order for maximum power transfer to occur. Looking in from the output terminals, this is also the image impedance.

Fig. 11.8 Three forms of constant-k low-pass filters.

Fig. 11.9 Characteristic curve of a constant-k low-pass filter.

By knowing the value of f_c and Z_o, it is possible to design the filter from the following equations which are given here without derivation:

$$L = \frac{Z_o}{3.14f_c} \qquad (11.7)$$

$$C = \frac{1}{3.14f_c Z_o} \qquad (11.8)$$

where L = inductance, H
$\quad\quad\;\; C$ = capacitance, F
$\quad\quad\;\; Z_o$ = load impedance, Ω

Once we have calculated the required values of L and C, we can verify that this is a constant-k filter by substituting the values into the equation

$$Z_o = k = \sqrt{\frac{L}{C}} \qquad (11.9)$$

where Z_o, L, C = values in Eqs. (11.8) and (11.7)
$\quad\quad\;\; k$ = constant k, and is equal to the image impedance for all frequencies that the filter is designed to pass

The following example illustrates the design of a low-pass constant-k filter. The cutoff frequency is given as 1000 Hz. All frequencies above 1000 Hz are attenuated. A filter of this characteristic might be used in narrow-band transmission of information.

example 11.3 Design a constant-k low-pass filter that will attenuate frequencies above 1000 Hz (f_c = 1000 Hz). The load across the filter output terminals, Z_o, is a pure resistance of 600 Ω. In order to obtain sharp cutoff characteristics, use a three-section ladder configuration.

solution The greater the number of filter sections used, the sharper the cutoff—hence the three-section ladder. Not all applications could justify the cost of components, and in many cases it is not necessary that the cutoff be sharp.

Figure 11.10 shows the circuit. The three-section filter can be made by combining two L sections with a T section, as shown in Fig. 11.10a. The first step is to calculate the required value of L from Eq. (11.7).

$$L = \frac{Z_o}{3.14f_c}$$

$$= \frac{600}{3.14 \times 1000} = 0.191 \text{ H} \qquad (11.7)$$

As shown in Fig. 11.10a, each inductor in this illustration actually has a value of $L/2$, or 0.095 H.

The next step is to calculate the required value of C from Eq. (11.8).

$$C = \frac{1}{3.14 f_c Z_o}$$

$$= \frac{1}{3.14 \times 1000 \text{ Hz} \times 600 \text{ }\Omega} = 0.53 \text{ }\mu\text{F} \tag{11.8}$$

Each capacitor in Fig. 11.10a actually has a value of $C/2$, or 0.265 μF.

Figure 11.10b shows the circuit with the calculated component values. When the sections are combined to produce the required ladder filter, the two halves of L are combined to produce $L(L/2 + L/2 = L)$ as shown in Fig. 11.10c.

Fig. 11.10 An example of a constant-k filter design: (a) Prototype. (b) Element values. (c) Composite filter design.

If this is truly a constant-k filter, then the characteristic impedance will be 600 Ω, as calculated from Eq. (11.9).

$$Z_o = k = \sqrt{\frac{L}{C}} = \sqrt{\frac{0.191}{0.53 \times 10^{-6}}}$$

$$\approx \sqrt{360\ 000} \approx 600 \text{ }\Omega$$

It will not be possible to obtain capacitors and inductors with the precise computed values for the filter. Any compromise in component values will also result in a compromise in shape factor and other characteristics.

The impedance of the filter must be identical with the load impedance because the circuit is the same when viewed from either end. In practice, an improved impedance match to the source and load can be obtained by adding an m-derived half-section at each end of the constant-k filter of Fig. 11.10c. These added half-sections would be designed to match the image impedance of the filter to the resistive terminations of the source and the load.

HIGH-PASS CONSTANT-k FILTER DESIGN   The arrangement of components for a high-pass constant-k filter design and the resulting characteristic curve are shown in Fig. 11.11. Figure 11.11a shows the three possible configurations for a constant-k high-pass filter. As with the low-pass filters, these configurations are named according to the letter of the alphabet that they are similar to: T, π, and L. The equations for calculating the value of L and C are given here without derivation.

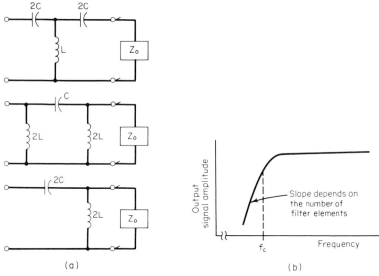

(a)   (b)

Fig. 11.11   Circuits and characteristic curves of constant-k high-pass filter: (a) Three forms of constant-k high-pass filter circuits.   (b) Characteristic curve.

$$L = \frac{Z_o}{12.6 f_c}$$ (11.10)

$$C = \frac{1}{12.6 f_c Z_o}$$ (11.11)

where L = inductance, H
       C = capacitance, F
       Z_o = load impedance, Ω

Components L and C are identified in the circuit of Fig. 11.11a.   The characteristic impedance is Z_o, which is also the value of k, and also the value of the load impedance. The cutoff frequency is f_c and its meaning is indicated in Fig. 11.11b.

BANDPASS FILTER DESIGN   The constant-k bandpass filter circuit and its characteristic curve are shown in Fig. 11.12.

The series circuits in Fig. 11.12a (comprised of L_1 and C_1) are resonant to the frequency being passed.   Actually, when L_1 and C_1 are tuned to the center frequency of the passband, there is very little opposition at that frequency.   The parallel circuit

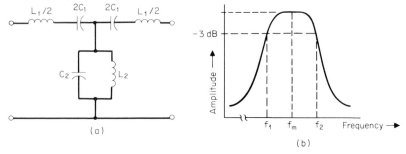

(a)   (b)

Fig. 11.12   Bandpass filter and its characteristic curve: (a) Constant-k bandpass filter.   (b) Bandpass filter characteristic curve.

(comprised of L_2 and C_2) is virtually an open circuit to the center frequency – which is also the parallel resonant frequency.

The characteristic curve of this filter circuit, shown in Fig. 11.12b, also defines the frequencies used in the calculations of inductance and capacitance. The equations for L_1, L_2, C_1, and C_2 are given here without derivation.

$$L_1 = \frac{Z_0}{3.14(f_2 - f_1)} \tag{11.12}$$

$$L_2 = \frac{(f_2 - f_1)Z_0}{12.6 f_1 f_2} \tag{11.13}$$

$$C_1 = \frac{f_2 - f_1}{12.6 f_1 f_2 Z_0} \tag{11.14}$$

$$C_2 = \frac{1}{3.14(f_2 - f_1)Z_0} \tag{11.15}$$

where L_1, L_2 = are identified in Fig. 11.12a and are measured in henrys
$\quad$ C_1, C_2 = are identified in Fig. 11.12a and are measured in farads
$\quad$ f_1, f_2 = limits of bandwidth at the −3-dB points
$\quad$ Z_0 = characteristic impedance, Ω

The constant k is equal to the image impedance, and is defined by the equation

$$k = \sqrt{\frac{L_1}{C_2}} \tag{11.16a}$$

$$= \sqrt{\frac{L_2}{C_1}} = Z_0 \tag{11.16b}$$

The geometric center frequency f_m is given by the equation

$$f_m = \sqrt{f_1 f_2}$$

$$f_m = \frac{1}{6.28 \sqrt{L_1 C_1}}$$

$$= \frac{1}{6.28 \sqrt{L_2 C_2}} \tag{11.17}$$

In the next example, the design of a bandpass filter is illustrated. Such a filter may find application in a spectrum analyzer, where a band of frequencies is to be viewed at the exclusion of all other frequencies.

example 11.4 (a) Design a bandpass filter for passing frequencies between 90 and 110 kHz. The load is 1000 Ω. (b) Calculate its center frequency.

solution (a) The bandpass filter circuit is shown in Fig. 11.12a. The load impedance is 1000 Ω, and this must equal the characteristic impedance Z_0. The lower frequency limit f_1 is 90 kHz, and the upper frequency limit is 110 kHz. With these values, the required inductances, L_1 and L_2, and the capacitance values, C_1 and C_2, can be calculated.

$$L_1 = \frac{Z_0}{3.14(f_2 - f_1)}$$

$$= \frac{1000 \ \Omega}{3.14(110 \ \text{kHz} - 90 \ \text{kHz})} = 15.9 \ \text{mH} \tag{11.12}$$

$$L_2 = \frac{(f_2 - f_1)Z_0}{12.6 f_1 f_2}$$

$$= \frac{(110 \ \text{kHz} - 90 \ \text{kHz})(1000 \ \Omega)}{12.6(90 \ \text{kHz})(110 \ \text{kHz})} = 160 \ \mu\text{H} \tag{11.13}$$

$$C_1 = \frac{f_2 - f_1}{12.6 f_1 f_2 Z_0}$$

$$\tag{11.14}$$

$$= \frac{110 \ \text{kHz} - 90 \ \text{kHz}}{12.6(90 \ \text{kHz})(110 \ \text{kHz})(1 \ \text{k}\Omega)} = 160 \ \text{pF}$$

$$C_2 = \frac{1}{3.14(f_2 - f_1)Z_o}$$

$$= \frac{1}{3.14(20 \text{ kHz})(1 \text{ k}\Omega)} = 0.0159 \ \mu\text{F} \qquad (11.15)$$

If these values are correct, then the value of the constant k will be equal to the characteristic impedance of the filter. This can be checked with Eq. (11.16a):

$$k = \sqrt{\frac{L_1}{C_2}} = \sqrt{\frac{15.9 \times 10^{-3}}{0.0159 \times 10^{-6}}}$$

$$k = Z_o = 1000 \ \Omega$$

The value of k can also be computed from Eq. (11.16b):

$$k = \sqrt{\frac{L_2}{C_1}} = \sqrt{\frac{160 \times 10^{-6}}{160 \times 10^{-12}}}$$

$$k = Z_o = 1000 \ \Omega$$

The value of k has been calculated in two different ways, using L_1 and C_2, and also L_2 and C_1. In both cases it is equal to the characteristic impedance required to match the filter to the load. This, in turn, is the image impedance as seen by the load.

(b) The center frequency of the band of frequencies passed by the filter is obtained from Eq. (11.3).

$$F_0 = \sqrt{f_1 \times f_2} = \sqrt{90 \times 110 \text{ kHz}} = 99.5 \text{ kHz}$$

This is the geometric center frequency which is sometimes designated f_m, the subscript m standing for midfrequency. The arithmetic frequency, when calculated from Eq. (11.2) is 100 kHz, which is close to the geometric mean.

BAND-REJECTION FILTER DESIGN Band-rejection filters are sometimes called *band-stop filters*, or *notch filters*.

Figure 11.13 shows a constant-k band-rejection filter circuit and its characteristic curve. The parallel-tuned circuits, L_1 and C_1, allow all frequencies to pass through the

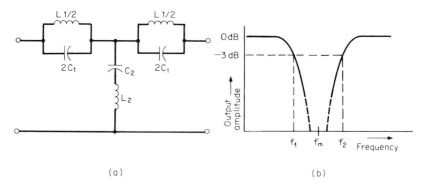

(a)   (b)

Fig. 11.13 Band-stop filter and its characteristic curve: (a) Band-rejection filter. (b) Filter response.

filter except the frequencies near resonance. The resonant frequency of these tuned circuits is the center frequency of the band being rejected. The series-resonant circuit, L_2 and C_2, short-circuits the unwanted frequency to ground. The width of the band being rejected is a direct function of the circuit Q. The slope of the curve — that is, the sharpness of tuning — is increased by adding filter sections.

The equations for L_1, L_2, C_1, and C_2 are given here without derivation.

$$L_1 = \frac{(f_2 - f_1)Z_o}{3.14 f_1 f_2} \qquad (11.18)$$

$$L_2 = \frac{Z_o}{12.6(f_2 - f_1)} \qquad (11.19)$$

$$C_1 = \frac{1}{12.6(f_2 - f_1)Z_o} \tag{11.20}$$

$$C_2 = \frac{f_2 - f_1}{3.14 f_1 f_2 Z_o} \tag{11.21}$$

where L_1, L_2 = are identified in Fig. 11.13a and are measured in henrys
C_1, C_2 = are identified in Fig. 11.13a and are measured in farads
f_1, f_2 = limits of the bandwidth being rejected as measured at the 3-dB points
Z_o = characteristic impedance, Ω

The constant k is equal to the image impedance and is defined by the equation

$$k = \sqrt{\frac{L_1}{C_2}} \tag{11.22a}$$

$$k = \sqrt{\frac{L_2}{C_1}} = Z_o \tag{11.22b}$$

The geometric center frequency f_m is given by the equation

$$f_m = \sqrt{f_1 f_2}$$
$$= \frac{1}{6.28 \sqrt{L_1 C_1}} = \frac{1}{6.28 \sqrt{L_2 C_2}} \tag{11.23}$$

11.8 SUMMARY OF CONSTANT-k FILTER CHARACTERISTICS

In each of the relationships given for the design of constant-k filters, it was necessary to know the value of the load impedance to be placed across the filter output terminals. The values of inductances and capacitances are always dependent upon (among other things) this impedance value. Use of the constant k was not necessary, and in our examples the only use of constant k was to show that it is equal to the image impedance. However, the derivations of the equations for L and C, which are rather tedious and were not given, *do* depend on use of the constant k.

The shapes of the characteristic curves for the constant-k filters cannot be controlled by the design engineer. He can, of course, add more sections to increase the selectivity, but the additional cost may not be justified. The sides of the selectivity curves of bandpass and band-stop filters are referred to as the *skirts*, and the shapes of these curves—as measured by the shape factor—are often referred to as the *skirt selectivity*. This is simply an indication of how sharply the circuit tunes, as is the shape factor. Constant-k filters are characterized by poor skirt selectivity when compared to other types of filters.

As we shall see in the next section, the first step in designing an m-derived filter is to design a constant-k filter. Then, the constant-k filter is modified by adding a reactance to either the series or shunt leg. Since the constant-k filter is designed first, it is often referred to as the *prototype*.

11.9 m-DERIVED FILTERS

Besides having improved skirt selectivity, the frequency of infinite attenuation can be readily selected in the m-derived filter. It still shares with the constant-k filter the disadvantage of not having a constant image impedance throughout its passband. The combination of constant-k and m sections, however, provides a flexibility in filter design not realized with the constant-k filter alone.

The constant-k filter has a rather gradual *roll-off*: that is, the amplitude of the signal does not change rapidly at the skirts. There are some applications where a sharper response is needed. This can be accomplished by adding reactance to either the series or parallel (shunt) branch. Figure 11.14 shows three examples with a reactance added to the shunt path. .Note that the same configurations (T, π, and L) are employed as were used for constant-k filters. In fact, these are identical with the constant-k low-pass filters except that an inductor has been added to resonate with the shunt capacitance. The series elements L_1 are unchanged, and, therefore, these configurations are referred to as series-type m-derived filters. The m-derived filter, then, is

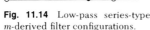

Fig. 11.14   Low-pass series-type *m*-derived filter configurations.

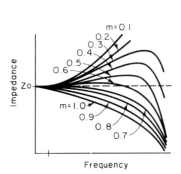

Fig. 11.15   The effect of *m* on filter impedance.

named for the branch that is *unchanged* from the constant-*k* type.   If the shunt paths C_2 are unchanged, and reactive components are added to the series elements, the configuration is then called a shunt-type *m*-derived filter.

Returning again to the circuits of Fig. 11.14, we see that the shunt path is now a resonant circuit.   There will be a resonant frequency at which the circuit presents a short across the line.   That frequency, called the *infinite attenuation frequency* f_x, and the cutoff frequency f_c do *not* have the same value.   Cutoff occurs at the point where the output signal amplitude is down 3 dB from the maximum value, just as in the constant-*k* filter.

The value of *m* in the *m*-derived filter is always less than 1, and may be obtained mathematically from one of the two equations given below.

For low-pass, *m*-derived filters:

$$m = \sqrt{1 - \frac{f_x}{(f_c)^2}} \qquad (11.24a)$$

For high-pass, *m*-derived filters:

$$m = \sqrt{1 - \frac{f_x}{f_c}} \qquad (11.24b)$$

where f_c = cutoff frequency Hz
f_x = infinite attenuation frequency, Hz
m = a constant used in determining the values of elements

A special case occurs for $m = 1.0$.   If the value of 1.0 is substituted into the equations for *m*-derived filters, the filter is converted into a constant-*k* filter.   This means that the constant-*k* filter may be considered to be a special type of *m*-derived filter for which $m = 1.0$.

Choice of the value of *m* is not entirely dependent on the relationship between f_c and f_x, as might be supposed from Eqs. (11.24a) and (11.24b).   Different values of *m* cause the filter to have different impedance characteristics.   The relationship between the value of *m* and the value of filter impedance is illustrated in the graph of Fig. 11.15.   In fact, this is one of the important advantages of *m*-derived filters: the designer can choose the particular impedance relationship that he wants.   In the constant-*k* filter the designer must accept the particular impedance relationship for that type of filter.

In Fig. 11.15, the constant-k filter impedance is represented by the curve marked $m = 1.0$.

In the graph of Fig. 11.15 the curve marked $m = 0.6$ provides the impedance versus frequency relationship which is nearest to being ideal. This curve is considered to be the best choice for many applications, and, therefore, $m = 0.6$ is a value that is frequently chosen in practical filter design. Of course, the greater the range of frequencies over which the impedance is constant, the greater the range of frequencies over which maximum power transfer can occur.

LOW-PASS m-DERIVED FILTER DESIGN (SERIES TYPE)   The equations for L_1, L_2, and C_2 are given here without derivation. In each case, it may be necessary to alter the calculated value in accordance with the specifications marked on Fig. 11.14. For example, suppose you are solving for L_1 in the T configuration, and you determine that $L_1 = 10$ mH from the equations below. According to the illustration, this value is divided by 2, so that the actual value of each of the two inductors will be $L_1/2 = 10/2 = 5$ mH.

$$L_1 = m\,\frac{Z_0}{3.14 f_c} \tag{11.25}$$

$$L_2 = \frac{1-m^2}{4m} \times \frac{Z_0}{3.14 f_c} \tag{11.26}$$

$$C_2 = m\,\frac{1}{3.14 Z_0 f_c} \tag{11.27}$$

where L_1, L_2 = series and shunt inductances, respectively, H
$\quad C_2$ = shunt capacitance, F
$\quad m$ = a constant obtained from the use of Eq. (11.24a)
$\quad f_c$ = cutoff frequency, Hz
$\quad Z_0$ = characteristic impedance of the filter circuit, Ω
If the value of m is chosen as unity ($m = 1.0$), then the three equations become

$$L_1 = \frac{Z_0}{3.14 f_c}$$

$$L_2 = 0$$

$$C_2 = \frac{1}{3.14 Z_a f_c}$$

The results mean that no inductance is used for the shunt leg ($L_2 = 0$), and the values of L_1 and C_2 are identical with those of the constant-k filter as given by Eqs. (11.7) and (11.8). This proves that a constant-k filter is a special case of an m-derived filter for which $m = 1.0$.

LOW-PASS m-DERIVED FILTER DESIGN (SHUNT TYPE)   Instead of the reactive component being placed in the shunt leg (as was done for the series-type m-derived filter), it may be placed in the series leg, as shown in Fig. 11.16. In this case a parallel circuit is formed by L_1 and C_1 which offers high impedance at the frequency to which they are tuned. If the inductor and capacitor were perfect elements—that is, if they contained no resistance—then the opposition to signal flow at the resonant frequency of L_1 and C_1 would be infinite. For this reason, the resonant frequency is called the infinite attenuation frequency, as it was in the series type of Fig. 11.14.

The equations for L_1, C_1, and C_2 are given here without derivation. As before, it may be necessary to modify the values of the elements as for the series type. Thus, L_1 becomes $L_1/2$ in the T type, C_2 becomes $C_2/2$ in the π type, etc.

$$L_1 = m\,\frac{Z_0}{3.14 f_c} \tag{11.28}$$

$$C_1 = \left(\frac{1-m^2}{4m}\right)\left(\frac{1}{3.14 Z_0 f_c}\right) \tag{11.29}$$

Fig. 11.16 Low-pass shunt-type *m*-derived filter configurations.

Fig. 11.17 High-pass series-type *m*-derived filter configurations.

$$C_2 = m \frac{1}{3.14 Z_o f_c} \qquad (11.30)$$

where L_1 = inductance in the series leg, H
$\quad\ C_1$ = capacitor that resonates with L_1, F
$\quad\ C_2$ = capacitance in the shunt leg
$\quad\ f_c$ = cutoff frequency, Hz
$\quad\ Z_o$ = characteristic impedance of the filter, Ω

HIGH-PASS m-DERIVED FILTER DESIGN (SERIES TYPE)   High-pass *m*-derived filters can assume the same three basic configurations as are used for low-pass filters (T, π, and L).   Also, each of these configurations may be designed as a series type or shunt type.   Like the low-pass filter, a high-pass *m*-derived filter is a series type when the shunt leg is modified, and it is a shunt type when the series leg is modified. Figure 11.17 shows the three configurations for the series-type high-pass *m*-derived filters.   The equations for C_1, C_2, and L_2 are given without derivation.

$$C_1 = \frac{1}{12.6 f_c Z_o m} \qquad (11.31)$$

$$C_2 = \left(\frac{4m}{1 - m^2} \right) \left(\frac{1}{12.6 f_c Z_o} \right) \qquad (11.32)$$

$$L_2 = \frac{Z_o}{12.6 f_c m} \qquad (11.33)$$

where C_1, C_2 = series and shunt capacitance values, F
$\quad\ L_2$ = inductance in the shunt leg that resonates with C_2, H
$\quad\ Z_o$ = characteristic impedance of the filter circuit, Ω
$\quad\ f_c$ = center frequency, Hz
$\quad\ m$ = a constant that determines the rate at which the filter characteristic rolls off.

In the circuits of Fig. 11.17 the series-resonant circuit comprised of C_2 and L_2 provides a short circuit to the frequency of infinite attenuation, thus preventing any signal at this frequency from passing through the filter.

HIGH-PASS m-DERIVED FILTER DESIGN (SHUNT TYPE)   Figure 11.18 shows the three configurations for an *m*-derived shunt-type high-pass filter. In this case a parallel-tuned circuit comprised of L_1 and C_1 is tuned to the frequency of infinite impedance. The parallel-tuned circuit prevents any signal at that frequency from passing. The equations for the values of elements C_1, L_1, and L_2 are given here without derivation.

$$C_1 = \frac{1}{12.6 f_c Z_0} \tag{11.34}$$

$$L_1 = \left(\frac{4m}{1 - m^2}\right)\left(\frac{Z_0}{12.6 f_c}\right) \tag{11.35}$$

$$L_2 = \frac{Z_0}{12.6 f_c m} \tag{11.36}$$

where L_1, L_2 = series and shunt inductances, respectively, H
C_1 = capacitance that resonates with L_1, F
f_c = cutoff frequency, Hz
Z_0 = characteristic impedance of the filter circuit, Ω
m = constant arrived at in the same manner as in other *m*-derived filters

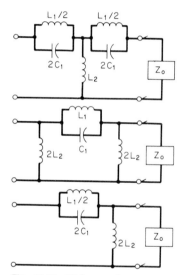

Fig. 11.18   High-pass shunt-type *m*-derived filter.

11.10 SUMMARY AND DESIGN EXAMPLE OF m-DERIVED FILTERS

A number of *m*-derived filter sections can be combined to form a ladder-type filter. In some cases an *m*-derived section may be combined with a constant-*k* section to obtain a particular overall characteristic. At the input and output ends of the filter, L sections may be used to obtain an impedance match.

Before an example of *m*-derived filter design is given, a few of the more important facts about *m*-derived filters will be summarized. The first step in designing an *m*-derived filter is to design the constant-*k* prototype. A value of *m* is decided upon next. If the circuit is to match a load impedance over a range of frequencies, then a value of $m = 0.6$ is generally used. The smaller the value of *m*, the sharper the cutoff characteristic of the filter and the greater the change in impedance with frequency. A constant-*k* filter section added to the filter will ensure that there is attenuation beyond the cutoff point of the filter.

Fig. 11.19 An example of m-derived filter design: (a) Filter prototype. (b) Filter sections with calculated values. (c) Composite filter.

example 11.5 Design an m-derived ladder-type low-pass filter. The cutoff frequency f_c is to be 1000 Hz and the load impedance is to be 600 Ω.

solution To get a sharp cutoff, we shall use two full T sections and two L sections. See Fig. 11.19. Figure 11.19a shows the circuit. The input and output L sections are series-type m-derived filters. The first T section (reading left-to-right) is a simple constant-k filter, and the other T section is an m-derived series section.

The first step is to find the value of L_1 and C_1 for the constant-k filter.

$$L_1 = \frac{Z_0}{3.14 f_c} = \frac{600}{3.14 \times 1000} \tag{11.7}$$

$$= 0.191 \text{ H (or 191 mH)}$$

$$C_1 = \frac{1}{3.14 f_c Z_0}$$

$$= \frac{1}{3.14 \times 1000 \times 600} = 0.531 \ \mu\text{F} \tag{11.8}$$

The values of L_1 and C_1 for the constant-k filter are shown in Fig. 11.19b. As required by the specification of Fig. 11.19a, the value of L_1 has been divided by two. The next step is to determine the values of L_2, L_3, and C_2 of the m-derived T section. The subscripts are changed from the original equations so that the various sections will not be confused.

$$L_2 = m \frac{Z_0}{3.14 f_c} \tag{11-25}$$

in which the term $(Z_0/3.14 f_c)$ is equal to the constant-k prototype value of L_1.

A value of $m = 0.866$ will be used for the m-derived T section in order to get adequate attenuation. A value of $m = 0.6$ will be used for the end sections in order to get an impedance match over a range of frequencies.

$$L_2 = \frac{0.866 \times 600}{3.14 \times 1000} = 0.166 \text{ H (or 166 mH)}$$

$$L_2 = \frac{0.3 \times 0.191}{2} = 29 \text{ mH}$$

$$L_3 = \frac{1 - m^2}{4m} \left(\frac{Z_o}{3.14 f_c} \right) \tag{11.26}$$

$$= \frac{1 - 0.75}{3.46} (0.19) = 0.014 \text{ H (or 14 mH)}$$

$$C_2 = \frac{m}{3.14 Z_o f_c} \tag{11.27}$$

$$= \frac{0.866}{1.884 \times 10^6} = 0.459 \ \mu\text{F}$$

The values of L_2 L_3, and C_2 are shown in Fig. 11.19b. The values of L_4, L_5, and C_3 will now be calculated.

$$L_4 = \frac{m Z_o}{3.14 f_c} \tag{11.28}$$

$$= \frac{0.6 \times 600}{3.14 \times 1000} = 0.114 \text{ H (or 114 mH)}$$

$$L_5 = \left(\frac{1 - m^2}{4m} \right) \left(\frac{Z_o}{3.14 f_c} \right) \tag{11.26}$$

$$= \frac{(1 - 0.36) \times 0.190}{2.4} = 0.051 \text{ H (or 51 mH)}$$

$$C_3 = \frac{m}{3.14 Z_o f_c} \tag{11.27}$$

$$= 0.6 \times 0.531 \times 10^{-6} = 0.319 \ \mu\text{F}$$

The end sections are shown with the calculated values in Fig. 11.19b. The value of C_3 is divided by two, and the value of L_5 is multiplied by 2, as required by the specification shown in Fig. 11.19a.

In combining the four sections of Fig. 11.19b, the series inductances are added to make one single inductance value, thus reducing the number of components required. For example, L_4 is added to $L_1/2$ in combining the first and second sections. The final composite filter design is shown in Fig. 11.19c.

Although these filters are easy to design, they suffer from certain shortcomings. The most important is that the terminations must be such that at the center frequency of the filter the source and load must be pure resistances equal to the image impedance. As the frequency increases, the impedance must decrease to zero at the cutoff frequency. This cannot be done with lumped circuits, and so the actual band shape with resistive terminations is never exactly as shown on the graphs.

11.11 SOME ADDITIONAL TYPES OF FILTERS

Constant-k and m-derived filters are easy to design (in comparison to other types), and they have reasonably good roll-off characteristics. Many other methods of designing LC filters are also possible, each with its own advantages and disadvantages.

Filter manufacturers may make use of a computer to design a variety of filters having different attenuation and phase characteristics. The designers choose the filter with the characteristic most nearly equal to their required characteristics. This is comparable to the situation with tubes or transistors in which case the designer purchases the tube or transistor most nearly suited to his needs.

Since the computer has been made available to filter designers, there is a trend toward obtaining "tailor-made" filters. The designer specifies the characteristics he wants, and the computer performs the computation. What used to amount to hours and hours of tedious equation solving can now be accomplished in minutes. Also, more complex designs can now be obtained—designs that were not economically feasible when calculated without computers.

The designer of today, then, must know the characteristics, advantages, and disadvantages of the various types of filters. This enables him to specify filters so as to get the best possible design for his needs.

Fig. 11.20 The configuration of a typical Butterworth or Chebyshev filter.

Fig. 11.21 Typical attenuation response curves for Butterworth and Chebyshev filters: (a) Butterworth filter. (b) Chebyshev filter.

BUTTERWORTH FILTER This type of filter is also called a *maximally flat filter*. It is a form of *LC* filter that has a relatively flat response—that is, it offers a relatively constant impedance to signals with frequencies in its passband range.

The configuration of a typical Butterworth filter is shown in Fig. 11.20. You will note that it is like a constant-*k* filter. However, the values of the components are different. This is to be expected because they are arrived at by a different method.

The Butterworth filter has a good phase response. This is a measure of the amount of phase shift that occurs when a signal passes through the filter. It also has good amplitude response, which is another way of saying that its response is flat. However, it has a rather gradual roll-off, which makes it unsuitable in applications where it is necessary to sharply cut off adjacent frequencies.

CHEBYSHEV FILTER The Chebyshev filter is also a form of *LC* filter in the same configuration as the Butterworth filter. The design of the Chebyshev filter is based on the assumption that all the frequencies to be passed are equally important. In this respect, it is different from the filters previously discussed which are designed on the assumption that zero frequency (dc) is the most important factor in design.

The cutoff characteristic of a Chebyshev filter is much sharper than that of the Butterworth or other types discussed (constant-*k* and *m*-derived). An important disadvantage of the Chebyshev is that there is a ripple in its passband response. This is another way of saying that its response curve is not flat in the passband region. Figure 11.21 compares the attenuation response curves of the Butterworth filter (which has no ripple) with that of the Chebyshev filter (which has ripple).

CRYSTAL FILTER (SOMETIMES CALLED "QUARTZ FILTER") Figure 11.22 shows two equivalent circuits for crystals in their holders. The equivalent circuit of a piezo-

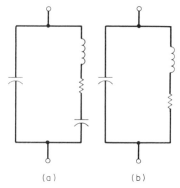

Fig. 11.22 Equivalent circuits for a crystal: (a) This is the equivalent circuit that is usually shown. (b) Mathematically, this circuit can be obtained from the one shown in (a).

electric crystal that is usually shown is in Fig. 11.22*a*. The extremely high Q of a crystal causes it to behave as a bandpass filter with a very narrow bandwidth.

Mathematically, it is possible to convert the equivalent circuit of Fig. 11.22*a* into another equivalent circuit like the one shown in Fig. 11.22*b*. This one is easier to visualize, and calculations for this type of circuit were discussed in Chap. 10 of this book.

Figure 11.23 shows a crystal in filter circuit configurations. If a coil L_1 with a resistance R_1 is placed in series with the crystal, as shown in Fig. 11.23*a*, the circuit behaves as a simple parallel-resonant tuned circuit. It will pass all frequencies except the one to which it is tuned. Because of the high circuit Q, a resistor or coil may be placed in parallel with the circuit. This will lower the Q and broaden the frequency response. Also, a variable capacitor may be placed across the crystal circuit to tune it through a narrow range of frequencies. Figure 11.23*b* shows the practical circuit for a crystal filter, and Fig. 11.23*c* shows a typical response curve.

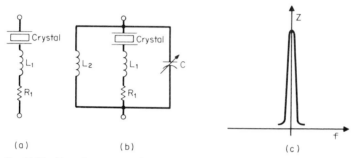

Fig. 11.23 Crystal circuits used for filters: (*a*) A coil L_1 and its dc resistance R_1 in series with the crystal makes it behave as a simple parallel-resonant circuit with a high Q. (*b*) In this circuit the variable capacitor C tunes the circuit through a narrow range, and coil L_2 lowers the circuit Q for a broader response.

Crystal filters are used in applications where a parallel-resonant LC circuit response curve is applicable. For example, crystal filters are used in i-f amplifier coupling circuits to provide a narrow passband with sharp skirt selectivity. A number of crystals may be used in a stagger tuning arrangement to obtain a wider response than is normally obtainable with one crystal circuit. The very narrow bandwidth of a crystal filter is useful in CW receivers because their sharp selectivity makes possible separate signals of adjacent carrier frequencies. Q *multipliers* in the i-f stage may employ crystals for this purpose.

Disadvantages of crystal filters include their cost, which may be high compared to that of an LC filter. (However, if the LC filter was designed to produce exactly the same resonant curve as the crystal filter, it would be much bulkier and far more expensive.) Crystal filters cannot be tuned over a wide range of frequencies. They are bulkier than the ceramic filters that will be described next.

CERAMIC FILTER Ceramic filters have characteristics similar to those of quartz filters, but are much smaller in size. Instead of using a piece of quartz for the vibrating element, the ceramic filter uses the piezoelectric properties of lead zirconate–lead titanate resonators.

Figure 11.24 shows the construction of a simple three-terminal ceramic filter. The dot electrode terminal (in the center of the ceramic disk) is the one to which the input signal is delivered. The input signal causes the ceramic disk to vibrate at its mechanical resonant frequency. If the input signal frequency is other than the mechanical resonant frequency of the ceramic disk, no vibration will take place. When the ceramic disk is vibrating, voltages are generated around its edges which are picked off by the ring electrode terminal. The common terminal, which is simply a silver plate across the bottom of the ceramic disk, is normally operated at ground potential. As an indication of the very-small physical size of the ceramic filter, a ceramic disk 0.2 in. in diameter by 0.1 to 0.4 in thick will resonate at a frequency of 455 kHz.

Fig. 11.24 Construction details of a three-terminal ceramic filter.

A two-terminal ceramic filter is made by placing a silver plate on both sides of the piezoelectric ceramic disk. There is no common electrode terminal. In this type of filter, the signal is fed to one of the plates and taken from the other.

To get a sharper bandpass characteristic, a number of the two-terminal disks can be stacked to produce what is known as a *ceramic ladder filter*. Figure 11.25 presents this type of filter. Figure 11.25a shows its physical appearance, and Fig. 11.25b indicates how this type of filter is constructed by stacking ceramic wafers. Figure 11.25c gives the response curve of a ladder filter compared with that of an ideal filter (shown in dotted lines). Note that the shape of the ceramic ladder filter response curve is very near to the shape of the ideal curve.

Figure 11.26 demonstrates an i-f amplifier stage Q_1 coupled to another i-f amplifier stage Q_2 with ceramic filter coupling. Note that this is a three-terminal type filter, but it may also be a stacked filter. In other words, it is possible to stack three-terminal ceramic disks just as it is possible to stack the two-terminal type in order to get the desired bandpass characteristics.

The input signal to the filter in Fig. 11.26 is taken from collector-load resistor R_3, passed through the filter. The output signal is delivered to the base of Q_2 at the junction of its base-bias voltage divider comprised of R_5 and R_6. Although the circuit looks like a resistance-capacitance amplifier, remember that the filter is equivalent to an LC filter circuit.

(a)   (b)   (c)

Fig. 11.25 The ladder-type ceramic filter and its characteristic curve: (a) Physical appearance of a ceramic ladder filter. (b) Construction of a ceramic ladder filter. (c) Response of a ceramic ladder filter.

Fig. 11.26 An i-f stage that uses a ceramic filter for coupling.

MECHANICAL FILTERS (ALSO CALLED "MAGNETOSTRICTIVE FILTERS") Both the quartz crystal and ceramic filters employ the piezoelectric effect to produce a mechanical vibration in the filter material. This mechanical vibration, which determines the frequency of the filter, is directly related to its physical size and can be held to a very close frequency tolerance.

Instead of the combination of piezoelectric principle and mechanical resonance, it is also possible to use the magnetostrictive principle. The two magnetostrictive effects that are related to the use of mechanical filters are the *Joule effect* and the *Vallari effect*. The Joule effect relates to the change in physical length of a ferromagnetic material in the presence of an applied magnetic field. The Vallari effect relates to the change in magnetization of a material when an external stress is applied to that material. Both these effects are usually combined and referred to as the magnetostrictive effect.

Figure 11.27 presents a mechanical filter and its equivalent *LC* circuit. The construction details of a mechanical filter are shown in Fig. 11.27*a*. The electric input signal produces a varying magnetic field around a magnetostrictive driving rod. This varying field causes the driving rod to change in length at a rate that is directly equal to the rate at which the field varies. The varying length of the rod produces a mechanical vibration in the resonant mechanical section. Note that the resonant section is comprised of metal disks which are manufactured to be precisely resonant at a certain frequency. These disks are coupled together by rigid coupling rods.

Any frequency other than the resonant frequency cannot pass through the mechanical filter because the resonant frequency of the disks is quite precise—that is, they are manufactured to very close physical dimensions. Thus, the mechanical filter cannot be set into vibration by frequencies other than the mechanical frequencies. The mechanical oscillations are converted back to electric oscillations by a transducer output coil.

The bias magnets at each end produce a restraining force on the motion of the disks in order to make the filter vibrate only once for each input cycle. Remember that one complete cycle of ac signal consists of two peak currents. Each of these peaks will magnetize the rod and cause it to change its length. Thus, two vibrations would occur for each cycle if it were not for the biasing magnet. The magnetic field from the bias magnet holds the resonant disks so that they can move easily in one direction, but cannot move easily in the opposite direction. With this arrangement the output frequency is the same as the input frequency.

(a)

(b)

Fig. 11.27 Construction details of a mechanical filter and its equivalent circuit: (*a*) Elements of a mechanical filter. (*b*) Electrical analogy of a mechanical filter.

The electric equivalent circuit in Fig. 11.27*b* shows that the mechanical filter is comparable to a ladder-type *LC* filter.

A variation in design of the mechanical filter, which is used at much lower frequencies, employs a tuning fork instead of vibrating mechanical metal disks. The input signal is electromagnetically coupled to one prong of the tuning fork which sets it into vibration. The output signal is taken from the other prong. Mechanical filters of this type are used at audio frequencies.

11.12 ACTIVE FILTERS

As mentioned in the beginning of the chapter, an active filter contains, in addition to op amps, resistors and capacitors. Inductors, a component in *LC* filters, are thereby eliminated. Because of their relative large size and cost, the elimination of inductors in filters is indeed a considerable achievement. This is especially true for filters having cutoff frequencies of 100 Hz or less.

There are many methods one may use for the realization of active filters. In this section we shall concentrate on an active filter that provides the Butterworth response of Fig. 11.21. Two basic low-pass circuits for realizing Butterworth active filters are illustrated in Fig. 11.28.

(a)   (b)

Fig. 11.28 Basic low-pass Butterworth filters: (*a*) First-order section. (*b*) Second-order section.

A first-order low-pass section is shown in Fig. 11.28*a* and a second-order low-pass section in Fig. 11.28*b*. The voltage gain, A_r, of each section, from Chap. 13, is determined by resistors R_1 and R_2 connected to the inverting (−) terminal of the op amp:

$$A_r = 1 + \frac{R_2}{R_1} \tag{11.37}$$

The half-power cutoff frequency f_c for each section is determined from the following expression,

$$f_c = \frac{0.159}{RC} \tag{11.38}$$

where f_c is in hertz, R in ohms, and C in farads. The basic sections of Fig. 11.28, as

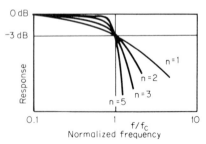

Fig. 11.29 Frequency-response curves of a Butterworth low-pass filter for different orders, *n*.

will be explained shortly, may also be used to realize high-pass, bandpass, and band-reject filters.

Figure 11.29 provides plots of frequency response of a low-pass Butterworth filter for different orders, n. A particular value of n is obtained by cascading the basic sections of Fig. 11.28. As n increases, the shape factor improves. The frequency scale is normalized by plotting the ratio of f/f_c. At $f = f_c$ ($f/f_c = 1$), the response of the filter is down by -3 dB.

For a first-order ($n = 1$) Butterworth low-pass filter, the first-order section of Fig. 11.28a is used. For a second-order ($n = 2$) filter, the second-order section of Fig. 11.28b is employed. If a third-order ($n = 3$) filter is necessary, the first-order and second-order sections of Fig. 11.28 are cascaded. A fourth-order ($n = 4$) filter is realized by cascading two second-order sections, and so on.

The gain of the first-order section is arbitrary. The gain of a second-order section, however, is defined. It is related by coefficient a, listed in Table 11.1, for different values of n. The relationship between A_r and a is

$$A_r = 3 - a \qquad (11.39)$$

Example 11.6 illustrates the use of the preceding equations and Table 11.1 in a filter design problem.

TABLE 11.1 Coefficients for a Second-Order Butterworth Section

n	a
2	1.414
3	1
4	0.765, 1.848
5	0.618, 1.618
6	0.518, 1.414, 1.932

patch p. 11-27

example 11.6   Design a fifth-order ($n = 5$) Butterworth low-pass filter having a cutoff frequency of 1 kHz. For each section, $R = 1$ kΩ and $R_1 = 5$ kΩ. The first-order section has a gain of 10.

solution   The filter consists of a cascaded connection of a first-order and two second-order sections, as illustrated in Fig. 11.30. By Eq. (11.28), for $R = 1$ kΩ,

$$C = \frac{0.159}{f_c R}$$

$$= \frac{0.159}{10^3 \times 10^3} = 0.159 \times 10^{-6} \text{ F} = 0.159 \ \mu\text{F}$$

For a gain of 10 for the first-order section, by Eq. (11.37),

$$10 = 1 + \frac{R_2}{5 \text{ k}}$$

Solving, $R_2 = 45$ kΩ.

Fig. 11.30   An example of a fifth-order ($n = 5$) 1-kHz low-pass Butterworth active filter. (See Example 11.6.)

Fig. 11.31   Basic high-pass Butterworth filters: (a) First-order section. (b) Second-order section.

Referring to Table 11.1, for the first second-order section in Fig. 11.30, $a = 0.618$.   By Eq. (11.39),

$$A_{rl} = 3 - 0.618 = 2.382$$

By Eq. (11.37) then,

$$2.382 = 1 + \frac{R_2}{5 \text{ k}}$$

and $R_2 = 6.91$ kΩ.

For the remaining second-order section, from Table 11.1, $a = 1.618$.   Hence,

$$A_{r2} = 3 - 1.618 = 1.382$$

$$1.382 = 1 + \frac{R_2}{5 \text{ k}}$$

Solving, $R_2 = 1.91$ kΩ.

If, for example, a sixth-order ($n = 6$) filter were required, three second-order sections of Fig. 11.28b would be cascaded.   In calculating the values of R_2, from Table 11.1, $a = 0.518$ would be used for the first section, 1.414 for the second, and 1.932 for the last section.

High-pass filter   If the R's and C's in Fig. 11.28 are interchanged, first-order and second-order high-pass sections are obtained.   These are illustrated in Fig. 11.31. Table 11.1 and Eqs. (11.37) to (11.39) also apply to the high-pass sections.   This is illustrated in the following example.

example 11.7   Design a fifth-order ($n = 5$) Butterworth high-pass filter having a cutoff frequency of 1 kHz.   For each section, $R = 1$ kΩ and $R_1 = 5$ kΩ.   The first stage has a voltage gain of 10.

solution   The filter, shown in Fig. 11.32, uses the basic high-pass sections of Fig. 11.31. The values of C and R_2 were obtained in the manner illustrated in Example 11.6 for the fifth-order low-pass filter.

Fig. 11.32   An example of a fifth-order 1-kHz high-pass Butterworth active filter.   (See Example 11.7.)

Bandpass filter An active bandpass filter may be obtained by cascading high- and low-pass active filters, as illustrated in the block diagram of Fig. 11.33. The cutoff frequency of the low-pass filter must be greater than the cutoff frequency of the high-pass filter. If the cutoff frequencies for the high-pass and low-pass filters are f_H and f_L, respectively, the bandwidth of the bandpass filter is $f_L - f_H$.

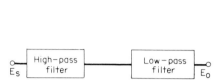

Fig. 11.33 Block diagram of a bandpass active filter.

Fig. 11.34 Realizing an active band-reject filter by connecting a low-pass and a high-pass filter to a summer.

Band-reject filter An active band-reject filter may be realized by connecting together the inputs of a low-pass and a high-pass filter and their respective outputs to a summing amplifier. This is illustrated in Fig. 11.34. For this filter, f_L must be less than f_H. The stop band is then equal to $f_H - f_L$.

Chapter 12

Transistor Amplifiers and Oscillators

12.1 INTRODUCTION

An amplifier increases the magnitude of, or *amplifies*, an electric signal. The signal may be derived, for example, from a TV antenna, a phono cartridge, or a strain gage. Of all components, the amplifier is by far the most widely used building block in electronics systems.

An oscillator generates a waveform without the presence of an external signal. If the waveform is a sine wave, the oscillator is referred to as a *sinusoidal* oscillator. For a waveform other than a sine wave, such as a square or sawtooth wave, the oscillator is called a *nonsinusoidal*, or *relaxation*, oscillator. Sinusoidal oscillators are used, for example, to test the performance of circuits, to generate radio frequencies, and in receivers.

In this chapter we shall consider how amplifiers are classified, their characteristics, biasing, how to calculate their gain and frequency response, and the effects of utilizing more than one amplifier stage. The various parameters used for characterizing transistors are defined in Chap. 8. Feedback, where a portion or all of the output signal is returned to the input of an amplifier, and the operation of sinusoidal oscillators, conclude the chapter.

12.2 CLASSIFICATION OF AMPLIFIERS

An amplifier may be represented by the simple block diagram of Fig. 12.1. The input signal voltage is denoted by E_g and the input signal current by I_g. Across the output terminals is load resistor R_L. Resistance R_L may represent, for example, the resistance of a loudspeaker coil, a motor winding, or the input of a connected amplifier stage. Voltage E_o is the output (load) voltage, and current I_o the output (load) current. Symbols P_i and P_o are the input and output signal powers, respectively.

If the amplifier is optimized to amplify voltage signals, it is called a *voltage* amplifier. For current signals, it is referred to as a *current* amplifier. If it is to develop output power of one watt or more, it is generally classified as a *power* amplifier.

Another method of classification is based on the amplitude of signals. For small signals, in the order typically of microvolts and millivolts, the amplifier may be classified as *small-signal*, and for larger signals as a *large-signal* amplifier. Simple algebra is all that is required for calculating the performance of small-signal amplifiers. Graphical methods, however, are generally used in the analysis of large-signal amplifiers.

Amplifiers are also classified in terms of load current flow. Figure 12.2a shows one cycle (360°) of a sine wave input signal to an amplifier. If the load current also flows

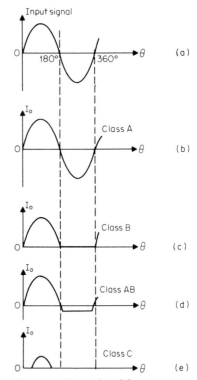

Fig. 12.1   Block diagram of an amplifier.

Fig. 12.2   Classes of amplifier operation: (a) Input signal. (b) Load current flows 360° for class A operation. (c) Load current flows 180° for class B operation. (d) In class AB operation, load current flow is greater than 180° but less than 360°. (e) Class C operation, where load current flows for less than 180°.

for 360°, as indicated in Fig. 12.2b, the amplifier is operating as *class A*. *Class B* operation, illustrated in Fig. 12.2c, exists when the load current flows for only 180°. If the load current flows for more than 180 but less than 360°, *class AB* operation is realized (Fig. 12.2d). *Class C* operation is obtained if the load current flows for less than 180° (Fig. 12.2e).

Class A operation is used for small-signal and single-transistor (referred to as *single-ended*) power amplifiers. If class B, AB, or C operation is used in a single-ended amplifier, the output waveform is badly distorted and the amplifier is worthless. For this reason, for class B or AB operation the amplifier contains two transistors, referred to as a *push-pull* amplifier. Class C operation is often used in the amplification of r-f signals in a circuit containing a tuned load.

Amplifiers are also classified according to their intended operation. Examples include *audio*, *r-f*, *video*, *microwave*, and *pulse* amplifiers.

12.3 CHARACTERISTICS OF AMPLIFIERS

The performance of amplifiers is characterized by the following terms:

Voltage gain
Current gain
Power gain
Input resistance

Output resistance
Bandwidth
Distortion
Slewing rate

VOLTAGE GAIN   The voltage gain of an amplifier A_v is defined as the ratio of the output signal voltage E_o to the input signal voltage E_s:

$$A_v = \frac{E_o}{E_s} \tag{12.1}$$

CURRENT GAIN   The current gain of an amplifier A_i is defined as the ratio of the output signal current I_o to the input signal current I_s:

$$A_i = \frac{I_o}{I_s} \tag{12.2}$$

POWER GAIN   The power gain of an amplifier A_p is defined as the ratio of the output signal power P_o to the input signal power P_i:

$$A_p = \frac{P_o}{P_i} \tag{12.3a}$$

An alternative definition is that the power gain is equal to the product of the voltage and current gains:

$$A_p = A_v A_i \tag{12.3b}$$

INPUT RESISTANCE   The input resistance (or impedance) of an amplifier R_i (or Z_i) is the resistance (or impedance) across the input terminals of an amplifier.

OUTPUT RESISTANCE   The output resistance (or impedance) of an amplifier R_o (or Z_o) is the resistance (or impedance) across the output terminals of an amplifier with the signal source set to zero (see Chap. 6).

BANDWIDTH   The bandwidth of an amplifier BW is a band of frequencies that is amplified with modest attenuation in gain. A typical bandwidth curve is illustrated in Fig. 12.3. Gain, expressed in decibels (dB), is generally plotted along the y axis, and frequency along the x axis. To cover conveniently a large range of frequencies, a logarithmic scale is used for the frequency axis.

Fig. 12.3 An example of bandwidth characteristics of an amplifier.

The nominal gain over a large range of frequencies is expressed as 0 dB. At frequencies $f_L = 20$ Hz and $f_H = 20$ kHz, the gain is down by -3 dB. Therefore, frequency f_L is referred to as the *lower* -3-dB frequency, and f_H as the *upper* -3-dB frequency. Other terms used include the lower and upper *break, cutoff, corner,* and *half-power* frequencies. The bandwidth is equal to the difference between the upper and lower -3-dB frequencies:

$$BW = f_H - f_L \tag{12.4}$$

From Fig. 12.3, BW $= 20\,000 - 20 \simeq 20\,000$ Hz $= 20$ kHz.

DISTORTION Because the characteristics of the BJT and FET are not strictly linear (see Chap. 8), the output of an amplifier contains frequencies that were not present in the input signal. Assume, for example, that a 1-kHz sine wave is impressed across the input terminals of an amplifier. The output will contain, in addition to a 1-kHz frequency, referred to as the *fundamental frequency*, integrally related frequencies of 2, 3, 4, . . . , kHz. These frequencies are referred to as *harmonics:* 2 kHz is the *second harmonic*, 3 kHz is the *third harmonic*, and so on.

Harmonic distortion Harmonic distortion is defined as the ratio of the amplitude of a harmonic component to the amplitude of the fundamental frequency. For example, if the 1-kHz output is 10 V and the 2-kHz output is 0.5 V, the *second-harmonic distortion* D_2 is $0.5/10 = 0.05 = 5$ percent.

Intermodulation distortion Assume that a 1000-Hz and a 400-Hz signal are fed into an amplifier. The output, in addition to the 1000- and 400-Hz signals, will contain frequencies equal to their *difference* $(1000 - 400 = 600 \text{ Hz})$ and to their *sum* $(1000 + 400 = 1400 \text{ Hz})$. The sum and difference frequencies is the result of intermodulation distortion in an amplifier.

SLEWING RATE The slewing rate of an amplifier indicates how well it responds to a rapidly changing waveform. To measure the response, a rectangular pulse is applied to the input of the amplifier (Fig. 12.4). Because of internal capacitances of

Fig. 12.4 Determining the slewing rate of an amplifier. The value of the slewing rate equals $\Delta E/\Delta t$.

the amplifier, it takes a finite time for the output to reach its maximum value. How fast it reaches it is the slewing rate of the amplifier.

As indicated in Fig. 12.4, an increment of time Δt (Δ is the Greek letter *delta*) is required for the output voltage to increase by the increment ΔE along its linear ascent. The ratio $\Delta E/\Delta t$ is equal to the slewing rate. Its unit is generally expressed in volts per microsecond.

12.4 BIASING AND STABILIZATION

Because of its wide use, in this section we consider how the BJT and FET amplifiers are biased and stabilized for class A operation.

BIASING THE COMMON-EMITTER AMPLIFIER A typical common-emitter (CE) junction transistor amplifier is illustrated in Fig. 12.5. A single bias source, E_{CC}, sup-

Fig. 12.5 Basic class A common-emitter amplifier biased from a single dc source, E_{CC}.

plies the necessary collector and base currents for class A operation. Capacitor C_c, called a *coupling* capacitor, blocks any direct current from the base that may be in signal source E_s. Output signal voltage E_o is taken across the collector and ground.

A simple method of finding the *quiescent operating point* Q is by graphical analysis. First, summing the dc voltages in the collector circuit of Fig. 12.5, we have

$$E_{CC} = R_c I_C + V_{CE} \qquad (12.5a)$$

Solving Eq. (12.5a) for I_C,

$$I_C = -\frac{V_{CE}}{R_C} + \frac{E_{CC}}{R_C} \qquad (12.5b)$$

Expression (12.5b) is an equation of a straight line of the form $y = mx + b$, where m is the slope and b is the y intercept. Comparing the equation of a straight line with Eq. (12.5b), we see that $y = I_C$, $m = -1/R_C$, $x = V_{CE}$, and $b = E_{CC}/R_C$. Because only two points determine a straight line, it is convenient to superimpose Eq. (12.5b) on the collector characteristics of a transistor. The superimposed line is referred to as a *load line*.

To obtain the two points, Eq. (12.5b) is first solved for $I_C = 0$; then

$$V_{CE} = E_{CC} \qquad (12.6a)$$

This is one point needed to plot the load line. The second point is obtained by setting $V_{CE} = 0$. For this condition, Eq. (12.5b) reduces to

$$I_C = \frac{E_{CC}}{R_C} \qquad (12.6b)$$

The two points and the load line drawn between them on the collector characteristics are illustrated in Fig. 12.6. The slope of the load line is $-1/R_C$. Location of the Q

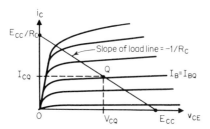

Fig. 12.6 Superimposing load line on collector characteristics for location of the Q point.

point is at the intersection of the load line and base current curve I_{BQ}. For class-A operation the Q point is located to be approximately at the center of the load line. The quiescent collector voltage and current are denoted by V_{CQ} and I_{CQ}, respectively.

Returning to Fig. 12.5, the equation for the dc quantities in the input circuit is

$$E_{CC} = I_B R_1 + V_{BE} \qquad (12.7a)$$

where, at room temperature, $V_{BE} = 0.7$ V for a silicon transistor and 0.3 V for a germanium transistor (see Chap. 8).

Because the bipolar junction transistor is a *current-operated* device, it is biased by a base current I_B. Solving Eq. (12.7a) for I_B, we obtain

$$I_B = \left(\frac{E_{CC} - V_{BE}}{R_1}\right) \qquad (12.7b)$$

The application of this and the other expressions derived in the section is illustrated in Example 12.1.

Fig. 12.7 Locating the Q point in Example 12.1.

example 12.1 In the circuit of Fig. 12.5, $E_{CC} = 10$ V, $R_C = 5$ Ω, $I_B = 100$ mA, and $V_{BE} = 0.7$ V. Assume that the collector characteristics of Fig. 12.7 are appropriate for the transistor. (a) Write the equation for the load line. (b) Superimpose the load line on the collector characteristics. (c) Locate the Q point. (d) Determine the value of R_1.

solution (a) Substitution of $R_C = 5$ Ω and $E_{CC} = 10$ V in Eq. (12.5b) yields $I_C = -V_{CE}/5 + 10/5 = -V_{CE}/5 + 2$, which is the equation of the load line.
   (b) For $I_C = 0$, by Eq. (12.6a), $V_{CE} = 10$ V. For the second point, letting $V_{CE} = 0$, by (12.6b), $I_C = {}^{10}\!/_5 = 2$ A. The load line drawn between the two points is shown in Fig. 12.7.
   (c) From Fig. 12.7, the Q point is located at the intersection of the load line and the $I_B = 100$ mA curve: $V_{CQ} \approx 4.7$ V and $I_{CQ} \approx 1.1$ A.
   (d) Solving Eq. (12.7b) for R_1,

$$R_1 = \frac{E_{CC} - V_{BE}}{I_B}$$

Substitution of the given values in the equation yields

$$R_1 = \frac{10 - 0.7}{0.1} = 93 \ \Omega$$

If collector characteristics are not available, the Q point can be estimated. The procedure is illustrated in the next example.

example 12.2 Using the values given in Example 12.1 and the fact that the dc current gain, $h_{FE} = 10$, find the Q point of the amplifier of Fig. 12.5.

solution Collector current I_C equals the product of base current I_B and the dc current gain h_{FE}: $I_C = I_B h_{FE}$. From Example 12.1, $I_B = 100$ mA $= 0.1$ A; hence, $I_C = I_{CQ} = 0.1 \times 10 = 1$ A. Solving Eq. (12.5a) for V_{CE}, we obtain

$$V_{CE} = E_{CC} - R_C I_C$$

Therefore, $V_{CE} = V_{CQ} = 10 - 5 \times 1 = 5$ V. The Q point is located at $V_{CQ} = 5$ V and $I_C = 1$ A. These values are reasonably close to those obtained graphically in Example 12.1.

BIASING THE COMMON-BASE AMPLIFIER Generally, two dc sources are used for biasing the common-base (CB) amplifier. Required are E_{CC} for the output circuit and E_{BB} for the input circuit, as shown in Fig. 12.8a. Because the current gain of a transistor in the common-base configuration is close to unity, the common-base collector

(a)   (b)

Fig. 12.8 Biasing a common-base amplifier: (a) Using two dc sources, E_{CC} and E_{BB}. (b) Using a single source, E_{CC}.

characteristics are not very informative. The Q point is therefore calculated in a manner similar to that illustrated in Example 12.2. The dc current gain h_{FB} of a CB amplifier is taken to be equal to one: $h_{FB} = 1$.

A method for biasing a common-base amplifier from a single dc source E_{CC} is shown in Fig. 12.8b. *Bypass capacitor* C_2 is chosen so that its reactance to the lowest signal frequency is at least one-tenth the value of the resistance of R_2. This ensures that the *base is common to the input and output signals.* For direct current, however, the capacitor acts like an "open circuit," and the transistor is biased as a common-emitter amplifier. Resistor R provides a dc return path between the emitter and ground.

BIASING THE EMITTER FOLLOWER   A basic emitter follower, also referred to as a *common-collector* (CC) amplifier, is illustrated in Fig. 12.9. The collector is connected directly to dc bias source E_{CC}. Output signal voltage E_o is taken across the emitter resistor R_E. Because the emitter and collector currents are approximately the same in a transistor, the common-emitter collector characteristics can be used in the analysis of the emitter follower.

Fig. 12.9 Basic emitter follower circuit.

Writing an equation for the dc voltages around the input circuit, we have

$$E_{CC} = R_1 I_B + V_{BE} + R_E I_C$$

assuming that I_C is approximately equal to the emitter current I_E. Solving for I_C,

$$I_C = \frac{E_{CC} - R_1 I_B - V_{BE}}{R_E} \qquad (12.8)$$

The dc voltage across the emitter resistor $E_{o(dc)}$ is

$$E_{o(dc)} = R_E I_C \qquad (12.9)$$

STABILIZING THE COMMON-EMITTER AMPLIFIER   Because of variations in reverse saturation current, current gain, and the base-emitter voltage with temperature and aging, it is necessary to *stabilize* the Q point of most common-emitter amplifiers. (The common-base and emitter follower amplifiers may be considered to be inherently stable.) A number of stabilization techniques exist and are examined in this section.

Current-feedback stabilization   An example of current-feedback stabilization is illustrated in Fig. 12.10a. Resistors R_1 and R_2 establish the Q point. Emitter resistor R_E, connected between the emitter of the transistor and ground, provides stabilization. In analyzing the circuit, it is convenient to replace resistors R_1 and R_2 connected to E_{CC} with a Thévenin equivalent circuit (see Chap. 6). Thévenin resistance R_B is

$$R_B = R_1 \| R_2 = \frac{R_1 R_2}{R_1 + R_2} \qquad (12.10a)$$

The Thévenin voltage E_B is

$$E_B = \frac{E_{CC} R_2}{R_1 + R_2} \qquad (12.10b)$$

The Thévenin circuit is given in Fig. 12.10b.

Assume that the dc current gain increases beyond its initial value. As a result, the collector current tends to rise, and the voltage across the emitter resistor E_E increases.

(a)   (b)

Fig. 12.10   Common-emitter amplifier with current-feedback stabilization: (a) Circuit. (b) Biasing network replaced by its Thévenin
equivalent circuit.

Because voltage E_E opposes E_B, the base-emitter voltage tends to decrease. Less base current flows, and the collector current is reduced to approximately its initial value.

If, for some reason, the collector current tends to decrease, voltage E_E also decreases. The base-emitter voltage increases, and more base (and collector) current flows.

From the previous description, current-feedback stabilization provides "automatic regulation" of the location of the Q point. A useful criterion for the effectiveness of current-feedback stabilization is the *current stability factor* S_i. It relates changes in collector current to changes in the reverse saturation current. A conservative approximate expression for S_i is

$$S_i \simeq 1 + \frac{R_B}{R_E} \qquad (12.11)$$

where, referring to Fig. 12.10b, R_B is the resistance of the parallel combination of R_1 and R_2 and R_E is the emitter resistance.

It is noted that by Eq. (12.11), if R_B is zero or R_E is infinite, $S_i = 1$, which represents perfect stabilization. For either of these values, however, the signal gain would be zero. Practical values for S_i are in the range of 4 to 10.

example 12.3   If, in Fig. 12.10a, $R_1 = 60$, $R_2 = 12$, and $R_E = 2$ kΩ, calculate the value of S_i.

solution   By Eq. (12.10a), $R_B = R_1 \| R_2 = (60 \times 12)/(60 + 12) = 10$ kΩ. By Eq. (12.11), $S_i = 1 + {}^{10}\!/_2 = 1 + 5 = 6$.

Voltage-feedback stabilization   Another stabilization technique, voltage-feedback stabilization, is illustrated in Fig. 12.11. The feedback resistor R_F connected between the collector and base of the transistor is the key element.

Fig. 12.11 Common-emitter amplifier with voltage-feedback stabilization.

Fig. 12.12 Offsetting the reduction in signal gain of an amplifier with current- or voltage-feedback stabilization: (*a*) Bypassing emitter resistance. (*b*) Decoupling feedback network.

Assume that, because of high-temperature operation, the base-emitter voltage has been reduced. Base current and, as a result, collector current increase. Consequently, the collector-emitter voltage V_{CE} is reduced. Because the feedback current I_F in the feedback resistor is approximately equal to the collector-emitter voltage divided by the feedback resistance ($I_F \simeq V_{CE}/R_F$), the feedback current falls. This, in turn, causes the base and collector currents to decrease. The Q point, therefore, tends to return to its initial value. Changes in reverse saturation current or dc current gain, whether increasing or decreasing, are stabilized in a similar manner.

The current stability factor for voltage-feedback stabilization is given by

$$S_i \simeq 1 + \frac{R_F}{R_C}$$ (12.12)

For good stability R_F should be low and R_C high in value. A low value of feedback resistance, however, reduces the signal gain of the amplifier.

Bypass and decoupling capacitors To realize a low value of stability factor without appreciable deterioration in signal gain, *bypass* and *decoupling* capacitors are employed. In Fig. 12.12*a*, bypass capacitor C_E is in parallel with emitter resistor R_E. If the reactance of C_E is one-tenth or less of the value of R_E at the lowest signal frequency, the signal gain of the amplifier is hardly affected.

In Fig. 12.12*b*, resistor R_F is replaced by two resistors, R_A and R_B. Their sum is made equal to R_F ($R_F = R_A + R_B$). Often, R_A and R_B are chosen to be equal to one-half the value of R_F. Decoupling capacitor C_F is connected between the junction of R_A, R_B and ground. For minimum effect on signal gain, C_F is chosen so its reactance is one-tenth the value of R_A or R_B, whichever is smaller, at the lowest signal frequency.

Diode stabilization A method employing a junction diode for stabilizing a common-emitter amplifier is illustrated in Fig. 12.13. The cathode of junction diode D is connected to the emitter of the transistor. Source E_D forward-biases the diode. It is reasonable to expect that variation in the diode junction voltage will track and oppose variations in the base-emitter voltage of the transistor. Consequently, changes in base-emitter voltage are canceled. In a silicon transistor, the most commonly used BJT, changes in the Q point are very dependent on the base-emitter voltage.

BIASING JFET AND DEPLETION-TYPE MOSFET AMPLIFIERS Because the FET is a voltage-operated device, *self-biasing* may be used. This is the prime method for biasing the JFET or depletion-type MOSFET. An example of an n-channel FET, self-biased by source resistor R_S connected between the source terminal of the FET and ground, is shown in Fig. 12.14. Drain current I_D results in a voltage drop across R_S which is equal to $I_D R_S$. The polarity of the voltage drop is such that the source is positive with respect to the ground.

Resistor R_G (typical value is 100 kΩ) is connected between the gate terminal and ground. It provides a dc return path for the voltage across the source resistor. As a

Fig. 12.13 Example of diode stabilization for a common-emitter amplifier.

Fig. 12.14 An n-channel common-source FET amplifier which is self-biased by source resistor R_S.

result, the gate is biased *negatively* with respect to the source by $-I_D R_S$ volts. For a p-channel device, the polarity of E_{DD} is reversed and the gate-source voltage is $+I_D R_S$ volts.

In addition to providing a bias voltage, the source resistor also stabilizes the operation of the amplifier. Its behavior is analogous to the emitter resistor used in current feedback stabilization of a BJT amplifier. To minimize the decrease in signal gain, R_S is bypassed with a capacitor, as in the common-emitter amplifier.

Writing an equation for the dc voltages in the output side of Fig. 12.14, we obtain

$$E_{DD} = I_D(R_D + R_S) + V_{DS}$$

Solving for I_D,

$$I_D = -\frac{V_{DS}}{R_D + R_S} + \frac{E_{DD}}{R_D + R_S} \tag{12.13}$$

Equation (12.13), similar to Eq. (12.5b), is the equation of a load line. Its use is illustrated in the next example.

example 12.4   Assume that in Fig. 12.14 the sum of R_D and R_S is 40 Ω and $E_{DD} = 20$ V. If the Q point is located at $V_{DQ} = 10$ V, determine (a) the quiescent current and the gate-source voltage, and (b) the values of R_D and R_S. Use the drain characteristics of Fig. 12.15.

Fig. 12.15 Determining the Q point of a FET amplifier. (See Example 12.4.)

solution   (a) Substitution of the given values in Eq. (12.13) yields $I_D = -V_{DS}/40 + {}^{20}\!/_{40} = -V_{DS}/40 + 0.5$. For $I_D = 0$, $V_{DS} = 20$ V, which is one point on the load line. For $V_{DS} = 0$, $I_D = 0.5$ A $= 500$ mA, the second point on the load line. The load line is drawn between the two points in Fig. 12.15.

For $V_{DQ} = 10$ V, the load line intersects the -3-volt bias curve. Therefore, the quiescent current I_{DQ} is approximately 230 mA, and the gate-source bias voltage is -3 V.

(b) Because the gate-source voltage is -3 V, $-I_D R_S = -3$. The quiescent drain current was found in (a) to be 230 mA $= 0.23$ A. Solving for R_S, we have $R_S = {}^3\!/_{0.23} \approx 13$ Ω. The value of R_D is $40 - 13 = 27$ Ω.

BIASING THE ENHANCEMENT-TYPE MOSFET   An example of biasing an n-channel enhancement-type MOSFET is illustrated in Fig. 12.16. Feedback resistor R_F is

Fig. 12.16 Biasing an n-channel enhancement-type MOSFET.

connected between the drain and gate terminals of the transistor. By voltage division, the gate-source voltage V_{GS} equals the product of the drain-source voltage V_{DS} and the ratio $R_G/(R_G + R_F)$:

$$V_{GS} = \frac{V_{DS}R_G}{R_G + R_F} \qquad (12.14)$$

For a p-channel enhancement-type MOSFET, the polarity of E_{DD} is reversed.

In addition to providing a bias voltage, the feedback resistor also stabilizes the Q point of the transistor. To minimize the attenuation in signal gain, R_F may be decoupled as in the common-emitter amplifier using voltage-feedback stabilization.

12.5 SMALL-SIGNAL AMPLIFIERS

The performance of small-signal amplifiers is determined with the aid of a few simple equations. These equations are based on *models*, or *equivalent circuits*, of the amplifier in question. In this section the performance of small-signal BJT and FET amplifiers is considered. The basic parameters for these devices are defined in Chap. 8.

In the analysis of amplifiers it is desirous to consider specific frequency ranges of signals. This allows the development of simple models and provides insight into amplifier performance. In this section it is assumed that the range of frequencies is less than 5000 Hz, referred to the *midfrequency range*. At midfrequencies, the inherent transistor and stray wiring capacitances may be neglected.

ANALYZING THE BJT AMPLIFIER   The use of hybrid (h) parameters in the analysis of the small-signal BJT amplifier leads to a simple and generalized model. Let the transistor be represented by a circle with three leads numbered 1, 2, and 3, as shown in Fig. 12.17a. For example, lead 1 may be the base, lead 2 the collector, and lead 3 the emitter. A *hybrid model* of the BJT, suitable for midfrequencies, is shown in Fig. 12.17b.

Parameter h_i is the short-circuit input resistance; h_r is the reverse transfer ratio; h_f is the short-circuit forward current gain; and h_o is the output admittance. The unit for h_i is ohms and for h_o it is mhos. Parameters h_f and h_r have no units; that is, they are

Fig. 12.17 Developing a small-signal model of the BJT: (a) General representation of a transistor. (b) Hybrid model.

numerics. Sources $h_r e_2$ and $h_f i_1$ are examples of *dependent*, or *controlled*, voltage and current sources, respectively (see Chap. 6).

The three useful configurations for the BJT are the common-emitter (CE), common-base (CB), and common-collector (CC), or emitter follower amplifiers. To distinguish the configuration, a second letter is added to each subscript. For example, if the transistor is used in the common-emitter configuration, h_i becomes h_{ie}, h_r becomes h_{re}, and so on. The notation is summarized in Table 12.1.

TABLE 12.1   Hybrid Parameter Notation

	Configuration		
h parameter	CE	CB	CC
h_i	h_{ie}	h_{ib}	h_{ic}
h_r	h_{re}	h_{rb}	h_{rc}
h_f	h_{fe}	h_{fb}	h_{fc}
h_o	h_{oe}	h_{ob}	h_{oc}

Because the CE configuration is widely used, the manufacturer normally provides only common-emitter hybrid parameters on his data sheets. If the common-base or common-collector parameters are needed, they can be obtained by the approximate equations summarized in Table 12.2. In the table the common-base and common-collector parameters are expressed in terms of the common-emitter parameters.

TABLE 12.2   Approximate Hybrid Parameter Conversions

$h_{ib} \approx h_{ie}/(1 + h_{fe})$	$h_{ic} \approx h_{ie}$
$h_{rb} \approx h_{ie}h_{oe}/(1 + h_{fe}) - h_{re}$	$h_{rc} \approx 1$
$h_{fb} \approx -h_{fe}/(1 + h_{fe})$	$h_{fc} \approx -(1 + h_{fe})$
$h_{ob} \approx h_{oe}/(1 + h_{fe})$	$h_{oc} \approx h_{oe}$

example 12.5   The common-emitter h parameters of a particular transistor are: $h_{ie} = 1000$ Ω, and $h_{re} = 10^{-4}$, $h_{fe} = 49$, $h_{oe} = 10^{-5}$ mho. Calculate the h parameters for the transistor in the (a) common-base and (b) common-collector configurations.

solution   (a) By the equations of Table 12.2, $h_{ib} = 1000/(1 + 49) = 20$ Ω; $h_{rb} = 10^3 \times 10^{-5}/(1 + 49) - 10^{-4} = 2 \times 10^{-4} - 10^{-4} = 10^{-4}$; $h_{fb} = -49/(1 + 49) = -49/50 = -0.98$; and $h_{ob} = 10^{-5}/(1 + 49) = 0.2 \times 10^{-6}$ mho.

(b) From Table 12.2, $h_{ic} = 1000$ Ω; $h_{rc} = 1$; $h_{fc} = -(1 + 49) = -50$; and $h_{oc} = 10^{-5}$ mho.

Simplified hybrid model   For most small-signal amplifiers, the simplified model of Fig. 12.18 may be used. Comparing Fig. 12.18 with Fig. 12.17b, $h_r e_2$ and h_o have been eliminated in the simplified hybrid model. The model for each of the three configurations is illustrated in Fig. 12.19. In each case the numerals have been replaced by letters E, B, and C. It is interesting to note that the same basic model is used for all three configurations.

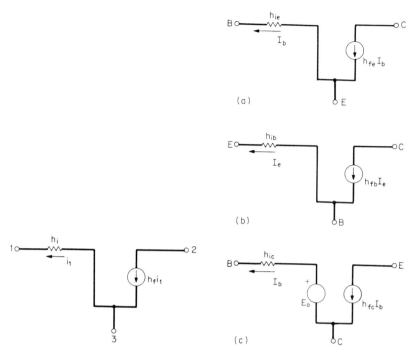

Fig. 12.18 Simplified hybrid model of a BJT.

Fig. 12.19 Hybrid models for the three transistor configurations: (a) Common emitter. (b) Common base. (c) Common collector (emitter follower).

Application of the simplified hybrid model   We consider first the common-emitter amplifier of Fig. 12.20a. Resistance R_s is the resistance of the source (source resistance). It is reasonable to assume that the reactance of capacitors C_c and C_E are zero: that is, they act like short circuits. Because our concern here is with the signal behavior of an amplifier, the dc sources (E_{CC} in this case) are set to zero. The resultant model of the amplifier is given in Fig. 12.20b. Resistance R_B is equal to the parallel combination of bias resistors R_1 and R_2.

Figure 12.21a shows a common-base amplifier, and its model is given in Fig. 12.21b. The emitter follower and its model are illustrated in Fig. 12.22. In deriving the models for these two configuration, the dc sources were set to zero and the coupling capacitor was assumed to act as a short circuit.

Fig. 12.20   An example of a common-emitter amplifier: (a) Circuit. (b) Model.

Fig. 12.21   An example of a common-base amplifier: (a) Circuit. (b) Model.

Fig. 12.22   An example of an emitter follower: (a) Circuit. (b) Model.

Equations for current gain A_i, input resistance R_i, voltage gain A_v, and output resistance R_o, for each of the three configurations are summarized in Table 12.3. The equations, expressed in terms of the common-emitter hybrid parameters, are based on:
   1. The simplified hybrid model.
   2. It is assumed that R_B is much greater than the input resistance and that R_C is equal to or less than 5000 Ω.
The power gain of an amplifier may be taken as the product of the voltage and current gains.

TABLE 12.3   Expressions for Evaluating the Performance of a BJT Amplifier

	CE	CB	CC
A_i	h_{fe}	$-h_{fe}/(1 + h_{fe})$	$-(1 + h_{fe})$
R_i	h_{ie}	$h_{ie}/(1 + h_{fe})$	$h_{ie} + (1 + h_{fe})R_E$
A_v	$-h_{fe}R_C/(R_s + R_i)$	$R_C/(R_s + R_i)$	1
R_o	∞	∞	$(R_s + h_{ie})/(1 + h_{fe})$

   example 12.6   The transistor used in the amplifiers of Figs. 12.20 to 12.22 have an h_{fe} of 49 and an h_{ie} of 900 Ω. Assume that $R_C = 5000$, $R_E = 1000$, and $R_s = 100$ Ω. With the aid of Table 12.3, calculate A_i, R_i, A_v, and R_o for each amplifier.

   solution   For the CE amplifier of Fig. 12.20, $A_i = 49$; $R_i = 900$ Ω; $A_v = -49 \times 5000/(100 + 900) = -245$; and $R_o = \infty$.
   For the CB amplifier of Fig. 12.21, $A_i = -^{49}/_{50} = -0.98$; $R_i = 900/(1 + 49) = ^{900}/_{50} = 18$ Ω; $A_v = 5000/(100 + 18) = 5000/118 = 42.5$; and $R_o = \infty$.
   For the CC (emitter follower) amplifier of Fig. 12.22, $A_i = -(1 + 49) = -50$; $R_i = 900 + (1 + 49) \times 1000 = 50\,900$ Ω, $A_v = 1$; and $R_o = (100 + 900)/(1 + 49) = ^{1000}/_{50} = 20$ Ω.

Practical considerations   The negative sign before an expression for voltage or current gain indicates that the output signal is 180° *out of phase* with respect to the input signal. Referring to Fig. 12.23, during the positive half-cycle of the input signal,

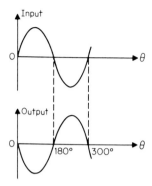

Fig. 12.23 Input and output signal waveforms 180° out of phase.

Fig. 12.24 Small-signal admittance (y) model of a FET operating at midfrequencies.

the output is a negative half-cycle. As the input signal goes through its negative half-cycle, the output is positive-going.

Although the voltage gain of an emitter follower never exceeds unity, its usefulness lies in its relatively *high input resistance* and *low output resistance*. (In Example 12.6 we found for the emitter follower that $R_i = 50\ 900$ and $R_o = 20\ \Omega$.) For example, the emitter follower is used if a voltage source having a large source resistance is required to deliver power to a low-resistance load. If an emitter follower is interposed between the source and load, virtually full power is delivered to the load.

The equations summarized in Table 12.3 assume that R_B is much greater than the input resistance of the amplifier. If this is not true, because of *current division,* a portion of the input signal current is drained through R_B, and less base current flows. The method for considering current division is covered in Chap. 6.

If, in Fig. 12.20a, emitter resistor R_E is not suitably bypassed, the equations for the CE amplifier of Table 12.3 must be modified. For an unbypassed emitter resistor, the input resistance becomes

$$R_i = h_{ie} + (1 + h_{fe})R_E \qquad (12.15)$$

Substitution of Eq. (12.15) for R_i in the equations for the CE amplifier of Table 12.3 yields the correct results for an unbypassed emitter resistor.

ANALYZING THE FET AMPLIFIER In a manner analogous to the development of the small-signal model for the BJT, the basic model for a FET, or MOSFET, operating at midfrequencies is illustrated in Fig. 12.24. For the field-effect device, *admittance* (y) *parameters* are used. The equivalent circuit is referred to as a y, or an admittance, model. Table 12.4 lists the y parameters for midfrequency operation for the three configurations of the FET or MOSFET: common source (CS), common gate (CG), and common drain (CD), or source follower, amplifiers. Parameter g_{fs} is the common-source forward transconductance and g_d the drain-source admittance. Their unit is the *mho.*

The three amplifier configurations and their models for the field-effect transistor are

TABLE 12.4 Small-Signal y Parameters for the Field-Effect Transistor

Parameter	CS	CG	CD
y_{11}	0	$g_{fs} + g_d$	0
y_{12}	0	$-g_d$	0
y_{21}	g_{fs}	$-(g_{fs} + g_d)$	$-g_{fs}$
y_{22}	g_d	g_d	$g_{fs} + g_d$

Fig. 12.25 Three basic FET amplifiers and their small-signal models: (a) Common source. (b) Common gate. (c) Common drain (source follower).

given in Fig. 12.25. Table 12.5 summarizes the equations for voltage gain, etc., for each circuit. The equations are based on the following assumptions:
1. $R_G \gg R_s$.
2. $g_{fs} \gg g_d$.
3. Capacitors C_c and C_S act like short circuits.
4. Input resistance does not include R_G.
5. Midfrequency operation.

example 12.7 A FET has a $g_{fs} = 0.1$ mho and $g_d = 10^{-6}$ mho. If $R_D = 10$ kΩ and $R_S = 1$ KΩ, calculate the values of A_v and R_o the common-source, common-gate, and common-drain amplifiers of Fig. 12.25.

solution Substituting the given values in the appropriate expressions of Table 12.5, we obtain

TABLE 12.5 Expressions for Evaluating the Performance of a FET (or MOSFET) Amplifier

	CS	CG	CD
A_v	$-g_{fs}/(1/R_L + g_d)$	$g_{fs}/(1/R_L + g_d)$	$g_{fs}/(1/R_L + g_{fs})$
A_i	∞	-1	∞
R_i	∞	$\dfrac{1}{g_{fs} - g_d g_{fs}(1/R_L + g_d)}$	∞
R_o	$1/g_d$	$1/g_d$	$1/g_{fs}$

CS: $A_v = -g_{fs}/(1/R_L + g_d) = -0.1/(1/10^4 + 10^{-6}) = -0.1/(10^{-4} + 10^{-6}) \simeq -0.1/10^{-4} = -1000.$ $R_o = 1/g_d = 1/10^{-6} = 1$ MΩ.

CG: $A_v = g_{fs}/(1/R_L + g_d) = 0.1/(1/10^4 + 10^{-6}) \simeq 1000.$ $R_o = 1/g_d = 1/10^{-6} = 1$ MΩ.

CD: $A_v = g_{fs}/(1/R_L + g_{fs}) = 0.1/(1/10^4 + 0.1) \simeq 1.$ $R_o = 1/g_{fs} = 1/0.1 = 10$ Ω.

Like the emitter follower, the source follower has a high (infinite) input resistance and a low (10 Ω in this example) output resistance. Its voltage gain can never exceed unity.

If the emitter resistance of the common-source amplifier (Fig. 12.25a) is not suitably by-passed, the equation for A_v in Table 12.5 cannot be used. Instead, the voltage gain is ex-pressed by

$$A_v = \frac{-g_{fs}}{1/R_L + g_d + g_{fs}R_S/R_L}$$

12.6 FREQUENCY RESPONSE OF AN AMPLIFIER

In the previous section, effects of the inherent transistor capacitances and the coupling capacitor on amplifier performance were neglected. Transistor capacitance attenuates high-frequency signals and the coupling capacitor attenuates low-frequency signals. These effects, for the common-emitter and common-source amplifiers, are examined in this section.

HIGH-FREQUENCY MODEL OF COMMON-EMITTER AMPLIFIER The widely used *hybrid-pi model* of a transistor in the CE configuration at high-frequency operation is provided in Fig. 12.26a. Resistance $r_{bb'}$, the *base-spreading resistance*, is the ohmic

(a)

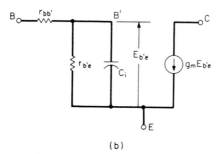

(b)

Fig. 12.26 Hybrid-pi model of a BJT in the common-emitter configuration for high-frequency operation: (a) Actual. (b) Simplified.

resistance of the base region. Its value is in the order of 100 Ω. *Base-emitter resistance* $r_{b'e}$ varies with the collector current: the higher the current, the lower its value, and vice versa. The sum of $r_{bb'}$ and $r_{b'e}$ is equal to h_{ie}:

$$h_{ie} = r_{bb'} + r_{b'e} \tag{12.16}$$

The dependent source is $g_m E_{b'e}$, where g_m is the *forward transconductance* of the transistor, and $E_{b'e}$ is the voltage across points B' and E. The transconductance is related to h_{fe} and $r_{b'e}$ by

$$g_m = \frac{h_{fe}}{r_{b'e}} \tag{12.17a}$$

In general, $r_{bb'}$ is much less than $r_{b'e}$; hence, $r_{b'e} \simeq h_{ie}$, and

$$g_m \simeq \frac{h_{fe}}{h_{ie}} \tag{12.17b}$$

The transconductance also depends on the dc collector current I_C. At room temperature (25°C), the relationship is

$$g_m = \frac{I_C}{26} \text{ millimhos} \tag{12.18}$$

where I_C is expressed in milliamperes.

The *collector-emitter resistance* r_{ce} is equal to one divided by h_{oe}:

$$r_{ce} = \frac{1}{h_{oe}} \tag{12.19}$$

It is interesting to note that the noncapacitive elements in the hybrid-pi model are related to the hybrid parameters.

Capacitances $C_{b'e}$ and $C_{b'c}$ are referred to the *diffusion* and *depletion*, or *barrier, capacitances*. Values of $C_{b'e}$ range from 100 to 1000 pF (picofarad $= 10^{-12}$ F) and $C_{b'c}$ is typically a few picofarads.

Simplified hybrid-pi model   Owing to a phenomenon, called the *Miller effect*, capacitance $C_{b'c}$ suitably modified may be placed in parallel with $C_{b'e}$. Capacitance $C_{b'c}$ is multiplied by one plus the product of g_m and R_C. The result C_M, referred to as the *Miller capacitance*, is

$$C_M = (1 + g_m R_C) C_{b'c}$$

Addition of C_M to $C_{b'e}$ yields the effective input capacitance C_i:

$$C_i = C_{b'e} + (1 + g_m R_C) C_{b'c} \tag{12.20}$$

The *simplified hybrid-pi model* containing C_i is shown in Fig. 12.26b. Because r_{ce} is usually much greater than R_C, it is omitted from the figure. The high-frequency model of the common-emitter amplifier of Fig. 12.20a appears as in Fig. 12.27.

Because at increasing frequencies the reactance of C_i decreases, the voltage across it, $E_{b'e}$, also drops. As a result, the collector signal current I_c, which equals the product of g_m and $E_{b'e}$, and the output voltage E_o are reduced. The voltage gain, therefore, falls with increasing frequency.

The upper -3-dB frequency f_H at which the voltage gain falls to approximately 70 percent of its value at midfrequencies (usually taken at 1 kHz) is determined as follows. Signal source E_s is set to zero. (As explained in Chap. 6, in setting a voltage source to zero, it is short-circuited; in setting a current source to zero, it is open-circuited.) A

Fig. 12.27   High-frequency model of a common-emitter amplifier.

reasonable assumption is that R_B is much greater than source resistance R_s. Therefore, neglecting R_B, the equivalent resistance R_{eq} across C_i is $(R_s + r_{bb'})$ in parallel with $r_{b'e}$ (Fig. 12.28). Hence,

$$R_{eq} = \frac{(R_s + r_{bb'})r_{b'e}}{R_s + r_{bb'} + r_{b'e}} \tag{12.21}$$

At the upper −3-dB frequency f_H, R_{eq} is equal to the reactance of $C_i = 1/2\pi f_H C_i$. Therefore,

$$R_{eq} = \frac{1}{2\pi f_H C_i}$$

Solving for f_H,

$$f_H = \frac{1}{2\pi R_{eq} C_i} = \frac{0.159}{R_{eq} C_i} \tag{12.22}$$

since $1/2\pi = 0.159$.

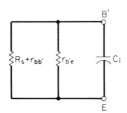

Fig. 12.28 Model of input circuit for determining the upper −3-dB frequency, f_H.

example 12.8 In Fig. 12.27, $g_m = 0.1$ mho, $C_{b'e} = 300$ pF, $C_{b'c} = 4$ pF, $R_C = 1000\ \Omega$, $R_s = 900\ \Omega$, $r_{bb'} = 100\ \Omega$, and $r_{b'e} = 1000\ \Omega$. Neglecting R_B, calculate (a) C_i, (b) R_{eq}, and (c) f_H.

solution (a) Substituting the given values, by Eq. (12.20), $C_i = 300 + (1 + 0.1 \times 1000) \times 4 = 300 + 101 \times 4 = 704$ pF.
(b) By Eq. (12.21), $R_{eq} = (900 + 100) \times 1000/(900 + 100 + 1000) = 500\ \Omega$.
(c) By Eq. (12.22), $f_H = 0.159/(500 \times 704 \times 10^{-12}) = 452,000$ Hz $= 452$ kHz $= 0.452$ MHz.

Figures of merit Three commonly used figures of merit for comparing the high-frequency performance of bipolar transistors are:
1. *Short-circuit beta cutoff frequency* f_β is the frequency at which h_{fe} is equal to approximately 70 percent of its value at midfrequencies. It is expressed by

$$f_\beta = \frac{0.159}{r_{b'e}(C_{b'e} + C_{b'c})} \tag{12.23}$$

The lower the values of $r_{b'e}$, $C_{b'e}$, and $C_{b'c}$, the higher is f_β.
2. *Short-circuit gain-bandwidth product* f_T is the frequency at which h_{fe} is equal to one. It is given by

$$f_T = \frac{0.159 g_m}{C_{b'e} + C_{b'c}} \tag{12.24}$$

The relationship of h_{fe}, f_T, and f_β is given by

$$f_T = h_{fe} f_\beta \tag{12.25}$$

3. *Short-circuit alpha cutoff frequency* f_α is the frequency at which h_{fb} is equal to approximately 70 percent of its value at midfrequencies. It is expressed by

$$f_\alpha = \frac{0.159 g_m}{C_{b'e}} \tag{12.26}$$

Because $C_{b'c}$ is generally much less than $C_{b'e}$, comparing Eqs. (12.24) and (12.26), $f_\alpha \approx f_T$.

HIGH-FREQUENCY MODEL OF COMMON-SOURCE AMPLIFIER A high-frequency model of a FET in the common-source configuration is given in Fig. 12.29a. Capacitances C_{gs} and C_{gd} are depletion capacitances between gate and source, and gate and drain, respectively. Capacitance C_{ds} is the drain-source capacitance of the channel.

By using the Miller effect, as was done for the common-emitter amplifier, the high-

Fig. 12.29 High-frequency model of a FET (or MOSFET) in the common-source configuration: (a) Actual. (b) Simplified.

frequency model is simplified, as shown in Fig. 12.29b. Input capacitance C_i is expressed by

$$C_i = C_{gs} + (1 + g_m R_D) \qquad (12.27)$$

The value of $1/g_d$ is generally much greater than practical values of load resistance R_D; therefore it is omitted from the simplified model. Because C_i is the influencing capacitance in determining the high-frequency response, C_{ds} is also omitted from the figure.

The high-frequency model of the common-source amplifier of Fig. 12.25a is illustrated in Fig. 12.30. Similar to the analysis of the CE amplifier, the upper -3-dB frequency f_H is determined by equating source resistance R_s to the reactance of C_i:

$$R_s = \frac{0.159}{f_H C_i}$$

Fig. 12.30 High-frequency model of a common-source amplifier.

Fig. 12.31 Input circuits at low frequency for (a) a CE amplifier and (b) a CS amplifier.

Solving for f_H,

$$f_H = \frac{0.159}{R_s C_i} \qquad (12.28)$$

LOW-FREQUENCY RESPONSE   Assume that in a common-emitter amplifier the emitter resistor is suitably bypassed. (For a common-source amplifier, it is assumed that the source resistor is suitably bypassed.) The low-frequency response is then determined by coupling capacitor C_c. This is illustrated in Fig. 12.31 where the input circuits for the CE and CS amplifiers at low-frequency operation are shown.

The reactance of a capacitor, as mentioned earlier, decreases with increasing frequencies and increases with decreasing frequencies. Thus, at high frequencies, the reactance is so much less than the series resistance in the circuit that it acts like a short circuit. As the frequency is reduced, however, the reactance is increased. As a result, the voltage across the base and emitter (or gate and source) terminals is reduced and the gain of the amplifier falls.

The lower -3-dB frequency f_L is determined by setting the total series resistance in the input circuit equal to the reactance of the coupling capacitor at f_L. For the CE amplifier,

$$R_s + h_{ie} = \frac{0.159}{f_L C_c}$$

Solving for f_L,

$$f_L = \frac{0.159}{(R_s + h_{ie})C_c} \qquad (12.29a)$$

For the CS amplifier,

$$R_s + R_G = \frac{0.159}{f_L C_c}$$

and

$$f_L = \frac{0.159}{(R_s + R_G)C_c} \qquad (12.29b)$$

A similar approach is used in calculating the low-frequency response of the other amplifier configurations.

12.7 CASCADING AMPLIFIER STAGES

To realize greater signal gain than obtainable from a single amplifier stage, a number of stages are *cascaded*, as illustrated in the block diagram of Fig. 12.32. In a cascaded amplifier, the output of one amplifier stage is *coupled* to the input of the following stage. If the stages are *noninteracting*, that is, if there is no appreciable change in the gain or other parameters of an individual stage when cascaded, the overall gain A_t is equal to the product of the individual gains,

$$A_t = A_1 \times A_2 \times \cdots \times A_n \qquad (12.30)$$

where A_1 is the gain of the first stage, A_2 the gain of the second stage, and A_n the gain of the nth stage.

Owing to its nearly infinite input resistance, the FET amplifier may be considered as being noninteracting. The BJT, because its input resistance is relatively low, interacts and Eq. (12.30) cannot be used. Instead, working from the output toward the input stage, the gain of each stage must be considered separately, taking into account loading effects of the cascaded stages.

COUPLING CASCADED STAGES   Two basic methods are used for coupling cascaded stages:
1. Resistance-capacitance (*RC*) coupling.
2. Direct coupling (*DC*).

Fig. 12.32 Block diagram of a cascaded amplifier.

Examples of resistance-capacitance and direct-coupled BJT stages are illustrated in Fig. 12.33. (If a FET is substituted for a BJT, the circuits are identical.)

In *RC* coupling, the dc collector voltage of the preceding stage is blocked by the coupling capacitor from the base of the following stage. In direct coupling, the collector of one transistor is coupled directly to the base of the following transistor. Because the dc operating point of a transistor tends to drift, special circuit precautions are taken in *DC* amplifier design. The greatest use of direct coupling is in integrated-circuit amplifiers (see Chap. 13). Unless direct current or slowly varying signals are to be amplified, *RC* coupling is invariably used in discrete amplifiers.

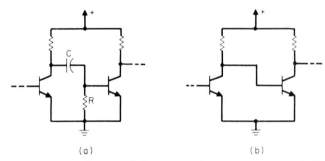

(a)   (b)

Fig. 12.33 Coupling cascaded stages: (*a*) Resistance-capacitance (*RC*) coupling. (*b*) Direct coupling (*DC*).

FREQUENCY RESPONSE OF A CASCADED AMPLIFIER The overall frequency response of a cascaded amplifier is less than that for an individual stage. For *noninteracting identical stages*, the upper −3-dB frequency of an n-stage cascaded amplifier f'_H is

$$f'_H = f_H \sqrt{2^{1/n} - 1} \qquad (12.31a)$$

The lower −3-dB frequency f'_L is

$$f'_L = f_L / \sqrt{2^{1/n} - 1} \qquad (12.31b)$$

where f_H and f_L are the upper and lower −3-dB frequencies of an individual stage.

example 12.9 An individual stage of a two-stage *RC*-coupled amplifier has $f_H = 20$ kHz and $f_L = 200$ Hz. Its bandwidth, therefore, is $20\,000 - 200 = 19\,800$ Hz $= 19.8$ kHz. Assuming that the stages are identical and noninteracting, calculate the bandwidth of the cascaded amplifier.

solution By Eq. (12.31*a*), $f'_H = 20\sqrt{2^{1/2} - 1}$. The quantity $2^{1/2} = \sqrt{2} = 1.414$; hence, $\sqrt{1.414 - 1} = \sqrt{0.414} = 0.64$. Therefore, $f'_H = 20 \times 0.64 = 12.8$ kHz. By Eq. (12.31*b*), $f'_L = 200/0.64 = 313$ Hz.

The bandwidth of the cascaded amplifier is $f'_H - f'_L = 12\,800 - 313 \approx 12.5$ kHz. With respect to an individual stage, the bandwidth of the cascaded amplifier was reduced in excess of 60 percent.

12.8 POWER AMPLIFIERS

A useful criterion for comparing power amplifiers is the conversion efficiency η (Greek letter *eta*). The conversion efficiency is defined as the signal power delivered to the load P_{ac} to the dc power P_{dc} supplied to the collector or drain circuit of a transistor. Expressed as a percentage,

$$\eta = \frac{P_{ac}}{P_{dc}} \times 100\% \qquad (12.32)$$

For example, the maximum conversion efficiency realized with class A operation is 50 percent and with class B operation 78.4 percent. Class A and B power amplifiers are examined in the following sections.

CLASS A POWER AMPLIFIER A typical class A power amplifier is illustrated in Fig. 12.34. Load R_L, which may be the coil of a loudspeaker, is coupled to the collector circuit of the transistor by *coupling transformer T*. The turns ratio of the transformer n is the ratio of N_1 to N_2 turns:

$$n = \frac{N_1}{N_2} \qquad (12.33)$$

The value of R_L "seen" by the transistor is modified. The modified, or *reflected*, value R_L' is given by the product of n squared and R_L:

$$R_L' = n^2 R_L \qquad (12.34)$$

The output power is generally determined graphically. Two load lines are drawn on the collector (or drain) family of curves. The dc, or *static*, load line represents the dc resistance of transformer winding N_1. Because its resistance is extremely low, it is taken to be zero. A resistance value of zero ohms appears as a *vertical line* perpendicular to the voltage axis at E_{CC} (or E_{DD}), as shown in Fig. 12.35.

Fig. 12.34 Typical class A power amplifier.   **Fig. 12.35** Superimposing the dc and ac load lines on the collector characteristics.

The Q point is determined by the intersection of the dc load line and base current I_{BQ}. The value of the base current depends on the values of R_1 and R_2 of the biasing network (Fig. 12.34). Intersecting the Q point and having a slope of $-1/R_L'$ is the ac, or *dynamic*, load line.

Assume that the input sine wave signal has a positive peak value of I_{B2} and a negative peak value of I_{B1}. The corresponding maximum and minimum values of collector current and voltage are found by the intersection of the ac load line and the I_{B1} and I_{B2} curves. As indicated in Fig. 12.35, these values are denoted by I_{max}, E_{max}, I_{min}, and E_{min}, respectively. The output power P_{ac} in watts is

$$P_{ac} = \frac{(E_{max} - E_{min})(I_{max} - I_{min})}{8} \qquad (12.35)$$

COLLECTOR DISSIPATION The power dissipated as heat in the collector is referred to as collector dissipation P_C. (For a FET, the dissipated power in the drain is called the *drain dissipation P_D*.) The value of P_C is

$$P_C = E_{CC} I_{CQ} - P_{ac} \qquad (12.36a)$$

A class A amplifier operating without a signal is said to be *idling*. When idling, $P_{ac} = 0$, and the amplifier is *dissipating maximum power*, $P_{C,max}$:

$$P_{C,max} = E_{CC} I_{CQ} \qquad (12.36b)$$

It is important that the maximum dissipated power never exceed the value given by the manufacturer. Further, from Fig. 12.35, it is seen that the maximum collector signal voltage E_{max} exceeds E_{CC}. The value of E_{max} must always be less than the collector (or drain) breakdown voltage of the transistor specified by the manufacturer.

SECOND-HARMONIC DISTORTION   Second-harmonic distortion of an amplifier provides an indication of how faithful the amplified output signal is with respect to the input signal.   Using the values found graphically in determining the output power, second-harmonic distortion D_2 is

$$D_2 = \frac{|I_{max} + I_{min} - 2I_{CQ}|}{2|I_{max} - I_{min}|} \times 100\% \qquad (12.37)$$

The vertical lines indicate the *magnitude of the quantity.*

example 12.10   The dc and ac load lines are superimposed on the collector characteristics of Fig. 12.36. The Q point for class A operation is located at approximately $E_{CC} = 4$ V and $I_{CQ} = 1$ A.   For an input signal swing between $I_B = 200$ mA and 10 mA, the following values are estimated from the intersection of the ac load line and the base current curves: $E_{max} \simeq 7.3$ V, $E_{min} \simeq 2.7$ V, $I_{max} \simeq 1.4$ A, and $I_{min} \simeq 0.2$ A.   Find (a) P_{ac}; (b) η; (c) P_C and $P_{C.max}$; (d) D_2.

solution   (a) By Eq. (12.35), $P_{ac} = (7.3 - 2.7)(1.4 - 0.2)/8 = 4.6 \times 1.2/8 \simeq 0.69$ W.
(b) $P_{dc} = E_{CC}I_{CQ} = 4 \times 1 = 4$ W.   By Eq. (12.32), $\eta = (0.69/4) \times 100\% = 17.2\%$.   This is considerably less than the maximum conversion efficiency of 50 percent for class A operation.
(c) By Eqs. (12.36a) and (12.36b), $P_C = 4 - 0.69 = 3.31$ W; $P_{C.max} = 4$ W.
(d) By Eq. (12.37), $D_2 = \dfrac{|1.4 + 0.2 - 2 \times 1|}{2|1.4 - 0.2|} \times 100\% = 0.4/2.4 \times 100\% = 16.6\%$.

PUSH-PULL POWER AMPLIFIERS   Examples of push-pull amplifier circuits are illustrated in Fig. 12.37.   An advantage of push-pull over single-ended operation is that,

(a)

(b)

Fig. 12.36 Determining output power of a class A single-ended amplifier. (See Example 12.10.)

Fig. 12.37 Examples of class B push-pull power amplifiers: (a) Conventional type. (b) Complementary symmetry type.

assuming identical transistors, all even harmonics of distortion are canceled. If the transistors are operated as class B, conversion efficiencies as high as 78.4 percent are realized. Further, when the amplifier is idling, the collector dissipation is zero. The discussion to follow is limited to class B operation.

The circuit of Fig. 12.37a requires two center-tapped transformers, T_1 and T_2. Transformer T_1 is referred to as a *phase-inverting input transformer.* The voltage polarity at the dotted terminals is always the same, either plus or minus. If the base of transistor Q_1 is positive, the base of transistor Q_2 is negative, and vice versa. Output transformer T_2 couples load resistance R_L across the collectors of Q_1 and Q_2.

Assume that the signal is a sine wave. On the positive half-cycle, the base of Q_1 is positive and the base of Q_2 negative. Transistor Q_1 conducts and Q_2 is cut off. During the negative half-cycle, the polarities are reversed. Now, Q_1 is cut off and Q_2 conducts current. During a cycle, therefore, each transistor in a class B push-pull amplifier conducts current half the time.

The circuit of Fig. 12.37b, referred to as a *complementary symmetry amplifier,* does not require input and output transformers. (It may be regarded as a *transformerless* power amplifier.) Transistor Q_1 is npn and Q_2 is pnp. For good operation, Q_1 and Q_2 should be closely matched in their electrical characteristics. During the positive half-cycle, Q_1 conducts and Q_2 is cut off. On the negative half-cycle the situation is reversed: Q_2 conducts and Q_1 is cut off.

CROSSOVER DISTORTION The input resistance of an amplifier increases with decreasing collector current and decreases with increasing collector current. Consequently, in class B operation, the sine wave is flattened, as illustrated in Fig. 12.38. The flattening is referred to as *crossover distortion.*

To reduce crossover distortion, power amplifiers are generally operated as class AB. Under this mode of operation, a small collector current flows for zero signal current. The conversion efficiency of a class AB push-pull amplifier is in the order of 60 to 70 percent.

Fig. 12.38 Illustration of crossover distortion.

Fig. 12.39 Thermal equivalent circuit of a transistor mounted on a heat sink.

HEAT SINKS Because of possible large collector dissipation, power transistors (as well as power diodes and other semiconductor devices) are often mounted on heat sinks. A heat sink is a metal plate or structure, typically aluminum, which aids in dissipating the heat away from the device.

One can draw a *thermal equivalent circuit* of a transistor mounted on a heat sink, as shown in Fig. 12.39. This circuit is similar to a series circuit of resistors where temperature T is analogous to voltage; thermal resistance Θ (Greek letter *theta*) is analogous to electrical resistance; and dissipated power P_C is analogous to current. By Ohm's law, $I = E/R$; for the thermal equivalent circuit, one can write $P_C = T/\Theta$.

The temperature of the collector-base junction is designated by T_J; T_C is the temperature of the transistor case; T_S is the temperature of the heat sink; and T_A is the ambient (environment) temperature. The unit of temperature is degrees Celsius.

Thermal resistance Θ_{JC} is the resistance to the flow of heat between the junction and case of the device; Θ_{CS} is the thermal resistance between the case and heat sink; and Θ_{SA} is the thermal resistance between the heat sink and the ambient. The unit for Θ is degrees Celsius per watt and P_C is in watts.

Applying Ohm's law to the thermal equivalent circuit, the allowable dissipated power is equal to the difference in the junction and ambient temperature divided by the sum of the thermal resistances:

$$P_C = \frac{T_J - T_A}{\Theta_{JC} + \Theta_{CS} + \Theta_{SA}} \tag{12.38}$$

The application of Eq. (12.38) is illustrated in the following example.

example 12.11 A silicon transistor is mounted on a heat sink. If $T_J = 125°C$ and $T_A = 25°C$, determine the thermal resistance of the heat sink needed in order for the transistor to dissipate 50 W. Assume that $\Theta_{JC} = 0.5°C/W$ and $\Theta_{CS} = 0.3°C/W$.

solution Substitution of the given values in Eq. (12.38) yields $50 = (125 - 25)/(0.5 + 0.3 + \Theta_{SA}) = 100/(0.8 + \Theta_{SA})$. Solving, $0.8 + \Theta_{SA} = \frac{100}{50} = 2$. Hence, $\Theta_{SA} = 2 - 0.8 = 1.2°C/W$.

12.9 FEEDBACK AMPLIFIERS

Examples of feedback have already been considered in this chapter. It was seen that current- and voltage-feedback stabilization helped maintain the Q point of an amplifier relatively fixed, regardless of changes in temperature or other parameters. In general, an amplifier with feedback (*feedback amplifier*) provides gain stability, increases frequency response, and decreases distortion.

The operation of a feedback amplifier is influenced by its output signal, in addition to its input signal. This is realized by returning, or feeding back, a portion or all of the output signal to the input of the amplifier. In *voltage feedback*, the feedback signal is proportional to the load voltage; in *current feedback*, the returned signal is proportional to the load current.

If the feedback signal subtracts from the input signal, *negative feedback* is said to exist. For *positive feedback*, the feedback signal adds to the input signal. Negative feedback is used in amplifiers; positive feedback is the basis for the operation of sinusoidal oscillators.

PROPERTIES OF NEGATIVE FEEDBACK To highlight some important properties of negative feedback, reference is made to the block diagram of a feedback amplifier in Fig. 12.40. Output voltage E_o, which is across load resistor R_L, is also impressed across the input terminals of the feedback network. The output voltage of the feedback network E_f is connected in series with signal source E_s. The gain of the amplifier block is A_v, and the gain of the feedback network is B.

Because the feedback voltage is proportional to the output voltage, voltage feedback is obtained. Furthermore, the feedback voltage is in series with the input signal. For these reasons, the configuration of Fig. 12.40 is referred to as a *voltage-feedback series-input* feedback amplifier.

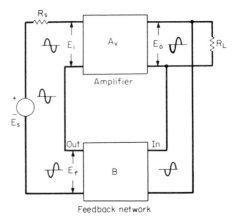

Fig. 12.40 Block diagram of a feedback amplifier.

For the verification that negative feedback indeed exists, voltage waveforms at the input and output terminals of the amplifier and feedback network are indicated in Fig. 12.40. Assume that the input sine wave to the amplifier E_i is positive-going. Also, the numerical value of A_v is negative, indicating that the output is 180° out of phase with respect to the input. As a result, output E_o is negative-going.

The input to the feedback network, which is equal to E_o, is also negative-going. Because the feedback network is generally comprised of resistors, there is no phase inversion, and its output E_f is negative-going. Negative feedback exists since E_f opposes E_s.

We now derive an expression for voltage gain with feedback A_{vf}. Referring back to Fig. 12.40, input voltage E_i is equal to the algebraic sum of the signal voltage E_s and feedback voltage E_f:

$$E_i = E_s + E_f \qquad (12.39a)$$

The feedback voltage is equal to the product of output voltage E_o and the gain of the feedback network B:

$$E_f = BE_o \qquad (12.39b)$$

Substitution of Eq. (12.39b) in Eq. (12.39a) yields

$$E_i = E_s + BE_o \qquad (12.39c)$$

The gain of the amplifier block A_v is equal to the ratio of its output to input voltages

$$A_v = \frac{E_o}{E_i} \qquad (12.39d)$$

Substituting Eq. (12.39c) for E_i in Eq. (12.39d) results in

$$A_v = \frac{E_o}{E_s + BE_o} \qquad (12.39e)$$

Solving Eq. (12.39e) for A_{vf}, which is equal to output voltage E_o divided by the signal voltage E_s, we obtain

$$A_{vf} = \frac{A_v}{1 - BA_v} \qquad (12.40)$$

It is common practice to refer to A_{vf} as the *closed-loop gain*, A_v as the *open-loop gain*, and the product BA_v as the *loop gain*.

Equation (12.40) illustrates some very important facts about feedback. If $B = 0$ (no feedback present), the closed-loop and open-loop gains are equal ($A_{vf} = A_v$). For negative feedback, the product BA_v is negative. Consequently, the denominator is greater than one and the voltage gain with negative feedback is less than without feedback. This is a very important property of negative feedback.

If BA_v is much greater than one, the one can be neglected in the denominator of Eq. (12.40). Consequently,

$$A_{vf} \simeq \frac{A_v}{-BA_v}$$

The A_v terms cancel and

$$A_{vf} \simeq -\frac{1}{B} \qquad (12.41)$$

Equation (12.41) demonstrates that if $BA_v \gg 1$, the voltage gain depends essentially on the gain of the feedback network. This means that, as long as B is constant, even if the gain of the amplifier changes owing to variation in temperature or in device parameters, the gain of the feedback amplifier is hardly affected. The gain of the feedback network, because it is usually comprised of resistors, is easy to stabilize. Another important property of negative feedback, therefore, is *gain stability*.

example 12.12 Negative feedback is to be introduced around an amplifier with an open-loop voltage gain of $A_v = -200$. It is required that the voltage gain with feedback also equal

-200 ($A_{vf} = -200$). Because of negative feedback, the open-loop gain must be increased to realize $A_{vf} = -200$. If $B = 4 \times 10^{-3}$, determine the new value of A_v.

solution   Solving Eq. (12.40) for A_v yields

$$A_v = \frac{A_{vf}}{1 + BA_{vf}} \qquad (12.42)$$

Substituting the given values in Eq. (12.42), we have $A_v = -200/(1 - 4 \times 10^{-3} \times 200) = -200/0.2 = -1000$.

Bandwidth   Another advantage of negative feedback is that it extends the bandwidth of an amplifier. Letting the upper -3-dB frequency with feedback be f_{HF} and the lower -3-dB frequency be f_{LF}, we have

$$f_{HF} = f_H(1 - BA_v) \qquad (12.43a)$$

$$f_{LF} = \frac{f_L}{1 - BA_v} \qquad (12.43b)$$

For example, if $f_H = 10$ kHz and $f_L = 100$ Hz, then for $1 - BA_v = 10$, $f_{HF} = 10 \times 10$ kHz $= 100$ kHz and $f_{LF} = {}^{100}/_{10} = 10$ Hz. The bandwidth has been extended by a factor slightly greater than ten.

Distortion   Negative feedback also reduces distortion in an amplifier. Letting D_F equal the distortion with feedback, and D the distortion without feedback, then

$$D_F = \frac{D}{1 - BA_v} \qquad (12.44)$$

For example, if the second-harmonic distortion without feedback D_2 is 10 percent, then the second-harmonic distortion with feedback $D_{F2} = {}^{10}/_{10}$ is 1 percent. The distortion has been reduced by a factor of ten.

Instability   At very low and high frequencies, the loop gain BA_v may become equal to, or close to, one. If a one is substituted for BA_v in Eq. (12.40), we obtain

$$A_{vf} = \frac{A_v}{1 - 1} = \frac{A_v}{0} = \infty$$

The gain of the feedback amplifier, therefore, approaches infinity. What this means in practical terms is that the amplifier is *unstable* and sine-wave oscillations are generated. This is undesirable for amplifiers although, as we shall see, a necessary condition for a sine-wave oscillator.

Whenever negative feedback is introduced in an amplifier, an opportunity exists for the feedback amplifier to become unstable. There are criteria, such as the *Nyquist criterion*, for determining the stability of a feedback amplifier. *Compensation networks* can be used to convert an unstable feedback amplifier to a stable one.

EFFECTS ON INPUT AND OUTPUT RESISTANCE   In addition to providing voltage and current gain stability, increased bandwidth, and reduced distortion, negative feedback also modifies the input and output resistances of an amplifier. Depending on how feedback is introduced, it can either decrease or increase the input and output resistances of a feedback amplifier. There are four basic types of feedback, as illustrated in the examples of Fig. 12.41.

  a. *Voltage feedback, series input.* Input resistance, R_{iF}, is *greater* than input resistance without feedback, R_i: $R_{iF} > R_i$. Output resistance, R_{oF}, is *less* than output resistance without feedback, R_o: $R_{oF} < R_o$.

  b. *Voltage feedback, shunt input:* $R_{iF} < R_i$; $R_{oF} < R_o$.
  c. *Current feedback, series input:* $R_{iF} > R_i$; $R_{oF} > R_o$.
  d. *Current feedback, shunt input:* $R_{iF} < R_i$; $R_{oF} > R_o$.

12.10 SINUSOIDAL OSCILLATORS

In the discussion of instability in feedback amplifiers, it was pointed out that if the loop gain $BA_v = 1$, the amplifier oscillates. The condition that the loop gain be unity for

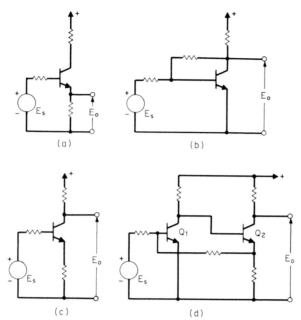

Fig. 12.41 Four basic types of feedback: (*a*) Voltage feedback, series input. (*b*) Voltage feedback, shunt input. (*c*) Current feedback, series input. (*d*) Current feedback, shunt input.

sine-wave oscillations is called the *Barkhausen criterion*. In this section a number of commonly used sine-wave oscillators are examined.

PHASE-SHIFT OSCILLATOR   An example of a phase-shift oscillator using a FET is illustrated in Fig. 12.42. Biased for class A operation, the output of the FET is connected to three cascaded *RC* sections. In turn, the output of the three *RC* sections, which serves as the feedback network, is connected to the input (gate) of the amplifier.

The amplifier and the *RC* network each contributes 180° phase shift at the frequency of oscillation. Total phase shift, therefore, is $180 + 180 = 360°$. This ensures that positive feedback exists and, if A_v and B are chosen properly, the Barkhausen criterion is satisfied. The magnitude of the amplifier gain must be at least 29 and the frequency of oscillation f in hertz is

$$f = \frac{0.159}{(\sqrt{6}RC)} \tag{12.45}$$

Fig. 12.42 An example of a phase-shift oscillator.

Fig. 12.43   Examples of tuned-circuit oscillators: (*a*) Hartley.   (*b*) Colpitts.

TUNED-CIRCUIT OSCILLATORS   Two examples of tuned-circuit, or *resonant circuit*, oscillators using a BJT are illustrated in Fig. 12.43.   In the *Hartley* oscillator of Fig. 12.43*a*, appropriate phase shift to satisfy the Barkhausen criterion is obtained by the mutual coupling M between inductors L_1 and L_2.   The output of the amplifier is coupled to the input side by coupling capacitor C_c.   Assuming small mutual coupling, an approximate expression for frequency of oscillation is

$$f = \frac{0.159}{\sqrt{C(L_1 + L_2)}} \qquad (12.46)$$

In the *Colpitts* oscillator of Fig. 12-43*b*, suitable phase shift is obtained by capacitors C_1, C_2, and inductor L.   As in the Hartley oscillator, C_c couples the output voltage of the BJT to the input side.   Because capacitor C_1 is across the base-emitter junction of the transistor, the input capacitance C_i of the transistor should be included.   Letting $C_1' = C_1 + C_i$, the frequency of oscillation is given by

$$f = \frac{0.159}{\sqrt{LC_1'C_2/(C_1' + C_2)}} \qquad (12.47)$$

FREQUENCY STABILITY   The frequency of oscillation tends to vary, or *drift*, with variations in temperature or aging.   *Frequency stability* refers to the ability of an oscillator to maintain its frequency constant over long periods of time.   To ensure frequency stability, various methods are employed.   One method uses a capacitor with a negative-temperature coefficient (see Chaps. 2 and 10) to cancel the positive temperature coefficient of the inductor.   Another approach is to use a very high-Q tuned circuit.   An example of a high-Q circuit is the quartz crystal, which is the basis of the crystal oscillator.

Fig. 12.44   Quartz crystal: (*a*) Electrical symbol.   (*b*) Approximate model.

Fig. 12.45   An example of a crystal oscillator.

CRYSTAL OSCILLATOR   Quartz crystal has the property of becoming mechanically deformed when a voltage is impressed across its faces.   Or, if mechanically strained, it produces a voltage.   This phenomenon is referred to as the *piezoelectric effect*.   The electrical symbol and approximate model of a quartz crystal are shown in Fig. 12.44. An examination of the model shows that the quartz crystal acts like a parallel LC tuned circuit.   Its Q may be as high as 10 000, thereby ensuring excellent frequency stability.

An example of a FET crystal oscillator is illustrated in Fig. 12.45.   The crystal is connected to the input circuit of the amplifier, and an $L_D C_D$ circuit tuned to the frequency of the crystal is in the output circuit.   Suitable coupling to satisfy the Barkhausen criterion is obtained through the internal drain-gate capacitance C_{gd}, and stray capacitance.   The frequency of oscillation is determined by the crystal.   If, for example, a 2.5-MHz oscillator is needed, a 2.5-MHz crystal is inserted in the circuit.

Chapter 13

Operational Amplifiers

13.1 INTRODUCTION

The phenomenal development of monolithic *IC* technology has made the operational amplifier (*op amp*) perhaps the most versatile component in electronics. With proper selection of feedback elements, the op amp may be used as a precision gain-voltage amplifier, buffer, summer, current source, converter, oscillator, and in many more applications. Its field is generally limited by the user's ingenuity and imagination.

Actually, the operational amplifier is not new. It was, and still is, used in analog computers. (An analog computer is employed to simulate physical systems.) Because of IC technology, it is possible to obtain an op amp, containing a dozen transistors, in a TO-5 package, which is used for packaging a single discrete transistor. Also, the cost of some IC operational amplifiers are less than what one paid for a single transistor just a few years ago.

13.2 THE BASIC OP AMP

In this section we treat the op amp as a *black box* with ideal characteristics. By considering the op amp as an ideal component, we can easily derive results of its application in different circuits. Fortunately, the derivations also hold, with an error of a few percent, for the majority of commercially available op amps.

The basic op amp is represented by the symbol of Fig. 13.1. The input denoted by the negative sign is referred to as the *inverting input;* the plus sign denotes the *noninverting input.* For simplicity, ground, power-supply, and frequency-compensation (to be considered later) connections are omitted from the figure.

If a signal is impressed across the inverting terminal with respect to ground, the output signal is 180° out of phase with respect to the input signal, as in the common-emitter amplifier. For a signal applied across the noninverting terminal with respect to ground, the output is in phase with the input. If signals are impressed across both the inverting and noninverting terminals with respect to ground, the output is proportional to the difference of the input signals.

IDEAL CHARACTERISTICS The attributes of the ideal op amp, which many physical op amps approach, are:
1. Voltage gain is infinite.
2. Input impedance is infinite.
3. Output impedance is zero.
4. The frequency response is flat (constant) for all frequencies.
5. The output signal is zero if the input signal is zero.

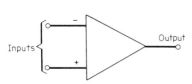

Fig. 13.1 Symbol for the basic op amp.

Fig. 13.2 Model of an ideal op amp.

Based on these attributes, a model of the ideal op amp is drawn in Fig. 13.2. The output voltage E_o equals the product of the gain A and the net input signal across the inverting and noninverting terminals E_i.

FEEDBACK Physical op amps can exhibit voltage gains as high as 10^9. If the input voltage is finite, but close to zero, the output voltage of such a high-gain amplifier would soar to infinity. Actually, the output would be limited by the power-supply voltages. Under this condition, the amplifier is said to be *saturated*, and useless. To make the op amp useful and versatile, *feedback* is introduced around the amplifier.

Consider the operational circuit, called an *inverting amplifier*, of Fig. 13.3. The noninverting terminal is returned to ground and a feedback resistor R_F is connected between the inverting and output terminals. A resistor R_1 is in series with the signal source E_s and the inverting terminal.

Fig. 13.3 An inverting amplifier.

Fig. 13.4 The concept of the virtual ground.

As mentioned earlier, the ideal op amp exhibits infinite input impedance and voltage gain. The latter property implies that the voltage across the inverting and noninverting terminals is zero. Because the input impedance is infinite, no current flows into the inverting terminal. This condition is referred to as a *virtual ground* (Fig. 13.4). From the figure it is seen that I_1, the current flowing in R_1, is equal to I_F, the current flowing in R_F. Current $I_1 = (E_s - 0)/R_1$ and $I_F = (0 - E_o)/R_F$. Hence,

$$\frac{E_s}{R_1} = \frac{-E_o}{R_F}$$

Solving for the voltage gain with feedback, $A_f = E_o/E_s$, we have

$$A_f = -\frac{R_F}{R_1} \tag{13.1}$$

What is striking and significant about Eq. (13.1) is that the voltage gain is only a function of resistors R_F, R_1, and is independent of the amplifier. For example, if $R_F = 100$ and $R_1 = 1 \text{ k}\Omega$, $A_f = {}^{-100}/_1 = -100$. If R_F and R_1 are precision resistors, the gain of -100 will hold, regardless of variations in amplifier characteristics, power-supply voltage, etc. This behavior is typical of high-gain commercial op amps with suitable feedback.

13.3   APPLICATIONS

The inverting amplifier has already been analyzed. In this section, examples of other commonly used applications of the op amp are considered.

NONINVERTING AMPLIFIER   The noninverting amplifier is shown in Fig. 13.5. In this circuit, the signal is fed directly to the noninverting terminal of the op amp. Resistors R_F and R_2 comprise the feedback network. Because of the virtual ground, the voltage at the inverting and noninverting terminals is equal to E_s. Owing to the

Fig. 13.5   A noninverting amplifier.

Fig. 13.6   A difference amplifier.

input impedance being infinite, $I_F = I_2$. Current $I_F = (E_o - E_s)/R_F$ and current $I_2 = E_s/R_2$. Hence,

$$\frac{E_o - E_s}{R_F} = \frac{E_s}{R_2}$$

Solving for $A_f = E_o/E_s$,

$$A_f = \frac{(R_2 + R_F)}{R_2} = 1 + \frac{R_F}{R_2} \tag{13.2}$$

For the noninverting amplifier, the output and input voltages are in phase; for the inverting amplifier, they are 180° out of phase.

DIFFERENCE AMPLIFIER   The difference amplifier of Fig. 13.6 provides an output that is proportional to the difference of the input signals, $E_{s1} - E_{s2}$. Using the concept of the virtual ground, the voltage at the inverting and noninverting terminals is equal to voltage E_3. Current $I_1 = (E_{s1} - E_3)/R_1$ and $I_F = (E_3 - E_o)/R_F$. Because $I_1 = I_F$,

$$\frac{E_{s1} - E_3}{R_1} = \frac{E_3 - E_o}{R_F}$$

By voltage division, $E_3 = E_{s2}R_F/(R_1 + R_F)$. Substitution of this expression in the preceding equation yields

$$\frac{E_{s1} - E_{s2}R_F/(R_1 + R_F)}{R_1} = \frac{E_{s2}R_F/(R_1 + R_F) - E_o}{R_F}$$

Solution for the last equation for the output voltage E_o gives

$$E_o = -\frac{R_F}{R_1}(E_{s1} - E_{s2}) \tag{13.3}$$

Thus the output voltage is proportional to the difference in the input voltages.

example 13.1   An input signal feeding an inverting amplifier has 5 mV of 60-Hz hum. How can a difference amplifier be used to cancel the hum?

solution   The circuit using a difference amplifier is shown in Fig. 13.7. The variable-voltage 60-Hz source is adjusted until no 60-Hz hum appears in the output E_o. This condition is monitored on a scope connected across the output terminal and ground of the op amp.

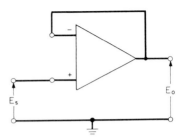

Fig. 13.7 Cancellation of 60-Hz hum present in a signal.

Fig. 13.8 A voltage follower.

VOLTAGE FOLLOWER An op amp circuit which has unity voltage gain, called a voltage follower, is shown in Fig. 13.8. Voltage followers, like emitter and source followers, are used for matching a high-impedance source to a low-impedance load.

The output terminal of the op amp is connected directly to the inverting terminal; the signal is impressed between the noninverting terminal and ground. Because of the virtual ground, the voltage at the inverting terminal is equal to the voltage at the noninverting terminal E_s. Since the output is connected to the inverting terminal, $E_o = E_s$, and

$$A_f = \frac{E_o}{E_s} = 1 \tag{13.4}$$

SUMMING AMPLIFIER The summing amplifier of Fig. 13.9 provides an output voltage proportional to the sum of the input voltages, $E_1, E_2, \ldots, E_n$. If resistors $R_1 = R_2 = \cdots = R_F$, then the output is equal to the negative sum of the inputs. If the resistors are unequal, then each input is summed with a *scale factor*.

Owing to the virtual ground, $I_1 = E_1/R_1$, $I_2 = E_2/R_2, \ldots, I_n = E_n/R_n$ and $I_F = -E_o/R_F$. Applying Kirchhoff's current law (KCL), the algebraic sum of the currents at the inverting terminal equals zero. Hence,

$$\frac{E_1}{R_1} + \frac{E_2}{R_2} + \cdots + \frac{E_n}{R_n} = \frac{-E_o}{R_F}$$

Solving for E_o, we obtain

$$E_o = -\left(\frac{R_F}{R_1} E_1 + \frac{R_F}{R_2} E_2 + \cdots + \frac{R_F}{R_n} E_n\right) \tag{13.5}$$

Fig. 13.9 A summing amplifier.

Fig. 13.10 An example of a multirange dc voltmeter. (See Example 13.3.)

example 13.2 Three input voltages are to be summed in the summing amplifier of Fig. 13.9. The feedback resistor $R_f = 100$ kΩ, $E_1 = 0.1$ V, $E_2 = 0.2$ V, and $E_3 = 0.05$ V. Determine E_o if (a) $R_1 = R_2 = R_3 = 100$ kΩ and (b) $R_1 = 20$, $R_2 = 50$, $R_3 = 10$ kΩ.

solution (a) For equal resistors, Eq. (13.5) reduces to $E_o = -(E_1 + E_2 + E_3) = -(0.1 + 0.2 + 0.05) = -0.35$ V.

(b) For unequal input resistors, each input voltage, from Eq. (13.5), is multiplied by a scale factor: R_f/R_1, R_f/R_2, and R_f/R_c. Hence,

$$E_o = -(^{100}\!/_{20} \times 0.1 + ^{100}\!/_{50} \times 0.2 + ^{100}\!/_{10} \times 0.05)$$
$$= -(5 \times 0.1 + 2 \times 0.2 + 10 \times 0.05)$$
$$= -(0.5 + 0.4 + 0.5) = -1.4 \text{ V}$$

example 13.3 A precision multirange dc voltmeter is to be constructed. The voltage ranges desired are zero to 0.1, 1, 2, and 5 V full scale. Using an op amp and a 0- to 5-V dc meter, design the circuit.

solution A design of such a meter is shown in Fig. 13.10. The unknown voltage to be measured, E_x, is impressed across the noninverting terminal and ground. To realize the four voltage ranges, a four-position switch and four resistors, $R_{F1}, \ldots, R_{F4}$, are connected in the feedback path. Resistor R_2, connected to the inverting terminal and the switch, completes the feedback circuit. A reasonable value for R_2 is 5 kΩ.

When $E_x = 0.1$ V, the 0- to 5-V meter must read 0.1 V full scale. The gain of the noninverting amplifier is, therefore, equal to $5/0.1 = 50$. From Eq. (13.2), $A_f = 1 + R_f/R_2$; hence, $50 = 1 + R_{F1}/5$. Solving, $R_{F1} = (50 - 1) \times 5 = 245$ kΩ.

Following the same procedure, for the 0- to 1-V scale, $A_f = 5/1 = 5$. Therefore, $R_{F2} = (5 - 1) \times 5 = 20$ kΩ. For the 0- to 2-V scale, $A_f = 5/2 = 2.5$; $R_{F3} = (2.5 - 1) \times 5 = 7.5$ kΩ. Finally, for the 0- to 5-V scale, $A_f = 5/5 = 1$; $R_{F4} = (1 - 1) \times 5 = 0$.

LOGARITHMIC AMPLIFIER This circuit provides an output voltage that is proportional to the logarithm of the input voltage. A simple version of a logarithmic amplifier

Fig. 13.11 An elementary logarithmic amplifier.

is obtained by replacing the feedback resistor R_F in the inverting amplifier of Fig. 13.3 by a diode (Fig. 13.11). The diode current I_D may be approximated by

$$I_D = I_o \epsilon^{+E_D/k} \tag{13.6}$$

where I_o is the diode reverse saturation current
$\quad E_D$ is the voltage across the diode
$\quad k$ is a term that includes physical constants and temperature
Because of the virtual ground, $I_1 = E_s/R_1 = I_D$. Also, $E_D = 0 - E_o = -E_o$ V. Making these substitutions in Eq. (13.6), we obtain

$$\frac{E_s}{R_1} = I_o \epsilon^{-E_o/k} \tag{13.7}$$

Taking the natural logarithm (ln) of Eq. (13.7) yields

$$\ln \frac{E_s}{R_1} = \ln I_o - \frac{E_o}{k}$$

Therefore,

$$E_o = -k\left(\ln \frac{E_s}{R_1} - \ln I_o\right) \tag{13.8}$$

Equation (13.8) demonstrates that the output voltage is indeed proportional to the natural log of the input voltage.

CURRENT-TO-VOLTAGE CONVERTER   Devices, such as photomultipliers and photocells, act like an ideal current source. The output current of these devices is essentially constant, independent of the load. The current-to-voltage converter provides an output voltage which is proportional to the current.

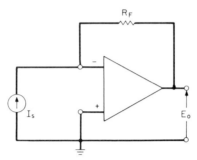

Fig. 13.12   A current-to-voltage converter.

An example of such a converter is illustrated in Fig. 13.12. The device is represented by the ideal current source I_s. Because of the virtual ground,

$$I_s = \frac{-E_o}{R_F}$$

or
$$E_o = -I_s R_F \qquad (13.9)$$

The output voltage is, therefore, proportional to the input current.

13.4   PRACTICAL CONSIDERATIONS

To this point we represented the op amp as having only three terminals: the inverting and noninverting input terminals and an output terminal. In addition to these, there are terminals for ground, power supply, and compensating network connections (Fig. 13.13).

Generally negative $(-E)$ and positive $(+E)$ power supplies are required. There are at least two terminals to which are connected a compensating network. Compensation is required to ensure stable amplifier operation. The compensating network usually consists of a small capacitor in series with a resistor.

FREQUENCY RESPONSE AND COMPENSATION   A frequency response plot of an *uncompensated* operational amplifier may take the form shown in Fig. 13.14 (solid lines). Along the vertical axis is plotted the magnitude of the voltage gain without feedback, or the *open-loop gain*, in decibels (dB). A voltage gain of 100 dB corre-

Fig. 13.13   Symbol for a physical op amp.

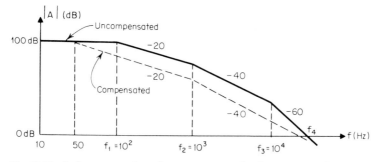

Fig. 13.14   Bode response plots of an uncompensated and a compensated op amp.

sponds to an amplifier gain of 100 000. Frequency is plotted to a logarithmic scale along the horizontal axis. This permits a large range of frequencies to be plotted in a compact manner.

The actual curve is approximated by straight-line segments. Such a representation is referred to as a *Bode*, or an *asymptotic*, plot. The Bode plot is easy to draw and provides useful information.

In Fig. 13.14 frequency $f_1 = 100$ Hz is called a *break frequency*, where the gain is actually 3 dB less than at lower frequencies. For frequencies between f_1 and f_2, the gain decreases, or *rolls off*, −20 dB for a decade of frequency. This means that for a 10-to-1 increase in frequency, the voltage gain falls by 20 dB. An *equivalent expression*, −6 dB/octave, means that as the frequency is *doubled*, the gain decreases by 6 dB. At frequency f_2 the gain begins to roll off at −40 dB/decade (−12 dB/octave), and at f_3 it rolls off at −60 dB/decade (−18 dB/octave). Frequency f_4, which intersects the frequency axis at unity voltage gain (0 dB), is the *crossover frequency*.

An amplifier that exhibits an open-loop response of −60 dB/decade (or −18 dB/octave) at the crossover frequency is *unstable*. It can break into oscillation and generate sine waves. By the proper selection of a compensating network, the amplifier is made stable. For this condition, indicated by the dashed lines in Fig. 13.14, the slope is −40 dB/decade (−12 dB/octave) at the crossover frequency.

The specific compensation network to use is recommended by the manufacturer of the op amp. It may, for example, be a 2200-pF capacitor connected in series with a 10-kΩ resistor.

From Fig. 13.14 the dominant frequency of the compensated amplifier is only 50 Hz. It might appear, therefore, that the op amp is really not too useful. Because of the use of feedback, which increases the bandwidth of an amplifier, this is not true.

Consider an op amp with an open-loop gain of 100 000 (100 dB). If, for an inverting amplifier, $A_f = 100$ (which equals 20 log 100 = 40 dB), the difference between the open-loop gain and the gain with feedback, referred to as the *closed-loop gain*, is 100 − 40 = 60 dB at low frequencies. This is illustrated in Fig. 13.15. Note that the

Fig. 13.15   Feedback increases the bandwidth of an amplifier.

bandwidth is much greater than 10 Hz; in this example it is 10 kHz. For frequencies greater than 10 kHz, the open- and closed-loop gains are equal.

OFFSET   For a zero input, an ideal op amp has exactly zero output. In physical op amps, because of variations in transistor characteristics, there is a nonzero output for a zero signal. To compensate for this, a dc voltage is applied to the amplifier such that, for a zero input signal, the output is also zero. Before we see how this is accomplished, some terms used in the literature and on data sheets are defined.

Input offset current   The input offset current I_{io} is the difference in dc bias currents of the inverting and noninverting transistors in an op amp.

Input offset voltage   The input offset voltage E_{io} is the voltage impressed across the input of an op amp to ensure that the output is zero when the input signal is also zero.

Input offset current drift   The input offset current drift is the ratio of input offset current change ΔI_{io} to a change in temperature ΔT. It is expressed by $\Delta I_{io}/\Delta T$.

Input offset voltage drift   The input offset voltage drift is the ratio of input offset voltage change ΔE_{io} to a change in temperature ΔT. It is expressed by $\Delta E_{io}/\Delta T$.

OFFSET CORRECTION   A circuit for offset correction in an inverting amplifier is illustrated in Fig. 13.16. Connected to the noninverting input is a voltage divider

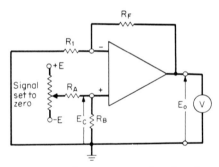

Fig. 13.16   Offset correction of an inverting amplifier.

consisting of resistors R_A and R_B. One end of resistor R_A is connected to the wiper arm of the potentiometer (pot). A positive voltage (+E) is connected to one end of the pot, and the other end is connected to a negative (−E) voltage. This arrangement makes available both positive and negative voltages for offset correction. The value of the pot is in the order of 20 kΩ.

Offset correction is done as follows: With the signal set to zero, a sensitive dc voltmeter is connected across the output to monitor output voltage E_o. The pot is then adjusted until $E_o = 0$. It must be emphasized that, because offset voltage and current drift with temperature, the correction obtained is only valid for a *given operating temperature*. Variations in offset current and voltage with temperature are illustrated in Fig. 13.17.

example 13.4   Referring to Fig. 13.16, assume that $\pm E = \pm 20$ V, $R_A = 250$ kΩ, and $R_B = 100$ Ω. Determine the maximum value of the offset correction voltage E_C.

solution   By voltage division, $E_C = \pm 20 \times 100/(250\ 000 + 100) \approx \pm 2000/250\ 000 = \pm 8$ mV.

A circuit for correcting offset in a noninverting amplifier is given in Fig. 13.18. As for the inverting amplifier, with a dc voltmeter connected across the output, the pot is adjusted until $E_o = 0$.

SLEWING RATE   An important characteristic of an op amp is how well it follows a rapidly changing waveform. Assume that the input to a noninverting amplifier, with a

Fig. 13.17 Examples of how input offset current and voltage vary with temperature. (*Courtesy RCA.*)

gain of 10, is the step input signal of Fig. 13.19a. The output is shown in Fig. 13.19b. Note that the output voltage of 10 V is not reached instantaneously. The rate of change of the rising voltage ΔE_o with respect to time Δt is the *slewing rate:*

$$\text{Slewing rate} = \frac{\Delta E_o}{\Delta t} \qquad (13.10)$$

The unit of the slewing rate is usually expressed as volts per microsecond. A high slewing rate indicates an op amp with a good response to rapidly changing waveforms.

Fig. 13.18 Offset correction of a noninverting amplifier.

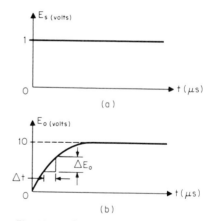

Fig. 13.19 Slewing rate of an op amp: (*a*) Step input to a noninverting amplifier. (*b*) Output of amplifier.

13.5 OP AMP CIRCUITRY

An operational amplifier contains a number of stages, each of which performs a specific function. This is illustrated in the block diagram (Fig. 13.20) of a typical op amp. Each of the circuits will be first briefly described and then examined in detail.

The differential amplifier in the first stage provides the inverting and noninverting inputs, gain, and is very stable in operation. The constant-current sink may be thought of as simulating a high emitter resistance with good efficiency. For additional gain, a second amplifier stage is required. Often, the second stage is another differential amplifier.

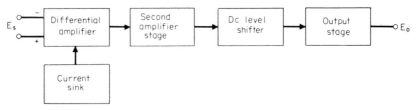

Fig. 13.20   General block diagram of an operational amplifier.

Because the circuits in an op amp are direct-coupled, as the signal progresses from the input to the output there is a buildup in dc voltage. The dc level shifter cancels out the built-up dc voltage. This ensures that, neglecting offset, the output is zero for zero input signal.

The output stage may be an emitter follower. For high-level signals, however, a push-pull type of output stage is used.

DIFFERENTIAL AMPLIFIER   A basic differential amplifier is illustrated in Fig. 13.21a. A striking feature of the circuit is its *symmetry*. Because IC processing yields devices that are closely matched, the differential amplifier may be viewed as a Wheatstone bridge shown in Fig. 13.21b. Two power supplies are generally required for its operation: $+E_{CC}$ and $-E_{EE}$ volts. Input signals E_1 and E_2, or either one alone, are applied to the transistors, Q_1 and Q_2, respectively.

Assume that the input signals are equal: $E_1 = E_2$. For this condition, the output voltages are also equal ($E_{o1} = E_{o2}$) and, viewed as a bridge, the bridge is said to be balanced. If input E_1 is greater than E_2, more collector current flows in Q_1 and less collector current in Q_2. Hence, $E_{o1} < E_{o2}$. On the other hand, if $E_2 > E_1$, more collector current now flows in Q_2 than in Q_1, and $E_{o1} > E_{o2}$.

Practical signals may be thought of as containing two components: the *common-mode* (E_c) and *difference-mode* (E_d) signals. The common-mode signal is *common to both inputs*. Examples of this kind of signal include noise, hum from a power supply, and effects of temperature change. The difference-mode signal is the signal that is to be amplified. For an ideal differential amplifier, therefore, the gain to the common-mode signal would be zero, and the gain to the differential-mode signal would be very high.

Neglecting h_{re} and h_{oe} of the transistors, the common-mode signal gain A_c is given by

$$A_c \approx \frac{-R_C}{2R_E} \tag{13.11}$$

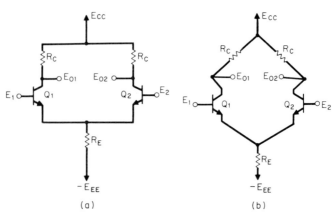

(a)   (b)

Fig. 13.21   Basic differential amplifier: (*a*) Circuit. (*b*) Amplifier viewed as a bridge.

and the difference-mode signal gain A_d is

$$A_d \simeq \frac{-h_{fe}R_C}{2h_{ie}} \qquad (13.12)$$

From Eq. (13.11), for a given value of R_C, a small common-mode gain requires that R_E should be large. For very practical reasons, R_E cannot be made too large. For one, the practical limit of diffused resistors is 30 kΩ. Furthermore, for a large value of R_E, excessive power is wasted as heat. For these reasons a current sink, which simulates electronically a large value of resistance, is used in place of R_E. (The current sink is treated later in this chapter.)

Referring to Eq. (13.12), for a high value of difference-mode gain, h_{fe} and R_C should be large and h_{ie} low. High-gain transistors can be fabricated by making the bases exceedingly thin. One such device, called the *super beta transistor*, has an h_{fe} of a few thousand. For practical reasons, the maximum value of R_C is 30 kΩ. The value of h_{ie} varies inversely with collector bias current. Because the transistors in a differential amplifier are biased at a low quiescent current, a typical value of h_{ie} is a few kilohms.

A figure of merit for differential (and operational) amplifiers is the *common-mode rejection ratio*, CMRR. It is defined as the ratio of the difference-mode gain to the common-mode gain:

$$CMRR = \frac{A_d}{A_c} \qquad (13.13)$$

The CMRR is generally expressed in decibels. The greater the value of CMRR, the better is the differential (or operational) amplifier.

example 13.5 Referring to the differential amplifier of Fig. 13.21a, assume that $R_C = R_E = 10$ kΩ. $h_{fe} = 100$, and $h_{ie} = 4$ kΩ; h_{re} and h_{oe} are negligible. Determine (a) A_c, (b) A_d, and (c) CMRR.

solution (a) From Eq. (13.11), $A_c \simeq -R_C/2R_E = -10/(2 \times 10) = -0.5$.
(b) From Eq. (13.12), $A_d \simeq -h_{fe}R_C/2h_{ie} = -100 \times 10/(2 \times 4) = -125$.
(c) CMRR $= A_d/A_c = -125/(-0.5) = 250$. In decibels, CMRR $= 20 \log (250) \simeq 48$ dB.

CURRENT SINK In the previous section it was seen that for a low value of common-mode gain, resistor R_E should be large. An efficient method for simulating a high-value resistance is by use of a *current sink*. An example of such a circuit connected to the emitters of a differential amplifier is shown in Fig. 13.22a.

Transistor Q_3 acts like a constant-current source. Resistors R_1 and R_2 bias the transistor and R_3 stabilizes its quiescent operating point. The circuit of Fig. 13.22a simulates a resistance of a few hundred kilohms. A general symbol for a current sink, often found in the literature, is illustrated in Fig. 13.22b.

An example of a commercially integrated differential amplifier, the Fairchild μA730, is shown in Fig. 13.23. Transistors Q_1 and Q_2 comprise the differential amplifier; Q_5 is the current sink. For low output impedance, Q_3 and Q_4 are connected as emitter followers.

Fig. 13.22 Constant-current sink: (a) Circuit. (b) Symbol.

Fig. 13.23 Example of a commercial IC (μA730) differential amplifier. (*Courtesy Fairchild Semiconductor.*)

CASCADING STAGES To achieve high gain, an additional amplifier stage is cascaded to the differential amplifier. An example is the cascaded pair of differential amplifiers shown in Fig. 13.24. Transistors Q_1 and Q_2 and the constant current sink comprise the first differential amplifier stage. The second differential-amplifier stage contains transistors Q_3 and Q_4. Because only a single-ended output E_o from Q_4 is required, no collector resistor is necessary for Q_3. Also, an emitter resistor R_E, instead of a current sink, is adequate for the second stage.

Another circuit for increasing gain is illustrated in Fig. 13.25. Transistors Q_1 and Q_2 constitute the differential amplifier; Q_3 and Q_4 provide increased gain and convert the double-ended output of the differential amplifier to a single-ended output. The bases of Q_3 and Q_4 are fed from a common voltage point P through equal resistors R_{C1}. Transistor Q_3 inverts the output of Q_1 and compares it with the output of Q_2. The full differential output of the differential amplifier, therefore, is impressed across Q_4. Transistor Q_4 amplifies the differential signal and provides a single-ended output.

Another interesting feature of the circuit is that it is relatively insensitive to changes in the positive supply voltage E_{CC}. If, for example, E_{CC} should rise, the collector cur-

Fig. 13.24 A cascaded pair of differential amplifiers for increased gain.

rents in Q_3 and Q_4 also rise. Consequently, the collector voltages of Q_3 and Q_4 tend to decrease. As their voltage falls, point P becomes less positive, thereby reducing the base current to Q_4. If the base current to a transistor is reduced, its collector voltage increases. Thus we have a regulating (feedback) mechanism in operation that ensures the collector voltage of Q_4 remaining essentially constant.

DC LEVEL SHIFTER   In monolithic integrated circuits, coupling capacitors are not practical. Instead, direct coupling is used and, as the signal travels from the input to the output stage, there is a buildup in dc voltage. If this built-up voltage is not canceled, a nonzero output is obtained for zero signal.

A circuit for canceling the built-up dc voltage in an op amp is called a *dc level shifter*. There are many circuits for accomplishing this; a basic dc level shifter is shown in Fig. 13.26.

Fig. 13.25 Transistors Q_3 and Q_4 provide additional gain and convert the double-ended output of the differential amplifier (Q_1, Q_2) into a single-ended output voltage, E_o.

Fig. 13.26 A basic dc level shifter.

In this circuit, transistor Q is connected as an emitter follower. The emitter resistance consists of two resistors: R_1 in series with R_2. The dc base-emitter voltage E_{BE} is nominally 0.7 V at room temperature. Let E_I be the dc voltage at the base of Q and E_O the output dc voltage. The shifted voltage is equal to $E_I - E_O$ and is given by

$$E_I - E_O = E_{BE} + I_E R_1 \qquad (13.14)$$

For example, if $I_E = 2$ mA and $R_1 = 600\ \Omega$, then the dc voltage shift is, by Eq. (13.14), $0.7 + 2 \times 0.6 = 1.9$ V.

OUTPUT STAGE   The output stage of an op amp must exhibit low output impedance and be capable of providing the required load power, and its output signal should be limited only by the supply voltages for a maximum input signal. For low output power, an emitter follower is often used as the output stage in an op amp. If the power requirements are large, however, excessive power is dissipated in the emitter resistor. For this case, the circuits of Fig. 13.27 are generally employed as the output stage of an op amp.

In the circuit of Fig. 13.27a, transistor Q_1 operates as a *phase splitter*. If $R_C = R_E$, the signals are equal and 180 out of phase at the collector and in phase at the emitter with respect to input signal E_i. Transistors Q_2 and Q_3 operate as a push-pull amplifier, often referred to as a *totem-pole amplifier*.

Assume, for example, that the base voltage to Q_2 is decreasing and the base voltage to Q_3 is increasing. Transistor Q_3, therefore, conducts more current than Q_2. The difference in transistor currents is the load current, which flows in load resistor R_L. In many integrated circuits, emitter resistor R_E of Q_1 is replaced by a diode.

The circuit of Fig. 13.27b, called a *complementary pair* output stage, eliminates the

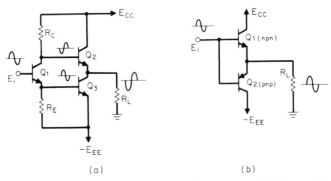

Fig. 13.27 Examples of output stages used in op amps: (*a*) Totem pole. (*b*) Complementary pair.

need for a phase splitter. Transistors Q_1 and Q_2 are of opposite types: Q_1 is npn and Q_2 is pnp. For efficient operation, Q_1 and Q_2 should be matched.

On the positive half-cycle of input signal E_i, Q_1 is conducting (because it is npn) and Q_2 is nonconducting (because it is pnp). Current, therefore, flows in the positive direction through Q_1 to load resistor R_L. On the negative half-cycle, Q_1 is nonconducting and Q_2 is conducting. Current flows in the negative direction through the pnp transistor Q_2 to R_L, thereby completing the output waveform.

13.6 EXAMPLES OF PRACTICAL OP AMPS

Two examples of commercially available op amps will now be analyzed. The first is the Motorola MC-1530 op amp of Fig. 13.28. Transistors Q_2 and Q_4 comprise the first differential amplifier stage, which is cascaded to a second differential amplifier containing Q_6 and Q_7. A current sink, Q_3, is used for the first stage and an emitter resistor, R_6, for the second stage.

Dc level shifting is achieved primarily by transistors Q_8 and Q_{10}. A phase splitter, Q_{11}, drives the totem-pole amplifier containing transistors Q_{12} and Q_{13}. Note that a diode is used instead of an emitter resistor in the phase splitter.

Fig. 13.28 Circuit of the Motorola MC-1530 op amp. (*Courtesy Motorola Semiconductor Products, Inc.*)

Fig. 13.29   Circuit diagram of the RCA CA 3008 op amp.   (*Courtesy RCA.*)

As a second example, consider the RCA CA 3008 op amp of Fig. 13.29.   Transistors Q_1 and Q_2 constitute the first differential amplifier stage which is coupled to a second differential amplifier (Q_3, Q_4).   Current sinks are used for both stages (Q_6, Q_7), and diodes D_1 and D_2 provide thermal stabilization for their operating points.

Transistor Q_5 compensates the effect of common-mode signals.   Assume, for example, that the dc power-supply voltage decreases.   As a result, the voltage at the emitters of Q_3 and Q_4 also decrease.   Less collector current flows in Q_5, as well as in Q_7 and Q_9.   Because of the decrease in collector current of Q_7, which acts as a sink, less current flows in Q_3 and Q_4.   This results in an increase in their collector voltage, thereby canceling the decrease in the supply voltage.

Transistors Q_8 and Q_9 provide dc level shifting.   The output stage is an emitter follower.   Note that emitter resistor R_{18} of the follower is returned to the junction of resistors R_{16} and R_{17}, instead of to the negative supply voltage.   This connection provides *positive feedback* and increases the gain of the emitter follower from less than 1 to about 1.5.

The current flowing in R_{18} is added to the current in R_{16}.   The voltage drop across R_{16} is thereby increased.   Because the power-supply voltages are fixed, the voltage across R_{14} is reduced.   This results in the collector voltage of Q_9, which is connected to the base of the emitter follower, to increase.   The greater input voltage to the emitter follower results in a higher output voltage of the op amp.   Resistor R_{15} limits the signal load current in case output terminal 12 is short-circuited to terminal 8.

13.7   UNDERSTANDING THE DATA SHEET FOR AN OP AMP

In selecting an op amp for a particular application, the user must consult the data (specs) sheet provided by the manufacturer of the device.   Representative data sheets for an op amp, the Fairchild μA709C, are reproduced in Fig. 13.30.   Invariably, the data included in the sheets are:

1. General description   A cogent statement that stresses special features and application areas of the op amp under consideration.   For example, the μA709C features

Fig. 13.30   Example of data sheets for an op amp provided by a manufacturer. (*Courtesy Fairchild Semiconductor.*)

low offsets, high input impedance, etc.   The amplifier is intended for such applications as dc servos, analog computers, low-level instrumentation, and the generation of special functions.

2. Absolute maximum ratings   Included under this heading are such items as supply voltage, internal power dissipation, and operating temperature range.   For reliable operation and long life, the absolute maximum ratings must never be exceeded.

3. Physical dimensions   Provides detailed dimensions of the package for mounting. Additional examples of IC packages are illustrated in Fig. 13.31.

4. Schematic diagram   The schematic diagram is a circuit description of the op amp.   The μA709C has some interesting features.   The first differential stage (Q_1, Q_2)

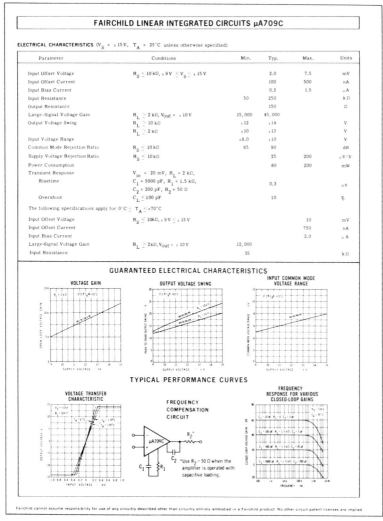

FAIRCHILD LINEAR INTEGRATED CIRCUITS μA709C

ELECTRICAL CHARACTERISTICS (V$_S$ = ± 15 V, T$_A$ = 25°C unless otherwise specified)

Parameter	Conditions	Min.	Typ.	Max.	Units
Input Offset Voltage	R$_S$ ≤ 10 kΩ, ± 9 V ≤ V$_S$ ≤ ± 15 V		2.0	7.5	mV
Input Offset Current			100	500	nA
Input Bias Current			0.3	1.5	μA
Input Resistance		50	250		kΩ
Output Resistance			150		Ω
Large-Signal Voltage Gain	R$_L$ ≥ 2 kΩ, V$_{out}$ = ± 10 V	15,000	45,000		
Output Voltage Swing	R$_L$ ≥ 10 kΩ		± 12	± 14	V
	R$_L$ ≥ 2 kΩ		± 10	± 13	V
Input Voltage Range		± 8.0	± 10		V
Common Mode Rejection Ratio	R$_S$ ≤ 10 kΩ	65	90		dB
Supply Voltage Rejection Ratio	R$_S$ ≤ 10 kΩ		25	200	μV/V
Power Consumption			80	200	mW
Transient Response	V$_{in}$ = 20 mV, R$_L$ = 2 kΩ,				
Risetime	C$_1$ = 5000 pF, R$_1$ = 1.5 kΩ,		0.3		μs
	C$_2$ = 200 pF, R$_2$ = 50 Ω				
Overshoot	C$_L$ ≤ 100 pF		10		%
The following specifications apply for 0°C ≤ T$_A$ ≤ +70°C					
Input Offset Voltage	R$_S$ ≤ 10kΩ, ± 9 V ≤ ± 15 V			10	mV
Input Offset Current				750	nA
Input Bias Current				2.0	μA
Large-Signal Voltage Gain	R$_L$ ≥ 2kΩ, V$_{out}$ = ± 10 V	12,000			
Input Resistance		35			kΩ

GUARANTEED ELECTRICAL CHARACTERISTICS

VOLTAGE GAIN   OUTPUT VOLTAGE SWING   INPUT COMMON MODE VOLTAGE RANGE

TYPICAL PERFORMANCE CURVES

VOLTAGE TRANSFER CHARACTERISTIC   FREQUENCY COMPENSATION CIRCUIT   FREQUENCY RESPONSE FOR VARIOUS CLOSED-LOOP GAINS

μA709C

R$_2$*

C$_2$ *Use R$_2$ = 50 Ω when the amplifier is operated with capacitive loading.

C$_1$ R$_1$

Fairchild cannot assume responsibility for use of any circuitry described other than circuitry entirely embodied in a Fairchild product. No other circuit patent licenses are implied

Fig. 13.30 Continued.

is cascaded to a second differential stage which uses *Darlington pairs* (Q_3, Q_5 and Q_4, Q_6) for the transistors. A Darlington pair provides a very high input impedance. (In some op amps, field-effect transistors are used in the first differential amplifier stage to achieve a high input impedance.) The output stage (Q_{13}, Q_{14}) is a complementary pair circuit.

5. Connection diagram   Identifies the terminals on the package.

6. Electrical characteristics   Provides ranges of parameter values such as input offset voltage, gain, and power consumption. Test conditions for the parameters must always be explicit. The *supply-voltage rejection ratio* is the ratio of a change in offset voltage for a corresponding change in a supply voltage.

7. Guaranteed electrical characteristics   Curves show how voltage gain, output voltage swing, and input common-mode voltage range vary with the supply voltage.

To–5 style
12–terminal
package

To–5 style
10–terminal
package

To–5 style
8–terminal
package

Dual–in–line
ceramic package

Dual–in–line
plastic package

Ceramic flat pack

Fig. 13.31 Examples of commonly used IC packages. (*Courtesy RCA.*)

The *common-mode voltage range* is the maximum input voltage that may be applied to either the inverting or noninverting terminal of an op amp for safe operation.

8. Typical performance curves The *voltage-transfer characteristic* relates the output voltage to the input voltage over the operating temperature range of the op amp. The frequency compensation circuit and response curves are extremely important. As mentioned earlier, frequency compensation is required to ensure stable operation of the amplifier. It is noted that the closed-loop gain is dependent on the values of the capacitors and resistors used in the compensation network.

Chapter 14

Digital Circuit Fundamentals

14.1 INTRODUCTION

The development of the junction diode and transistor has been the chief catalyst in the rapid growth of computers and digital systems in general. Semiconductor devices can be made to act like nearly perfect switches, making them ideal for digital circuits. By the use of a few elementary circuits over and over again, all the necessary functions in the operation of a digital system are realized.

Because a switch is either ON or OFF, it is said to exhibit two stable states. (A switch, for this reason, is sometimes referred to as a *bistable device*.) For semiconductor switches to perform arithmetic and logic operations, it therefore becomes necessary to represent numbers by two digits, or *bits*, 0 and 1. In this manner, a 0 may be represented by a closed switch and a 1 by an open switch. Such a number system using only 0s and 1s, called the *binary number system*, is employed in digital systems.

In this chapter we consider first how numbers are expressed in the binary number system. It is also necessary to know how to convert a decimal number into a binary number and back again to decimal. Basic logic functions, such as AND, OR, NOT, NAND, and NOR, are examined next. These functions are described completely by the *truth table*, which is discussed. Operation of multivibrators is also considered. The chapter concludes with a study of the response of switching devices to rapidly changing waveforms, referred to as *dynamic response*.

14.2 BINARY NUMBERS

Because we are born with ten fingers, it is by sheer accident that we count by 10s. The *base*, or *radix*, of our number system is ten, referred to as the *decimal number system*. A zero and nine different symbols are needed to represent decimal numbers: 0, 1, 2, 3, 4, 5, 6, 7, 8, 9. In some past societies people used both their fingers and toes to count. It is not surprising to learn that the base of their number system was twenty. If we were born with only two hands and no fingers, it is a safe bet that we would count by 2s.

Number systems are *positional*. The value of a number depends not only on its symbols but also on their position in the number. For example, 367 and 673 are two numbers having the same symbols but different positions. One may regard a decimal number, such as 367, as a shorthand for

$$3 \times 10^2 + 6 \times 10^1 + 7 \times 10^0 = 300 + 60 + 7 = 367$$

where $10^0 = 1$, $10^1 = 10$, and $10^2 = 100$. Thus, 367 is equal to 7 units (10^0), 6 tens (10^1), and 3 hundreds (10^2). All decimal numbers are expressed in this manner, regardless of their magnitude.

The binary number system is also positional. For example, 1101 is equal to

$$1 \times 2^3 + 1 \times 2^2 + 0 \times 2^1 + 1 \times 2^0 = 8 + 4 + 0 + 1 = 13$$

where $2^0 = 1$, $2^1 = 2$, $2^2 = 4$, and $2^3 = 8$. Thus, binary number 1101 is equal to 1 unit (2^0), 0 twos (2^1), 1 four (2^2), and 1 eight (2^3).

In comparing numbers having different base systems, it is customary to place a subscript, equal to the number base, after the number. Hence, one would write

$$13_{10} = 1101_2$$

DECIMAL-TO-BINARY CONVERSION To convert a whole decimal number to a binary number, the following procedure is used:

1. Divide the number by 2; the remainder is either a 0 or a 1.
2. Place the remainder to the right of the partial quotient obtained in step 1.
3. Divide the partial quotient of step 1 by 2, placing the remainder to the right of the new partial quotient.
4. Repeat the preceding steps until a quotient of zero is obtained.
5. The binary number is equal to the remainders arranged so that the first remainder is the least significant bit (LSB) and the last remainder is the most significant bit (MSB) of the binary number.

example 14.1 Convert the following decimal numbers to binary numbers: (a) 13; (b) 103.

solution (a)

$$
\begin{array}{r}
2)\underline{13} \\
2)\underline{6} \\
2)\underline{3} \\
2)\underline{1} \\
0
\end{array}
\qquad
\left.
\begin{array}{l}
1 \\
0 \\
1 \\
1
\end{array}
\right\} \text{Remainders}
$$

Rearranging remainders, $13_{10} = 1101_2$. (This equality was verified in the preceding section.)

(b)

$$
\begin{array}{rl}
2)\underline{103} & \\
2)\underline{51} & 1 \\
2)\underline{25} & 1 \\
2)\underline{12} & 1 \\
2)\underline{6} & 0 \\
2)\underline{3} & 0 \\
2)\underline{1} & 1 \\
0 & 1
\end{array}
$$

Hence, $103_{10} = 1100111_2$. To verify the answer, the binary number is expanded into its constituent parts: $1100111_2 = 1 \times 2^6 + 1 \times 2^5 + 0 \times 2^4 + 0 \times 2^3 + 1 \times 2^2 + 1 \times 2^1 + 1 \times 2^0 = 1 \times 64 + 1 \times 32 + 0 \times 16 + 0 \times 8 + 1 \times 4 + 1 \times 2 + 1 \times 1 = 64 + 32 + 4 + 2 + 1 = 103_{10}$.

DECIMAL-TO-BINARY FRACTION CONVERSION To convert a decimal fraction to a binary fraction, the following procedure is used:

1. Multiply the decimal fraction by 2.
2. If, as a result of step 1, a number equal to or greater than 1 is obtained, the 1 is placed to the right of the partial product. For example, multiplying 0.6 by 2 yields 1.2, which is a number greater than 1. The 1 is placed to the right of the partial product 0.2. If the result obtained is less than one, a zero is placed to the right of the partial product.
3. The partial product obtained in step 2 is multiplied by 2. The process is repeated until the partial product is 0 or the resulting binary fraction is to the required places of the binary point (which corresponds to the decimal point of a decimal fraction).
4. The 1s and 0s in the order obtained are equal to the binary fraction.

example 14.2 Convert the following decimal fractions to binary fractions: (a) 0.375; (b) 0.68.

solution (a)

$$
\begin{array}{ll}
2 \times 0.375 = 0.750 & 0 \\
2 \times 0.750 = 1.500 & 1 \\
2 \times 0.500 = 1.000 & 1 \\
2 \times 0.000 = 0.000 &
\end{array}
$$

Hence, $0.375_{10} = 0.011_2$. To verify the answer, 0.011 is expanded into its constituent parts: $0.011 = 0 \times 2^{-1} + 1 \times 2^{-2} + 1 \times 2^{-3} = 0/2 + 1/4 + 1/8 = 0.25 + 0.125 = 0.375$.

(b)
$$2 \times 0.68 = 1.36 \qquad 1$$
$$2 \times 0.36 = 0.72 \qquad 0$$
$$2 \times 0.72 = 1.44 \qquad 1$$
$$2 \times 0.44 = 0.88 \qquad 0$$
$$2 \times 0.88 = 1.76 \qquad 1$$

and so on. In this example the binary fraction is unending. To five *binary places*, $0.68_{10} \approx 0.10101_2$.

BINARY-TO-DECIMAL CONVERSION   To convert a binary whole number to a decimal number, the following procedure is used:

1. Multiply the most significant bit (MSB) by 2.
2. If the next to the MSB is a one, add a 1 to the partial product obtained in step 1; if a zero, add a 0.
3. Multiply the result obtained in step 2 by 2. Continue the process until the least significant bit (LSB) is included in the conversion.

example 14.3   Convert the following binary numbers to decimal numbers: *(a)* 11010; *(b)* 110101.

solution *(a)*   MSB   LSB

$$2 \times 1 + 1 = 3 \qquad 2 \times 3 + 0 = 6 \qquad 2 \times 6 + 1 = 13 \qquad 2 \times 13 + 0 = 26$$

Therefore, $11010_2 = 26_{10}$.

(b)   1   1   0   1   0   1

$$2 \times 1 + 1 = 3 \quad 2 \times 3 + 0 = 6 \quad 2 \times 6 + 1 = 13 \quad 2 \times 13 + 0 = 26 \quad 2 \times 26 + 1 = 53$$

Hence, $110101_2 = 53_{10}$.

BINARY-TO-DECIMAL FRACTION CONVERSION   To convert a binary fraction to a decimal fraction, the following procedure is used:

1. Divide the LSB by 2.
2. If the next to the LSB is a 1, add it to the result of step 1.
3. Divide the result of step 2 by 2. If the next bit is a 1, add it to the result.
4. Continue the process until the binary point is reached.

example 14.4   Convert the following binary fractions to decimal fractions: *(a)* 0.0111; *(b)* 0.01001.

solution   *(a)* Dividing the LSB of 0.0111 by 2 yields $\frac{1}{2} = 0.5$. Because the next bit is a 1, it is added to 0.5. Hence, the first ratio, R_1, is

$$R_1 = 0.5 + 1 = 1.5$$

Dividing 1.5 by 2 and adding 1 yields

$$R_2 = \frac{1.5}{2} + 1 = 1.75$$

Dividing 1.75 by 2 and, because the next bit is a zero, we have

$$R_3 = \frac{1.75}{2} + 0 = 0.875$$

Finally, the binary point is reached and $0.875/2 = 0.4375$. Hence, $0.0111_2 = 0.4375_{10}$.

(b)
$$R_1 = 0.5 + 0 = 0.5$$
$$R_2 = \frac{0.5}{2} + 0 = 0.25$$
$$R_3 = \frac{0.25}{2} + 1 = 1.125$$
$$R_4 = \frac{1.125}{2} + 0 = 0.5625$$
$$R_5 = \frac{0.5625}{2} = 0.28125$$

Hence, $0.01001_2 = 0.28125_{10}$.

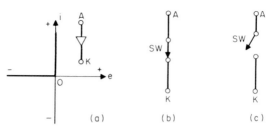

Fig. 14.1 Ideal diode: (*a*) Current-voltage characteristics. (*b*) Closed-switch model for a forward-biased ideal diode. (*c*) Open-switch model for a reverse-biased ideal diode.

14.3 DIODE AS A SWITCH

As explained in Chap. 8, a junction diode when forward-biased conducts current and when reverse-biased it does not conduct current. *It behaves like a switch.*

Consider the characteristics of an ideal diode, shown in Fig. 14.1*a*. When forward-biased (*A* positive with respect to *K*), the voltage across the ideal diode is zero, regardless of the current flowing. The forward-biased ideal diode, therefore, may be represented by the closed-switch model of Fig. 14.1*b*. On the other hand, when reverse-biased, the current is zero regardless of the magnitude of the reverse voltage. A reverse-biased ideal diode, therefore, can be represented by the open-switch model of Fig. 14.1*c*.

Fig. 14.2 Physical junction diode: (*a*) Current-voltage characteristics. (*b*) Model for a forward-biased physical diode. (*c*) Model for a reverse-biased physical diode.

The characteristics of a physical junction diode are illustrated in Fig. 14.2*a*. When forward-biased, before appreciable current flows, the forward voltage must exceed the threshold voltage V_t (0.6 V for silicon at room temperature). In conduction, the forward resistance of the diode, which can be as low as a few ohms, is designated by R_F. A forward-biased physical diode, therefore, acts like a closed switch in series with a battery of V_t volts and a resistance of R_F ohms, as illustrated in Fig. 14.2*b*.

When reverse-biased (the reverse voltage must not exceed the breakdown voltage of the diode), the reverse resistance is extremely high, in the order of megohms. Designating the reverse resistance by R_R, the reverse-biased diode is represented by a switch in parallel with R_R shown in Fig. 14.2*c*. In practice, the models for the ideal diode are used often to represent a physical diode.

14.4 TRANSISTOR AS A SWITCH

Figure 14.3*a* is an elementary circuit of an npn junction transistor operating as a switch. (Field-effect transistor switches are considered in Chap. 15.) The configuration is that

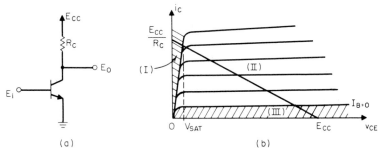

Fig. 14.3 Transistor operating as a switch: (*a*) Elementary circuit employing a npn transistor in the common-emitter configuration. (If a pnp transistor is used, the polarity of E_{CC} is reversed.) (*b*) Load line superimposed on the collector characteristics. Regions I, II, and III are the saturation, active, and cutoff regions, respectively.

of a common-emitter amplifier. Input E_i is applied to the base of the transistor. The collector is connected to resistance R_C in series with the collector supply voltage E_{CC}. Output voltage E_o is taken across the collector and ground.

The load line superimposed on the collector characteristics is shown in Fig. 14.3*b*. One may define three regions of operation, indicated by Roman numerals I, II, and III.

I. Saturation region In the saturation region, both the collector-base and base-emitter junctions are *forward-biased*. The voltage across the collector and emitter is referred to as the *saturation voltage*, V_{sat}. Typically, V_{sat} for most switching transistors is 0.1 V or less.

When a transistor is in the saturation region, it is considered to be ON. It may be represented by the model of a closed switch in series with a battery of V_{sat} volts, as in Fig. 14.4*a*. A transistor that when ON is in saturation is referred to as a *saturated*

Fig. 14.4 Models for a common-emitter transistor switch: (*a*) For the ON state. (*b*) For the OFF state.

switch. If a transistor is ON and not in the saturated region, it is referred to as a *non-saturated switch*. Because of greater efficiency, transistors are generally operated as saturated switches.

II. Active region In the active region, the collector-base junction is *reverse-biased* and the base-emitter junction is *forward-biased*. The Q point of an amplifier, for example, is located in the active region of the collector characteristics.

III. Cutoff region In the cutoff region, both the collector-base and base-emitter junctions are *reverse-biased*. When in the cutoff region, the transistor is considered to be OFF. It may be represented by the model of an open switch of Fig. 14.4*b*.

A transistor operated as a saturated switch is either in the saturation (ON) or cutoff (OFF) regions. In switching a transistor from ON to OFF (or from OFF to ON), it makes a transition through the active region. A *fast* switching transistor makes the transition rapidly.

14.5 BASIC LOGIC CIRCUITS

There are three basic logic circuits which, when used in various combinations, perform the arithmetic and control functions in a digital computer or system. These circuits are the OR gate, AND gate, and the INVERTER (NOT).

OR GATE The logic symbol for a two-input OR gate is given in Fig. 14.5. Output X is obtained when A *or* B, *or* both A and B, are present at the input. The preceding statement can be expressed by a *logic equation* as

$$X = A + B \qquad (14.1)$$

where the plus sign (+) is interpreted as OR. Because the binary number system is used in logic systems, inputs A and B, and output X can assume only the values of 0 and 1.

An elementary example of an OR gate is the two parallel manual switches, A and B of Fig. 14.6. The input to both switches is voltage source E; the output X is a glowing lamp. The lamp glows if switch A *or* B *or* both A and B are closed.

Fig. 14.5 Logic symbol for a two-input OR gate ($X = A + B$). OR gates having more than two inputs are also possible.

Fig. 14.6 An example of an OR gate consisting of two parallel manual switches. The lamp glows ($X = 1$) if switch A, or B, or A and B are closed.

The operation of a logic circuit can be described by a *truth table*. In a truth table, all possible combinations of inputs and their corresponding outputs are displayed in tabular form. Inputs A and B can assume only four possible combinations of values: 00, 01, 10, and 11. These values and their corresponding output X are listed in the truth table for the two-input OR gate in Fig. 14.7. It is read as an output X exists when A = 0 or B = 1; A = 1 or B = 0; or A = B = 1. No output exists when A = B = 0.

The number of different possible combinations assumed by n inputs is expressed by

$$\text{Combinations} = 2^n \qquad (14.2)$$

As we saw in the preceding paragraph, for two inputs A and B four combinations were possible ($2^n = 2^2 = 4$).

example 14.5 Draw the truth table for a three-input (A, B, C) OR gate.

A	B	C	X
0	0	0	0
0	0	1	1
0	1	0	1
0	1	1	1
1	0	0	1
1	0	1	1
1	1	0	1
1	1	1	1

A	B	X
0	0	0
0	1	1
1	0	1
1	1	1

Fig. 14.7 Truth table for a two-input OR gate.

Fig. 14.8 Truth table for a three-input OR gate. (See Example 14.5.)

solution   Because $n = 3$, by Eq. (14.2), $2^3 = 8$ combinations are possible. These are: 000, 001, 010, 011, 100, 101, 110, and 111. (A simple way of listing the combinations is to write the binary numbers corresponding to decimal numbers 0 to $2^n - 1$.) The combinations and the corresponding values of X for the three-input OR gate are listed in the truth table of Fig. 14.8.

Positive and negative logic   In *positive logic*, the voltage that represents a 1 is greater than the voltage that represents a 0. The reverse is true in *negative logic*: A 1 is represented by a voltage less than the voltage denoting a 0. Letting E_1 be the voltage for a 1 and E_0 be the voltage for a 0, examples of positive and negative logic are illustrated in Fig. 14.9. The 1 and 0 are commonly referred to as *logic 1* and *logic 0*.

OR gate circuit   An example of a two-input OR gate (for positive logic) using diodes is illustrated in Fig. 14.10. (In negative logic the diodes are reversed.) The circuit is an example of *diode logic*.

Fig. 14.9   Example of positive and negative logic levels: (*a*) Positive logic: $E_1 > E_0$. (*b*) Negative logic: $E_1 < E_0$.

Fig. 14.10   An example of a two-input OR gate using diodes for positive logic. (In negative logic, the diode connections are reversed.)

Let $E_1 = 5$ V and $E_0 = 0$ V. Assuming ideal diodes, if $E_A = E_B = 0$ V, both diodes are nonconducting, and the output $= 0$ V ($X = 0$). If $E_A = 5$ V and $E_B = 0$ V, diode D_A conducts and diode D_B is reverse-biased; the output, therefore, equals 5 V ($X = 1$). If $E_B = 5$ V and $E_A = 0$ V, diode D_B now conducts and D_A is reverse-biased; output equals 5 V. For $E_A = E_B = 5$ V, both diodes conduct and the output is at 5 V ($X = 1$).

Because of noise in digital systems, it is possible that voltages E_A and E_B are unequal. Suppose that $E_A = 6$ V and $E_B = 5$ V for a logic 1. If $E_A = 6$ V and $E_B = 0$ V, the output equals 6 V. If $E_A = 5$ V and $E_B = 0$ V, the output will be at 5 V. If both inputs are applied simultaneously, $E_A = 6$ V and $E_B = 5$ V, the output is equal to the *larger of the inputs*.

Assume that indeed $E_A = 6$ V and $E_B = 5$ V. The output is 6 V. Because the anode of diode D_B is at 5 V and the cathode at 6 V, diode D_B is reverse-biased and does not conduct.

Fig. 14.11   Logic symbol for a two-input AND gate ($X = AB$). AND gates having more than two inputs are also possible.

AND GATE   The logic symbol for a two-input AND gate is shown in Fig. 14.11. Output $X = 1$ when inputs A *and* B are present; otherwise, $X = 0$. Expressed by a logic equation,

$$X = A \times B \tag{14.3a}$$

where the x is interpreted as AND. As in algebra, the x may be eliminated and Eq. (14.3a) expressed by

$$X = AB \tag{14.3b}$$

A	B	X
0	0	0
0	1	0
1	0	0
1	1	1

Fig. 14.12
Truth table
for a two-
input AND
gate.

Fig. 14.13 An example of an AND
operation. The lamp glows $(X = 1)$
only if switches A and B are closed.

The truth table for a two-input AND gate is given in Fig. 14.12. It is seen that an output exists $(X = 1)$ when *both* inputs equal 1; otherwise, $X = 0$. For a three-input AND gate, $X = 1$ only when all three inputs equal logic 1, and so on.

An elementary example of an AND operation is illustrated in Fig. 14.13. Switches A and B are connected in series with source E and the lamp. The lamp glows $(X = 1)$ when switches A *and* B are closed. If either or both switches are open, the lamp does not glow $(X = 0)$.

AND gate circuit An example of a two-input AND gate for positive logic using diodes is shown in Fig. 14.14. Resistance R is returned to voltage E_D which is equal to or greater than E_1. (For negative logic, the diodes and polarity of E_D are reversed.)

Fig. 14.14 An example of a two-input
AND gate for positive logic. (In negative
logic, the diode connections and supply
voltage E_D are reversed.)

Assume ideal diodes, $E_1 = 5$ V, $E_0 = 0$ V, and $E_D = 5$ V. If either E_A or $E_B = 0$ V, diode D_A or D_B conducts; the output is zero volts $(X = 0)$. If $E_A = E_B = 5$ V, both diodes are reverse-biased and the output equals $E_D = 5$ V $(X = 1)$.

When E_A and E_B for logic 1 are unequal, the output is equal to the *lesser* voltage. Assume that $E_A = 5$ V and $E_B = 3$ V. Diode D_B becomes forward-biased because its anode was initially at $E_D = 5$ V and its cathode at $E_B = 3$ V. The output is equal to 3 V $(X = 1)$. Since the anode of diode D_A is now at 3 V and its cathode at $E_A = 5$ V, diode D_A is reverse-biased.

INVERTER The logic symbol for the INVERTER, also referred to as a NOT, is given by either of the symbols of Fig. 14.15. Output X is equal to the *not* of A. Expressed by a logic equation,

$$X = \overline{A} \qquad (14.4)$$

where the bar over A indicates a NOT operation. If $A = 1$, output $X = 0$; if $A = 0$, $X = 1$. The truth table for the inverter is given in Fig. 14.16.

INVERTER circuit An example of an inverter using an npn junction transistor is illustrated in Fig. 14.17. Resistance R_1 is in series with the transistor base and input A.

Fig. 14.15 Logic symbols for the INVERTER ($X = \overline{A}$).

A	X
0	1
1	0

Fig. 14.16
Truth table
for an IN-
VERTER.

Resistance R_2 is connected between the base and the negative-voltage supply, $-E_{BB}$ volts. This ensures that for $A = E_0$ (logic 0), the base-emitter junction is reverse-biased and the output $= E_{CC}$ volts (logic 1). When $A = E_1$ (logic 1), transistor Q conducts and the output is equal to V_{sat} volts (logic 0).

Fig. 14.17 An example of an INVERTER circuit using a npn transistor. For a pnp transistor the polarities of E_{CC} and E_{BB} are reversed.

Fig. 14.18 Models for the input (base) circuit of a transistor switch: (a) For $E_0 = 0$ V. (b) for $E_1 = 10$ V. (See Example 14.6.)

example 14.6 For the INVERTER of Fig. 14.17, $E_{CC} = 10$ V, $-E_{BB} = -10$ V, $R_C = 1$ kΩ, $R_1 = 10$ kΩ, and $R_2 = 90$ kΩ. Logic signals $E_0 = 0$ V and $E_1 = 10$ V. (a) Verify that the circuit behaves as an INVERTER. (b) Determine the minimum dc current gain for the transistor, $h_{FE, min}$ to ensure that the transistor operates as a saturated switch. Assume that $V_{sat} = 0.1$ V and that the base-emitter voltage is 0.7 V when the transistor is ON.

solution (a) For $E_0 = 0$ V, the base-emitter junction is reverse-biased and is represented by the model of an open switch, shown in Fig. 14.18a. The voltage across the base-emitter junction, by voltage division, is $-10 \times 10/(10 + 90) = -1$ V. A reverse bias of -1 V is more than necessary for a transistor to be OFF (nonconducting).

For $E_1 = 10$ V, the model for the base circuit is illustrated in Fig. 14.18b. The base-emitter junction is represented by a 0.7-V battery. To calculate the base current I_B, superposition is used. Setting E_{BB} to zero, the current owing to E_1 is $I_1 = (10 - 0.7)/10 = 0.93$ mA. Setting E_1 to zero, the current due to E_{BB} is $I_2 = (10 + 0.7)/90 = 0.12$ mA. The base current is equal to the difference of I_1 and I_2: $I_B = I_1 - I_2$. Hence, $I_B = 0.93 - 0.12 = 0.81$ mA.

From Fig. 14.17, collector current I_C is equal to the difference of E_{CC} and V_{sat} divided by R_C: $I_C = (E_{CC} - V_{sat})/R_C$. Substitution of the given values in the equation yields $I_C = (10 - 0.1)/1 = 9.9$ mA. When input $A = E_1$ (10 V), the output equals V_{sat} (logic 0) if $I_B = 0.81$ mA.

(b) The minimum dc current gain is expressed by $h_{FE.min} = I_C/I_B$. From part (a), $h_{FE.min} = 9.9/0.81 = 12.2$. Transistor Q, therefore, must have a dc current gain of at least 12.2 to ensure that it operates as a saturated switch.

14.6 HALF-ADDER

To demonstrate the usefulness of the AND, OR, and INVERTER circuits, these building blocks are used to construct a basic component of a binary adder, the half-adder (HA). Besides addition, the binary adder can be used for subtraction, multiplication, and division. In subtraction, the subtrahend is inverted and added to the minuend. Multiplication may be performed by repeated addition, and division by repeated subtraction. The binary adder is indeed a universal circuit.

The addition table for binary numbers is extremely simple, as illustrated in Fig. 14.19. We see that $0 + 0 = 0$; $0 + 1 = 1$; $1 + 0 = 1$; and $1 + 1 = 10$ (decimal 2). The heart of the binary adder is the half-adder, shown symbolically in Fig. 14.20. It has two inputs A and B, and two outputs C (carry) and S (sum).

Fig. 14.19 Addition table for binary numbers.

Fig. 14.20 Symbol for a half-adder. Bits to be added are A and B. Output S is the sum bit and C is the carry bit.

A	B	C	S
0	0	0	0
0	1	0	1
1	0	0	1
1	1	1	0

Fig. 14.21 Truth table for a half-adder.

Fig. 14.22 A half-adder realized with AND and OR gates and INVERTERS.

The truth table for the half-adder is given in Fig. 14.21. A sum bit results when $A = 0$ and $B = 1$, or $A = 1$ and $B = 0$. When $A = 1$ and $B = 1$, a 0 results in the sum column and a 1 in the carry column.

The preceding results may be summarized by two logic equations. Sum $S = 1$ when $A = 0$ *and* $B = 1$ *or* $A = 1$ *and* $B = 0$. Carry $C = 1$ when A *and* $B = 1$. Hence,

$$S = \overline{A}B + A\overline{B} \tag{14.5a}$$

$$C = AB \tag{14.5b}$$

Using AND and OR gates, and INVERTERS, the half-adder may be realized as shown in Fig. 14.22.

14.7 INHIBIT AND EXCLUSIVE OR GATES

The logic symbol for a three-input INHIBIT gate is illustrated in Fig. 14.23a. It is basically an AND gate with one input (C) negated by an INVERTER, as shown in Fig. 14.23b. This is indicated on the symbol by a circle drawn at input C.

(a)   (b)

Fig. 14.23 An INHIBIT gate: (a) Logic symbol ($X = AB\overline{C}$). (b) It consists of an AND gate with one input negated by an INVERTER. INHIBIT gates having more than three inputs, as well as only two inputs, are possible.

The operation of an INHIBIT gate is such that $X = 1$ when $A = B = 1$ and $C = 0$; if C also equals 1, $X = 0$. Input C therefore acts to *inhibit* the gate. If $C = 0$, it *enables* the gate, permitting it to perform an AND operation. The logic equation for the INHIBIT gate is

$$X = AB\overline{C} \tag{14.6}$$

The truth table is given in Fig. 14.24. Because three inputs are present, $n = 3$. By Eq. (14.2), $2^3 =$ eight possible combinations of A, B, and C exist. It is seen that $X = 1$ only if $A = 1$, $B = 1$, and $C = 0$.

The logic symbol for an EXCLUSIVE OR gate is shown in Fig. 14.25. For this gate, $X = 1$ if $A = 1$, $B = 0$, or $A = 0$, $B = 1$. If $A = B = 0$ or $A = B = 1$, the output is zero. Expressed by a logic equation,

$$X = \overline{A}B + A\overline{B} \tag{14.7}$$

A	B	C	X
0	0	0	0
0	0	1	0
0	1	0	0
0	1	1	0
1	0	0	0
1	0	1	0
1	1	0	1
1	1	1	0

Fig.   14.24
Truth table for a three-input INHIBIT gate.

Fig. 14.25 Logic symbol for an EXCLUSIVE OR gate ($X = \overline{A}B + A\overline{B}$).

Equation (14.7) is identical with Eq. (14.5a) for the sum S of a half-adder. The circuit for S in Fig. 14.22 can therefore be used for realizing an EXCLUSIVE OR gate.

The truth table is illustrated in Fig. 14.26. Comparing this table with the truth table for the OR gate of Fig. 14.7, it is seen that in the OR gate, $X = 1$ even if $A = B = 1$. The OR gate, therefore, is sometimes referred to as an INCLUSIVE OR, because it provides an output when all inputs are at a logic 1.

A	B	X
0	0	0
0	1	1
1	0	1
1	1	0

Fig. 14.26
Truth table for an EX-CLUSIVE OR gate.

14.8 NAND AND NOR GATES

Two universal gates, the NAND and NOR, are covered in this section. With either of these gates, the OR, AND, and NOT functions can be realized.

NAND GATE Logic symbols for the NAND gate are illustrated in Fig. 14.27. The NAND gate may be regarded as an AND gate followed by an INVERTER, as shown in

(a)   (b)

Fig. 14.27 Logic symbols for a two-input NAND gate ($X = \overline{AB} = \overline{A} + \overline{B}$). NAND gates with more than two inputs are also possible.

Fig. 14.28. Output $X = 1$ when $A = B = 0$; otherwise $X = 0$. Expressed by a logic equation,

$$X = \overline{AB} \tag{14.8a}$$

which is read as "X is equal to the *not* of A *and* B."

By a form of a very useful theorem, De Morgan's theorem, the *not* of A *and* B is equal to the *not* of A *or* the *not* of B: $\overline{AB} = \overline{A} + \overline{B}$. The logic equation for NAND operation may therefore also be expressed by

$$X = \overline{A} + \overline{B} \tag{14.8b}$$

Fig. 14.28 The NAND gate viewed as an AND gate followed by an INVERTER.

A	B	X
0	0	1
0	1	1
1	0	1
1	1	0

Fig. 14.29
Truth table for a two-input NAND gate.

Based on Eq. (14.8b), an alternative symbol for the NAND gate is given in Fig. 14.27b, which shows negated inputs to an OR gate.

The truth table for a two-input NAND gate for positive logic is provided in Fig. 14.29. It is seen that $X = 0$ when $A = B = 1$; for all other inputs, $X = 1$.

If both inputs A and B are connected together, the NAND gate functions as an IN-VERTER. An INVERTER following a NAND yields an AND gate, as illustrated in Fig. 14.30. A *not of a not* ($\overline{\overline{AB}}$), indicated by two horizontal bars above AB, yields the function itself, AB. For negative logic, where a logic 1 is represented by a voltage that is less positive than for a logic 0, the positive NAND gate behaves as an OR gate.

example 14.7   Verify that for negative logic the positive NAND gate behaves as an OR gate.

solution   In the truth table of Fig. 14.29, a 0 becomes a 1 and a 1 becomes a 0 for inputs A and B in negative logic. The new truth table, therefore, appears as in Fig. 14.31. This is identical with the truth table for the OR gate of Fig. 14.7.

A	B	X
1	1	1
1	0	1
0	1	1
0	0	0

Fig. 14.30   A two-input AND gate realized by a NAND gate followed by a second NAND gate acting as an INVERTER.

Fig. 14.31 Truth table for a two-input posi-tive NAND gate func-tioning as an OR gate for nega-tive logic. (See Exam-ple 14.7.)

NOR GATE   Logic symbols for the NOR gate are given in Fig. 14.32. The NOR gate may be viewed as an OR gate followed by an INVERTER, illustrated in Fig. 14.33. Output $X = 1$ only when $A = B = 0$; otherwise $X = 0$. Expressed by a logic equation,

$$X = \overline{A + B} \qquad (14.9a)$$

which is read as "X is equal to the *not* of A or B."

By another form of De Morgan's theorem, $\overline{A + B}$ is equal to the *not* of A *and* the *not* of B. Therefore X also equals $\overline{A} \times \overline{B}$:

$$X = \overline{A} \times \overline{B} \qquad (14.9b)$$

Based on Eq. (14.9b), an alternative symbol for the NOR gate is given in Fig. 14.32b, which shows negated inputs to an AND gate.

The truth table for a two-input NOR gate for positive logic is illustrated in Fig. 14.34. It is seen that output $X = 1$ only for $A = B = 0$; for all other pairs of inputs, $X = 0$.

If both inputs are tied together, the NOR gate acts as an INVERTER. An INVERTER

(a)   (b)

Fig. 14.32   Logic symbols for a two-input NOR gate ($X = \overline{A + B} = \overline{A} \times \overline{B}$). NOR gates with more than two inputs are also possible.

Fig. 14.33 A NOR gate viewed as an OR gate followed by an INVERTER.

A	B	X
0	0	1
0	1	0
0	1	0
1	1	0

Fig. 14.34 Truth table for a two-input NOR gate.

following a NOR yields an OR gate, as shown in Fig. 14.35. For negative logic, the positive NOR gate acts as an AND gate.

example 14.8 Verify that for negative logic the positive NOR gate acts as an AND gate.

solution In the truth table of Fig. 14.34, a 0 becomes a 1 and a 1 becomes a 0 for inputs A and B in negative logic. The resulting truth table appears in Fig. 14.36. This is identical with the truth table for the AND gate of Fig. 14.12.

Fig. 14.35 A two-input OR gate realized by a NOR gate followed by a second NOR gate acting as an INVERTER.

A	B	X
1	1	1
1	0	0
0	1	0
0	0	0

Fig. 14.36 Truth table for a two-input positive NOR gate functioning as an AND gate for negative logic. (See Example 14.8.)

14.9 MULTIVIBRATORS

There are three kinds of multivibrators: the *bistable, monostable,* and *astable.* Their fundamental behavior and properties are explained initially with the aid of the block diagrams shown in Fig. 14.37.

   a. The block diagram of a bistable multivibrator (MV), also referred to as a *flipflop, binary,* and an *Eccles-Jordan circuit,* is illustrated in Fig. 14.37a. The bistable MV has two inputs, labeled 1 and 2, and two outputs, X and $\overline{X}$. Output $\overline{X}$ (*not X*) is said to be the *complement* of X. If X is a logic 1, $\overline{X}$ is a logic 0; if $X = 0$, then $\overline{X} = 1$.

Fig. 14.37 Block diagrams of multivibrators: (*a*) Bistable. (*b*) Monostable. (*c*) Astable.

Assume that $X = 1$. The bistable MV remains in this state until a trigger is applied to, say, input 1; then $X = 0$. Output $X = 0$ until a trigger is applied to terminal 2; then $X = 1$ once again. Thus, the bistable MV exhibits two stable dc states: $X = 1$ and $X = 0$. It remains in one of these two states until a trigger is applied to either terminal 1 or 2 to change its state.

The bistable MV exhibits *memory*. Its output depends not only on the present input, but also on which input terminal the previous trigger was applied. Such a circuit is referred to as *sequential*. The OR, AND, etc., gates depend only on their present inputs and are said to be *combinational*.

The bistable MV is used in *shift registers* and *counters*. These are two important components employed in digital computers and systems. Shift registers are used to store information, such as data or an instruction. The counter, as its name implies, counts pulses.

b. The block diagram of a monostable MV, commonly referred to as a *one-shot*, is illustrated in Fig. 14.37*b*. It has one input and one output terminal. With no trigger applied, the output voltage is approximately 0 V (logic 0). It remains in this stable dc state until a trigger is applied to the input terminal. As a result of the trigger, a well-defined rectangular pulse appears at X. At the termination of the rectangular pulse, the output returns to its stable state, logic 0.

In addition to generating rectangular pulses, the one-shot is also used as a delay. If, for example, the output pulse of a one-shot is applied to terminal C of the INHIBIT gate of Fig. 14.23*a*, the output of the gate is delayed for a time equal to the length of the rectangular pulse.

c. The block diagram of an astable, or *free-running*, MV, is given in Fig. 14.37*c*. There are two output terminals, X and $\overline{X}$, and no input terminals. At the outputs are rectangular-type waveforms. The astable MV is an example of a *nonsinusoidal*, or *relaxation*, oscillator. It is used as a *clock* in digital computers to ensure that operations, such as addition, are synchronized with other operations in the system.

BISTABLE MV   The basic circuit of a *collector-coupled* bistable MV employing npn transistors is illustrated in Fig. 14.38. Note the symmetrical nature of the circuit. The

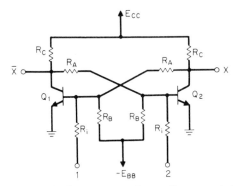

Fig. 14.38 Basic circuit of a collector-coupled bistable MV employing npn transistors. (For pnp transistors, the polarities of E_{CC} and E_{BB} for this and other multivibrator circuits are reversed.)

collector of one transistor is coupled to the base of the other transistor by resistor R_A. Resistor R_B is connected to the base and supply source, $-E_{BB}$ volts. Output X is taken at the collector of transistor Q_2 and $\overline{X}$ at the collector of transistor Q_1.

Assume that $X = 1$ and, therefore, $\overline{X} = 0$. In this state, Q_1 is conducting (ON) and Q_2 is nonconducting (OFF). The base-emitter voltage of Q_1 is approximately 0.7 V and the base-emitter voltage of Q_2 is negative, ensuring that it is cut off.

To change the state of the flipflop, a negative trigger is applied to terminal 1. (For best triggering of multivibrators, the trigger is selected to *turn off* the ON transistor.)

Transistor Q_1 is turned off and its collector voltage rises toward E_{CC} volts. The increasing collector voltage is coupled by resistor R_4 to the base of Q_2. Its base-emitter voltage, therefore, becomes less negative and is finally equal to 0.7 V. Transistor Q_2 is now ON ($X = 0$), and Q_1 is cut off ($\overline{X} = 1$).

To obtain $X = 1$ and $\overline{X} = 0$, a negative trigger is applied to terminal 2. The chain of events described in the preceding paragraph occurs, forcing Q_1 to go ON and Q_2 to go OFF. Now, $X = 1$ and $\overline{X} = 0$.

MONOSTABLE MV   The basic circuit of a collector-coupled monostable MV (one-shot) using npn transistors is shown in Fig. 14.39. As in the bistable MV, the collector of Q_2 is coupled to the base of Q_1 by resistor R_4. The collector of Q_1, however, is coupled to the base of Q_2 by coupling capacitor C.

In its dc stable state, Q_2 is ON ($X = 0$). To ensure that indeed Q_2 is ON, the maximum value of R, R_{max}, is equal to the net voltage across it (E_{CC} − base-emitter voltage E_{BE2}) divided by base current I_{BS} required for Q_2 to be in saturation:

$$R_{max} = \frac{E_{CC} - E_{BE2}}{I_{BS}} \qquad (14.10)$$

To aid us in understanding the operation of the one-shot, waveforms at the bases and collectors of Q_1 and Q_2 are shown in Fig. 14.40. For time less than t_1, the one-shot is in

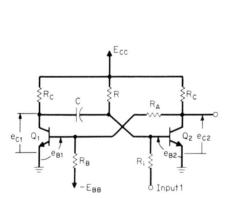

Fig. 14.39 Basic circuit of a collector-coupled monostable MV employing npn transistors.

Fig. 14.40 Waveforms at the bases and collectors of transistors in a collector-coupled monostable MV: (a) Base of Q_2. (b) Collector of Q_2. (c) Base of Q_1. (d) Collector of Q_1.

its stable state where Q_1 is OFF and Q_2 is ON. The base-emitter voltage of Q_2, e_{B2}, is at approximately 0.7 V. Assuming that $V_{sat} = 0$, the collector-emitter voltage of Q_2 is, therefore, $e_{C2} = 0$. Because Q_1 is OFF, its base-emitter voltage is negative ($-E_{B1}$ volts) and its collector-emitter voltage, e_{C2}, is equal to the supply voltage, E_{CC} volts.

At $t = t_1$ assume that a negative trigger is applied to the input terminal. Transistor Q_1 turns ON and Q_2 turns OFF. For $V_{sat} = 0$, the voltage at the collector of Q_1 drops by E_{CC} volts. Because the voltage across a capacitor cannot change instantaneously, the drop in E_{CC} volts is transmitted to the base of Q_2. Hence, at $t = t_1$, $e_{B2} = -(E_{CC} - 0.7)$ volts, and Q_2 is cut off.

Voltage e_{B2} begins to rise exponentially toward E_{CC} volts. When it is equal to the threshold voltage V_t, transistor Q_2 turns ON and Q_1 turns OFF. At the base of Q_2 an *overshoot* of E' volts occurs. Because of the coupling capacitor e_{C1} rises instantaneously by E' volts. During the time Q_2 is OFF, its collector-emitter voltage, e_{C2}, is equal to E_{C2} volts which is slightly less than the supply voltage, E_{CC} volts.

The length, or width, of the rectangular pulse T generated at the collector of Q_2 is given by

$$T \simeq 0.69RC \qquad (14.11)$$

where T is in seconds, R in ohms, and C in farads. Equation (14.11) is an excellent approximation for T if E_{CC} is much greater than 0.7 V. The length of the negative trigger that initiates the one-shot is generally a small fraction of T.

ASTABLE MV The basic circuit of a collector-coupled astable MV using npn transistors is shown in Fig. 14.41. Note that the collector of each transistor is coupled by capacitor C to the base of the other transistor. The astable MV, therefore, has no dc stable states.

Fig. 14.41 Basic circuit of a collector-coupled astable MV employing npn transistors.

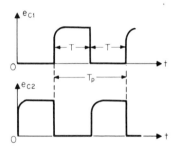

Fig. 14.42 Waveforms at the collectors of Q_1 and Q_2 of a collector-coupled astable MV.

Referring to the waveforms at the collectors of Q_1 and Q_2, e_{C1} and e_{C2}, respectively (Fig. 14.42), it is seen that each transistor is always generating a waveform. The shape of the waveforms is similar to that generated at the collector of Q_1 in the one-shot. The length T indicated in Fig. 14.42 is equal to 0.69RC, as for the one-shot. The total period T_p is therefore equal to $2T$:

$$T_p = 1.38RC \qquad (14.12a)$$

The frequency f is one divided by the total period. Hence the frequency of the generated square wave, in hertz, is

$$f = \frac{1}{1.38RC} \qquad (14.12b)$$

14.10   DYNAMIC RESPONSE OF DIODE AND TRANSISTOR SWITCHES

Because of various factors, to be examined in this section, the response of a diode or transistor operating as a switch is delayed and the waveshape distorted. As a result, the rate at which a diode or transistor can be switched is limited.

RESPONSE OF A DIODE SWITCH Assume that the waveform of Fig. 14.43a is applied to a diode, such as a diode in a positive OR gate. During the interval of t_1 seconds, the diode is forward-biased and a current I_{D1} flows, as indicated in Fig. 14.43b.

At $t = t_1$, the input voltage is $-E$ volts, and the diode becomes reverse-biased. What one expects is that the current in a reverse-biased diode should be zero. Instead, an appreciable current, $-I_{D2}$, flows for a time t_s seconds, which eventually goes to zero.

To explain this anomalous behavior, it is necessary to examine the action of a pn junction (see Chap. 8). When a diode is forward-biased, there is a huge buildup of electrons in the p region and holes in the n region in the vicinity of the pn junction. When the diode is reverse-biased, it takes time for the electrons and holes to depart from the junction region. Consequently the diode conducts in the reverse direction for a time t_s, referred to as the *storage time*.

After t_s has elapsed, the reverse current gradually decreases and ultimately reaches zero. The reason for the gradual decay in reverse current is that it takes time for the diode depletion capacitance to charge to the reverse voltage, $-E$ volts. The time from t_1, when the diode is first reverse-biased, to the time when the reverse current equals a specified value, $-I_R$, is the *reverse-recovery time* t_{rr} of the diode. For good switching diodes, t_{rr} is in the order of nanoseconds (10^{-9} s).

RESPONSE OF A TRANSISTOR SWITCH   Assume that the rectangular pulse (length $= t_p$) of Fig. 14.44a is impressed across the input terminals of an INVERTER.

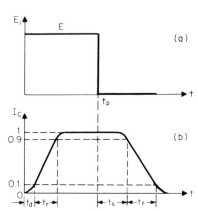

Fig. 14.43 Response of a diode switch: (a) Voltage waveform applied to diode. (b) Diode current flow in response to impressed voltage waveform.

Fig. 14.44 Response of a transistor switch: (a) Pulse applied to the input of an INVERTER. (b) Collector current waveform in response to the input pulse.

The collector current appears as shown in Fig. 14.44b. The output waveshape is stretched and distorted. For convenience, the maximum value of collector current is normalized to one. Four specific times are indicated in the figure. These are:

1. *Delay time* t_d is defined as the time it takes the collector current to reach 0.1 (10 percent) of its maximum value. It corresponds to the time required to charge the junction capacitance plus the time it takes for the carriers (electrons or holes) to reach the collector region.

2. *Rise time* t_r is defined as the time it takes the collector current to rise from 0.1 (10 percent) to 0.9 (90 percent) of its maximum value. The rise time is the time needed to charge additional transistor capacitances.

3. *Storage time* t_s is defined as the time, measured from t_p to where the collector

current falls to 90 percent of its maximum value.   In saturation, both the base-emitter and collector-base junctions are forward-biased.   Consequently the collector, in addition to the emitter, injects electrons (or holes) into the base region.   Similar to turning off a forward-biased diode, it takes time for the carriers to leave the base region.   This phenomenon results in a storage time delay.

4. *Fall time t_f* is defined as the time required for the collector current to fall from its 90 to 10 percent maximum values.   During this interval, the transistor capacitances are being discharged.

The sum of the delay and rise times, $t_d + t_r$, is equal to the *turn-on time t_{on}* of the transistor:

$$t_{on} = t_d + t_r \qquad (14.13a)$$

Similarly, the sum of the storage and fall times, $t_s + t_f$, is equal to the *turn-off time*, t_{off}, of the transistor:

$$t_{off} = t_s + t_f \qquad (14.13b)$$

To decrease the turn-on time, the base is driven hard.   In this condition, more base current is supplied to the base than required for the transistor to saturate.   Because of the excess base current the storage time, and hence the turn-off time, increases.   For high-speed switching, the transistor may be operated as a nonsaturated switch.   An example of a nonsaturating circuit is emitter-coupled logic (ECL), considered in Chap. 15.

Fig. 14.45   The use of a speed-up (commutating) capacitor C_s across the series input resistor R_1 to improve the switching speed of a transistor.

A method used often in discrete circuits to decrease the turn-on and turn-off times is to place a *speed-up*, or *commutating, capacitor* across the resistor in series with the transistor base (Fig. 14.45).   The value of the speed-up capacitor C_s is in the order of 50 pF.   Initially, when the input rises from 0 to E volts, the capacitor acts as a short circuit.   Maximum base current therefore flows and the turn-on time is reduced.   At the termination of the input pulse, the capacitor also acts as a short circuit and aids in removing the excess carriers from the base region.   As a result, the storage and turn-off times are also reduced.

Chapter 15

Digital Integrated Circuits

15.1 INTRODUCTION

In the development of the digital integrated circuit (DIC), families of logic evolved which exhibit well-defined properties. Three important families that enjoy wide use are:

1. *Transistor-transistor logic* (TTL, T^2L) is the most widely used logic. It has good speed, reasonably low power dissipation per gate, and is relatively low in cost.

2. *Emitter-coupled logic* (ECL), or *current-mode logic* (CML), is the fastest logic available today. Operating as a nonsaturated switch, it can switch frequencies as high as 500 MHz. Its power dissipation per gate, as well as its cost, is relatively high.

3. *Complementary metal-oxide semiconductor logic* (CMOS, COS/MOS) exhibits the lowest dissipation per gate. It is slower than T^2L or ECL, and its cost is moderate. Whereas T^2L and ECL employ the bipolar junction transistor in their circuits, CMOS uses the enhancement-type MOSFET. A comparison of the three major families of logic is provided in Table 15.1.

Owing to rapid advances in integrated circuit processing, logic circuits are available with an excess of 1000 logic gates on a single chip of silicon. One can define four levels of complexity in terms of the number of gates on a chip:

1. *Small-scale integration* (SSI): A DIC containing less than 12 gates on a chip

2. *Medium-scale integration* (MSI): A DIC containing more than 12, but less than 100 gates on a chip

3. *Large-scale integration* (LSI): A DIC containing more than 100, but less than 1000 gates on a chip

4. *Grand-scale integration* (GSI): A DIC containing more than 1000 gates on a chip

This chapter defines the terms used to characterize a DIC; discusses the operation of T^2L, ECL, and CMOS logic; and considers their application as flipflops in shift registers and counters. The chapter concludes with a discussion of semiconductor memories, digital-to-analog and analog-to-digital converters, and the microprocessor.

15.2 DIC TERMS AND PARAMETERS

In addition to the specification of required power-supply voltages and operating temperature range, a number of terms and parameters are used in characterizing the performance of a digital integrated circuit. Important specifications include:

FAN IN   The maximum number of inputs to a gate.

TABLE 15.1 Comparison of IC Logic Families

Parameter	T²L				ECL				CMOS
	Standard	Low-power	High-speed	Schottky	1-nS	2-nS	4-nS	8-nS	
Logic function	NAND	NAND	NAND	NAND	OR/NOR	OR/NOR	OR/NOR	OR/NOR	NOR/NAND
Supply voltage, V	5	5	5	5	-5.2	-5.2	-5.2	-5.2	3 to 18
Power dissipated/gate, mW	12	1	22	18	60	25	22	31	10^{-5} [°]
Propagation delay/gate, nS	10	33	6	3	1	2	4	8	70
Speed, MHz	35	3	50	125	500	200	165	30	5
Noise margin	Very good	Very good	Very good	Good	Fair	Fair	Fair	Fair	Excellent
Fan out (typical)	10	10	10	10	10	25	25	25	>50

[°] This is the *static* value. In switching, the power dissipation per gate at 1 MHz is approximately 1 mW and increases with frequency.

FAN OUT   The maximum number of gates that may be connected to the output terminal of a single gate.

PROPAGATION DELAY   The propagation delay t_{dp} is the difference in time between the application of an input signal and its presence at the output of a DIC. Its unit is generally nanoseconds.

SPEED   This parameter indicates how fast a flipflop can change states. Its unit is megahertz.

GATE DISSIPATION   The average dc power dissipated in a gate. Its unit is milliwatts or microwatts.

SPEED-POWER PRODUCT   The product of propagation delay (in nanoseconds) and gate dissipation (in milliwatts). Because the product of power and time yields energy, the unit for the speed-power product is picojoules, pJ ($1 \text{ pJ} = 10^{-12}$ J).

NOISE MARGIN   Noise margin NM is the maximum extraneous voltage that causes a gate to change its state. Its unit is volts or millivolts.

15.3 T²L

A basic T²L circuit having a fan-in of two is illustrated in Fig. 15.1. Transistor Q_1 is unique; it has multiple emitters. Each base-emitter junction behaves like a diode.

Fig. 15.1 Basic T²L gate shown with a fan-in of two. The circuit performs positive NAND logic.

The two base-emitter junctions, therefore, act like diodes of an AND gate to which are connected inputs A and B. The base-collector junction of Q_1 serves as a diode in series with the base of Q_2, which is connected as an INVERTER.
   The logic performed by T²L is positive NAND logic. Assume that either input A or B is at logic 0. One of the base-emitter junctions becomes forward-biased. The base of Q_1 is therefore near ground, and the collector-base junction is reverse-biased. No base current flows in Q_2, and the transistor is OFF: $X = 1$.
   If both inputs are at logic 0, the same condition exists as though either one is at logic 0 and output $X = 1$. When both A and B are at logic 1, the base-emitter diodes are reverse-biased. The collector-base junction of Q_1 is now forward-biased, and base current flows in Q_2. Output X, therefore, is at logic 0.
   There exists a number of members in the T²L family. They include the standard, low-power, high-speed, and Schottky-clamped series. These circuits are examined in the following discussion.

STANDARD T²L   The circuit for a two-input standard T²L NAND gate is shown in Fig. 15.2. Transistor Q_1 serves the same function as in the basic circuit of Fig. 15.1. Transistor Q_2 is connected as a *phase splitter*. When a logic 0 is at its base, its collector is at a logic 1 and its emitter at a logic 0. If the input is equal to a logic 1, the reverse is true. Now, the collector is at a logic 0 and the emitter at a logic 1.

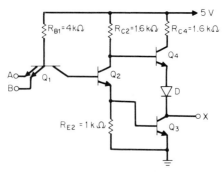

Fig. 15.2 Circuit for a two-input standard T²L NAND gate.

Transistors Q_3 and Q_4 are connected as a *totem-pole amplifier*. This configuration minimizes the effects of any capacitance present, such as stray, across the output. One serious effect of capacitance across the output of a gate is to reduce its switching speed. The function of diode D is to limit the collector current to a reasonable value when Q_3 is in the ON state.

Assume that inputs A and B are at logic 0. The base-emitter junctions of Q_1 are forward-biased, and a logic 0 appears at the base of Q_2. Owing to the operation of the phase splitter, at the collector of Q_2 there appears a logic 1 and at its emitter a logic 0. Transistor Q_4 is therefore ON and Q_3 is OFF. Output X is at a potential equal to E_{CC} less the voltage drops across R_{C4}, the collector-emitter of Q_4, and the diode; hence, $X = 1$.

If either input is at logic 0, the same condition described in the preceding paragraph prevails, and $X = 1$. Assume that both inputs are at logic 1. A logic 1 therefore appears at the base of Q_2. As a result, a logic 0 is at the collector and a logic 1 at the emitter of Q_2. Transistor Q_4 is now OFF, and Q_3 is ON; $X = 0$.

LOW-POWER T²L A version of T²L which dissipates less than one-tenth the power of standard T²L is illustrated in Fig. 15.3. Referred to as low-power T²L, the NAND gate is very similar to the standard circuit. To ensure low-power dissipation, the resistance values are increased appreciably. For example, $R_{B1} = 40$ kΩ, and $R_{C2} = 20$ kΩ; in the standard T²L, their corresponding values are 4 and 1.6 kΩ, respectively.

Transistors Q_3 and Q_4 constitute a Darlington pair. In addition to providing greater current gain than a single transistor, the Darlington pair also increases the switching speed of the gate.

Fig. 15.3 An example of a low-power T²L gate. Transistors Q_3 and Q_4 are connected as a Darlington pair which provides more current gain than a single transistor.

Fig. 15.4 High-speed T²L gate. It features low values of resistances. Diodes D_A and D_B protect the base-emitter junctions of transistor Q_1 against voltage breakdown.

HIGH-SPEED T²L Similar in configuration to low-power T²L, a high-speed T²L NAND gate is shown in Fig. 15.4. To achieve high-speed operation, the resistance values in the circuit are kept low. Because at high switching speeds stray inductance, in addition to stray capacitance, enters the picture, oscillations may be superimposed on the pulses. This phenomenon is referred to as *ringing*. The purpose of diodes D_A and D_B, referred to as *clamping diodes*, connected between each input and ground is to limit negative signal swings because of ringing. In this manner the base-emitter junctions are protected from avalanche or zener breakdown.

(a)   (b)

Fig. 15.5 Schottky-clamped transistor: (*a*) Schottky diode connected across the base and collector terminals of a junction transistor. (*b*) Symbol for a Schottky-clamped transistor, referred to as a Schottky transistor.

SCHOTTKY-CLAMPED T²L The transistors in the previous examples of transistor-transistor logic all operate in saturation when in the ON state. As explained in Chap. 14, a saturated switch gives rise to storage time which reduces its speed of operation. To increase switching speed, the transistor is operated as a nonsaturated switch. In this mode of operation, the transistor in the ON state is in the active region close to, but never in, saturation.

To achieve efficient nonsaturated switching, a Schottky diode (see Chap. 8) is connected across the collector and base of a transistor, as illustrated in Fig. 15.5a. When the transistor is turned on, some of the base current is diverted by the diode from the base. Consequently, less base current flows, and the transistor is prevented from saturating. The symbol for a Schottky-clamped transistor, also referred to as a *Schottky transistor*, is given in Fig. 15.5b.

An example of a Schottky-clamped T²L NAND gate is illustrated in Fig. 15.6. The circuit is similar to the high-speed T²L gate of Fig. 15.4. A low-power Schottky-clamped T²L gate is also available. As in the low-power T²L gate, high-resistance values are used in this version.

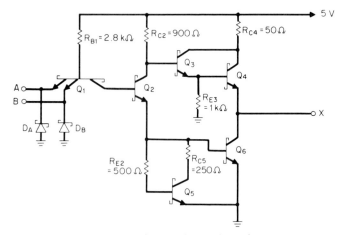

Fig. 15.6 An example of a Schottky-clamped T²L gate.

15.4 ECL

With its transistors operating as nonsaturated switches, emitter-coupled logic (ECL), also referred to as current-mode logic (CML), is the fastest family of logic circuits. It is available in four basic types having propagation delays of 8, 4, 2, and 1 ns (see Table 15.1). Emitter-coupled logic provides two logic functions: the NOR and OR. Its logic symbol is illustrated in Fig. 15.7.

Fig. 15.7 Logic symbol for an ECL gate.

A recent member of the family is the 2-nS ECL gate. This version optimizes switching speed and gate dissipation (see Table 15.1). An example of a 2-nS gate having a fan-in of two is provided in Fig. 15.8. It is powered by a -5.2-V dc power supply. A logic 0 is approximately -1.7 V, and a logic 1 is equal to -0.9 V. Because a logic 1 is less negative than a logic 0, their values correspond to positive logic.

A basic circuit of ECL is the differential amplifier comprised of transistors Q_2 and Q_3

Fig. 15.8 An example of a 2-nS ECL gate.

(see Chap. 13). Emitter current I_E is essentially constant. The base of Q_3 is held at a constant dc reference voltage of 1.29 V. The reference voltage is stabilized with respect to temperature by transistor Q_4 and diodes D_1 and D_2. Switching occurs when the input to either transistor, Q_1 or Q_2, is approximately 0.1 V greater or less than the reference voltage. Emitter followers Q_5 and Q_6 provide dc level shifting to ensure that a logic 0 corresponds to -1.7 V and a logic 1 to -0.9 V.

Assume that inputs A and B are at logic 0. For this condition, Q_1 and Q_2 are OFF and Q_3 is ON (logic 0). The output of Q_5, therefore, is $X = 0$ (OR) and at Q_6, $\overline{X} = 1$ (NOR).

Suppose that A is at logic 0 and B at logic 1. Now Q_3 is OFF, Q_2 is ON; and $X = 1$, $\overline{X} = 0$. If both inputs are at logic 1, as in the preceding case, $X = 1$ and $\overline{X} = 0$. We see therefore that ECL provides an OR and a NOR output.

15.5   CMOS

Complementary metal-oxide semiconductor (CMOS) logic dissipates minute power and can operate over a power-supply range of 3 to 18 V. It is, however, slower than T^{2}L or ECL (see Table 15.1). In this logic family, no resistors are used. Instead, p-channel and n-channel enhancement-type MOSFETs are employed in complementary pairs. As a result, a much greater packing density is realized than with the bipolar junction transistor in monolithic integrated circuits.

INVERTER   A CMOS inverter is illustrated in Fig. 15.9. Source S_2 of the p-channel device is returned to the supply voltage (E_{DD} volts), and source S_1 of the n-channel device is returned to ground. Output X is obtained at the junction of the drain terminals, D_1 and D_2. Both gates, which are connected together, constitute input terminal A. The substrate (Sub) is connected to the source of each transistor.

A logic 1 is equal to E_{DD} volts and a logic 0 to 0 V. Assume that A is at logic 1. Because G_1 is positive with respect to S_1, the n-channel MOSFET (Q_1) is ON. The gate-source voltage of the p-channel device (Q_2), however, is at 0 volts ($E_{DD} - E_{DD} = 0$); hence, Q_2 is OFF. Output $X = 0$.

If A is at logic 0, the n-channel transistor is OFF. The p-channel device, whose gate is now negative with respect to its source ($0 - E_{DD} = -E_{DD}$), is ON. The output in this condition may be thought of as being shifted to E_{DD} volts and $X = 1$.

In either case, one transistor is always ON and the other is always OFF. Since the transistors are connected in series, the drain current flowing is equal to their leakage current. Because of its extremely small value, the dissipation is in the order of 0.01 μW. (In switching, however, the gate dissipation is approximately 1 mW at 1 MHz and increases with frequency.)

NOR GATE   An example of a two-input NOR gate employing CMOS logic is shown in Fig. 15.10. Two p-channel MOSFETs are connected in series and two n-channel

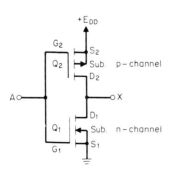

Fig. 15.9 A CMOS INVERTER. Transistor Q_1 is an n-channel enhancement-type MOSFET and Q_2 a p-channel device.

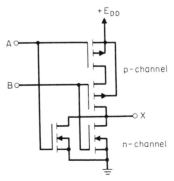

Fig. 15.10 A CMOS two-input NOR gate.

MOSFETs are connected in parallel. The gate of each n-channel transistor is connected to the gate of a p-channel transistor. The substrates of the p-channel devices are connected to E_{DD} and the substrates of the n-channel devices are returned to ground.

Assume that A and B are at logic 0. In this case both p-channel devices are ON, and the n-channel devices are OFF. Output $X = 1$ (the NOT of a zero is one). If A or B, or both, are at logic 1, one or both p-channel transistors are OFF. Also, one or both n-channel transistors are ON. Output $X = 0$ (the NOT of a one is zero).

> **example 15.1**   Show how the two-input NOR gate of Fig. 15.10 may be expanded into a three-input NOR gate.
>
> **solution**   This is accomplished by adding a p-channel MOSFET in series and an n-channel MOSFET in parallel (Fig. 15.11). Their gates are connected together for the third input C. In such fashion, one may develop a four-, or greater, input NOR gate.

NAND GATE   A two-input CMOS NAND gate is illustrated in Fig. 15.12. In contrast with the NOR gate of Fig. 15.10, the p-channel devices are in parallel and the n-channel devices are in series in the NAND gate. For each additional input, a p-channel MOSFET is connected in parallel and an n-channel MOSFET in series. Their gates are then joined together for the new input.

Fig. 15.11   A CMOS three-input NOR gate. (See Example 15.1.)

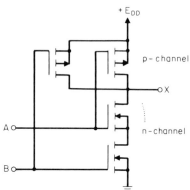

Fig. 15.12   A CMOS two-input NAND gate.

Assume that both A and B are at logic 1. The p-channel devices are OFF, and the n-channel devices are ON. Output $X = 0$ (the NOT of a one is zero). If either or both A and B are at logic 0, one or both p-channel transistors are ON and one or both n-channel transistors are OFF. Output $X = 1$ (the NOT of a zero is one).

15.6   FLIPFLOPS

The bistable MV, described in Chap. 14, is the basic circuit in a number of flipflops used in shift registers and counters. In this section, their behavior is illustrated by truth tables. Because the gates of logic families are generally either NAND or NOR (see Table 15.1), the implementation of flipflops is based on these configurations.

R-S FLIPFLOP (LATCH)   An R-S (reset-set) flipflop, also referred to as a *latch*, implemented with NAND gates is illustrated in Fig. 15.13a. The output of each gate is connected to the input of the other gate. In the truth table of Fig. 15.13b, subscript $n + 1$ indicates the state of the flipflop after being triggered by a suitable pulse at either the R or S input terminals. Subscript n denotes the state of the flipflop before being triggered. The logic symbol for an R-S flipflop is provided in Fig. 15.14.

S	R	Q_{n+1}
0	0	Not permitted
0	1	1
1	0	0
1	1	Q_n

(a)   (b)

Fig. 15.13 An R-S flipflop (latch) using NAND gates: (a) Circuit. (b) Truth table.

Fig.15.14 Logic symbol for an R-S flipflop.

For no input trigger ($S = R = 0$), both outputs tend to go to a logic 1 (the NOT of a zero is one). Because this is an unstable state, this condition is not permitted.

Assume that $Q_n = 0$ and, therefore, $\overline{Q}_n = 1$. Because output $\overline{Q}_n$ is an input to NAND gate 1, one of its inputs, therefore, is a logic 1. If $S = 0$, the output of gate 1 is a one (the NOT of a zero is a one); hence, $Q_{n+1} = 1$. Similarly, if $Q_n = 1$ and $R = 0$, Q_{n+1} becomes a 0. For $S = R = 1$, the state of the flipflop is unchanged.

example 15.2 Show how an R-S flipflop may be implemented using NOR gates. Draw a truth table for the flipflop.

solution The circuit is shown in Fig. 15.15a and the truth table in Fig. 15.15b. For $S = R = 0$, there is no change in state ($Q_{n+1} = Q_n$). Assume that $Q_n = 1$; therefore $\overline{Q}_n = 0$. Output $\overline{Q}_n$ is one of the inputs to NOR gate 1. If $R = 1$, the NOT of $1 + 0 = 0$; hence $Q_{n+1} = 0$ and $\overline{Q}_{n+1} = 1$.

Now assume that $Q_n = 0$ ($\overline{Q}_n = 1$). Output Q_n is one of the inputs to NOR gate 2. If $S = 1$, the NOT of $0 + 1 = 0$; hence $Q_{n+1} = 1$ and $\overline{Q}_{n+1} = 0$. The condition $S = R = 1$ is not permitted.

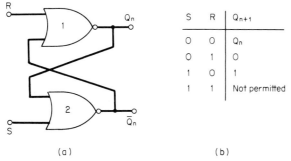

S	R	Q_{n+1}
0	0	Q_n
0	1	0
1	0	1
1	1	Not permitted

(a)   (b)

Fig. 15.15 An R-S flipflop using NOR gates: (a) Circuit. (b) Truth table. (See Example 15.2.)

CLOCKED R-S FLIPFLOP In many applications, such as shift registers, it is necessary that the operation of flipflops be synchronized by a master clock in the system. Such a flipflop is referred to as a clocked or *synchronous* flipflop. An example of a clocked R-S flipflop employing NAND gates and its truth table are illustrated in Fig. 15.16. The logic symbol for a clocked R-S flipflop is shown in Fig. 15.17.

NAND gates 1 and 2 are referred to as *steering gates*. A clock pulse C is one of the inputs to each steering gate. NAND gates 3 and 4 constitute the R-S flipflop of Fig.

S	R	Q_{n+1}
0	0	Q_n
0	1	0
1	0	1
1	1	Not permitted

(a)   (b)

Fig. 15.16   A clocked R-S flipflop using NAND gates: (a) Circuit. (b) Truth table.

Fig.   15.17 Logic symbol for a clocked R-S flipflop.

15.13a. If $S = R = 0$, the flipflop remains in its original state. The condition $S = R = 1$ is not permitted.

Assume that $Q_n = 0$ ($\overline{Q}_n = 1$). A trigger at terminal S and a clock pulse to gate 1 result in a zero at its output. At the input of gate 3, therefore, there is a 0 from gate 1, and a 1 owing to the output of gate 4. The output of gate 3 is now a logic 1 ($Q_{n+1} = 1$). The flipflop has changed its state.

To return the flipflop to its original state ($Q_n = 0$), a trigger is applied to terminal R and a clock pulse to the other terminal of gate 2. The inputs to gate 4 are now a 0 from gate 2 and a 1 owing to the output of gate 3. Hence, $Q_{n+1} = 0$ and $\overline{Q}_{n+1} = 1$.

example 15.3 Show how a clocked R-S flipflop may be implemented using NOR and AND gates. Draw a truth table for the flipflop.

solution The circuit is given in Fig. 15.18; the truth table is identical with that of Fig. 15.16b.

Fig. 15.18 A clocked R-S flipflop employing NOR and AND gates. (See Example 15.3.)

J-K FLIPFLOP The circuit of a clocked J-K flipflop and its truth table are provided in Fig. 15.19. The input terminals, in addition to the C terminal, are denoted by J and K. The basic circuit is the clocked R-S flipflop considered in the preceding section. The outputs of AND gates 1 and 2 are connected to the S and R terminals, respectively.

The first three entries in the truth table of Fig. 15.19b are identical with the clocked R-S flipflop. Although $S = R = 1$ is not permitted in the R-S flipflop, $J = K = 1$ is allowed in the J-K flipflop. An examination of the truth table reveals that for this input condition the output is complemented ($\overline{Q}_n$). That is, if $Q_n = 1$, $Q_{n+1} = 0$ and if $Q_n = 0$, $Q_{n+1} = 1$. The logic symbol for the J-K flipflop is given in Fig. 15.20.

example 15.4 Show how a J-K flipflop may be implemented using NAND gates.

solution This is accomplished by replacing the two-input NAND gates (1 and 2) of Fig. 15.16a by three-input NAND gates, as illustrated in Fig. 15.21.

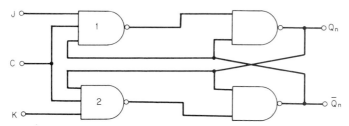

Fig. **15.19** The *J-K* flipflop: (*a*) Circuit. (*b*) Truth table.

Fig. **15.20** Logic symbol for a *J-K* flipflop.

Fig. **15.21** A *J-K* flipflop implemented with NAND gates. (See Example 15.4.)

T FLIPFLOP This flipflop changes its state with each application of a trigger to a single input terminal *T*. Also referred to as a *toggle*, the *T* flipflop is used often in counters. An example of a clocked *T* flipflop implemented by using a *J-K* flipflop, and its truth table, are provided in Fig. 15.22. It is seen that the *T* input terminal is formed by joining together the *J* and *K* terminals. The logic symbol for the *T* flipflop is shown in Fig. 15.23.

example 15.5 Show how a *T* flipflop may be realized using AND gates and an *R-S* flipflop.

solution The circuit is given in Fig. 15.24.

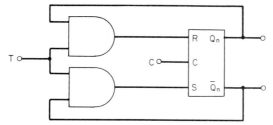

Fig. **15.22** A *T* (toggle) flipflop implemented with a *J-K* flipflop: (*a*) Circuit. (*b*) Truth table.

Fig. **15.23** Logic symbol for a *T* flipflop.

Fig. **15.24** A *T* flipflop realized by using AND gates and an *R-S* flipflop. (See Example 15.5.)

Fig. 15.25 A delay (D) flipflop implemented with a J-K flipflop and an INVERTER: (a) Circuit. (b) Truth table.

Fig.   15.26 Logic symbol for a D flip-flop.

D FLIPFLOP An example of a clocked D, or *delay*, flipflop using a J-K flipflop and an INVERTER is given in Fig. 15.25a. Its truth table is provided in Fig. 15.25b. An R-S flipflop may be used instead of the J-K type. In this case the input to the INVERTER is connected to the S terminal and the output to the R terminal. The D terminal is also connected to S. The logic symbol for the D flipflop is shown in Fig. 15.26.

The D flipflop is used as a one-bit delay. When $D = 1$, $J = 1$ and, because of the INVERTER, $K = 0$. Referring to the truth table of Fig. 15.19b for the J-K flipflop, $Q_{n+1} = 1$. Also, if $D = 0$, $J = 0$, and $K = 1$; hence, from the truth table, $Q_{n+1} = 0$. Output Q_{n+1} therefore equals D delayed by one clock pulse, or bit.

CLEAR AND PRESET It is often necessary to *clear* a flipflop to its zero $(Q = 0)$ or *preset* it to its one $(Q = 1)$ state. An example of how this is realized is illustrated in Fig. 15.27. All four NAND gates have three inputs each. One input to gate 3 is the *preset*, Pr, input; and Cl to gate 4 is the *clear* input. The logic symbol for a clear-preset J-K flipflop is illustrated in Fig. 15.28.

In either clearing or presetting a flipflop, the clock pulse is at logic zero $(C = 0)$. Assume that the flipflop is to be preset to one $(Q = 1)$. Inputs $\text{Pr} = 0$, $\text{Cl} = 1$ and, as mentioned previously, $C = 0$. Because $\text{Pr} = 0$, even if the other two inputs to gate 3 are

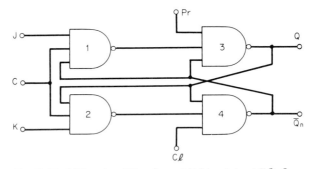

Fig. 15.27 Adding clear (Cl) and preset (Pr) inputs to a J-K flipflop.

Fig.   15.28 Logic symbol for a clear-preset J-K flipflop.

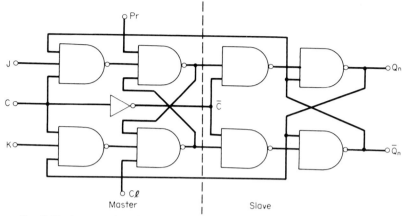

Fig. 15.29 A master-slave flipflop. The master is a *J-K*, and the slave an *R-S* flipflop.

at logic 1, the output $Q = 1$. Why? Because the NOT of 0 *and* 1 *and* 1 equals the NOT of zero which is a logic 1.

If the flipflop is cleared, $Q = 0$. For this condition, Cl = 0, Pr = 1, and $C = 0$. Because Cl = 0, the output of gate 4 is a logic 1; hence $\bar{Q} = 1$ and $Q = 0$. Once a flipflop has been cleared or preset, both the Cl and Pr inputs are maintained at a logic 1 before the next input bit arrives at the *J* or *K* terminal.

MASTER-SLAVE FLIPFLOP If the length of a trigger pulse is greater than the propagation delay of a gate, it may not be possible to change its state reliably. This problem is referred to as a *race condition*. Because propagation delays are generally very small (see Table 15.1), there is a reasonable chance of a race condition existing in digital integrated circuits. One technique for avoiding this is to use a *master-slave* flipflop.

An example of a master-slave flipflop is shown in Fig. 15.29. The first flipflop is called the *master* and the second the *slave*. The master is a *J-K* and the slave an *R-S* flipflop. An INVERTER is connected between the clock terminals of the *J-K* and *R-S* flipflops. In operation, Pr = Cl = 1.

When a clock pulse is present, $C = 1$ and, because of the INVERTER, $\bar{C} = 0$. The master is operative (enabled), and the slave is inoperative (inhibited). Consequently, during the clock pulse, the *J-K* flipflop can change state, but not the *R-S* flipflop.

In the interval when the clock pulse returns to zero, $C = 0$ and $\bar{C} = 1$. The *R-S* flipflop is now enabled, and the *J-K* is inhibited. During this interval, when the clock pulse is zero, the content of the *J-K* flipflop is transferred to the slave. Output *Q* corresponds to the output of the *J-K* flipflop.

15.7 COUNTERS

Flipflops connected to count pulses are referred to as counters. Counters have many applications in digital systems. They are used, for example, to count clock pulses (CP) in controlling the sequence of operations in a digital computer.

A basic counter, referred to as a *ripple counter*, is illustrated in Fig. 15.30. In this

Fig. 15.30 An example of a three-stage ripple counter employing *T* flipflops.

Fig. 15.31   Timing diagram for the ripple counter of Fig. 15.30.

example, a three-stage ripple counter implemented with T flipflops is shown. The J and K terminals of each flipflop are connected and returned to a logic 1. This connection converts the J-K to a T flipflop (see Fig. 15.22a). Because the J-K terminals are returned to a logic 1, output Q of each flipflop is at logic zero. The outputs of FF_0, FF_1, and FF_2 are designated as Q_0, Q_1, and Q_2, respectively. Output Q_0 is the LSB and Q_2 the MSB.

Assume that each flipflop changes its state on the falling (trailing), or *negative-going edge*, of a pulse. A timing diagram for the ripple counter of Fig. 15.30 is given in Fig. 15.31. On the falling edge of CP_1, Q_0 goes to a logic 1. Because each flipflop changes its state on the falling edge, FF_1 and FF_2 are unaffected; that is, $Q_1 = Q_2 = 0$. Thus, after one clock pulse, we have a binary count of 1:

$$\frac{Q_2 \; Q_1 \; Q_0}{0 \quad 0 \quad 1}$$

At the end of CP_2, Q_0 in going to zero triggers Q_1; Q_2 remains unchanged. Hence, we now have a binary count of 2:

$$\frac{Q_2 \; Q_1 \; Q_0}{0 \quad 1 \quad 0}$$

Proceeding in this manner, at the end of CP_7, we have a binary count of 7:

$$\frac{Q_2 \; Q_1 \; Q_0}{1 \quad 1 \quad 1}$$

At the end of CP_8, the counter resets itself: that is, $Q_0 = Q_1 = Q_2 = 0$. For a four-stage counter, the counter resets at the end of the 16th pulse, and so on.

It is interesting to note from the timing diagram that the output of FF_0 changes its state once for every two clock pulses; the output of FF_1 changes its state once for every four clock pulses, and so on. A ripple counter, therefore, may be used as a *divide-by* 2, 4, 8, etc., circuit.

OTHER TYPES OF COUNTERS   The ripple counter of Fig. 15.30 is an example of an *asynchronous counter*. In an asynchronous counter, the individual flipflops are not clocked simultaneously. In a *synchronous counter*, all flipflops are clocked and change their states simultaneously. Definitions of other counters include:

Up counter   An up, or *forward*, counter adds each input pulse to the count. The ripple counter of Fig. 15.30 is an example of an up counter.

Down counter   A down, or *reverse*, counter subtracts 1 from a preset number during each input pulse.

Up-down counter   An up-down, or *bidirectional*, counter can operate as either an up or a down counter.

Modulo counter   A counter that counts other than binary multiples is referred to as a

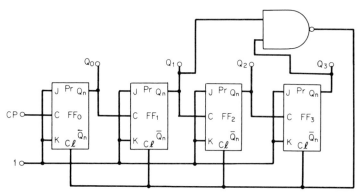

Fig. 15.32 An example of a decade counter using J-K flipflops. (See Example 15.6.)

modulo counter. For example, a counter that counts 1, 2, 3, 4 and resets itself to 0 on the 5th pulse is a modulo-5 counter. A *decade counter* that counts 1, . . . , 9 and resets itself to 0 on the 10th pulse is a modulo-10 counter.

example 15.6 Using J-K flipflops, design a decade counter.

solution A decade (modulo-10) counter employing J-K flipflops is illustrated in Fig. 15.32. Four flipflops are required. If three flipflops are used, a maximum count of only 7 (111) is possible, which is inadequate for a decade counter.

On the 10th input pulse, the decade counter must reset to zero. Decimal 10 equals binary 1010. Therefore, on the 10th pulse, $Q_3 = 1$ (MSB), $Q_2 = 0$, $Q_1 = 1$, and $Q_0 = 0$ (LSB). Outputs Q_3 and Q_1 are fed to the two-input NAND gate. When, on the count of 10, $Q_1 = Q_3 = 1$, the output of the NAND gate equals zero (the NOT of 1 *and* 1 = 0). The zero is returned to the clear (Cl) inputs of each flipflop, thereby resetting the counter to zero.

15.8 SHIFT REGISTERS

A shift register may be regarded as a component for storing a binary number or an instruction. For example, if a binary number or instruction has n bits, it is referred to as an n-bit word. An n-bit shift register consists of n-cascaded flipflops, indicated symbolically in Fig. 15.33. Flipflop FF_0 contains the least significant bit (LSB) and FF_{n-1} the most significant bit (MSB) of the word stored in the register.

MSB   LSB

Fig. 15.33 A symbolic representation of an n-bit shift register.

There are a number of varieties of shift registers, as illustrated symbolically in Fig. 15.34. For simplicity, four-bit registers are shown.

Serial register (Fig. 15.34a) In this register the word is stored and read out serially, that is, in sequence. Because the word is shifted to the *right*, it is referred to as a *right-shift* register. A register that shifts to the *left* is called a *left-shift* register.

Serial-in, parallel-out register (Fig. 15.34b) In this register, the word is stored serially and read out in parallel. The serial-in, parallel-out register is also referred to as a *serial-to-parallel converter*.

Parallel-in, serial-out register (Fig. 15.34c) In this register the word is stored in parallel (each bit being stored simultaneously) and read out serially. The parallel-in, serial-out register is also referred to as a *parallel-to-serial converter*.

Parallel-in, parallel-out register (Fig. 15.34d) In this register the word is stored and read out in parallel.

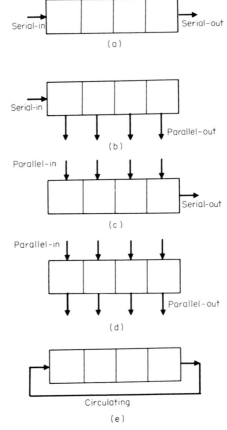

Fig. 15.34 Different types of shift registers. For simplicity, four-bit registers are illustrated: (*a*) Serial. (*b*) Serial-in, parallel-out (serial-to-parallel converter). (*c*) Parallel-in, serial-out (parallel-to-serial converter). (*d*) Parallel-in, parallel-out. (*e*) Circulating (dynamic shift register; shift-right read-only memory).

Circulating register (Fig. 15.34*e*)   In this register the word circulates continuously. The circulating register is also referred to as a *dynamic shift register* and a *shift-right read-only memory.*

An example of a four-bit serial shift register employing *J-K* flipflops is illustrated in Fig. 15.35. During a clock pulse, each flipflop assumes the state of its preceding flip-

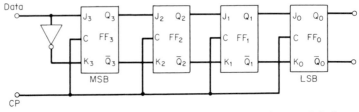

Fig. 15.35 An example of a four-bit serial shift register employing *J-K* flipflops.

flop. Assume that the word to be stored in the register is 1011. The sequence of operations is illustrated in Table 15.2.

Before CP_1, the state of each flipflop is at a logic zero ($Q_3 = Q_2 = Q_1 = Q_0 = 0$). At the termination of CP_1, the LSB, 1, is stored in FF_3; hence, $Q_3 = 1$ and $Q_2 = Q_1 = Q_0 = 0$. At the end of CP_2, the LSB has been shifted to FF_2 and the next bit, 1, is shifted to FF_3; hence, $Q_3 = Q_2 = 1$ and $Q_1 = Q_0 = 0$.

TABLE 15.2   Storing the Word 1011 in a Serial Shift Register

CP	Bit	Flipflop status			
		Q_3	Q_2	Q_1	Q_0
		0	0	0	0
1	1	1	0	0	0
2	1	1	1	0	0
3	0	0	0	1	0
4	1	1	0	1	1

Upon the termination of CP_3, the contents of FF_3 and FF_2 are shifted to FF_1 and FF_0, respectively; now, $Q_3 = 0$, $Q_2 = Q_1 = 1$, and $Q_0 = 0$. Finally, at the end of CP_4, the word 1011 is stored in the register with $Q_3 = 1$, $Q_2 = 0$, and $Q_1 = Q_0 = 1$.

To read out the word, four clock pulses are applied to the register. At the end of the fourth clock pulse, $Q_3 = Q_2 = Q_1 = Q_0 = 0$. Because the stored word is destroyed in reading out the shift register, the process is called *destructive readout* (DRO).

Fig. 15.36   A four-bit serial-in, parallel-out shift register. (See Example 15.7.)

example 15.7   Show how the serial shift register of Fig. 15.35 may be converted to a serial-in, parallel-out shift register (serial-to-parallel converter).

solution   The circuit of a serial-in, parallel-out shift register is given in Fig. 15.36. Four AND gates are connected to the circuit. One input to each AND gate is connected to the output of a flipflop. The second input is connected to an *enable* signal. When the enable signal equals a logic 1, the content of each flipflop in the register is read simultaneously. Because the word in the shift register is not destroyed in reading out, the process is called *nondestructive readout* (NDR).

15.9 SEMICONDUCTOR MEMORIES

One important building block found in digital computers and systems is the memory. A memory is required for the storage of instructions (computer program), data, and results obtained in the processing of data. Whereas in the past ferrite cores were widely used, semiconductor memories have replaced cores in most digital systems. In a semiconductor memory, flipflops are used to store a 0 or a 1 bit. The transistors in the flipflop may be bipolar (BJT) or field-effect (MOSFET).

Memories are organized to store W words, each word being B bits in length. The storage capacity of a memory, therefore, is $W \times B$ bits. A typical capacity of a semiconductor memory chip is 4096 bits (1024 words $\times$ 4 bits/word). A group of bits is often referred to as a *byte*. Generally a byte is taken to be equal to 8 bits. Most semiconductor memories exhibit nondestructive readout.

TYPES OF MEMORIES   Semiconductor memories may be categorized as:
  1. Read-only memory (ROM).
  2. Random-access memory (RAM).
In the ROM, the data are written in once and thereafter cannot be changed. This type of memory, for example, may be used to store a computer program. Information stored in the RAM can be readily changed in the processing of data.

A variation of the ROM is the *programmable* ROM (PROM). In the PROM, each memory element is in series with an aluminum or Nichrome strip that acts as a fuse link. The user can open the links by passing a specified current through the element. In this manner elements may be removed to represent a 0 or a 1 bit.

Memories are also categorized as being either *static* or *dynamic*. A static memory requires no clock pulses; a dynamic memory needs clock pulses. Generally, dynamic memories are less costly and consume less power than static memories.

ACCESS AND CYCLE TIMES   Two important terms used in characterizing memories are *access* and *cycle time*. Access time is the time needed to read out a word from the memory. Cycle time is equal to one over the rate at which a word may be selected for reading or writing. Typical access and cycle times are 400 and 600 ns, respectively.

ROM   An example of a ROM using p-channel depletion-type MOSFETs is illustrated in Fig. 15.37. For simplicity, only three-bit (B_0, B_1, B_2) and three-word (W_0, W_1,

Fig. 15.37   An example of a ROM employing p-channel depletion-type MOSFETs. For simplicity, only three bit and word lines are shown.

W_2) lines are shown. Typically, a semiconductor ROM has 1024 word lines, each word being four bits in length.

Transistors Q_A, Q_B, and Q_C have their drains connected to the gate terminals. These transistors serve as the load resistors for the memory transistors. For example, Q_A is the load transistor for Q_{01} and Q_{02}, Q_B is the load resistor for Q_{10}, and Q_C is the load resistor for Q_{22}. Each drain of memory transistors Q_{01}, Q_{12}, etc., is connected to a bit line; their gates are connected to a word line. The presence of a transistor, such as Q_{10}, indicates a stored bit of data. An absence of a transistor indicates no stored bit of data.

For example, if word W_2 is to be read, a negative voltage is applied to word line W_2. Because the transistors are p-channel devices, Q_{02} and Q_{22} are turned on. Bit lines B_0 and B_2, therefore, read approximately 0 volts. Where a transistor is absent, a negative voltage is read. If a conducting transistor represents a 1 and an absent transistor a 0, the binary number read is 101.

RAM An example of a RAM cell using bipolar junction transistors is illustrated in Fig. 15.38. (Random-access memories using MOSFETs are also available.) Transistors Q_1 and Q_2 have multiple emitters, similar to that used in T^2L. The circuit is a bistable MV in which the collector of one transistor is directly coupled to the base of the other transistor.

Fig. 15.38 A typical memory cell using bipolar junction transistors found in random-access memories. In a four-bit, 1024-word RAM, for example, 4096 of these cells are used in the memory.

Bit lines B_1 and B_2 are connected to an emitter of each transistor. The *word-select line* is connected to the remaining emitters. Assume that when Q_1 is conducting (Q_2 is therefore OFF), a binary 1 is stored in the cell. If Q_2 is ON (Q_1 is now OFF), a binary 0 is stored.

In its quiescent state, the word-select line is returned to ground or a low voltage. If a binary 1 is stored in the memory cell, transistor Q_1 is ON and Q_2 is OFF. Current, therefore, flows through emitter E_{12}.

To read the contents of the cell an appropriate positive voltage is applied to the word-select line. Emitter E_{12} becomes reverse-biased with respect to the base of Q_1. Emitter E_{11}, however, is forward-biased owing to a small potential present at bit line B_1. As a result, emitter current flows in bit line B_1, indicating that a binary 1 was stored in the cell. Because Q_2 is OFF, no current flows in bit line B_2 when it is sensed.

To write or store data in the cell, the word-select line is raised to a suitable positive potential. If a small voltage is applied to bit line B_2, transistor Q_2 is turned ON and Q_1 is turned OFF. A binary 0 is thereby stored in the cell. If a large positive voltage is applied to bit line B_2, the reverse is true: Q_1 is turned ON and Q_2 is turned OFF. This corresponds to a binary 1.

15.10 DIGITAL-TO-ANALOG AND ANALOG-TO-DIGITAL CONVERTERS

In data processing systems, there is the need to convert a digital signal to its analog equivalent or an analog signal to its digital equivalent. To cite an example, consider the digital voltmeter (DVM). To read a voltage, which is an analog quantity, the voltage is converted to a digital quantity by an analog-to-digital (A/D) converter. If the digital voltage is to be converted to an analog voltage, a digital-to-analog (D/A) converter is required. In this section, both types of converters are examined.

D/A CONVERTER   An example of a D/A converter that converts a three-bit binary quantity to an equivalent analog quantity is illustrated in Fig. 15.39. The basic circuit of the D/A converter is a summing amplifier using an op amp (see Chap. 13). Output voltage E_o is equal to

$$E_o = \frac{-R'}{R} E_1 - \frac{R'}{2R} E_2 - \frac{R'}{4R} E_3 \qquad (15.1)$$

For simplicity, mechanical single-pole, double-throw (SPDT) switches are shown in Fig. 15.39. When SW_1 is in position A, it reads the most significant bit (MSB). Switch SW_3 in position A reads the least significant bit (LSB). Any switch in position B is returned to ground.

Fig. 15.39   An example of a weighted-resistor D/A converter. For simplicity, mechanical switches are shown.

Assume that the binary quantity N to be converted is $N = 001$. Let the feedback resistor R' be equal to $R/2.5$ and $E_R = -10$ V. For $N = 001$, switches 1 and 2 are returned to ground and switch 3 to $E_R = -10$ V. By Eq. (15.1),

$$E_{o(1)} = \frac{-R/2.5}{4R} (-10) = 1 \text{ V}$$

Suppose that $N = 011$ (decimal 3). Then, switches 2 and 3 are connected to $E_R = -10$ V, and SW_1 is returned to ground. By Eq. (15.1),

$$E_{o(3)} = \frac{-R/2.5}{2R} (-10) - \frac{-R/2.5}{4R} (-10) = 3 \text{ V}$$

Consider $N = 111$ (decimal 7). All three switches are now connected to $E_R = -10$ V, and

$$E_{o(7)} = \frac{-R/2.5}{R} (-10) - \frac{R/2.5}{2R} (-10) - \frac{R/2.5}{4R} (-10) = 7 \text{ V}$$

The preceding results indicate that the output voltage always corresponded to the binary input, and, indeed, a digital quantity was converted to an analog equivalent. For a four-bit quantity, another resistor equal to $8R$ in series with a switch is connected to point P. In general, for an n-bit quantity, n resistors are required. The resistance value for the least significant bit is equal to $2^{n-1} R$.

The D/A converter of Fig. 15.39 is referred to as a *weighted-resistor converter*, because its output voltage depends on the values of the input summing resistors. These resistances have different values (R, $2R$, and so on) and the resistor for the LSB can become excessively large ($2^{n-1} R$).

Fig. 15.40 A ladder-type D/A converter for a three-bit word. The values of resistors in this kind of converter are either R or $2R$.

Owing to these facts, it is difficult for all the resistors to track with temperature. If good tracking is not realized, errors result when the converter is operated at different temperatures. To overcome these difficulties, the *ladder-type D/A* converter of Fig. 15.40 has been developed. Although it uses twice as many resistors as the weighted-resistor type D/A, the values of the resistors are only either R or $2R$.

Assume that $N = 100$ (decimal 4). Switch SW_1 is connected to reference voltage E_R, and switches 2 and 3 are returned to ground. The resistance at node A with respect to ground is $2R\|2R = R$. Addition of this resistance to $R_A = R$ yields $2R$. The resistance at node B with respect to ground is therefore $2R\|2R = R$. Addition of this resistance to $R_B = R$ yields $2R$. Hence, looking to the left from node C, there is an equivalent resistance of $2R$ to ground, as shown in Fig. 15.41a. Because the input voltage to the op amp is approximately zero, point C also "sees" an equivalent resistance of $2R$ to its right.

Current I flowing in the $2R$ resistor connected to E_R is, therefore,

$$I = \frac{10}{2R + 2R\|2R} = \frac{10}{3R}$$

The voltage at node C is current I multiplied by $2R\|2R$:

$$\frac{10}{3R} \times 2R\|2R = \frac{10}{3R} \times R = \frac{10}{3} \text{ V}$$

Using these results and letting $R' = 2.4R$, the model of Fig. 15.41b is drawn. By Eq. (15.1),

$$E_o = \frac{10}{3} \times \frac{2.4R}{2R} = 4 \text{ V}$$

which is the decimal equivalent of binary 100.

Fig. 15.41 A ladder-type D/A converter processing the binary number $N = 100$: (a) Resistance to the left or right of node C is equal to $2R$. (b) Model for determining output voltage, E_o.

Fig. 15.42 Ladder-type D/A converter processing $N = 001$: (a) Circuit. (b) Model for determining E_o. (See Example 15.8.)

example 15.8 In the ladder-type D/A converter of Fig. 15.40, determine E_o when $N = 001$. Assume that $R' = 2.4R$.

solution For $N = 001$, switches 1 and 2 are returned to ground and SW_3 is connected to E_R. Referring to Fig. 15.42a, node A "sees" a resistance to its left of $2R$ and an equivalent resistance of $2R$ to its right. Because $2R\|2R = R$, the current in SW_3 is $10/(2R + R) = 10/3R$ amperes.

At node A, half of current I ($I/2$) flows to the left, and the other half flows to the right. Reaching point B, the current is divided again by 2, and $I/4$ flows toward node C. At C it is split again, and $I/8$ flows through $2R$ in series with SW_1.

The voltage at node C with respect to ground is equal to the product of $I/8$ and $2R$:

$$\frac{I}{8} \times 2R = \frac{1}{8} \times \frac{(-10)}{3R} \times 2R = \frac{-5}{6} \text{ V}$$

This is illustrated in the model of Fig. 15.42b. By Eq. (15.1)

$$E_o = \left(\frac{-2.4R}{2R}\right)\left(\frac{-5}{6}\right) = 1 \text{ V}$$

Electronic switches are used in place of mechanical switches. To gain some idea how this is accomplished, consider the transistor switch of Fig. 15.43. As in the mechanical switch, point A goes to an input resistor and point B to ground.

Fig. 15.43 An elementary pnp transistor switch used in place of mechanical switches in D/A converters.

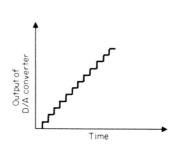

Fig. 15.44 An example of an *A/D* converter. The digital (binary) output is obtained from the binary counter.

Fig. 15.45 A ramp (staircase) wave-form at the output of the *D/A* converter of Fig. 15.44.

When a bit is present at the Bit terminal, assume that a positive voltage is impressed across the base of the transistor. Because the transistor is pnp, it is cut off, and the input resistor is connected to reference voltage $-E_{CC}$ through resistance R_C. If a bit is not present, a negative is impressed across the base. The transistor is turned ON, and the input resistor is returned to ground.

Because of collector resistor R_C, some error is introduced in the converter. Other configurations, containing a pair of bipolar or field-effect transistors, are used for mini-mizing the error.

A/D CONVERTER   One example of a commonly used *A/D* converter is provided in Fig. 15.44. One input to the AND gate is derived from the clock, and the other input is obtained from the output of a difference amplifier. The difference amplifier compares the output of the *D/A* converter and the analog input signal. For this type of operation the difference amplifier is referred to as a *comparator*.

If the output of the *D/A* converter is less than the analog signal, the AND gate remains operative, and the binary counter continues counting clock pulses. When their differ-ence is zero, the AND gate becomes inoperative, and the binary counter stops counting. The output of the counter is a binary number equal to the analog input signal.

The output of the *D/A* converter continuously increases by one bit as the binary counter counts. The resultant waveform, referred to as a *ramp*, or *staircase*, waveform is illustrated in Fig. 15.45. For this reason, the circuit of Fig. 15.44 is referred to as a ramp *A/D* converter. Although other methods are used in analog-to-digital conversion, the ramp type is fairly common.

15.11 THE MICROPROCESSOR

A microprocessor may be regarded as a general-purpose digital computer that is avail-able on a few silicon chips. Currently, over a dozen IC manufacturers are producing more than two dozen different types of microprocessors. Although most microproces-sors use MOSFETs, a few employ the BJT. Applications of the microprocessor include traffic control systems, numerically controlled machine tools, and the automatic testing of engines.

For illustrative purposes, the microprocessor chips manufactured by National Semi-conductor Corporation are considered. National produces two basic microprocessor chips. These are the Register, Arithmetic, and Logic Unit (RALU); and the Control Read-Only Memory (CROM).

The RAUL performs the basic control and arithmetic functions in processing data. The CROM contains the instructions (program) that direct the execution of data by the RALU. In addition, National provides software, that is, programs for a given applica-tion of the microprocessor. By using a number of RALU and CROM chips, it is possible to realize a 16-bit word computer, referred to as a *microcomputer*.

Chapter 16

Power Supplies

16.1 INTRODUCTION

The function of an electronic power supply is to convert the available prime power source into a form required by a particular system. Typical prime power sources are rated at: 115 V, 60 Hz, single phase; 208 V, 60 Hz, three phase; 115 V, 400 Hz, single phase; and 12 or 28 V, dc.

The voltage requirements will vary depending on the type of system and also the portion of the system. Solid-state systems require considerably lower dc voltages than vacuum-tube systems. Some systems require regulated dc voltages, while others do not. In a television receiver, the r-f and video portions may require low to moderate values of dc voltages, while the picture tube may require high voltages in the order of 30 000 V. In computers, low dc voltages may be required for solid-state logic circuits, while intermediate values of dc voltages may be required for solid-state analog circuits.

Regardless of the type of electronic power supply involved, the basis of the supply is its rectifier system. We will begin our study of power supplies with a review of the basic types of rectifier systems.

16.2 RECTIFIER SYSTEMS

A block diagram of an electronic power supply is shown in Fig. 16.1. The ac power source is generally the power delivered by the power company, but in some applications other sources may be used. One example is in automotive electric systems that employ alternators. An alternator is basically an ac power source, and the ac power is converted to dc power by diodes which are mounted on the metal frame of the alternator. In this particular example the diodes are semiconductor types, but in other applications vacuum-tube diodes may be used.

Returning to the block diagram, the ac power is delivered to the primary of a power transformer. This transformer may be a step-up or a step-down type, depending on how much voltage is required by the load. Some electronic power-supply circuits operate without a transformer, and are called *transformerless power supplies*. Such supplies are less expensive to build, and this is considered to be an important advantage. However, the amount of voltage delivered to the load is a function of the voltage at the power source, and therefore, the use of such transformerless supplies is limited.

The output of the power-supply transformer is delivered to a section marked "diode rectifier circuit." The purpose of the rectifiers is to convert the ac voltage input to a pulsating dc voltage. Diodes function as rectifiers because they will allow current flow in only one direction—that is, from cathode to plate or anode. When an ac voltage is placed across the diode, it conducts during the half-cycles when the plate (or anode) is

Fig. 16.1   Block diagram of an electronic power supply.

positive with respect to the cathode, but it will not conduct on the alternate half-cycles when the plate is negative with respect to the cathode.

The output of the diode rectifier circuit is a pulsating voltage. It cannot be used for operating transistor and vacuum-tube circuits because these circuits normally require an unvarying dc voltage. Therefore, a filter circuit is used to smooth out the voltage ripples. Usually filter circuits are *passive,* and are comprised of resistors and capacitors, or inductors and capacitors. (A passive filter is one that simply operates as a smoothing circuit.) There is always some loss of power in such a circuit, but this is not considered to be a serious disadvantage since the input stages of the power supply can be designed to deliver a sufficient amount of power for the filter loss, as well as for the load.

In the block diagram there is a dotted section marked *regulator* between the filter circuit and the load. Not all power supplies have regulators. Their purpose is to maintain a constant value of voltage across the load regardless of changes in the load resistance or ac input voltage. The regulator is usually an electronic circuit that accomplishes its function by sensing any change in the voltage across the load, and then feeding this change into the regulator circuit for use as a control to offset the change.

The last stage in the block diagram of Fig. 16.1 is the *load.* The load is the component or circuit that receives the dc power. In a radio or television set the load will be the vacuum tubes or transistors operating in the electronic circuitry. In other applications the load may be a dc motor, a battery that is being charged, or any other circuit or component requiring a dc source of power.

From the discussion of the block diagram, it can be seen that the term "electronic power supply" is somewhat misleading. This circuit does not *supply* the power, but rather, converts it from one form (ac) to another form (dc). However, the name is widely used in electronics. Transformers and filter circuits are covered in other chapters in this book. Our concern here will be primarily with various rectifier configurations, how the rectifier circuits operate, and the advantages and disadvantages of each circuit.

HALF-WAVE RECTIFIERS   The simplest rectifier circuit is the one-diode configuration shown in Fig. 16.2. This is a half-wave rectifier, so named because the output voltage waveform for the circuit is only one-half of the ac input voltage waveform. It is an example of a transformerless power supply, and it is very popular in low-cost home-entertainment electronic systems such as table-model radios and record players.

The input voltage to the circuit is an ac waveform. On one-half cycle, which we have designated the *positive half-cycle,* the anode of the diode is made positive and electron current can flow through the diode. The path of electron current flow is shown by the arrows in Fig. 16.2*a.* It will be noted that the current is flowing upward through the load resistor R_L. This means that the voltage drop across R_L will be such that the negative polarity of the voltage is at the bottom and the positive polarity is at the top. The output waveform during the positive half-cycle is an exact reproduction of the positive half-cycle of input waveform.

During the next half-cycle, the input voltage is negative. (See Fig. 16.2*b.*) This causes a negative voltage to be applied to the anode of D_1. No electron current can flow through the diode during this half-cycle, and the output voltage will be zero, as shown by the waveform on the diagram. Figure 16.2*c* shows the waveform of the voltage across the load resistance.

The waveform of Fig. 16.2*c* is referred to as a *pulsating dc voltage.* It is a dc voltage because its polarity is always positive (or zero). It is pulsating because the amplitude changes from moment to moment. Such a waveform is not useful for operating tube or transistor circuits unless it is first filtered. The purpose of filtering is to take out the pulsations. Because the voltage rises from zero to its maximum value and back to zero, and then stays at zero for a certain period of time, makes this waveform difficult to filter.

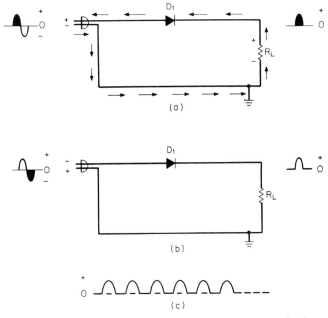

Fig. 16.2 The half-wave rectifier circuit with current flow and voltage waveforms: (*a*) Current flows during the positive half-cycle. (*b*) No current flows during the negative half-cycle. (*c*) Output voltage waveform of a half-wave rectifier circuit.

If this waveform is going to be converted to a steady dc voltage, then all the periods during which the voltage is zero must be filled in.

FILTERING The procedure for filtering the voltage usually involves charging a capacitor to the *peak* ac input voltage during a time when the diode is conducting and then having the capacitor discharge through the circuit in order to maintain the voltage across the load. To understand this, we will look at a simple filter circuit consisting only of a capacitor, as shown in Fig. 16.3. The only difference between this half-wave rectifier circuit and the one shown in the previous illustration is that a filter capacitor C has been added. This filter capacitor is an electrolytic type with a high capacitance value which enables it to store a considerable amount of energy during each half-cycle. The positive sign beside the capacitor symbol indicates that the capacitor is of the electrolytic type.

During the positive half-cycle of input voltage, the diode conducts as before. In this case, however, only a small part of the conducting current initially flows through the load, producing the output voltage waveform. The rest of the current flows into one plate of the capacitor and out of the other, thus charging the capacitor. This is shown in Fig. 16.3*a*. During the negative half-cycle, as shown in Fig. 16.3*b*, a negative input voltage is delivered to the plate of the diode, and it can no longer conduct. At this time the voltage across the capacitor causes current to flow through the load resistance. The capacitor discharge current is indicated by the arrow in Fig. 16.3*b*. This current prevents the voltage from dropping to zero, as it normally would without the presence of the capacitor.

The filter capacitor charges to the peak value of the ac input voltage. When the value of input voltage begins to decrease below the voltage across the capacitor, then the capacitor begins to discharge through the resistor. Figure 16.3*c* shows the filtered output waveform from the power supply in solid lines. The unfiltered output waveform (without the filter capacitor) is shown with dotted lines. Although the voltage drops

Fig. 16.3 The half-wave rectifier with a simple filter: (*a*) During the first positive half-cycle the capacitor charges. (*b*) During the next half-cycle the capacitor discharges. (*c*) The filtered output waveform.

somewhat during the negative half-cycles of input voltage, it does not drop all of the way to zero.

The operation of the simple capacitor filter shown in Fig. 16.3 is based on the fact that a capacitor can store energy. Basic theory tells us that a capacitor stores energy in the form of an *electrostatic field*, and an inductor (another device that can store energy) stores it in the form of an *electromagnetic field*. Passive filter circuits always rely on the ability of either a capacitor or an inductor (or both) to store energy and return the energy to the circuit at the proper time. Figure 16.4 shows a half-wave rectifier with a more elaborate filter. This circuit is popular with low-cost electronic circuits. The switch (SW) energizes or deenergizes the circuit. The pi filter, comprised of R_1, C_1, and C_2, smooths the dc pulsations from the half-wave rectifier. This circuit is used mainly for low-current applications, unless the capacitors are very large in value. In the latter case, the resistor can be kept to a reasonably low value, to minimize the voltage drop across it.

FULL-WAVE RECTIFIERS   One of the most important disadvantages of the half-wave rectifier circuit is the difficulty in filtering its output waveform. The relatively

Fig. 16.4   Half-wave rectifier with a filter.

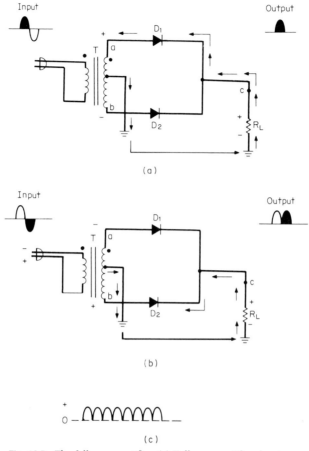

Fig. 16.5 The full-wave rectifier: (*a*) Full-wave rectifier showing current flow during one half-cycle. (*b*) Full-wave rectifier showing current flow during the next half-cycle. (*c*) Output waveform of the full-wave rectifier.

long period between positive voltage peaks makes it necessary for the filter circuit to supply a considerable amount of current flow through the load for a long period of time — that is, the discharge time is long compared to the charge time of the filter components. In spite of this disadvantage, the half-wave rectifier is a popular circuit because it is inexpensive and relatively trouble-free.

Figure 16.5 shows a full-wave rectifier circuit. This type of circuit *requires* a transformer T_1 with a center-tapped secondary winding. In addition to the need for the transformer, the circuit also requires the use of two diodes — shown as D_1 and D_2 in the circuit.

With an ac voltage across the primary of the transformer, the polarity of the voltage across the secondary will periodically reverse. The dot notation on the transformer is a standard way of indicating points of identical phase. When the voltage on the primary winding goes positive at the point where the dot is located, then at that instant the voltage at the secondary also goes positive at the point where the dot is located.

On one half-cycle, point *a* will be positive; and on the next half-cycle, point *a* will be negative. The voltage at point *b* will always have the opposite polarity of the voltage at point *a*. The center tap which is connected to the common, or *ground point*, will always be maintained at zero volts.

Figure 16.5*a* shows the operating condition during the half-cycle when point *a* is positive and point *b* is negative. The negative voltage on the plate of D_2 prevents it from conducting. The positive voltage on the plate of D_1, however, allows that tube to conduct. The conduction path is shown by arrows. If point *a* is positive, and if the center tap is zero, then it stands to reason that the center tap is negative with respect to point *a*. Starting at the negative terminal of the *voltage source* (which in this case is one-half the secondary winding) the electron current flows out of the center tap into the common circuit. It then flows out of the common circuit, through R_L, and to the junction of the two cathodes. The electron current cannot flow through D_2 because it is cut off with a negative voltage on its anode. Therefore, all the current flows through diode D_1. This completes the circuit.

During the first half-cycle, the current is seen to be flowing upward through the load resistor, making the voltage drop across it positive at point *c*. The output waveform shown on the illustration is seen to be a reproduction of the input waveform during this half-cycle. The amplitude of the output may be either smaller or larger than the input, depending on the turns ratio of the power transformer. If the turns ratio is 1:1, it means that the number of turns on the primary and on *one-half* of the secondary are equal. In such a case, and assuming there is no filter circuit, the output amplitude will be approximately equal to the amplitude to the input signal. It is necessary to say "approximately" because there is always some voltage drop across the diode itself. The actual value of this drop can be obtained by multiplying the current through the diode by the forward resistance of the diode. Although this voltage drop may be small enough to be neglected in most cases, it does, nevertheless, reduce the output amplitude to something less than the input voltage (measured across one-half of the transformer secondary winding). There is also a voltage drop across the diode in the half-wave rectifier circuit previously discussed.

In Fig. 16.5*b* the circuit is shown with the voltage polarity across the secondary reversed. In this case, the anode of D_1 is negative (with respect to ground) and the anode of D_2 is positive. Therefore, only D_2 can conduct during this half-cycle. Again, electron current leaves the center tap, flows into the common connection, and then through the load resistor R_L. The current flow is shown by arrows. It is important to note that the electron current is again flowing in the same direction as before, through R_L, making point *c* again positive with respect to ground. This means that a positive-going voltage will appear across the load for *both* half-cycles of input voltage.

Figure 16.5*c* shows the output waveform of a full-wave rectifier. Since there are no long periods during which zero voltage is generated across the load, this waveform is easier to filter than the half-wave rectifier output. That is one of the important advantages of the full-wave rectifier circuit.

TUBE RECTIFIERS   Some full-wave rectifier power supplies are designed with diode tubes having filament emitters. Filament emitters are more efficient, in general,

Fig. 16.6   A full-wave rectifier circuit using tubes with filament emitters.

than cathode-type emitters. Cathodes are used whenever it is necessary to isolate the cathode emitter from the heater circuit. Figure 16.6 shows a full-wave rectifier with the filament emitters. The filament voltage is obtained from a separate winding on the transformer. This is a common practice with full-wave rectifier circuits. In other rectifier circuits, there may be a separate filament transformer, or the filament may be in series with a number of other filaments which produce (when in series) a total voltage drop equal to the line voltage. When the latter method is used, only tubes with cathode emitters can be used, so that the ac filament circuit is isolated from the dc supply voltage. (The only purpose of the filament winding in the circuit of Fig. 16.6 is to provide current for heating the filament. It does not enter into the operation of a rectifier power supply in any other way.)

The circuit shows the current flow for one-half cycle, during which point a is positive and point b is negative. As before, only the plate with the positive voltage can attract electrons. Therefore, only that half of the tube conducts during this half-cycle. Electron current leaves the center tap, flows upward through the load resistor into the filament, and then from the filament to the plate of the upper diode. During the next half-cycle the upper diode will have a negative voltage on its plate, and the lower diode will have the positive voltage. Therefore, the current flow will be through the lower diode.

An important consideration here is the fact that the filament winding of the transformer will be above ground by the amount of voltage drop across R_L because of the common connection at point c. In some power supplies this may be 300 or 400 V (or more) above ground. In choosing a transformer for this type of circuit, it is important to determine that the transformer secondary can withstand the resulting voltage stress. If the filament winding is above ground by 400 V, and point b is negative by 400 V during one-half cycle, then the voltage difference between the filament winding and that half of the winding is 800 V. Although this is not an excessively high voltage, it does, nevertheless, put a voltage stress between windings within the transformer itself and the transformer core, which is grounded.

Although the output waveform of the full-wave rectifier is easier to filter, this circuit is not without disadvantages. The fact that the secondary winding of the transformer *must be center-tapped* means that only one-half of the total secondary voltage will appear across the load. As an example, in the circuit of Fig. 16.6 assume that the voltage between point a and point b is 1000 V. This would be the total secondary winding voltage. Under this condition the voltage across R_L would be only 500 V. (Again, we are neglecting the drop across the diode itself.)

Another disadvantage that must be taken into consideration is that two diodes are needed for the operation of the full-wave rectifier. This means an added expense and somewhat reduced circuit reliability.

BRIDGE RECTIFIERS   A full-wave rectifier circuit that makes use of the *full* transformer secondary-winding voltage, and does not require the center tap, is shown in Fig. 16.7. It is called a bridge rectifier because of its similarity to the configuration of the Wheatstone bridge. Although a transformer is shown in the circuit, it may also be connected as a transformerless power supply.

On the first half-cycle, shown in Fig. 16.7a, point a becomes positive with respect to point b. If we consider the secondary winding of the transformer as the voltage source, we can trace the electron current flow (starting at point b) for this half-cycle. The arrows show the current path flowing from b through D_1, up through R_L, and through D_2 back to point a on the secondary winding. Diodes D_3 and D_4 are cut off during this half-cycle, because D_3 has a positive voltage on its cathode, and D_4 has a negative voltage on its anode. In order for a diode to conduct, it is necessary for the anode to be positive with respect to the cathode. If the cathode of a diode is made highly positive with respect to the anode, then this is the same as saying that the anode is negative with respect to the cathode. That is why the positive voltage on the cathode of D_3 prevents it from conducting.

The next half-cycle of operation is shown in Fig. 16.7b. The polarity of voltage across the secondary winding has now been reversed, making point a negative with respect to point b. Starting at the negative terminal of the source (which is now point a on the secondary winding), current flows through D_3, through R_L, through D_4, and back to point b on the secondary winding. Diode D_1 cannot conduct because of the positive

Fig. 16.7 A bridge rectifier circuit: (a) Current flow in the bridge rectifier circuit for the first half-cycle of operation. (b) Current flow during the second half-cycle.

voltage on its cathode, and D_2 cannot conduct because of the anode during this half-cycle.

The diodes in the bridge rectifier circuit may be thought of as switches which are operated by voltage polarities. During both half-cycles of operation, the current is seen to flow in the same direction through R_L. This means that the circuit produces a full-wave pattern like the one shown in Fig. 16.7b.

When semiconductor diodes are used for bridge rectifiers, they are often mounted in the same package for convenience.

The bridge rectifier circuit is found in many applications in the communications field and elsewhere. It is also used in some instrument circuits. Most meter move-

Fig. 16.8 The half-wave voltage doubler circuit: (a) This half-cycle charges C_1 to 170 V. (b) During this half-cycle, the 170 V across C_1 is in series with the 170-V input, making a total of 340 V; capacitor C_2 charges to 340 V.

ments on voltmeters respond only to a dc voltage. In order to measure an ac voltage, it is necessary to convert the alternating current into a direct current. A half-wave rectifier circuit *could* be used, but the bridge rectifier is generally employed because it is a full-wave rectifier. The input voltage – that is, the voltage being measured – is applied directly across the bridge at points *a* and *b*, and the meter movement is connected into the circuit at the position of R_L in the circuit of Fig. 16.7.

HALF-WAVE VOLTAGE DOUBLERS The half-wave voltage doubler circuit of Fig. 16.8 will provide an output voltage across R_L that is approximately twice the peak input ac voltage to the circuit. Thus we have a transformerless circuit which steps up the line voltage and converts it to a dc potential.

During one-half cycle of input power, the voltage polarity will be as shown in Fig. 16.8a. The positive voltage on the anode of D_2 causes it to conduct, and the conduction path is shown by the arrows. This charges C_1 to the peak voltage (170 V in this example). At the same time, diode D_1 cannot conduct because its anode is not positive with respect to its cathode.

The next half-cycle of operation is illustrated in Fig. 16.8b. This shows that the voltage across capacitor C_1 and the input voltage are in series. In other words, they are connected plus to minus as are the two batteries in series shown at the inset. The input voltage to the circuit during this half-cycle is equal to the voltage across C_1 (170 V) plus the peak voltage across the line, making a total of 340 V. The negative side of the line is connected to the anode of D_2 which prevents that diode from conducting during this half-cycle. The positive voltage on capacitor C_1 is applied to the anode of D_1, and therefore D_1 *can* conduct during this half-cycle. The conduction path is shown by the arrows.

It will be noted that the current in Fig. 16.8b divides: part of it flowing into capacitor C_2, and the remainder flowing through the load. Capacitor C_2 will charge to the full 340 V during this half-cycle (minus, of course, the drop across the diode, which can usually be neglected in this type of circuit).

On the next half-cycle, the condition of Fig. 16.8a again prevails. The input voltage recharges capacitor C_1 by its conduction through D_2. At the same time D_1 is cut off during this half-cycle. It will be remembered that capacitor C_2 was charged to the 340-V input during the half-cycle shown in Fig. 16.8b. Now capacitor C_2 will discharge through the load, as shown by the dotted arrow. This prevents the voltage across the load from dropping to zero, as it would without the filter capacitor.

The output voltage waveform is a filtered half-wave voltage like the one shown in Fig. 16.3c. If the value of load resistance in this circuit is small, then capacitor C_2 will discharge rapidly through it, causing the voltage across R_L to go to zero during some portion of the half-cycle when D_1 is not conducting. Therefore, this type of power supply cannot be used for circuits that require a high current flow—that is, circuits that load the supply. Both C_1 and C_2 are normally quite large in capacitance value, and are usually electrolytic capacitors. However, even with a large capacitance value for C_2, the power-supply circuit cannot sustain a heavy load current.

If R_L were a variable resistor, it would be found that the smaller the value of resistance, the lower the output voltage for the circuit. What is more important is that when the load current through R_L varies from moment to moment, the output voltage of the power supply also varies. An ideal power supply would be one that maintains a constant output voltage, regardless of the amount of load current being drawn. Such a power supply is not realizable in practice, but can be approached closely under certain conditions. The *regulation* of the power supply is a measure of how well it maintains its output voltage under variable load conditions. The disadvantage of the half-wave voltage doubler shown in Fig. 16.8 is that it has poor regulation. Therefore a half-wave voltage doubler is not generally used where the load varies widely.

FULL-WAVE VOLTAGE DOUBLERS   The distinguishing feature of the half-wave voltage doubler just described is that it delivers power to the load only during one-half cycle of input alternating current. This, in part, accounts for the relatively poor regulation of the circuit. A full-wave doubler provides better regulation and is easier to filter.

A full-wave voltage doubler is shown in Fig. 16.9. On the first half-cycle, the input voltage will have a polarity as marked at the plug in Fig. 16.9a. A positive voltage, under this condition, is delivered to the cathode of D_1, and that diode is cut off. At the same time, the anode of D_2 receives a positive voltage, and that diode *will* conduct. The conduction path is shown by the arrows. Capacitor C_1 charges to the peak line voltage (170 V) during this period of time.

On the second half-cycle, the polarity of the voltage at the plug reverses—placing a negative voltage on the anode of D_2, and also a negative voltage on the cathode of D_1. (See Fig. 16.9b.) Under this condition, only D_1 will conduct. The conduction path is again shown by arrows. Capacitor C_2 charges to the peak line voltage of 170 V. The voltage across C_2 is added in series to the voltage across C_1 (which is obtained during the previous half-cycle). The two voltages, in series, combine to produce a 340-V output voltage. The load resistance is connected across the two series capacitors.

It might be argued that this is not really a full-wave rectifier system because the total voltage is not developed at the output capacitor circuit on *both* cycles of input. Instead, the voltage is delivered to one-half of the output circuit each time the input voltage reverses polarity. However, at least *some* voltage is delivered to the output capacitor circuit each half-cycle, and this makes it more of a full-wave rectifier than the one shown in Fig. 16.8.

Although the regulation of the full-wave voltage doubler circuit is somewhat improved over that of the half-wave doubler, the circuit is still unable to sustain a heavy load current without a serious drop in the output voltage.

VOLTAGE TRIPLERS AND OTHER VOLTAGE MULTIPLIERS   In the voltage doubler circuits just discussed, the principle of operation is based on charging a capacitor in one-half cycle, and then adding the voltage across that capacitor to the circuit voltage (or to the voltage across another capacitor) during the next half-cycle. It should be obvious that any number of capacitors can be charged, and their voltages added. However, as the amount of voltage multiplication goes up, the amount of regulation of the power-supply circuit becomes poorer and poorer.

Figure 16.10 shows another example of a voltage multiplier. (Voltage multipliers are diode rectifier circuits that increase the input voltage. The *half-wave doubler*, and

Fig. 16.9 The full-wave voltage doubler: (*a*) First half-cycle of operation. (*b*) Operation during the second half-cycle.

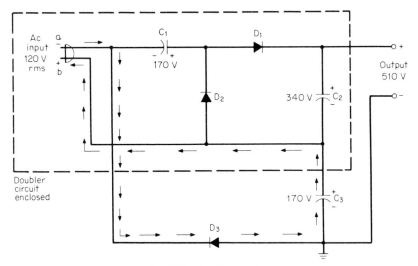

Fig. 16.10 A voltage tripler.

the *full-wave doubler* are examples. Other diode circuits that multiply the input volt-age are triplers, quadruplers, quintuplers, etc.) The circuit of Fig. 16.10 is a voltage tripler. The portion of the circuit enclosed in dotted lines is exactly the same as for the half-wave voltage doubler previously discussed, and illustrated in Fig. 16.8. Capacitor C_2 is charged to twice the peak input line voltage in this part of the circuit. Diode D_3 and capacitor C_3 have been added to the doubler circuit to convert it into a tripler.

On one-half cycle, point a on the plug will be negative with respect to point b, and this places a negative voltage on the cathode of D_3. The electron current flow through D_3 starts at point a, flows through the diode up into capacitor C_3, and then out of the capacitor and back to point b. This electron current path is indicated by solid arrows on Fig. 16.10. During the same half-cycle when C_3 is charged, D_2 is also conducting and charging capacitor C_1.

On the next half-cycle point, a is positive with respect to point b, and the diode D_3 is cut off. Diode D_1 conducts, and charges capacitor C_2 to twice the peak line volt-age, as discussed previously for the doubler circuit.

The overall result is that the voltage across C_2 is twice the peak line voltage, and the voltage across C_3 is equal to the peak line voltage. Assume, as an example, that the line voltage is 120 V rms. This will result in a voltage across C_2 of 340 V, and a voltage across C_3 of 170 V. The total voltage across the two capacitors (in series) is 510 V — which is three times the peak input line voltage. This voltage appears across the out-put terminals of the supply. The output voltage of the tripler is high, but the circuit cannot sustain a heavy load current.

If a negative output voltage is required, the diodes in any of the circuits can all be reversed. If this is done, the connections for the electrolytic capacitors must also be reversed because these capacitors are *polarized*—that is, they can only be connected into a circuit with the correct polarity.

LOAD VERSUS LOAD RESISTANCE   The terms load and load resistance have been used interchangeably in discussions up to now, but these terms actually mean different things. The load of a power supply always refers to the amount of *current* that the power supply must deliver. Therefore, when it is written that a power supply is under a *heavy load*, it means that the power supply must deliver a heavy current to the load. The load resistance, on the other hand, is the value of load resistance across the power-supply output terminals. The higher the load resistance value, the lower the amount of current that must be delivered by the power supply. Therefore, a high load resistance means a low power-supply load.

COMBINATION POSITIVE AND NEGATIVE SUPPLIES   In some circuits it is de-sirable to obtain both a positive and a negative output voltage from the same power supply. For example, an amplifier may use both pnp and npn transistors, and a vacuum-tube amplifier requires a positive plate voltage and a negative grid voltage. Trans-former-type power supplies can be readily adapted to provide both polarities from the same circuit. Such a circuit is shown in Fig. 16.11. Instead of using a common tie point at the center tap of the transformer secondary winding, this point is connected to one side of a series resistance branch comprised of R_1 and R_2. The junction of the two resistors is tied to the common point of the circuit.

Since this is a full-wave rectifier circuit, the current will flow through the resistors on both half-cycles of input signal. The arrows show the paths of current flow. The polarities of the resulting voltage drops across the resistors are marked on the diagram. Since the current is flowing from point b toward ground, point b must be a negative voltage with respect to ground. At the same time current is flowing from the ground point toward point a, making point a positive with respect to ground. Both voltage polarities are obtained from the same power-supply circuit. The output voltages are usually filtered by two different filter circuits — one for each polarity of voltage.

If the circuit to which the power-supply voltage is being delivered requires a highly positive voltage for operation of the plates of a tube and a relatively low negative volt-age for operating the grid, a separate transformer may be used to obtain the negative voltage. For transformerless power supplies, the positive voltage may be obtained with a half-wave rectifier or other transformerless supply, and the negative dc voltage

Fig. 16.11 This circuit will supply a positive and a negative output voltage.

may be obtained by a separate transformer supply. The advantage of using two separate supplies over that of the circuit shown in Fig. 16.11 is that the negative supply can be more efficiently designed. Normally the grid circuits of vacuum-tube amplifiers do not draw current, and therefore the load on such a supply is small. A small, inexpensive transformer may be used for the negative grid circuit supply. The identical power-supply principles are equally applicable to power supplies for solid-state circuits.

16.3 REGULATED POWER SUPPLIES

An ideal regulated dc power supply is one that maintains a constant dc output, regardless of changes in external load requirements or changes in input voltage. The degree to which the performance of an actual power supply approaches this ideal is defined by its specifications, which will be discussed in depth later in this chapter. Figure 16.12 shows a general-purpose regulated power supply used in laboratories.

Fig. 16.12 A general-purpose lab power supply. (*Courtesy Lambda Electronics Corp.*)

In general, dc power supplies may be divided into two functional categories: constant-current and constant-voltage types. Constant-current supplies maintain a constant current through a load resistor, regardless of the value of the load resistance, within a certain voltage range. Constant-voltage supplies maintain a fixed voltage across the load resistor as long as the current through the load resistor stays within a specified range. Equations (16.1 and 16.2) give the maximum and minimum values of load resistance for a constant current power supply.

$$R_{max} = \frac{E_{max}}{I} \qquad\qquad (16.1)$$

$$R_{min} = \frac{E_{min}}{I} \qquad\qquad (16.2)$$

where R_{max} = maximum value of load resistance, Ω
$\quad\quad R_{min}$ = minimum value of load resistance, Ω
$\quad\quad E_{max}$ = maximum value of output voltage, V
$\quad\quad E_{min}$ = minimum value of output voltage, V
$\quad\quad\quad I$ = output current, A

example 16.1   A constant-current power supply has an output current of one ampere and a voltage range (called the *compliance*) of 0 to 50 V. What are the maximum and minimum values of load resistance?

solution   Using Eq. (16.1), we may find the maximum load resistance:

$$R_{max} = \frac{E_{max}}{I} = \frac{50\ V}{1\ A} = 50\ \Omega \qquad\qquad (16.1)$$

Using Eq. (16.2), we may find the minimum load resistance:

$$R_{min} = \frac{E_{min}}{I} = \frac{0\ V}{1\ A} = 0\ \Omega \qquad\qquad (16.2)$$

In this case, the minimum load resistance is zero ohms, which is a short circuit. Equations (16.3) and (16.4) give the maximum and minimum values of load resistance for a constant-voltage power supply.

$$R_{max} = \frac{E}{I_{min}} \qquad\qquad (16.3)$$

$$R_{min} = \frac{E}{I_{max}} \qquad\qquad (16.4)$$

where R_{max}, R_{min} = same values as stated above
$\quad\quad\quad\quad E$ = output voltage, V
$\quad\quad\quad I_{min}$ = minimum value of output current, A
$\quad\quad\quad I_{max}$ = maximum value of output current, A

In most supplies, $I_{min} = 0$; the maximum load resistance for a constant-voltage supply is then infinity, or an open circuit.

example 16.2   A constant-voltage power supply has an output voltage of 50 V and an output current range from 50 mA to one ampere. What are the maximum and minimum values of load resistance?

solution   Using Eq. (16.3) to find the maximum load resistance,

$$R_{max} = \frac{E}{I_{min}} = \frac{50\ V}{50\ mA} = 1000\ \Omega$$

Using Eq. (16.4) to find the minimum load resistance,

$$R_{min} = \frac{E}{I_{max}} = \frac{50\ V}{1\ A} = 50\ \Omega$$

If this supply had been capable of going to a no-load condition, zero amperes output current, the maximum load resistance would be infinite or an open circuit.

Many power supplies are designed to operate in either a constant-voltage mode or in a constant-current mode. They are usually equipped with *automatic crossover*, a feature that allows the supply to go directly from one mode of operation to the other as the load resistance changes.

Many supplies are designed only for constant-voltage operation over a narrow range. These are frequently referred to as *slot-range* supplies. They generally are equipped with a *foldback* current-limiter circuit, which functions to decrease the output voltage and current simultaneously when the load resistance is too low; i.e., the output current rating of the supply would be exceeded if the output voltage remained at its normal value.

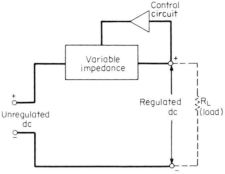

Fig. 16.13 Block diagram of a series-regulator power supply.

BASIC TYPES Regulated dc power supplies may also be classified by the type of input supply they utilize: ac or dc. The ac type usually employs a transformer to isolate the input from the output and to either raise or lower the input voltage to the level required. The ac output of the transformer is then rectified and filtered. The filter output is an unregulated dc voltage. From this point on, the regulator which follows may be considered a "dc-to-dc regulator." Thus, once voltage changing, rectification, and filtering are accomplished, the ac-to-dc and the dc-to-dc regulators may be treated alike.

There are three basic methods of regulating an unregulated dc source. Many power systems utilize a combination of two or in some cases all three of these.

Series type The most common system employed is the series regulator, which is shown in block form in Fig. 16.13. The series regulator is essentially a variable impedance placed between the unregulated input and the regulated output. By varying this impedance, the output voltage (or current) is maintained constant. The variable impedance is an active device, usually a power transistor in modern equipment, but vacuum tubes are utilized in older supplies.

Shunt type Another frequently encountered regulation system is the shunt regulator, shown in block form in Fig. 16.14. Here a fixed-source impedance is placed between the unregulated source and the regulated output. A variable impedance, called a "shunt" or "parallel" impedance, is placed across the output. The shunt impedance changes to maintain the output voltage (or current) constant. As in the case of the series regulator, the variable impedance is an active device, either a transistor or a vacuum tube.

Switching type In either of these two above approaches, all energy not delivered to the load must be dissipated in the regulating elements. More efficient systems involve switching circuits which utilize some form of *duty-cycle* control. The details

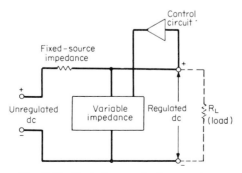

Fig. 16.14 Block diagram of a shunt regulator.

of this *nondissipative* type of regulator will be discussed in more detail later in this chapter. Some systems use only a nondissipative circuit, while others employ this circuitry as a coarse preregulator followed by a fine regulator of either the shunt or series type. This produces the finer regulation of the dissipative control, while having the efficiency advantage of a "nondissipative" circuit.

16.4 POWER-SUPPLY REGULATOR CIRCUITS

Figure 16.15 shows a simple power-supply regulator consisting of a series resistor R_s and a zener diode D_1. A gaseous regulator tube would serve the same purpose as the zener diode in this circuit, but it is not available with low-voltage ratings. It usually takes more than 70 V to "fire" a gaseous regulator tube, while the zener has the advantage that it is available in a wide selection of voltage values. Another advantage of the zener diode over the gaseous regulator is that it is available in a much wider range of current-handling abilities.

Fig. 16.15   A simple power-supply regulator.

The value of resistance for the series resistor in the simple shunt regulator of Fig. 16.15 can be calculated from the equation

$$R_s = \frac{E_i - E_o}{I_m + I_z} \qquad (16.5)$$

where E_i = unregulated input voltage, V
  E_o = regulated output voltage, V
  I_m = maximum load current, A
  I_z = current through the zener diode, A
As a general rule, the zener diode current I_z should be made equal to about 10 percent of the value of the maximum load current. Mathematically,

$$I_z = 0.1 I_m \qquad (16.6)$$

where I_m and I_z have the same meaning as for Eq. (16.5).
  The power rating of R_s can be determined from the equation

$$P_s = 2(I_m + I_z)^2 R_s \qquad (16.7)$$

Once the zener current rating is obtained from Eq. (16.7), then its power rating can be determined from the equation

$$P_z = E_o I_z \qquad (16.8)$$

Manufacturer's specifications can be consulted in order to choose the zener that meets the particular requirements of a power supply. A sample problem will show how the regulated zener supply can be designed.

  example 16.3 Design a 14-V regulated supply that will provide a maximum load current of 100 mA from an unregulated power supply that delivers 25 V.

  solution A 14-V zener diode will be needed.
The zener current rating:

$$I_z = 0.1 I_m$$
$$= 0.1 \times 100 \text{ mA} = 10 \text{ mA} \qquad (16.6)$$

Resistance value of R_s from Eq. (16.5):

$$R_s = \frac{E_i - E_o}{I_m + I_z} = \frac{25 - 14}{0.1 + 0.01} = 100 \ \Omega$$

Power rating of R_s from Eq. (16.9):

$$P_s = 2(I_m + I_z)^2 R_s = 2(0.1 + 0.01)^2 \times 100 = 2.42 \text{ W}$$
$$P_z = E_o I_z = 14 \times 0.01 = 0.14 \text{ W}$$

The series regulator will have a resistance of 100 Ω and a power rating of over 2.4 W. The 14-V zener will have a current rating of 10 mA and a power rating of over 0.28 W.

Figure 16.16 shows a closed-loop power-supply regulator circuit. At the unregulated input side of the regulator, there is a series circuit comprised of R_1, Q_1, and zener diode D_1. At the regulated output terminals there is a series resistor circuit comprised of R_2, R_3, and R_4.

Resistor R_2 is a variable resistor that sets the base voltage of Q_1. The series resistor string is sometimes called the *sense circuit* because it delivers any change in the output voltage to the base of transistor Q_1.

In order to understand the operation of the power-supply regulator shown in Fig. 16.16, it will be assumed that the regulated output voltage begins to rise for some reason (such as a decrease in load current). The emitter voltage of Q_1 is a fixed value set by the zener diode. Therefore, when the output voltage rises, it causes the base of Q_1 to become more positive, causing Q_1 to conduct harder through resistor R_1. The base voltage for control amplifier transistor Q_2 is obtained from the drop across R_1. Increasing the current through R_1 causes the base of Q_2 to become less positive. This decreases the current through Q_2, which controls the base current of Q_3. Decreasing the base current of Q_3, the power regulator transistor, has the same effect as increasing the resistance between the emitter and collector of the transistor.

To summarize, a rise in the voltage at the output terminals is accompanied by an increase in resistance at Q_2, causing an increased drop across Q_3 and bringing the voltage at the output terminals back to its regulated value.

A decrease in the output voltage causes the base at Q_1 to become less positive. Therefore, it conducts less through resistor R_1. That causes the base of Q_2 to go in a positive direction, increasing the current through Q_2 and also through Q_3. This is equivalent to decreasing the resistance of Q_3 and decreasing the voltage drop across it.

To summarize, a decrease in voltage at the output terminals is accompanied by a decrease in drop across Q_3, and the output voltage remains constant. In this closed-loop feedback regulator circuit, Q_3 serves as a variable resistor which is automatically adjusted to compensate for changes in load voltage. The voltage across Q_3 increases or decreases, as required, in order to maintain the voltage across the output terminals at a constant.

The closed-loop regulator of Fig. 16.16 will also regulate against changes in input voltage. Suppose, for example, that the input voltage – that is, the unregulated voltage

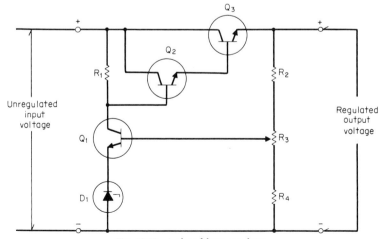

Fig. 16.16   A closed-loop regulator.

— increases. This makes the collector voltage of Q_1 greater and causes it to conduct harder. The base of transistor Q_2 becomes less positive, decreasing the current through that transistor and hence decreasing the current through Q_3. This is the same as increasing the emitter-to-collector resistance of Q_3 and increasing the voltage drop across it. Thus, an increase in the input voltage on the unregulated side is accompanied by an increase in the voltage drop across Q_3, and the output voltage at the regulated side does not change.

A similar chain of events occurs when the output voltage decreases. The result is that when the output voltage decreases, the feedback loop sets the base of Q_3 so that it conducts harder. This is the same as lowering the resistance of Q_3 and decreasing the voltage drop across it. Thus, when the input voltage decreases, the decrease also occurs across Q_3, and the voltage remains unchanged.

16.5 CHARACTERISTICS OF FILTERING CIRCUITS FOR UNREGULATED SUPPLIES

Regulated power supplies often have only a simple capacitive filter between the unregulated output and the regulator circuitry. The regulator itself serves to filter out the ripple, and may be referred to as an *electronic filter*. For unregulated supplies the use of filters is common practice.

The subject of filters was covered in Chap. 11, but the relationship between the filter and the power supply will be taken up here. The three most commonly encountered power-supply filter configurations are shown in Fig. 16.17. In practice, there may be several identical filter sections, but the basic principle is the same. They are *low-pass filters*, which means that they will pass direct current but reject ripple frequencies. The capacitors bypass the ripple frequencies to ground and the inductors (or resistors) oppose the passage of ripple frequencies.

Ideally the filter would pass only direct current — the lowest possible frequency — and reject all ripple frequencies. However, when using lumped components as shown in

(a)

(b)

(c)

Fig. 16.17 Popular power-supply filters: (a) Capacitive input filter. (b) RC filter. (c) Choke-input filter.

Fig. 16.17, it is not possible to achieve these ideal conditions, although they can be approached. For a half-wave rectifier on a 60-Hz power line the ripple frequency is 60 Hz, and the cutoff frequency of the filter must be below this value. For a full-wave rectifier on a 60-Hz power line the ripple frequency is 120 Hz, and it is easier to achieve a cutoff frequency below this than for a half-wave supply.

Figure 16.17a shows a *capacitor input filter*, so designated because the first component following the supply is C_1. The filter configurations of Fig. 16.17a and b are also called *pi filters* because of their resemblance to the Greek letter π. Capacitor C_1 will charge to the peak of the ripple frequency, and an important characteristic of capacitive-input filters is that they provide a higher voltage at the output terminals when compared to choke-input filters. However, this type of filter must never be used with a gaseous rectifier because the initial input capacitor charging current could damage the tube.

The "peak-inverse" voltage across the rectifier may be twice the peak-input voltage. Figure 16.18 shows why. The input capacitor has charged to the peak voltage, as

Fig. 16.18 This illustration shows why the peak-inverse voltage across a diode is larger than applied peak voltage, when a capacitor-input filter is used.

shown in Fig. 16.18. (The drop across the rectifier has been disregarded in this illustration.) On the half-cycle that puts the rectifier plate at a potential of -170 V, the inverse voltage across the diode is clearly seen to be 340 V. This may be excessive for some rectifiers. Both tube and semiconductor devices are rated by the maximum peak-inverse voltage (p.i.v.) they can withstand, and this must be taken into consideration when making a replacement during servicing.

The percent regulation of a power supply is given as the equation

$$\text{Percent regulation} = 100\,\frac{E_1 - E_2}{E_2} \tag{16.9}$$

where E_1 = power-supply output voltage when no load is connected

E_2 = power-supply output voltage when delivering the maximum load current

Supply regulation is affected by the filter since it has resistance and it is between the supply and the load. This is more readily apparent from the pi filter of Fig. 16.17b. Any change in load current will cause a change in the voltage drop across the filter resistor R and will, therefore, directly affect the output voltage. The advantage of using a choke instead of a resistor is that the supply regulation will be better. The disadvantage is that a choke costs considerably more than a power resistor. The advantage of better regulation does not justify the cost of the power supply if used in a system where the load is relatively constant. For example, in a radio the load consists of a fixed number of tubes or transistors in operation, and regulation of the supply is not an important factor.

The filter resistor (Fig. 16.17b) must dissipate power in the form of heat. If the power amplifiers of a system conduct through the filter resistor, its power rating must be quite high. However, power amplifiers are relatively insensitive to small changes in voltage on their anodes, so they can be connected to the power-supply side – that is, to point a in Fig. 16.17b.

The choke-input filter of Fig. 16.17c provides better regulation than a capacitor-input filter, but it does not have as high an output voltage for a given input. Choke-input filters are used with gaseous rectifiers.

For choke-input filters the inductance of the choke must be equal to or greater than a certain minimum value called the *critical inductance*. This inductance is defined by the equation

$$\text{Critical inductance (in henrys)} = \frac{E \text{ (in volts)}}{I \text{ (in milliamperes)}} \qquad (16.10)$$

where E = output voltage of the supply

I = current flow through the choke

All the filters of Fig. 16.17 show a bleeder resistor R_B across the output terminals. An important function of this resistor is to discharge the filter capacitors when the supply is turned off. If properly designed, regulation is improved by use of a bleeder.

16.6 SPECIFICATIONS

Each power-supply manufacturer provides his customers with performance specifications for his product. Although the actual details of the specifications vary from one manufacturer to another, a basic understanding of the parameters will enable one to interpret most "spec" sheets.

REGULATION   The basic purpose of a regulated power supply is to provide a constant output (either voltage or current, depending on the type of supply), independent of changes in either the specified input line or output load. The specifications will enable the user to determine the maximum deviation of the output voltage or current that will occur owing to changes in line or load.

Line regulation is defined as the change in output voltage, or current, for a given line change, with the load and the ambient temperature held constant. It is usually denoted as a fixed change plus a percentage change for any line-voltage change within the operating range of the supply: for example, 0.01% + 1 mV for a constant-voltage supply, or 0.05% + 2 mA for a constant-current supply. Equations (16.11a) and (16.11b) describe the line regulation for a constant-voltage and constant-current power supply, respectively.

$$\Delta E_o = (\text{percentage regulation})E_o + \text{fixed error} \qquad (16.11a)$$

where ΔE_o = change in output voltage due to an input line change

E_o = value of output voltage before the line is changed

$$\Delta I_o = (\text{percentage regulation})I_o + \text{fixed error} \qquad (16.11b)$$

where ΔI_o = change in output current due to an input line change

I_o = value of the output current before the line is changed

Load regulation is defined as the change in output voltage, or output current, for any given load change within the specified limits with the input line voltage and ambient temperature held constant. In a constant-voltage supply, this means that the load impedance may be changed in such a manner that the output load current remains within the current range of the supply. In a constant-current supply, this means that the load impedance may change between the limits defined by Eqs. (16.1) and (16.2). As with line regulation, load regulation is usually specified as a percentage of the output plus a fixed change, for example, 0.01% + 1 mV or 0.02% + 2 mA.

Equations (16.11a) and (16.11b) are valid for both line and load regulation.

TEMPERATURE COEFFICIENT   The temperature coefficient of a power supply is the change in output voltage (or output current) due to changes in the external ambient temperature, with the input line and output load held constant. This change is due to the temperature coefficient of the reference element, usually a zener diode, and of the amplifier offset and gain. The temperature coefficient is most often specified as percentage of the output plus a fixed change per degree Celsius, e.g., (0.01% + 1 mV)/°C or (0.05% + 1 mA)/°C. Equations (16.5c) and (16.5d) may be used to determine the change in output for a given temperature change.

$$\Delta E_o = \Delta T_x (\text{percentage temperature coefficient } E_o + \text{fixed change})/°C \quad (16.11c)$$

$$\Delta I_o = \Delta T (\text{percentage temperature coefficient } I_o + \text{fixed change})/°C \quad (16.11d)$$

where ΔE_o = change in output voltage of a constant-voltage supply for a change in ambient temperature ΔT

ΔI_o = change in output current of a constant-current supply for a change in ambient temperature ΔT

ΔT = change in ambient temperature in degrees Celsius

STABILITY Line regulation, load regulation, and temperature coefficient describe the changes in the power-supply output due to external variations. They imply that if the line, load, and ambient temperature are held constant, the output will not vary. In general, this is not true. There are output changes due to internal component changes caused by temperature changes within the supply. These temperature shifts are the result of localized internal heating and external air currents which are rarely completely eliminated.

If all the effects of temperature gradients are eliminated, the output will still change somewhat because of component aging changes. These changes affect the "stability" of a power supply. Stability is defined as the change in output, either voltage or current, with line, load, and ambient temperature held constant. The measurement is taken over a fixed time period, for example, over 30 days. The time period is usually either eight hours or thirty days.

Most power supplies do not specify stability. In order to obtain a guaranteed degree of stability, it is necessary to mount sensitive components in such a way that they will not be affected by normal air flows and, therefore, will not be subject to the resulting temperature variations. In addition to these precautions, it is necessary to use components whose stability is controlled. This control is usually obtained through an aging process during which all unstable parts are culled out. This is an expensive and time-consuming procedure. These additional expenses make these techniques uneconomical in all but the most expensive laboratory supplies where knowledge of the supply stability is essential. In units of this type, the reference and parts of the amplifier are usually placed in a constant temperature oven in order to reduce the effects of external ambient changes.

TRANSIENT RESPONSE Line and load regulation define the steady-state limits of the output of the supply for line or load changes. It is often of interest to know what the output does during the transition from one state to the other. This transition period is known as the transient response.

Methods of specifying transient response vary. In most cases, load transient response is more important than line transient response. The load transient response is specified for a fixed step load change, usually from 10 percent of full load to full load, or from full load to 10 percent of load. There are two important factors: the maximum amplitude of the transient response, and the time required for the output to return to some specified point. The specified point is usually within 50 mV of the final value.

Figure 16.19 shows the transient which occurs when a constant-voltage power supply

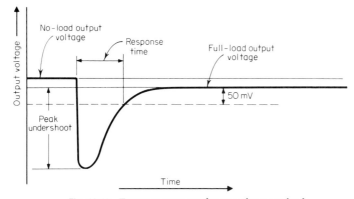

Fig. 16.19 Transient response due to a change in load.

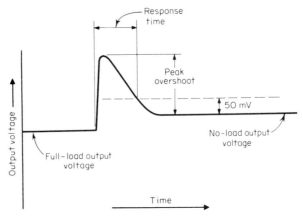

Fig. 16.20   Load off transient response.

goes from a light-load to a full-load condition. The rapid drop in output voltage represents the discharge of the output capacitor of the supply. The rate of discharge is limited only by the wiring resistance and inductance. During this time the control circuitry has not reacted, and the supply is still providing the original load current. Until the control circuitry reacts, the output capacitor is supplying the current to the load. As the control circuit turns on, the rate of decay of the output voltage decreases. When the power circuit is completely on, the power supply then provides load current and recharges the output capacitor. The output voltage then increases to its new value within the regulation band.

There is also a transient that occurs when the load is suddenly removed. This overshoot is shown in Fig. 16.20. The main source of overshoot is that it is impossible to locate the output capacitor directly at the point of regulation. Therefore, there is a voltage drop due to wiring resistance from the output terminal of the supply. When the load current is decreased, this drop decreases. The voltage on the capacitor cannot change instantaneously; the voltage difference must appear at the output terminals. This is illustrated in Fig. 16.21. For a one-ampere supply, 50 mΩ of resistance will provide 50 mV of overshoot.

RIPPLE   The final parameter that is usually specified is output ripple. This is the ac component which is superimposed on the dc output. The ripple is generally specified as an rms (root mean square) value and a peak-to-peak value. A typical specification might read:

Output ripple and noise, 0.250 mV rms, 1 mV p-p

Fig. 16.21   Relation of output capacitor location and internal wiring resistance to transient response.

The amount of ripple that may be tolerated varies from user to user. High ripple, particularly any switching spikes that might be present, would be unacceptable in logic circuits where the power-supply noise could appear as logic inputs and cause an error. In transmitting and receiving equipment, it is usually necessary to have a fairly low ripple to avoid hum, especially in the audio sections.

Constant-current power supplies also have ripple specifications which are similar to those of constant-voltage supplies except that they are in terms of current instead of voltage. For example, a supply may be rated as 0.5 mA rms, 2 mA p-p.

16.7 SPECIAL DESIGN FEATURES

Four important power-supply design features will now be discussed. They are found as either standard or optional features on most high-quality power supplies.

REMOTE SENSING   Modern power supplies are capable of load or line regulation in the order of tens or hundreds of microvolts. If this regulation is to be utilized, it is necessary to have the load at the point of regulation.

Figure 16.22 shows the output terminals of a 1-A, 10-V power supply with the load connected by 20 ft of no. 24 wire which has a resistance of 0.026 Ω/ft. If the load is changed by one ampere, the voltage change at the output of the supply is 2 mV, but the change at the load is 2 mV + 1 A × 0.52 Ω, the total wiring resistance.

Fig. 16.22 Local sensing when the load is away from the power-supply output terminals.

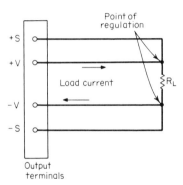

Fig. 16.23 Remote sensing to avoid loss of load regulation due to power-line voltage drops.

To eliminate this degeneration of load regulation, most modern power supplies are equipped with two additional terminals, the sensing terminals, +S and −S, for remote sensing. Wires may be run from these terminals to the point where regulation is desired. Since no load current flows through these "sensing" leads, the effects of the line drops in the power leads are compensated for. It is generally advisable to twist the leads from +S and −S together to eliminate the pickup of any stray fields. In applications where the load is relatively constant, it is possible to connect the +V terminal to the +S terminal and the −S terminal to the −V terminal at the power-supply barrier strip. Figure 16.23 shows a power supply hooked up in a remote sensing mode.

REMOTE PROGRAMMING   Most power supplies come equipped with an internal potentiometer which may be used to control the output voltage. In many applications it is desirable to have the output voltage control at some point remote from the power supply itself. The ability to have such remote programming is found in most modern power supplies.

Remote programming is usually specified in terms of resistive programming and voltage programming. The former specifies a certain number of ohms per volt which

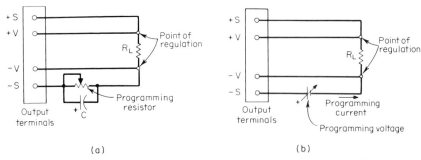

Fig. 16.24 Remote-programming systems: (a) Resistive remote programming. (b) Voltage remote programming.

gives the required external programming resistance change for a one-volt change in output voltage. The latter specifies a certain number of volts per volt, usually one, which gives the required external programming voltage change for a one-volt change in output voltage.

Figure 16.24 shows a voltage remote-programming system. Figure 16.24a shows a typical hookup for remote resistance programming. In this case, the internal control is set to zero and all control is maintained by the external resistor. The programming resistance is usually between 200 and 1000 Ω/V. The capacitor across the programming resistor is used to keep the output ripple to a minimum.

In Fig. 16.24b a power supply is shown being programmed by an external voltage source. It is necessary that the external voltage source be able to draw a reverse current equal to the programming current of the power supply, usually between 1 and 5 mA. If the programming voltage cannot do this, it should be shunted by a fixed resistor which will draw the required programming current at the operating voltage of the programming supply.

OVERVOLTAGE PROTECTION In many applications, particularly those involving integrated circuits, an abnormally high voltage at the output of the power supply could be catastrophic. In applications such as these, it is usually recommended that some type of overvoltage protection be employed.

There are two basic types of overvoltage protectors, or "O.V.'s" as they are frequently called. The first compares the output voltage of the supply to a preset level. When the output voltage exceeds this level, the O.V. *crowbars* (or short-circuits) the output by firing an SCR (silicon-controlled rectifier), which is directly across the output terminals of the supply. If the overvoltage condition has been caused by a failure in the power supply, either a fuse will open or a breaker will trip to disable the set. If the overvoltage condition was caused by an operator inadvertently adjusting the output voltage beyond the set point of the O.V., then the supply will go into current limiting. It is then necessary to turn the supply off and lower the voltage adjust setting to reset the overvoltage protector.

Figure 16.25 shows two block diagrams used for protection purposes. Figure 16.25a is a block diagram of an overvoltage protector. The comparator compares the output voltage of the power supply to a reference voltage. If the output voltage exceeds the reference, the comparator output actuates a triggering circuit, which in turn fires the SCR which crowbars the output of the supply. The output voltage is reduced to the forward conduction voltage of the SCR which is usually about one volt.

The other basic type is described as a *tracking O.V.* This type compares the reference voltage of the power supply to the output voltage. It will fire only when the supply has malfunctioned and there is too large a difference between the output voltage and the reference voltage. This type will not protect the external circuitry from operator error.

The overvoltage protector is a safety device placed between the supply and the load. As such, its circuitry should be independent of the power-supply control circuitry. This is not true of the tracking overvoltage protector.

(a)

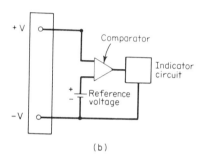

(b)

Fig. 16.25   Two block diagrams of circuits used for protection purposes: (*a*) Block diagram of an overvoltage protection circuit. (*b*) Block diagram of an undervoltage warning circuit.

Fig. 16.26   Example of a dual-output power supply, capable of remote sensing, remote programming, and series and parallel operation. (*Courtesy Lambda Electronics Corp.*)

UNDERVOLTAGE PROTECTION In some circuits, especially memory circuits, it is undesirable for the output voltage of the power supply to fall below a certain level. Then some form of undervoltage protection is used.

The simplest undervoltage protection merely compares the supply voltage to a preselected value. When the output falls below this value, some type of signal is actuated to alert the user to this condition.

Figure 16.25*b* is a block diagram of an undervoltage protection circuit. The comparator compares the output voltage of the power supply to the reference voltage. When the output falls below the reference voltage, the comparator output actuates the indicator circuit. The indicator circuit may be a light, a bell, or any other type of signal that would call the operator's attention to this condition.

A more sophisticated approach is for the undervoltage protector to switch in an auxiliary power source.

Undervoltage protection is generally not found as a standard feature on most supplies. One reason is that most undervoltage conditions are due to power-line (input ac) failure. In this case, the power source for any indicating system or an auxiliary power supply is also disabled. Thus, an undervoltage protection system would have to be tailored to a particular installation.

Figure 16.26 illustrates a Lambda LCD-2-55 dual-output power supply. This type of supply is used primarily in systems applications. The supply is capable of remote sensing, remote programming, and series and parallel operation. The two appendages are overvoltage protectors.

Chapter 17

Batteries

17.1 INTRODUCTION

The use of chemical reactions to produce electricity dates from March 1800 when the Italian physicist, Alessandro Volta (1745–1827), described two batteries to the Royal Society of London. The first of these was called a "Crown of Cups" since it consisted of a group of cups arranged in a circular pattern. Each cup was filled with salt solution (the electrolyte) and contained two strips (electrodes) of dissimilar metals (silver and zinc), with each of the two strips externally connected to its adjoining mate. These, of course, were series-connected and were the forerunner of our various modern cylindrical cell systems. In the other design that Volta presented, the dissimilar metals were stacked in the form of a "pile," very much like coins in a dispenser, with a piece of paper soaked in salt separating alternate layers of metal. This was the forerunner of our modern-day wafer or flat-plate cells.

For the most part, batteries representing variations of the Crown of Cups concept were used until about 1860 when Gaston Plante described a battery using lead and lead oxide electrodes and an electrolyte of sulfuric acid to the French Academy of Science. This, of course, ultimately developed into our familiar automobile battery, and has by now evolved into a myriad of sizes and engineering types to serve a very large number of applications. Plante's contribution was particularly noteworthy in that it achieved high-power levels.

Working at about the same time, another Frenchman, Georges LeClanche, conceived the idea of having all the battery's electrolyte absorbed by its electrodes so as to essentially function as though it were dry. In 1868 LeClanche received a patent for such a "dry" cell and undertook its manufacture. As one of the electrodes in his cell, LeClanche used manganese dioxide and a carbon-rod current collector. A cup of zinc not only formed the cell container, but also served as the other electrode. LeClanche's cell, therefore, was the direct lineal forebear of our familiar flashlight dry cells, and was a most significant development since it provided a truly mobile source of electric energy, thereby triggering the growth of the battery industry and a host of battery-using devices.

Today we have a broad spectrum of devices that yield electric energy from a controlled chemical reaction. In size, these range from tiny mercury "button" cells such as are used to power hearing aids, to the large lead-acid cells used to provide all power requirements, including propulsion of submerged submarines. A small button cell (i.e., the M5 size) will be less than $\frac{5}{16}$ in. in diameter by a little more than $\frac{1}{8}$ in high and will weigh about 0.02 oz. A perfectly conventional submarine cell will be about 18 in wide, 18 in long, $5\frac{1}{2}$ ft high, and will weigh over one ton. The little M5 has a capacity of 36 mAh whereas its giant submarine cell cousin has a capacity measured in kiloampere-hours.°

° The abbreviations "h" for "hour" and "s" for "second" follow the SI system recommended by the IEEE.

Not only is there an almost continuum of sizes between these two extremes, but there are now, as regularly available commercial products, a number of different cell systems, each having a combination of properties against which specific needs can be matched. In addition to Plante's cell and LeClanche's cell, we can consider nickel-cadmium batteries for extended cycle life, nickel-iron cells for ruggedness, silver-zinc cells for very high power, mercury cells for very high capacities, alkaline cells for outstanding versatility, magnesium dry cells for stability in hot, humid climates, and so on. In addition to sizes and types, additional variety of operating properties is obtained from different basic physical forms. Cells are available in cylindrical designs, rectilinear forms (using flat electrodes or tubular electrodes). Other forms include wafer cells (both round or rectangular) for easy series stacking for high voltages, and now even in pile form for still easier stacking to still higher voltages. In spite of this variety, the underlying principles whereby chemical reactions are used to provide electric energy — or the process is reversed so that electric energy can be used to bring about chemical reactions — can be readily understood.

GENERAL CONSIDERATIONS Of the various kinds of chemical reactions that can take place, one type generally called a redox reaction (i.e., reduction-oxidation) is of particular interest to us. A redox reaction not only involves a chemical reaction between two substances, but also a transfer of electrons from one of the substances (the reducing agent) to other substances (the oxidizing agent). As might be expected, if a redox reaction can be set up in such a manner as to cause the transfer of electrons to take place through a wire, we would in effect have a means of generating an electric current. Just such conditions are set up in a cell or battery which thereby acts as a source of electricity. (A battery is a group of cells connected in series and parallel to provide the desired voltage and power. By common usage, single cells are also commonly referred to as batteries.)

There are three essential components in a battery: an anode, a cathode, and an electrolyte. The anode and cathode together are referred to as the electrodes and are made up of the chemicals which will ultimately react to cause the electron transfer. In addition, the electrodes must provide a means for physically supporting the reacting chemicals, and a suitable conductive path for facilitating the flow of the electrons. This latter structure is referred to as the grid of the electrode. The anode is the electrode which will release electrons (and become oxidized) during the discharge of the battery. Similarly the cathode is the electrode which will absorb electrons (and be reduced) during discharge of the battery. The electrolyte is a liquid, usually an aqueous solution made conductive by the dissolution of an acid, a base, or a salt. The electrolyte completes the conductive path between the electrodes of opposite polarity. We can summarize and illustrate this with a conventional primary dry cell.

PRIMARY BATTERY In the common dry cell (also referred to as a carbon-zinc or LeClanche cell), manganese dioxide (MnO_2) is used as a cathode and zinc (Zn) as an anode. The electrolyte is a solution of ammonium chloride and/or zinc chloride. When the anode and cathode are connected through a wire so as to permit electrons to flow from anode to cathode, the reaction between MnO_2 and Zn can take place, and an electric current is set up in the external circuit and the battery is discharging, as illustrated in Fig. 17.1.

In operation, zinc is converted to zinc oxide (ZnO), releasing two electrons which then flow through the external circuit to become available to the MnO_2. These electrons are then absorbed by the MnO_2, thereby changing the latter to a lower oxide of manganese (such as Mn_2O_3). It should be noted that hydroxyl ions (OH) and water also enter into the reactions, but their net participation is canceled out. (Hydroxyl ions are particles in solution made up of hydrogen and oxygen atoms and carry a single negative charge.) As can be seen, the battery can continue to set up an electron flow until one of the other chemical reactants is exhausted, in which case the battery is said to be discharged.

In summary, a battery stores chemicals (not electricity) and can bring about a reaction between these chemicals in such a manner as to cause an electron flow through an external circuit.

The cell we have just described is classified as a primary battery since the chemical

Fig. 17.1 Fundamental operation of a dry (primary) battery.

Anode reaction:

$$Zn + 2OH^- \rightarrow ZnO + H_2O + 2e$$

Cathode reaction:

$$2MnO_2 + H_2O + 2e \rightarrow Mn_2O_3 + 2OH^-$$

Overall reaction (obtained by adding anode and cathode reactions):

$$Zn + 2MnO_2 \rightarrow ZnO + Mn_2O_3$$

reactions that gave rise to the electric current are not readily reversible. By convention, the electric current flow is opposite to that of the electron flow, and the cathode is therefore designated as the positive electrode, and the anode as the negative.

SECONDARY BATTERY Once the active materials are used up, a primary battery is not capable of yielding any more electric energy and is then discarded. Other battery systems use materials that can be restored to their original chemical state by reversing the current flow, i.e., by providing electric energy to the cell from some external source. This process is known as charging, and a battery that is capable of undergoing a number of discharge-charge cycles is called a secondary battery. An example of a secondary battery is the lead-acid system, illustrated in Fig. 17.2. By definition, the anode on discharge becomes the cathode on charge. Similarly, the cathode on discharge becomes the anode on charge. However, the terminal designations of positive and negative do not change since the current flow reverses when the roles of the electrodes of secondary batteries are most often referred to as positives and negatives, and while this designation can also be used for primaries, the electrodes of the latter are more usually referred to as anodes and cathodes.

In their most common forms both primary and secondary batteries are entirely operative when they are made. Because of this, slow deteriorations can occur owing to a small amount of spontaneous reaction or "local action" that goes on all the time in activated batteries.

RESERVE BATTERIES Where extra long shelf life (i.e., ability to yield its electric energy after long storage) is desired, the electrolyte is withheld from the battery, and steps are taken to exclude moisture since water, in some form, is usually required for the local action reactions. Such batteries are extremely stable in storage, and are activated by the addition of the appropriate electrolyte. Since the power of such batteries is thereby kept in reserve, they are frequently referred to as "reserve batteries." A special type of reserve battery is sometimes also referred to as a "dry charged" battery. In this case the electrodes of the battery are brought to a state of full charge at the time they are made, and then deactivated by the removal of electrolyte. The restoration of electrolyte at the time and place the battery needed results in a virtually

Conventional lead-acid cell

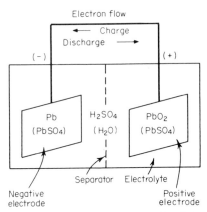

Fig. 17.2   Fundamental operation of a lead-acid (secondary) battery.

Discharge:

$$Pb + SO_4^{--} \rightarrow PbSO_4 + 2e$$
$$PbO_2 + SO_4^{--} + 4H^+ + 2e \rightarrow PbSO_4 + 2H_2O$$

Charge:

$$PbSO_4 + 2e \rightarrow Pb + SO_4^{--}$$
$$PbSO_4 + 2H_2O \rightarrow PbO_2 + SO_4^{--} + 4H^+ + 2e$$

factory-fresh battery. A great many automobile batteries are now shipped in the dry charged condition with the station attendant adding the (sulfuric acid) electrolyte at the time of installation.

There is a special class of reserve batteries that use nonaqueous or molten salt electrolytes. Such electrolytes contain no water and utilize organic liquids, or salts that become conductive in the molten condition. This latter kind of electrolyte must be liquefied before it becomes functional, and the battery can therefore be activated by heat. This is usually accomplished by a heat squib, which melts the electrolyte to activate the battery. Such batteries are known as thermal-type reserve batteries, heat-activated reserve batteries, or simply thermal batteries. A typical thermal battery will be in the form of a small button cell with a calcium or magnesium anode, a silver chromate cathode, and an electrolyte consisting of a mixture of potassium, sodium, and lithium chlorides. Such batteries have an indefinite storage capability, but discharge times of only a few minutes at best. However, they have found special uses such as power for proximity fuses. Further descriptions of reserve batteries will be considered later in this chapter along with other special types.

17.2 CHARACTERISTICS OF IMPORTANT COMMERCIAL PRIMARY BATTERIES

A large number of different primary-cell types are now being produced commercially, but only relatively few have reached the point where they are in volume production and widely distributed. Actually, each cell system has physical and electrical properties different from all the others, so that the user now has an extensive choice from which to select the type most adaptable to his particular application. Table 17.1 shows the comparative properties of the different primary-cell systems that are most widely distributed and of commercial importance.

TABLE 17.1   Characteristics of Commercially Important Primary Cells

Type	Open-circuit voltage, V	Average operating voltage, V	Wh/lb	Wh/in³
LeClanche	1.50–1.65	1.25	5–40	0.5–3
Alkaline	1.52	1.20	20–50	2.0–3.5
Mercury	1.35–1.40	1.25	25–55	5–8
Magnesium dry	1.90–1.95	1.50	15–50	1–4

Of these, the most important is the LeClanche system, also known as carbon-zinc cells or simply "dry cells." This latter designation has come into widespread use although the cells are certainly not dry. As a matter of fact, they contain electrolyte just as any other battery does, but this electrolyte is absorbed into the cathode active material so that the cell appears to be dry.

BASIC TYPES   The dry-cell industry has standardized on three basic engineering types: round cells, flat cells, and wafer cells. The industry has further standardized dimensions for 14 round cell sizes, 4 flat cell sizes, and 5 wafer sizes. Virtually all batteries are made up from these standard sizes, detailed in Table 17.2. Many manufacturers, however, issue their own catalog information showing the exact dimensions of their finished cells and batteries. Dry cells are called upon to provide a wide variety of electrical needs in terms of capacity, voltage, and current. For example, a flashlight may require 0.25 A at 3 V, while a small transistor radio may require only 5 mA at 9 V. To satisfy an almost endless variety of capacity-voltage-current requirements, the industry has developed these standard sizes and types, which are then combined in series-parallel arrangements to satisfy almost any desired power requirement.

POLARIZATION OF DRY CELLS   During the discharge of a dry cell, the electrolyte actively enters into the electrode reactions. In the vicinity of the actual reaction sites, the electrolyte at one electrode becomes depleted of its reactants, while the electrolyte at the other electrode becomes concentrated with the products of the reaction. This condition, known as polarization, slows down (or may even block) any further reaction. This has the effect of lowering the voltage at which the battery is actually discharging, or restricting the magnitude of the current that can be taken from the battery. When the battery is allowed to stand on open circuit, the natural diffusion of the components of the electrolyte permits the reactants and products to redistribute themselves more uniformly, thereby reducing (or even entirely overcoming) the effects of polarization. After this occurs, the battery can continue its discharge. Thus, dry cells can yield

TABLE 17.2 Standard Commercial LeClanche (Dry) Cell Sizes

Cell designation		Nominal dimensions°, in	
American National Standards Institute	International Electrochemical Commission	Diameter	Height of can
ROUND CELLS			
N	R1	$^{27}/_{64}$	$1^1/_{64}$
N	R1	$^7/_{16}$	$1^3/_{32}$
R	R4	$^{17}/_{52}$	$1^3/_8$
AAA	R03	$^{25}/_{64}$	$1^{11}/_{16}$
AA	R6	$^{17}/_{32}$	$1^7/_8$
A	R8	$^5/_8$	$1^7/_8$
B	R12	$^3/_4$	$2^1/_8$
C	R14	$^{15}/_{16}$	$1^{13}/_{16}$
CD	R18	1	$3^3/_{16}$
D	R20	$1^1/_4$	$2^1/_4$
E	R22	$1^1/_4$	$2^7/_8$
F	R25	$1^1/_4$	$3^7/_{16}$
G	R26	$1^1/_4$	$3^{13}/_{16}$
6	R40	$2^1/_2$	6
		Length Width Thickness		
FLAT CELLS			
F40	F40	$1^1/_4$	$^{27}/_{32}$	$^7/_{32}$
F70	F70	2	$1^1/_4$	$^7/_{32}$
F90	F90	$1^5/_8$	$1^9/_{16}$	$^{13}/_{16}$
F100	F100	$2^{13}/_{32}$	$1^{13}/_{16}$	$1^{27}/_{32}$
WAFER CELLS			
F15	F15	$^9/_{16}$	$^9/_{16}$	$^1/_8$
F20	F20	$^{15}/_{16}$	$^{17}/_{32}$	$^1/_8$
F30	F30	$1^1/_4$	$^{27}/_{32}$	$^1/_8$
F25	F25	$^{15}/_{16}$	$^{15}/_{16}$	$^7/_{32}$
F40	F40	$1^1/_4$	$^{27}/_{32}$	$^9/_{32}$

° Bare cells, i.e., less manufacturer's finishing and labeling.

more energy on an intermittent discharge routine than on a continuous one. As can be expected from the foregoing, a normal discharge of a dry cell shows a continuously decreasing voltage until the desired cutoff voltage (usually 0.9 V) is reached. Since diffusion is working against polarization, cells discharging at low rates (where diffusion can keep pace with polarization) show very slowly decreasing voltages, dropping off rapidly, however, as the battery is exhausted. Cells discharging at high rates (where polarization effects dominate) show very rapidly decreasing voltages during discharge, but good recovery after a "rest" period. This is illustrated in Fig. 17.3.

PERFORMANCE DATA Because of the wide variety of circumstances under which dry cells are discharged, performance data are difficult to summarize. Nevertheless, some fairly good methods of correlating the anticipated performance with the use conditions have been developed. One fairly widely used procedure is to correlate the capacity, expressed as operating time, with the initial drain rate at the start of the discharge, i.e., after initial stabilization of battery voltage. Such discharges are based on the load having a constant resistance. A graphical presentation usually gives the service life in hours as the ordinate, against an abscissa of initial drain rate in milliamperes. For convenience, a log-log plot is used. As indicated earlier, minimum serviceable voltage also has an effect, and such plots usually show a family of curves for several end of discharge voltages. Figures 17.4 and 17.5 present representative rating curves for a round cell and a wafer cell, operating continuously and intermittently.

Fig. 17.3 Discharge characteristics of LeClanche dry cells; D-size cells showing continuous discharge (solid line) and high and low rates of intermittent discharges for 4 h/day (dashed lines).

Fig. 17.4 Rating curves for typical round LeClanche cell. Solid line shows continuous discharge; dashed line intermittent discharge (4 h/day).

Fig. 17.5 Rating curves for typical wafer dry cells. Solid lines show continuous discharge; dashed lines show intermittent discharge (4 h/day).

example 17.1 How long can a D cell be expected to continuously operate a resistive load of 25 Ω to an end point of 0.9 V?

To approximate the initial drain, note that the initial stable cell voltage will be about 1.5 V. (See Fig. 17.3.)

$$I = \frac{E}{R} = \frac{1.5}{25} = 0.06 \text{ A}$$
$$= 60 \text{ mA}$$

On Fig. 17.4, locate an initial drain of 60 mA on the abscissa (horizontal axis). Follow this point vertically upward until it intersects the topmost solid curve (which represents a 0.9-V end point) and read 53 h as the ordinate (vertical value) of that point.

example 17.2 How long can the cell of Example 17.1 be expected to operate if used intermittently for four hours per day to the same end point?

For this problem, follow the 60-mA initial drain line until it intersects the topmost dashed line, and read its ordinate as 90 h.

The manufacturers of dry batteries have considerable control over the performance of their batteries through control of the mix formulations. Those batteries intended for essentially low drain use, but requiring high capacity (as for example in powering transistor radios) use mixes designed to give these properties; others needing only low capacity, but at high drain rates (as for example in photoflash applications) use mix formulations that give this combination. This is illustrated in Fig. 17.6 showing how discharge characteristics can be controlled even in a single physical size.

example 17.3 How much longer will a flashlight having an initial drain of 200 mA operate on a heavy-duty D than on a general-service flashlight battery if both are operated to a 1.0-V/cell end point?

Fig. 17.6   Rating curves illustrating effect of mix composition; D-size cells showing intermittent discharge (4 h/day), low voltage percent.

Using Fig. 17.6, find the initial drain of 200 mA on the abscissa, and follow it vertically upward until it intersects the flashlight battery line and read 15 h as the ordinate. Then continue vertically until the 200-mA line intersects the heavy-duty battery line and read 20 h as the ordinate. The heavy-duty battery will therefore yield $20 - 5 = 5$ h of additional service.

Another means of specifying capacity is in terms of the amount of electricity that can be drained under the conditions of discharge. This is usually stated in ampere-hours, milliampere-hours, or ampere-minutes. Capacity expressed in this way, however, is sometimes difficult to arrive at since with a constant load the current will vary with time as a result of the battery voltage varying with time. Under such conditions, the ampere-hour capacity can be taken as the product of the average current and the discharge time.

AMERICAN NATIONAL STANDARDS INSTITUTE REQUIREMENTS   For many purposes, the American National Standards Institute has developed test procedures and minimum specified requirements. They are described in detail in National Bureau of Standards Handbook 71, but some of the more important test procedures and specifications are abstracted in Table 17.3.

example 17.4   A radio A battery uses 3F-size dry cells in series. How shall it be tested for compliance with ANSI Standards? Does this battery meet ANSI Standards if it runs 41 days on this test?

Using Table 17.3, the test shown for radio A service is set up to operate the battery daily for one four-hour period each day. The load will be $3 \times 25 = 75 \ \Omega$ to an end point of $3 \times 1.0 = 3.0$ V. If the battery runs 46 days, it will have yielded $41 \times 4 = 164$ h. Since the ANSI requirement is 140 h, this battery is well over the minimum requirement.

TABLE 17.3   ANSI Test Procedures and Minimum Performance Specifications

Test name	Daily discharge schedule	Resistance per $1\frac{1}{2}$-V unit, Ω	End point per $1\frac{1}{2}$-V unit, Ω	Cell size	ANSI initial requirement
General-purpose flashlight to represent 0.5-A lamp	1.5-min period	2.25	0.65	D general purpose	400 min
General-purpose flashlight to represent 0.3-A lamp	1.5-min period	4	0.75	C AA	325 min 80 min
General-purpose flashlight to represent 0.22-A lamp	1.5-min period	5	0.75	AAA	50 min
Heavy industrial flashlight to represent 0.3-A lamp	32 4-min periods	4	0.9	D industrial	800 min
Light industrial flashlight to represent 0.3-A lamp	8 4-min periods	4	0.9	D general purpose D industrial	600 min 950 min
Railroad lantern to represent 0.15-A lamp	8 $\frac{1}{2}$-h periods	8	0.9	F	45 h
Photoflash test	60 1-s periods (1 discharge/min for 1 h/day)	0.15	0.5 D size, 0.25 C & AA sizes	D photoflash C photoflash AA photoflash	800 s 700 s 150 s
Radio A	1 4-h period	25	1.0	F G	140 h 170 h
Radio B	1 4-h period	$166\frac{2}{3}$	1.0	F40 A F90	30 h 130 h 225 h
Heavy intermittent	2 1-h periods	$\frac{2}{3}$	0.85	F (4 cells in parallel) #6 general purpose #6 industrial	70 h

SHELF LIFE   Among the special characteristics of dry batteries, a very important one is shelf life. As previously discussed, batteries produce their electric energy as a result of reactions that take place simultaneously at two otherwise independent electrodes. Under certain circumstances it is possible for each electrode reaction to take place independently of the other, thus not providing useful electric energy to an external circuit. Such reactions are referred to as "local action" phenomena, and have the effect of using up the materials without providing useful electric energy. In a battery, local action predominately affects the anode or negative electrode. Usually one of the by-products of local action is the generation of hydrogen gas which can result in a buildup of internal pressure, and some provision must therefore be made to allow hydrogen to escape. This is usually done by making the central carbon electrode porous so that hydrogen can slowly diffuse safely from the cell. It should be noted that this represents a design difficulty, since water which must be kept in the cell in order for the cell to operate properly, can also escape by this route (particularly at elevated temperatures).

Through developmental efforts over the years, the shelf life of quality dry batteries in casual storage (i.e., at normal temperatures and humidities) has reached the 18 to 21 months level, at which time the battery can still give up to two-thirds of its original service life.

In considering the storage of LeClanche cells, low temperatures are very beneficial. For example, a dry cell stored at 32°F will retain virtually all its original capacity for as long as two years. It should be noted, however, that the battery should be allowed to warm up to room temperature before it can yield its full capacity. This should be done slowly to prevent condensed moisture from physically disrupting the battery assembly.

The LeClanche system is also capable of designs that produce high voltages in small spaces. For such batteries, the flat-type cell or wafer-type cell is most frequently used. These batteries find application where high capacities are not required, but high voltages are. Since the flat cell can be readily stacked in rectangular spaces, the efficiency of space utilization for such batteries is high. In addition, the series connection of a large number of cells to produce the required voltages can be efficiently handled. Flat cells or wafer-cell batteries have therefore found extensive use in instrumentation, as radio B batteries and even radio A batteries (particularly where the power requirements are not excessive).

The special electrical properties and variety of available designs and shapes have resulted in the very wide use of LeClanche batteries in flashlights, radio receivers and transmitters, photographic equipment, portable meters, and other (generally low-power) devices.

ALKALINE BATTERIES At first appearance, alkaline-MnO_2 primary cells, sometimes called alkaline dry cells, or simply alkaline cells, seem to be very closely related to LeClanche cells. In point of fact, there are a great many similarities such as the electrochemically active materials which are the same (MnO_2 cathode, zinc anode) a illustrated in Fig. 17.7. In addition, the similarity is continued in available physica

Fig. 17.7 Schematic diagram of a typical alkaline primary cell.

sizes and shapes. However, in the alkaline cell, a solution of a strong alkali, potassium hydroxide, is used as the electrolyte in place of the $ZnCl_2$ and NH_4Cl of the conventional LeClanche system. In addition, the zinc exists in granular form so that it presents a very extensive surface to the electrolyte. These factors greatly alter the electrical and other characteristics of the battery. Polarization effects in alkaline cells, for example, are much less pronounced than in conventional dry cells (under equivalent discharge conditions), and much of the energy can be withdrawn at higher power levels. Along related lines, it is also found that their discharge characteristics are more easily defined and catalogued since the voltage variation during discharge is likely to be somewhat less than in a LeClanche cell. Alkaline dry batteries are, therefore, rated according to their capacity, usually cited at a particular drain rate. For convenience the initial drain rate is used, and the capacity is usually expressed in milliampere-hours or ampere-hours at this rate. Table 17.4 summarizes the sizes and electrical characteristics of the commercially important alkaline cell sizes.

TABLE 17.4   Sizes and Characteristics of Alkaline Primary Cells

Size designation	Capacity		Diameter,° in	Height,° in
	mAh	At initial drain of mA		
N	580	18	0.47	1.12
AAA	750	18	0.41	1.75
AA	1,800	60	0.55	1.94
C	5,000	100	1.31	1.93
D	10,000	300	1.31	2.37

° Finished cells; i.e., including manufacturer's labels, etc.

example 17.5 How long can an alkaline AA cell be expected to operate at a drain of (a) 60 mA, (b) 40 mA, (c) 100 mA?
solution From Table 17.4, the capacity of an AA alkaline cell is 1800 mAh at a drain of 60 mA. The operating time therefore will be 1800/60 = 30 h. At 40 mA the capacity will be slightly higher, but for estimating purposes would still be taken as 1800 mAh: 1800/40 = 45 h. At 100 mA the capacity will be slightly lower, but for estimating purposes can still be taken as 1800 mAh: 1800/100 = 18 h. (At higher current drains, the capacity should be determined for the specific use and circumstances, the capacity given in the table being used only for approximations.)

Like LeClanche cells, alkaline cells have generally good shelf-life characteristics. As a matter of fact, alkaline cells are actually sealed since local action reactions in healthy cells are not serious. As a consequence, hydrogen production is very much slower than in LeClanche cells, and in fact by proper care during manufacture can be made negligible. Since alkaline cells are sealed, there is also no loss of moisture during storage, and alkaline cells therefore accept prolonged storage very well. As a matter of fact, a shelf life of two to three years (to 80 percent of original capacity) is common. Also, because they are sealed, alkaline cells can be stored at higher temperatures than LeClanche cells. Storage up to 130°F for extended periods of time is entirely feasible. However, since elevated temperatures accelerate local action reactions, it is not recommended that they be exposed to temperatures above 130°F for any length of time. In the other direction, as the temperature of the cells is reduced, there is only a slow falloff in capacity for alkaline cells, making them well-suited for operation in moderately low-temperature ambients. Quantitatively the effects of temperature are summarized in Table 17.5.

Although the alkaline dry cell is generally not troubled with the evolution of hydrogen gas, under adverse conditions such as overdischarging or excessive temperatures some gassing can occur, and most alkaline cells are therefore provided with a safety blowoff which permits hydrogen to be vented at a certain pressure so that the hydrogen can be harmlessly relieved.

Because of their basic and special properties and ability to provide relatively high power levels, alkaline dry batteries are now extensively used to power portable radios (particularly where high power levels are required), in photographic equipment, and

TABLE 17.5 Effect of Ambient Temperature on Capacities of Alkaline Primary Cells

Discharge temperature, °F	Approximate Service, %°		
	Light drain	Medium drain	Heavy drain
113	100	100	100
70	100	100	25
30	70	40	25
−4	25	20	5

Note. When discharge times (at normal ambient temperatures) exceed approximately 30 h, drains are considered light; when less than approximately 15 h, drains are considered heavy.
° Cells cooled to ambient temperature before discharge; service at 70°F to 0.8-V end point taken as 100%.

for driving fractional horsepower electric motors, such as are used in electric shavers, hobbycraft, and the like.

MERCURY DRY BATTERIES Another cell system using an alkaline electrolyte that has achieved a significant commercial position is the mercury dry cell. The outstanding characteristics of mercury cells is that they represent the highest available capacities for a given volume of any of the primary batteries. As can be seen in Fig. 17.8, one type of mercury battery is very similar in construction to the alkaline battery. Physically the major difference is that mercuric oxide (HgO) is substituted for the manganese dioxide, but this brings about some profound changes in cell characteristics. In addition to having very high capacities, mercury cells also have an outstanding shelf life since mercury or mercury compounds have a strong effect in reducing local action. As a matter of fact, the shelf-life problem as far as mercury cells are concerned is the ability to thoroughly seal the cell rather than any chemical action that might take place.

Because of this, the practical shelf life seems to be limited to about two or three years on larger sizes, and perhaps to one or two years on the smaller sizes. Mercury cells are marketed in essentially the same sizes as alkaline and LeClanche dry cells, but in addition are available in extremely small sizes called button cells, schematically illustrated in Fig. 17.8. Table 17.6 gives the general sizes and weights of the more important commercial sizes.

The temperature characteristics of mercury cells are very similar to those of alkaline cells, except that at low temperatures capacities fall off a little more rapidly, so that the lowest temperature suitable for mercury-cell operation would be about 30°F, as shown in Table 17.7.

Fig. 17.8 Schematic diagram of typical mercury button cell. Note that cylindrical mercury cells closely resemble alkaline primary cells, shown in Fig. 17.7, differing principally in the composition of the cathode mix.

TABLE 17.6   Sizes of Commercially Important Mercury Cells

Size designation, ANSI	Rated capacity		Nominal dimensions, in°	
	mAh	At drain of mA	Diameter	Height
M5	36	2	0.31	0.14
M8	100	3	0.46	0.13
M15	160	5	0.46	0.21
M25	350	3	0.46	0.57
M40	1000	35	0.63	0.65
M55(AA)	2400	50	0.55	1.97
M100(D)	14000	250	1.28	2.39

° Bare cells, i.e., less manufacturer's finishing and labeling.

TABLE 17.7   Temperature Characteristics of Mercury Cells

Discharge temperature, °F	Approximate service, %°	
	Light drain	Heavy drain
113	100	100
70	100	100
40	95	7
30	6	2

Note. When discharge times (at normal ambient temperatures) exceed approximately 25 h, drains are considered light; when less than approximately 25 h, drains are considered heavy.
° Cells cooled to ambient temperature before discharge; service at 70°F to 0.9-V end point taken as 100%.

example 17.6 A mercury battery can operate a particular radio receiver for 160 h at normal ambients. How long can it be expected to operate the receiver in an ambient of 35°F.? Table 17.7 indicates that the battery would operate the radio (i.e., a light drain) for 0.95(160 = 152 h at 40°F, and only 0.06(160) = 9.6 h at 30°F. To estimate performance at 35°F, it is necessary to extrapolate between these values: 0.5(152 − 9.6) + 9.6 = 80.8 h.

In addition to their extremely good shelf life and high capacity, mercury cells have another outstanding characteristic: namely, uniformity of discharge voltage. As can be seen from Fig. 17.9, mercury cells have an almost uniform discharge voltage, except for a short time at the very beginning and of course at the very end of the discharge.

Fig. 17.9 Typical discharge characteristics of mercury cells.

This permits designers of instrumentation and electronic equipment to use circuits that need not accommodate voltage fluctuations. Not only is the voltage of mercury cells stable, but it is also highly reproduceable. Mercury cells, therefore, find wide application in instrumentation as stable voltage working cells (to be periodically matched against standard cells for very high precision and long service). As a consequence of their high capacity, availability in very small sizes, good shelf life, and stable discharge voltage, mercury cells have found a wide range of applications, as, for example, in portable voice recorders, radio transceivers, and the like.

MAGNESIUM DRY BATTERIES   No discussion of primary batteries would be complete without significant mention of the new magnesium dry cell. Fundamentally, the magnesium dry cell is an exact counterpart of the LeClanche system in which the outer can is made of magnesium alloy instead of zinc.

This rather "small" change is accompanied by a number of more subtle changes that result in an entirely new system with a set of characteristics all its own. The outstanding properties of the magnesium dry battery are its ability to withstand high-temperature, high-humidity storage, and extremely long shelf life under casual storage. In addition, the magnesium dry battery operates at voltage levels considerably higher than conventional LeClanche or mercury dry cells. This is due to the fact that magnesium is a far more electropositive (i.e., active) metal than zinc. However, because of its reactivity, one would ordinarily expect a magnesium cell to be unstable, but what actually happens is that magnesium forms a very thin but very impervious oxide coating, thereby protecting the underlying metal from local action. Although special electrolytes must be used (generally solutions of magnesium bromide or magnesium perchlorate) and the cell must be closed by very special techniques in order to retain moisture and still permit hydrogen to escape (since hydrogen is liberated as a by-product of the discharge), it is really the protective film which forms on the magnesium that has made the magnesium dry cell possible. At their present stage of development, these batteries can be successfully stored at 160°F for 30 days without significant loss of capacity (i.e., less than 10 percent).

As yet, the total production from all suppliers is being funneled into military applications, but it is anticipated that units will soon be available for normal commercial applications. Cell sizes have not as yet been standardized, and only cursory performance data can be presented at this time. What general information is available is presented in Table 17.8 and Figure 17.10.

TABLE 17.8   Sizes of Typical Magnesium Cells

Cell designation, ANSI	Nominal dimensions, in°	
	Diameter	Height
N	0.42	1.09
R	0.53	1.38
A	0.63	1.88
CD	1.00	3.19
D	1.25	2.25

° Bare cells, i.e., less manufacturer's finishing and labeling.

As can be seen from Table 17.1, magnesium dry batteries are fully in a position to compete with mercury and alkaline batteries in terms of power capability per unit weight and capacity per unit volume. Ultimately, magnesium dry cells should also be more favorably priced than their mercury or alkaline counterparts. At least for the immediate future, however, on the basis of costs, magnesium dry cells are not expected to make serious inroads into conventional dry battery areas, except for those applications where extremely long shelf life (even up to five years) is required, or where the batteries are to be used in hot, humid areas where the logistics of getting fresh LeClanche cells to the user are poor. Magnesium batteries, since they can withstand extremely adverse storage conditions for extended periods of time, can be delivered

Fig. 17.10 Typical discharge characteristics of magnesium dry cells.

to the ultimate user in tropical and semitropical countries at full or almost full capacity. The outstanding shelf life of the system (as compared to that of conventional dry cells) is illustrated by Fig. 17.11. Here both the carbon-zinc and magnesium types show very little capacity loss after three months' storage at room temperature, but for storage at elevated temperatures the superiority of the magnesium system quickly asserts itself. At really high storage temperatures, the carbon-zinc type fails rather quickly, while the magnesium type retains most of its original capability.

There are other primary batteries that have achieved a degree of commercial value, such as silver-zinc, air-depolarized, and others, but these are not yet in sufficiently general use to warrant discussion at this time. However, it should be mentioned that the silver-zinc primary does have an outstanding high rate capability and a capacity as much as 80 percent greater than its secondary counterpart, described later in this chapter. Physically it is very similar to the dry charged form of the secondary version (shown in Fig. 17.15) except that it usually does not have an ion-restructive diaphragm. The higher capacity is due to the fact that the silver electrode can be oxidized to a higher level before cell assembly than can conveniently be done after cell assembly and activation (as would be the case for secondary operation).

Before going on to rechargeable systems, Fig. 17.12 is included to illustrate the actual physical forms of some of the primary systems we have considered.

Fig. 17.11 Comparison of high-temperature shelf-life capabilities of carbon-zinc and magnesium dry batteries.

Fig. 17.12   Sample display of primary batteries. (*Courtesy Burgess Battery Company*)

17.3 CHARACTERISTICS OF IMPORTANT COMMERCIAL SECONDARY TYPES

Although there are a large number of types among the secondary, or rechargeable batteries, only three have achieved any really significant commercial importance. These are the lead-acid, nickel-cadmium, and silver-zinc systems.

LEAD-ACID BATTERY   Of these, the lead-acid system is the most important from a commercial point of view and is widely distributed. It is available in a very large variety of sizes and shapes to satisfy applications requiring a broad range of capacities and power levels. A summary of commercially important types is given in Table 17.9. It should be noted that while the battery sizes are arranged in order of increasing capacity, some are for 6-V (three-cell), 12-V (six-cell), or 2-V (single-cell) units so that the table covers a much broader range than at first appearance.

An interesting point should be noted concerning the 50- and 100-Ah units which are approximately the same size and weight in spite of the twofold capacity difference. This is explained by the fact that they both have the same energy content, 600 Wh (obtained by multiplying the capacity in ampere-hours by the battery voltage, as indicated above). It is important to realize that the capacity and power capability of a rechargeable battery depend in very large measure on its prior history. A typical situation, particularly in lead-acid batteries, is that the capacity builds up during early cycle life,

TABLE 17.9   Representative Sizes and Weights of Lead-Acid Batteries

Nominal size Ah capacity @ 20-h rate	Number of cells	Nominal dimensions, in°			Approximate weight
		Length	Width	Height	
4	3	$2^{3}/_{4}$	$1^{7}/_{8}$	4	$1^{3}/_{4}$
8	3	$5^{1}/_{2}$	2	$4^{3}/_{4}$	4
30	6	8	$5^{3}/_{8}$	$9^{1}/_{8}$	27
50	6	$11^{1}/_{2}$	$7^{1}/_{8}$	$9^{1}/_{8}$	50
100	3	$10^{3}/_{8}$	$7^{1}/_{8}$	$9^{1}/_{8}$	46
200	6	$21^{5}/_{8}$	$11^{1}/_{8}$	$10^{1}/_{4}$	147
280	1	$5^{1}/_{8}$	$6^{1}/_{4}$	$15^{1}/_{4}$	60
420	1	$5^{1}/_{8}$	$6^{1}/_{4}$	$21^{3}/_{8}$	84

° Overall dimensions, i.e., including terminals and vent plugs.

then maintains a substantially uniform value for most of the battery's useful life, and finally falls off (more and more rapidly) toward the end of life. Unless some accidental life-terminating event occurs, such as a short circuit or loss of electrolyte through a cracked case, etc., the end of life of the battery will be reached when its capacity falls below that which is useful for the application. For a great many applications, this will be at about 60 percent of the original capacity.

As for the rating of new batteries, different manufacturers hold to somewhat different views. Some, for example, might rate a battery on what it will do on the very first discharge cycle in the user's hands, whereas others may take advantage of the fact that the capacity normally builds up during early cycle life, and will rate a battery after it has been "conditioned" in use. Where the capacity of the battery is critical, the user should determine which of these systems is the basis for rating the battery.

The conditions of discharge also have a very profound effect on the capacity and power capability of rechargeable batteries. As a general rule, a battery will display its greatest capacity when discharged at very low rates. For lead-acid batteries this rate is generally chosen to be the 20-h rate. A great deal of confusion has arisen as a result of the industry method of expressing the discharge rate of a battery. The phrase "the 20-h rate" means that the battery is being discharged at such a current as will exhaust the capacity of the battery in 20 h. The confusion results from the fact that it is necessary to know what the capacity of the battery is at the particular current drain in order to determine the current at which it should be discharged.

To simplify this situation, the discharge current is now generally related to a nominal capacity of the battery. For example, if the nominal capacity of a battery is 10 Ah, a discharge rate of 0.5 A would be called the "20-h rate." Most leading producers now cite capacity at the 20-h rate and at normal (78°F) ambient temperatures. For lead-acid batteries, there is usually very little difference in capacity as the rate of discharge is increased from the 20-h rate to perhaps the 10-h rate. But thereafter, as the discharge rate is further increased, there is an increasingly rapid falloff of capacity. Except for very special designs, the half-hour rate is generally considered to be the maximum sustained rate at which lead-acid batteries can be used with any degree of efficiency.

The lead-acid system has a number of characteristics that are particularly noteworthy. Although these batteries do not withstand high-temperature storage very well (anything above 130°F being somewhat deleterious), they do withstand low-temperature storage extremely well and are capable of yielding considerable power and operation at low temperatures. As a typical example, the capacity of a lead-acid battery is likely to drop off only about 30 percent as the temperature is decreased from room temperature to 0°F. Additional information is given in Table 17.10. In determining the capacity of lead-acid batteries from Table 17.10, it should be noted that the discharge rate has considerable significance. In using the table, the column calling for the discharge rate at which the battery is to be operated as measured at room temperature must first be found, and this column is then followed downward to the appropriate temperature. The figure cited is the capacity (expressed as a percent of the 20-h rate capacity at room temperature) at the same current drain.

TABLE 17.10   Effect of Temperature on Capacity of Lead-Acid Batteries

Temperature, °F	Running time, hours°					
	1	5	10	50	100	500
	Drain current, amperes					
100	118	21	10	1.4	0.6	0.06
80	114	20	9.0	1.3	0.5	0.05
60	106	19	8.3	1.1	0.5	0.05
40	97	17	7.5	1.0	0.4	0.04
20	90	15	6.7	0.9	0.3	0.03
10	82	13	5.6	0.8	0.2	0.02

° Capacity at 80°F and 20-h rate = 100 Ah.

example 17.7 A battery has a capacity of 100 Ah at the 20-h rate at room temperature. What is its capacity at the 10-h rate at room temperature and at 20°F?

solution For a 100-Ah battery the current drain at the 20-h rate is 5A. From Table 17.10, its capacity at room temperature at the 10-h rate will be 91 percent of 100 Ah, or 91 Ah. This is the nominal capacity of the battery at room temperature and at the 10-h rate. For this battery, therefore, the discharge current at room temperature and the 10-h rate is 9.1 A. Following the 10-h rate column downward to 20°F, the capacity at a drain of 9.1 A will be 66 percent of what it was at room temperature, or 60 Ah. This table may seem a little difficult to use, but it is a very serviceable one in that it contains a great deal of information since interpolations for other rates and temperatures give entirely acceptable results.

In a rechargeable battery, charge retention, i.e., the ability to maintain capacity after charge, corresponds to shelf life in a primary battery. In a conventional lead-acid battery (sometimes referred to as an antimonial alloy lead battery), the charge retention varies considerably from the early stages of life to the latter stages, falling off slowly but continuously as the battery is cycled. When new, the battery is very likely to be capable of losses of less than 10 percent per month in (casual) storage, but this figure could easily approach 100 percent for an old unit nearing the end of its useful life. A newer type of lead-acid battery is now available, however (i.e., the calcium alloy lead battery), which has a remarkable degree of charge retention that does not fall off to any significant extent as the battery is used. In a calcium alloy type, the charge retention is likely to be better than 85 percent after as long as one year in casual storage, and this will be fairly uniform throughout its life. Generally speaking, this latter type is used primarily for standby power since it has only a limited cycle life capability.

Because of the variety of properties that can be developed by design and/or electrical treatment, lead-acid batteries have achieved an amazing variety of applications. Apart from the popular ones of providing starting, lighting, and ignition power for automobiles, operating electric fork-lift trucks and similar commercial applications, they have now found significant applications in standby power (for telephones, industrial purposes, hospitals, emergency lighting, etc.), as well as in a variety of special applications such as portable TV, electric power tools, and ordnance guidance systems, to name but a few of the more important ones.

NICKEL-CADMIUM BATTERIES A second type of rechargeable battery of great and continuously growing commercial importance is the nickel-cadmium system. Within the basic system, there are several different engineering types and designs, each design reflecting the development of specific properties for a specific field of application. The pocket-type nickel-cadmium battery, for example, is an extremely long-life unit capable of thousands of very deep discharges. Because it can also moderate rates of discharge, it is used principally where motive power is desired, such as in electric wagons and the like. Of more recent development, the sintered-plate-type nickel-cadmium battery is available in both the sealed and vented forms. The sintered-plate system is by far the fastest growing type in the nickel-cadmium field and will be discussed in more detail.

Basically, the sintered-plate type utilizes a very porous nickel to support the active materials and serve as a current path. This matrix is made by sintering finely divided particles of carbonyl nickel powder (a special grade of "macelike" particles). The resulting plaques can be as much as 80 percent porous and are impregnated with the active materials, nickelic hydroxide in the positive electrode, and metallic cadmium in the negative. As is common in all nickel-cadmium types, the sintered-plate nickel-cadmium battery utilizes a potassium hydroxide electrolyte. A schematic diagram of a sintered-plate nickel-cadmium battery is shown in Fig. 17.13, and common commercially available sizes are given in Table 17.11.

Vented cells are usually encased in plastic containers, while sealed cells are usually in thin-walled steel containers. Both are extremely rugged, and the battery can take a good deal of physical abuse, as well as electrical abuse, without permanently deleterious effects. Nickel-cadmium batteries are capable of very high rates of discharge, and as a matter of fact many applications will discharge such batteries at the one, and even the one-half hour rate at good efficiencies. The high rate capability of these batteries also makes them useful for "pulsed" discharges (i.e., very high rate, very short duration discharges).

In addition to their fine physical characteristics, nickel-cadmium batteries perform

Fig. 17.13 Schematic diagram of sintered-plate nickel-cadmium cell.

TABLE 17.11   Sizes of Common Nickel-Cadmium Cells

| ANSI size designation | Nominal capacity | | Size, in | |
	mAh	At drain of mA	Diameter	Height
		SEALED, ROUND		
	50	5	0.61	0.24
	150	15	0.98	0.28
	450	45	1.70	0.30
AA	450	45	0.56	1.95
C	1200	120	0.88	1.67
D	4000	400	1.30	2.43

| | Capacity | | Size, in | | |
	Nominal Ah	At drain of amperes	Length	Width	Height
		SEALED, PRISMATIC			
	1.5	0.15	1.36	0.75	3.36
	4.5	0.45	1.93	1.67	3.13
	7.5	0.75	1.93	1.67	4.25
	15	1.50	3.56	1.34	4.90
		VENTED, PRISMATIC			
	6.5	0.65	0.96	2.18	4.06
	24	2.4	1.08	3.16	8.28
	40	4.0	1.39	3.16	9.40
	120	12.0	2.17	6.81	8.73
	230	23.0	3.17	8.14	9.40

well at both high and low temperatures, but capacity does fall off if the temperature is raised or lowered from ambient room temperatures. The temperature characteristics of typical nickel-cadmium cells are shown in Table 17.12. As can be seen from Table 17.12, the falloff in capacity as the temperature is lowered is about the same as for lead-acid types for the temperature range covered. However, as the temperature is lowered beyond the range covered in Tables 17.10 and 17.12, the performance of the nickel-cadmium battery falls off more slowly than its lead-acid counterpart. Here only qualitative data can be cited (since we are in an area where specific designs play a most important role), but a nickel-cadmium battery can be quite useful down to −40°F (i.e., yielding about 5 to 10 percent of its normal capacity), whereas a lead-acid battery would probably have a practical limit of −20°F. There is another very important factor to consider as regards high rate capability at low temperatures, in that the current capability per unit weight or per unit volume, for short-duration discharges, is very favorable for nickel-cadmium sintered-plate batteries. For these reasons this type of battery is favored for low-temperature engine-starting applications.

TABLE 17.12 Temperature Characteristics of Nickel-Cadmium Cells

Discharge temperature, °F	% of room-temperature capacity at 10-h rate
115	93
78	100
40	93
32	90
0	60
−5	50

The charge retention characteristics of a nickel-cadmium battery are not too far different from those of a lead-acid battery, but in at least one respect the nickel-cadmium battery does have an outstanding characteristic: it can withstand extremely long periods of idle storage (up to 7 to 10 years being not at all uncommon) without permanent, deleterious effects. In fact, when it is desired to store a battery, it should be completely discharged all the way down to zero volts (by even short-circuiting the battery at low rates when its voltage has dropped to a point where this can be done safely). By contrast, a lead-acid battery must be electrically maintained during idle storage (by a monthly low-rate trickle, or some other routine-freshening charge). A nickel-cadmium battery needs no such electrical maintenance, and when placed into service after long idle storage will, after a few cycles, pretty much return to its proper capacity.

Perhaps one of the most outstanding characteristics of the nickel-cadmium battery is its ability to accept a recharge very quickly. Since the back emf of the battery rises relatively slowly during charge, as indicated on Fig. 17.14, most of the charge can be completed at high rates before the back emf begins to seriously limit the charge acceptance. However, since the battery is capable of accepting charges at very high rates, it is not overly important to control the charge to a great degree, particularly at the early stages. This is true for vented cells to a much greater degree than for sealed cells since the latter can also accept high overcharge rates. During overcharge of vented cells, however, water is electrolyzed into hydrogen and oxygen, using up water and electricity, but otherwise having no further deleterious effect on the battery.

The nickel-cadmium system is thus a sophisticated combination of virtues and limitations. In spite of its high initial cost, when used under proper circumstances it can be quite economical since it provides the closest approach the battery industry has yet made toward the goal of long-life low (or no)-maintenance batteries. These batteries have therefore found extensive use in jet aircraft starting, radio and telephone communications, subcutaneous low-power medical applications, cold weather starting, low power starting (as in lawnmowers, chain saws, etc.), and, of course, in cordless appliances (such as toothbrushes, shavers, etc).

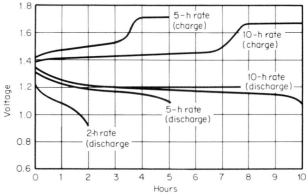

Fig. 17.14 Constant-current charge-discharge characteristics of nickel-cadmium batteries.

SILVER-ZINC BATTERIES   In the spectrum of battery power sources, the silver-zinc system has achieved a firmly entrenched position principally because of its power capability. Whereas nickel-cadmium batteries do quite well at discharge rates as high as the one-hour rate, silver-zinc primaries and secondaries easily go to the ten-minute rate, and for pulse power even to the one-second rate with relatively good efficiencies. The general construction of silver-zinc secondaries is shown in Fig. 17.15. Essentially the primary version is made in about the same way, but the ion-restrictive diaphragm need not be used, nonwoven fabrics being used instead (to provide lower resistance and more rapid activation). Apart from stability in the dry condition (making it suitable for reserve battery applications), one of the principal virtues of the primary version is that it can take advantage of higher degrees of oxidation of the silver (since the positive electrode is "formed" before cell assembly under conditions favorable to silver oxidation). This manifests itself as a greatly increased discharge capacity.

Typical primary and secondary characteristics are shown in Fig. 17.16. This figure shows the voltage-time curves for a charge and discharge cycle of a silver-zinc secondary, and for the discharge of a primary. It is interesting to note that Fig. 17.16 makes

Fig. 17.15 Schematic diagram of typical silver-zinc secondary cell (end view to show separator configuration).

Fig. 17.16 Charge-discharge characteristics of typical silver-zinc secondary cell.

no reference to size (i.e., capacity) since the general shape of the curves fit all sizes with remarkable uniformity when each is operated at its appropriate current value. It is also interesting to note that while secondary batteries are usually furnished dry and uncharged, they can be furnished dry charged, in which case the first discharge will have the characteristics of a typical primary, reverting back to normal secondary characteristics for subsequent cycles. Of course, because of the presence of an ion-restrictive barrier, which usually is slow wetting, the activation time of a dry charged secondary will be as long as for any other silver-zinc secondary.

There are, of course, too many physical forms that have been developed for secondary batteries, but a representative sampling is shown in Fig. 17.17.

Fig. 17.17 Sample display of secondary batteries. (*Courtesy Burgess Battery Company*)

Referring now to the secondary version, the silver-zinc system shows a marked buildup in capacity during early (perhaps 20 percent of) life, followed by a long period (perhaps 70 percent of life) of reasonably stable capacity. Toward the end of life there is a rapid decline in capacity to a normal end point, usually about 60 percent of rated capacity. In operation the cycle life is highly variable, depending on use circumstances, but generally the range is from 20 for deep, rapidly repeated cycles, to perhaps 200 cycles of moderate (50 percent) depth, provided that sufficient time is allowed between discharge-charge-discharge to avoid temperature buildup.

High temperatures can be particularly harmful since the solubility of the negative electrode in the electrolyte is greatly increased, and deleterious degradation of the diaphragm also occurs. Overcharge also has seriously deleterious effects, but as in calcium-alloy batteries, a pronounced increase in back emf at the approach of the full-charge condition permits convenient control of the charge so that serious overcharging can be readily avoided.

Since silver oxide is one of the active materials, the cost of silver-zinc batteries is obviously high, limiting its applications to date to essentially military gear and special communication devices. Nevertheless it has an extremely good low-temperature capability, a modest charge retention (although this has been and is being improved), and a reasonable cycle life. For those applications requiring very high power levels, or extremely high bursts of energy (pulse loads) silver-zinc batteries indeed provide a convenient, reliable power source in small, lightweight packages.

17.4 MAINTENANCE

Unlike primary batteries which require no special attention, the maintenance of secondary batteries is of extreme importance, particularly where maximum performance and life are to be obtained. Charging has a profound effect in this regard, and several good methods have been developed. The constant-current method simply returns the capacity that has been discharged at a fixed current for a fixed period of time, or until the back emf (i.e., the countervoltage) of the battery rises to some predetermined value. It is therefore sometimes called the time-controlled or voltage-controlled method of charging. This is perhaps the simplest method in terms of the required control equipment, but does not account for, or take advantage of the fact that the battery has the ability to accept a charge very rapidly at the early stages of recharge (particularly after a deep discharge) and that this situation is reversed as the battery approaches the fully charged condition.

As the battery approaches full charge, the efficiency with which the battery accepts the charge falls off very rapidly (reaching perhaps only 4 to 5 percent at about 95 percent of full charge). This results in an undesirable condition in constant-current charging, since it is too slow at the beginning of the recharge (when the battery can accept charge efficiently) and too fast toward the end of recharge (when the battery can accept charge only slowly). A wasteful recharge not only manifests itself as an undesirable cost, but also as excessive positive grid corrosion (which will result in short life) and excessive watering frequency.

To improve this situation, we can take advantage of the fact that the back emf of the battery rises as the state of charge increases. Many manufacturers now market charging devices which use a modified constant-potential procedure in which the initial stages of the recharge are carried out at a high, but constant current, the value of which is essentially dependent on the power capability of the charging source. As the back emf of the battery reaches a predetermined point, which is toward the latter part of the recharge, automatic switching circuits change the system to a fixed potential recharge. This potential is set so that it is a little higher than the end of charge back emf of the battery. As the back emf of the battery approaches the potential of the charging circuit, the charging current falls off as the ability of the battery to accept the charge falls off. This turns out to be efficient, not only in terms of time and cost, but also in the avoidance of overcharge and in reducing watering frequency.

LEAD-ACID BATTERY MAINTENANCE   Apart from the selection of a suitable charging procedure, the electrolyte in lead-acid batteries must be maintained at the proper concentration and level. Negative electrodes exposed to the atmosphere as a result of low electrolyte level are likely to oxidize very rapidly, a process that is only partially reversible by charging, and results in varying degrees of permanent damage to the electrode. When the concentration of sulfuric acid is allowed to fall, the capacity of the battery also suffers rather quickly since sulfate ions are a participant in the discharge reactions. In addition, the negative plates tend to sulfate (i.e., form an irreducible lead sulfate) in low-gravity acid.

On the other hand, if the electrolyte concentrates in the battery, the positive grids are likely to corrode more rapidly, especially during overcharge periods, and again the total life of the battery will be impaired. Since sulfuric acid is not lost from the battery as a result of the direct electrolytic reactions, only water must be replaced periodically to maintain the electrolyte composition and level. During overcharge, gases are evolved as a result of the decomposition of water, and this not only represents a loss of water, but also the gases produce an electrolyte spray above the upper portions of the cell. This electrolyte spray is extremely fine, and so quickly finds its way out of the cell as entrainment with the gases being evolved.

If this should occur, the top of the battery is very likely to get wet with sulfuric acid, which, being a very good electric conductor, permits stray currents to be set up, not only from cell to cell, but also from each cell to ground. With poor top-of-battery cleanliness, these currents can indeed reach very significant values, and the battery can be rather quickly discharged or even destroyed. To avoid this, many manufacturers have developed ingenious spray traps and have incorporated these in their vent plugs. However, careful and proper charging can go a long way toward eliminating this difficulty. Where electrolyte has come out of the cell, it is necessary to neutralize it (boric acid or bicarbonate of soda being recommended) followed by a thorough washing and drying of the top of the battery. This procedure will minimize the destructive effects of corrosion on external battery parts, terminals, and holders.

The loss of sulfuric acid as a result of spray entrainment will, of course, also impair the capacity of the battery. However, this sulfuric acid can be replaced, using chemically pure sulfuric acid, to bring the concentration (or specific gravity) of the electrolyte up to the manufacturer's recommended value.

In doing this, it is best to remove as much of the free electrolyte to which the additional acid is to be mixed to an external vessel. The new acid mixture is properly adjusted for concentration and can then be returned to the battery. If fresh acid is simply added to the battery, the dense, fresh acid will simply settle to the bottom and will probably attack the separators on the way down. The addition of fresh sulfuric acid also generates quite a bit of heat, and, of course, concentrated sulfuric acid is quite dangerous to handle. This operation should, therefore, be conducted only with proper protective clothing and facilities, and under qualified supervision.

SEALED UNITS   To minimize maintenance, a number of brands of lead-acid batteries have been put on the market as "sealed" or "closed" units. Such batteries are generally made using the calcium alloy system which has an extremely high back emf (of the order of 2.8 to 3.0 V/cell when approaching the full charge condition). For this reason inexpensive automatic charge control equipment can be used to limit the amount of overcharge. This, coupled with the fact that local-action shelf losses are extremely small for calcium alloy batteries, has resulted in a condition in which not a great deal of water is used up during the normal use of the battery, nor is there any serious gassing on standby. Under these conditions the cell can essentially be "closed," since the need to replace water will have been reduced to the point where the cell can go without rewatering for its useful life. The fact that only very little gas is produced and even this will be principally hydrogen (a gas that is very difficult to contain) permits the gas to slowly but surely leak out via even very minute flaws in the seal. Since such batteries can, therefore, go for very extended periods without any maintenance, their use, particularly in small power applications and automobiles, is increasing.

Since many small power applications require a degree of portability (as, for example, power tools, etc.) it is also desirable to immobilize the electrolyte. This is usually done by forming a "gel" in the electrolyte by the addition of silica gel (finely divided SiO_2) to the electrolyte. This in effect absorbs the electrolyte, permitting the battery to be turned even upside down without reorientation of the electrolyte.

However, in immobilized electrolyte-sealed types, it is very desirable to avoid disruption of the gel, such as would occur with excessive gassing. Consequently such batteries must be recharged with perhaps a little more care than those having free electrolyte, although the usual rules still apply.

CHARGING NICKEL-CADMIUM BATTERIES   The ability to be recharged quickly, and here we are talking of within an hour or so from a full discharge, is also true for the sealed version. However, in the sealed version, the danger of disrupting the battery physically during or even approaching the overcharge period is greater than in the vented type. The sealed nickel-cadmium battery is made possible by the fact that the negative electrode is capable of recombining the oxygen liberated by the positive electrode during the overcharge period, while at the same time it is possible to arrange the relative capacities of the electrodes so that the liberation of hydrogen at the negative electrode during the overcharge period is entirely avoided.

In addition, local-action losses during storage are such as to avoid the production of hydrogen, still further enhancing the conditions that permit the sealing of such cells. Thus, while a sealed cell can also be recharged rapidly, greater precautions (such as exercise of greater electric control) must be taken to reduce the rate as the cell approaches gassing conditions. Generally speaking, a sealed nickel-cadmium cell will be operated at somewhere in the vicinity of 20 to 30 percent lower available capacity than a similarly sized vented cell. However, the convenience of no physical maintenance, and the cleanliness of external components (i.e., freedom from spray and spillage) have made the sealed nickel-cadmium cell very popular in small power devices where extended cycle life is particularly important.

Although nickel-cadmium batteries are indeed rugged, there is an electrical condition that can quickly lead to the demise of either vented or sealed types. This, known as "runaway," occurs when the battery is charged in such a manner as to create a great deal of internal heat. This can result from excessive charge rates, or from having the battery in an environment where the heat of the charge (particularly during the overcharge period) cannot be readily dissipated. As the temperature of the battery rises, its internal resistance drops very sharply. When the charge is being carried out from a "system-controlled" power source (i.e., essentially a constant-potential charge), a lowering of the internal resistance will result in an increase in the charging current. This increased current in turn results in a lowering of the efficiency of the charge acceptance, still further increasing the heat generation, still further raising the temperature of the cell. The cell can thus very quickly be brought to a condition where the heat becomes physically disruptive. This runaway condition is prevented by providing for better electric controls and adequate ventilation of the cells and the compartment in which it is housed.

An interesting characteristic of nickel-cadmium batteries is the so-called "memory effect" in which the battery seems to "remember" the manner in which it has been cycled. The memory effect shows up only in a highly repetitive cycling routine, and manifests itself as an apparently permanent loss of reserve capacity. Thus, if a 10-Ah battery is cycled on a routine procedure to a depth of only 4 Ah, with the intention to keep the remaining 6 Ah in reserve against emergencies or to enhance long cycle life, the actual available capacity of the battery falls to 4 Ah after a fairly short time on this routine. To avoid this, batteries in such service are periodically given conditioning cycles in which the depth of charge and discharge, and/or the rates at which these are performed, are varied from the routine. This treatment completely avoids the memory effect, and in fact can restore the capacity of a battery so affected, although the latter is more difficult and requires a greater number and variety of conditioning cycles.

CHARGING SILVER-ZINC BATTERIES   All the general rules for charging secondary batteries apply to silver-zinc types as well, only perhaps more so. Since such batteries are frequently used in higher-voltage systems, in fact up to 150 to 200 V, top cleanliness to avoid leakage currents and stray currents to ground becomes most important. The deleterious effects of excessive overcharge, and particularly high end-of-charge rates, are also magnified in this kind of battery. In overcharge resulting in gassing, the loss of electrolyte can be quite serious since (unlike lead-acid batteries, for example) there is very little free electrolyte (i.e., that which is not absorbed by the separators and electrodes). Any loss of free electrolyte causes drastic changes in the electrolyte level, and its replacement is difficult, especially if the distribution of the electrolyte among the various electrochemical components has been disturbed. Then too, since the zinc electrode is partially soluble in the electrolyte, a loss of the latter also actually means a loss of electrode material.

A still more serious effect of overcharging can occur where heat is also allowed to build up inside the cell. This is rather easy to do since silver-zinc batteries are almost always housed in plastic cell containers which are poor thermal conductors. A number of close-packed cells, charged and overcharged at excessive rates, will rapidly result in high temperatures, particularly at the center of the battery. If we remember that silver-zinc cells have organic separators and diaphragms under contact pressure with a strong oxidant such as the silver oxide of the positive electrode (see Fig. 17.15), there is little doubt that high temperatures result in the failure of these components.

In charging silver-zinc secondaries, therefore, sufficient time should be allowed between the end of discharge and the start of charge to allow the battery to cool down. A similar rest would also be very desirable between the end of charge and subsequent discharge. Since very little heat gets out around the sides of the battery, a good practice is to blow air gently over the terminals and intercell connectors. These, being of silver or silver-plated copper, can conduct a good deal of heat from inside the battery to where it can be conveniently dissipated.

While voltage-actuated control systems can be used, Fig. 17.16 indicates that the largest voltage rise as a silver-zinc cell is charged occurs when the battery is only about two-thirds charged, and that the rise at full charge is not very pronounced. Also, voltage is not a clear-cut means of establishing the advent of overcharge. Although not shown in Fig. 17.16, the battery voltage at the end of charge is not at all fixed, slowly increasing from about 1.88 V for a fresh battery to even as high as 2.1 V for one that is nearing end of life. For these reasons the most common practice is to recharge silver-zinc batteries at constant current for a fixed period of time, and to do so for such a time as to return 105 to 125 percent of the ampere-hours removed during the previous discharge. A slightly more sophisticated system takes advantage of the fact that silver-zinc batteries will accept a charge at very high rates at the start of a charge (after a deep discharge), and in fact will do so up to the first sharp voltage rise at about two-thirds full charge. This charging system simply uses a high constant-current charge period with a time control to switch to a lower constant-current level for completion of the charge. This two-step system has a great deal of merit, and many manufacturers now recommend such a charging procedure.

17.5 SPECIAL TYPES

A considerable amount of research and development is going on in the battery field today relating to new electrode combinations. Since practically any redox reaction can be conducted electrolytically in a cell from which electric energy can be withdrawn, a great many couples (electrode combinations) have been proposed. A few of these have found specific applications. Among the more important are:

1. Magnesium dry cells
2. Organic depolarized cells
3. Cuprous chloride–magnesium cells
4. Silver chloride–magnesium cells
5. Air-depolarized cells
6. Air-zinc cells

The general properties of these batteries are outlined in Table 17.13.

Each of these has at least one and sometimes even several special properties that has been a key factor in its growth. Magnesium dry cells are capable of storage under high-temperature high-humidity conditions; organic depolarized dry cells are potentially capable of very high capacities with light weights; cuprous chloride–magnesium types are capable of operation in very low pressure and low temperature ambients; silver chloride–magnesium batteries are quickly activated by dilute saline solutions; air-depolarized cells require practically no maintenance; air-zinc cells are capable of high energy yields per unit weight; and so on. It is important, therefore, to consider these properties in greater detail, and also in combination with other properties.

TABLE 17.13   Characteristics of Special Cell Systems

Type	Open-circuit voltage, V	Average discharge voltage, V	Capacity, Wh Per lb	Capacity, Wh Per in^3
Magnesium dry	1.95	1.60	45	3
Organic depolarizer (magnesium anode)	1.65	1.20	64	3
Cuprous chloride– magnesium	1.5	1.15	31	2
Silver chloride– magnesium	1.9	1.5	75	8
Air-depolarized	1.45	1.05	71	2
Air-zinc	1.45	1.2	52	6

MAGNESIUM BATTERIES   At the head of the list, in terms of commercial production, is the magnesium dry cell which we have already introduced under primary batteries. This particular battery very much resembles a conventional LeClanche dry cell in which a magnesium can has simply been substituted for the conventional zinc can. This accomplishes several things including raising the voltage of the cell, increasing its power capability, and (as a result of other design modifications) also its capacity.

However, it is the nature of the magnesium anode that it very quickly forms a very protective film when in contact with the battery electrolyte, and it is this film that gives the magnesium dry cell its outstanding property. This film protects the underlying metal and permits the battery to have an extremely long shelf life. It also helps it to withstand very adverse storage conditions such as high temperatures and high humidities. Not a great deal of reliable data are yet pertinent to the shelf life of magnesium dry cells, but there are the indications that they can easily be stored for as long as five to seven years with only modest losses (i.e., 10 to 20 percent for cells of current designs). In terms of adverse storage conditions, magnesium dry batteries can be stored continuously for 30 days (and many for 60 and even 90 days) in ambients of 160°F and 90 to 95 percent humidity with losses of only about 10 percent in capacity; and even this is largely due to loss of moisture instead of anode deterioration. This latter feature makes the magnesium dry cell extremely attractive for use in those areas of the world which are subject to hot, humid climates, and where distribution logistics require a fairly long time interval between manufacturer and user.

ORGANIC DEPOLARIZED BATTERIES   The organic depolarized cell is physically similar to the magnesium dry cell and, in fact, may even employ the same magnesium anode can, but in this case reducible organic material takes the place of the usual MnO_2 depolarizer used in magnesium or LeClanche dry cells. Organic depolarizers, in addition to being synthetically produced materials (and, therefore, of more uniform and continuing supply than conventional depolarizers), have the outstanding characteristic of being able to absorb a large number of electrons during the course of their reduction. This produces batteries of very high capacity for a given space and weight. However, because the organic depolarizers are completely nonconductive electrically, the battery must have a substantial grid structure (usually a carbon "chain" as in other dry cells), and this tends to reduce its space advantage. Even so, organic depolarizer cells show a fairly high internal resistance, but where the drain rates are low, as in radio receivers, it warrants, and is receiving, a great deal of attention.

CUPROUS CHLORIDE–MAGNESIUM BATTERIES   The cuprous chloride-magnesium (CuCl-Mg) cell is again a very specialized type which has found considerable use in radiosonde work. Since it is stored in a dry condition, the CuCl-Mg battery is classified as a reserve type, activated by immersion in water (either fresh or saline). For radiosonde work, the CuCl-Mg battery uses a separator which is highly absorbent so that the electrolyte remains within the cell after it is withdrawn from the activating liquid. In this condition the cell can be sent to very high altitudes where pressures and temperatures are extremely low. The special characteristics of the cuprous chloride–magnesium battery which permits this to be done is the fact that as soon as

Fig. 17.18 Cuprous chloride–magnesium radiosonde battery. (*Courtesy Clevite Corp.*)

the battery starts to discharge, copper is transferred from the cathode to the anode where it catalyzes the local action reactions. The resulting high local action rate produces a great deal of heat which keeps the battery at a suitable operating temperature, in spite of the fact that it may be operating in extremely low ambients.

Figure 17.18 shows a typical cuprous chloride–magnesium battery that powers a radiosonde for high-altitude weather observations. This unit supplies both A and B power, the latter being provided by 78 series-connected cells in three parallel stacks. The peculiar spatial arrangement provides the proper balance between the heat generated and the heat lost to the outside necessary to keep the battery at a proper operating temperature.

SILVER CHLORIDE–MAGNESIUM BATTERIES The silver chloride–magnesium (AgCl-Mg) battery is, in a general way, similar to the cuprous chloride–magnesium battery, and in fact has been used for radiosonde work where higher power requirements are needed than can be supplied by a CuCl-Mg battery. The principal application for silver chloride–magnesium batteries, however, is to power sonobuoys and torpedoes. The AgCl-Mg system has an outstanding characteristic in being extremely stable in storage without any special preparation or care, and units that have been stored for more than ten years have given full performance when activated and discharged.

In sonobuoy or torpedo batteries, the saline water of the ocean itself forms the electrolyte. Because this type (dilute salt) of electrolyte is a fairly poor conductor, the electrodes must be spaced very closely together, thus achieving the low internal resistance which makes possible the high power levels at which this type of battery is generally used. Here again local action at the magnesium anode produces a good deal of heat, so that means are usually provided for continuously flushing the battery with sea water. This not only keeps the internal water temperature down, but also keeps the electrodes free of sludge resulting from the discharge. The silver chloride–magnesium battery, the characteristics of which are given in Fig. 17.19, is capable of an extremely wide range of rates, performing well even at very high power levels.

AIR-DEPOLARIZED BATTERIES Since the air-depolarized cell actually uses atmospheric oxygen as its depolarizer, it can have very high capacities within a given weight. However, it does not have very high or even medium rate capabilities, and although it has been on the market for a number of years, it has found only limited appli-

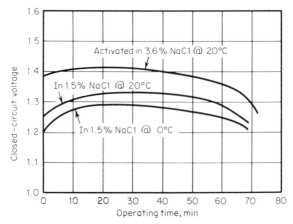

Fig. 17.19   Characteristics of (sonobuoy type) silver chloride–magnesium cell.

cation in low-power devices at inaccessible locations.   The air-depolarized cell uses a cast block of zinc as its anode, and a porous carbon electrode open to the air as its cathode.   The electrolyte is potassium hydroxide, and is usually present in large excess of operating requirements so as to practically eliminate watering and other maintenance procedures.

AIR-ZINC BATTERIES   Recently more sophisticated versions of the air-depolarized cell have been under intensive development, and such cells (renamed air-zinc cells to differentiate them from the earlier versions) are now made in a variety of sizes.   These are beginning to find one application in powering radio transceivers.   Still further applications are envisioned in the form of motive power (and even automobile) systems, since the cells receive their depolarizer from the air as needed.

Fig. 17.20   Air-zinc battery with replaceable anodes.   (*Courtesy Clevite Corp.*)

In the modern air-zinc cell, special low-cost cathode catalysts are employed to achieve respectably high rates of discharge. However, no really standardized sizes have yet been presented for general commercial use, since most batteries developed to date have been against highly specific military uses. In a broad sense, air-zinc cells can also be considered to be reserve batteries since they are packaged in dry form and are activated by addition of water. They carry solid potassium hydroxide (usually in the pores of the anode), which by combining with the water forms the electrolyte of the cell. After activation, the shelf life of the battery is fairly poor, and pretty nearly all its capacity will be dissipated in a matter of weeks. However, some considerable work is now being done to improve this situation, principally through amalgamation of the zinc with mercury, a process that greatly reduces the rate at which the zinc will spontaneously react with the electrolyte.

In addition to improvement of the primary version, work is also being directed toward recharging the zinc electrodes in air-zinc cells in a more or less conventional manner, thus producing a secondary battery. At the present time such batteries have not reached any really commercial performance, but an alternate development, the so-called mechanically rechargeable battery, has become available. In this type of battery, illustrated in Fig. 17.20, the zinc anodes are designed in such a way as to be easily removed from the battery upon completion of their discharge, and new ones are slipped into their place. During this process the electrolyte is also replaced, and the result is a battery virtually capable of brand-new performance. The positive, or air electrode, has been developed to a current life expectancy equivalent to about 50 anodes, making this a fairly economical battery to operate.

17.6 FUEL CELLS

The discussion of air-zinc cells, where one of the reactants (oxygen) enters the cell as needed, leads directly into a group of electrochemical generators usually referred to as fuel cells. As indicated earlier, many electrochemical reactions are reversible. From this the concept was evolved that if water can be broken down electrochemically into hydrogen and oxygen, it should be possible to recombine hydrogen and oxygen in an electrochemical cell (i.e., on suitable electrode surfaces) so as to yield electric energy. This indeed proved to be true, and cells were developed as early as the 1920s which accomplished the pertinent reactions at reasonably good efficiencies. In a great many respects, such cells had the characteristics of a prime mover–generator combination since the electrochemical materials (hydrogen and oxygen) were gases, and could be stored outside the cell to be metered into the cell as needed. Within limits it was found that the greater the rate at which the gases were fed, the greater was the electric power generated by the cell. Furthermore these gases acted very much as they might have as a fuel and an oxidant, and hence the popular term fuel cell.

In more recent times, a great variety of "fuels" have been found to be suitable, as have a fairly large number of oxidants, so that by now there is a variety of fuel-cell systems. Also, the concept has been enlarged to embrace other than gaseous fuels, and typically, in addition to hydrogen, such materials as hydrocarbons, hydrazine (N_2H_2), and molten metals (such as sodium, lithium, potassium, and other reactive easily fused metals) have been used. Oxidants such as air, chlorine, bromine, and others, have also provided interesting cell characteristics. As a matter of fact, the various fuel-cell combinations of oxidants and reductants that have been investigated have resulted in fuel-cell systems with highly specialized characteristics for very specific applications.

Basically, however, pretty nearly all fuel cells are variations of the fundamental design shown schematically in Fig. 17.21. This figure shows a hydrogen-oxygen cell and the auxiliary equipment needed to support it. This auxiliary equipment turns out to be extremely important for the extended operation of a fuel cell. In the cell illustrated, the overall reaction involves the oxidation of hydrogen to water. This water enters the electrolyte and, thereby, dilutes the latter. Some system must be provided, therefore, to remove the water in order to maintain the electrolyte concentration at appropriate levels. If the hydrogen or oxygen contains any impurities, these will enter the cell and ultimately build up to the point where they will seriously impair the performance of the cell, and the fuel and oxidant must therefore be brought to a high level of purity before entering the cell. These turn out to be very difficult engineering problems, and a good

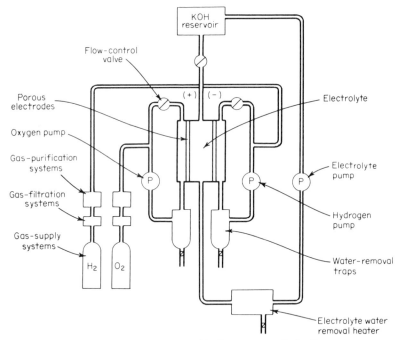

Fig. 17.21   Schematic diagram of hydrogen-oxygen fuel-cell system.

Fig. 17.22   Hydrogen-oxygen fuel-cell power system. (*Courtesy Union Carbide Corp.*)

TABLE 17.14   Characteristics of Typical Fuel-Cell Systems

Power capability, W	Application	Fuel	Oxidant	Unattended life, h	Power capability	
					Kw/lb	Kw/ft³
200	Radio communication	Hydrogen	Oxygen	1500	0.006	0.4
500	General power	Hydrazine	Air	1000	0.005	0.2
10 000	Space power	Alcohol	Air	2500	0.000	0.3
25 000	Marine	Hydrocarbon	Air	7500	0.01	1.3

deal of the progress that has been made has resulted from careful development of the auxiliary equipment. A photograph of an actual fuel-cell system is shown in Fig. 17.22, with general operating characteristics of a number of fuel-cell systems detailed in Table 17.14.

Fuel cells are, therefore, no longer simple electrochemical power generators in that the movement of fuel and oxidant into the cell and the products of the oxidation out of the cell have become a fairly complicated materials movement problem, and we find that modern fuel cells have a great deal of pumping, filtering, throttling, and control equipment associated with the electrochemical generator itself. In many instances, as for the H_2O_2 cell, it has been necessary to go to very high pressures and temperatures (to achieve high power outputs) further complicating the equipment. Others, such as the lithium-fluorine system, require extremely high temperatures in order to operate properly.

Many fuel cells are now available with fairly large power capabilities and have found use as the power source for manned space vehicles, as for example in the Apollo program. Fuel cells using alcohol have been produced in smaller sizes and have been promoted for military communication. To date, fuel-cell designs are developed against very specific purposes, and there are no generalized sizes, or standardization of components, so that tabularized information of this nature is not currently available. Nevertheless the wide variety of properties that can be achieved, coupled with long-life catalysts which greatly increase the rate of reaction possible at the electrode surfaces (and thereby increase the power capability of the cell), have accounted for the tremendous current interest being shown in these devices. In addition, a great many organizations, both government and private, are supporting major projects to increase still further the efficiency and the ease of the operation of these units, and it is very likely that standard forms will be evolved over the next few years.

At the moment, the indications are that these forms will be directed particularly toward motive power applications (small automobiles and the like), stationary generators (particularly for standby service), and radio communication. Fuel cells have indeed shown themselves to be capable of producing usable quantities of energy directly from conventional fuel sources and have a long life with relatively little maintenance. As a matter of fact, it is now entirely feasible to have completely unattended remote fuel-cell power plants, or to utilize waste fuels (such as hydrogen from caustic production) to produce "convenient" power. At the present time, however, the cost of fuel cells and their operation are high, but there is every expectation that as their use becomes more widespread and improvements continue, and as some degree of standardization is achieved, fuel cells will occupy an important position as electrochemical power sources.

Chapter **18**

Microprocessors and Microcomputers

Gary C. Meyer

Central Region Service Manager
Philips Test & Measuring Instruments, Inc.
Minnetonka, Minnesota

18.1 INTRODUCTION

Our world is exploding with technology that has made computer a household word. Semiconductor evolution has advanced from the discrete transistor to the integrated circuit (IC). The heart of modern computers is an IC known as a microprocessor. The microprocessor (sometimes referred to as a central processing unit, or CPU) is also being used in many different applications. This device is used in "on-board" automotive computers, sewing machines, process control instrumentation, burglar alarms, oscilloscopes, programmable temperature controllers, personal computers, video games, and numerous other electronic systems.

Microprocessors integrate registers, limited memory, an address bus, a data bus, a control bus, timing control circuits, and arithmetic and logic circuits all in a single chip. In addition to the microprocessor, a computer requires peripheral devices plus additional memory to store programs and data. Examples of peripheral devices include digital test instruments, printers, cathode-ray tube (CRT) displays, floppy disks, and cassette tape.

Several manufacturers produce microprocessors that offer a diversity of operational techniques and specifications. Leading manufacturers include Intel Corporation, Motorola, Rockwell, and Zilog. Documentation available from manufacturers includes some types of programming manual and specification sheets for each individual device.

This chapter discusses the Intel 8085, as a model, including hardware and software (the written instructions) to provide a basis for the study of other microprocessors. The chapter concludes with an examination of such popular CPUs as the Zilog Z80, Motorola 6800, and the Synertek 6502.

18.2 TRANSPORTING DIGITAL WORDS

A microprocessor is a relatively unintelligent device that requires precise instructions to function properly. The only language it understands is "machine language" consisting of 0s and 1s (binary numbers). Combinations of 0s and 1s comprise a digital word. Refer to Fig. 18.1. Each box represents a *bit*. And eight bits is referred to as a *byte*, and 4 bits is called a

Fig. 18.1   An example of an 8-bit (1-byte) word.

Fig. 18.2   Dividing the word in Fig. 18.1 into two 4-bit fields.

nibble. Digital words are simply a group of bits treated as a unit. A 16-bit word, for example, is 2 bytes wide.

The digital word in Fig. 18.1 represents a decimal number which may be determined as follows:

$$10011010_2 = (1 \times 2^7) + (0 \times 2^6) + (0 \times 2^5) + (1 \times 2^4) + (1 \times 2^3) + (0 \times 2^2)$$
$$+ (1 \times 2^1) + (0 \times 2^0)$$
$$= 128 + 16 + 8 + 2 = 154_{10}$$

where subscript 2 indicates a binary number and subscript 10 a decimal number. This binary number (or decimal number) can be converted to hexadecimal number 9AH.

Digital words used in microprocessor applications are divided into groups of four and referred to as a *4-bit field*. Each 4-bit field also can be represented by binary-coded decimal (BCD) notation. BCD should not be confused with binary numbers. For example, dividing the digital word of Fig. 18.1 into two 4-bit fields is illustrated in Fig. 18.2. Each field can be reduced to hexadecimal notation by finding the equivalent BCD in Table 18.1 and writing out the corresponding number in hexadecimal notation.

Consider the first field of our example. A BCD of 1001 appears opposite number 9 in Table 18.1. The second field of 1010 is opposite number A; hence, the hexadecimal is 9AH, or simply 9A if hexadecimal notation is understood.

Referring to Fig. 18.3, we see the voltmeter can be switched to any line, thus monitoring whether the line is high or low. As switches S1 through S4 are changed, decimal numbers 1 through 15 or hexadecimal numbers 0 through F can be represented. The lines are also identified by the least significant bit (LSB) and most significant bit (MSB). This identification does have value in relating the BCD notation to the hardware of Fig. 18.3. LSB refers to the far right bit of the BCD, while MSB refers to the far left bit.

example 18.1   For the circuit of Fig. 18.3, which switches must be closed and which open to produce the 4-bit field 1010? What are the LSB and MSB values? What hexadecimal number is represented?

solution   In general, the data field takes on the format

TABLE 18.1   Decimal, BCD, and Hexadecimal Equivalents

Decimal	BCD	Hexadecimal
0	0000	0
1	0001	1
2	0010	2
3	0011	3
4	0100	4
5	0101	5
6	0110	6
7	0111	7
8	1000	8
9	1001	9
10	1010	A
11	1011	B
12	1100	C
13	1101	D
14	1110	E
15	1111	F

Fig. 18.3   Relationship of BCD to hardware.

For 1010, D4 = 1, D3 = 0, D2 = 1, D1 = 0. Therefore, S4 = closed, S3 = open, S2 = closed, S1 = open. Also LSB = 0, MSB = 1, and hexadecimal number = A.

The connecting wires of Fig. 18.3 are called a *bus*, or *data bus*. In this example, it is called a four-wire bus. Since data is flowing in only one direction, the term *unidirectional*, or one-way, bus is applied. Obviously eight wires can be used to represent two 4-bit data fields. This is the case in many microprocessor systems.

If several devices are to be tied together in a system, the same bus may be used to communicate between devices. It is also convenient to show a bus as two parallel lines instead of drawing 4, 8, or 16 wires on a diagram.

Refer to Fig. 18.4. Device 1 is using a one-way, or unidirectional, bus to the main data bus. Devices 2 and 3 are shown with double arrows, indicating they can output, or receive, data from the main bus. This is called a *bidirectional*, or two-way, bus. Devices 2 and 3 can receive data from device 1 or communicate with each other.

Note that only two circuits can "talk" to each other on the bus at a given time; there-fore, precise timing and control functions are accomplished by adding two additional buses (control and address), as shown in Fig. 18.5.

Fig. 18.4   Example of unidirectional and bidirectional buses.

Each device is assigned a fixed number, specified as an address, just as houses on a city block can be identified by a house number and street. The timing and control circuit controls which devices are connected to the data bus first by giving the address, second by controlling the direction of data flow, and third by providing exact timing via the control bus.

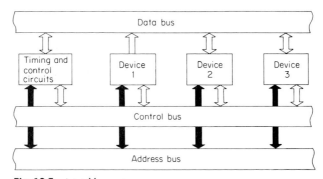

Fig. 18.5   A total bus system.

18.3 INTRODUCTION TO MICROPROCESSOR SYSTEMS

Before we examine a microprocessor system, a checkbook analogy is presented. An individual keeping a checkbook current is required to enter the amount of a check and subtract that amount from the previous balance. Referring to Fig. 18.6, assume the individual (operator) has never used a checkbook or a calculator. Very detailed instructions, therefore, must be given. The operator can only read instructions, perform instructions, read numbers, and write down numbers. All instructions must be carried out in the exact sequence specified to achieve correct results. Note that every instruction is identified by a fixed number. Specific line numbers (or check numbers) in the checkbook also identify a location on the page. Microprocessor systems work in much the same way; Table 18.2 summarizes the comparison.

A CPU requires a detailed set of instructions called a *program*. The program sequence is controlled by assigning a number called an *address* to each line. Programs for a CPU can be written in assembly language by using shorthand words called *mnemonics*. Programming is covered later in this chapter.

A block diagram of a typical microprocessor system is shown in Fig. 18.7. The read-only memory (ROM) and/or part of the random-access memory (RAM) contains the set of instructions to run the system in a given sequence. Each location in these memories is given a specific address with data at each address. The CPU initiates the program in the following manner:

1. The CPU outputs the starting memory address on the address bus.
2. The starting address located in memory is recognized and outputs its corresponding data onto the data bus.

Fig. 18.6 Checkbook-microprocessor analogy.

TABLE 18.2   Summary of Checkbook-Microprocessor Analogy

Real-life terms (see Fig. 18.6)	Microprocessor system terms
Set of instructions	Program or software—stored in memory
Instruction line numbers	Program addresses
Checkbook	Data stored in RAM memory
Check numbers	Address where data is stored
Read instruction	Instruction register
Understand instruction	Instruction decoder
Perform correct subsequence	Timing and control
Keep track of sequence	Program counter
Writing and reading	Write data to/read data from RAM
First number/second number/	Input-output ports
answer	Temporary registers
Calculator	Arithmetic logic unit (ALU)

3. The CPU inputs the data from the data bus, interprets the data, performs an operation, and places the next address on the address bus.

During this procedure, the control bus receives information from the CPU to enable the corresponding external block [that is, input-output (I/O), ROM, or RAM] at the appropriate time. Other control lines output signals to I/O or memory to enable a read or write function. *Read* means that data is being received from the memory by the CPU; *write* means that the CPU is sending data to memory. All operations must take place at a specified time determined automatically by the CPU. Since the processor performs operations very quickly (usually in a few microseconds), sometimes the control lines are used to delay the CPU until slower peripheral devices complete their operations.

An I/O device shown in Fig. 18.7 extends the system capability to input data from, or output data to, the outside world. This term refers to peripheral devices such as CRT displays, control circuits, tape recorders, and printers. Also I/O ports have addresses to identify their location.

MICROCOMPUTERS   The microprocessor system shown in Fig. 18.7 is the heart of a microcomputer. Some type of keyboard is connected to one I/O input port for programming, and a CRT readout is interfaced with another I/O output port. A printer also may be driven from another I/O output port to produce *hard copy*, a name given to the printed page. Microcomputer ROMs contain programs to read the keyboard and control the display.

Fig. 18.7 Block diagram of a basic microprocessor system.

Often system RAMs are used to store a program inputted from the keyboard, magnetic tape, or disks which are then used to run the microcomputer.

Computer specifications include the amount of storage available in terms of 1K, 4K, 16K, 32K, or 64K bytes of memory. These notations are rounded-off numbers specifying the approximate number of memory locations (1K byte = 1024).

Additional hardware outside the microprocessor system is often required to complete a microcomputer. For example, the CRT must have additional control circuitry and a character generator to produce a display. Keyboards also have associated circuitry to decode the switches during typing. If BASIC computer language is used to program the system, an additional ROM program is supplied to assemble or change the keyboard entries to digital words of 1s and 0s which a microprocessor understands.

In its simplest form, a microcomputer is a microprocessor system interfaced to external input-output peripheral devices for programming and readout.

18.4   THE 8085 MICROPROCESSOR

A detailed discussion of the block diagram (Fig. 18.8) will give a better understanding of how the microprocessor is able to control digital words (addresses, data, or control) on the three main bus lines. Obviously, measurements cannot be made inside the chip; however, internal data can be placed on the bus lines and examined with a logic analyzer or a development system. Logic analyzers are covered in Chap. 19.

REGISTERS   Examination of Fig. 18.8 shows several registers used for temporary storage of data during the execution of a program. The register array consists of six 8-bit registers designated by the letters B, C, D, E, H, and L and three 16-bit registers named stack pointer, program counter, and address latch incrementer-decrementer. All 8-bit registers in the register array can be combined in specific pairs to form 16-bit registers if desired. Registers B

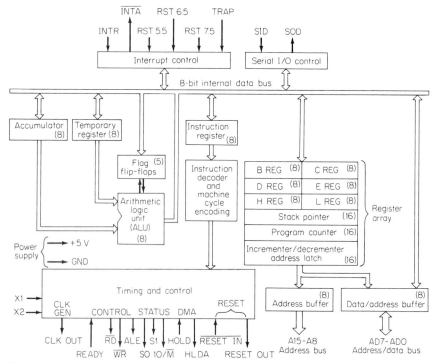

Fig. 18.8   Functional block diagram of the 8085 microprocessor. *(Courtesy Intel Corporation.)*

and C, D and E, and H and L are the specified pairs. The lettered registers are sometimes called *scratchpad* registers since they are used to store data for short intervals. The HL register pair can be used as a special memory pointer to specify an address in memory which becomes the source or destination of data.

The *accumulator* is an 8-bit register serving some special functions:

1. All arithmetic and logic operations performed in the ALU are based on data in the temporary register and accumulator. Note that the temporary register stores data captured from the internal data bus.

2. Answers from the ALU are returned to the accumulator.

3. Some special load, store, input, and output instructions use the accumulator (see Sec. 18.5).

4. Data bits in the accumulator can be rotated left or right by software or internally in some arithmetic operations. Rotation shifts data bits one position to left or right as specified.

Another 8-bit register is the *flag flipflops* consisting of 5 useful bits and 3 undefined bits. Each bit is either *set* (logic 1) or *reset* (logic 0), depending on the result of a given operation. Data bits in the flag register can be pictured as shown in Fig. 18.9. The five defined flag bits are used to indicate some operation or status of the accumulator register. Bits that are defined indicate the following:

D_7	D_6	D_5	D_4	D_3	D_2	D_1	D_0
S	Z	–	AC	–	P	–	CY

Fig. 18.9   Example of a flag register.

S   Sign flag D7 (MSB). The sign bit duplicates the accumulator sign bit and stores the result for future reference. This permits negative or positive numbers to be handled by the accumulator from -128 to $+127$. Changes in this bit occur after each execution of arithmetic or logic operations.

Z   Zero flag D6. If certain operations leave all 0s in the accumulator, the zero flag will be set (logic 1).

AC   Auxiliary Carry (AC) flag D4. The AC flag is set (logic 1) by an overflow out of bit 3 in the accumulator.

P   Parity flag D2. If the number of 1 bits in the accumulator is even (even parity), this flag is set. If odd parity occurs, the flag is reset (logic 0).

CY   Carry flag D0 (LSB). Arithmetic operations affect this flag. An overflow out of accumulator bit D7 (MSB) sets the flag.

Many of the arithmetic and logic operations affect one or more flags. A complete description of this interaction is given in the Intel 8080/8085 Assembly Language Programming Manual.[°]

Sometimes the accumulator and flag registers are considered together as a 16-bit register and given the name *program status word* (PSW). Under certain program conditions, it is desirable to save the accumulator and flag register contents on the stack. This is accomplished by storing the PSW (16 bits) or register pair in a portion of RAM called *stack*, discussed in the next section.

STACK POINTER   Consider Fig. 18.10 where information from the CPU register pairs can be saved on the stack. The stack pointer (SP) keeps track of the address in RAM where the last register information is saved. Each time data is "pushed" onto the stack, the SP register is decremented by 2 (subtracts 2). When data is "popped" off the stack back to the CPU, the SP is always incremented by 2. A programmer must initialize the SP in the program, giving the starting stack address.

Fig. 18.10   Operation of the stack pointer (SP).

PROGRAM COUNTER   A 16-bit register called the *program counter* (PC) is used to keep track of the program instruction sequence. It points to the memory location containing the next instruction to be executed.

° *Intel 8080/8085 Assembly Language Programming Manual*, Intel Corp., Santa Clara, Calif.

ARITHMETIC LOGIC UNIT   The ALU, as mentioned previously, inputs the accumulator and temporary register, performs an arithmetic or logic operation, and then returns the result to the accumulator.

ADDRESS LATCH INCREMENTER-DECREMENTER   The address latch outputs an address for a fixed period of time to the address buffers. This 16-bit address may be generated by the program counter, stack pointer, or any 16-bit register. The incrementer-decrementer section provides a way to increase or decrease the register contents by one count at a time.

DATA OR ADDRESS BUFFER   Addresses from the 16-bit address latch are separated into two bytes called the *high byte* and *low byte*. The high byte (A15–A8) is sent to the 8-bit buffer which is connected to the high-order address bus. The low byte (A7–A0) uses a separate 8-bit buffer which shares data output into the address or data bus at different points in time. This sharing is called *multiplexing* and is discussed later.

INSTRUCTION REGISTER AND DECODER   The first instruction fetch in a program places a coded instruction on the internal data bus which is transferred to the instruction register. An instruction decoder examines the code and transforms the bits to a language understood by the timing and control circuits. Depending on the instruction, the decoder outputs control the registers, ALU, data buffer, and address buffers.

TIMING AND CONTROL   Timing signals originate in an internal clock generator, and the clock frequency is determined by an external crystal. Several inputs and outputs accessible to the user provide a means of control or a readout of the CPU status. Figure 18.8 shows the following inputs and outputs:

CLK OUT	Clock out.
$\overline{\text{READY}}$	Ready input used for external synchronization of slower peripherals.
$\overline{\text{RD}}$	Read output present when reading from memory or I/O port, active low (as indicated by bar over letters).
$\overline{\text{WR}}$	Write output, active low, present when writing information to memory or I/O port.
ALE	Address latch enable, used to latch the low byte (A7–A0) signal.
S0, S1	Status 0 and status 1 lines indicate machine cycle operations taking place inside the CPU.
IO/$\overline{\text{M}}$	Input-output or memory status line; a high indicates I/O, a low indicates a memory read or write. IO/$\overline{\text{M}}$, S0, and S1 are often used together to indicate other status conditions. (See Fig. 18.13.)
HOLD	An external device is permitted to take over the bus lines; an input signal disconnects the CPU from bus lines.
HLDA	Hold acknowledge indicates the CPU is in hold state.
$\overline{\text{RESET IN}}$	Reset in (active low) sets the program counter to zero and resets interrupt enable and HLDA flipflops. Also it may affect internal register contents.
RESET OUT	Reset out indicates CPU is being reset; it may be used for system reset signal.

INTERRUPT CONTROL   The interrupt control block inputs provide a means of interrupting normal program flow. Five interrupt codes are given different priorities, as shown in Table 18.3.

TABLE 18.3   Priorities of Interrupt Codes

Priority	Symbol	Vectored address
1	TRAP	0024H
2	RST7.5	003CH
3	RST6.5	0034H
4	RST5.5	002CH
5	INTR	Specify

TRAP is a hardware-controlled interrupt. A signal applied to the TRAP interrupt causes the PC to jump to address 0024H. If this interrupt is used in a system, the interrupt service program routine must be written at this address.

Interrupts RST7.5, RST6.5, and RST5.5 are software-controlled. These interrupts are also vectored to specific addresses as listed and are maskable (can be turned on or off by software). One interrupt can also interrupt another, depending on priorities. This is called *nesting*.

The fifth input, INTR, is also enabled or disabled by software. Bytes 2 and 3 of this instruction allow the programmer to specify the interrupt service routine address.

SERIAL I/O CONTROL   Serial input data (SID) from the outside world is converted to parallel data in this block. The resultant parallel data is fed to the internal data bus. This provides a useful conversion for data transmitted on a two-wire communication system, such as a telephone line, or outputted from a cassette recorder to be used by the microprocessor. Serial output data (SOD) provides a means of feeding the parallel bus data into a two-wire system. Essentially, this block is a communications terminal.

BASIC SYSTEM TIMING   Each device in a microprocessor system has a unique timing diagram. Clock signals generated by the processor control the exact time that data is placed on, and removed from, the bus. Data on a bus must be valid before it is stored in memory or sent to an I/O port. Similarly, data from memory or an I/O port must be valid on the bus before it is read by the processor. The time required to fetch and execute an instruction is called an *instruction cycle*. Several terms associated with processor timing are shown in Fig. 18.11.

Instruction cycles are subdivided into machine cycles (M). Every machine cycle completes a read or write operation. Read and write operations provide the necessary communication between the processor and other devices to execute any instruction. One to five machine cycles may be required to complete an instruction cycle, depending on the instruction being executed. Figure 18.11 shows an instruction cycle consisting of three machine cycles. The number of states within a machine cycle varies with the operation being carried out.

example 18.2   If the clock frequency is 3 MHz, what is the total time required for the instruction cycle shown in Fig. 18.11?

solution

$$T_1 = \frac{1}{3 \times 10^6} = 0.33 \ \mu s$$

A machine cycle is M = number of states × time per state. Hence,

$$M1 = 4 \times 0.33 = 1.32$$
$$M2 = 3 \times 0.33 = 0.99$$
$$\underline{M3 = 3 \times 0.33 = 0.99}$$
$$\text{Instruction cycle} = 3.30 \ \mu s$$

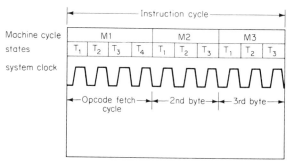

Fig. 18.11   Example of an instruction cycle.

A   B   C   D

Address (low byte)   Data (8 bits)

Time ⟶   Data not valid

Fig. 18.12   Illustration of multiplexing.

MULTIPLEXING   Multiplexing is a process of sending more than one set of data over the same bus at the same time. The multiplexed microprocessor uses a time sharing technique made possible by the short time required to execute instructions. As previously discussed, the address/data bus uses this method to place the address low byte on the bus first, followed by data; Fig. 8.12 illustrates the process. Note that for a short time the data is invalid. If the bus is sampled or strobed between A and B, the address (low byte) is seen. If the sampling is performed between C and D, data is extracted. By using the symbology of Fig. 18.12, a timing diagram for the 8085 can be constructed (Fig. 18.13).

The system clock generates timing pulses in which each cycle constitutes one timing state. Processor output A8–A15 is the high-order address which changes with every machine cycle. A0–A7 is the low-order address of the program counter placed on the AD bus. To capture 16 bits (2 bytes) of address information before the AD bus is multiplexed, the address latch receives a signal called the *address latch enable* (ALE). When the ALE signal produces a negative slope (the trailing edge of the pulse), bus lines A8–A15 and AD0–AD7 are sampled and latched. Now data can be placed on the AD bus.

$\overline{\text{RD}}$ is the read signal (active low) which is used to enable the memory when data from the latched address location is present on the bus. The microprocessor must have valid data to feed the instruction register; therefore, it waits for the third state, T3, before sampling the data. Once the data is in a register, the read pulse goes high.

$\overline{\text{WR}}$ is the write signal (active low) which is used to enable the memory to receive data from the microprocessor.

$\text{IO}/\overline{\text{M}}$ is a single line used to indicate an input-output operation when the signal level is high or a memory operation when the level is low. In conjunction with the $\overline{\text{RD}}$ and $\overline{\text{WR}}$ signals, the $\text{IO}/\overline{\text{M}}$ ensures proper selection of devices and operations.

The bottom line of Fig. 18.13, labeled STATUS, is the control bus consisting of S0, S1, $\text{IO}/\overline{\text{M}}$, $\overline{\text{RD}}$, $\overline{\text{WR}}$, and $\overline{\text{INTA}}$. Table 18.4 shows the logic levels present on the bus for seven possible types of machine cycles.

example 18.3   The following data bits are observed on the bus lines during M1. (*a*) What is the 16-bit address? (*b*) What type of operation is being executed? (*c*) What is the data expressed in hexadecimal numbers?

Fig. 18.13   Timing diagram for the 8085 microprocessor.

solution   (*a*) Bits A15–A8 represent the high-order byte of the program counter (PC$_H$ of Fig. 18.13). Bits A7–A0 are the low-order byte (PC$_L$). The complete address is determined as follows:

First, the bits are divided into 4-bit fields. Next, the 4-bit fields are converted to hexadecimal numbers; hence, the address is 4519H.

(*b*) Table 18.4 is used to find the operation. Comparison of the data bits shown above with this chart indicates an OPCODE FETCH.

(*c*) The solution procedure is the same as in (*a*):

```
D7                 D0
┌─────────┬─────────┐
│ 0 0 1 1 │ 1 1 1 0 │    or    3EH
└─────────┴─────────┘
    3         E
```

TYPES OF ADDRESSING   Addressing is the process of pointing to the location (address) of the instruction to be executed or data to be processed. Four basic types of addressing are employed in microprocessors:

1. Immediate addressing
2. Direct addressing
3. Indirect addressing
4. Indexed addressing

TABLE 18.4   Machine Cycle Chart for the 8085 Chip

Machine cycle operation		Status			Control		
		IO/$\overline{\text{M}}$	S1	S0	$\overline{\text{RD}}$	$\overline{\text{WR}}$	$\overline{\text{INTA}}$
Op code fetch	(OF)	0°	1†	1	0	1	1
Memory read	(MR)	0	1	0	0	1	1
Memory write	(MW)	0	0	1	1	0	1
I/O read	(IOR)	1	1	0	0	1	1
I/O write	(IOW)	1	0	1	1	0	1
Interrupt acknowledge	(INA)	1	1	1	1	1	0
Bus idle	(BI): DAD	0	1	0	1	1	1
	INA(RST/TRAP)	1	1	1	1	1	1
	HALT	TS‡	0	0	TS	TS	1

° 0 = logic 0
† 1 = logic 1
‡ TS = high impedance

The 8085 uses the first three, which are discussed here. Indexed addressing is covered in Sec. 18.8.

Immediate addressing means that the data word, called an *operand*, is located in the memory location immediately following the instruction code. As an example of this type of addressing, assume that the number FF is to be placed in the accumulator during a program. The instruction code (also called operational code, or *op code* for short) for moving data immediately to the accumulator is 3E hexadecimal (00111110). Instruction 3E placed at address 2002 must be followed by the operand or data of FF (11111111) at the next address, 2003. (See Fig. 18.14.) Addressing of this nature is often used to set the contents of a register.

Hex address	Memory	
2000H	8 bits	
2001		
2002	Instruction code	⎤
2003	Operand	⎬ Immediate
2004		⎦
2005	Instruction code	⎤
2006	Low address byte	⎬ Direct
2007	High address byte	⎦

Fig. 18.14   Immediate and direct addressing.

Direct addressing uses a different technique. Immediately following the instruction code in memory is the address of the operand. An example of this technique is the jump instruction. Address 2005 in Fig. 18.14 could be the code for an unconditional jump instruction, or C3 hexadecimal (11000011). Then address 2006 must contain the low byte and 2007 the high byte of the address specified by the programmer. This sequence would cause the program counter to jump to the new specified address.

Indirect addressing utilizes the processor's 16-bit registers to point out an address in memory. First, the desired address is stored in the CPU register. Second, an instruction code, including a reference to the proper register, is stored in memory. If an operand is needed several times in a program, this method saves program length and memory space. For example, consider Fig. 18.15. Moving register B contents to a memory location (M) in an 8085 system can be accomplished as follows:

1. Load the 16-bit register pair HL with the address 4000. The HL register pair is used as a memory pointer.

2. When the processor comes to 300A in the program, it senses the op code for a register B to memory move (70).

3. The processor interrogates the HL register for an address (4000) and duplicates B register contents into that memory location.

example 18.4   Refer to Fig. 18.15. If the B register contents are to be placed at address 4001 (by using indirect addressing), what address location and what data must be changed?

solution   Because 4001 instead of 4000 is the new address, the low-order address byte located at memory address 3001 must be changed from 00 to 01. No other change is required.

Fig. 18.15   Example of indirect addressing.

18.5   THE 8085 INSTRUCTION SET

Every microprocessor has an instruction set which defines all the operations possible. Instruction sets are copyrighted by the manufacturer; therefore, similar instructions for various processors will have different names. A thorough understanding of the instruction set is required to write a program.

The instruction set for the 8085 is divided into four groups:

1. *Data transfer group.* Encompasses all the register and memory moves, plus load and store operations.

2. *Arithmetic and logic group.* Addition, subtraction, incrementing, decrementing, rotate, and all logic operations are grouped under this heading. A few specials are also included.

3. *Branch control group.* Includes jumps, calls, returns, and restarts.

4. *I/O and machine control.* Stack operations, I/O, interrupt controls, no operation, and halt operations are covered.

This section lists instructions for the four groups, defines the symbology, and provides some examples to illustrate the required sequence of operation.

DATA TRANSFER GROUP   This group contains move and transfer instructions used to transfer data internally between registers or externally between registers and memory:

Mnemonic	Meaning
MOV	Move
MVI	Move immediate
LDA	Load accumulator directly from memory
STA	Load accumulator directly in memory
LHLD	Load H and L registers directly from memory
SHLD	Store H and L registers directly in memory
LXI	Load register pair with immediate data
LDAX	Load accumulator from address in register pair
STAX	Store accumulator in address in register pair
XCHG	Exchange H and L with D and E
XTHL	Exchange top of stack with HL

The addition of an X indicates usage of an "extended" 16-bit register, also called a register pair (RP).

In addition to these mnemonics, each transfer instruction must specify a source and a destination register for the data to be moved. Symbolically, the destination register is named first, followed by the source register. For example, a register-to-register move would be shown as MOV (destination),(source). If data in accumulator A is to be moved into the B register, this becomes MOV B,A. This operation does not alter the data in the A register, but merely copies the 8 bits of data into the B register. If the letters are reversed (MOV A,B), the direction of transfer is reversed. Then data is moved from register B to register A.

Data also can be moved to and from memory M with a MOV instruction. If the data byte in register D is to be moved to memory, the instruction is MOV M,D. The destination is memory, and the location in memory is determined by the address programmed into the HL register.

The basic MOV instruction is called a *1-byte* instruction because it occupies 1 byte of memory space. Examples of 1-, 2-, and 3-byte instructions are shown in Fig. 18.16. From Fig. 18.16b, the move immediate instruction (MVI) must be followed by the byte of data to be placed in the specified register. For example, MVIA in the first byte and FF in the second byte would put all 1s in the 8 accumulator bits (FF = 11111111).

Figure 18.16c shows a 3-byte instruction LXI, or load an extended (16-bit) register with

←One byte→	←One byte→	←One byte→
MOV D,E	MVI	LXI
	Data	Low byte
		High byte
(a)	(b)	(c)

Fig. 18.16   Examples of (a) 1-, (b) 2-, and (c) 3-byte instructions.

TABLE 18.5 The 8080A/8085A Instruction Set (Courtesy Intel Corporation)

Data transfer group

Move

MOV:
A,A	7F		B,A	47
A,B	78		B,B	40
A,C	79		B,C	41
A,D	7A		B,D	42
A,E	7B		B,E	43
A,H	7C		B,H	44
A,L	7D		B,L	45
A,M	7E		B,M	46

C,A	4F
C,B	48
C,C	49
C,D	4A
C,E	4B
C,H	4C
C,L	4D
C,M	4E

Move (cont)

MOV:
E,A	5F	H,A	67	L,A	6F
E,B	58	H,B	60	L,B	68
E,C	59	H,C	61	L,C	69
E,D	5A	H,D	62	L,D	6A
E,E	5B	H,E	63	L,E	6B
E,H	5C	H,H	64	L,H	6C
E,L	5D	H,L	65	L,L	6D
E,M	5E	H,M	66	L,M	6E

Move Immediate

MVI:
A, byte	3E
B, byte	06
C, byte	0E
D, byte	16
E, byte	1E
H, byte	26
L, byte	2E
M, byte	36

Load Immediate

LXI:
B, dble	01
D, dble	11
H, dble	21
SP, dble	31

Load/Store

LDAX B	0A
LDAX D	1A
LHLD adr	2A
LDA adr	3A
STAX B	02
STAX D	12
SHLD adr	22
STA adr	32

Arithmetic and logic group

Add*

ADD:
A	87		A	8F
B	80		B	88
C	81		C	89
D	82	ADC:	D	8A
E	83		E	8B
H	84		H	8C
L	85		L	8D
M	86		M	8E

Subtract*

SUB:
A	97
B	90
C	91
D	92
E	93
H	94
L	95

Increment†

INR:
A	3C		B	03
B	04		D	13
C	0C	INX:	H	23
D	14		SP	33
E	1C			
H	24			
L	2C			
M	34			

Decrement‡

DCR:
A	3D		B	0B
B	05		D	1B
C	0D	DCX:	H	2B
D	15			
E	1D			
H	25			
L	2D			
M	35			

Logical*

ANA:
A	A7	A	AF	A	B7
B	A0	B	A8	B	B0
C	A1	C	A9	C	B1
D	A2	D	AA	D	B2
E	A3	E	AB	E	B3
H	A4	H	AC	H	B4
L	A5	L	AD	L	B5
M	A6	M	AE	M	B6

(XRA middle column, ORA right column)

Branch control group

Jump
JMP adr	C3
JNZ adr	C3
JNZ adr	C2
JZ adr	CA
JNC adr	D2
JC adr	DA
JPO adr	E2
JPE adr	EA
JP adr	F2
JM adr	FA
PCHL	E9

Call
CALL adr	CD
CNZ adr	C4
CZ adr	CC
CNC adr	D4
CC adr	DC
CPO adr	E4
CPE adr	EC
CP adr	F4
CM adr	FC

Return
RET	C9
RNZ	C0
RZ	C8

I/O and machine control

Stack Ops

PUSH:
B	C5
D	D5
H	E5
PSW	F5

POP:
B	C1
D	D1
H	E1
PSW*	F1

XTHL	E5
SPHL	F9

Input/Output
OUT byte	D3
IN byte	D8

Control
DI	F3
EI	FB
NOP	00
HLT	76

New Instructions (8085 Only)
RIM	20

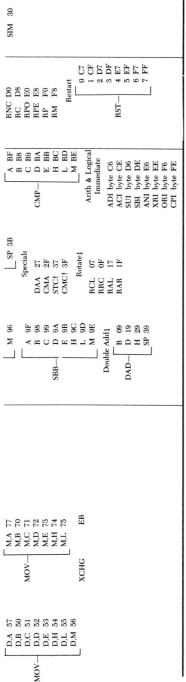

SIM 30

MOV
- D,A 57
- D,B 50
- D,C 51
- D,D 52
- D,E 53
- D,H 54
- D,L 55
- D,M 56

MOV
- M,A 77
- M,B 70
- M,C 71
- M,D 72
- M,E 73
- M,H 74
- M,L 75

XCHG EB

SBB
- M 96
- A 9F
- B 98
- C 99
- D 9A
- E 9B
- H 9C
- L 9D
- M 9E

Double Add‡
DAD
- B 09
- D 19
- H 29
- SP 39

SP 3B

Specials
- DAA 27
- CMA 2F
- STC† 37
- CMC† 3F

Rotate‡
- RLC 07
- RRC 0F
- RAL 17
- RAR 1F

CMP
- A BF
- B B8
- C B9
- D BA
- E BB
- H BC
- L BD
- M BE

Arith & Logical Immediate
- ADI byte C6
- ACI byte CE
- SUI byte D6
- SBI byte DE
- ANI byte E6
- XRI byte EE
- ORI byte F6
- CPI byte FE

RNC D0
RC D8
RPO E0
RPE E8
RP F0
RM F8

Restart
RST
- 0 C7
- 1 CF
- 2 D7
- 3 DF
- 4 E7
- 5 EF
- 6 F7
- 7 FF

byte = constant, or logic/arithmetic expression that evaluates to an 8-bit data quantity (second byte of 2-byte instructions).
dble = constant, or logic/arithmetic expression that evaluates to a 16-bit data quantity (second and third bytes of 3-byte instructions).
adr = 16-bit address (second and third bytes of 3-byte instructions).
° = all flags (C, Z, S, P, AC) affected.
† = all flags except CARRY affected (exception INX and DCX affect no flags).
‡ = only CARRY affected.

immediate 2 bytes of data. An example is LXIH, which loads the register pair HL with data (or an address) specified in the next 2 bytes. The low byte for register L must appear in the byte immediately after the LXI instruction, followed by the high byte which is loaded into register H.

Space does not permit complete coverage of the entire instruction set. The manufacturer's programming manual for each microprocessor must be consulted. A few examples are covered for each group of instructions. Table 18.5 summarizes the entire instruction set and op codes needed to write and understand programs for 8085 chip.

example 18.5 Given the following information in system memory, what bits reside in the D register? Consider address 8000 as the first step of a program.

Address	Data (hexadecimal)	Status
8000H	3E	Op Code
8001	1F	Data
8002	57	Op code

solution Here 3E appears at address 8000H. The first program step is always an op code fetch. From Table 18.5, 3E is an op code for MVIA (move immediate data to accumulator). Figure 18.16b specifies an MVI instruction as 2 bytes; therefore, at 8001 (second byte), the digital number 1F is moved into the accumulator. At 8002, op code 57 from Table 18.5 is MOV D,A. This instruction copies the number 1F from the accumulator into the D register. Therefore,

$$1F \text{ (hexadecimal)} = \underbrace{00011111}_{\text{bits in D register}}$$

example 18.6 Given the 3-byte instruction LXID, what hexadecimal number must be placed in the 3 memory bytes to store 55 in the D register and 44 in the E register?

solution LXID from the description under data transfer group is "load register pair with immediate data." This loads the extended register pair D,E with data specified in the next 2 bytes.

Memory

First byte	11	Op code for LXID from Table 18.5
Second byte	44	Low-order byte in E
Third byte	55	High-order byte in D

ARITHMETIC AND LOGIC GROUP This group can be subdivided into two subgroups, A and B.

A. Arithmetic Add, subtract, increment, or decrement memory or register data.

ADD   Add to accumulator
ADI   Add immediate data to accumulator
ADC   Add to accumulator, using carry flag
ACI   Add immediate data to accumulator, using carry flag
SUB   Subtract from accumulator
SUI   Subtract immediate data from accumulator
SBB   Subtract from accumulator, using borrow (carry) flag
SBI   Subtract immediate from accumulator, using borrow
INR   Increment specified byte by 1
DCR   Decrement specified byte by 1
INX   Increment register pair by 1
DCX   Decrement register pair by 1
DAD   Double register add: Add contents of register pair to H,L register pair

example 18.7 The number 1A is stored in the accumulator. Using an arithmetic instruction, make the accumulator contents equal 00H.

solution   One solution would use SUB A, which subtracts the contents of register A from register A. The result is placed in the accumulator: 1A − 1A = 00H. If the difference were not zero, the accumulator bits would be presented in a 2's-complement format.

Another solution is SUI, which subtracts the next immediate byte from the accumulator. Instruction SUI works only with the accumulator.

Byte 1	D6	op code for SUI
Byte 2	1A	Data to be subtracted

This solution uses 2 bytes, whereas the first solution requires only 1 byte. Both solutions accomplish the objective; however, the first saves time and memory space.

B. Logic   Perform logical, compare, and rotate operations on the data.

ANA   Logic AND with accumulator
ANI   Logic AND with accumulator, using immediate data
ORA   Logic OR with accumulator
ORI   Logic OR with accumulator, using immediate data
XRA   Exclusive logic OR with accumulator
XRI   Exclusive OR, using immediate data
CMP   Compare 8-bit value with accumulator
CPI   Compare, using immediate data
RLC   Rotate accumulator left
RRC   Rotate accumulator right
RAL   Rotate left through carry
RAR   Rotate right through carry
CMA   Complement accumulator
CMC   Complement carry flag
STC   Set carry flag

example 18.8   The ANA instruction performs a logic AND operation with a specified register. Given the ANAB instruction with the following bits in the accumulator and register B, list the resulting bits in the accumulator.

Accumulator =   D7   D0
1 0 1 0 1 0 1 0   (AAH)

Register B =   1 1 0 0 1 1 0 0   (CCH)

solution   Refer to Chap. 14. An AND gate operation yields a 1 output if both inputs are 1 and a 0 output if the inputs are different. All the bits in registers A and B are ANDed one at a time:

$$A = 10101010 \quad (AAH)$$
$$B = \underline{11001100} \quad (CCH)$$
$$\text{Result in A} = 10001000 \quad (88H)$$

example 18.9   The CMP instruction compares the accumulator bits with a specified register and sets flags to indicate the result, as shown in Table 18.6. Given a CMPC instruction with CFH in the accumulator and CDH in register C, will the zero (Z) and carry (CY) flags be set?

TABLE 18.6   Results of Compare Operation

	Flag	Condition	Meaning
If sign flag is 1 (positive number)	Zero	Set (1)	Equal values in A and register
	Zero	Not set (0)	Unequal values in A and register
	Carry	Not set (0)	A ≥ register value
	Carry	Set (1)	A < register value
If sign flag is 0 (negative number) or one value is complemented	Zero	Set (1)	Equal values in A and register
	Zero	Not set (0)	Unequal values in A and register
	Carry	Not set (0)	A < register value
	Carry	Set (1)	A ≥ register value

solution   Since A > register C (CF > CD), from Table 18.6 Z = 0 (not set) and CY = 0 (not set).

example 18.10   Assume the accumulator contains the bit pattern 10001000. If you use an RLC instruction, how many times must this instruction be executed to produce 22H in the accumulator? Note: RLC rotates A contents 1 bit to the left with the MSB transferring to the LSB.

solution   The transfer sequence after each execution is

Initial value:	10001000	(88H)
First execution:	00010001	(11H)
Second execution:	00100010	(22H)

Therefore, two executions are required.

Examination of the arithmetic and logic group instructions shows the accumulator to be central in the execution of these instructions. Remember the ALU in the CPU always uses data supplied from the accumulator. Compare instructions compare an 8-bit value with the accumulator. Rotate instructions shift the accumulator contents 1 bit position to the left or right. Instructions included in this group, not limited to the accumulator, are the decrement and increment instructions which can be applied to other registers as well. Most of these instructions affect different combinations of the five flag bits often used as indicators of the operational results. The programming manual should be consulted for flag status information stemming from many possible combinations.

BRANCH CONTROL GROUP   Changes in the normal program sequence flow are initiated with branch control instructions. Two categories, unconditional and conditional, make up this group.

Unconditional   Decision to jump is not based on flag conditions.

JMP   Jump to specified address
CALL   Call subroutine at given address
RET   Return to previous program

Conditional   Decision to jump is based on flag conditions. Terminology is as follows with the effect on condition flags shown in parentheses:

NZ   Not zero (Z = 0)
Z   Zero (Z = 1)
NC   No carry (C = 0)
C   Carry (C = 1)
PO   Parity odd (P = 1)
PE   Parity even (P = 1)
P   Plus (S = 0)
M   Minus (S = 1)

Using this terminology, we have

Jump	Call	Return	
JC	CC	RC	(Carry)
JNC	CNC	RNC	(No carry)
JZ	JNC	RZ	(Zero)
JNZ	CNZ	RNZ	(Not zero)
JP	CP	RP	(Plus)
JM	CM	RM	(Minus)
JPE	CPE	RPE	(Parity even)
JPO	CPO	RPO	(Parity odd)

Two additional branching instructions change the program counter:

PCHL   Moves H and L to PC.
RST   Special restart instruction used with interrupts. Restart is a special-purpose call.

Refer to Sec. 18.6 for an example of an unconditional jump. Examples of CALL and RET instructions are given under the section "Subroutines" (see Fig. 18.20).

STACK, I/O, AND MACHINE CONTROL GROUP   The fourth group of instructions includes those for stack and I/O operations, and for machine controls.

Stack Affects stack and stack pointer.
  PUSH Push 2 bytes of data onto the stack
  POP Pop 2 bytes of data off the stack
  XTHL Exchange top of stack with H and L
  SPHL Move contents of H and L to stack pointer
I/O
  IN Initiate input operation (from data bus to the accumulator)
  OUT Initiate output operation (from accumulator to data bus)
Machine controls
  EI Enable interrupt system
  DI Disable interrupt system
  HLT Halt
  NOP No operation

18.6 SOFTWARE—WRITING A PROGRAM

Writing programs for microprocessor systems is an art learned by doing rather than reading. A few tools and general techniques are discussed in this section; however, the same end result may be accomplished by using different instructions. If memory space is at a premium or time is essential, the shortest possible program must be developed.

STEPS IN WRITING A PROGRAM

Step 1: Description Write down in plain language what the program is to accomplish. A detailed description at the beginning will greatly facilitate the mechanics involved in writing the program. This also assists others in understanding the program, in addition to serving as a future reference to the programmer.

Step 2: Flowchart A flowchart is a pictorial diagram of the overall program flow. Figure 18.17 shows symbols often used to draw charts. Consider a program designed to input two numbers from a digital voltmeter (DVM), subtract them, output the result to a controller, and repeat the process (Fig. 18.18). The circle indicates the starting point. The parallelogram block inputs the first voltage from the DVM to be stored (a process depicted by the

Fig. 18.17 Flowchart symbols.

Fig. 18.18 Example of a flowchart.

rectangle) in the microprocessor. Next, V2 is inputted to the accumulator, V1 is subtracted from V2 in the ALU unit, and the difference is sent to an output port for use by a controller. By jumping back to start, the program runs in a continuous loop.

Step 3: Examine system addresses First, the memory map of the processor system must be examined to determine the addresses available for programming and data storage.

Address   Memory   I/O

Fig. 18.19 A memory-mapping example.

A typical memory map example is given in Fig. 18.19. If we assume a keyboard type of system is being used, addresses 0000 through 10FF are specified as the monitor program. This section in ROM would be used for items such as reading the keyboard, system reset, and control of the display. Addresses 2000 through 20FF, specified as RAM, provide space to store the program. In a system serving a fixed function such as the program of Fig. 18.18, usually the program is stored in ROM which is nonvolatile; however, additional RAM is sometimes provided for storage of variable data. This example also illustrates memory-mapped I/O, which means the I/O ports are identified by port addresses. IN and OUT instructions are followed by the port address.

Step 4: Instruction set With the flowchart as a guide, instructions from the instruction set can be selected to accomplish the program objectives. An assembly language program can be written now by using the mnemonics for the microprocessor. The mnemonics are written opposite each address along with a remark concerning the purpose of the instruction.

Step 5: Coding Once the mnemonics are selected, the corresponding op codes from the manufacturer's programming sheet are written by the mnemonic.

Table 18.7 illustrates how steps 1 to 5 are put into practice for the program of Fig. 18.18.

Each IN (or OUT) instruction is followed by the port address at program addresses 2001, 2004, and 2007. At address 2001, V1 is in the accumulator. Since the next IN instruction (2003) will input V2 into the accumulator, V1 must be stored in another location. This is accomplished by MOV B,A, which moves V1 from the accumulator to register B (any other register could be used). After V2 is in the accumulator, the SUBB instruction subtracts the B register contents (V1) from the new accumulator value (V2), writes the result back into

TABLE 18.7 Program for Fig. 18.18

Step 1

Program to input V1 from port 20 and V2 from port 21; subtract two voltages; output difference to port 80; loop.

Step 2

Step 2 is the flowchart in Fig. 18.18.

Step 3, Address	Step 4, Mnemonic	Step 5, Op code	Remarks
2000H	IN	DB	Input V1 from port
2001	20	20	Port 20
2002	MOV B,A	78	Store V1 in B register
2003	IN	DB	Input V2 from port
2004	21	21	Port 21
2005	SUB B	90	V2 − V1
2006	OUT	D3	Output difference to
2007	80	80	Port 80
2008	JMP	C3	Jump back to start
2009	00	00	Adr low byte
200A	20	20	Adr high byte

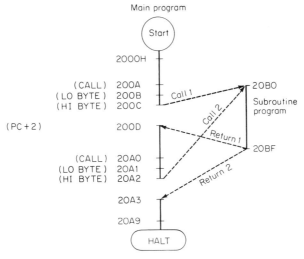

Fig. 18.20 Execution of a subroutine.

the accumulator, and outputs the difference to port 80. JMP at address 2008 unconditionally jumps the program back to the starting address 2000 (low byte 00, high byte 20), causing the program to run in a continuous loop.

SUBROUTINES Subroutines are used in programs requiring a certain operation to be performed several times during the main program. A subroutine is assigned addresses outside the main line program and called when needed.

Figure 18.20 shows a main program starting at address 2000H. At 200A a CALL instruction appears followed by the subroutine address B0 (low byte first) and 20 (high byte second). This 3-byte instruction saves the next program counter address following the CALL plus 2 on the stack (200D) as a return address. Programming RET at the end of the subroutine forces (pops) the saved address off the stack, thus jumping the PC back to the main program. A return instruction does not require a specified address because the stack takes care of this task automatically. The same procedure is used at 20A0 with the same subroutine. Advantages of using subroutines are that they save memory space, simplify writing programs, and encourage a modular program structure.

It is also possible to CALL a subroutine within another subroutine, as illustrated in Fig. 18.21. When subroutines are called within subroutines, it is called *nesting*. Return addresses are saved on the stack in the correct order to keep the correct program sequence. The stack is sometimes referred to as a LIFO (last in, first out); hence, an address at point A on the diagram is placed on the stack first, and B is pushed onto the stack last. After completion of subroutine 2, address B is popped off the stack first, allowing subroutine 1 to resume. At the end of subroutine 1, address A is popped off the stack, and there is return to the main program.

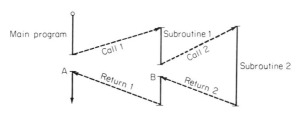

Fig. 18.21 Nesting of subroutines.

18.7 SUPPORT DEVICES

Development of a microcomputer requires selection of support devices for the microprocessor. The number of devices used in a system will depend on the functions that the system must provide. A few devices are mentioned in this section.

DECODERS When several memory or I/O chips are used in a system, a means of activating each device must be provided. Decoders act as a switch to enable one of several devices in a system. Figure 18.22 shows the logic symbol for a "1 out of 8" binary decoder. Signals on A0–A2 (address inputs) and E1–E3 (enable inputs) will produce an output on only one of the decoded output lines (O0–O7). The outputs are used as chip enable (CE) or chip select (CS) signals to activate other devices. Manufacturers publish truth tables showing input versus output data.

I/O PORTS Refer to Fig. 18.23. This device is enabled by selecting a digital control word consisting of STB (strobe), DS1 and DS2 (device select), and MD (mode). Terminals STB, DS1, DS2, and MD comprise the control logic used to set up device selection, data latching, output buffers, and/or an interrupt request mode. Normally an I/O port is initialized in the beginning of a program to configure the circuitry as an input or output. Then the initialized device will latch into this configuration and will not have to be reprogrammed unless the same port is used for the opposite function later in the routine. Manufacturers supply truth tables in their specifications for setup configurations.

PROGRAMMABLE PERIPHERAL INTERFACE (PPI) Support devices like the PPI offer a system great flexibility in a small package. Software controls I/O ports in various combinations. A block diagram of the Intel 8255A PPI is shown in Fig. 18.24. Processors are interfaced to peripheral devices via the three-state bidirectional data bus buffer. Transmission and reception of data are controlled by the read-write control logic block programmed by the CPU. As a result, the three 8-bit ports (A, B, C) can be selected as inputs or outputs for the microprocessor. Nomenclature for Fig. 18.24 is shown in Table 18.8.

OTHER SUPPORT DEVICES Many other support chips are used in different systems. The following list summarizes the characteristics of some chips.

1. Programmable DMA (direct memory access) controller. Certain peripheral devices operate at higher speeds than the CPU or may require use of the bus system independent of the CPU. A DMA controller provides signals to tristate the CPU from the bus via a HOLD function, generate timing and control signals, direct data transfer between peripherals, and return control to the CPU at completion of DMA activity.

2. Programmable interrupt controller. This device permits several vectored interrupts to be programmed if the microprocessor does not have the necessary capability.

3. Programmable CRT controller. Systems requiring a CRT display use this device to interface and control CRT circuitry via the CPU.

4. Programmable keyboard interface. A keyboard interface will scan a keyboard used to program a processor system and send the result to the CPU.

Fig. 18.22 A "1 out of 8" binary dec.
(Courtesy Intel Corporation.)

Fig. 18.23 Programmable 8-bit I/O port.
(Courtesy Intel Corporation.)

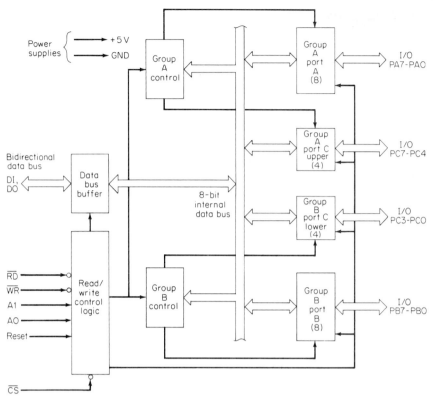

Fig. 18.24 Block diagram of an 8255A programmable peripheral interface. (*Courtesy Intel Corporation.*)

The manufacturer's data must be consulted to understand the individual support devices, to troubleshoot systems, or to design interfaces.

18.8 MICROPROCESSOR SURVEY

Our discussion now shifts from the Intel 8085 to a brief description of other types. Space does not permit exhaustive coverage of every processor; however, an introduction to the programming structure, including registers and instructions, is presented. Before delving into a survey, we discuss addressing modes used by various CPUs.

Three addressing modes were covered earlier (immediate, direct, indirect). Three additional categories of indirect addressing—extended, implied, and accumulator—are described now along with indexed addressing and its special case, called *relative addressing*.

EXTENDED ADDRESSING   Extended addressing is a type of indirect addressing that uses a 16-bit register and a 3-byte instruction. Byte 1 contains the instruction, byte 2 contains 1

TABLE 18.8   Nomenclature for Fig. 18.24

D7–D0	Data bus	A0, A1	Port address
Reset	Reset input	PA7–PA0	Port A bits
$\overline{CS}$	Chip select	PB7–PB0	Port B bits
$\overline{RD}$	Read input	PC7–PC0	Port C bits
$\overline{WR}$	Write input	Vcc	+5 V

byte of the address, and byte 3 contains the second byte of the address. Low and high address bytes must be placed in the proper sequence as determined by the type of microprocessor being used.

IMPLIED ADDRESSING   Implied addressing is another type of indirect addressing mode in which the instruction op code contains the operand address. An example would be the 8085 MOV instructions or STC affecting only the carry flag.

ACCUMULATOR ADDRESSING   Accumulator addressing is a special case of implied addressing. The instruction relates directly to an accumulator which contains the operand. Examples from the 8085 instructions include decimal adjust accumulator (DAA) and rotate instructions.

INDEXED ADDRESSING   Indexed addressing employs an "offset number" plus a specified address which point to another address storing the operand. The process works as follows:

1. The index register contains an address. Assume 3000H as an example.
2. An offset number is specified in the second byte of the instruction, for example, 2AH.
3. The program counter jumps to the sum 3000H + 2AH, or 302AH.

The offset, sometimes called *displacement*, is an 8-bit value limited in addresses from 00H to FFH. If a number of operands are stored sequentially in memory, indexed addressing provides an easy method of advancing the program by incrementing the index register.

RELATIVE ADDRESSING   Relative addressing is a special case of indexed addressing in which the program counter is used as the index register. The program counter points to the displacement stored in memory, adds that offset to its own value, and points to the resulting address storing the operand.

COMPARISON OF MICROPROCESSORS   The 8080, Z80, 6800, and 6502 provide a cross section of popular 8-bit processors using different programming and hardware configurations. A simple block diagram showing registers available to a programmer along with a brief description of operation is given for each type.

Intel 8080 (Fig. 18.25)   In previous sections of this chapter we described the 8085 in detail; the 8080 was the forerunner of the 8085. The basic differences between the two devices are shown in Table 18.9. Programs written for the 8080 use the same instructions as the 8085 except for the additional 8085 features of interrupts and serial data I/O.

Fig. 18.25 Simplified block diagram of the 8080 microprocessor.

Z80   See Fig. 18.26. Examination of the register configuration shows some similarity to the 8080/8085; however, additional registers include duplicate accumulators, flags, extended register pairs B'–D'–H', and two index registers. The added capability permits indexed addressing as well as numerous data moves between registers. Data words also can be rotated

TABLE 18.9   Intel 8080 Compared with Intel 8085

8080	8085
One built-in interrupt	Five priority interrupts available
No SOD or SID	Serial I/O ports included
Needs external two-phase clock chip	Built-in clock generator
+12-, +5-, −12-V power supplies	Single +5-V power supply
2-μs machine cycle	Approximately 50% faster
78 instructions	Added instructions for interrupts and serial I/O
Needs external controller chip	Built-in system controller

in any register or memory location. Extensive memory-block move and search commands provide flexible data transfers. Seven flag bits are available for testing program status. The instruction set of Z80 has about 158 instructions, which include the 8080 instructions under different mnemonics.

6800 An absence of scratchpad registers is noticeable in Fig. 18.27. Instead of storing and moving data inside the processor, the 6800 stores and moves data in RAM memory with various instructions. This CPU is an example of a *memory-oriented* type of processor. Short instructions place the first 256 memory locations within easy access of the processor. Seven addressing modes previously described are available. Many instructions require a selection of the addressing

Fig. 18.26 Simplified block diagram of the Z80 microprocessor.

mode as well as the operation to be performed. Index and relative addressing provides efficient movement of data between memory locations.

All data in the internal registers and accumulators are 8 bits wide. A condition code register of 6 bits serves the same function as the 8085 flags. The bus structure of the 6800 contains a 16-bit address bus and 8-bit data and control buses. Three software interrupts and one hardware interrupt also are supplied; however, this is not a vectored interrupt as in the 8085. A two-phase external clock chip must be used with the 6800. The newer type 6802 has an on-chip clock generator and an on-chip RAM of 128 locations 8 bits wide.

6502 See Fig. 18.28. As in the case of the 6800 family, the 6502 is memory-oriented. Only one 16-bit register, the program counter, is present. All other registers are 8 bits wide, including an accumulator, status register (flags), stack pointer, and two index registers.

An outstanding feature of the 6502 is the 13 possible addressing modes. The modes that can be specified for the 56 instructions of the instruction set include the addressing modes discussed earlier—immediate, accumulator, implied, relative, and indirect. In addition to these, the following addressing modes are also available:

1. Absolute. Similar to extended addressing on the 6800. Bytes 2 and 3 contain the 16-bit memory address where the operand is located.

2. Zero page. Similar to direct addressing in the 6800, sometimes called *base-page* addressing. The CPU categorizes 256 bytes of memory into a page. Zero page refers to memory addresses 00H to FFH, the first 256 memory bytes. So 1K byte of memory would have four pages.

3. Indexed indirect X. A zero page address in the second byte is the offset value added to index register X resulting in another zero page address. This address holds yet another address pointing to the operand anywhere in memory. Another term used for this mode is *preindexing*.

4. Indirect indexed Y. The second instruction byte specifies a zero page address. Immediately following the zero page address byte is another address byte (16-bit indirect address) used to fetch yet another 8-bit address which is added to the Y index register, resulting in a pointer for the operand. This is also referred to as *postindexing*.

5. Zero page X and zero page Y. Both modes are types of indexed addressing. The second instruction byte address is added to the index register (X or Y) to get the effective address. This sum must be less than the zero page limit of FFH.

Fig. 18.27 Simplified block diagram of the 6800 microprocessor.

Fig. 18.28 Simplified block diagram of the 6502 microprocessor.

6. Absolute X and absolute Y. Another type of indexed addressing is used. The index register (X or Y) is added to the address in the second and third instruction bytes, providing a fetch address for the operand. The sum must be on zero page.

18.9 SUMMARY

This chapter provides a basic introduction to microprocessor operation. Technological advancement continues at an astonishing pace with the development of 16-bit microprocessors, multi-microprocessor systems, and single-chip microcomputers. The study of any microprocessor requires the user to do the following:

1. Examine the block diagram (architecture).
2. Study the instruction set.
3. Relate the instruction set to the architecture.
4. Review the timing diagrams and control signals.

Development systems supplied by manufacturers provide a tremendous learning tool for hardware and software. These systems come in many sizes, shapes, and prices. Another valuable learning tool is a logic analyzer (Chap. 19) which presents a visual readout of programs, control signals, and timing relationships.

Chapter 19

Logic Analyzers, Logic Probes, and Signature Analysis

"Logic Analyzers" by Pat Zagaria
Product Marketing Manager
Philips Test & Measuring Instruments, Inc.
Mahwah, New Jersey

"Signature Analysis" by Kenneth Jessen
Material Engineer
Hewlett-Packard Co.
Loveland, Colorado

19.1 INTRODUCTION TO LOGIC ANALYZERS

More and more in our electronic world, digital solutions are used to realize sophisticated functions. The exploding application field of microcomputers is just one example of this fast-growing innovation. But analog test and measuring instruments are not always adapted to the measuring requirements of digital environments. So a new type of instrument, a specialized tool for this digital environment called the *logic analyzer*, has been developed.

In this section we explain why we need a logic analyzer and how it works. We cover the following topics in turn:
- Brief review of digital circuitry
- Fault conditions in digital circuits
- Basic principles of logic analyzers
- Main functions of logic analyzers
- Applications of logic analyzers
- Glossary of logic analyzer terms

19.2 REVIEW OF DIGITAL CIRCUITRY

First, let us look at an analog circuit. Figure 19.1 shows a very simple, straightforward radio receiver. The signal reaching the input is processed sequentially (we could also say serially in time) by all the stages in turn, finally arriving at the output. Transfer of information tends to be over one line. Fault finding in such a system is simple. It can be executed by just studying every stage separately with a simple measuring tool or an oscilloscope. For example, if the signal at the input of the detector stage is correct but that at the output of this stage is incorrect, then the detector stage must be defective. This *sequential* processing of the signal is the most significant property of most analog circuits.

Now let us look at a simple digital circuit. Figure 19.2a shows an AND gate with three

Fig. 19.1   Sequential processing of analog signals.

inputs and one output. The three input conditions a, b, and c of this gate together determine the output condition; so for fault finding on this gate, we need an instrument with four inputs, which will measure and/or display all four signals at the same time.

The second AND gate in Fig. 19.2b adds three more input conditions to the circuit. Combining these two AND gates via an OR gate, we get a basic circuit often found in digital applications. If we want to trace the signals in such a circuit to find a fault, we need to be able to trace at least seven signal lines; and if we want to see the intermediate signals as well, we need two extra inputs. Only the capability to trace these nine signal lines at the same time enables us to find the fault. This is what we call *combinatorial* circuitry. Digital circuits have multinodal signal points. This is a major difference between analog and digital circuits. In practical applications however, not only are the digital circuits combinatorial; also the signal is sequentially processed.

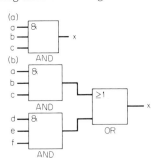

Fig. 19.2   Combinatorial logic circuit.

This combination of sequential and combinatorial signal processing is clearly illustrated in Fig. 19.3. On the left, we see several input conditions. Only if these input conditions are fulfilled during a certain time will the resulting output condition be correct.

A more complicated example of such a system is the microcomputer system of Fig. 19.4. The signals (data) are sequentially transported over the bus lines. Functions such as the central processing unit and the memories are realized by combinatorial logic. Clearly many

Fig. 19.3   Combination of sequential and combinatorial signal processing in digital system.

Fig. 19.4 Microcomputer system.

conditions will have to be examined in combination and in sequence to trace a fault in such a system.

We have the data bus with data present on 8 lines, the 6-line control bus, and the 16-line address bus. In addition there are various output, control, and input lines. All together, there is much more than we will ever be able to examine with an oscilloscope. We need a special, new kind of tool for analyzing all these signals. This is the logic analyzer.

19.3 TYPES OF FAULTS IN DIGITAL SYSTEMS

The common faults that can occur in digital circuitry may be due to the hardware, the software, or both; software faults occur, of course, only if the application is controlled by a software program. The following hardware faults may be distinguished:
- No data or wrong data
- Glitches
- Spikes
- Races
- Timing error
- Ringing
- Wrong level

These faults can occur in analog circuits also; however, the large number of signal lines in combinatorial circuits can greatly complicate the task of fault finding.

Software faults are mainly program errors from different sources, or memory defects.

HARDWARE FAULTS

No data Little need be said about this point. No data or wrong data can be due to a broken line, short circuit, or power failure in some circuits.

Glitches Figure 19.5 shows one way in which a glitch can be generated. The circuit is a 4-bit counter gated in such a way that it counts only up to 3. At count 3, both outputs *a* and *b* are high; the AND gate is then enabled, and both flipflops are reset to zero. This means that the *a* output is high only for a very short time here before it is reset. The resulting short pulse (shown in the figure) is called a *glitch*. This type of unwanted signal normally is found only in the design stage. Correct design should eliminate all glitches.

Fig. 19.5 Glitches.

Spikes A spike is an unwanted signal very similar to a glitch. Figure 19.6 shows an example of spike generation. The sharp rising edge of a signal transition present on one or

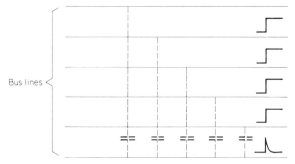

Fig. 19.6   Generation of (parasitic) spikes.

more of the lines in a bus system can cause a parasitic signal to appear on one or more bus lines owing to capacitive coupling. This parasitic signal is called a *spike*. It can create problems in a logic circuit connected to the line involved, for the circuit would receive the unwanted signal for a short time.

Races   Races are quite similar to glitches and spikes. They generally occur when signals of different speed are combined in one logic circuit (see Fig. 19.7). If both signals *a* and *b*

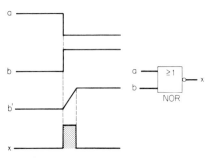

Fig. 19.7   Races.

have sharp edges, as shown in the top half of the figure, there will be no signal at the output of the NOR gate. However, heavy loading of the *b* signal line can make the rising edge of the *b* signal somewhat slower than the falling edge of the *a* signal. Consequently, it will take longer than usual for *b* input to reach its switching threshold. For a short time, therefore, both *a* and *b* will be within the range characterized as low, giving a high output. As soon as the *b* signal reaches the threshold level, the output goes low, which results in a short unwanted output pulse.

Timing errors   Another hardware fault condition can be caused by timing errors. Figure 19.8 shows data on a data bus together with the corresponding sampling pulse. Suppose the data are taken from the bus and sent to another device at the falling edge of the sampling pulse. If for some reason the data sampling pulse is delayed and appears at the moment when the data are changing (indicated by the crossing of the data lines), then undefined data are sampled and the result may be false.

Ringing   Ringing is another hardware fault. Figure 19.9 shows a two-input AND gate. If the inductance of one of the input lines of the AND gate is excessive, the sharp edge of

Fig. 19.8   Timing error.   **Fig. 19.9**   Ringing.

Fig. 19.10   Logic levels.

an input signal could cause oscillations of such an amplitude that the signal passes the gate threshold several times, causing the output shown.

Wrong level   Each family of logic circuits has its own defined range of switching levels. Figure 19.10 shows the logic ranges of some popular digital circuits: TTL, CMOS, and ECL.

Let us take a TTL circuit as an example. A TTL output will deliver an output level of less than 400 mV when the signal is logic 0 and of at least 2.4 V when the input signal is logic 1. At the input, however, the circuit will still recognize 800 mV as a 0 and signals from 2 V onward as 1. If the input level is between 0.8 and 2.0 V, the logic condition is undefined. The output signal then can be either a 1 or 0, so a faulty condition can occur. These wrong levels can be caused by excessive loading of the logic gates or by interference.

SOFTWARE FAULTS   Software errors are mainly program errors. These can be subdivided into three main groups: wrong instructions, latent faults, and timing faults.

Wrong instructions   The first is an error caused by wrong instructions, so the program is not executed as planned. This could also be due to missing instructions, wrong addresses, etc.

Latent faults   Figure 19.11 shows a small part of a program where an add operation is followed by an unconditional jump instruction. If now the jump instruction is changed by mistake into a "jump nonzero" instruction, the program will run as planned as long as the result of the preceding "add b" operation is not a zero. However, if the result is a zero, the jump instruction will not be executed and the program will continue with the output routine, which certainly was not the intention. Errors of this type may be due to wrong instructions in the program or even to hardware defects. As we can see from Fig. 19.11, the codes in this example for the jump and the jump nonzero instructions differ by only 1 bit, the last.

Fig. 19.11   Latent software fault.

Timing faults   In some programs for communication between processor and peripherals, the transmission rate is software-controlled. If the peripheral unit is too slow in executing its function and the software is not programmed to wait for completion of the operation in question, the program will continue, which means that various faults can occur.

Memory defects   Another type of error, sometimes wrongly regarded as a software fault, is a memory defect that is due, for example, to an open circuit or a short circuit in the memory lines. Such a memory defect also could be the cause of the mix-up between the unconditional jump and jump nonzero instructions mentioned above. In fact, these are hardware defects, and we need hardware tools to find them.

Our short review of digital circuits and the possible faults makes it quite clear how difficult it can be to trace a fault in such circuits, because of the complexity and speed of the hardware and the intricate structure of the software. This is particularly true if more than one fault is present.

We should now be able to specify the requirements of the ideal tool for these applications, the logic analyzer.

19.4 REQUIREMENTS ON LOGIC ANALYZER

First we list some general requirements:
- We should have the capability to trace and display many input signals, say 16, at the same time. This will permit measurements on combinatorial logic circuitry.
- We require an optimum sampling rate, to be able to work with logic circuits of all kinds.
- The input level (threshold) should be adaptable to different logic families.
- Storage facilities are needed, to capture sequences of data occurring only once or at very low repetition rates.
- We should have extensive trigger facilities, enabling us to capture specific blocks of data in a data stream.
- And we need a variety of convenient display modes to enable us to analyze the problem on hand simply.

A more specific requirement is the capability to measure in the data domain as well as in the time domain. The *time domain* is the field for work in the hardware environment.
- Here we work with a high clock rate, normally asynchronously, for optimum resolution in high-speed circuitry.
- We should have detection facilities for hardware faults such as glitches, spikes, races, and ringing.
- We should also be able to make accurate time measurements.
- In the *data domain*, which is mainly in the software environment, we normally work with a synchronous system clock taken from the system under test.
- We also should be able to compare the incoming data with reference data stored in additional memory.
- Selection of specific data in a data stream should be possible via selective sampling.
- It should be possible to present captured data in a state table in various formats: binary, octal, or hexadecimal (hex). Presentation in the map mode is also useful for solving special problems.

As an additional feature, logic analyzers should be able to offer facilities for analog measurement of quantities such as level and time with an oscilloscope (which is still the only tool for detailed real-time measurements).

After this brief discussion of why we need logic analyzers and what they should be able to do, it is time to consider what this invaluable measuring tool really is and how it does what we need it to do.

19.5 BASIC PRINCIPLES OF THE LOGIC ANALYZER

As may be seen from Fig. 19.12a, a logic analyzer is basically nothing more than a *memory*, which functions here as a multiple shift register for at least 16 bits in parallel, where we store digital *data*, taken from a system under test. The operation of the logic analyzer is controlled by a *clock signal* (the clock together with the data input is the data acquisition unit). A sample of the data present is taken each time a clock pulse occurs and transferred to the memory. How does this work?

After the first clock pulse, one data word is stored in the first memory location; see Fig. 19.12b. On receipt of the next clock pulse, this data word is shifted one place further, and the next data word is transferred to the first memory location. Each subsequent clock pulse causes the string of data words to be shifted one place further in this way. Figure 19.12c is a photograph of a logic analyzer.

MEMORY A prime consideration for defining the memory of a logic analyzer is the width of our data word, or how many bits in parallel can be taken at every clock pulse.

Fig. 19.12 Basic design of logic analyzer.

In accordance with the requirements mentioned previously, Philips logic analyzers have 16 or more inputs (up to 96 channels) so the data word we can take from the system under test is at least 16 bits wide. The number of words we can store at a time is a second consideration. It may be 64; then the memory will have 64 locations for the data words—but any other size is possible in principle. Sizes up to 2000 are not uncommon.

We have seen how data are taken in every clock pulse and shifted through the memory until they "disappear" at the other end. To visualize the data of interest, we need to be able to freeze the data flow at a required spot. This is done with the aid of the trigger function, a very important part of every logic analyzer.

TRIGGER FUNCTION   The trigger function tells the logic analyzer to stop data acquisition and to freeze the data present in the memory at that instant. At the appropriate moment we take the data stored, format them, and transfer them to the display—the third main function of the logic analyzer.

Fig. 19.13 (a) Use of direct leads for data sampling from a micro-computer system. (b) Actual hookup to a logic analyzer.

DATA ACQUISITION   The sampled data should, of course, be acquired correctly from the system under test. That is done via probes, with color-coded leads being used to connect the probes to the system under test.

Depending on the type of analyzer, we also can use *individual* probes for connecting each data input to the circuit of interest, e.g., for sampling the data present on the data bus and address bus of a microcomputer system as shown in Fig. 19.13a. Figure 19.13b shows an actual hookup to a logic analyzer.

DISPLAY   The data are stored in the memory in the normal binary code; however, the display can be formatted in various different ways, as shown in Fig. 19.14. In most logic analyzers the data displayed are preceded by a line number (related to the trigger word) for the sake of convenience. We now discuss the various possible display formats.

Time format   One possibility is to convert the 1s and 0s of the data stored in the memory to pulses; then the display has the form shown in Fig. 19.14a. Up to 16 waveforms can be displayed, since we have 16 inputs in this example. This display looks very much like that of a normal multitrace oscilloscope; however, it is obtained in quite a different way (as a sampled pseudo-timing waveform).

The pulses for a given data word are vertically under one another; from left to right, first

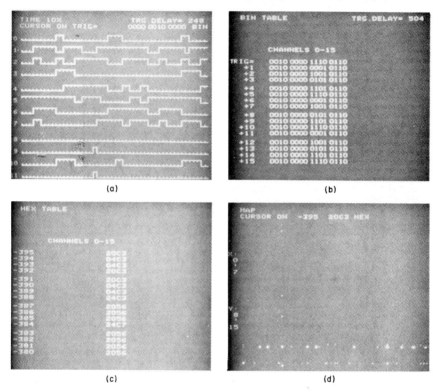

Fig. 19.14 Display formats.

we get all the bits for word number 0, then those for word number 1, word number 2, and so on.

Since the memory capacity is normally larger than the display capacity, the display shown on the screen represents a selection from the captured data in our memory, as seen through a "display window," which in principle can be given any desired width by the designer of the logic analyzer. In our example, the display window is 16 data words wide. The time format is used mainly for hardware analysis.

Binary format In the data domain, we often use other data representations. The binary format (Fig. 19.14b), for example, takes the same binary-coded data from the memory as for the time mode and represents them directly in 1s and 0s to give what is known as a *state table*. Here the individual data words are displayed horizontally, not vertically as in the time format. The display window is again 16 data words wide.

Hexadecimal format A more compact, and often more convenient, state table is the hexadecimal representation shown in Fig. 19.14c, which is simply a translation of the binary format of Fig. 19.14b to the hexadecimal notation (where the numbers from 0 to 15 are represented by 0, 1, 2, 3, 4, 5, 6, 7, 8, 9, A, B, C, D, E, and F, successively). For example, D7 in hexadecimal = binary 11010111 = 215 in normal decimal notation.

Hexadecimal is widely used by people in the microcomputer world as well as by many minicomputer users—so why not use it in the analyzer too? A variant of this is the octal format, applied by some minicomputer users.

Map format A completely different display mode is the map format illustrated in Fig. 19.14d. This has many interesting applications as a kind of routing indicator for analyzing program flow. Once again, we take the same data from the analyzer memory, but convert them in another way. We come back to this map display later. The various display formats are generally pushbutton- or switch-selected in a logic analyzer.

Summarizing our brief account of the operation of the logic analyzer, we see that this instrument takes in multichannel data (in our example, one new 16-bit data word every clock pulse) and stores the data in the memory. The trigger function stops data flow at the desired point. Then the data are displayed in a convenient way.

It is important to realize that the three main functions—acquisition, triggering, and display—are completely independent of one another.

19.6   MAIN FUNCTIONS OF A LOGIC ANALYZER

The three main functions of the logic analyzer, grouped around the memory where the sampled data are stored, are data acquisition, trigger function, and display. We now discuss each of the functions in greater detail.

DATA ACQUISITION

The clock   A more detailed diagram of the data input (acquisition) unit is given in Fig. 19.15. We see here once again the memory, the data input and the *clock*, introduced above. The data input and the *clock* belong together: a new data word is picked up each time a clock pulse is given. We consider first the case where the clock belongs to the system under test—the physical source of input for the analyzer. This is denoted by "system clock" in Fig. 19.15. The data naturally run synchronously with this clock.

The system clock need not be the master clock of the system under test; indeed, this is seldom the case. It is often a derived clock or a control line (read-write or chip select line, for example). In any case, the system clock we select in our test object must be synchronous with the data we want to analyze. This type of operation is the *synchronous mode*.

We can use the analyzer's internal clock also. This is completely independent of the system clock and thus runs asynchronously. The internal clock is a free-running signal produced by the analyzer, having no time relation with the incoming data. We explain later the reasons for using this asynchronous mode.

Data input   Before we can transfer data from the real world via the input to the analyzer's memory, we must choose the appropriate range of input levels by means of the threshold level selection (TLS) unit, shown in Fig. 19.15. The threshold level is determined by the logic family in the system under test (for example, TTL, ECL, or MOS).

Figure 19.16 shows the situation for a single data input line. The switching threshold for

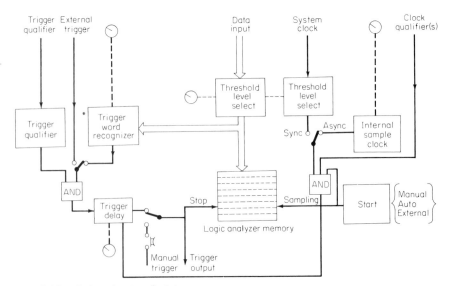

Fig. 19.15 Clocking data into the LA memory.

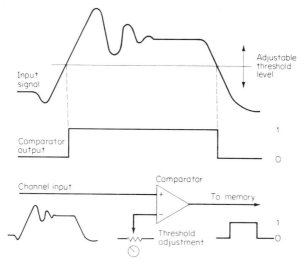

Fig. 19.16   Input with threshold detection.

measurements on TTL circuitry is about 1.5 V. If we set the TLS unit to this value, every incoming voltage higher than 1.5 V is regarded as being a logic 1 and every input voltage below 1.5 V as a logic 0.

The comparator output signal indicated in Fig. 19.16 is sent to the memory for sampling.

Setup and hold times   Figure 19.17 shows the timing diagram for one input channel, with data at the input and a clock pulse. Now, it is not the clock pulse as a whole but rather its leading edge that determines the moment of data sampling.

If we take a positive-going clock pulse, the positive-going *edge* is important. It is also possible to use the negative-going edge of a negative clock pulse, as shown in Fig. 19.17. The choice between these two types of clock pulses is user-selectable.

Now, the data must be present for a certain time (called the *setup time*) before the clock edge appears, to make sure that the data at the entry to the analyzer memory are completely stable when the clock edge does appear. The setup time is of the order of a few nanoseconds. The maximum permissible clock rate is also an important parameter, and it is related to the setup time. It is generally between 10 and 100 MHz (corresponding to clock pulse rates between 100 and 10 ns).

The hold time is another important parameter in this connection. The data should remain stable for a short time after the sampling edge, to ensure correct data sampling. This hold time is generally very near to zero.

Another important factor is the duration of the clock pulse itself, even though data sam-

Fig. 19.17   Definition of setup and hold times.

pling is basically controlled by the leading edge of the clock pulse. This should generally be of the order of a couple of nanoseconds.

Clock qualifiers   The synchronous mode is used mainly in the data domain. For selection of data in a data stream, wide use is made of the clock qualifier. The word *qualifier* sometimes gives rise to confusion. There is also another qualifier, called the *trigger qualifier*, which has a completely different function and is discussed later.

A clock qualifier is used for selective sampling of the data words wanted from the data stream. Application of the clock qualifier comes down to gating the clock with an additional condition. This is illustrated in Fig. 19.18, which shows part of a system program from which we would like to extract and store in the memory the data words marked "yes" in the program. This is done with the aid of a clock signal taken from the system under test, together with a control signal from the system as clock qualifier (in this example, the read-write line is used).

Fig. 19.18   Clock and clock qualifiers (selective sampling).

We adopt the convention that a 1 (high) on the read-write line means that we have a read operation, so data are sampled on receipt of a clock edge. If this signal is not a 1, we have a write operation and we do not sample.

As can be seen, the first data word (at A in the program) coincides with a "true" clock qualifier, so this data word is sampled and shifted into the memory. One clock pulse later, the clock qualifier is still true, so the data word at B is also taken and transferred into the memory after the previous one. At C, however, the clock qualifier is false (0), so this data word is not taken; and so on. The overall result is that we store the four data words marked "yes" out of the total of eight data words in the memory.

In another application, we might study the write operation. For this purpose the logic assignation of the clock qualifier is reversed, and we take the high level as false and the low level as true.

One main advantage of the clock qualifier is that it permits us to make more efficient use of the memory capacity of the logic analyzer. A disadvantage is that the time relation in our captured data is lost, since only the selected data words are stored. However, data acquisition is still synchronous with the enabled clock pulses. As long as we are not interested in time-related data sequences, this will not matter.

In the data domain, we are interested in related sequences of data, so a clock qualifier is of great importance. Use of a number of clock qualifiers permits selection of the data words in even greater detail or depth.

Start function (analyzer begins to gather data)   One function not mentioned so far is the *start* function used to initiate data acquisition. This may be manual—we press the button and sampling starts. It also may be automatic with starts at regular intervals of, say, 1 s for automatic pseudo-real-time measurements in the data stream or via the external input with an electric signal (e.g., a TTL signal).

TRIGGER FUNCTION   As long as data acquisition is not stopped, the data flow through the memory continues and no data are displayed. Data acquisition is stopped with the aid of the trigger function, which is a very important part of every logic analyzer. It permits us to capture only those blocks of data from the data stream in which we are interested. Unlike the trigger in an oscilloscope (which *starts* the time base), the trigger in a logic analyzer *stops* data acquisition. We discuss the trigger function with reference to Fig. 19.19, where the crosshatching represents the trigger section. Triggering can be realized in various ways: externally, internally, or manually.

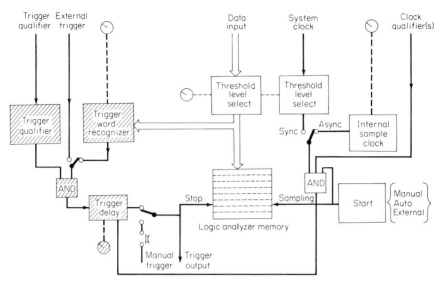

Fig. 19.19   Trigger function.

Manual triggering   Here we just press a button to stop further data acquisition. This manual trigger is, in fact, a manual stop. It is not normally used, but we see later that the other trigger facilities do not always give us correct triggering under all circumstances. For example, if the measuring conditions are such that data acquisition keeps on continuously and the "not triggered" signal appears on the screen, then we can use the manual stop to freeze the data passing by at that moment, to get some idea of what is going on. We could compare this with a beam finder on an oscilloscope, which also gives an indication of what is the matter with the measurement.

Internal triggering   Internal triggering (also called *combinatorial* triggering) is performed with the aid of the *trigger word recognizer*. This is the most important trigger facility, and it is used in the majority of measurements. Here we select the trigger word with the operator-controlled trigger word selector, and we stop the data flow as soon as the last data word sampled from the data stream coincides with the trigger word. This trigger word can be selected with mechanical switches, but there are other means, depending on the analyzer. Philips PM 3500 logic analyzer uses three-way switches, which can be set to 1, 0, or "don't care" (immaterial) at each position. This enables us to select one unique trigger word from 2^{16} possibilities (we have 16 inputs). In Fig. 19.20a, we see such a selected trigger word, with a match between the trigger word and incoming data at the second captured data word. Only then will the trigger circuit generate the stop pulse.

Analyzers now use what is known as *sequential*, or *nested*, triggering. This means that not one trigger word is selected but several are one after another. All trigger words should be matched in a selected sequence before the stop pulse can be given. This is illustrated in Fig. 19.21, which shows part of a program where both trigger words 0 and 1 have to be matched before a stop pulse is given and the data can be displayed. This permits data analysis before and during this conditional call as well as the subroutine. If the subroutine were called from another place in the main program, we would not get a stop pulse.

Further additional conditions, e.g., that a certain time should elapse or a certain event should occur between trigger words 0 and 1, another between trigger words 1 and 2, and so on, are sometimes built in. The *trigger qualifier* is an extra input, an extra qualification for triggering. If we combine this trigger qualifier with the trigger word recognizer, we could also speak of a "17th input channel." Although there is no storage or display of data for this channel, it is taken into account for triggering. Not only should the trigger word be matched, but also the trigger qualifier condition should be fulfilled before a trigger pulse is given to stop data acquisition. We see in Fig. 19.20b that both these conditions are not

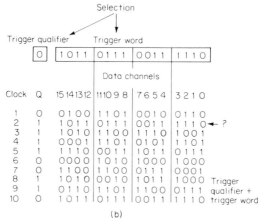

Fig. 19.20 (a) Trigger word. (b) Trigger word and trigger qualifier.

Fig. 19.21 Nested triggering.

Logic analyzer

Trigger qualifier — 17th input

16 bit — Trigger word

Trigger word extender — Trigger selection

1
x
0

Pods

Triggering
on
16 bits

24 bits

32 bits

40 bits

Fig. 19.22 Trigger word expansion.

fulfilled by the second data word in the table, so data acquisition will not be stopped here after all; we have to wait for the last data word in the table before this occurs. This 17th channel can be enlarged in multiples of 8 by means of trigger pods, as illustrated in Fig. 19.22. This enables us to select one unique trigger word out of a much larger data word. Switches are provided on the trigger probes to select a 0, a 1, or a "don't care" for each additional trigger input. This is called a *trigger word expander*.

External triggering The logic analyzer also can be triggered externally via an input which accepts an electric signal from such sources as an interrupt line, a timer, or a specific control line of the system under test. In all three trigger modes—external, internal, and manual—the trigger function works independently and can be used in exactly the same way in synchronous and asynchronous sampling.

Trigger delay Trigger delay allows us to select from the data stream a data block that comes a preselected number of clock pulses after the selected trigger word. This is a delay by a specific number of samples, not by a specific time. The delay can be user-selected to any integer number between 0 and 10 000, say. It also can be switched off, which is functionally the same as zero delay. If a certain delay is selected, for example, 63, an internal down counter in the delay circuit will count down, using the same clock pulse as for clocking new data in the analyzer memory. We would like to explain the use of the trigger delay with reference to the simple program of Fig. 19.23. On receipt of every sample pulse, one new data word is captured and is shifted through the analyzer memory in the data stream.

If no trigger delay is selected (Fig. 19.23a), then as soon as the trigger word recognizer finds a match between data and selected trigger word, it produces an immediate stop. The data are displayed, and new data are no longer sampled. The last data word stored is our selected trigger word. We have captured historical data, i.e., data occurring *before* the

Fig. 19.23   Trigger delay.

selected trigger word. This can be very useful. If, for example, the trigger word itself represents a fault condition in the test object, we can study what occurred before it. Display data of this type are called *pretriggered* data.

To simplify the explanation, a memory capacity of only eight words—less than the real memory capacity of a logic analyzer—is shown here. Let us now select a certain trigger delay (seven in the example of Fig. 19.23b). After the selected trigger word matches a data word, the down counter is started and counts down one step for every sample pulse; only at the point where this counter reaches zero will a stop pulse will be given. This means that after the match, seven new data words are sampled, and the display shows the trigger word followed by these seven data words. Such data are called *post-trigger* (or *future*). Historical data and future data (always with respect to the trigger word) are easy to recognize from their line numbers. The minus sign stands for historical data; a positive sign or no indication indicates future data. If we take a delay of four, we see three words of historical data and four words of future data, with the trigger word in between (see Fig. 19.23d). This could be called *center triggering*, since the trigger word is found in the middle of the memory.

Figure 19.23c shows a case where a delay of 1000 has been chosen. After the trigger word has been found, the delay counter again starts counting down and now allows 1000 new data words to be sampled before the stop pulse is given. Since our memory capacity is eight in this example, the trigger word itself is not displayed (it has been shifted through the analyzer memory long ago), though its value is, of course, still set on the trigger word selector. The trigger delay enables us to cover a range of historical data as large as the memory capacity and of future data as large as the delay (1000 data words in this example); thus we effectively stretch the memory.

The trigger delay function can be regarded as a means of trigger positioning at an appropriate point in our data stream. The unique trigger word provides a reference point, and the trigger delay positions the required block of data from the data stream relative to this trigger word.

The stop pulse can be used for other purposes apart from stopping data acquisition. For example, we can trigger another instrument such as a counter or an oscilloscope with it, via the trigger output. This instrument can then carry out a supplementary measurement, related to a specific point in the data stream.

DISPLAY   The display function links the captured data in the analyzer memory to the user's interpretation of the data display. It takes the data as they are stored and presents them to the user in a convenient format.

Let us start by recapitulating some of the facts about the display mentioned previously (see Fig. 19.24). The *binary* representation (Fig. 19.24*b*) may be regarded as the most straightforward. It provides a state table giving the data in 0 and 1 form for all 16 inputs.

The *hexadecimal* representation (Fig. 19.24*c*) also provides a state table, but here the binary data are combined in groups of 4 bits, each of which is replaced by a hexadecimal character. A 16-bit word length thus corresponds to four hexadecimal characters. The octal state table compresses the data in a similar way, by groups of 3 bits. Other formats such as ASCII and DECI as well as binary and hexadecimal are available.

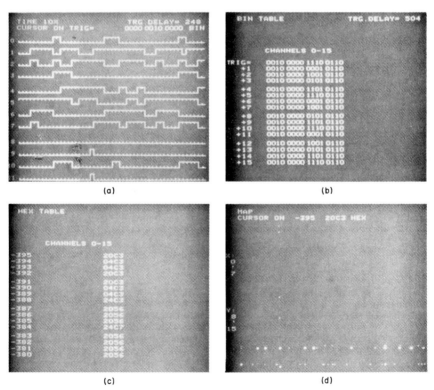

(a) (b) (c) (d)

Fig. 19.24   Display formats (see also Fig. 19.14).

Another useful feature in formatting the display is the possibility of decoding the words in the data stream into mnemonics representing disassembled microprocessor instructions, e.g., for the 8080/85 family,

-2	7E	MOV AM
-1	80	ADD B
0	C9	Return

or a combination of IEC bus commands and data:

-3	LAD 4
-2	DATA
-1	DATA
0	GET
1	TAD 7

With the aid of the *cursor control*, we select the block of stored data we want to display. In Fig. 19.25, this block contains 16 lines of 16-bit data words out of a total memory capacity of 504 × 16 bits. This cursor-controlled part is called the *display window*. By turning the cursor control we can, as it were, move the display window past the stored data until we bring to light the data block of interest. Additional control is provided by the *blanking* facility. If we are interested in only a limited number of channels (for example, only 5 bits), we can blank the other ones. Remember that this is only a display control; for example, the trigger word still has to be set up in full on the trigger word selector, and the full data word remains in the logic analyzer memory.

In the *compare mode*, we compare the new data sampled in a given measurement with reference data already stored in an additional memory. Each data bit that is not equal to the corresponding reference bit is intensified in the display. Every bit of the entire contents of the memory is compared with the corresponding reference bit, but only a selected part is shown in the display window. Additional information is, therefore, given above the display, in the form of a message *equal* (the data in the memory are completely identical with the reference data) or *unequal* (one or more bits out of the total memory contents are different). We can then move the display window up and down to trace all the different—intensified—bits.

In the *time* format (Fig. 19.24a), the display waveforms look just like those from oscilloscopes. However, as we mentioned above, these waveforms are not real-time as in an oscilloscope, but just represent the stored data from the analyzer memory reconfigured into a pseudo-time display. The time format is used mostly for measurements in the hardware environment.

The time mode can be used to display the entire contents of the memory (by setting the magnification control to "×1"). A part of this display (the *cursor block*) is intensified and can be moved along the memory with the aid of the cursor control. The first data word of this cursor block is also displayed, together with the line number relative to the trigger word and the contents of this data word in binary, hexadecimal or octal notation. The same cursor block will stay on display when we select only a part of the memory contents (magnification ×10) or an even smaller part (magnification ×25). Blanking also is possible here; the remaining channels will be regrouped to fill the entire display for user convenience.

A completely different type of display formatting is provided by the *map mode* (see Fig.

Fig. 19.25   Display window and blanking.

Fig. 19.26 Map display.

19.26). This gives us an overview of all captured data words (504 in the present examples) in what we might call a geographic way. The map format is used mostly in data-domain measurements. For this purpose, each data word in the analyzer memory is split into two halves, the least significant byte and the most significant byte. The contents of each byte is processed by a separate digital-analog converter (DAC), to yield a deflection of the spot in the horizontal direction for the least significant byte and in the vertical direction for the most significant byte. Each data word is thus converted to a point on the display screen, characterized by its own xy coordinates.

The values of the 2 bytes determine the unique position of the spot for each different data word. If we have 504 locations in the memory, 504 different spots can be displayed at the *same* time. If we now take the address lines of the system under test as data input for the analyzer (see Fig. 19.27), we can display the flow of a captured part of the program in map mode. Remember, the value of the program counter, present on the address bus in some form, always indicates the *next* program step.

The example of Fig. 19.28 shows the map display of a program structure with several subroutines stored at different address locations. The addresses of this program are sampled and stored in the analyzer's memory. The specific pattern in which the map-mode dots appear on the display characterizes our captured program.

The first addresses sampled from the main program appear here as the dots in the bottom straight line of Fig. 19.28, where successive dots represent successive addresses. At a certain point in the program there is a call for the first subroutine; the start address for this subroutine is thus the next address sampled. This is clearly indicated on the display, since the next spot is located at a different place (1 in Fig. 19.28).

At the end of this subroutine we return to the main program and continue the straight line at the bottom of the display. At the second call for this first subroutine, the next address is again the start address of the subroutine. The spots on this subroutine are more intensified

Fig. 19.27 Application of map display mode to a microcomputer system.

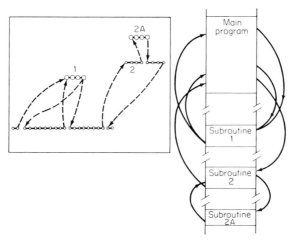

Fig. 19.28   Map display of program flow.

than the ones from the main program because the subroutine is executed twice, the main program once. This gives an easy way of recognizing parts of the program that are executed more often than others. The remaining part of Fig. 19.28 can be understood in much the same way.

SYNCHRONOUS AND ASYNCHRONOUS SAMPLING   Let us now return to the data acquisition process. We can take a system clock as sample pulse, synchronous with the sampled data for data-domain measurements. For the hardware environment, however, this does not always satisfy the measuring requirements. There are often conflicting timing relationships or short, unwanted signals such as spikes and glitches which could disturb the correct functioning of a digital system. We find that not all the data are captured in the synchronous mode (see Fig. 19.29). The solution is to select a faster clock. If a clock with higher repetition rate is not available in the system under test, we take the logic analyzer's internal clock with (normally) a rate that is roughly 5 to 10 times faster. The repetition time of this clock can be user-selected between, say, 10 ns and 20 ms, but this clock is completely independent of, or asynchronous with, the data.

Fig. 19.29   Synchronous data sampling.

Figure 19.30 shows an example of asynchronous sampling of the data, at a much higher clock rate. For the sake of convenience, Philips' PM 3500 logic analyzer shows the clock ticks, which are given for easy time reference in the time display mode.

Fig. 19.30   Asynchronous data sampling.

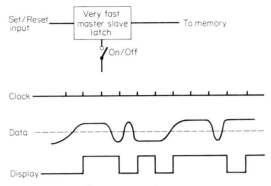

Fig. 19.31 Glitch-capturing mode.

GLITCH-CAPTURING MODE So far, we have considered only data that are taken in at clock-pulse edges (synchronously or asynchronously). This is called the *sample mode*. It is not possible to see data transitions in between clock pulses in the sample mode. However, this is possible in the *latch* or the *glitch-capturing* mode (Fig. 19.31). In this mode an additional, very fast master-slave latch located immediately after each input is switched on. We also could say that an extra memory capacity of one 16-bit data word is used to capture signal transitions in between clock pulses.

Each initial transition of the data signal from its original state to the opposite one (1 to 0, or 0 to 1) will be stored in the fast latch, and this information will be transferred to the normal working memory. Even glitches down to 3 ns can be captured in this mode.

19.7 APPLICATIONS OF THE LOGIC ANALYZER

EXERCISES IN MONITORING DATA TRANSPORT VIA BUS LINES In this section we give some elementary exercises that could be used to familiarize the trainee electronics engineer (or anyone new to the logic analyzer) with the working of this instrument.

In every microprocessor system, the data are transferred over the buses to various peripherals such as memories and I/O devices. A logic analyzer must be capable of acquiring and monitoring this information in the most convenient way. We give below simple exercises to illustrate some ways in which the logic analyzer functions to this end.

All our examples refer to the basic microprocessor system of Fig. 19.32, consisting of a 8085 microprocessor, 8355 PROM/IO (programmable read-only memory with associated input-output devices), 8155 RAM/IO (random-access memory with associated input-output devices), and 8205 address decoder.

Fig. 19.32 Block diagram of microcomputer setup used as a basis of the exercises in this section.

TABLE 19.1   Program Used in Exercises

Address	User label	Instruction mnemonic	Operand or address	Description
0575	XMPL4	LXI	H, 2000H	Load 2000H in register pair HL
0578		MVI	A, 0DH	Set port A to write, port B to read
057A		OUT	20H	Write to RAM
057C	XM41:	INR	B	Increment counter (add 1 to contents of B register)
057D		MVI	A, 0CH	Move 0C$_H$ (hexadecimal notation for decimal 12, or digital 0000 1100) to accumulator
057F		SIM		Mask interrupt 7.5
0580		OUT	23H	Write 0C to port A
0582		MOV	A, B	Move contents of B register into A
0583		OUT	21H	Write to port A
0585		IN	22H	Read port B
0587		MVI	M, 0FFH	Write FF (hexadecimal notation for decimal 255, or digital 1111 1111) to address 2000
0589		JMP	057C	Go and get next number

We assume that the program shown in Table 19.1 has been loaded into the PROM. It would take us too long to explain this program completely to someone unfamiliar with the workings of a microcomputer, but the following information may prove useful.

Address.   Addresses are the locations in the memory in which the data and instructions are written (have been stored). The addresses are given in hexadecimal notation, in which the digits 0 to 9 have their normal decimal significance and the letters A to F are used to represent decimal 10 to 15. Thus, e.g., the address 057F corresponds to $5 \times 16^2 + 7 \times 16^1 + 15 \times 16^0 = 1280 + 112 + 15 = 1407$ in decimal notation. It is customary to write hexadecimal numbers with a subscript H (for example, 2000$_H$) to distinguish them from decimal numbers of the same form.

User Label.   These codes have the following significance here: XMPL4 = example 4 and XM41 = first label in example 4.

Instruction Mnemonic.   These mnemonics (also known as *op codes*, short for "operational codes") are more or less comprehensible representations of the relevant instructions taken from the instruction set of the 8085 microprocessor. The mnemonics used in this program have the following significance:

LXI   Load immediate register pair
MVI   Move immediate register
IN   Input
OUT   Output
INR   Increment register (add 1 to current contents of register)
SIM   Set interrupt mask
JMP   Jump unconditional

In fact, each of these mnemonics is stored in the PROM in the form of an 8-bit "machine code" which would be very difficult for the programmer to work with directly. The correspondence between the mnemonics and the machine codes comprising the instruction set may be found in the data sheet for the microprocessor in question. For example, the mnemonic SIM represents the machine code 00110000 (which corresponds to 030 in hexadecimal notation, or decimal 48 in the instruction set for the 8085).

Operand.   Operands are also stored in the PROM in the form of 8-bit bytes. Each mnemonic may be followed by no data or address bytes, one data or address byte (if there is no comma in the entry in this column) or two such bytes (if there *is* a comma in the entry).

Description.   Assembler programs used to translate from machine language to higher-level languages such as the instruction mnemonics given here include routines for printing out very brief descriptions of the various instructions, in order to guide the electronics engineer working on the development of a given microcomputer system. However, the descriptions given by the assembler are so cryptic as to be hardly intelligible to any but the insider; we have, therefore, expanded the description somewhat here.

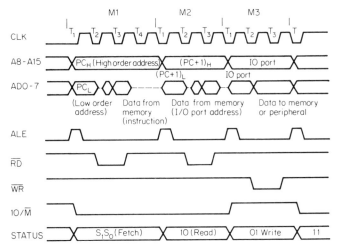

Fig. 19.33   Basic timing diagram of 8085 microprocessor.

The overall effect of the program listed above is to ensure that the sequence of (decimal) digits 0, 1, 2, 3, 4, etc. is written to input-output (I/O) port A of the 8155 RAM, and port B is read immediately afterward. Also, each time one digit from the above sequence is handled by ports A and B, the hexadecimal combination 0C is written to port C and the combination FF to memory location 2000H. We thus get the data sequence 00, FF, 0C, 01, FF, 0C, 02, FF, 0C, 03, FF, 0C, etc. passing along the data bus lines.

Figure 19.33 shows the basic timing diagram of the 8085 microprocessor. A timing diagram is a fundamental tool for logic analysis, for it gives all sorts of information such as types of data, control and status information, parameter setting, clock qualifiers, etc. at the appropriate points with the PROM addresses for the RAM (2000H) and I/O (2121, be used for all these exercises.

Exercise 1   First we want to check all the addresses used in this program.

Connections

Clock input of logic analyzer to ALE (address latch enable) lines

Eight data channels of PM3540 to most significant address bus lines A8–A15

Eight data channels of PM3540 to least significant address bus lines AD0–AD7

Settings   (see heading of Fig. 19.34)

It may be seen from the timing diagram that all addresses are stable on the buses during the negative-going edge of the ALE pulse.

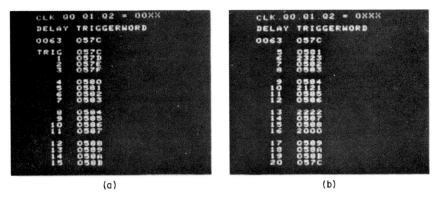

(a)   (b)

Fig. 19.34   (a) Checking the program addresses. (b) Display is "rolled up" five addresses in order to see entire program.

A falling edge (CLK = 0) has, therefore, been chosen as clock parameter for the logic analyzer.

The trigger word can be chosen at any convenient point of the program flow; we took 057C here.

Post-triggering can be chosen in order to capture *all* the information after the trigger word; a delay of 63 is chosen to get all the information plus trigger word in the logic analyzer's 64-word memory.

Results. In Fig. 19.34a we see a photograph of the display of the PM3540, showing all the addresses from 057C to 0588 at which the various instructions are written, interspersed at the appropriate points with the PROM addresses for the RAM (2000H) and I/O (2121, 2323, etc.).

By rolling or paging through the memory contents the whole program can be displayed; Fig. 19.34b shows the display rolled up to reveal the addresses from 0581 to 0588.

Exercise 2: Checking a selected part of the program flow (addresses)   An address decoder decodes addresses so as to activate enable lines for the various devices connected to the system, thus selecting which chip is to be used in the system at a given time (here either the PROM or the RAM). To select the data, we use the logic analyzer's clock qualifier, which gates the clock pulses.

Connections

   Clock (CLK) input of logic analyzer to ALE line

   Clock qualifier Q_0 to CS (or CE) of PROM

   Data channels as in Exercise 1

Settings

   CLK = 0, that is, negative-going edge of ALE is active

   Q_0 = 0, since CS signal is active when low

   Trigger word: any word in 05xx series or xxxx can be chosen (here we choose 057C again)

   Delay: 63

Results. As may be seen from Fig. 19.35, only the addresses of the program as such (057C–058B) are displayed now; those of RAM (2000) and I/O ports (2121, 2323) are skipped.

Exercise 3: Analyzing the data flow   If we want to see the data, we cannot use the ALE signal to clock the logic analyzer. As may be seen from the timing diagram of Fig. 19.33, either the $\overline{WR}$ (write) or $\overline{RD}$ (read) pulse can, in general, be used as clock to pick out the data flow. In this case, we are interested in seeing the data written to the memory and the I/O ports; the write pulse is thus the appropriate choice here.

Connections

   Clock input of logic analyzer to $\overline{WR}$ line

   Data channels as in Exercise 1

Settings

   CLK = 1, since $\overline{WR}$ is active on positive-going edge (see timing diagram of Fig. 19.33)

   Since the 8085 is an 8-bit multiplexed microprocessor, only bus lines AD0–AD7 transfer data. The most significant half of the address, which is also clocked into the analyzer, thus is not relevant information for this exercise and can be blanked. Hence, we choose the trigger word xxoo (or ??oo in hexadecimal notation, where ? replaces the binary "x" or "don't care" sign).

   Delay: 63

Results   The data flow (00, FF, 0C, 01, etc. . . .) in the programmed sequence is displayed in Fig. 19.36.

Exercise 4: Selecting I/O data only   As the timing diagram of Fig. 19.33 indicates, the signal IO/$\overline{M}$ determines whether data are transferred to I/O ports or to the memory. We can select, e.g., the data to I/O ports by feeding this IO/$\overline{M}$ signal to the clock qualifier input of the logic analyzer.

Connections

   Clock input of logic analyzer to $\overline{WR}$

   Q_0 to IO/$\overline{M}$

   Data channels as Exercise 3

Settings

   CLK = 1

Fig. 19.35 Program address listing that skips the addresses of the RAM and I/O ports.

Fig. 19.36 Analyzing data flow.

$Q_0 = 1$, because data are sent to I/O ports when IO/$\overline{\text{M}}$ is high (see timing diagram of Fig. 19.33)

Trigger word: xxoo

Delay: 63

Results. Only the information passed to the I/O lines is displayed, as shown in Fig. 19.37. If we had selected $Q_0 = 0$, then only the data written to the memory would be shown, i.e., an uninterrupted sequence of FF, FF, FF, ...

Exercise 5: Selecting only data written to port A   If we call up the most significant half of the addresses, other settings and connections being as for Exercise 4, we get the display shown in Fig. 19.38.

It can be seen that the most significant half of the address of port A is 21_{H}, and that of port C is 23_{H}. In binary terms this means that these ports differ only in the second bit of their address, so this bit gives the vital information on what port is called. A second clock qualifier input receiving this bit can thus be used to select data only from port A—or only from port C.

Connections

Clock of LA to $\overline{\text{WR}}$

Q_0 to IO/$\overline{\text{M}}$

Q to A8

Data channels as in Exercise 4

Settings

CLK = 1

Fig. 19.37 Data flow through the I/O ports only.

Fig. 19.38 As in Fig. 19.37, but with the most significant half of the addresses displayed (recalled).

Fig. 19.39   Data flow to port A, selected on basis discussed in caption to Fig. 19.38.

$Q_0 = 1$
$Q_1 = 0$ for selecting port A
Trigger word: xxoo
Delay: 63

Results.   Figure 19.39 shows only the data written to port A. If we had selected $Q_1 = 1$, only the data written to port C (i.e., an uninterrupted sequence of 0C, 0C, 0C, etc.) would have been displayed. Note that we would have needed a new trigger word—xx0C—for this purpose.

19.8 GLOSSARY OF LOGIC ANALYZER TERMS

The following "dictionary" of terms commonly associated with the logic analyzer (generally abbreviated LA below) has been split into a number of sections dealing with triggering, sampling, display functions, memory considerations, and general topics. A final section dealing with common microprocessor terminology has been included as general information.

Triggering

Center Triggering.   When the memory contains data leading up to and immediately following the trigger event.
Delayed Trigger.   When the memory contains data that occur a present digital delay after the trigger event.
Edge Triggering.   Trigger word recognition occurring on a clock edge which controls the sampling of an external input.
External Trigger.   An additional input channel, usually without data storage, which acts as a single channel trigger.
Glitch Triggering.   The ability of an analyzer to trigger on extremely short pulses which it would not normally sample into its memory.
Manual Triggering.   User triggers the logic analyzer by pushing a button.
Pass Counter.   A selectable counter used in sequential triggering. Its action is to delay the arming of the second word until a preset number of passes of the first have occurred.
Post-triggering.   When the memory contains the trigger event and the data immediately following it.
Pretrigger.   When the memory contains the data that occurred up to and including the trigger event.
Sequential or Nested Triggering.   Triggering of the LA which only takes place in a certain sequential logic flow; for example, word A must be followed by word B before triggering occurs. Two or more trigger words must be selectable in the logic analyzer. It is used to isolate a particular conditional entry to a subroutine at different points.
Trigger Qualifier.   An additional input that must be "true" before the word recognizer can recognize the trigger word.
Trigger Word.   User-selected logic pattern on which the analyzer triggers.
Word Recognizer.   The part of the LA that identifies the trigger word in the logic flow.

Sampling

Asynchronous Sampling. The sampling rate of the logic analyzer is controlled by its own internal high-speed clock.

Clock Qualifier. A logic state input that is ANDed with the sampling clock edge. Only when this state input is true at the moment of the active clock edge does sampling occur.

Clock Width. Minimum clock high time or low time for sampling.

Dual Clocking. System enabling a logic analyzer to sample information from two clock edges (usually two separate systems in one box). Application area in multiplexed bus microprocessors where one clock determines the address and a second identifies the data at that address.

Hold Time. Time during which data must be stable after the active clock edge.

Latching or Capture or Glitch-Detector Mode. Logic analyzer operating mode based on use of fast input circuits detecting the presence of transient pulses that would not be sampled by the clock edge.

Sample Interval. The time between samples (= reciprocal of the speed).

Setup Time. Time during which data must be stable prior to the active clock edge.

State Analyzers. Those analyzers that operate with an external clock provided by the system under test.

Synchronous or State Sampling. The sampling rate of the logic analyzer is controlled by a selected edge in the system being monitored.

Timing Analyzers. Those analyzers that have the ability to operate asynchronously with the system.

Timing Analyzer Speed. The repetition rate of the sampling edge—*not* the bandwidth.

Display

ASCII (American Standard Code for Information Interchange), Equivalent to the ISO 7-Bit Code) Mode. A display mode which converts a seven-digit binary value to alphanumeric characters. See Table 19.2.

BIN (Binary Display Mode). Each channel is shown as series of 1s and 0s.

Blanking. The removal of unwanted channels from the display.

Data Field. That area of the display which details the captured data.

Display Window. The area of the captured data that is displayed in the data field.

Graph. Display mode that plots the serial number of each captured item of data on the horizontal axis against the magnitude of the data at that location on the vertical axis.

HEX (Hexadecimal Display Mode). The binary value of four binary channels expressed as a hexadecimal value (0 to F).

IEC 625 (IEEE 488, GPIB) Mode. A display mode that decodes the binary information being transferred across the IEC interface bus.

Inverted-Logic Display. Logic high is a 0, and logic low is a 1. In this display mode, all values are inverted in state displays; but voltage levels remain true in the timing mode.

Mapping. A display mode that shows the full-memory depth in a single display. The high-order 8 bits control a digital-to-analog converter on the X axis, and the low 8 bits control a digital-to-analog converter on the Y axis, giving each word a unique location in the display.

Menu. Listing of parameter settings in software-operated logic analyzers.

Mnemonics. Display mode used to monitor microprocessors. Display shows the hexadecimal value plus the mnemonic code (abbreviated description of op code).

OCT (Octal Display Mode). The binary value of three binary channels expressed as an octal figure (0 to 7).

Parameter Field. That area of the display which details the parameter.

Memory

Compare Mode. Comparison of new data with reference data previously stored in a second memory within the LA.

Formattable Memory. On some LAs it is possible to change the width and depth of the memory, e.g., halve the width and double the depth.

TABLE 19.2   ISO 7-Bit Code (646-1972)

b_4	b_3	b_2	b_1	$b_7\,b_6\,b_5$ / C/R	0 0 0 / 0	Hexa-decimal	Dec-imal	0 0 1 / 1	Hexa-decimal	Dec-imal	0 1 0 / 2	Hexa-decimal	Dec-imal
0	0	0	0	0	NUL	00	0	DLE	10	16	SP	20	32
0	0	0	1	1	SOH	01	1	DC1	11	17	!	21	33
0	0	1	0	2	STX	02	2	DC2	12	18	"	22	34
0	0	1	1	3	ETX	03	3	DC3	13	19	#	23	35
0	1	0	0	4	EOT	04	4	DC4	14	20	$	24	36
0	1	0	1	5	ENQ	05	5	NAK	15	21	%	25	37
0	1	1	0	6	ACK	06	6	SYN	16	22	&	26	38
0	1	1	1	7	BEL	07	7	ETB	17	23	'	27	39
1	0	0	0	8	BS	08	8	CAN	18	24	(	28	40
1	0	0	1	9	HT	09	9	EM	19	25	)	29	41
1	0	1	0	10	LF	0A	10	SUB	1A	26	°	2A	42
1	0	1	1	11	VT	0B	11	ESC	1B	27	+	2B	43
1	1	0	0	12	FF	0C	12	FS	1C	28	'	2C	44
1	1	0	1	13	CR	0D	13	GS	1D	29	-	2D	45
1	1	1	0	14	SO	0E	14	RS	1E	30	.	2E	46
1	1	1	1	15	SI	0F	15	US	1F	31	/	2F	47

List of codes used:
NUL   Null/idle
SOH   Start of heading (CC)
STX   Start of text (CC)
ETX   End of text (CC)
EOT   End of transmission (CC)
ENQ   Enquiry (CC)
ACK   Acknowledge (CC)
BEI   Audible or attention signal
BS   Backspace (FE)

HT   Horizontal tabulation (punch card skip) (FE)
LF   Line feed (FE)
VT   Vertical tabulation (FE)
FF   Form feed (FE)
CR   Carriage return (FE)
SO   Shift out
SI   Shift in
DLE   Data link escape (CC)
DC1   Device control

Memory Depth.   The number of samples that the memory can contain.
Memory Width.   The number of channel inputs that can store data.
Search Mode.   Philips—a very high-speed comparison mode. Others—the memory is searched for an operator-chosen word. Its location is then highlighted.

General

Bit.   Single unit of binary information, either 1 or 0.
Byte.   A group of bits considered and processed together, generally 8 bits.
CMOS.   +5-V logic; low is <1.5 V, high is >3.5 V
         +10-V logic; low is <3.0 V, high is >7.0 V
         +15-V logic; low is <4.5 V, high is >10.5 V
Computer Word.   Usually 2 bytes (16 bits) but depending on system can be 8, 16, 32 bits, etc.
ECL.   −3-V logic; low is <−1.6 V, high is >−0.75 V
Tristate.   Used to describe a system having an intermediate logic state (high-impedance level) in addition to low and high.
TTL.   +5-V logic; low is <0.8 V, high is >2.0 V
Minimum Logic Swing.   Dead band around the selected threshold which the input signal must exceed to change the polarity of the comparator circuit.

Common microprocessor terminology

Address.   The location of a data word in the memory. The memory commences at address 0000 (hexadecimal) and continues up to FFFF (hexadecimal) with 16-bit address capabilities.

0 1 1	Hexa-decimal	Dec-imal	1 0 0	Hexa-decimal	Dec-imal	1 0 1	Hexa-decimal	Dec-imal	1 1 0	Hexa-decimal	Dec-imal	1 1 1	Hexa-decimal	Dec-imal
3			4			5			6			7		
0	30	48	@	40	64	P	50	80	`	60	96	p	70	112
1	31	49	A	41	65	Q	51	81	a	61	97	q	71	113
2	32	50	B	42	66	R	52	82	b	62	98	r	72	114
3	33	51	C	43	67	S	53	83	c	63	99	s	73	115
4	34	52	D	44	68	T	54	84	d	64	100	t	74	116
5	35	53	E	45	69	U	55	85	e	65	101	u	75	117
6	36	54	F	46	70	V	56	86	f	66	102	v	76	118
7	37	55	G	47	71	W	57	87	g	67	103	w	77	119
8	38	56	H	48	72	X	58	88	h	68	104	x	78	120
9	39	57	I	49	73	Y	59	89	i	69	105	y	79	121
:	3A	58	J	4A	74	Z	5A	90	j	6A	106	z	7A	122
;	3B	59	K	4B	75	[	5B	91	k	6B	107	{	7B	123
<	3C	60	L	4C	76	\	5C	92	l	6C	108	|	7C	124
=	3D	61	M	4D	77	]	5D	93	m	6D	109	}	7D	125
>	3E	62	N	4E	78	↑	5E	94	n	6E	110	~	7E	126
?	3F	63	O	4F	79	←o	5F	95	o	6F	111	DEL	7F	127

DC2	Device control	EM	End of medium
DC3	Device control	SUB	Start of special sequence
DC4	Device control (stop)	ESC	Escape
NAK	Negative acknowledge (CC)	FS	File separator (IS)
SYN	Synchronous idle (CC)	GS	Group separator (IS)
ETB	End of transmission block (CC)	RS	Record separator (IS)
		US	Unit separator (IS)
CAN	Cancel	DEL	Delete

Address Bus. Usually 16 lines on which a binary code is placed by the microprocessor to identify a single address location.

Assembler. Enables programmers to write in a language one level above machine code, using mnemonic codes. The assembler software translates the mnemonics to machine code.

Control Lines. Logic levels on these lines signify to the rest of the system what sort of action is occurring. They are the conductors and traffic lights in the system.

Data Bus. Usually eight lines upon which the data being transported to or from the microprocessor are placed.

Disassembler. Translates instructions from machine code to mnemonics so the program can be more easily read by humans.

Instruction Cycle. The time required to complete one complete instruction.

I/O. Input-output parts of the system. Used for communication with the outside world: data are written to them and read from them in a similar way as to and from memory.

Machine Cycle. The time required to complete one processor action, e.g., to fetch the op code or write data to a RAM location.

Mnemonics. Programmers' shorthand, used when programs are written in assembly language rather than machine code.

Multiplexed Bus Microprocessor. The address lines are time-shared with the data lines. First, these lines contain address information that is latched into a decoder by a control line; then, they carry data information that is read with another control line.

Program. The program flows sequentially through the addresses unless otherwise instructed, e.g., when a certain condition arises and needs servicing by a special subprogram contained in another part of the memory.

Program Counter. Contains the next memory location from which to read data.

RAM (Random-Access Memory). Contains data and sometimes the program. Data lost if power removed.

ROM (Read-Only Memory). Contains the system program.

19.9   LOGIC PROBE

A logic probe is a small, hand-held "instrument on a chip" that is used for functional testing and troubleshooting of digital circuits. Physically, a logic probe looks like an oversized pen with a needle point at one end. The needle point is placed on various digital circuit terminals to detect the presence of logic highs or logic lows. From the opposite end of the logic probe, a strain-relieved power cord extends. This cord is connected to the power supply of the circuit under test.

A logic probe can detect the logic level at any point in a digital circuit. The display is by means of a band of light near the tip of the probe as follows: (1) a logic high is indicated by a bright light, (2) a logic low is indicated by no light, (3) an incorrect signal level produces a dim light, and (4) pulsing produces a flashing light.

A logic probe is extremely useful in tracing logic levels and pulses through IC circuitry. It can instantly show whether the mode probed is high, low, "bad level," open-circuited, or pulsing.

19.10   LOGIC PULSER

A logic pulser is a small, hand-held instrument on a chip. A logic pulser injects digital pulses between gates without requiring the unsoldering of components. Physically, a logic pulser looks like a logic probe. It can automatically drive low nodes high or high nodes low.

A logic pulser is a pulse generator with high-output current capability. It has sufficient current to override IC outputs in either the high or low state. It is a valuable instrument used for fault isolation or circuit design.

19.11   SIGNATURE ANALYSIS FOR DIGITAL TROUBLESHOOTING

Troubleshooting digital circuitry presents new challenges to the test technician. Not only have digital devices become more complex, but also the large amount of digital data generated by many of the newer designs has become difficult for the technician to interpret. The trend in the electronics industry is to replace traditional analog functions with equivalent digital circuitry. Many devices operate only at higher speeds which precludes the use of single-step testing. To detect faults of a dynamic nature, testing must be done "at speed."

A number of approaches are used to test and troubleshoot digital circuitry. Although it is not portable, the use of a computer-based system is a logical choice. The computer is capable of operating at high speeds (given the appropriate drivers and receivers to interface with the board under test) and is able to interpret the large amount of digital data often involved in testing. Using techniques to simulate the operation of the board under test, the computer can provide the user with a measure of the effectiveness of a given test. It can even be used to determine how a board will react to a hypothetical failure. Computer-based systems, however, are expensive and require programming skills and a great deal of data on the board to be tested.

A logic analyzer is a second tool used to look at digital information. Each bit of binary data is viewed on a cathode-ray tube (CRT), and in many ways a logic analyzer is equivalent to looking at waveforms on an oscilloscope. Providing the user can interpret the data, the logic analyzer provides exact information on the binary state of each bit of information for a number of parallel input lines.

19.12   BASIS OF SIGNATURE ANALYSIS

Signature analysis (SA) is yet another tool, and in many cases it offers a practical approach to digital troubleshooting. Arbitrarily long strings of binary data on a single node are compressed into a single, manageable signature. Signature analysis is based on the time-honored technique of signal tracing. In contrast to other digital troubleshooting methods, SA produces a signature at a node which is either good or bad. Good signatures are derived from looking at the nodes on a known good board, and all subsequent testing is done against these signatures. No analytical information is provided as to which bit in a string of binary data might be incorrect, but the advantages of SA are its simplicity and ease of operation.

Signature analysis requires that the circuitry under test be stimulated by a repetitive pattern. As a general rule, the stimulus should cause the state of every node on the board under test to change state at least once. However, SA lacks the capability to compute how effective a given stimulus is in finding faults. In summary, the effectiveness of SA depends heavily on the quality of the stimulus to cause faults to propagate outward to one of the nodes where a measurement is made.

19.13 TEST ROUTINE IN ROMs

For many microprocessor-based products, manufacturers have embedded a test routine in one of the ROMs. The microprocessor and its associated memory elements can be placed into the test routine via a jumper wire. During troubleshooting, the microprocessor uses the test routine to provide a repetitive stimulus. In some cases, the microprocessor can be forced into a free-run mode by disconnecting the data bus (an NOP instruction). This forces the microprocessor into a counting routine which repeats itself. Where the manufacturer has not provided an internal stimulus, the test routine must be supplied externally by the user.

The signature is observed over a preselected window in the binary string of data at a given node. The SA is turned on at the starting point in the sequence and terminated at the ending point. The signature is derived from the digital activity which takes place during the window.

TESTING CONTENTS OF SEVERAL ROMs Proper selection of the window allows the user to test various portions of digital circuitry from the same node. A good example of this would be to independently test the contents of several ROMs tied to common address and data buses, as shown in Fig. 19.40. By selecting the beginning and ending addresses for a specific ROM, its contents may be viewed as signatures by probing each output node on the data bus. Once that ROM has been tested, shifting the window to include the addresses of another ROM will allow it to be tested by using the same stimulus. The stimulus need only contain the addresses of all the ROMs in the circuit to be tested (often as simple as a binary counting routine). To test U7 (in Fig. 19.40), the signature analyzer is turned on at the starting address A000 and turned off at the ending address AFFF. The contents of U8 also can be tested by prodding the same nodes on the data bus but shifting the window to a starting address of B000 and an ending address of BFFF.

The signature analysis is synchronized to the clock rate of the board under test, and testing is accomplished dynamically at or near normal operating speed. Many SA units allow detection of unstable signatures to pinpoint timing problems. At times, unstable signatures are due to propagation delays normal to the circuit under test, and the clock must be "deskewed" by using a tapped delay line in series with the clock to the signature analyzer.

TESTING UNKNOWN BOARDS The entire SA technique is predicated on having good signatures from a known good board. Testing on unknown boards is accomplished by using the signatures derived from the good board. The danger in the SA approach is that small differences between boards could cause "bad" signatures on a good board. Such cases are

Fig. 19.40 Individual ROMs can be tested by using SA.

Fig. 19.41   Manufacturers using SA as a troubleshooting tool will typically list signatures in their operating and service manuals on a node-by-node basis.

the exception, however. As illustrated in Fig. 19.41, manufacturers list signatures on a node-by-node basis in their operating and service manuals.

TROUBLESHOOTING TECHNIQUE   Troubleshooting is carried out by the test technician who uses conventional back-tracing methods. If a bad signature is detected, the fault can be traced to the failing node. Because of the bus structure so common in modern digital equipment in which many devices are tied to a single set of nodes, fault isolation to a single defective component may not be possible. In other cases, it may be possible to view both signatures at the input and output nodes of a device.

19.14   THE DESIGN OF A SIGNATURE ANALYZER

As illustrated in Fig. 19.42, the design of a signature analyzer is conceptually simple. The unit consists of shift registers with multiple feedback taps summed with the input data. The incoming binary data are sent though the shift registers and are modified by the feedback loops in such a way that the probability of having the same signature for two varying binary sequences is extremely low.

At the end of the window, the process of shifting data is stopped. The remainder left in the registers is the unique signature for that binary sequence. The signature itself is typically displayed as four hexadecimal characters. If 1 bit in the string is changed, the entire signature is changed.

In summary, SA is a powerful digital troubleshooting tool adaptable for field use. It does require some planning and designed-in features on the part of the manufacturer for the

Fig. 19.42   Design of signature analyzer.

product to the tested. The technique relies on previously recorded signatures derived from a known good board. From the technician's standpoint, conventional back-tracing methods may be applied to isolate faults on digital circuitry.

REVIEW QUESTIONS AND ANSWERS ABOUT SIGNATURE ANALYSIS

1. What does signature analysis demand in terms of the stimulus? *Answer:* The stimulus must be repetitive and should cause the state of every node to change at least once during the test sequence.

2. Can the preselected window be changed to test other portions of the circuitry by using the same stimulus? *Answer:* Yes. Often this technique is used to test the contents of ROMs connected to a common bus.

3. Does signature analysis give the user any diagnostic information as to what bit in a binary sequence might be missing to cause an incorrect signature? *Answer:* No. Signature analysis tests on a go–no-go basis.

Chapter **20**

Microwaves

Richard J. Pankow
Senior Microwave Engineer
Norden Systems
Norwalk, Connecticut

20.1 BASIC CONCEPTS

The fact that electricity does not travel instantaneously greatly affects circuits and devices in the area generally known as microwaves. The velocity of electricity, although very fast $(3 \times 10^{10}$ cm/s), results in voltages and currents being different at physically separated points. Although this phenomenon always holds in low-frequency circuits, the effects are trivial; but when the frequency of operation is sufficiently high, these effects cannot be ignored.

WAVELENGTH *Wavelength* is that distance traveled by electricity during the time of one complete cycle. In free space,

$$\lambda = \frac{c}{f} \tag{20.1}$$

where λ = wavelength
c = speed of light
f = frequency

example 20.1 What is the wavelength in air of a 1-GHz (10^9 Hz) wave?

solution The velocity in air is essentially the same as in free space, 2.9979×10^{10} cm/s. So

$$\lambda = \frac{c}{f} = \frac{2.998 \times 10^{10}}{10^9} = 29.98 \text{ cm} \qquad \textit{Answer}$$

$$\frac{29.98}{2.54} = 11.8 \text{ in}$$

Physical devices that may occupy several inches of length must account for the delays resulting from the finite velocity of propagation.

MICROWAVE REGION The boundaries of the microwave region are somewhat arbitrary. It is generally agreed, however, that signals lying between 1 and 250 GHz are considered to be in the microwave region. This corresponds to wavelengths of 12 to 0.05 in (or

30.5 cm to 1.27 mm), respectively. The region is divided by convention into bands for which letters have been assigned. (See Fig. 20.1.)

Fig. 20.1   Microwave band letter designations and frequency in gigahertz.

POWER AS A FUNDAMENTAL QUANTITY   Often in the measurements of microwave circuits, the concepts of voltage and current tend to lose their importance. Power is more readily measurable since it can be converted to heat. Some of the earliest microwave instruments were those designed to measure power by means of a resistive element which heated up when microwave energy impinged on it. The resulting change in resistance was sensed by a balanced Wheatstone bridge. Another early power-measuring instrument detected the temperature rise of the resistor element by means of flowing water. There were, at the time, very few instruments that could measure voltage or current directly. Furthermore, they were so large physically that they could not be used to measure a circuit without upsetting it. The evolutionary process thus begun results in many microwave components to this day being described in terms of power.

DECIBELS   Because of the large dynamic range of power levels encountered in microwave work, it is often convenient to use the mathematical convenience of decibels. *Decibel* literally means one-tenth of a bel, a unit named after Alexander Graham Bell who first applied the logarithm principle to measurements he performed during his experiments on deafness. A typical microwave receiver may have an input power level of 1 μW and an output power of 1 W, which is a gain of 1,000,000. To avoid writing large numbers, Bell conceived of using the logarithm of the number. The definition of the decibel (dB) is

$$dB = 10 \log \frac{P_2}{P_1} \tag{20.2}$$

where P_1 and P_2 are the powers being compared.

Some important relations which are immediately apparent are as follows:

When P_2 equals P_1, dB = 0.
When P_2 is larger than P_1, the decibels are positive.
When P_2 is smaller than P_1, the decibels are negative.
When P_2 is 10 times (100 times) P_1, dB = 10 (dB = 20).
When P_2 is 0.1 times (0.01 times) P_1, dB = −10 (dB = −20).

Another useful relation ensues because the logarithm of 2 is 0.3010. When the powers P_2 and P_1 are related by factors of 2, 4, 0.5, and 0.25, the values are +3, +6, −3, and −6 dB, respectively. In other words, when you multiply by 2, add 3 dB; when you divide by 2, subtract 3 dB.

The decibel is not an absolute unit. Rather, it indicates the relation between two powers. Decibels cannot be converted to watts except when P_1 (denominator) is held constant by mutual agreement. A special name is usually assigned in such cases. The most common example is the dB_m, where the denominator is always 1 mW. A signal strength of 1 mW is 0 dB_m, 2 mW is 3 dB_m, etc. One can use dB and dB_m for solving problems.

example 20.2   An amplifier has an input power of 8 μW and an output power of 4 mW. What is the gain in decibels?

solution   Gain compares output relative to input:

$$dB = 10 \log \frac{P_2}{P_1} = 10 \log \frac{4 \times 10^{-3}}{8 \times 10^{-6}}$$
$$= 10 \log 500 = 10(2.7) = 27 \text{ dB} \qquad \textit{Answer}$$

example 20.3   What is the output of a 27-dB gain amplifier whose input power is 25 μW?

solution

$$dB_m = 10 \log \frac{P_2}{1 \text{ mW}} = 10 \log \frac{25 \times 10^{-6}}{1 \times 10^{-3}}$$
$$= 10 \log 0.025 = 10(-1.6) = -16 \text{ dB}_m$$

So -16 dB$_m$ is the input power, because output power equals input power times gain when you use dB.

$$\text{Output power (dB}_m) = \text{input power (dB}_m) + \text{gain (dB)}$$
$$\text{Po} = -16 + 27 = +11 \text{ dB}_m$$

example 20.4   What power in watts is equivalent to $+11$ dB$_m$?

solution   Here we use the antilogarithm. Since

$$dB = 10 \log \frac{P_2}{P_1}$$

we know

$$P_2 = P_1(10^{dB/10})$$

In our case, $P_1 = 1$ mW since we are dealing with dB$_m$. So

$$P_2 = 1 \times 10^{-3} \times 10^{11/10} = 10^{-3} \times 10^{1.1}$$
$$= 10^{-3} \times 12.6 \text{ W}$$
or
$$= 12.6 \text{ mW} \qquad \textit{Answer}$$

When you use dB$_w$, it is understood that the denominator P_1 is always 1 W. This is, we are always comparing the measured power to 1 W; it may be greater than, less than, or equal to 1 W.

Generally dB$_c$ is applied when several signals coexist and one is considered the prime signal (carrier). Here we compare the relative value of the modulation, the sidebands, or any interfering signals to the desired one and express it as in decibels.

20.2   TRANSMISSION LINES

Any conductor or wire can be considered a transmission line. However, when we are dealing with frequencies whose wavelengths are comparable to the dimensions of the circuit, special conductors with known properties have been devised. These special conductors take several generic forms. Coaxial cable and parallel wire (such as TV lead-in) are probably the most widely known conventional types. At microwave frequencies these can be quite lossy, and other types of lines come into use. Among these are waveguides and dielectric cylinders (fiber optics). The microwave band, letter designations, and frequencies in gigahertz are given in Fig. 20.1.

In general, we discuss lossless lines. By this we mean that there are no dissipative elements, namely resistance. Obviously, real conductors have resistance and are lossy. However, the solutions to those problems are very complex, and as long as losses are kept low, the idealized lossless solutions give excellent results.

CHARACTERISTIC IMPEDANCE   It is useful at this point to introduce the term *characteristic impedance* Z_0. This impedance is the ratio of the voltage to the current of a wave traveling in either direction on a transmission line. A forward and backward wave may exist simultaneously in a transmission line. At points on the line where the two waves exist simultaneously, they combine as follows:

1. Voltage waves traveling in opposite directions add.
2. Current waves traveling in opposite directions subtract.

The impedance Z_0 is characteristic of the physical geometry of the transmission line. Transmission lines made of perfectly conducting wires (resistance = 0) and perfect insulating material (resistance $= \infty$) are known as lossless lines. Many practical lines act practically like lossless lines.

COAXIAL CABLE Coaxial cable is made of two concentric conductors separated by an insulator. The velocity of propagation of an electric wavefront is given by $V = c/\sqrt{\epsilon_r}$, where ϵ_r is the permittivity (dielectric constant) of the insulator. This velocity is always less than c and is independent of frequency. The characteristic impedance Z_0 for a coaxial transmission line is

$$Z_0 = \frac{60}{\sqrt{\epsilon_r}} \ln \frac{b}{a} = \frac{138}{\sqrt{\epsilon_r}} \log \frac{b}{a} \qquad (20.3)$$

where b = radius of outer conductor
  a = radius of inner conductor
  ϵ_r = relative permittivity
An abbreviated table of permittivities is shown in Table 20.1.

example 20.5 Find the characteristic impedance of a Teflon-filled coaxial line shown below.

solution

$a = 0.031$ in
$b = 0.250$ in

$$Z_0 = \frac{138}{\sqrt{\epsilon_r}} \log \frac{b}{a}$$

$$= \frac{138}{\sqrt{2.1}} \log \frac{0.250}{0.031}$$

$$= 86.3 \ \Omega \qquad\qquad Answer$$

example 20.6 For the air-filled coaxial line shown, what should the inner conductor diameter be so that $Z_0 = 50 \ \Omega$?

solution We know that $\epsilon_r = 1$ for air. We want to solve for a. Thus

$a = ?$
$b = 0.250$ in

$$Z_0 = 138 \log \frac{b}{a}$$

$$\log \frac{b}{a} = \frac{Z_0}{138}$$

$$\frac{b}{a} = 10^{Z_0/138}$$

$$a = \frac{b}{10^{Z_0/138}}$$

TABLE 20.1   An Abbreviated Table of Relative Permittivities

Material	Relative dielectric constant	
	@ 1 MHz	@ 10 GHz
Free Space	1.00	1.00
Teflon	2.10	2.10
Polyethylene	2.25	2.25
Polystyrene	2.56	2.54
Nylon	3.40	3.00
Fused Quartz	3.78	3.78
Glass, Corning 7070	4.00	4.00
Bakelite	4.40	3.70
Micarta	4.50	3.20
Water	78.00	55.00

Substituting values yields

$$a = \frac{0.250}{10^{50/138}} = 0.109 \text{ in} \qquad \qquad Answer$$

Coaxial cable comes in many physical forms depending on application requirements. For example, power handling, voltage breakdown, mechanical flexibility, and loss per foot must all be considered. Figure 20.2 depicts some typical forms. Table 20.2 is an abbreviated list of frequently used types of coaxial cable.

Coaxial cable is used at VHF, UHF, and the lower microwave frequencies.

PARALLEL-WIRE TRANSMISSION LINE As its name implies, the parallel-wire transmission line consists of two wires held at a constant spacing either by spacers placed at regular intervals or by a continuous film of insulator (TV twin lead). The characteristic impedance for an air dielectric line is

$$Z_0 = 120 \ln \frac{D}{a} = 276 \log \frac{D}{a}$$

where a is the conductor radius and D is the center-to-center spacing. The parallel-wire transmission line is used at VHF and UHF frequencies but not at microwave frequencies.

STRIPLINE AND MICROSTRIP Important types of transmission lines in microwave circuits besides coaxial cable are stripline and microstrip. Stripline consists of a narrow, flat conductor sandwiched between two dielectric boards whose outside surfaces are coated with conductor. Microstrip is a narrow, flat conductor on a single dielectric board whose opposite face is metallized, much like an ordinary printed-circuit board. The velocity of propagation and characteristic impedance of these types of lines are not expressible in closed form. Many handbooks contain charts for calculating these parameters from mechanical dimensions, similar to the graphs shown in Figs. 20.3 and 20.4.

OTHER SHAPES Occasionally, one encounters odd shapes, such as in an experimental setup, where it is desired to calculate the characteristic impedance. Figure 20.5 illustrates some of the more common ones and the characteristic impedance formulas.

WAVEGUIDES Waveguides come in many different shapes. The most widely used is the rectangular type, although circular, elliptical, and double-ridged shapes are available for special purposes. See Fig. 20.6. The waveguide may be made of copper, brass, or aluminum. It may have polished interior walls and sometimes may be silver-flashed. Waveguides function on the principle of propagating a wave by bouncing it between the walls of the waveguide at a skew angle, like images in a house of mirrors. The concepts of voltage, current,

(a)   (b)   (c)   (d)

Fig. 20.2 Typical forms of coaxial cable. (*a*) Semirigid, thin-walled copper outer conductor with solid dielectric insulator. (*b*) Flexible braided outer conductor with vinyl jacket. (*c*) Helical spacer for inner conductor. (*d*) Radial spacer for inner conductor.

TABLE 20.2 Coaxial Line Design Data *(courtesy Alpha Wire Corp., Elizabeth, N.J.)*

RG cable type	ALPHA product number	Velocity of prop., %	Nominal impedance, Ω	Attenuation (db/100 ft) frequency in Megahertz						
				10	50	100	200	400	1000	3000
6A/U	9006A, 9807 +	66.0	75.0	0.80	1.40	2.90	4.30	6.40	11.00	—
6/U	9843, 9804	78.0	75.0	0.62	1.50	2.10	3.10	4.40	7.10	—
8/U	9008	66.0	52.0	0.66	1.50	2.20	3.20	4.60	9.00	19.00
8A/U	9008A	66.0	52.0	0.66	1.50	2.20	3.20	4.60	9.00	19.00
8/U	9846, 9802	78.0	50.0	0.49	1.20	1.80	2.60	3.80	6.20	11.30
9B/U	9009B	66.0	50.0	0.66	1.50	2.20	3.20	4.60	9.00	19.00
11/U	9011	66.0	75.0	0.66	1.50	2.20	3.20	4.60	9.00	19.00
11A/U	9011A	66.0	75.0	0.66	1.50	2.20	3.20	4.60	9.00	19.00
11/U	9847	78.0	75.0	0.41	1.00	1.50	2.20	3.30	5.60	10.90
12A/U	9012A	66.0	75.0	0.65	1.50	2.15	3.20	4.70	8.20	18.00
22B/U	9816 + +	66.0	95.0	1.20	2.80	4.20	6.30	9.50	—	—
55/U	9055	66.0	53.5	1.35	3.00	4.30	6.00	8.80	16.50	—
55B/U	9055B	66.0	53.5	1.35	3.00	4.30	6.00	8.80	16.50	—
58/U	9058	66.0	53.5	1.20	3.10	4.60	7.00	10.00	17.50	—
58A/U	9058A	66.0	50.0	1.40	3.30	4.90	7.30	11.00	20.00	—
58C/U	9058C, 9158^x	66.0	50.0	1.40	3.30	4.90	7.30	11.00	20.00	—
58/U	9848, 9801	78.0	50.0	1.20	3.10	4.80	6.90	10.10	18.00	—
59/U	9059, 9830, 9159^x	66.0	75.0	1.10	2.30	3.30	4.70	6.70	11.50	—
59B/U	9059B, 9845	66.0	75.0	1.10	2.30	3.30	4.70	6.70	11.50	—
59/U	9840, 9844, 9803 9062, 9162^x, 9805	78.0	78.0	0.95	2.10	2.90	4.10	6.60	11.00	—

62/U	9863, 9864, 9865	84.0	93.0	0.90	1.90	2.80	3.70	5.80	8.50	—
62A/U	9062A	84.0	93.0	0.90	1.90	2.80	3.70	5.80	8.50	—
63B/U	9063B	84.0	125.0	0.50	1.10	1.50	2.30	3.40	5.70	—
71B/U	9071B	84.0	93.0	0.90	1.90	2.80	3.70	5.80	8.50	18.60
174/U	9174	66.0	50.0	3.80	6.50	8.90	12.00	17.50	31.00	—
178B/U	9178B	70.0	50.0	5.30	10.00	13.30	20.00	27.50	45.00	—
179B/U	9179B	70.0	75.0	5.00	7.90	9.80	12.70	15.80	25.00	—
180B/U	9180B	70.0	93.0	3.10	4.20	5.10	7.30	10.40	16.50	—
187A/U	9187A	70.0	75.0	5.00	7.90	9.80	12.70	15.80	25.00	—
188A/U	9188A	70.0	50.0	3.80	7.90	11.50	15.00	20.00	30.00	58.00
195A/U	9195A	70.0	95.0	3.10	4.20	5.10	7.30	10.40	16.50	—
196A/U	9196A	70.0	50.0	5.30	10.00	13.30	20.00	27.50	45.00	—
212	9212	66.0	50.0	0.65	1.60	2.40	3.60	5.20	8.80	16.70
213	9213	66.0	50.0	0.66	1.50	2.20	3.20	4.60	9.00	19.00
214	9214	66.0	50.0	0.66	1.50	2.20	3.20	4.60	9.00	19.00
215	9215	66.0	50.0	0.66	1.50	2.20	3.20	4.60	9.00	19.00
217	9217	66.0	50.0	0.41	1.00	1.40	2.10	3.10	5.80	13.00
218	9218	66.0	50.0	0.23	0.56	0.81	1.20	1.90	3.80	9.00
219	9219	66.0	50.0	0.23	0.56	0.81	1.20	1.90	3.80	9.00
223/U	9223	66.0	50.0	1.35	3.00	4.30	6.00	8.80	16.50	—
AlphaMini	9450	70.0	50.0	6.50	13.00	16.00	25.50	33.00	—	—
AlphaMini	9806	78.0	93.0	3.10	4.20	5.10	7.30	10.40	—	—
AlphaMini	9475	80.0	75.0	5.00	7.90	9.80	12.70	15.80	—	—
AlphaMini	9500	80.0	100.0	4.50	7.40	9.60	12.50	15.70	—	—

Note: + = precision video cable.
     ++ = twinaxial construction.
     × = plenum cable.

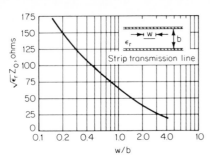

Fig. 20.3 Characteristic impedance of strip transmission line for the special case of a very thin center strip. (After S. B. Cohn, "Problems in Strip Transmission Lines," *IRE Transactions on Microwave Theory and Techniques*, vol. 3, no. 2, March 1955.)

Fig. 20.4 Characteristic impedance and λ_0/λ_m ratio for microstrip transmission line. (From the 1969 *Microwave Engineers' Handbook and Buyers' Guide*, as presented by Burke, Gelnovatch, and Chase, based on Wheeler's work.)

Fig. 20.5 Unusual shapes for transmission lines and their characteristic impedance formulas.

Fig. 20.6 Waveguides: (*a*) Rectangular, (*b*) elliptical, (*c*) circular, (*d*) double-ridged, (*e*) method of propagation.

characteristic impedance, and velocity of propagation are far less useful in understanding the waveguide than in the previously discussed transmission lines. Waveguides are less lossy at a given frequency than an equal length of coaxial cable. Waveguide is more restricted in its range of frequency usefulness than is coaxial cable. For example, RG-8B/U, a popular coaxial cable, has a loss of 17 dB per 100 ft at 3 GHz whereas WR-284 waveguide loss per 100 ft is about 0.8 dB at the same frequency, a reduction of 20 to 1. However, the recommended frequency range for the waveguide is 2.60 to 3.95 GHz, as shown in Table 20.3 while RG-8B/U cable is useful from direct current to 10 GHz. The reason for the narrowband nature of the waveguide lies in the method of energy propagation. When the free-space wavelength is too large, i.e., the frequency is too low, the wave does not "fit" inside the guide. As an analogy, compare AM and FM reception when you are driving through a short tunnel:

Wavelength of AM   980 ft
Wavelength of FM   9.8 ft
Diameter of tunnel   37 ft

Specifically, a wave will not propagate in a rectangular waveguide if the frequency is less than the cutoff frequency

$$f_{co} = \frac{c}{2a}$$

where c is speed of light in air (air-filled waveguide) and a is the larger inside dimension of the waveguide. At the high-frequency end, the wavelength becomes sufficiently small that more than one wavelength can fit into the waveguide. This is known as higher-order modes and is undesirable.

TABLE 20.3   Reference Table of Rigid Rectangular Waveguide Data and Fittings (Courtesy Microwave Development Laboratories, Inc., Natick, Mass.)

EIA WG designation WR-( )	Recommended operating ranges for TE$_{10}$ mode Frequency (GHz)	Recommended operating ranges for TE$_{10}$ mode Wavelength (cm)	Cutoff for TE$_{10}$ mode Frequency (GHz)	Cutoff for TE$_{10}$ mode Wavelength (cm)	Range $\frac{2\lambda}{\lambda_c}$	Range $\frac{\lambda_g}{\lambda}$	Theoretical CW power rating lowest to highest frequency (MW)	Theoretical attenuation lowest to highest frequency (db/100ft)
2300	0.32–0.49	93.68–61.18	0.256	116.84	1.60–1.05	1.68–1.17	153.0–212.0	.051–.031
2100	0.35–0.53	85.65–56.56	0.281	106.68	1.62–1.06	1.68–1.18	120.0–173.0	.054–.034
1800	0.41–0.625	73.11–47.96	0.328	91.44	1.60–1.05	1.67–1.18	93.4–131.9	.056–.038
1500	0.49–0.75	61.18–39.97	0.393	76.20	1.61–1.05	1.62–1.17	67.6–93.3	.069–.050
1150	0.64–0.96	46.84–31.23	0.513	58.42	1.60–1.07	1.82–1.18	35.0–53.8	.128–.075
975	0.75–1.12	39.95–26.76	0.605	49.53	1.61–1.08	1.70–1.19	27.0–38.5	.137–.095
770	0.96–1.45	31.23–20.67	0.766	39.12	1.60–1.06	1.66–1.18	17.2–24.1	.201–.136
650	1.12–1.70	26.76–17.63	0.908	33.02	1.62–1.07	1.70–1.18	11.9–17.2	.317–.212 .269–.178
510	1.45–2.20	20.67–13.62	1.157	25.91	1.60–1.05	1.67–1.18	7.5–10.7	
430	1.70–2.60	17.63–11.53	1.372	21.84	1.61–1.06	1.70–1.18	5.2–7.5	.588–.385 .501–.330
340	2.20–3.30	13.63–9.08	1.736	17.27	1.58–1.05	1.78–1.22	3.1–4.5	.877–.572 .751–.492
284	2.60–3.95	11.53–7.59	2.078	14.43	1.60–1.05	1.67–1.17	2.2–3.2	1.102–.752 .940–.641
229	3.30–4.90	9.08–6.12	2.577	11.63	1.56–1.05	1.62–1.17	1.6–2.2	
187	3.95–5.85	7.59–5.12	3.152	9.510	1.60–1.08	1.67–1.19	1.4–2.0	2.08–1.44 1.77–1.12
159	4.90–7.05	6.12–4.25	3.711	8.078	1.51–1.05	1.52–1.19	0.79–1.0	
137	5.85–8.20	5.12–3.66	4.301	6.970	1.47–1.05	1.48–1.17	0.56–0.71	2.87–2.30 2.45–1.94
112	7.05–10.00	4.25–2.99	5.259	5.700	1.49–1.05	1.51–1.17	0.35–0.46	4.12–3.21 3.50–2.74
90	8.20–12.40	3.66–2.42	6.557	4.572	1.60–1.06	1.68–1.18	0.20–0.29	6.45–4.48 5.49–3.83
75	10.00–15.00	2.99–2.00	7.868	3.810	1.57–1.05	1.64–1.17	0.17–0.23	
62	12.4–18.00	2.42–1.66	9.486	3.160	1.53–1.05	1.55–1.18	0.12–0.16	9.51–8.31 6.14–5.36
51	15.00–22.00	2.00–1.36	11.574	2.590	1.54–1.05	1.58–1.18	0.080–0.107	
42	18.00–26.50	1.66–1.13	14.047	2.134	1.56–1.06	1.60–1.18	0.043–0.058	20.7–14.8 17.6–12.6 13.3–9.5
34	22.00–33.00	1.36–0.91	17.328	1.730	1.57–1.05	1.62–1.18	0.034–0.048	
28	26.50–40.00	1.13–0.75	21.081	1.422	1.59–1.05	1.65–1.17	0.022–0.031	21.9–15.0
22	33.00–50.00	0.91–0.60	26.342	1.138	1.60–1.05	1.67–1.17	0.014–0.020	31.0–20.9
19	40.00–60.00	0.75–0.50	31.357	0.956	1.57–1.05	1.63–1.16	0.011–0.015	
15	50.00–75.00	0.60–0.40	39.863	0.752	1.60–1.06	1.67–1.17	0.0063–0.0090	52.9–39.1
12	60.00–90.00	0.50–0.33	48.350	0.620	1.61–1.06	1.68–1.18	0.0042–0.0060	93.3–52.2
10	75.00–110.00	0.40–0.27	59.010	0.508	1.57–1.06	1.61–1.18	0.0030–0.0041	
8	90.00–140.00	0.333–0.214	73.840	.406	1.64–1.05	1.75–1.17	0.0018–0.0026	152–99
7	110.00–170.00	0.272–0.176	90.840	.330	1.64–1.06	1.77–1.18	0.0012–0.0017	163–137
5	140.00–220.00	0.214–0.136	115.750	.259	1.65–1.05	1.78–1.17	0.00071–0.00107	308–193
4	170.00–260.00	0.176–0.115	137.520	.218	1.61–1.05	1.69–1.17	0.00052–0.00075	384–254
3	220.00–325.00	0.136–0.092	173.280	.173	1.57–1.06	1.62–1.18	0.00035–0.00047	512–348

Material alloy	JAN WG designation RG-( )/U	JAN flange designation Choke UG( )/U	Cover UG( )/U	EIA WG designation WR-( )	Dimensions (in) Inside	Tolerance	Outside	Tolerance	Wall thickness nominal
Alum.	290			2300	23.000–11.500	±.020	23.250–11.750	±.020	0.125
Alum.	291			2100	21.000–10.500	±.020	21.250–10.750	±.020	0.125
Alum.	201			1800	18.000–9.000	±.020	18.250–9.250	±.020	0.125
Alum.	202			1500	15.000–7.500	±.015	15.250–7.750	±0.15	0.125
Alum.	203			1150	11.500–5.750	±.015	11.750–6.000	±.015	0.125
Alum.	204			975	9.750–4.875	±.010	10.000–5.125	±.010	0.125
Alum.	205			770	7.700–3.850	±.005	7.950–4.100	±.005	0.125
Brass	69		417A	650	6.500–3.250	±.005	6.660–3.410	±.005	0.080
Alum.	103		418A						
Brass	337			510	5.100–2.550	±.005	5.260–2.710	±.005	0.080
Brass	104		435A	430	4.300–2.150	±.005	4.460–2.310	±.005	0.080
Alum.	105		437A						
Brass	112		553	340	3.400–1.700	±.005	3.560–1.860	±.005	0.080
Alum.	113		554						
Brass	48	54A	53	284	2.840–1.340	±.005	3.000–1.500	±.005	0.080
Alum.	75	585	584						
Brass	340			229	2.290–1.145	±.005	2.418–1.273	±.005	0.064
Brass	49	148B	149A	187	1.872–0.872	±.005	2.000–1.000	±.005	0.064
Alum.	95	406A	407						
Brass	343			159	1.590–0.795	±.004	1.718–0.923	±.004	0.064
Brass	50	343A	344	137	1.372–0.622	±.004	1.500–0.750	±.004	0.064
Alum.	106	440A	441						
Brass	51	52A	51	112	1.122–0.497	±.004	1.250–0.625	±.004	0.064
Alum.	68	137A	138						
Brass	52	40A	39	90	0.900–0.400	±.003	1.000–0.500	±.003	0.050
Alum.	67	136A	135						
Brass	346			75	0.750–0.375	±.003	0.850–0.475	±.003	0.050
Brass	91	541	419	62	0.622–0.311	±.0025	0.702–0.391	±.003	0.040
Alum.	349	—	—						
Silver	107	—	—						
Brass	352			51	0.510–0.255	±.0025	0.590–0.335	±.003	0.040
Brass	53	596	595	42	0.420–0.170	±.0020	0.500–0.250	±.003	0.040
Alum.	121	598	597						
Silver	66	—	—						
Brass	354	600	599	34	0.340–0.170	±.0020	0.420–0.250	±.003	0.040
Alum.	355	—	—						
Silver	96	—		28	0.280–0.140	±.0015	0.360–0.220	±.002	0.040
Brass	271		383						
Silver	97		—	22	0.224–0.112	±.0010	0.304–0.192	±.002	0.040
Brass	358			19	0.188–0.094	±.0010	0.268–0.174	±.002	0.040
Brass	273		385	15	0.148–0.074	±.0010	0.228–0.154	±.002	0.040
Silver	98		—						
Brass	274		37	12	0.122–0.061	±.0005	0.202–0.141	±.002	0.040
Silver	99		—						
Brass	359			10	0.100–0.050	±.0005	0.180–0.130	±.002	0.040
Silver	138	—	—	8	0.080–0.040	±.0003	0.156 DIA	±.001	—
Silver	136	—	—	7	0.065–0.0325	±0.00025	0.156 DIA	±.001	—
Silver	135	—	—	5	0.051–0.0255	±0.00025	0.156 DIA	±.001	—
Silver	137	—	—	4	0.043–0.0215	±0.00020	0.156 DIA	±.001	—
Silver	139	—	—	3	0.034–0.0170	±0.00020	0.156 DIA	±.001	—

	General
Cutoff frequency (air-filled)	$f_{co} = \dfrac{c}{2} \sqrt{\left(\dfrac{m}{a}\right)^2 + \left(\dfrac{n}{b}\right)^2}$
	For TE modes $\quad m = 0, 1, 2, \ldots$ $\quad n = 0, 1, 2, \ldots$ but $m \neq n$ if either $= 0$ For TM modes $\quad m = 1, 2, 3, \ldots$ $\quad n = 1, 2, 3, \ldots$
	TE$_{10}$ mode
Cutoff frequency	$f_{co} = \dfrac{c}{2a}$
Velocity of axial propagation (group velocity)	$v_p = c \sqrt{1 - \left(\dfrac{f_{co}}{f}\right)^2}$
Phase velocity	$v_\phi = \dfrac{c}{\sqrt{1 - (f_{co}/f)^2}}$
Guide wavelength	$\lambda_g = \dfrac{\lambda_0}{\sqrt{1 - (f_{co}/f)^2}}$

Fig. 20.7   Key rectangular waveguide equations.

The velocity of propagation through a waveguide varies with frequency while in coaxial cable it does not. Again, this is due to the nature of the propagation method. Just above the cutoff frequency, energy begins to progress down the guide, and the velocity approaches the speed of light as frequency increases. As previously stated, waveguides are generally not used above the next higher-order mode frequency. See Figs. 20.7 and 20.8.

example 20.7   Waveguide type WR-62 is propagating a 16.5-GHz signal in the TE$_{10}$ mode. To make slotted-line measurements, we need to find the guide wavelength.

solution   The free-space wavelength is

$$\lambda_o = \frac{c}{f} = \frac{11.8}{16.5} = 0.715 \text{ in}$$

$$c = 3 \times 10^{10} \text{ cm/s}$$

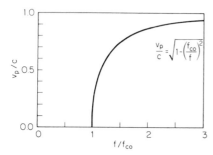

Fig. 20.8   Velocity of propagation in waveguide TE$_{10}$ mode.

or $$c = 11.8 \text{ in/ns}$$

and the cutoff frequency is

$$f_{co} = \frac{c}{2a} = \frac{11.8 \times 10^9}{2(0.622)} = 9.486 \text{ GHz}$$

So $$\lambda_g = \frac{\lambda_0}{\sqrt{1 - (f_{co}/f)^2}} = \frac{0.715}{\sqrt{1 - (9.486/16.5)^2}} = 0.874 \text{ in}$$

What is the next higher-order mode frequency?

$$f_{co} = \frac{c}{2} \sqrt{\left(\frac{m}{a}\right)^2 + \left(\frac{n}{b}\right)^2}$$

For TE$_{20}$: $$f_{co} = \frac{11.8 \times 10^9}{2} \sqrt{\left(\frac{2}{0.622}\right)^2 + 0} = 18.9 \text{ GHz}$$

For TE$_{11}$: $$f_{co} = \frac{11.8 \times 10^9}{2} \sqrt{\left(\frac{1}{0.622}\right)^2 + \left(\frac{1}{0.311}\right)^2} = 21.21 \text{ GHz}$$

For a piece of this waveguide 100 ft long, what delay is caused, compared to free space?

$$v_P = c \sqrt{1 - \left(\frac{f_{co}}{f}\right)^2} = 11.8 \times 10^9 \sqrt{1 - \left(\frac{9.486}{16.5}\right)^2} \quad 9.655 \times 10^9 \text{ in/s}$$

$$t = \frac{d}{v_P} = \frac{100 \times 12}{9.655 \times 10^9} = 124.3 \text{ ns} \qquad \text{through waveguide}$$

$$t = \frac{d}{c} = \frac{100 \times 12}{11.8 \times 10^9} = 101.7 \text{ ns} \qquad \text{free space}$$

DIELECTRIC TRANSMISSION LINES   Dielectric transmission line (fiber optics, inverse guide) is similar to waveguide in its method of transporting energy; the difference lies in the surface used to bounce the waves within the guide. Although in waveguide conducting metals are used as boundaries, here total internal reflection (as in a prism) is used to confine energy. These transmission lines most often are reserved for frequencies above 100 GHz and quasi-optical energy. An exception is the commonly used dielectric rod or polyrod antenna, where because of the lower frequency the internal reflections are not total and energy is radiated into space.

REFLECTION COEFFICIENT   The mechanism by which a load at the end of transmission line communicates its presence to the source is the reflection coefficient. In Fig. 20.9, when the switch closes at time zero, a voltage and current wave start down the transmission line

Fig. 20.9   Illustration of the reflection coefficient.

toward the load. As stated previously, they are related by the characteristic impedance and have no idea what load they will encounter. If, as in the figure, the line is terminated with a resistor, the voltage and current waves must satisfy two conditions simultaneously at the resistor: first, Ohm's law, which says that the voltage divided by the current must equal the value R; second, the quotient of a voltage-to-current wave must equal Z_0. For both conditions to be satisfied simultaneously, one might hypothesize the generation of a backward-going voltage and current wave. While this may seem like a giant mental leap, it satisfies the mathematical criteria of solution in a simple way and does not contradict any physical observations. The backward-traveling wave, sometimes called the *reflected* wave, is related to the forward or *incident* wave by the factor Γ_v as derived in Fig. 20.9. Almost always one speaks of the voltage reflection coefficient. It is easy to show that the relation of the forward and backward current waves is the negative of the forward and backward voltage waves: $\Gamma_i = -\Gamma_v$

example 20.8   Find the voltage reflection coefficient Γ_v for a 50 $\Omega = Z_o$ line terminated

solution   For part (a) of the example,

$$\Gamma_v = \frac{R - Z_0}{R + Z_0} = \frac{100 - 50}{100 + 50} = \frac{1}{3} \qquad\qquad Answer$$

For part (b) of the example,

$$\Gamma_v = \frac{R - Z_0}{R + Z_0} = \frac{25 - 50}{25 + 50} = \frac{-25}{75} = -\frac{1}{3} \qquad\qquad Answer$$

It is easy to show that for any passive value of resistance (a positive value) the reflection coefficient ranges from -1 at $R = 0$ to $+1$ at $R = \infty$.

When the excitation is a sinusoid, the principle of the reflection coefficient still applies. If the load has reactive components, then Γ_v takes on complex values.

example 20.9   Find $\tilde{\Gamma}_v$ for a load of $\tilde{Z}_L = 50 + j50$ on a line 50 $= Z_0$.

solution

$$\tilde{\Gamma}_v = \frac{\tilde{Z}_L - Z_0}{\tilde{Z}_L + Z_0} = \frac{50 + j50 - 50}{50 + j50 + 50} = \frac{j50}{100 + j50} = \frac{50\underline{/90°}}{111.8\underline{/26.6°}}$$

$$= 0.45\underline{/63.4°} = \tilde{\Gamma}_v \qquad\qquad Answer$$

$$= 0.20 + j0.40 = \tilde{\Gamma}_v \qquad\qquad Answer$$

VOLTAGE STANDING WAVE RATIO   When forward and backward sinusoidal traveling waves exist simultaneously, they create an interference pattern, because at different locations on the line the sine waves will be in phase, out of phase, or somewhere in between. The ratio of the largest magnitude voltage to the smallest magnitude voltage is called the VSWR. Note that VSWR relates magnitudes only and so can never be a complex number like the reflection coefficient. One can, however, establish a relationship between the magnitude of Γ_v and VSWR, as shown in Example 20.10.

example 20.10   What is the VSWR on the 50-Ω line terminated as shown?

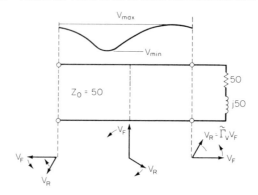

solution From Example 20.9,

$$\tilde{\Gamma}_v = 0.20 + j0.40 \qquad \text{or} \qquad \tilde{\Gamma}_v = 0.45\underline{/63.4°}$$

Since $\qquad V_{max} = |V_F| + |V_R| = |V_F| + |\tilde{\Gamma}_v||V_F| = |V_F|(1 + |\tilde{\Gamma}_v|)$

similarly, $\qquad V_{min} = |V_F|(1 - |\tilde{\Gamma}_v|)$

So $\qquad\qquad\qquad\qquad \text{VSWR} = \dfrac{V_{max}}{V_{min}} = \dfrac{1 + |\tilde{\Gamma}_v|}{1 - |\tilde{\Gamma}_v|}$

Now $|\tilde{\Gamma}_v| = 0.45$, so $\qquad\qquad \text{VSWR} = \dfrac{1 + 0.45}{1 - 0.45} = 2.64 \qquad\qquad\qquad \textit{Answer}$

Historically, measurements of VSWR were the first to be done because the apparatus required is quite simple. By measuring the VSWR and the location of the nulls of the interference pattern, one could determine algebraically the nature and value of an unknown reflection coefficient.

SMITH CHART Phillip H. Smith devised a transmission line chart in 1939 which bears his name. It was an improvement on the charts that existed previously for microwave work. One of its primary uses is to quickly translate VSWR measurements to values of reflection coefficient and load impedance. The chart also finds great application in the area of filter design. Remember when you use the chart that all impedances are expressed in a normalized fashion, that is, the actual impedance is divided by Z_0. Position on a transmission line is determined by rotating a point about the center of the chart; once around corresponds to a one-half wavelength distance. Several examples best illustrate its use.

example 20.11 See Fig. 20.10. An impedance of $Z_L = 25 - j100$ terminates a 50-Ω transmission line. What is the reflection coefficient Γ_v? What is the VSWR?

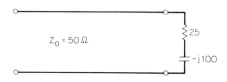

solution
Step 1. Normalize to the line impedance:

$$Z_L' = \frac{Z_L}{Z_0} = 0.5 - j2$$

Step 2. Plot the point P on the chart and draw circle (A) through the point centered on the chart.

Step 3. Where circle A crosses the diameter, project down and read $|\Gamma_v|$ and VSWR:

$$|\Gamma_v| = 0.82$$
$$\text{VSWR} = 10$$

Step 4. To determine the angle of Γ_v, draw radius B through point P and read the angle:

$$\angle\, \Gamma_v = -51°$$

example 20.12 See Fig. 20.11. By putting an element in shunt with the transmission line at some point, we may cause the input impedance of the line to appear real and equal to 50 Ω. Find the value and location of the shunt impedance required, which is closest to the load end of the line.

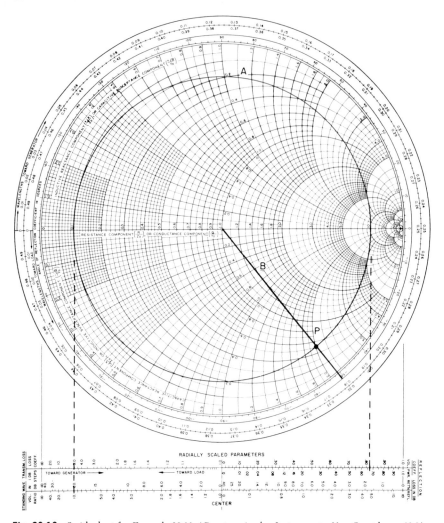

Fig. 20.10   Smith chart for Example 20.11. (*Courtesy Analog Instruments, New Providence, N.J.*)

solution

Step 1. Since we are going to use a shunt element, convert to admittance. Plot the admittance point Q and draw a circle through it:

$$Z_L = 25 - j100 \qquad \frac{Z_L}{Z_0} = 0.5 - j2 = Z'_L \qquad Y'_L = \frac{1}{Z'_L} = \frac{1}{0.5 - j2} = 0.12 + j0.47$$

(Note: Q is diametrically opposed to $Z'_L = 0.5 - j2$.)

Step 2. The first point toward generator (CW) where the circle intersects the circle $G = 1$ is point W. At this point on the line, the input admittance is $1 + j2.9$. The distance along the line away from the load is read off the chart as $0.202 - 0.07 = 0.132$ wavelength.

Step 3. By placing a shunt susceptance of $-j2.9$ at this point, the net input admittance is

$$Y' = 1 + j2.9 - j2.9 = 1 \qquad \therefore Z' = 1 \qquad \text{or} \qquad Z = 50 \ \Omega$$

Step 4. A susceptance of $-j2.9$ equals a reactance of $1/(-j2.9) = j.345$, which is inductive reactance.

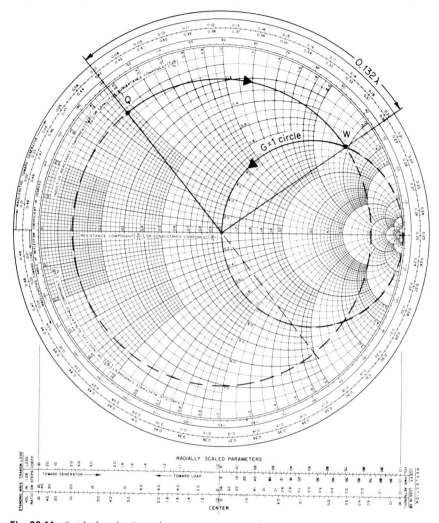

Fig. 20.11 Smith chart for Example 20.12. (*Courtesy Analog Instruments, New Providence, N.J.*)

Step 5. To unnormalize, multiply by Z_0:

$$X = j17.2$$

example 20.13 See Fig. 20.12. Slotted-line measurements on a WR-90 waveguide slotted line at 10 GHz show a minimum voltage at 4.64 cm from the reference plane. A VSWR of 3.0 was measured also. What is the impedance at the reference plane?

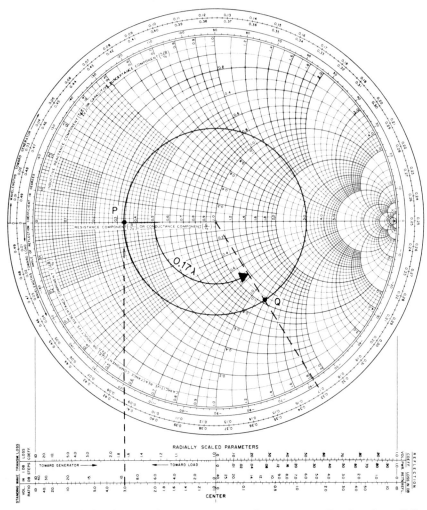

Fig. 20.12   Smith chart for Example 20.13. (*Courtesy Analog Instruments, New Providence, N.J.*)

solution
Step 1. Calculate the guide wavelength λ_g:

$$\lambda_g = \frac{\lambda_0}{\sqrt{1 - (f_{co}/f)^2}}$$

From Table 17.3 f_{co} = 6.557 GHz, and

$$\lambda_0 = \frac{3 \times 10^{10}}{10 \times 10^9} = 3 \text{ cm}$$

so
$$\lambda_g = \frac{3}{\sqrt{1 - (6.557/10)^2}} = 3.97 \text{ cm}$$

Step 2. Locate the VSWR on the scale of the Smith chart, and draw the circle of this radius.

$\tilde{a}$ = forward or incident wave
$\tilde{b}$ = backward or reflected wave

$$\tilde{S}_{nm} = \frac{\tilde{b}_n}{\tilde{a}_m} \bigg|\ \text{all other } \tilde{a}\text{'s} = 0$$

Fig. 20.13   A generalized S parameter concept.

Step 3. The distance of 4.64 cm corresponds to $4.64/3.97 = 1.17$ wavelengths. Starting at the minimum voltage point P, move toward load (CCW) by 1.17 wavelengths. That is twice around the chart plus 0.17 wavelength.

Step 4. Read off the normalized impedance at Q:

$$Z_L' = 1.05 - j1.2$$

POWER TRANSFER   Most of us are already familiar with the principle of matching load impedance to source impedance for maximum power delivered to the load. At microwave frequencies when a transmission line is placed between source and load, that principle is easily extended. For maximum power to be delivered to the load, the source impedance, transmission line impedance, and load impedance should all be equal. Since the reflection coefficient for a resistor of value Z_0 is 0, all power is absorbed by it and none is reflected. At the input end to such a line, an impedance $Z_0 = R$ is presented to the source.

S PARAMETERS   With the advent of more sophisticated measuring equipment for microwave work, the S parameter method of characterizing circuits has gained popularity. Microwave circuits are categorized by the number of ports they have. A port might be a coaxial connector or a waveguide opening. To characterize a circuit means to quantify the effects of all other ports on a particular one until all have been quantified. A one-port circuit needs only one such parameter to describe it completely, a two-port circuit requires four, a three-port nine, etc. The four S parameters of a two-port circuit completely describe in both magnitude and phase how that circuit will react to any combination of input and output signals. The simplest example of an S parameter is the case of a one-port circuit where the complex reflection coefficient Γ_v is the only S parameter required.

The S parameters are the magnitude and phase relations between voltage waves. They are of two types:

1. Backward-to-forward voltage wave ratio at a particular port when all other ports have no input

2. Backward wave at a port due to forward wave at another port

Because we are dealing with sinusoidal voltages, the S parameters are usually complex quantities. A generalized S parameter concept drawing is shown in Fig. 20.13. For a two-port network, the S parameters have familiar common names:

S_{11} input reflection coefficient
S_{12} reverse insertion loss
S_{21} forward insertion loss
S_{22} output reflection coefficient

example 20.14   For a two-port network, solve for the S parameters.

solution   Our hypothesis is that the backward wave at any port is a linear combination of all incident waves, at least over some limited range of operation.

$$b_1 = S_{11}a_1 + S_{12}a_2 \qquad b_2 = S_{21}a_1 + S_{22}a_2$$

If we terminate port 2 in a matched load, then a_2 is 0. Thus

$$b_1 = S_{11}a_1 + S_{12}(0) \qquad S_{11} = \frac{b_1}{a_1}\Big|_{a_2=0} \qquad S_{21} = \frac{b_2}{a_1}\Big|_{a_2=0}$$

$$b_2 = S_{21}a_1 + S_{22}(0)$$

Similarly,

$$S_{12} = \frac{b_1}{a_2}\Big|_{a_1=0} \qquad S_{22} = \frac{b_2}{a_2}\Big|_{a_1=0}$$

example 20.15   When a_2 is not equal to 0, what is the input reflection coefficient b_1/a_1? (This is the case if port 2 is terminated in a nonmatched load.)

solution

$$Z \neq Z_0 \qquad \therefore \Gamma_v \neq 0$$

and

$$b_1 = S_{11}a_1 + S_{12}a_2$$

but

$$a_2 = \Gamma_v b_2 = \Gamma_v(S_{21}a_1 + S_{22}a_2)$$

Solving for a_2 in terms of a_1 yields

$$a_2 = \frac{\Gamma_v S_{21}a_1}{1 - \Gamma_v S_{22}}\, a_1$$

Substituting gives

$$b_1 = S_{11}a_1 + S_{12}\frac{\Gamma_v S_{21}a_1}{1 - \Gamma_v S_{22}}\, a_1$$

Finally,

$$\frac{b_1}{a_1} = S_{11} + \frac{S_{12}S_{21}\Gamma_v}{1 - \Gamma_v S_{22}}$$

SKIN EFFECT   Loss of signal power in a microwave transmission circuit is a very important consideration. A coaxial line has its loss expressed in units of decibels per 100 ft. The amount of this loss is a function of the materials used in manufacturing the cable and is related also to the frequency of operation. Here we depart again from ordinary circuit theory in that resistance (loss) depends on frequency, which has no basis in conventional theory. The explanation of this phenomenon is called the *skin effect*. When signals of high frequency pass through transmission line, the currents do not penetrate the entire metallic wires but travel near the surface of the wires. This results in an apparent increase of resistance as the frequency is raised and the current-carrying area decreases. For example, coils in high-power transmitters sometimes are made of hollow tubing which is silver-coated. The hollow interior does not increase the resistance of the tubing for that frequency of operation. Often the interior of copper waveguide is flashed with silver since most of the energy barely penetrates the surface. A detailed explanation would require extensive use of electromagnetic theory.

Current decays within a conductor in an exponential manner. The skin depth is defined as that thickness in which the current is reduced by a factor $1/\epsilon$ (note: this is analogous to the time constant). The skin depth δ is related to frequency f, dc conductivity σ, and magnetic permeability μ. For nonferrous metals, the value of μ is the same as the free-space value μ_0.

example 20.16   Compare the skin depth at 60 Hz and 6 GHz in copper.

solution

$$\delta = \frac{1}{\sqrt{\pi f \mu \sigma}} = \frac{1}{\sqrt{\pi (60)(4\pi \times 10^{-7})(58 \times 10^{6})}}$$
$$\delta \text{ (at 60 Hz)} = 8.53 \times 10^{-3} \text{ m} = 0.336 \text{ in}$$
$$\delta \text{ (at 6 GHz)} = 33.6 \text{ } \mu\text{in}$$

20.3   CIRCUIT COMPONENTS

In this section we describe many of the most frequently encountered microwave components. A few large categories are established first to arrange the components in a systematic manner. Active versus passive components are a first distinction. By *active* we mean any component that requires energy to function other than that supplied by the microwave signal itself. Otherwise, a component is passive. As an example, an amplifier is considered active because it requires dc input power to function, whereas an attenuator is a passive component, in general. We include in the active category components such as diode switches which do not produce more output power than is inputted but nonetheless require an external power source.

The next major subdivision is whether a component is linear or nonlinear. Linearity here implies the same meaning as in ordinary circuit theory: (1) the output signal is in direct proportion to the input signal, and (2) a combination of input signals will produce the same combination of responses at the output in the same proportions with no new signals created.

Further distinctions can be made based on the number of ports a component possesses. In the group of components having two or more ports, a further distinction is whether that component is reciprocal. A reciprocal two-port network would measure the same insertion loss and insertion phase regardless of the direction of the microwave energy. Clearly an amplifier is a nonreciprocal two-port device. During the era of microwave technology over the past 50 years, an enormous quantity of components have been developed. Many generic types of components exist simultaneously in different technologies. For example, a coupler may be made as a coaxial component or a waveguide component; similarly, an amplifier may be a traveling wave tube which is a thermionic device or a solid-state GaAs FET amplifier. In the family tree of Fig. 20.14, the components we will be discussing are arranged in their appropriate positions.

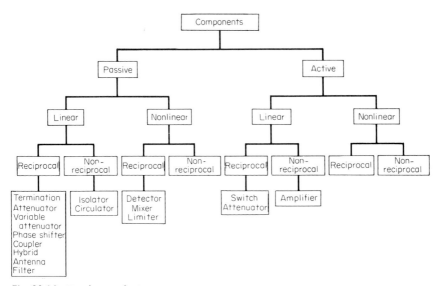

Fig. 20.14   Family tree of microwave circuit components.

(a)   (b)

Fig. 20.15 Waveguide terminations.

PASSIVE LINEAR RECIPROCAL COMPONENTS

Terminations Terminations are merely resistors designed to perform at microwave frequencies. They usually take a physical shape that is compatible to the transmission medium being used. Ordinary resistors with leads are not generally suitable for high-frequency application. The most frequently used terminations are 50 Ω which correspond to the most widely used coaxial line impedance.

In waveguide components, the termination fits into the waveguide opening. It is made from a block of graphite and, as shown in Fig. 20.15, is shaped as a pyramid or a staircase. In all cases, the purpose of a "matched" termination is to provide a reflectionless ending for a transmission line. For certain laboratory calibration purposes, deliberately mismatched terminations are sold. The termination is a one-port network.

Attenuator The purpose of an attenuator is to reduce the magnitude of a signal. Unlike the potentiometer in ordinary circuit analysis, the signal reduction must occur while a matched impedance is maintained at input and output ports. Early attenuators were designed following the principles of the well-known T pad and Pi pad, as shown in Fig. 20.16. Because there are three resistor values to manipulate, one can see that it is possible to provide signal reduction and maintain a 50-Ω input and output impedance.

In waveguide systems, an attenuator takes the form of a dissipative (carbonized) card which is partially inserted in the waveguide opening. Many specialized attenuators have been developed for high-power applications where the dissipation of heat is the primary design difficulty.

Variable attenuator Various clever mechanizations for mechanically adjustable attenuators have been devised. More recently, voltage-controlled variable attenuators that have no moving parts have made large inroads, and we will discuss these devices later.

The stepped variable attenuator is simply a selector switch with a number of fixed atten-

Fig. 20.16 Examples of Pi and T pads which attenuate the stated value and present a 50-Ω impedance to the source when the output is terminated in 50 Ω.

uators chosen individually by rotating the switch. This device finds wide use at the lower microwave frequencies, i.e., below 2 GHz.

The continuously variable attenuator is made by ganging several potentiometers such that they produce a matched T or Pi pad at all positions of rotation. These devices also are best suited to the lower microwave frequencies.

Continuously variable attenuators of stripline or microstrip construction are suited for the higher microwave frequencies. See Fig. 20.17. Typically a dissipative element is slid, by means of a lead screw, across the conductor of the transmission line. As the absorptive material approaches the line, more energy is removed from the transmission path.

Waveguide variable attenuators control insertion loss by adjusting the position of an absorbing card within the waveguide opening. See Fig. 20.18. Several standard methods are frequently used:

Fig. 20.17   Microstrip variable attenuator.

1. Through a slit in the broad wall of a waveguide a card is gradually inserted to vary the attenuation.

2. The card moves from the short wall of the waveguide gradually to the center of the waveguide for maximum attenuation.

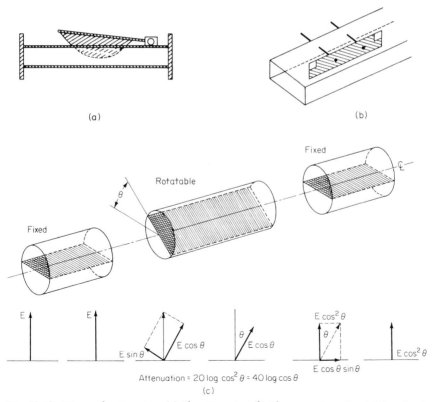

Fig. 20.18   Waveguide attenuators: (*a*) Flap attenuator. (*b*) Glass-vane attenuator. (*c*) Functional drawing including operating principle of the rotary-vane type of attenuator.

3. An absorbing card is rotated in a circular waveguide where its inclination relative to the electric field precisely determines the insertion loss. The rotary-vane attenuator is an extremely precise type of mechanical attenuator favored for use as a laboratory grade instrument.

Phase shifter   The phase shifter is a device used to vary the phase of a signal passing through it in a nearly lossless process. Most simply, a phase shifter may be a coaxial line manufactured in such a manner that it can be lengthened physically. A stripline phase shifter is made by sliding a dielectric card over the conductor. The presence of the dielectric causes the velocity of propagation to decrease and thus increases the number of electric degrees of phase shift per linear dimension. More recently, semiconductor diodes are employed. Waveguide phase shifters are made very similarly to waveguide attenuators except that the absorbing material is replaced with dielectric in the case of side wall or top wall phase shifters. The rotary-vane phase shifter, like the rotary-vane attenuator, is a precise laboratory instrument.

Coupler   The coupler is usually a three- or four-port device. Its function is to divide a signal entering a port into a fixed proportion among the other ports. This function can be accomplished with very small loss to the total power. All couplers of the low-loss type possess directional properties; for example, in Fig. 20.19a, the signal entering at port 1 produces 90 percent of its power at port 2 and 10 percent at port 3. However, a signal entering at port 2 produces 90 percent of its power at port 1 and ideally nothing at port 3. This type of coupler, as are all lossless couplers, is actually a four-port device with the fourth port terminated internally. Indeed, some couplers are sold as four-port couplers, as shown in Fig. 20.19b. A signal entering port 1 couples to port 3; a signal entering port 2 couples to port 4. None of a signal entering at port 2 should emerge at port 3; however, in real devices, some small amount does. That fraction is called *isolation*. The difference between isolation and coupling values is called *directivity*. Frequently, directional couples are available in waveguide or coaxial versions with standard coupling values of 6, 10, 20, and 40 dB.

Hybrids   Hybrids are, generally speaking, specialized 3-dB four-port couplers. Because they, like couplers, operate by proper phasing of signals, they are virtually lossless. The two major divisions of hybrids are based on the number of electric degrees which the coupler shifts the signals passing through it. The 90° hybrid, sometimes called the *short-slot hybrid* or *quadrature hybrid*, is shown in Fig. 20.20a with the phase relations. The 180° hybrid, often called the *magic T*, is shown in Fig. 20.20b. These devices are enormously useful for applications where signals need to be combined or canceled.

Antennas   The antenna can be thought of as a device for matching the transmission line impedance to the impedance of free space. In that sense it is merely a special type of transmission line. In Sec. 20.4 the subject is discussed more fully.

Filters   Filters are devices for selecting frequencies from all that are present; some are passed while others are attenuated. Most filters are constructed in a symmetric fashion and so are reciprocal. Because most filters contain only reactive and lossy elements, they are passive.

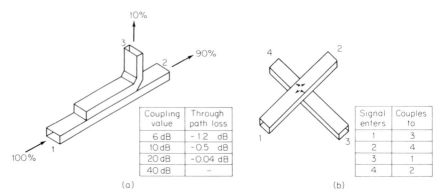

Coupling value	Through path loss
6 dB	− 1.2 dB
10 dB	− 0.5 dB
20 dB	− 0.04 dB
40 dB	−

Signal enters	Couples to
1	3
2	4
3	1
4	2

(a)   (b)

Fig. 20.19   (*a*) Broadwall coupler. (*b*) Crossguide coupler.

90° Hybrid
Signal enters port

		1	2	3	4
Response at port (mag/phase)	1	–	0	-3dB/0°	-3dB/-90°
	2	0	–	-3dB/-90°	-3dB/0°
	3	-3dB/0°	-3dB/-90°	–	0
	4	-3dB/-90°	-3dB/0°	0	–

(a)

180° Hybrid
Signal enters port

		1	2	3	4
Response at port (mag/phase)	1	–	0	-3dB/0°	-3dB/0°
	2	0	–	-3dB/0°	-3dB/180°
	3	-3dB/0°	-3dB/0°	–	0
	4	-3dB/0°	-3dB/180°	0	–

(b)

Fig. 20.20   Hybrid 3-dB four-port couplers.

PASSIVE LINEAR NONRECIPROCAL COMPONENTS

Isolator   An isolator is a two-port device that allows a microwave signal to pass in one direction only. A small loss is incurred by the signal in the forward direction, and a substantial loss occurs in the reverse direction. This behavior is somewhat analogous to the action of a diode as it relates to the passage of current. The mechanism which produces this effect depends on the interaction of the microwave signal with a static magnetic field. Typical modern isolators at X band have insertion losses of 0.7 dB and reverse insertion loss of 20 dB or more. Both coaxial and waveguide versions are available over the entire microwave band.

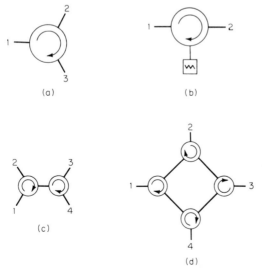

Fig. 20.21   (a) Basic three-port circulator. (b) Circulator connected as an isolator. (c) Four-port circulator, two junctions. (d) Four-port circulator, four junctions.

Circulator The nonreciprocal interaction of a microwave signal with a static magnetic field results in passive devices which sort signals based on their direction of travel. The three-port circulator in Fig. 20.21a is probably the most fundamental of these devices since all others can be derived from it. A signal entering at port 1 emerges at port 2, a signal entering at port 2 emerges at port 3, and a signal entering port 3 emerges at port 1. The circulator with one port terminated becomes an isolator (Fig. 20.21b). When the magnetic field is supplied by a reversible electromagnet, a circulator becomes an electromagnetic switch.

Fig. 20.22 labels below the figure:

90° Hybrid   Nonreciprocal phase shifters   90° Hybrid

Fig. 20.22 One type of differential phase shift, four-port circulator.

When more than three ports are desired in the circulation loop, arrangements such as those shown in Fig. 20.21c and d are often used with the three-port circulator as the basic building block.

A widely used circulator mechanization is called the Y junction, which is available in coaxial and waveguide models. It is simple to build, may be made by using printed-circuit techniques, and is relatively inexpensive and physically compact. If exceptionally high power is to be handled, a device known as a differential phase shift circulator is often used. See Fig. 20.22. This device contains no circulators at all, but rather is made up of nonreciprocal phase shifters and hybrids. Another type of circulator worth mentioning is the waveguide Faraday rotational circulator. This device separates signals traveling in opposite directions by rotating them through 45° each as they interact with the magnetic field, but in the opposite sense of rotation.

PASSIVE NONLINEAR RECIPROCAL COMPONENTS

Detector The detector is a junction diode. Its tiny size and low capacity enable it to probe a microwave signal and rectify it. Rectified voltage is then applied across a resistor, which completes the circuit. The small voltage that appears across the resistor usually must be amplified greatly in order to produce a useful signal. Today, in addition to p-n junction diodes, Schottky diodes and back diodes, which have a higher ratio of voltage produced to power input, are available. Within limited ranges of operation detectors are considered "square law." This simply means that the voltage produced is in direct proportion to the power incident on the diode, as shown in Fig. 20.23.

Mixer A mixer is one or more semiconductor junctions utilized to combine signals in a nonlinear fashion. Because the V-I characteristic of a diode can be thought to contain high-order polynomial terms (Taylor series), two important results occur:

  1. A single-frequency signal will produce its harmonics at the mixer output.

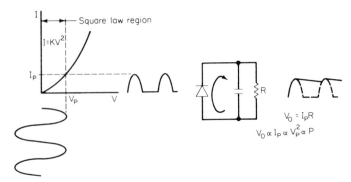

Fig. 20.23 Square-law operation of a detector diode.

$$i(t) = K_0 + K_1v + K_2v^2 + K_3v^3 + \text{----}$$

Single-tone intermodulation distortion

Harmonics of f_R	Harmonics of f_L: 0	1	2	3	4	5	
5	82>90>90	80 83 87	78>90 90	72 78 74	80>90>90	67 74 71	First order
	82>90>90	80 82 90	80>90>90	75 78 82	83>90>90	73 75 78	
4	89>90>90	86>90 >90	>90>90>90	86>90>90	95>90>90	82>90>90	
	89>90>90	82>90 >90	90>90>90	82>90>90	90>90>90	80>90>90	
3	58 65 69	56 62 74	63 68 76	54 54 57	68 72 76	50 53 53	Second order
	60 68 72	58 62 76	65 71 78	56 58 61	68 72 78	54 59 59	
2	71 76 77	65 78 76	74 80 80	67 75 77	74 85 82	65 75 70	
	71 78 80	64 77 78	74 83 82	65 75 78	74 86 82	63 74 71	
1	16 24 30	0 0 0	28 40 40	11 13 13	34 43 44	22 28 25	
	16 24 30	0 0 0	26 38 40	10 11 11	32 39 44	17 20 20	Third order
0	A B C	26 34 32	34 37 36	37 47 44	45 57 60	36 52 51	
		24 32 30	30 33 32	37 44 42	41 48 50	34 46 51	

A:
0.05 – 200 MHz
Class 1 mixer

B:
5 – 500 MHz
Class 1 mixer

C:
2 – 500 MHz
Class 1 mixer
with baluns

f_R = 49 MHz @ –10 dBm
f_L = 50 MHz

| A B C | f_L @ +10 dBm |
| A B C | f_L @ + 7 dBm |

Fig. 20.24 Mixer cross-product analysis. (*Courtesy Watkins-Johnson, Palo Alto, Calif.*)

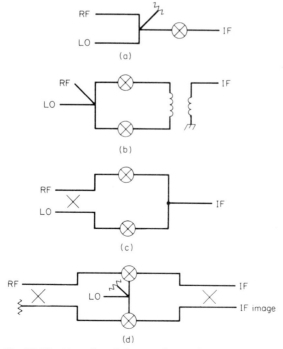

Fig. 20.25 (*a*) Single-ended mixer. (*b*) 180° balanced mixer. (*c*) 90° balanced mixer. (*d*) Image reject mixer.

TABLE 20.4   Mixer Comparison Guide

				Mixer comparison guide		
	Single-ended	Balanced (90°)	Balanced (180°)	Double-balanced	Image reject	Image recovery
Conversion loss	Good	Good	Good	Very good	Good	Excellent
VSWR LO, RF	Good, poor	Good, good	Fair, fair	Poor, poor	Good, good	Good, good
LO/RF isolation	Fair	Poor	Very good	Very good	Good	Very good
LO power required	+13	+5	+3	+10	+7	+7
Spurious rejection	Poor	Fair	Fair odd: fair	Good	Fair	Fair
Harmonic					Even: good	Even: good
suppression	Poor	Fair	Even: good	Very good	Odd: fair	Odd: Fair
Third-order intercept	—	+13 dB$_m$	+13	+18 dB$_m$	+15 dB$_m$	+15 dB$_m$

2. Two different frequency signals will produce sum and difference frequencies at the mixer output and sums and differences of harmonics. See Fig. 20.24, which shows the amplitude of the cross products in decibels relative to the desired term, $f_R \pm f_L$, for three typical double balanced mixers and two bias levels.

The frequency conversion made possible by mixers enables the building of the superheterodyne receiver, which is the mainstay of radio, TV, and radar.

When individual mixer diodes are combined with other passive components, namely hybrids, they are given special names. These new configurations possess features such as suppression of the second harmonic of one or both inputs or rejection of an undesired image which are not present in the basic single-diode mixer. See Fig. 20.25 and Table 20.4. Although mixer diodes resemble detectors in appearance, it must be remembered that the mixer deals only with RF signals and no detection occurs. The mixer's three ports are often referred to as:

RF   radio frequency
LO   local oscillator
IF   intermediate frequency

It is important that the semiconductor junction be operated in the proper nonlinear region of the *V-I* curve; this is called *biasing* the mixer. Bias is supplied by local oscillator signal power and/or dc bias.

Limiter   A limiter restricts the amount of power passing through it. Before the power reaches the limit value, the limiter has low loss. Two major types of limiters exist today: the diode limiter and the gas-tube limiter. A short-circuited diode placed in the transmission path will begin to conduct a current by detector action as the incident power is increased. The current that flows appears as a short circuit to the incident signal and reflects it. The gas tube is a short piece of waveguide filled with an inert gas. When the power level is reached that produces ionization of the enclosed gas, the plasma reflects the incident signal.

Fig. 20.26   (*a*) Series diode SPDT switch. (*b*) Shunt diode SPDT switch.

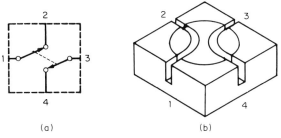

Fig. 20.27   *E* plane waveguide transfer switch. (*a*) Schematic diagram. (*b*) Cutaway view.

ACTIVE LINEAR RECIPROCAL COMPONENTS

Switch   The application of miniature semiconductor diodes as microwave circuit elements has enabled a profusion of active control devices. A reverse-bias p-n junction between the inner and outer conductors of a coaxial cable would nearly appear to be an open circuit. When forward-bias current from an external source is passed through the diode, it appears as a short circuit to an incident signal. Switches fabricated as shown in Fig. 20.26 that use diodes would allow extremely rapid switching between ports. Both series diode and shunt diode switches are possible. In general, a switch containing multiple diodes produces an isolation (off state) of 20 dB per junction. Switches using diodes as the control element are made in all transmission media. Usually an electronic circuit accompanies the device and drives current through the diodes in response to external logic commands. The bias current is much greater than any detected RF current so the RF signal has minimal effect on diode bias.

Mechanical coaxial and waveguide switches have long been used. Even those that have been motorized are extremely slow in comparison to the diode switch; however, insertion loss and isolation characteristics are superior. A type of mechanical switch frequently encountered in microwave work is called the *transfer switch*, as shown in Fig. 20.27. This type of switch is seldom encountered in electronic work.

Attenuators   The diode voltage-controlled attenuator is similar to the switch in its operation. Rather than a bilevel control of diode bias, as in the diode switch, a continuous control of bias current is utilized. This mode of operation permits the insertion loss of the device to be varied over a wide range. Multiple-diode attenuators are frequently produced. Often an electronic circuit that modifies the control-voltage-to-diode-current transfer function precedes the diode so that the insertion loss in decibels is linearly proportional to control voltage. Two major types of diode attenuators are currently used: the reflective and absorptive types. They differ in how the energy not passed is dissipated. The reflective type has a higher input VSWR because the energy off the diode goes directly backward. In the absorptive type, phase relationships are such that the reflected energy is directed to a load. See Fig. 20.28.

Another class of electronic attenuator consists of fixed pads and diode multithrow switches. This type is used when accuracy of attenuation and flatness of frequency response are desired. Fixed pads and diode switches also could be arranged in a binary circuit. See Fig. 20.29.

ACTIVE LINEAR NONRECIPROCAL COMPONENTS

Amplifiers   Today, nearly all low- to medium-power amplifiers use solid-state devices. For frequencies below 2 GHz, the bipolar transistor is the main device; at frequencies above 2 GHz, the gallium arsenide FET takes over. For amplifiers whose output power exceeds 1 W, thermionic devices, such as traveling wave tubes and crossed field amplifiers, dominate. The manufacture of solid-state microwave amplifiers is a highly competitive field and requires extremely fine photolithographic techniques at the higher frequencies. A "single-device" amplifier usually produces an 8-dB gain. Modern amplifiers do not require tuning and are inherently broadband. It is presently possible to purchase an octave band (8 to 16

Fig. 20.28 (*a*) Reflective-type attenuator. (*b*) Absorptive-type attenuator.

Fig. 20.29 (*a*) Multipad step attenuator. (*b*) 0- to 45-dB binary attenuator.

TABLE 20.5 Low Noise Wideband Amplifiers
Guaranteed Specifications at 25°C Case Temperature (courtesy Avantek, Santa Clara, Calif.)

Model	Frequency response (GHz) minimum	Gain (dB) minimum	Noise figure (dB) maximum	Power output for 1-dB gain compression (dBm) minimum	Gain flatness (±dB) maximum	Typical intercept point for IM products (dBm)	VSWR (50 Ω) maximum In	Out	Input power Volts DC	Current (mA)	Case drawing
4 to 8 GHz											
AMT-8032	4–8	17	4.5	+15	1.0	+25	2.0	2.0	+12	100	IC2
AMT-8033	4–8	27	4.5	+15	1.0	+25	2.0	2.0	+12	150	IC4
AMT-8034	4–8	36	4.5	+15	1.5	+25	2.0	2.0	+12	200	IC4
AMT-8035	4–8	45	4.5	+15	2.0	+25	2.0	2.0	+12	250	IC6
AMT-8052	4–8	10	8.5	+20	0.75	+30	2.0	2.0	+12	220	IC2
AMT-8053	4–8	19	6.0	+20	1.0	+30	2.0	2.0	+12	270	IC4
AMT-8054	4–8	28	5.0	+20	1.0	+30	2.0	2.0	+12	320	IC4
AMT-8055	4–8	37	4.5	+20	1.5	+30	2.0	2.0	+12	340	IC6
5 to 10 GHz											
AMT-10033	5–10	27	5.5	+13	1.0	+23	2.0	2.0	+12	160	IC4
AMT-10034	5–10	37	5.5	+13	1.5	+23	2.0	2.0	+12	210	IC4
AMT-10035	5–10	46	5.5	+13	2.0-	+23	2.0	2.0	+12	270	IC6
7 to 12 GHz											
AMT-12033	7–12	22	4.5	+15	1.5	+25	2.0	2.0	+12	150	IX4
AMT-12034	7–12	30	4.0	+15	1.5	+25	2.0	2.0	+12	200	IX4
AMT-12035	7–12	37	4.0	+15	2.0	+25	2.0	2.0	+12	250	IX6
AMT-12036	7–12	44	4.0	+15	2.0	+25	2.0	2.0	+12	300	IX6
8 to 12.4 GHz											
AMT-12433	8–12.4	22	4.5	+15	1.5	+25	2.0	2.0	+12	150	IX4
AMT-12434	8–12.4	30	4.0	+15	1.5	+25	2.0	2.0	+12	200	IX4
AMT-12435	8–12.4	37	4.0	+15	2.0	+25	2.0	2.0	+12	250	IX6
AMT-12436	8–12.4	44	4.0	+15	2.0	+25	2.0	2.0	+12	300	IX6

TABLE 20.5 Low Noise Wideband Amplifiers (continued)

Model	Frequency response (GHz) minimum	Gain (dB) minimum	Noise figure (dB) maximum	Power output for 1-dB gain compression (dBm) minimum	Gain flatness (±dB) maximum	Typical intercept point for IM products (dBm)	VSWR (50 Ω) In	VSWR (50 Ω) Out	Input power Volts DC	Input power Current (mA)	Case drawing
6 to 18 GHz											
AWT-18633	6–18	16	7.5	+12	1.0	+22	2.0	2.0	+12	160	IX4
AWT-18634	6–18	22	7.5	+12	1.0	+22	2.0	2.0	+12	200	IX4
AWT-18635	6–18	28	7.5	+12	1.5	+22	2.0	2.0	+12	250	IX6
AWT-18636	6–18	34	7.5	+12	1.5	+22	2.0	2.0	+12	300	IX8
AWT-18637	6–18	40	7.5	+12	2.0	+22	2.0	2.0	+12	350	IX8
AWT-18638	6–18	46	7.5	+12	2.0	+22	2.0	2.0	+12	400	IX8
AWT-18656	6–18	34	7.5	+20	1.5	+28	2.0	2.0	+12	350	IX6
8 to 18 GHz											
AWT-18033	8–18	15	7.5	+13	2.0	+22	2.0	2.0	+12	150	IX4
AWT-18034	8–18	20	7.5	+13	2.0	+22	2.0	2.0	+12	200	IX4
AWT-18035	8–18	25	7.5	+13	2.0	+22	2.0	2.0	+12	250	IX6
AWT-18036	8–18	20	7.5	+13	2.0	+22	2.0	2.0	+12	300	IX6
AWT-18037	8–18	35	7.5	+13	2.5	+22	2.0	2.0	+12	350	IX8
AWT-18038	8–18	40	7.5	+13	2.5	+22	2.0	2.0	+12	400	IX8
AWT-18039	8–18	45	7.5	+13	2.5	+22	2.0	2.0	+12	450	IX10
AWT-18057	8–18	35	7.5	+20	2.0	+28	2.0	2.0	+12	350	IX8
12 to 18 GHz											
AMT-18033	12–18	16	7.0	+13	2.0	+20	2.0	2.0	+12	150	IX4
AMT-18034	12–18	22	7.0	+13	2.0	+20	2.0	2.0	+12	200	IX4
AMT-18035	12–18	27	7.0	+13	2.0	+20	2.0	2.0	+12	250	IX6
AMT-18036	12–18	33	7.0	+13	2.0	+20	2.0	2.0	+12	300	IX6
AMT-18037	12–18	38	7.0	+13	2.0	+20	2.0	2.0	+12	350	IX8
AMT-18038	12–18	44	7.0	+13	2.0	+20	2.0	2.0	+12	400	IX8

GHz) GaAs FET amplifier with 60-dB gain and a saturated output power of $+20$ dBm which is unconditionally stable. An example of the diversity of available amplifiers is shown in Table 20.5.

20.4 ANTENNAS

The subject of antennas is in no way restricted to the microwave region, but is frequently introduced during the study of microwaves because of the profusion of antenna types in that frequency region. Some antennas associated with microwaves have counterparts at the lower frequencies while others have commonality with optical devices. Because the micro-wave wavelengths are of a convenient size, being neither too large nor too small, a great variety of antenna types have been developed.

In this discussion we present an overview of antenna principles and types, but avoid detailed discussion of electromagnetic field theory. We accept without resorting to Max-well's equations that radio energy can be made to propagate through space. The means by which the energy transitions into space or is collected for reception can be the general definition of an antenna.

RECIPROCITY   All antennas are reciprocal. We extend the definition of reciprocity at this time to include the directional properties of any antenna. For example, consider a dipole antenna connected to a radio transmitter. Current produced by the transmitter flows up and down the dipole antenna; energy radiates into space, and the signal strength is greatest broadside to the dipole. The principle of reciprocity now implies that when this antenna is used for receiving signals, those which emanate from a broadside direction will be received to a greater degree than those coming from other directions. A more succinct way of stating the same thing is that the directivity of an antenna is identical on transmission or reception.

ISOTROPIC RADIATOR   Before we define directivity and gain, the concept of isotropic radiation must be understood. The sun is an isotropic radiator of light, which is to say that the light intensity or the watts per square meter of light energy are directed equally in all directions. An isotropic antenna would produce concentric spheres of constant power intensity centered on the isotropic antenna. If one integrated the power intensity over a surface enclosing the antenna, one could calculate the total power radiated. This last fact is a consequence of the inverse-square law nature of propagating electromagnetic energy.

ANTENNA PATTERNS   An antenna pattern is a plot of the relative power intensity at a fixed distance from the antenna under test as a function of angular direction. Antenna patterns are measured at distances sufficiently removed that one can ignore the effect of radial distance on the relative pattern shape. This is called the *far field pattern*. The line of demarcation is generally taken as

$$R = \frac{2L^2}{\lambda}$$

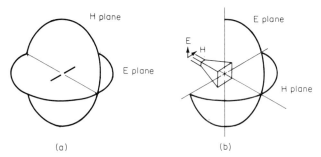

(a)   (b)

Fig. 20.30 Principal radiation planes for (a) dipole antenna, (b) horn antenna.

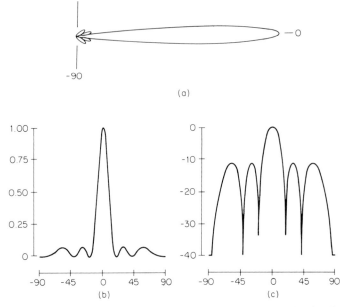

Fig. 20.31   Data formats for antenna patterns: (*a*) Direct power, polar plot. (*b*) Direct power, rectangular plot. (*c*) Logarithmic power (dB), rectangular plot.

where R is the minimum radial distance to the point of observation, L is the maximum linear dimension of the antenna under test, and λ is the operating wavelength.

Because we are measuring a three-dimensional characteristic on two-dimensional paper, many patterns are required to determine the relative power intensity at all points in space (see Fig. 20.30). The planes that contain only electric (E) field or magnetic (H) field at the antenna are usually orthogonal and determine the principal planes of the antenna pattern. Sometimes the principal planes are determined relative to mechanical references as, for example, with circularly polarized antennas. The principal planes are equivalently the equatorial plane and the plane of the prime meridian. When antennas are measured relative to terrestrial references, azimuth and elevation are often the principal planes.

The data may be plotted in polar or rectangular form, and the amplitude may be in direct power or decibels. See Fig. 20.31.

POLARIZATION   In describing linearly polarized antennas (the electric field is oriented in a constant direction), the plane parallel to the electric field is named for its orientation relative to a fixed reference. The antenna is then said to be polarized in that direction. Vertical and horizontal polarization refers to the direction of electric field relative to the earth. Parallel or perpendicular polarization refers to orientation relative to the plane of incidence, as in airborne radar sets. See Fig. 20.32.

DIRECTIVITY   Antennas are useful because we can increase the power intensity in one direction and reduce it in another. The total power radiated, of course, remains unchanged. The ratio of the maximum power intensity for a given antenna to that radiation intensity of an isotropic radiator of the same total power is called *directivity*.

GAIN   The gain of an antenna, while similar to directivity and often confused with it, has a different definition. *Gain* is the ratio of maximum power intensity to the maximum power intensity of a reference antenna having the same power input. Thus the gain characteristic includes losses between the antenna input terminals and free space. Recall that directivity makes no statement about the power input from a generator; thus the gain, relative to an isotropic radiator, of an antenna is less than its directivity.

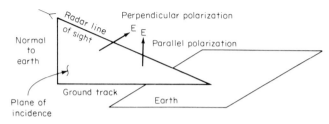

Fig. 20.32 Definition of polarization orientation.

EFFECTIVE AREA Effective area is most easily visualized by thinking of antennas in the receiving mode. The antenna sits in a flux of energy, collects it, and delivers it to a load. If we assume the process is lossless, then the power in the load is the intensity of the incident energy multiplied by the effective area of the antenna. All antennas possess an effective area even though their physical areas are small (e.g., a wire dipole antenna). The effective area does not always have a physical significance except in cases of horns or reflector-type antennas where the effective area and the physical area are nearly the same. Antenna effective area is simply another formulation for directivity, and a known fixed relation applies:

$$D = \frac{4\pi A}{\lambda^2}$$

where D is directivity, A is effective area, and λ is operating wavelength.

BEAMWIDTH Many antennas are designed to concentrate the energy in one or two directions. A measure of that concentration is called *beamwidth*. The radiation intensity decreases continuously from its maximum value as one moves away from the peak of the pattern and does not abruptly stop at the beamwidth angles, as might be construed. Normally the angular difference between the points where the radiation intensity is one-half of or 3 dB below its maximum value is called the beamwidth. See Fig. 20.33. As the beam becomes narrower, the directivity increases since the same energy is squeezed into a narrower solid angle.

ESTIMATING BEAMWIDTH AND DIRECTIVITY The relation between beamwidth and directivity is now approximated. See Fig. 20.34. If one assumes all power is contained within

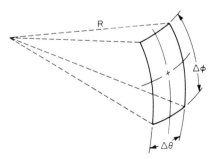

Fig. 20.33 Beamwidth of an antenna pattern.

Fig. 20.34 The principal plane beamwidths and solid-angle approximation.

the principal-plane beamwidths and is constant in that region and there is only one main lobe, then the power intensity over that area is the total power radiated divided by the area. Thus using radian angles

$$D = \frac{\text{maximum power intensity}}{\text{power intensity from isotropic source of equal total power}}$$

$$= \frac{W/A}{W/(4\pi R^2)} = \frac{W/(R\,\Delta\theta\,R\,\Delta\phi)}{W/(4\pi R^2)} = \frac{4\pi}{\Delta\theta\,\Delta\phi}$$

Converting to degrees yields

$$D = \frac{4\pi}{(\pi/180)\,\Delta\theta_1\,(\pi/180)\,\Delta\phi_1}$$

or

$$D = \frac{41\ 252}{\Delta\theta_1\,\Delta\phi_1}$$

To make this formulation more correctly reflect real-world patterns, i.e., to account for the approximations made, the number in the numerator is reduced to about 30 000.

Another useful rule of thumb, though not derived from the above, is mentioned at this point. When the size and frequency of an antenna are known, the beamwidth (and thus directivity) can be calculated:

$$\theta = \frac{60}{L_\lambda}$$

where θ is the half-power beamwidth in degrees and L_λ is the physical dimension, expressed as a number of wavelengths.

example 20.17   It is desired to take antenna pattern measurements at 20 GHz on a 3-in by 5-in aperture horn. Determine the pertinent parameters.

solution   The minimum distance for accurate measurements is

$$R = \frac{2L^2}{\lambda} = \frac{2(5)^2}{11.8/20} = \frac{50}{0.59} = 84.7 \text{ in}$$

The beamwidths are (interfering objects are kept clear of this region)

$$\theta = \frac{60}{5/0.59} = 7.08° \qquad \phi = \frac{60}{3/0.59} = 11.8°$$

The estimated directivity is

$$D = \frac{30\ 000}{7.08 \times 11.8} = 359, \text{ or } 25.5 \text{ dB}$$

At 84.7 in the size of the beam is

$$84.7 \times 7.08 \times \frac{\pi}{180} = 10.5 \text{ in}$$

$$84.7 \times 11.8 \times \frac{\pi}{180} = 17.4 \text{ in}$$

ANTENNA ARRAYS   The earliest technique used to shape the radiation pattern was to arrange antennas in geometric figures and control the phase angle and amplitude at each antenna so that the signals would add in phase in one direction and cancel at other directions. This technique is the basis of all antenna pattern analysis. The individual antennas making up the array are called *elements*. They may be vertical towers standing in a field for a broadcast station, slots cut periodically into a waveguide, or coplanar parallel metal rods, as in a TV antenna. These arrays are called *driven* arrays because each element is connected to the others and to the electronic device by a transmission line. In a *parasitic* array only one element is connected to the transmission line and the other elements are excited by mutual coupling to the driven element. We discuss the principles of driven arrays.

example 20.18   For two isotropic in-phase equal-amplitude radiators spaced 1 wavelength apart, find an expression for the far-field pattern.

solution   The path-length difference to the point of observation depends on the separation and angle of observation

$$l = d \sin \theta$$

The electric phase difference at that observation point is

$$360 \frac{d}{\lambda} \sin \theta = \psi$$

In the figure the voltage from radiator 2 lags that from radiator 1 by ψ degrees. Summing the voltages at P yields

$$\overset{(1)}{\tilde{V}} = 1\underline{/0} + \overset{(2)}{1\underline{/-\psi}}$$

$$= 1 + e^{-j\psi}$$

$$= (e^{j\psi/2} + e^{-j\psi/2})e^{-j\psi/2}$$

$$= \left(2 \cos \frac{\psi}{2}\right)(e^{-j\psi/2})$$

The magnitude is

$$|\tilde{V}| = 2 \cos \frac{\psi}{2}$$

The power is proportional to $|\tilde{V}|^2$, and normalizing for the number of elements yields

$$P = \cos^2 \frac{\psi}{2}$$

Now $\psi = 360 \dfrac{d}{\lambda} \sin \theta$ and $d = \lambda$

$$= 360 \sin \theta$$

Finally,

$$P = \cos^2 (180 \sin \theta)$$

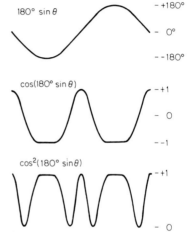

Notice that in Example 20.18 there are four lobes. The two at $\theta = 0°$ and $\theta = 180°$ are primary lobes, and the two at $\theta = +90°$ and $\theta = -90°$ are called *grating* lobes. Grating lobes cannot arise if separation between elements is less than approximately 0.6λ.

A similar analysis on N equal-amplitude uniform phase progression isotropic radiators yields the useful expression

$$P = \left[\frac{\sin (N\psi/2)}{N \sin (\psi/2)} \right]^2$$

where N is the number of elements, $\psi = \phi + 360(d/\lambda) \sin \theta$, and ϕ is the feed-driven electrical phase-shift between adjacent elements. The case of two elements is explored in Fig. 20.35.

PATTERN MULTIPLICATION   Sometimes the radiation patterns of complicated arrays can be generated by multiplying the array patterns of subdivisions of the original array by the pattern of isotropic radiators representing the overall array. As in Fig. 20.36, the subdivisions are called *element patterns*, and the overall representation is called an *array factor*. In Fig. 20.37, we demonstrate the binomial array by pattern multiplication. This array is said to have a taper; i.e., the central elements have greater amplitude than the edge elements.

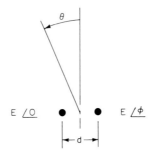

Fig. 20.35   Two equal-amplitude isotropic radiators.

MICROWAVE ANTENNAS

Dipoles   Although dipoles are used throughout the frequency spectrum, they are important to microwave antennas despite their small physical size. Dipoles can be printed on circuit boards along with their feed networks to form planar arrays. They also could be printed in diminishing lengths to make a log-periodic antenna (see Fig. 20.38). In the log-periodic antenna, the spacing between the Nth and $(N + 1)$st element is a fixed fraction of a wavelength of the Nth element. This is done so that each element when excited will "see" identical environments.

Spirals   The equiangular spiral shown in Fig. 20.38b is a logarithmic structure. The radius of the spiral and the width of the element increase in proportion to the angle subtended. Low frequencies excite the large-radius elements, and high frequencies excite the smaller-radius portions. The spiral antenna is a convenient means of producing circular polarization. Spirals are, in general, broad-beamed and low in gain. Another spiral antenna frequently encountered is called the *Archimedes spiral antenna*. In it the element width and pitch are constant.

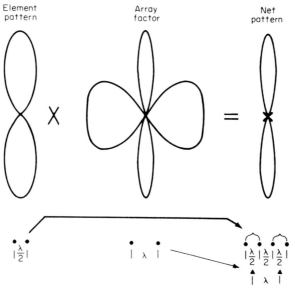

Fig. 20.36 Pattern generation for a four-element in-phase, equal-amplitude array by the method of pattern multiplication.

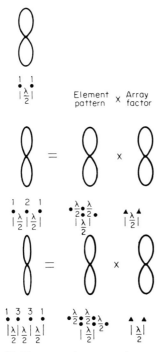

Fig. 20.37 Binomial array. Elements are spaced one-half wavelength apart, in phase, having amplitudes proportional to the binomial expansion coefficients. The overall pattern is derived by pattern multiplication in this case. The binomial array has no side lobes.

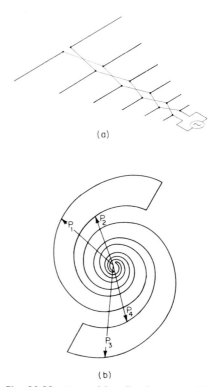

Fig. 20.38 Types of broadband antennas: (a) Log-periodic dipole array. (b) Equiangular spiral.

Horns   A horn antenna is a waveguide that is flared in one or two dimensions. Figure 20.39 illustrates the common rectangular and circular types. The gradual flaring creates a plane wave at the aperture of the horn. The larger the aperture, the narrower the beamwidth, and vice versa. The theory of horn antennas is so well understood that horns are often used as gain calibration standards. One can regard the emerging wavefront as a linear array of infinitely small point sources. See Fig. 20.40.

Focusing antennas   The antennas discussed next operate on quasi-optical principles. The analogy between reflective telescopes and reflector antennas is fairly clear. A metal surface takes the place of a polished mirror. The dielectric lens has the common glass lens as its analogy. The idea of collimating or focusing the sun's rays into a point by using shaving mirrors or a magnifying glass is a common experience shared by many experimenters.

Reflector Type.   It is possible to generate an aperture with an emerging plane wave by using the geometric properties of a parabola. A parabola is the locus of points equidistant from a point and a line. It follows that waves emitted from the focal point of a parabola will add in phase in a plane perpendicular to the parabolic axis, the aperture. See Fig. 20.41a and b.

Lens Type.   The planar wavefront also may be created by using a lens. Spherical wavefronts emitted from a source located at the lens' focus are straightened out as shown in Fig. 20.41c. The center of the lens, being thicker, will delay the wavefronts more than at the edges of the lens. In Fig. 20.41d, metal slats parallel to the electric field accelerate the wavefronts at the edges of the lens so that they catch up to the central portion. This is possible because the phase velocity is greater in a waveguide than in free space.

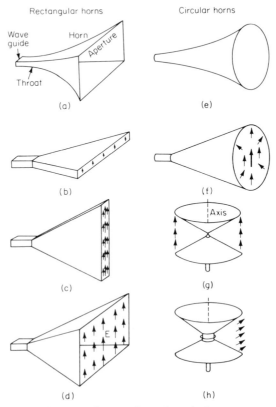

Fig. 20.39   Types of rectangular- and circular-horn antennas.

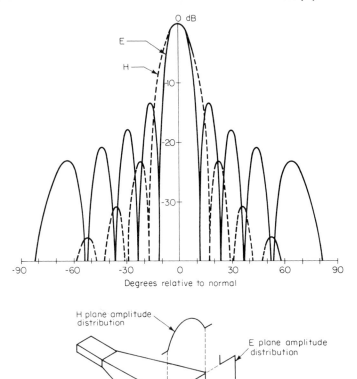

Fig. 20.40 Ideal E and H plane directivity patterns of a 5-wavelength square-horn antenna fed by a waveguide in the TE_{10} mode.

20.5 TEST EQUIPMENT

Because of the quantities measured and the technique used to measure them, microwave test equipment has become a separate specialty. The use of the equipment being restricted to those involved in microwave work, since it has little use to others, has caused a certain amount of "black magic" to be associated with it. To explain in detail all the equipment that currently exists requires far more space than is allotted. So in this section we present a few of the classical pieces of test equipment whose use never seems to diminish and two of the modern, sophisticated equipments available today.

POWER METER As its name implies, the power meter measures the absolute magnitude of the power absorbed by its sensing device. The sensing device consists of an appropriate waveguide or coaxial fitting and an element known as the *thermistor*. The thermistor changes its resistance because of the power it dissipates. The change in resistance is sensed (usually with a bridge circuit), amplified, and manipulated to produce an absolute power reading on the display device. All older power meters have a meter as the display device; however, some modern units are equipped with digital readouts (see Fig. 20.42a). Because temperature is involved in the measurement, all power meters are provided with a means of adjusting for the zero power point before a measurement is taken. Today's sensing ele-

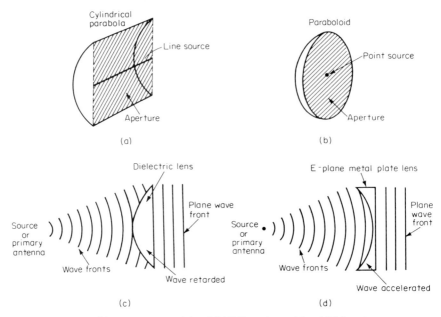

Fig. 20.41   Types of focusing antennas: (*a*) and (*b*) Reflector types. (*c*) and (*d*) Lens types.

ments can cover the range of 10 MHz to 26.5 GHz and the power range of -70 to $+20$ dB_m with just two sensors. It must be borne in mind that the VSWR of a sensor and any adaptors associated with it will affect the reading obtained. The power actually measured is reduced by that amount reflected or lost:

$$P_{meas} = P_{incid} - P_{refl} - P_{lost}$$
$$= P_{incid}(1 - |\Gamma|^2) - P_{lost}$$

SWR METER   An essential and extremely versatile microwave test equipment is the SWR meter. See Fig. 20.42*b*. This instrument measures relative power and depends on the square-law operation of the detector device. Frequently the detector is a crystal; however, a bolometer (a device with a large square-law region) also may be used. The meter scale is calibrated in decibels and has a direct-reading VSWR calibration, which simplifies the taking of measurements on the slotted line. The amplifier circuits in the meter are actually sharply tuned audio amplifiers operating at 1 kHz. See Fig. 20.43. For the meter to function, therefore, a 1-kHz amplitude modulation must be impressed on the microwave RF signal. The gain of the audio amplifier usually has two adjustments: a multiposition decade attenuator called *range* for 10-dB steps in gain and a continuously variable gain control so any input power may be set to a reference mark.

This scheme of detection provided a sensitive and convenient means of measurements at a time when microwave RF amplifiers were very bulky and expensive. Most signal generators for the microwave bands are equipped with the 1-kHz modulation capability.

SLOTTED-LINE   The slotted-line remains the foremost device for precise measurement of VSWR, insertion loss, and impedance. Slotted-lines are available in coaxial as well as waveguide versions. See Fig. 20.42*c*. In the waveguide type, a longitudinal slot is cut into the center of the broad wall of the waveguide. A thin electric probe and detector diode are held by a movable carriage which traverses the length of the slot. Adjustments for the probe penetration and a vernier for precise distance measurement are also part of the equipment assembly. Typically, the detected 1-kHz signal is then read by using an SWR meter. The distance vernier is a means of measuring frequency, but more importantly it enables the

Fig. 20.42 (*a*) Digital-reading power meter with sensor. (*b*) SWR meter. (*c*) Coaxial slotted-line meter. (*d*) Cavity-type frequency meter. *(Courtesy Hewlett-Packard Co.)*

Fig. 20.43 Operation of SWR meter.

determination of complex impedance when the data are evaluated with the aid of a Smith chart (see example 20.14).

example 20.19   Measure the VSWR of an unknown device by using a slotted-line meter and an SWR meter.

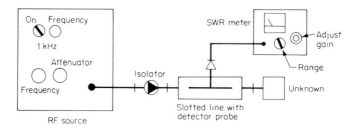

RF source

solution
1. Set the RF frequency to the value desired. Turn on 1-kHz modulation.
2. Adjust sequentially and repeat if necessary: 1-kHz frequency adjust, output attenuator, probe tuning (if so equipped), SWR meter range switch, and SWR meter continuous gain. Maximize the reading without exceeding square-law region of the detector. Use minimum probe penetration.
3. Rock slotted-line carriage to achieve a maximum indication on the SWR meter. Use gain control to set needle on VSWR = 1 (0 dB). A typical range position is 20 dB.

Range = 20 dB

4. Move the carriage until a minimum is reached. Read VSWR on the upper scale.

Range = 20 dB   VSWR = 1.7

5. If a minimum is not reached at VSWR = 3, increase the range control by 10 dB.

Range = 30 dB

6. Continue toward the minimum. Read VSWR on the lower scale.

Range = 30 dB   VSWR = 4.2

7. This process may continue. Alternate scales for each 10-dB increase, and add 10 to VSWR reading for each 20-dB increase in the range.

Range = 40 dB   VSWR = 12

FREQUENCY METER   The basic analog meter is made with an adjustable cavity. See Fig. 20.42d. The cavity resonance is varied by a precision thread arrangement which may be read directly in frequency. The unknown frequency is identified when the position of the frequency meter causes a dip in the power level transmitted through the meter. Electronic counters have substantially replaced cavity-type meters in CW applications; however, for pulse measurements the cavity meter has not yet been excelled.

RF SOURCES   RF sources fall into two general categories: single frequency and swept frequency. A single-frequency source such as a klystron usually incorporates a tuning mechanism to adjust its frequency over some limited range. The swept type of source is needed when one desires to measure the properties of a network at all frequencies between two limits. Swept sources are made from backward wave oscillators (BWOs) and recently Gunn diodes followed by swept yttrium iron garnet (YIG) filters. Most laboratory sources have power output levels below 50 MW ($+17$ dB$_m$).

SWEPT RETURN AND INSERTION LOSS TEST SET   At times it is desirable to use return loss (RL) rather than VSWR as a measure of the match of a network. The *return loss* is the relative power of the reflected wave compared to the incident wave, expressed in decibels. Clearly,

$$RL = 20 \log|\Gamma|$$

A relation exists among VSWR, return loss, and Γ which is tabulated for certain values in Table 20.6.

By using the same principle as in the SWR meter, i.e., detection of an amplitude-modulated microwave signal, amplification at audio frequencies, and subsequent audio detection, a swept frequency display can be generated on a CRT. See Fig. 20.44a. By arranging the AM frequency of the microwave signal to be considerably higher than the RF sweep rate (for example, 20 kHz versus 100 Hz), AM and FM can coexist and act independently. If the CRT horizontal sweep is derived from the RF sweep generator, a frequency response plot is obtained, providing the detector and generator have a flat response and adequate dynamic range. When detectors are made identically, two can be monitored at once and their outputs subtracted, in a logarithmic sense, to cancel variations in the generator output power as it sweeps. The CRT then displays the difference between input and output power (insertion

TABLE 20.6 Return Loss, VSWR, and Reflection Coefficient

Return loss (dB)	VSWR	Reflection coefficient
∞	1.00	0.000
26.44	1.10	0.048
20.82	1.20	0.091
17.69	1.30	0.130
15.56	1.40	0.167
13.98	1.50	0.200
12.74	1.60	0.231
11.72	1.70	0.259
10.88	1.80	0.286
10.16	1.90	0.310
9.54	2.00	0.333
9.00	2.10	0.355
8.52	2.20	0.375
8.09	2.30	0.394
7.71	2.40	0.412
7.36	2.50	0.429
7.04	2.60	0.444
6.75	2.70	0.459
6.49	2.80	0.474
6.25	2.90	0.487
6.02	3.00	0.500

Fig. 20.44 (a) Swept return and insertion loss test set shown with detectors and digital memory unit. (b) Microwave network analyzer. (Courtesy Hewlett-Packard Co.)

loss) or the difference between input and reflected power (return loss). These instruments also possess an absolute power measurement capability of somewhat less accuracy than a power meter.

This type of instrument lends itself to rapid production testing. With the addition of A/D converters and bus-controllable instruments, the entire measurement can be microprocessor-controlled. The disadvantage of this instrumentation is that no phase information is obtainable because detection occurs immediately.

MICROWAVE NETWORK ANALYZER   The most encompassing piece of microwave test equipment currently available is the network analyzer. See Fig. 20.44b. With it one can measure both the phase and the amplitude of insertion loss and return loss. The network analyzer consists of two similar superhetrodyne receivers with phase detection and amplitude comparison performed at a low intermediate frequency prior to detection. The local oscillator for the receivers is designed to follow a swept RF source of moderate slow rate over large frequency excursions. Depending on the display unit selected, one may have a meter indication, a decibel rectangular plot of amplitude and phase versus frequency, or a linear polar plot of relative amplitude and phase. Refinements to the basic network analyzer include calculator control of measurements which may also perform accuracy enhancement by vector algebra processing of measured data.

20.6 BIBLIOGRAPHY

Brown, R. G., et al.: *Lines, Waves and Antennas,* Ronald, New York, 1961.

Jasik, Henry: *Antenna Engineering Handbook,* McGraw-Hill, New York, 1961.

Jordan, Edward C.: *Electromagnetic Waves and Radiating Systems,* Prentice-Hall, Englewood Cliffs, N.J., 1950.

Kaufman, M., and A. Seidman: *Handbook of Electronics Calculations,* McGraw-Hill, New York, 1979.

Kraus, John D.: *Antennas,* McGraw-Hill, New York, 1950.

Lance, Algie L.: *Introduction to Microwave Theory and Measurements,* McGraw-Hill, New York, 1964.

Schelkunoff, S. A., and H. T. Friis: *Antennas—Theory and Practice,* Wiley, New York, 1952.

Chapter 21

Fiber-Optic Systems

Joseph R. Cleveland

Senior Manager
Transmission Technology Facility, MCI
Richardson, Texas

21.1 INTRODUCTION

Through the centuries people have been intrigued by the use of light to transmit information. Many quests have been initiated to realize optical communication systems which are practical and economical. One such example is the rapid progress made in fiber-optic technology. The large number of present and planned fiber-optic communication installations attests to how readily this technology has been adopted.

As with the established wire or radio communication technology, a basic fiber-optic system (Fig. 21.1) consists of a transmitter (light source with modulator or driver), propagation medium (optical fiber with connectors or splices), and a receiver (light detector and preamplifier). The light source may be either a light-emitting diode (LED), similar to that used in pocket calculators, or the more intense injection laser diode (ILD). Information is sent as intensity modulation (IM) of the light produced by the transmitter. In keeping with radio frequency (RF) technology, the optical signal is referred to as the *carrier*. Although frequency or phase modulation of the carrier can be accomplished, the technology is not yet available for practical implementation.

The position of the optical carrier in the RF spectrum is identified in Fig. 21.2. Note that optical communication frequencies range from 300 to 600 terahertz (THz). In the optical range it is common to specify the wavelength λ, given by

$$\lambda = \frac{c}{f} \tag{21.1}$$

where f is the frequency and c is the speed of light, which is 3×10^8 m/s in vacuum. Three units of wavelength commonly used are
  1. Angstrom (Å) $= 10^{-10}$ m
  2. Nanometer (nm) $= 10^{-9}$ m
  3. Micrometer (μm) $= 10^{-6}$ m

  example 21.1   An optical carrier has a frequency of 400 THz. What is its wavelength?

  solution

$$\lambda = \frac{3 \times 10^8 \text{ m/s}}{400 \times 10^{12} \text{ Hz}}$$
$$= 7.5 \times 10^{-7} \text{ m}$$
$$= 0.75 \ \mu\text{m} = 750 \text{ nm} = 7500 \text{ Å}$$

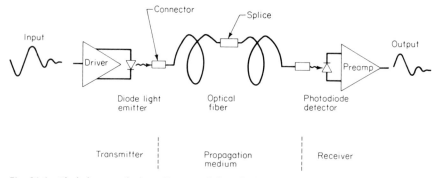

Fig. 21.1 Block diagram of a basic fiber-optic link. A duplex link requires an identical system but in reverse order.

Presently, fiber-optic systems operate in the range of 0.6 to 1.6 μm.

Fiber-optic systems offer a number of distinct advantages over wire or radio systems. Chief among these are

1. Low transmission losses
2. Greater bandwidths
3. Lower weight, less bulky cables

Fig. 21.2 Electromagnetic wave spectrum. The RF spectrum lies below 300 GHz, and the optical spectrum lies between 300 GHz and 30 000 THz. Optical transmitters used in fiber-optic systems emit light in the visible and near-infrared portions of the spectrum between 200 and 600 THz.

4. Immunity to inductive interference

5. Excellent signal confinement

These advantages relate to communications technology as follows:

1. Low transmission loss means that the light signal in the optical guide suffers little loss between the transmitter and receiver. Fibers are available now with an attenuation rating as low as 0.3 decibel per kilometer (dB/km) (a power reduction of 6.7 percent for each 1-km length). In contrast to coaxial cable characteristics, fiber attenuation is independent of modulation frequency, as shown in Fig. 21.3.

2. A greater bandwidth means that more information can be conveyed than with wire or radio communication systems. This property coupled with low fiber attenuation provides long-distance, high-data-rate transmission links. For example, transmission of 2×10^9 bits/s over a 44-km fiber link has been demonstrated.

3. Lower weight and less bulky cables permit increased transmission capacity in underground conduits. A 7-mm-diameter optical cable can replace a 7.6-cm (3-in) diameter, 900-pair copper cable (see Fig. 21.4) and provide the same information bandwidth. This size reduction permits the installation of four optical cables in a conduit occupied by the 900-pair wire cable. Together with the reduced size goes an enormous reduction in weight: 3.6 kg of optical cable can replace 94.3 kg of copper cable. The size and weight reduction make optical cables ideal for communication links in aircraft and ships. Of course, there are also cost savings associated with lower weight and less bulk.

4. Immunity to inductive interference exists because the fiber is a dielectric, so changing electric or magnetic fields do not induce interfering signals. That is, the optical fiber does not act as an antenna to pick up radio-frequency interference (RFI), electromagnetic interference (EMI), or electromagnetic pulses (EMP). Consequently, cross-talk, which is the coupling of information from one pair of copper wires to another nearby, is virtually nonexistent in optical fiber cables. The result is noise-free transmission.

5. Excellent signal confinement occurs because the optical guide is designed to trap the optical electromagnetic wave within the fiber. Hence, external electric and magnetic fields are virtually nonexistent. Such a phenomenon is important for secure communication links required by military organizations and desired by cable television services.

To employ these advantages, the designer must assemble a communication link which satisfies the system requirements for analog signal-to-noise ratio (SNR) or digital bit error rate (BER), bandwidth or data rate, transmission distance, and environment. To accomplish this goal, a number of factors must be considered. The amount of optical power coupled into the fiber and the receiver noise sets a limit to the maximum received signal quality. Losses of optical power in the fiber degrade this performance quantity. The longer the fiber, the worse these losses become.

Important to the bandwidth requirement are the contributions to the risetime and pulse broadening made by the system. The risetime T_r is the time required for a pulse to rise from

Fig. 21.3 Comparison of attenuation as a function of modulation frequency for several coaxial cables and an optical fiber. (*Courtesy AMP, Inc., Harrisburg, Pa.*)

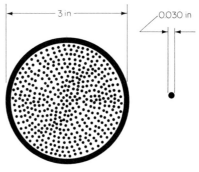

Fig. 21.4 Size comparison between a 900-pair copper telephone cable and an optical cable with the same information capacity.

Fig. 21.5 The meaning of risetime and pulse width of a pulse waveform.

10 to 90 percent of its maximum value, as indicated in Fig. 21.5. The pulse broadening $T_{1/2}$ refers to increases in the pulse width measured between the one-half maximum value. For a system with a transmitter, fiber, and receiver in cascade, the composite risetime is given by

$$T_r^2 \text{ (sys)} = (1.1)^2[T_r^2 \text{ (trans)} + T_r^2 \text{ (fiber)} + T_r^2 \text{ (rec)}] \qquad (21.2)$$

The pulse broadening for the system is determined in a similar manner. The corresponding equivalent bandwidth is approximately

$$B = \frac{0.32}{T_r} \qquad (21.3a)$$

or

$$B = \frac{0.44}{T_{1/2}} \qquad (21.3b)$$

in terms of pulse broadening. These equations are used frequently in optical system design.

21.2 OPTICAL FIBERS AND CABLES

Driving the development of fiber-optic systems has been the rapid advance in low-loss, wide-bandwidth optical fibers. These thin glass or plastic strands, with diameters between 0.003 and 0.5 mm, conduct the modulated optical carrier from the transmitter to the detector. The choice of the proper optical cable to accomplish this task depends on the fiber light-guiding ability, attenuation, bandwidth, cross-sectional structure (cabled fibers), and cable configuration (fiber connectors).

LIGHT GUIDING The simplest fiber type consists of a cylindrical core of glass (silica) or plastic with index of refraction n_1 surrounded by a concentric layer of glass or plastic of index n_2, as shown in Fig. 21.6. The index of refraction n is defined as

$$n = \frac{\text{speed of light in vacuum}}{\text{speed of light in material}} \qquad (21.4)$$

Fig. 21.6 Basic structure of a step-index fiber.

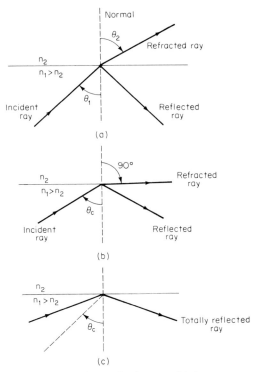

Fig. 21.7 Reflection and refraction of light rays at a boundary between two different materials for (a) light incident at an angle less than the critical angle, (b) light incident at the critical angle, and (c) light incident at an angle greater than the critical angle.

Its value ranges from 1.45 to 1.55 for silica glass. Hence, light travels approximately 33 percent slower in silica than in air. The second layer is called the *cladding*, and it has a refractive index that is less than that of the core by a few percent. Such step-index structures are identified as either all-plastic, plastic-clad silica (PCS) or glass-clad silica (CGS). Core cladding sizes range from 30 to 125 μm and 400 to 600 μm, respectively. Additional layers, called buffers, are added to protect the cladding and to strengthen the fiber.

Optical power coupled into the fiber core is prevented from escaping the core by being totally reflected at the core boundary. Therefore, the light is guided along the fiber core. The basis for light guiding is illustrated in Fig. 21.7. Notice in Fig. 21.7a the light ray directed toward a boundary at an angle θ_1. Some of the light is reflected at the same angle θ_1. The rest of the light crosses the boundary into the second medium as a refracted ray which travels at an angle θ_2, given by Snell's law:

$$n_1 \sin \theta_1 = n_2 \sin \theta_2 \tag{21.5}$$

For this illustration $n_2 > n_1$, so that $\theta_2 > \theta_1$.

example 21.2 If the core index is $n_1 = 1.5$, the cladding index $n_2 = 1.4$, and the angle of incidence $\theta_1 = 20°$, what is the angle of refraction?

solution By substituting the given values into Eq. (21.5), we get

$$1.5 \sin 20° = 1.4 \sin \theta_2$$

and
$$\theta_2 = 21.5°$$

Note that only part of the light is reflected at the boundary; the rest is lost owing to refraction. The objective is to eliminate the loss of light power owing to refraction.

For light rays that are directed into a material with a lower refractive index, there is a certain angle of incidence for which the refracted angle is 90°. This specific angle of incidence is called the *critical angle* θ_c and is illustrated in Fig. 21.6b.

example 21.3   For the index values given in Example 21.2, determine the critical angle.

solution

$$1.5 \sin \theta_1 = 1.4 \sin 90°$$

and

$$\theta_c = 69°$$

At angles of incidence greater than θ_c, total internal reflection of light occurs at the core-cladding boundary in a fiber, as shown in Fig. 21.7c.

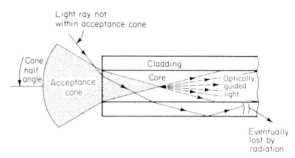

Fig. 21.8   Acceptance cone for a step-index fiber. Light that enters the fiber core at an angle greater than the cone half-angle is lost through refraction at the core-cladding boundary.

Therefore, light rays admitted to the fiber core only within a range of angles are guided by the fiber. This range defines an acceptance cone which is illustrated in Fig. 21.8. This cone is specified by the numerical aperture (N.A.) given by

$$\text{N.A.} = \sin \theta = \sqrt{n_1^2 - n_2^2} \tag{21.6}$$

where θ is the cone half-angle. For most fibers, the cladding index is only a few percent smaller than for the core, so Eq. (21.6) may be approximated as

$$\begin{aligned}\text{N.A.} &\approx \sqrt{2n_1(n_1 - n_2)} \\ &\approx \sqrt{2n_1 \, \Delta}\end{aligned} \tag{21.7}$$

where $\Delta = n_1 - n_2$. Step-index fibers have N.A. values between 0.20 and 0.60.

example 21.4   If the index difference is $\Delta = 0.05$ and $n_1 = 1.5$, find the numerical aperture.

solution

$$\begin{aligned}\text{N.A.} &= 2(1.5)(0.5) \\ &= 0.387\end{aligned}$$

This corresponds to an acceptance cone angle of 45.6°.

An important aspect of the numerical aperture is that it describes the light-capturing power of the fiber core. The larger the numerical aperture, the greater the optical power coupled into the fiber core, given approximately by

$$P_c = \frac{m + 1}{2} \frac{A_f}{A_e} (\text{N.A.})^2 P_o \tag{21.8}$$

where m = transmitter optical spatial distribution factor (discussed later)
   A_f = fiber core area
   A_e = light source emitting area
   P_o = total transmitter optical output power
If $A_f > A_e$, then $A_f/A_e = 1$.

example 21.5   For an LED with $P_o = 5$ mW, $m = 2$, and $A_e = 50 \times 150$ μm^2 coupled to a fiber with a 250-μm-diameter core and N.A. $= 0.45$, determine the coupled power coupled into the fiber.

solution

So
$$\frac{A_f}{A_e} = \frac{\pi(125 \ \mu m)^2}{50 \times 150 \ \mu m^2} = 6.5 > 1$$

$$P_c = \frac{3}{2}(1)(0.45)^2(5 \text{ mW}) = 1.52 \text{ mW}$$

The coupling loss in decibels is given by

$$\text{Loss} = 10 \log \frac{5 \text{ mW}}{1.01 \text{ mW}}$$
$$= 5.17 \text{ dB}$$

example 21.6   If the core diameter in Example 21.5 is reduced to 50 μm and the N.A. to 0.2, which is typical for telecommunications fibers, what is the coupled power?

solution

$$\frac{A_f}{A_e} = \frac{\pi(25 \ \mu m)^2}{50 \times 150 \ \mu m^2} = 0.26$$

then
$$P_c = \frac{2+1}{2}(0.26)(0.20)^2(5 \text{ mW}) = 80 \ \mu\text{W}$$

The coupling loss is 18 dB.

These examples illustrate the idea that as the N.A. decreases, the losses increase.

A problem with the step-index structure is that light rays may travel along several different paths, as shown in Fig. 21.9. For this reason, the step-index fiber is called a *multimode guide* and suffers from multimode dispersion. Since light energy traveling along two different paths propagates with different times, an input pulse of light is broadened, as shown in Fig. 21.10. Therefore, there is a limit to the pulse or data rate and to the fiber bandwidth. At present in commercially available step-index fibers, the bandwidth is limited to approximately 25 MHz for a 1-km length. This property is examined in more detail later.

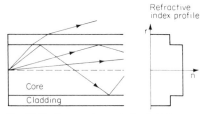

Fig. 21.9   Propagation paths for light in a step-index fiber.

To increase the bandwidth or pulse rate, the fiber core can be manufactured with the index of refraction gradually reduced as the core radius increases. The result is a graded-index fiber shown in Fig. 21.11. The effect is that light which travels farther from the core travels faster than that along the core center. For an appropriate index profile, the propagation time for the various modes is nearly equalized to yield a greatly reduced multimode dispersion term. Optical bandwidths for graded-index fibers range from 200 to over 1500 MHz for 1 km of fiber. A core buffer diameter of 50 to 125 μm is becoming a standard for telecommunications applications. Numerical aperture values range from 0.17 to 0.28.

If the core diameter is less than approximately 3.2 times the wavelength of the light signal in a step-index guide, then the light travels along one path, that is, in a single mode. This offers the potential for modulation bandwidths exceeding 1000 GHz. Single-mode fibers have a core diameter of 4 to 8 μm for 850-nm light. Because the core diameter of such fiber is so small, efficient light coupling between it and the optical transmitter or other fibers is difficult. For this reason, single-mode fibers are used primarily in laboratory demonstration applications. Single-mode fibers have cladding diameters in the range of 100 to 125 μm and numerical apertures of about 0.1.

ATTENUATION   Even with the best fibers available some of the light is lost by internal absorption and scattering, which attenuates the optical power. The optical intensity decreases geometrically as the signal travels from the source to the detector. For example,

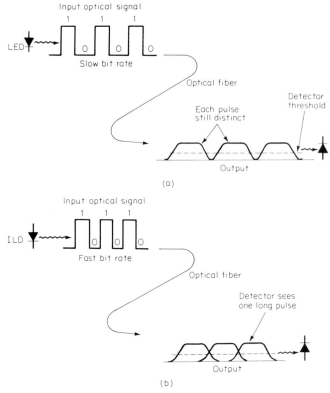

Fig. 21.10 Effect of multimode dispersion on digital pulse waveforms for (a) a low-dispersion fiber and (b) a high-dispersion fiber. *(Courtesy AMP, Inc., Harrisburg, Pa.)*

if the strength of the light falls to half its original value after the light has traveled 1 km, the intensity will fall to one-fourth of the original value at the end of the second kilometer, and so on. For long-distance transmission, the light source should be as powerful as possible and the cable losses or attenuation as low as possible, other things being equal.

Attenuation, or loss of light with distance, is expressed in decibels per kilometer (dB/km). Values range from less than 1 dB/km for low-loss telecommunications cables to 1000 dB/km for plastic fibers useful for short-distance transmission. Typical curves of attenuation versus optical wavelength for several fiber types are shown in Fig. 21.12. The attenuation is caused by four principal mechanisms: material absorption, material scattering, waveguide scattering, and leaky modes. The differences in the various curves represent the degree to which these imperfections have been reduced. For long fiber lengths, it is important to use an LED or laser diode which emits light with a wavelength where the fiber attenuation is a minimum. Properties of several fiber types are shown in Table 21.1.

Material absorption is caused by impurities in the core of the fiber which are exceedingly difficult to remove. The main impurities are the ions OH^-, Fe^{2+}, and Cr^{3+}, which in concentrations by weight of 1 part in 10^9 to 1.25 parts in 10^6 result in a peak loss of 1 dB/km. The absorption peaks near 950 and 1400 nm for silica-based fibers are due to the OH^- ions. Actual fibers con-

Fig. 21.11 Propagation paths for light in a graded-index fiber.

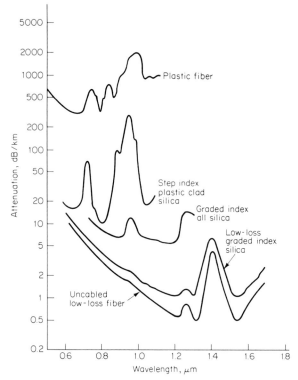

Fig. 21.12 Dependence of fiber attenuation on the optical wavelength for several fiber types.

tain varying amounts of these impurities depending on glass purification processes. Material or Rayleigh scattering is caused by frozen-in fluctuations in glass composition as well as glass-phase separations, which are small relative to the wavelength of light. This scattering varies as $1/\lambda^4$ and explains the decrease in attenuation with increasing wavelength for low-loss fibers.

Waveguide scattering results from geometric variations in the size of the core along its length, which in turn cause transfer of energy from one guided mode to another or from a guided mode to one that leaks into the cladding. In the former case, each mode may have different absorption and radiation coefficients. For a poorly made fiber, waveguide scattering losses are 10 dB/km.

The solution of Maxwell's equations which describe light propagation shows that even though light is totally reflected at the core boundary, the optical electromagnetic field extends into the cladding. (See Fig. 21.13.) This field is referred to as the *evanescent* field because its amplitude vanishes exponentially with radial distance from the core center. Attenuation or loss of the evanescent field owing to radiation or absorption from the cladding causes attenuation of the guided light in the core. Thus, the cladding thickness must be great enough that the evanescent field at the outer cladding surface is insignificant. Also, the cladding material should be almost as pure as that for the core. One of the reasons that plastic clad fibers have higher attenuation than all-glass fibers is that the plastic cladding is more lossy than glass cladding.

The importance of fiber attenuation in a fiber-optic link can be illustrated graphically. The optical power present in a fiber at a given distance L from the source can be computed as

$$P_f = P_c 10^{-\alpha L} \tag{21.9}$$

or as

$$P_f \, (\mathrm{dB_m}) = P_c \, (\mathrm{dB_m}) - \alpha L \tag{21.10}$$

TABLE 21.1 Properties of Selected Optical Fibers

Attenuation (dB/km @ λ in nm)	Bandwidth (MHz·km)	Numerical aperture	Core/clad diameter (μm)	Maximum length (km)	Manufacturer	Model no.
			Single mode (step index)			
0.7@1300	10^4		10/125	3	Fujikura Cable	SML 10/125–7
2.5@850	10^4		6/125	3	Fujikura Cable	SM 6/125–25
5@850	>1000	0.11	4/80	3	ITT, Electro-Optical	T 1601
3–20	≥80 000	0.12	5/100	1	Galite	7000
			Graded index			
1@1200	500	0.2	50/125	3	Valtec	ESP
1/2.5@	400–1000	0.2	50/125	1.1	Corning Glass Works	2504D2510D
2/3.5@	200–1000	0.2	50/125	1.1	Corning Glass Works	3502D–3510D
2–10	100–1000	≥0.21	63/125	3	Galite	6000 series
2.5@850	200–800	0.2	50/125	3	Valtec	
3@850	600–1500	0.21	50/125	1.1	Corning Glass Works	3006F–3015F
3@850	800	0.2	50/125	5–10	Sumitomo Electric	EG-5/4060
3@850	800	0.17	50/125	1	Dainichi-Nippon Cables	SM50G-SY-3L
3.5@850	400–800	0.21	50/125	1.1	Corning Glass Works	3504F–3508F
3.5@850	500	0.2	50/125	5–10	Sumitomo Electric	EG-5/4040
3.5@850	>1000	0.22	50/125	10	ITT, Electro-Optical	T 1211
4@850	200	0.2	50/125	3	Valtec	420
4–5@840	300	0.18	50/125	1.5	Fort	LG
4@840	500	0.17	60/125	2	Thomson CSF, LTT	GL 4
4@850	800	0.21	50/125	1.1	Corning Glass Works	4008F
4@850	300	0.2	50/125	3	Optronics	GIT
4@850	>1000	0.22	50/125	10	ITT, Electro-Optical	T 1209
5@850	400	0.21	50/125	1.1	Corning Glass Works	5004F
5–8@820	200–400	0.16	50/125	1	Times Fiber Commun	G-5000
5@850	400	0.25	62/125	1.0	General Fiber Optics	G-125/62-5
5–8@850	200–800	0.22	50/125	10	ITT, Electro-Optical	T 1200 series
5@850	800	0.17	60/150	1	Dainichi-Nippon Cables	SM60G-SY-5L
6–8	150	0.22	50/125	2	Standard Telephone	MA 3211-MA 2211
6@800	200	0.18	65/130	5	Fiberoptic Cable	G1-2/1-6UJ
6@850	200–400	0.21	50/125	1.1	Corning Glass Works	6002F–6004F
6@900	200	0.2	50/125	10	Fibronics	FGI-50-D
<7@820	400	0.21	50/125	1.1	Brand-Rex	Type II
			Step index			
2–10	10–20	≥0.21	63/125	3	Galite	5000 series
3–10	10–15	≥0.3	125/200	2	Galite	5020 series
4@820	30	0.21	90/125	20	Quartz & Silice	QSF 125AS-S
4@820	50	0.25	250/260	2	Fiberoptic Cable	Q1-LO-100J

TABLE 21.1 Properties of Selected Optical Fibers (*Continued*)

Attenuation (dB/km @ λ in nm)	Bandwidth (MHz·km)	Numerical aperture	Core/ clad diameter (μm)	Maximum length (km)	Manufacturer	Model no.
				Step index		
5 @ 820	30	0.4	125/300	25	Quartz & Silice	QSF 125A
5–20 @ 820	25	0.3	200/375	1	Times Fiber Commun	P-2000
5 @ 850	>30	0.25	50/125	10	ITT, Electro-Optical	T 1101
5 @ 850	30	0.17	100/150	1	Dainichi-Nippon Cables	SM 100-SY-5L
5 @ 850	30	0.19	200/400	3	Optronics	PC 200A
6 @ 820	40	0.25	125/150	5	Fiberoptic Cable	Q1-LO-5UJ
<7 @ 820	10	0.22	600	1.1	Brand-Rex	Type X
7 @ 700	20	0.3	100/140	1.0	General Fiber Optics	S-140/100-7
10 @ 850	15	0.30	100/140	1	ITT Components Group	SG5
10 @ 790	20	0.3	200/400	2	BOTec	EIS-200
10 @ 800	25	0.3	250/550	3	Valtec	PC 10
10 @ 820	25	0.24	200/300	4	Focom Systems	200/PSCI/10
10 @ 850	60	0.2	50/125	3	Optronics	SLT
12 @ 800	20	0.3	250/430	3.5	Fibronics	FSI-250
<15 @ 900	>10	0.25	100/150	1.5	Pilkington	15F150
15	5–15	≥0.4	204/305	2	Galite	4000LC
20 @ 790	25	0.23	200/400	3	Fiberguide Industries	Superguide-3
20 @ 820	25	0.25	250/260	2	Fiberoptic Cable	Q1-MD-10UJ
25 @ 820	22	0.24	250/380	2	Focom Systems	200/PCSI/25
25 @ 900	30	0.25	113/147	1	ITT Components Group	SG4
30 @ 820	35	0.4	125/300	25	Quartz & Silice	QSF 125C
30 @ 820		0.42	200/600		DuPont	S-120 type 30
<40 @ 900	6	0.48	200/250	10	Pilkington	40F250
40 @ 820	15	0.24	400/550	1.2	Focom Systems	400/PSCI/40
45 @ 670	20	0.25	250/310	1	ITT Components Group	PSI
<50 @ 820	25	0.22	200/400	1.1	Brand-Rex	Type VII
50 @ 660	20	0.23	200/400	3	Fiberguide Industries	Practiguide
50 @ 790	20	0.3	700/400	2	BOTec	EIS-200H
50 @ 820	20	0.4	400/550	5	Quartz & Silice	QSF 400B

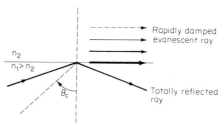

Fig. 21.13 Propagation of the evanescent light field for light which is totally reflected at a boundary between two different materials. None of the evanescent light is lost by radiation if the medium in which it travels extends to infinity.

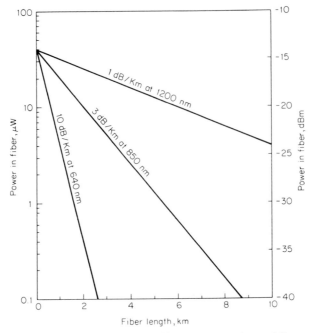

Fig. 21.14   The effect of different fiber attenuation values or different source wavelength values on the decrease of optical power in a fiber with increasing fiber length. The initial power level is 40 μW. For a receiver sensitivity of 0.7 μW (dashed horizontal line), the maximum fiber length can be estimated.

where α is the fiber attenuation in decibels per kilometer and dB_m is a unit for power referenced to 1 mW:

$$dB_m = 10 \log \frac{power}{1 \text{ mW}}$$

Note that Eq. (21.10) represents a straight line on a semilog graph of P_f versus L. Several graphs are shown in Fig. 21.14 for different fiber attenuation values. Here, the power coupled into the fiber is taken to be 40 μW (-14 dB_m). The three lines drawn on the graph indicate the optical power as a function of length of silica fiber for attenuation values at three different optical wavelengths. If the received optical power must exceed 0.7 μW (-31.5 dB_m), then the maximum fiber lengths possible are

    1.8 km for a 640-nm-wavelength source
    5.9 km for a 850-nm-wavelength source
    17.5 km (off scale) for a 1200-nm-wavelength source

The importance of matching the transmitter optical wavelength to the wavelength for minimum fiber attenuation should be apparent.

FIBER BANDWIDTH   In theory, an optical carrier provides a potential modulation bandwidth exceeding 10 THz. Practical optical guides, however, only permit the transmission of information at much lower rates, because of pulse-broadening effects. The bandwidth degradation may be described in terms of a pulse-broadening factor $\tau_{1/2}$ with units of nanoseconds per kilometer (ns/km), a risetime factor τ_r with units of nanoseconds per kilometer, or in terms of a bandwidth-length product (BLP) with units of megahertz-kilometers (MHz·km). To a first approximation, the pulse-broadening and risetime terms vary linearly with fiber length while the bandwidth varies inversely. For example, step index fiber with a bandwidth of 25 MHz at 1 km has a bandwidth of 6.25 MHz at 4-km length. These three terms are related by

$$\text{BLP} = \frac{0.32}{\tau_r} \tag{21.11}$$

$$\text{BLP} = \frac{0.44}{\tau_{1/2}} \tag{21.12}$$

Three effects contribute to this degradation of bandwidth: (1) multimode dispersion, (2) material dispersion, and (3) waveguide dispersion. Each is described below.

Recall from the discussion of fiber types that multimode dispersion may occur if the light travels along different paths in the fiber. For the step-index fiber the contribution to the pulse broadening varies as

$$\tau_{1/2} \backsim (NA)^2 \tag{21.13}$$

A large numerical aperture increases the optical power coupled into the fiber but reduces the bandwidth. The design of the step-index fiber represents a compromise between these two conflicting terms. For this reason the cladding index of refraction is only a few percent smaller than that for the core.

For graded-index fibers, the multimode dispersion is reduced significantly by equalizing the propagation time for each mode. However, this equalization depends critically on the index profile and on the transmitter wavelength. Although theory predicts that the pulse broadening for graded-index fibers may be 10 000 times smaller than that for step-index fibers, this performance level is not achieved in practice.

The second pulse-broadening effect is material dispersion which is important to wide-bandwidth applications. It arises from the wavelength dependence of the refractive index. Optical sources emit over a band of wavelengths—not just one. The result is that different optical frequencies in a pulse of light travel at varying relative speeds within the fiber. Hence, pulse broadening occurs as illustrated in Fig. 21.15. For LEDs and ILDs in the 800 to 900 nm range, the pulse broadening is approximately 5 and 0.5 ns/km, respectively. Hence, even if multimode dispersion were to be eliminated, material dispersion would limit the fiber bandwidth to 90 MHz·km with an LED transmitter and to 900 MHz·km for an ILD transmitter in this wavelength range.

Fortunately, fibers may be produced for which material dispersion is reduced to near zero near a wavelength of 1300 nm. Hence, there is much interest in graded-index fibers for applications between 1200 and 1300 nm where fiber attenuation is low also (refer to Fig. 21.12).

The remaining contribution to pulse broadening is the waveguide dispersion effect. Any carrier frequency f_c which is amplitude-modulated by a signal frequency f_s has two side-band frequencies produced: (1) the upper sideband at $f_c + f_s$ and (2) the lower sideband at $f_c - f_s$. The upper sideband travels in the optical guide with a slightly different speed than the lower sideband. Again, the result is pulse broadening. However, because the modulation or signal frequencies are much smaller than the carrier frequency, this effect may be ignored for all applications not involving single-mode fibers.

The combined pulse broadening for these fiber dispersion terms is given by

$$\tau_{1/2}^2 \text{ (fiber)} = \tau_{1/2}^2 \text{ (modal)} + \tau_{1/2}^2 \text{ (material)} + \tau_{1/2}^2 \text{ (waveguide)} \tag{21.14}$$

Fig. 21.15 Pulse broadening owing to material dispersion. The light energy at each wavelength may travel at a slightly different speed and arrive at the end of the fiber at a different time. *(Courtesy AMP, Inc., Harrisburg, Pa.)*

Generally, a system excited by an LED will have a greater pulse broadening and lower bandwidth owing to material dispersion than a system with an ILD source.

In general, caution should be exercised in system engineering with BLPs or pulse-broadening values from manufacturers' literature for the following reasons:

1. Many manufacturers measure the pulse-broadening effects with an ILD source which overestimates the fiber bandwidth for an LED-excited system.

2. The BLP is a function of fiber length. Many manufacturers measure its value with short links and extrapolate to 1 km.

3. The values obtained for τ_r and the BLP depend critically on the type of optical source and the source-to-fiber coupling technique. It is common to find that different laboratories will obtain values that vary by a factor of 2 or more for the same fiber type and length as a result of differences in optical coupling schemes.

For these reasons, a fiber dispersion factor should specify the measurement conditions including the fiber length and the source coupling scheme. Bandwidth-length products for several fiber types are given in Table 21.1.

example 21.7   An 820-nm LED is used as the source for a graded-index fiber which has a 200 MHz·km bandwidth-length product. Determine the effective BLP.

solution   The effective BLP is less because the 5-ns/km material dispersion term must be included. First determine the modal dispersion value from Eq. (21.12):

$$\tau_{1/2} = \frac{0.44}{200 \text{ MHz} \cdot \text{km}}$$
$$= 2.2 \text{ ns/km}$$

From Eq. (21.14) the effective pulse broadening becomes

$$\tau_{1/2}^2 = (2.2 \text{ ns/km})^2 + (5 \text{ ns/km})^2$$

and
$$\tau_{1/2} = 5.5 \text{ ns/km}$$

The bandwidth-length product becomes

$$\text{BLP} = \frac{0.44}{5.5 \text{ ns/km}}$$
$$= 80 \text{ MHz} \cdot \text{km}$$

CABLED FIBERS   A cladded, buffered fiber is not protected properly for the environments likely to be encountered in most applications. Provision must be made for the following:

1. Mechanical abuse from impact, crushing, flexure, and tension which arise from installation and service

2. Temperature extremes

3. Water ingress which may lead to stress corrosion or to stress owing to freezing

4. Chemical deterioration of cable materials

5. Radiation-induced attenuation

6. Damage from gnawing rodents

A variety of cable structures are available to accommodate or reduce the effect of these stresses. Examples are given in Figs. 21.16 and 21.17.

All cables have a maximum tensile load. Even though the tensile strength of individual fiber is in the range of 2×10^5 to 3×10^5 kg_f/m^2, the small cross-sectional area yields a maximum load of only 1.0 to 1.5 kg_f, or 2 to 3 lb. Therefore, optical fiber must be combined with strands of Kevlar°, fiber glass, or steel to provide the tensile strength required for installation. Cable structures for use with direct-bury, underground duct, plowed, or aerial installation techniques are available. Types of specific cable applications are listed in Table 21.2.

To prevent the introduction of permanent effects in the fiber properties during installation, the maximum tensible force should be limited to about 44 kg_f (100 lb). This should be adequate for cable installation for most applications. For example, a tensile load of 13 kg_f is normally required to pull a 1-km cable of 1.6-cm diameter through a multiply curved conduit. Some firms use a polyethylene tube or a lubricated metal duct tube to reduce friction in pulling cables through ducts.

° Kevlar is a registered trademark of DuPont, Inc.

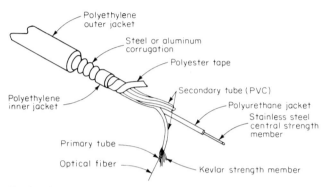

Fig. 21.16 Structures for a single-fiber cable and for a multiple-fiber cable.

Fig. 21.17 Cross sections of different structures for multiple-fiber cables. *(Courtesy General Cable Company, Woodbridge, N.J.)*

TABLE 21.2 Characteristics of Selected Fiber-Optic Cables

Attenuation (dB/km@λ in nm)	Bandwidth (MHz·km)	Number of fibers	Tensile strength (kg$_f$)	Maximum length (km)	Applications and features	Manufacturer	Typical product
0.7@1300	10	1–48	50–300	2	Interurban trunks	Fujikura Cable	NA/(5–60)
3–10@820	200–1500	5–60	300	1.1	General telecommunications	Compagnie Lyonnaise	SM50/6-LAP
3@850	30	6	150	1	All nonmetallic	Dainichi-Nippon	2-804350NM
3.5–8	200–800	4–12	400	2	Aerial	General Cable	T 2501
3.5–10@850	30–1000	1–19	400	4	Duct	ITT	T 2500
3.5–10@850	30–1000	1–19	400	4	Buried	ITT	T 2502
3.5–10@850	30–1000	1–19	400	1	Simplex	ITT	T 3000–3101
3.5–10@850	30–1000	1	30	4	Duplex	ITT	T 3001–3102
3.5–10@850	30–1000	2	60	4	General purpose	ITT	T 3500
5@850	200	1–7	100	4	High strength, lightweight	Valtec	MP-MG05
5@850	200	1–6	100	3	Military	Valtec	FC-MG05
5@850	200	6	50	3	Heavy-duty design	Valtec	HD-MG05
5@850	200	2	50	3	Low cost, intrabuilding	Valtec	GP-MG05-01
5@850	200	1	25	3	Telephone trunking	Valtec	TC-MG05
6@820	400	1–10	300–350	3	General purpose	Siecor	622
6@820	400@820	1–8	2–20	1	Oval, reinforced	Times Fiber	IN-6
6@820	400@820	2–12	2–45	1	Oval, reinforced air and duct	Times Fiber	AD-6
8@850	200	2	180–320	1	Communications & instrumentation	Belden	227002
9	20	2–18	113–317	1	Communications & instrumentation	Belden	226018
10@820	10	2	400	1	Data communications	Siecor	242
10@820	100	1	30	1	General purpose	Siecor	133
10@820	200	1–8	3–20	1	Oval, reinforced	Times Fiber	IN-10
10@820	200	1–18	113–317	1	Communications & instrumentation	Belden	220001
10@850	25	1	110	1	Communications & instrumentation	Belden	227001
12@800	200	1–2	50	1	Low cost, intrabuilding	Valtec	PC10
15@800	25	2	10	3	Small size, data communications	Valtec	DC-PCO8
25@820	25	1–8	100	3	Flame-retardant	Focom Systems	200/PCS8R/25
30@790	25	1–2	65–130	2	Resists radiation	DuPont, Pifax	S-120
35@850	5	1	150	1.0	Flame-resistant	Siecor	155

Cable bending can greatly affect fiber longevity. If a fiber with a cladding radius r is bent along a circumference of radius R, the outermost edge will be strained. The percentage strain is given by

$$\sigma_s = \left(\frac{R + 2r}{R + r} - 1 \right) \times 100\% \tag{21.15}$$

example 21.8 A fiber with a cladding radius of 125 μm is bent along a curve of radius $R = 2$ cm. Calculate the corresponding strain.

solution

$$\sigma_s = \left(\frac{2 \text{ cm} + 2 \times 0.0125 \text{ cm}}{2 \text{ cm} + 0.0125 \text{ cm}} - 1 \right) \times 100\%$$
$$= 0.62\%$$

To ensure fiber longevity the differential strain should not exceed 0.2 percent. For the fiber of Example 21.8, the bending radius R should be greater than 6.5 cm. When a company gives a minimum bending radius, it usually means the minimum radius that can be applied for a short time before the fiber breaks. Not given is the significant effect on fiber lifetime if the fiber is left in a curved condition with a radius greater than the minimum.

For protection against crushing or pests, the cables can be supplied with steel or aluminum armor, as in Fig. 21.16. However, the cable weight is increased greatly, and isolation from lightning is eliminated. Cables can be waterproofed, also, by the use of a petroleum-based jelly or a dry powder to fill cable voids.

Temperature extremes have an adverse effect on fiber attenuation. Plastic-clad and buffered fiber, on one hand, are usable to $-30°C$ where thermal expansion of the plastic causes stress that increases the attenuation markedly. Glass-clad fibers, on the other hand, are usable to below $-50°C$. At high temperatures, the main problem is the durability of the cable jacketing materials: Kevlar, fiber glass, polyethylene, or polyurethane. Cables that meet the IEEE-383 flame-retardance specification can be obtained on special order. In general, the choice of cable materials and structure is dictated by the specific temperature range of the application. The manufacturer should be consulted for cable viability in any application environment.

FIBER CONNECTORING Fiber-optic systems contain a number of individual components, source fibers, detectors, etc., which must be linked so that optical losses are minimal. There are three types of connectoring methods:

1. The splice joins two fibers or two fiber bundles.

2. The connector links one fiber to another, to repeaters, or to end devices and is demountable from the rest of the fiber-optic transmission system.

3. The coupler distributes optical power received from an input port to several output ports.

The key problem in any single-fiber splice or connector is, first, to align the central axis of one fiber with that of another fiber and, second, to minimize separation of the fiber ends. For example, a 76-μm (3-mil) misalignment of the axes for two 380-μm fiber cores leads to a 2-dB loss; a 76-μm end separation for the same fibers yields an additional 0.3-dB loss. The alignment should be done while fibers are accepted and joined with variations owing to manufacturing tolerances. For fibers with small core diameters (less than 50 μm), this process becomes critical.

Splices Several splicing techniques are illustrated for single fibers in Fig. 21.18 and for multiple fibers in Fig. 21.19. Of these, fusion or fiber welding with an electric arc has become widely implemented. At one time it was believed that fusion techniques were impractical for use in a nonlaboratory setting, particularly without highly trained workers. However, the recent success of cable splicers employed by United Telephone of Pennsylvania who had only 1 week of training and practice has dispelled this myth. Splices were made inside, outside, in breezy conditions, above ground, and in manholes.

In the fusion technique, the splicer works with the benefit of a 100X microscope, so the fibers appear about the size of pencils. The "weld" is made by melting the glass between electrodes. The key to a successful splice is to recognize the proper glass temperature by the color of the heated glass. As the glass fiber ends are heated, the color changes from orange

Fig. 21.18 Methods for splicing single optical fibers.

to pink to brilliant white. At that point the fibers are placed in contact for fusion. The average attenuation per splice ranges from 0.15 dB with a standard deviation of twice the mean value.

Another technique in Fig. 21.18 is the V groove which is preferentially etched into a silicon slice. A precision form is produced into which to lay the fibers. It also forms the basis for connecting multiple-fiber ribbons manufactured by Western Electric for the Bell system, as in Fig. 21.19a. Another, the precision sleeve splice, requires high degrees of accuracy and reproducibility of the inner sleeve dimensions and on the outer fiber diameter. For splices using index matching fluids and adhesives, precision tube splices have consistently achieved insertion losses less than 0.3 dB. The loose-tube splice uses a glass tube with a square cross section where the inner tube dimensions are purposely made larger than the outer fiber diameter. When the two fibers are bowed slightly, the loose tube rotates and holds the ends in place until the adhesive has cured. Various modifications of the loose-tube splice have been made with losses ranging from 0.3 to 0.5 dB.

Connectors   If repeated connections to a fiber are required, a demountable connector is used. An example, produced by AMP, Inc., is shown in Fig. 21.20. The key element is the ferrule which is made of resilient plastic. On being mated, a bushing compresses the ferrule to produce alignment of the fibers with the connector axis. Use of connectors of this type requires that the cable jacket materials be stripped to expose the fiber. Epoxy is used to fix the fiber in the ferrule and the jacket to the ferrule body. After the epoxy is cured, the exposed fiber end is lapped with polishing papers to a mirror finish. Other connectoring techniques, illustrated in Fig. 21.21, employ various alignment schemes to minimize losses. Regardless of the precision of connector components, fibers have minor geometric variations which add to potential misalignments even for perfectly mated male-female connector pieces. Commercially produced connectors have insertion losses ranging from 1.0 to 2.5 dB with costs ranging from $3 for large fibers (diameters greater than 240 μm) to $30 for precision single-fiber connectors (fiber diameter less than 140 μm).

Fig. 21.19 Structures for splicing multiple optical fibers.

Fig. 21.20 Example of a demountable connector. Fiber alignment is produced by compression of the resilient ferrule by the bushing. *(Courtesy AMP, Inc., Harrisburg, Pa.)*

Fig. 21.21 Structures for connectoring single optical fibers to minimize fiber alignment errors. The purpose is to maximize the coupling of light from one fiber into the other.

Couplers Couplers permit the distribution or combination of optical signals for any number of optical fiber lines or sources. Applications include data distribution networks, duplex communications with a single-fiber guide, and wavelength division multiplexing of several optical signals onto a single optical guide.

One common technique for producing a coupler is achieved by twisting lengths of unjacked fiber together in a bundle. While the fiber bundle is being fused, the fiber ends are pulled to form a biconical taper. Typically, the fused region is bonded to a substrate and encapsulated in a protective package, as illustrated in Fig. 21.22.

Fig. 21.22 Example of a biconical taper coupler.

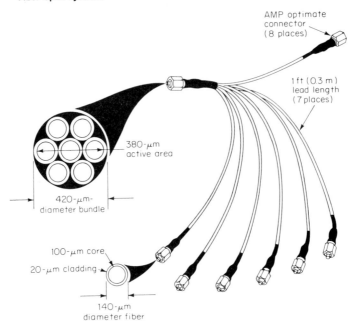

Fig. 21.23   A seven-way combiner-splitter. *(Courtesy AMP, Inc., Harrisburg, Pa.)*

A simple passive combiner or splitter can be configured with individual fibers, as shown in Fig. 21.23. Here seven fibers with a 140-μm cladding diameter are mounted in a standard fiber-optic connector to form a 400-μm-diameter bundle. The matching transmitter or receiver fiber should be a single fiber with a 400-μm cladding diameter. Two such units coupled back to back with a length of 400-μm fiber form a 14-arm star coupler.

Coupler technology is still in its infancy. Advances can be expected at a rapid pace which will increase the performance and reduce the cost of individual units.

Connectoring mismatch losses   Even if perfect fiber alignment and minimum core separation could be achieved with connectoring techniques, additional losses may arise from mismatches between the core ends. For example, changes in the core refractive index at the end boundaries lead to light reflection. This attenuation, called *Fresnel loss*, is given by

$$\alpha_{\text{Fresnel}} = -10 \log \frac{4n_1n_2}{n_1 + n_2} \tag{21.16}$$

where n_1 is the core refractive index and ni_2 is that for the medium outside the core end. For each silica-air boundary where $n_1 = 1.5$ and $n_2 = 1.00$, the loss is

$$\alpha_{\text{Fresnel}} = 10 \log \frac{(n_1 + n_2)^2}{4n_1n_2} = 10 \log \frac{2.5^2}{6} = 0.18 \text{ dB}$$

Because two such boundaries exist, the total Fresnel loss is approximately 0.35 dB per connection. Index-matching fluids or epoxy can reduce this loss term to near zero.

A second mismatch loss arises when light is coupled from a large-diameter core to a smaller-diameter receiving core. The resulting area mismatch loss α_A is

$$\alpha_A = 20 \log \frac{\text{larger core diameter}}{\text{smaller core diameter}} \quad \text{dB} \tag{21.17}$$

If the receiving fiber core is larger than the transmitting core, the area mismatch loss is zero.

example 21.9   A 200-μm fiber core couples optical power from an LED into a 50-μm graded-index fiber core. Calculate the area mismatch coupling loss.

solution Since the sending fiber is larger than the receiving fiber, the area mismatch coupling loss is

$$\alpha_A = 20 \log \frac{200 \ \mu m}{50 \ \mu m} \ dB$$
$$= 12 \ dB$$

A third mismatch loss term occurs if the sending core numerical aperture is larger than that for the receiving fiber core. This loss is given by

$$\alpha_{NA} = 20 \log \frac{\text{sending core N.A.}}{\text{receiving core N.A.}} \quad dB \qquad (21.18)$$

If the sending core N.A. is less than the receiving core N.A., then $\alpha_{NA} = 0$.

example 21.10 For the two fibers in Example 21.9, the 200-μm sending fiber has N.A. = 0.50, and the 50-μm receiving fiber has N.A. = 0.20. What is the coupling loss owing to mismatch of numerical apertures?

solution From Eq. (21.18) the coupling loss owing to mismatch of the numerical apertures is

$$\alpha_{N.A.} = 20 \log \frac{0.50}{0.20} \ dB$$
$$= 8 \ dB$$

The total loss in decibels experienced in connectoring is the sum of the effects resulting from the misalignment, fiber end separation, and mismatch terms. By combining the results of Examples 21.9 and 21.10 with Fresnel and alignment losses, the total connectoring attenuation is

Connectoring alignment	2 dB
$\alpha_{N.A.}$	8 dB
α_A	12 dB
Fresnel loss	0.3 dB
Total loss	22.3 dB

example 21.11 An LED couples 700 μW of optical power into a 200-μm-diameter, 0.50-NA fiber pigtail. This fiber is connected to a 50-μm-diameter, 0.20-N.A. telecommunications fiber. (See Example 21.10.) What is the power coupled into the optical transmission line?

solution

$$P_c = P_o \times \text{fraction lost}$$

or

$$P_c \ (dB_m) = P_o \ (dB_m) - \text{losses (dB)}$$

From Example 21.10 the coupling loss is 22.3 dB, so

$$P_c = -1.5 \ dB_m - 22.3 \ dB$$
$$= -23.8 \ dB_m$$

By converting from dB_m to milliwatts, the coupled power is only

$$P_c = 10^{-23.8/10} \ mW$$
$$= 0.042 \ mW = 4.2 \ \mu W$$

It should not be inferred that the manufacturer has an inferior product; the laws of physics apply to cause the coupling loss.

21.3. OPTICAL TRANSMITTERS AND SOURCES

The two types of optical sources which have been used widely are the light-emitting diode (LED) and the injection laser diode (ILD). These are solid-state pn junction devices that produce an increased light output with an increased forward-bias current. Electrons injected across the forward-biased pn junction combine with holes to produce photons of light which escape from the structure, as illustrated in Fig. 21.24. Various semiconductor compounds of gallium (Ga), arsenic (As), indium (In), and phosphorus (P) are used to produce devices which have a peak light emission at wavelengths from 600 nm in the visible part of the optical spectrum to 1600 nm in the infrared part. The range of light emission in the optical spectrum for each material type is given in Fig. 21.25. Included also are other gas and solid-

Fig. 21.24 Energy band diagram for a semiconductor light-emitting diode. Electrons injected into the conduction band of the n region recombine with holes in the valence band to produce photons of light. The larger the electron injection, the greater the production of photons.

state lasers used in optical communications. Many installations presently use aluminum gallium arsenide (AlGaAs) LEDs and ILDs which emit in the range of 780 to 900 nm. Future wide-bandwidth, long-distance links are expected to incorporate the InGaAsP devices, which transmit between 920 and 1650 nm.

The choice of an LED or ILD optical source rests on understanding the differences between the devices. For example, the ILD has a shorter risetime than the LED, which makes the ILD more suitable for wide-bandwidth and high-data-rate applications. In addition, more optical power can be coupled into a fiber with an ILD, which is important for long-distance transmission. However, the laser diode is strongly temperature-dependent and has a significantly higher cost. Which device type, LED or ILD, and what carrier wavelength to select depend on the transmission distance, data rate or bandwidth, fiber type, detector, and total system cost.

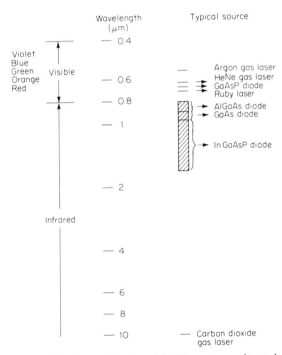

Fig. 21.25 The optical portion of the RF spectrum and typical sources at the various wavelengths. Solid-state, gas, and semiconductor optical emitters are given.

Fig. 21.26 (a) Directions of light emission for a light-emitting diode. (b) Structure and fiber coupling scheme for a heterojunction gallium aluminum arsenide surface emitting LED. (c) Structure and fiber alignment for an edge emitting LED.

LIGHT-EMITTING DIODES   The two most common LED structures are the surface emitter and the edge emitter, shown in Fig. 21.26. Surface emitters have a higher light output and efficiency than edge emitters because the light produced at the pn junction travels through only a thin absorbing layer of semiconductor material before it reaches the surface. For edge emitters, however, the photons travel the length of the structure parallel to the surface. Hence, more light is lost through absorption by the semiconductor material. Epitaxial layers of compound semiconductor materials form the p- and n-type regions. A heterojunction occurs if the pn junction is between two different types of semiconducting compounds, such as GaAs and AlGaAs.

The optical output characteristic versus current for a typical LED is illustrated in Fig. 21.27. For a drive current of 100 mA, the total available optical power for commercially available devices ranges between 1 and 10 mW. Amplitude modulation of the diode forward current produces a change in optical output intensity, as diagramed in Fig. 21.28. The maximum modulation index for analog transmission is typically on the order of 80 percent; for digital transmission, it is 100 percent. The output power for a given bias current may degrade up to 3 dB over the lifetime of the device, which for AlGaAs LEDs may exceed 100,000. Driver circuits for digital and analog modulation are found in Figs. 21.28 and 21.29.

Light-emitting diodes present the optical engineer or technician with several problems. One problem is that emitted light is spatially distributed as shown in Fig. 21.30. The optical intensity (power per unit area) at an angle θ from the normal to the emitting surface is given by

$$I(\theta) = I_0 \cos^m\theta \tag{21.19}$$

where I_0 is the intensity measured normal to the surface and m is an integer which describes how well the pattern is collimated. For $m = 1$, the output pattern is called a *Lambertian distribution*. For edge emitters, $m = 2$ or 3; for ILDs, $m = 3$ or 4. If the fiber is located too far from the emitting area, much of the light misses the fiber core.

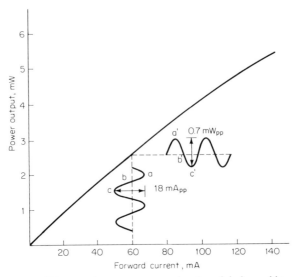

Fig. 21.27 Optical output power as a function of dc forward-bias current for a typical LED. A 9-mA peak modulation current applied at a 60-mA dc bias current produces a 14 percent modulation of the optical output power.

To minimize the loss of light power, a number of coupling schemes are used, as illustrated in Figs. 21.31 and 21.32. Devices with an attached optical fiber are identified as "pigtailed." Even with the most efficient coupling schemes, however, the power coupled into the optical fiber is 10 to 20 dB below that emitted by the LED. The amount of coupled power depends on the type of optical fiber as well as in the coupling scheme. How to determine the coupled power is discussed in the section on optical fibers.

Fig. 21.28 (*a*) A basic bias and modulation driver for an LED. (*b*) A better LED bias and driver circuit which has a 100-kHz bandwidth. (*c*) An LED bias and driver circuit with a 50-MHz bandwidth.

FIG. 21.28 *(Cont.)*

(c)

Q_1, Q_2: CTC   D3-28

$$I_B \cong \frac{V - 0.7\,V}{R_B}$$

$$I_A = \frac{10\,V - 0.7\,V}{R_A}$$

(a)

$$R_1 = \frac{V_{cc} - 3\,V}{I_F}$$

(b)

Fig. 21.29 Examples of digital driver circuits for digital applications (a) below 100 kilobits per second and (b) below 50 megabits per second.

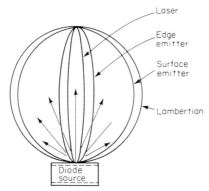

Fig. 21.30 Spatial emission pattern for light-emitting and laser diodes which shows the intensity as a function of angle from the normal to the emitting surface. A laser emits light within a narrow cone which results in efficient coupling of light into a fiber core.

Fig. 21.31 Mounting packages for light-emitting and laser diode pellets. *(Courtesy RCA, Solid State Division, Electro Optics and Devices, Lancaster, Pa.)*

Fiber light guide

TO-18 header

Highly polished
fiber tip

Index matching
medium

Semiconductor emitter
or detector

(a)

Active device bushing

Press-on
retention plate

Slots to
receive shielding

Active device module

(b)

Fig. 21.32 (a) Ferruled LED package and (b) the mounting bushing for use with the AMP connector illustrated in Fig. 21.20. (*Courtesy Motorola Semiconductor Products, Inc., Phoenix, Ariz., and AMP, Inc., Harrisburg, Pa.*)

A limitation of the LED is a relatively low-modulation bandwidth. GaAs and GaAsP devices with a risetime in the range of 600 to 800 ns have a bandwidth of 500 kHz or less. In contrast, edge-emitting AlGaAs diodes have risetimes between 2 and 20 ns and bandwidths greater than 200 MHz. For surface emitters, the inherent junction capacitance limits the maximum modulation bandwidth to less than 30 MHz. The modulation bandwidth varies with dc forward-bias current, as shown in Fig. 21.33.

The third problem is that the optical output power is spread over a band of wavelengths, as shown in Fig. 21.34. It is defined as the full width at half maximum (or FWHM), which is approximately 40 nm for most devices. Light rays with different wavelengths may travel down an optical guide with different velocities. Hence, a light pulse coupled into an optical fiber from an LED is broadened. The amount of pulse broadening plays a role in limiting the bandwidth of an optical link.

The characteristics of several LED com-

Fig. 21.33 Modulation bandwidth as a function of dc forward-bias current for an LED. (*Courtesy RCA, Solid State Division, Electro Optics and Devices, Lancaster, Pa.*)

Fig. 21.34   Typical spectral emission characteristics of light-emitting diodes fabricated from different semiconductor materials. *(Courtesy RCA, Solid State Division, Electro Optics and Devices, Lancaster, Pa.)*

ponents and transmitters are listed in Table 21.3. It is important that the selection of the optical emitter be made to match optimum performance properties of the fiber to be used as well as the maximum response of the photodetector.

INJECTION LASER DIODES   A laser diode is similar in structure to that for a heterojunction edge-emitting LED except that the end surfaces are cleaved or polished to produce a mirror finish. (Refer to Fig. 21.26c.) In addition, a channel or stripe geometry is used to

TABLE 21.3   LED Components and Transmitters

Wavelength (nm)	Bandwidth (MHz)	Power (μW)	Port/pigtail [diameter (μm)/N.A.]	Manufacturer and model
		Light-emitting diodes		
820	32	240	125/0.48	Laser Diode IRE 160FA
900	18	140	200/0.48	Motorola MFOE102F
820	38	700	200/0.50	Motorola MFOE106F
850	45	200	125/0.25	NEC NDL 4103 CP
820	150	60	100/0.21	RCA C86008E
1200	90	15	50/0.20	ITT T7632
1270	90	20	50/0.21	Laser Diode DE-1000F
		Analog LED transmitters		
820	30	140	300/0.22	LeCroy FAT-4H
820	20	20	62/0.22	MERET MDL 276-TV
840	20	100	—	ITT T 6211
		Digital LED transmitters		
	Bit rate (Mb/s)			
670	2	15		Burr Brown 3713T
660	5	15	370/0.53	LeCroy Diplink 1
820	10	100	100/0.30	Hewlett-Packard HFBR-1002
820	40	200	55/0.22	Laser Diode LDT-256
820	50	160	200/0.22	LeCroy FDT-5L
820	20	40	63/0.21	RCA C 86012E
820	10	275	200/0.48	Spectronics SPX 4140

reduce the region where electron injection occurs at the junction. This area reduction increases the injection current at the junction and increases the photon output. The mirrored end faces cause a large fraction of the light produced at the junction to be reflected into the channel region. Produced is an optical waveguide cavity which guides the photons to stimulate the generation of additional photons in phase with the stimulus. The result is light amplification by stimulated emission of radiation, from which the acronym *LASER* is obtained.

The output characteristic versus drive current for an ILD differs from that for the LED, as illustrated in Fig. 21.35. Significant light output does not occur unless the drive current exceeds a threshold; spontaneous emission similar to that for an LED occurs below threshold. Because small changes in drive current above the device threshold produce substantial changes in output optical power, the ILD is well suited for applications in digital communications systems. An example of a dc bias and modulation circuit for ILDs is given in Fig. 21.36.

A typical emission spectrum is shown in Figs. 21.37 and 21.38. At low power levels, the output power is distributed among several narrow emission lines or modes. As the drive current and optical power are increased, the power shifts to a dominant mode which is only a few nanometers wide. Note that the mode pattern and dominant emission wavelength are a function of the drive current. Such multimode operation contributes to high-frequency noise. Single-mode lasers have been developed that do not suffer from these characteristics.

The ILD has three main advantages over the LED. The first is its directionality. Because of the stimulated emission, the light emerges in a narrow beam, as indicated in Fig. 21.30. Hence, a large fraction of the optical power may be coupled into a small fiber core. Typically, the coupling loss is less than 3 dB. The second advantage is the very low junction

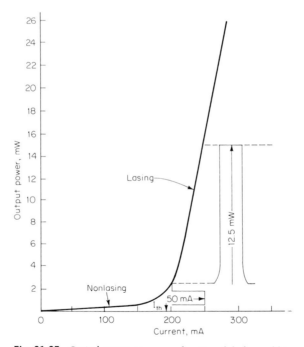

Fig. 21.35 Optical output power as a function of dc forward-bias current for an injection laser diode. Stimulated emission of light does not occur unless the threshold current is exceeded. Below threshold, spontaneous light emission occurs similar to that for an LED. The graph illustrates how a 50-mA current pulse yields a 12.5-mW change in the optical output power.

Fig. 21.36 Laser driver circuit with a modulation bandwidth between 30 Hz and 50 MHz. Included is a temperature control section. Diode bridge 1: 2 A, 50 V (peak inverse voltage). Diode bridge 2: 5 A, 50 V (peak inverse voltage). T: Thermistor Fenwall part GB41P2 (10 kΩ at 25°C) or equivalent. Thermo-electric module: Melcor part CP1417-10TL or equivalent. (*Courtesy RCA, Solid State Division, Electro Optics and Devices, Lancaster, Pa.*)

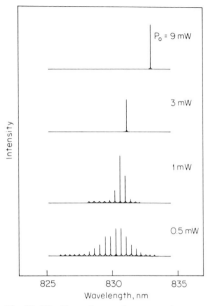

Fig. 21.37 Typical spectral emission characteristic for an aluminum gallium arsenide (AlGaAs)
laser diode as a function of optical output power.
At low power levels, multimode emission occurs.

capacitance, which permits a high-modulation bandwidth. Present AlGaAs devices are
available with bandwidths exceeding 1 GHz. The third advantage is the narrow wavelength
spread in the output spectrum, typically on the order of 1 to 2 nm. The result is a much
smaller material dispersion or pulse broadening. For these reasons laser sources can be used
to transmit data at higher rates over a given distance than is possible with LEDs.

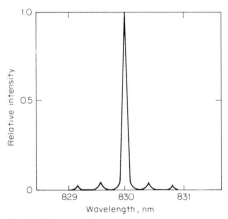

Fig. 21.38 Typical spectral characteristic of a single-mode laser emission line. The line width typically
is less than 0.5 nm, which reduces the effects of material dispersion.

A disadvantage of the ILD is the strong temperature dependence of the output characteristic curve, illustrated in Fig. 21.39. To control this dependence, special optical and electronic feedback drive circuits often are required. However, the result is a more complicated and costly transmitter. ILD sources may be either obtained as individual components or packaged as a complete transmitter which includes all driver feedback circuits. Several ILD components and transmitters are listed in Table 21.4.

Lasers also exhibit temporal instabilities which lead to "noise" in the optical output. One type of fluctuation occurs when changes in fiber-interface reflections perturb the laser optical cavity. A second noise category occurs as a train of discrete, narrow optical pulses even when the drive current is not modulated. These self-oscillations occur at frequencies above 200 MHz. A third category is modal noise, which arises from changes in the distribution of light among the fiber propagation modes. Such changes can occur as a result of wavelength changes when the laser is modulated. These noise characteristics have limited the application of the laser in broadband services such as CATV, which require high signal-to-noise ratios.

Fig. 21.39 Optical output power as a function of dc forward-bias current for a laser diode at different operating temperatures. A laser biased for proper operation at 25°C may cease at 40°C. Electronic regulation of the device temperature usually is required, as shown in Fig. 21.36.

21.4 OPTICAL DETECTORS AND RECEIVERS

The optical receiver must detect the received optical power and amplify the resulting electric signal with minimal distortion. It consists of a photodetector and a preamplifier, the choice of which greatly affects how well this requirement is satisfied. To appraise the performance of a receiver, it is necessary to evaluate the bandwidth, gain, noise, spectral response, sensitivity, and dynamic range, all of which affect the cost of a receiver system.

The two general classes of solid-state photodetectors are the reverse-biased pn diode and the phototransistor. One common type of diode has a PIN structure, illustrated in Fig. 21.40. A wide intrinsic region is placed between the p- and n-type materials to increase the junction width, which lowers capacitance and lowers bulk leakage currents that contribute to noise. Photons of light incident on the diode junction generate electron-hole pairs in the depletion region to produce an electric current proportional to the received optical power. The diode current ranges from picoamperes (10^{-12} A) to a few tenths of milliamperes. The efficiency of photon-to-carrier conversion is typically between 85 and 95 percent. Good efficiency and response times are obtained with reverse-bias voltages between 10 and 50 V.

Another diode detector type is the avalanche photodiode (APD) in which the photoexcited electrons trigger an avalanche of electrons through a high reverse-bias electric field. (Refer to Fig. 21.41.) This effect yields an apparent conversion gain and gives the detector its name. For this reason the APD is 10 to 15 dB more sensitive than the PIN diode. However, it is costly to produce, requires a reverse bias of 150 to 450 V, and exhibits a significant temperature dependence in the photoconversion gain. For many applications, these characteristics make APD devices less suitable than the PIN photodiodes.

In the phototransistor, the light detection diode becomes the base-collector junction of the transistor. (See Fig. 21.42.) The junction area is increased to improve the photon capture, but this causes higher base-collector capacitance and increases the Miller effect capacitance. Consequently, phototransistors are limited to applications requiring bandwidths less than 500 kHz. However, they can be manufactured at a cost lower than that for PIN diodes. A typical response efficiency is 80 percent for silicon devices for 850-nm light.

To amplify the small photocurrent, a low-noise preamplifier is required. Two commonly used receiver configurations are the transimpedance amplifier and the voltage amplifier,

TABLE 21.4 Laser Diodes and Transmitters

Wavelength (nm)	Bandwidth (MHz)	Power (μW)	Port/pigtail [diameter (μm)/N.A.]	Manufacturer and model
		Lasers		
780–850	>1200	7000	—	General Optronics GOLS
800–890	2000	1500	55/0.22	Laser Diode LCW-10F
840	500	500	55/0.25	ITT T 912
830	1000	2500	63/0.21	NEC NDL 3108 P
820	1000	2000	100/0.21	RCA C 86006 E
1300	1000	1000	50/0.20	RCA 86022 E
		Analog ILD transmitters		
780–850	1250	3000	55/0.25	General Optronics GO-ANA
830	500	1000	—	Times Fiber OTL-1101
840	150	2000	—	ITT T-6247
1300	500	3000	50/0.20	General Optronics GO-ANA
		Digital ILD transmitters		
	Bit rate (Mb/s)			
780–850	275	>1000	—	OIS OTX-5100
840	300	1000	—	ITT T6146
820	150	500	—	Meret MDL-4595

Fig. 21.40 Structure of a PIN photodiode which shows the generation of a hole-electron pair by the absorption of a photon. The applied reverse bias separates the carriers to yield the diode current. *(Courtesy RCA, Solid State Division, Electro Optics and Devices, Lancaster, Pa.)*

Fig. 21.41 Structure and electric field distribution for an avalanche photodiode. The large electric field causes impact ionization of additional carriers which multiplies the photogenerated carriers. *(Courtesy RCA, Solid State Division, Electro Optics and Devices, Lancaster, Pa.)*

shown in Fig. 21.43. The transimpedance amplifier is normally used for bandwidths below 100 MHz for which the output voltage is given by

$$V_{\text{out}} = -I_d R_f \qquad (21.20)$$

where R_f is the feedback resistance. It is attractive from a design standpoint because it provides better input and output impedance stability and yields reduced distortion. The voltage amplifier, however, is simpler and less costly to implement for large-bandwidth applications. It permits a larger bandwidth and better signal-to-noise ratio (SNR). However, because the RC network of the diode and load resistance acts as a differentiator, an output equalizer network is required. The RC product of the input circuit determines the amplifier bandwidth (BW), given by

$$\text{BW} = \frac{1}{2\pi RC} \qquad (21.21)$$

where R is the total input capacitance and C is the total device and amplifier input capacitance. For both PIN and APD photodetectors, the risetime typically is less than 3 ns. Hence, the bandwidth is limited usually by the preamplifier circuit.

Fig. 21.42 Structure of a phototransistor. *(Courtesy General Electric Company, Semiconductor Products Department, Auburn, N.Y.)*

example 21.12 A transimpedance amplifier has a feedback resistance of 20 kΩ and a feedback capacitance of 10 pF. If the PIN photodiode capacitance is 5 pF, calculate the bandwidth.

solution From Eq. (21.21) the bandwidth is

$$\text{BW} = \frac{1}{2\pi(20 \text{ k}\Omega)(10 \text{ pF} + 5 \text{ pF})}$$
$$= 530 \text{ kHz}$$

Just as in any communications system, the receiver configuration must detect and amplify the received signal with minimal distortion. The SNR must be great enough to maintain signal integrity and to keep errors below an acceptable minimum. The choice of detector and preamplifier circuit affects how well this is accomplished. For either amplifier configuration, FET stages yield lower noise figures below 10 MHz; above 10 MHz, bipolar devices are better. Which configuration to select depends on the circuit cost, required bandwidth, and available SNR. The detector and receiver specifications which determine the performance characteristics are described below.

RESPONSIVITY The optical-to-electrical conversion of a photodetector is given by the responsivity R_d:

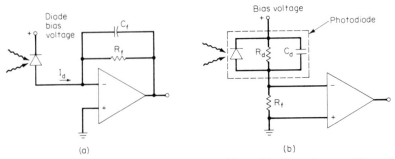

Fig. 21.43 Diagrams for (*a*) a transimpedance amplifier and for (*b*) a voltage amplifier used in optical receivers.

$$R_d = \frac{\text{photocurrent}}{\text{received optical power}} \tag{21.22}$$

The responsivity depends on the diode type, semiconductor material, bias voltage, and wavelength of the optical carrier. For the range of 650 to 950 nm, silicon photodiodes are used; for longer wavelengths the newly developed indium gallium arsenide (InGaAs) detectors are used. Typical responsivities for PIN detectors are shown in Fig. 21.44 and listed in Table 21.5. Peak values are 0.4 to 0.7 A/W for PIN diodes and 75 A/W for APD detectors.

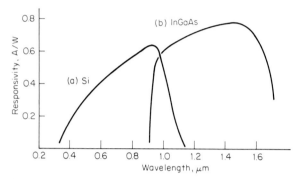

Fig. 21.44 Responsivity as a function of a wavelength for (a) silicon and (b) indium gallium arsenide PIN photodiodes.

The 100-fold increase in responsivity for the APD arises from the avalanche process. Germanium (Ge) detectors also may be used in the spectral range of 800 to 1800 nm; however, a large reverse current produces a high noise level.

An optical receiver also may have a responsivity defined in terms of the output voltage:

$$R_{amp} = \frac{\text{output voltage}}{\text{received optical power}} \tag{21.23}$$

For a transimpedance amplifier this equation may be rewritten by means of Eqs. (21.22) and (21.23) in the form

$$R_{amp} = R_d R_f \tag{21.24}$$

example 21.13 Suppose the amplifier circuit in Fig. 21.43a has $R_f = 20$ kΩ. For a PIN detector responsivity of 0.60 A/W, determine the amplifier responsivity.

TABLE 21.5 PIN and APD Detectors

λ_0 (nm)	R (A/W)	Risetime (ns)	NEP (10^{-13} W/$\sqrt{\text{Hz}}$)	Typical bias (V)	Manufacturer and model
			PIN diodes		
830	0.6	3	1.0	−45	RCA C30920E
820	0.3	3		−5	Spectronics SD-3322
900	0.4	25	0.05	−20	Motorola MFOD102F
940–1250	0.6	2	5	−5	ITT T7731
900–1650	0.6	1	5	−5	RCA C30979E
			APD diodes		
830	77	0.5	0.3	−225	RCA C30921E
800–900	85	1.2	0.08	−300	Centronic APD 05-4R

solution

$$R_{amp} = R_f R_d$$
$$= (20 \text{ k}\Omega) (0.6 \text{ A/W})$$
$$= 12\,000 \text{ V/W}$$

A number of manufacturers supply integrated photodetector and preamplifier packages. (Refer to Tables 21.6 and 21.7.)

Because the photocurrent is proportional to the received optical power, care must be exercised to interpret level changes and signal-to-signal noise values. Recall that the decibel is given by

$$dB = 10 \log \frac{P}{P_{ref}} \qquad (21.25)$$

or, in terms of a current ratio, as

$$dB = 20 \log \frac{I}{I_o} \qquad (21.26)$$

Therefore, a 50 percent decrease in *optical* power (-3-dB change) produces a -6-dB change in the output electric current. Hence, the optical and electric signal properties are related by

$$\text{Optical SNR (dB)} = \tfrac{1}{2}(\text{electric SNR) (dB)} \qquad (21.27)$$

This optical power-to-current conversion affects the interpretation of the bandwidth of an optical system. It is defined as the modulation frequency where the received optical power decreases by 3 dB. At this frequency, the electric output is down by 6 dB, however. Consequently, the electrical and optical bandwidths are related by

$$\text{Electrical bandwidth} = 0.707(\text{optical bandwidth}) \qquad (21.28)$$

example 21.14   If an optical fiber has an optical bandwidth-length product of 600 MHz·km, what is the equivalent electrical bandwidth-length product?

solution   By means of Eq. (21.28),

$$\text{Electrical BW} = 0.707(600 \text{ MHz·km})$$
$$= 424 \text{ MHz·km}$$

Failure to recognize the difference in the optical and electrical bandwidths easily may lead to a disappearance of system bandwidth.

NOISE EQUIVALENT POWER   In contrast to the normal communications receiver, the noise in the output of an optical receiver originates from two sources: the detector/amplifier circuit, and shot noise resulting from the random arrival of photons at the detector surface. (Photon-generated shot noise is sometimes called quantum noise.) This causes the noise level at the preamplifier output to depend on the level of the received optical power and the

TABLE 21.6   Fiber-Optic Receivers

Wavelength peak (nm)	Bandwidth (MHz)	TNEP (μW)	Manufacturer and model
	PIN receivers		
820	25	0.065	LeCroy FAR-4HS
820	5	0.5	Fibronics ALS-R1
840	6	1	ITT T 6358A
900	35	2.0	Motorola MFOD403F
900	10	0.3	Motorola MFOD404F
	APD receivers		
830	300	—	Times Fiber OR-2111A
840	6	0.06	ITT T 6358B
890	130	—	EG & G ODL-4A
905	1–100	0.02	Meret MDL-259

TABLE 21.7 Digital Fiber-Optic Receivers

Wavelength peak (nm)	Bit rate (Mb/s)	Sensitivity (μW)	Bit error rate (s^{-1})	Manufacturer and model
		PIN receivers		
650–1000	0.25	0.015	10^{-9}	Burr-Brown 3713R
670	1.0	0.25	10^{-8}	LeCroy DipLink 1
700–840	10	0.8	10^{-9}	Hewlett-Packard HFBR-2001
780–850	0.1–50 (150)	1 (5)	10^{-9}	OIS ORX-5000 (ORX-5100)
800–900	20	1.5	10^{-8}	LeCroy FDR-5AL
820	20 NRZ	0.5	10^{-9}	RCA C86012E
840	300	0.3	10^{-8}	ITT T6092
820	0.01–10	0.7	10^{-9}	Spectronics SPX 4141
		APD receivers		
780–850	0.1–50 (150)	0.05 (0.2)	10^{-9}	OIS ORX-5010 (5110)
840	107	0.01	10^{-8}	ITT T6196
905	100	0.5	10^{-12}	Meret MDL 4988

required SNR. The total output noise may be specified in terms of the total noise equivalent power (TNEP):

$$\text{TNEP} = \frac{\text{output noise current or voltage}}{\text{responsivity}} \qquad (21.29)$$

It represents the minimum detectable optical signal (SNR = 1). For example, the RCA C30807 PIN diode with a dark current of 10 nA and R_d of 0.6 A/W has a TNEP of 16.7 nW.

Recall that the rms noise voltage is proportional to the square root of the bandwidth BW. By dividing the TNEP by $\sqrt{BW}$, the noise equivalent power (NEP) is obtained:

$$\text{NEP} = \frac{\text{TNEP}}{\sqrt{BW}} \qquad (21.30)$$

It represents the noise in a unity bandwidth and provides a basis for comparing noise levels of different receivers, as in Fig. 21.45. At low optical intensities (low SNR), circuit noise

Fig. 21.45 Range of noise equivalent power (NEP) at unity bandwidth as a function of electric signal-to-noise ratio for optimum PIN and avalanche photodiode receivers. Actual receiver circuits may have higher NEP values than that indicated.

dominates the output of a PIN receiver but not an APD receiver; that is, the APD is limited by quantum noise. For this reason the APD detector is preferred for long-distance transmission of digital information since the required SNR rarely exceeds 20 dB. However, for many analog transmissions with high SNR values, shot noise dominates, so the PIN diode is the preferred detector in these applications.

SENSITIVITY   The sensitivity represents the optical power at the receiver input which yields a given electric SNR at the output. In terms of the TNEP or NEP, it is

$$P_r \text{ (dB}_m) = \text{TNEP (dB}_m) + \frac{1}{2}\text{SNR (dB)} - 10 \log m \qquad (21.31)$$
$$= \text{NEP (dB}_m) + 10 \log B + \frac{1}{2}\text{SNR (dB)} - 10 \log m$$

where m is the modulation index of the optical carrier. Since the optical power level is the desired quantity, one-half of the electric SNR is used. The value of the NEP may be read from one of the curves in Fig. 21.45. Two examples given below illustrate the calculation of the sensitivity.

example 21.15   A PIN diode receiver is used as part of a TV remoting link. The TV signal modulates the optical carrier with a modulation index of 0.7, a bandwidth of 4.5 MHz, and an SNR of 50 dB. What is the required receiver sensitivity?

solution   From Fig. 21.45 the NEP for a typical low-noise receiver for this application is approximately -118 dB$_m$. The sensitivity is computed from Eq. (21.31):

$$P_r = -118 \text{ dB}_m + 10 \log(4.5 \text{ MHz}) + 25 \text{ dB} - 10 \log 0.7$$
$$= -28 \text{ dB}_m = 1.6 \ \mu\text{W}$$

example 21.16   The same TV signal in Example 21.15 is digitized at the Nyquist rate (9 MHz) with an NRZ, 8-bit code. The bit rate is 72 Mbits/s, which requires a minimum bandwidth of 36 MHz. What is the required sensitivity of a PIN receiver and of an APD receiver?

solution   For a bit error rate of 10^{-9}, the electrical SNR is 19 dB. By means of Eq. (21.31), a PIN receiver has a sensitivity of

$$P_r \text{ (dB)} = -118 \text{ dB}_m + 10 \log(36 \text{ MHz}) + 9.5 \text{ dB} - 10 \log 0.7$$
$$= -34.5 \text{ dB}_m = 0.36 \ \mu\text{W}$$

For an APD receiver, the NEP is -132 dB$_m$, which results in a sensitivity of

$$P_r = -132 \text{ dB}_m + 10 \log(36 \text{ MHz}) + 9.5 \text{ dB} - 10 \log 0.7$$
$$= -48.4 \text{ dB}_m = 0.014 \ \mu\text{W}$$

The receiver sensitivity sets the minimum allowed optical power incident on the photodetector and has a direct bearing on the selecting of the transmitter and fiber. The difference between the power coupled into the fiber and the sensitivity gives the maximum loss allowed between the transmitter and the receiver. Characteristics of photodetectors and receivers are summarized in Tables 21.5 to 21.7.

21.5 SYSTEM DESIGN

The objective is to combine the right set of components to produce the desired signal-to-noise ratio or bit error rate, bandwidth or bit rate, and cost benefit at a specific link distance. Achievement of the desired system rests on appropriate choices of the following (with the least costly given first in each item):
   1. LED of ILD transmitter
   2. Step of graded-index fiber
   3. High or low fiber attenuation
   4. High or low fiber dispersion
   5. PIN- or APD-based receiver
   6. Connecting or splicing method
   7. Transmitter wavelength in the range of 550 to 700 nm, 800 to 920 nm, or 1100 to 1600 nm
Note that there are four possible types of transmitter-receiver combinations to consider. Which one to select is influenced by the choice of optical fiber. In these considerations, the primary constraint is the limited optical power that can be coupled into a fiber core. Unlike radio or wire transmission systems, a need for more power cannot be satisfied by using a more powerful transmitter.

TABLE 21.8   Bit Error Rate Versus Signal-to-Noise Ratio

BER	SNR (db)
10^{-1}	5.16
10^{-2}	10.34
10^{-3}	12.81
10^{-4}	14.42
10^{-5}	15.61
10^{-6}	16.55
10^{-7}	17.33
10^{-8}	17.99
10^{-9}	18.57
10^{-10}	19.08

The first step is to budget the power needed at the receiver to achieve the necessary signal-to-noise ratio. For this purpose, Eq. (21.31) is used. For digital systems, use Table 21.8 to convert the bit error rate (BER) to the equivalent electric signal-to-noise ratio. Next, select sources from Table 21.3 or 21.4 that are anticipated to couple adequate power into different fiber cores. The difference between the coupled power in dB_m and the receiver sensitivity in dB_m is the allowance for fiber and connecting losses as well as time, temperature, and noise degradation factors:

$$P_c - P_r \geq \alpha L + C + D \qquad (21.32)$$

where P_c = coupled optical power, dB_m
  α = fiber attenuation, dB/km
  L = fiber length, km
  C = connectoring losses, dB
  D = degradation factors, dB
Examples below illustrate how the power budget is used to select a fiber type.

Next determine the system risetime and bandwidth, if necessary. To use Eq. (21.2), the distinction between electrical bandwidth and optical bandwidth given by Eq. (21.28) must be maintained. Usually the calculations are made in optical terms. Therefore, in terms of an optical risetime T_{or}, Eq. (21.2) is written as

$$T_{or}^2 = (1.1)^2 [\tfrac{1}{2}T_{t-r}^2 + (\tau_f L)^2] \qquad (21.33)$$

where T_{t-r} = transmitter-receiver *electrical* risetime
   = $0.32/electrical$ bandwidth
  τ_f = fiber risetime, ns/km
  L = fiber length
The corresponding system bandwidths are given by

$$\text{Optical bandwith} = \frac{0.32}{T_{or}} \qquad (21.34a)$$

$$\text{Electrical bandwidth} = \frac{0.22}{T_{or}} \qquad (21.34b)$$

Equation (21.33) lends itself to an approximation method for estimating the maximum system bandwidth. The procedure is to draw two straight-line segments on log-log coordinate graph paper which represent bandwidth versus length tradeoffs. Figure 21.46 illustrates this technique for a transmitter-receiver risetime of 11.3 ns (28.3 MHz electrical and 40 MHz optical) coupled to a 100-MHz·km fiber. The estimated optical bandwidth at 4 km of fiber length is 25 MHz. This procedure is illustrated in Example 21.19.

example 21.17   A computer data link is to be designed for a 9600-bit/s, return-to-zero (RZ) data rate with BER = 10^{-9} at a length of 2 km. The modulation index is $m = 0.7$. Determine a suitable combination of optical transmitter, fiber, and receiver.

solution   For an RZ digital format, the system bandwidth equals the bit rate. For this low bandwidth, a viable system configuration is an infrared LED transmitter, a PIN diode receiver,

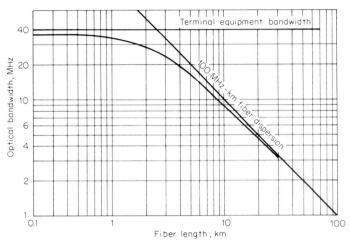

Fig. 21.46   The effect of remoting distance on system bandwidth for a 28-MHz transmitter-receiver pair (40-MHz optical bandwidth) linked to a 100-MHz·km fiber. The straight-line segments yield a first-order approximation to the bandwidth; the curved line gives the actual response.

and a step-index fiber. An infrared transmitter is selected because the fiber attenuation does not permit transmission over a few hundred meters with visible light sources. From Eq. (21.31) and Fig. 21.45 the optimum receiver sensitivity is

$$P_r = -117 \text{ dB}_m + 10 \log 9600 + \tfrac{1}{2}(18.6 \text{ dB}) - 10 \log 0.7$$
$$= -66.3 \text{ dB}_m = 0.23 \text{ nW}$$

As a trial transmitter, choose from Table 21.3 the Hewlett-Packard model HFBR-1002 with 100 μW (-10 dB$_m$) of power coupled into a 100-μm-diameter, 0.30-N.A. fiber pigtail. Take the step-index fiber to have a core diameter of 200 μm and an N.A. of 0.20. For these choices, the coupled power, by means of Eqs. (21.17) and (21.18), is

$$P_c = P_o - \alpha_A - \alpha_{NA} - \text{alignment loss} - \text{Fresnel loss}$$
$$= -10 \text{ dB}_m - 0 \text{ dB} - 3.5 \text{ dB} - 2 \text{ dB} - 0.3 \text{ dB}$$
$$= -15.8 \text{ dB}_m = 26.3 \ \mu\text{W}$$

The allowed loss budget if an optimum receiver is used becomes

$$\text{Budget} = -15.8 \text{ dB}_m - (-66.3 \text{ dB}_m)$$
$$= 50.5 \text{ dB}$$

This allowed loss may be allocated to fiber losses and system degradation terms:

Time degradation	3 dB
Fiber connectoring (one time)	3 dB
Fiber-detector connectoring	3 dB
Total	9 dB

Hence, 50.5 dB $-$ 9 dB is allowed for fiber attenuation. For a 2-km length, the maximum fiber attenuation is 20.8 dB/km. A number of optical cables are available with fibers which satisfy this requirement.

A bandwidth analysis is not necessary here since the bandwidth of all components exceeds greatly that required for the system.

example 21.18   An analog TV remoting link with a 4.5-MHz bandwidth, 50-dB SNR, and 2.5-km distance is to be designed. For an optical modulation index of $m = 0.7$, what combination of transmitter, fiber, and receiver is suitable?

solution   Consider the power budget available. The receiver sensitivity given by Eq. (21.31) and Fig. 21.45 is

$$P_r = -117 \text{ dB}_m + 10 \log(4.5 \times 10^6) + \tfrac{1}{2}(50 \text{ dB}) - 10 \log 0.7$$
$$= -23.9 \text{ dB}_m = 4.1 \ \mu\text{W}$$

Because of the high SNR required, the APD-based receiver provides no advantage. An examination of Table 21.3 reveals that an LED source in the range of either 800 to 900 nm or 1100 to 1300 nm does not couple sufficient power into a low-loss fiber. The loss margin in either case would be only 6 to 12 dB. An analog ILD transmitter is required. From Table 21.4, ILDs couple 1 mW (0 dB_m) into 50-μW, 0.20-N.A. fiber cores. The available loss margin becomes 23.9 dB. By using the same fiber as that for the ILD pigtail, the loss and degradation terms are

Source coupling

α_A	0 dB
$\alpha_N A$	0 dB
Alignment	3 dB
Link connectoring	
Fiber (2X fusion)	2 dB
Fiber detector	2 dB
Time degradation	3 dB
Temperature degradation	3 dB
Total	13 dB

The loss allowance for fiber attenuation is 23.9 dB $-$ 13 dB $=$ 10.9 dB. For a 2.5-km length, the fiber attenuation must be less than 4 dB/km. The fiber selected would be a graded-index type with a 200-MHz·km bandwidth-length product.

Since all components have a risetime or bandwidth that greatly exceeds the design requirements, a risetime calculation is not needed.

example 21.19 Estimate the maximum repeatorless transmission distance for a single-pulse coded modulation TV channel that is quantized with 2^8 levels and sent in a non-return-to-zero (NRZ) format.

solution To minimize sampling artifacts in the reconstructed video signal, choose a sampling frequency of 10.74 MHz, the third harmonic of the 3.58-MHz color subcarrier frequency. Each data word consists of an 8-bit vestigial sideband sample, 1 parity bit, and 1 frame-synchronizing bit for a 10-bit word. Thus, a 107.4-Mbits/s stream is required. The minimum electrical bandwidth for an NRZ format is one-half the bit rate, or 58.7 MHz, which has an equivalent optical bandwidth of 83 MHz. [Refer to Eq. (21.28).] This high modulation rate eliminates many LED transmitters and all step-index fibers.

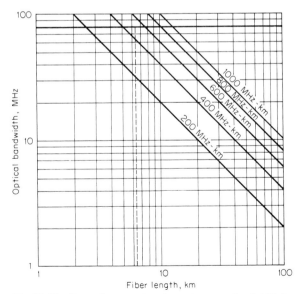

Fig. 21.47 First-order approximation to the system bandwidth for an analog system with an 83-MHz optical bandwidth. For a 6.3-km fiber length, a 600-MHz·km fiber would be required to yield the needed bandwidth.

Because of the maximum sensitivity requirement implied by the length specification, the system consists of an ILD transmitter and an APD receiver. Choose a 4-dB/km, graded-index fiber in order to compare Example 21.18 with the results here. From Table 21.4, a typical ILD couples 1 mW (0 dB$_m$) into a typical graded-index fiber.

The receiver sensitivity given by Eq. (21.31) and Fig. 21.45 is approximately (refer to Table 21.8 for the SNR)

$$P_r = -132 \text{ dB}_m + 10 \log(58.7 \times 10^6) + \frac{1}{2}(18.6 \text{ dB}) - 10 \log 0.7$$
$$= -43.5 \text{ dB}_m = 0.045 \ \mu\text{W}$$

Note that considerably less optical power is required at the receiver for this digital application:

$$\text{Maximum length} < \frac{(43.5 - 15) \text{ dB}}{(4 + 0.5) \text{ dB/km}}$$
$$L_{max} < 6.3 \text{ km}$$

which is much greater than the 2.5-km distance for the analog example.

To determine the fiber dispersion necessary, the graphical approach described at the beginning of this section is used in Fig. 21.47. To achieve an 83-MHz optical bandwidth at 6.3 km, the fiber bandwidth-length product must be greater than 400 MHz·km.

In the examples above, the electronic signal is applied directly to the driver to modulate the intensity of the optical carrier. Since the available power is limited, it may be necessary to precondition the electronic signal. For example, conversion of an analog signal to a digital format reduces the optical transmission SNR requirement, as shown in Example 21.19. Of course, bandwidth has been sacrificed. An alternative approach is to frequency-modulate a subcarrier which then amplitude-modulates the optical carrier. For a transmission SNR of 10 dB, a receiver postdetection SNR of 49 dB can be obtained since the FM process provides a 39-dB gain. A thorough treatment of these techniques is beyond the scope of the treatment presented here.

21.6 SELECTED BIBLIOGRAPHY

Aggarwal, I. D.: "Recent Developments in Optical Fibers," *Laser Focus*, vol. 18, no. 11, 1982, pp. 115–117.

Barnoski, M. K. (ed.): *Fundamentals of Optical Fiber Communications*, 2d ed., Academic, New York, 1981.

Barnoski, M. K.: "Fiber Systems for the Military Environment," *Proceedings of the IEEE*, vol. 68, no. 10, 1980, pp. 1315–1320.

Bliss, J., and D. W. Stevenson: "Fiber Optics: A Designer's Guide to System Budgeting," *Electro-Optical System Design*, vol. 13, no. 8, 1981, pp. 23–32.

Campbell, L., and A. Barnhart: "Fiber Optic Transmission Link Analysis," *Proceedings of the SPIE*, vol. 224, pp. 12–18 (April 1980 meeting in Washington).

Channin, D. J.: "Emitters for Fiberoptic Communications," *Laser Focus*, vol. 18, no. 11, 1982, pp. 105–113.

CSELT Technical Staff, *Optical Fibre Communications*, McGraw-Hill, New York, 1980.

Elion, G. R., and H. A. Elion: *Fiber Optics in Communications Systems*, Dekker, New York, 1978.

Forrest, S. R.: "Photodiodes for Long-Wavelength Communication Systems," *Laser Focus*, vol. 18, no. 12, 1982, pp. 81–90.

Giallorenzi, T. G.: "Fiber Optic Communications," *Proceedings of the IEEE*, vol. 66, no. 7, 1978, pp. 744–780.

Hewlett-Packard Application Note 1000: "Digital Data Transmission with the HP Fiber Optic System," Hewlett-Packard Components, 640 Page Mill Road, Palo Alto, Calif. 94304.

Howes, M. J., and D. V. Morgan (eds.): *Optical Fibre Communications: Devices, Circuits and Systems*, Wiley, New York, 1980.

Kao, C. K.: *Optical Fiber Systems*, McGraw-Hill, New York, 1982.

Lacy, E. A.: *Fiber Optics*, Prentice-Hall, Englewood Cliffs, N.J., 1982.

Miller, S. E., and A. G. Chynoweth (eds.): *Optical Fiber Telecommunications*, Academic, New York, 1979.

Personick, S. D.: *Optical Fiber Transmission Systems*, Plenum, New York, 1981.

Personick, S. D., N. L. Rhodes, D. C. Hanson, and K. H. Chan, "Contrasting Fiber-Optic Component-Design Requirements in Telecommunications, Analogue and Local Data Communications Applications," *Proceedings of the IEEE*, vol. 68, no. 10, 1980, pp. 1254–1262.

Seippel, R. G.: *Optoelectronics*, Reston Publishing, Reston, VA, 1981.

Sharma, A. B., S. J. Halme, and M. M. Butusov: *Optical Fiber Systems and Their Components*, Springer-Verlag, New York, 1981.

Tebo, A. R.: "Sensing with Optical Fibers: An Emerging Technology," *Electrooptical System Design*, vol. 14, no. 2, 1982, pp. 39–45.

Additional technical literature on fiber optics may be found in the following special journal issues:

The Western Electric Engineer, winter 1980.
Proceedings of the IEEE, vol. 68, no. 10, 1980.
IEEE Transactions on Microwave Theory and Techniques, vol. MTT-30, no. 4, April 1982; no. 10, October 1982.
IEEE Journal of Lightwave Technology.

Chapter 22

Active Filters

Roger C. Thielking
Professor of Electrical Technology
Onondaga Community College
Syracuse, New York

22.1 INTRODUCTION

An active filter has circuitry that contains one or more active components. In most cases, operational amplifiers (op amps) are used. Compared with passive filters, active filters have these potential advantages:

1. Voltage amplification is often possible.

2. Output impedance is very low, and input impedance is usually very high. Thus the filter performance is unlikely to be affected by the signal source or load characteristics.

3. Inductors are not needed. Any desired characteristics may be obtained by using only resistors, capacitors, and op amps.

4. Capacitors are usually small, even for low frequency applications.

At high frequencies, however, performance is limited by the gain-bandwidth product of the op amps. For frequencies approaching 1 MHz and above, passive filters may be better.

Passive filters usually include both capacitors and inductors, which, being large, are costly if low frequencies are involved. With active networks, small values of capacitance may be used for the same functions, as illustrated with two basic circuits. Figure 22.1a shows a capacitance multiplier circuit. The input impedance Z_{in} is equivalent to a resistor R' in series with a capacitor C', as shown in Fig. 22.1b. At low frequencies where $R' < X_{C'}$, the resistance can be neglected and the circuit acts as a capacitor to ground, with the value of C_1 multiplied by the ratio R_1/R_2. With the values shown, the multiplication factor is 1000, or 10^3. A simulated inductor is illustrated in Fig. 22.2a. Op amp OA_1 inverts the phase of the input current, which effectively shifts the phase angle of the input impedance by 180° so that it looks inductive rather than capacitive. The equivalent circuit is shown in Fig. 22.2b. Resistors R_1 and R_2 influence Q of the equivalent inductance L' as well as its value. For the values shown, $L' = 10$ H.

There are a very large number of possible active filter circuits. Complete analysis and derivation of formulas are usually rather complex and difficult. In this chapter, selected filter configurations that are representative of various types, and among the most useful, are shown. Design data and formulas are given without derivation. Phase shift data, which are usually of less interest, are not generally shown, but phase shift considerations are discussed briefly with all-pass filters, in Sec. 22.5.

Fig. 22.1   Capacitance multiplier. (*a*) Circuit. (*b*) Equivalent input impedance.

22.2 LOW-PASS AND HIGH-PASS FILTERS

Ideally, a low-pass filter will pass all frequencies within its passband (below the cutoff frequency f_c) without attenuation and will reject all higher frequencies. Conversely, in a high-pass filter, frequencies below f_c are rejected and all higher frequencies pass unchanged. (See Fig. 13.1*a* and *b*.) High-frequency response in active filters is, however, generally limited by the high-frequency response of the op amps used.

FIRST-ORDER FILTERS   These are perhaps the simplest of active filters, consisting of a single *pole*, or *RC* (resistor-capacitor) combination, and an op amp connected as a voltage follower for low-output impedance, to isolate the load.

The amplitude and phase characteristics of a filter are given by its *transfer function*, the ratio of output voltage and phase to input voltage and phase. The general transfer functions for first-order low- and high-pass filters are as follows:

Low-pass filter
$$\frac{V_{out}(s)}{V_{in}(s)} = \frac{A_l \omega_c}{s + \omega_c}$$
(22.1)

High-pass filter
$$\frac{V_{out}(s)}{V_{in}(s)} = \frac{A_h s}{s + \omega_c}$$
(22.2)

The operator is $s = j\omega$, where $j = \sqrt{-1}$ and ω is frequency in radians per second; ω_c is the cutoff radian frequency; and A_l and A_h are the passband voltage gains. Radian frequencies are related to frequencies in hertz by $\omega = 2\pi f$ and $\omega_c = 2\pi f_c$.

A low-pass filter is shown in Fig. 22.3*a*. For this circuit, passband gain $A_l = 1$, the cutoff radian frequency is $\omega_c = 1/(RC)$, and the cutoff frequency in hertz is

$$f_c = \frac{1}{2\pi RC}$$
(22.3)

Fig. 22.2   Simulated inductance. (*a*) Circuit. (*b*) Equivalent input impedance.

(a)

Normalized frequency f/f_c

(b)

Fig. 22.3   First order low-pass filter with unity gain. (a) Circuit. (b) Frequency response, normalized to $f_c = 1$ Hz.

Relatively high impedance values may be used for R and C, since the op amp OA_1 does not appreciably load the network. Frequency response is determined by Eq. (22.1) and is shown in Fig. 22.3b, with frequency normalized to f_c.

Figure 22.4a shows the high-pass filter. The passband gain is $A_h = 1$, the cutoff radian frequency is $\omega_c = 1/(RC)$, and f_c is given by Eq. (22.3). The transfer function, Eq. (22.2), produces the frequency response of Fig. 22.4b.

example 22.1   Design a low-pass first-order filter with a cutoff frequency $f_c = 800$ Hz. If the input V_{in} is a sine wave with frequency $f = 1.6$ kHz and 5.0 V peak-to-peak (p–p) amplitude, what is the amplitude of the output voltage V_{out}?

solution   From Eq. (22.3), $C = 1/(2\pi f_c R)$. Select a reasonable value for R, say 1.0 MΩ. Substituting values, we find $C = 1/(2\pi \cdot 800$ Hz $\cdot 1.0$ MΩ$) = 199$ pF. Use these values in the circuit of Fig. 22.3a.

At the signal frequency, $f/f_c = 1.6$ kHz$/800$ Hz $= 2.0$. From Fig. 22.3b, the attenuation is approximately -6.7 dB $= 20 \log (V_{out}/5.0$ V$)$. Solving yields $V_{out} = 5.0$ V$[\log^{-1} (-6.7$ dB$/20)] \approx 2.3$ V p–p.

example 22.2   A high-pass filter (Fig. 22.4a) has $R = 220$ kΩ and $C = 10$ nF. Find the cutoff frequency f_c and the frequency where the gain is reduced to 20 percent of the passband value.

solution   From Eq. (22.3), the cutoff frequency is $f_c = 1/(2\pi \cdot 220$ kΩ $\cdot 10$ nF$) = 72.3$ Hz. For 20 percent gain, $-20 \log 0.2 = -14.0$ dB. From Fig. 22.4b, $f/f_c = f/72.3$ Hz ≈ 0.20. Solving gives $f \approx 14.5$ Hz.

Amplifying filters   First-order filters may be modified to provide voltage amplification by adding two resistors in the op amp feedback, as shown in Fig. 22.5. In the transfer functions, Eqs. (22.1) and (22.2), the gain factors $A_l = A_h = (R_2 + R_3)/R_3$. Equation (22.3) becomes $f_c = 1/(2\pi R_1 C)$. The frequency responses, normalized to passband gain, are as shown in Figs. 22.3b and 22.4b.

Fig. 22.4 First-order high-pass filter with unity gain. (*a*) Circuit. (*b*) Frequency response, normalized to f_c = Hz.

Fig. 22.5 First-order low-pass filter, modified for variable gain (for high-pass, R_1 and C are interchanged).

Fig. 22.6 Relationship of natural frequency f_n and cutoff frequency f_c to passband (typical).

SECOND-ORDER FILTERS   Probably the most widely used circuit modules are those with two poles, referred to as *second-order* filters. Although only resistors and capacitors are used, the circuits are configured so that one of the poles acts like the inductive reactance of a second-order passive filter. The transfer functions are as follows:

Low-pass
$$\frac{V_{out}(s)}{V_{in}(s)} = \frac{A_l \omega_1 \omega_2}{(s + \omega_1)(s + \omega_2)} \tag{22.4}$$

High-pass
$$\frac{V_{out}(s)}{V_{in}(s)} = \frac{A_h s^2}{(s + \omega_1)(s + \omega_2)} \tag{22.5}$$

where ω_1 and ω_2 are the two pole frequencies (in radians per second). The *natural frequency*, or resonant frequency, ω_n is given by

$$\omega_n = \sqrt{\omega_1 \omega_2} \tag{22.6}$$

the transfer functions also may be expressed as follows:

Low-pass
$$\frac{V_{out}(s)}{V_{in}(s)} = \frac{A_l \omega_n^2}{s^2 + (\omega_n/Q)s + \omega_n^2} \tag{22.7}$$

$$\tag{22.8}$$

High-pass
$$\frac{V_{out}(s)}{V_{in}(s)} = \frac{A_h s^2}{s^2 + (\omega_n/Q)s + \omega_n^2}$$

where Q is the *quality factor* of the circuit. Sometimes the *damping factor* ζ is used instead. It is related to Q by $\zeta = 1/(2Q)$.

The natural frequency in hertz is $f_n = \omega_n/(2\pi)$. The passband may extend somewhat beyond f_n (see Fig. 22.6). Its edge corresponds to the *cutoff frequency* f_c (hertz) or ω_c (radians per second). The natural and cutoff frequencies are related by the passband extension factor k_c as follows:

Low-pass $\omega_c = k_c \omega_n$ or $f_c = k_c f_n$ (22.9)

High-pass $\omega_c = \dfrac{\omega_n}{k_c}$ or $f_c = \dfrac{f_n}{k_c}$ (22.10)

Now k_c is related to Q as shown in Table 22.1 and depends on circuit values.

Effect of Q   All second-order low- and high-pass filters have a flat response at frequencies well within the passband and attenuate at the rate of 40 dB/decade at frequencies well beyond cutoff. Near cutoff, however, the response varies with different values of the quality factor Q. Response types for selected Q values are listed in Table 22.1. Chebyshev characteristics, which include a range of Q values, have a peak in their frequency response near cutoff, where the passband amplitude is exceeded. The decibel values in Table 22.1 are for these peaks.

The Bessel characteristic has the most nearly constant signal delay for all passband frequencies. With the Butterworth characteristic, the amplitude response has maximum flatness within the passband. The Chebyshev characteristics have a more abrupt passband edge and slightly greater attenuation beyond, together with the response peak. Response curves for the Bessel, Butterworth, and 2-dB Chebyshev filters are shown in Fig. 22.7. These cor-

TABLE 22.1   Filter Response Types

Filter response type	Quality factor Q	Passband extension factor k_c
Bessel	0.5774	0.7851
Butterworth	0.7070	1.000
Chebyshev 0.5-dB peak	0.8637	1.128
Chebyshev 1.0-dB peak	0.9564	1.159
Chebyshev 2.0-dB peak	1.128	1.184

Fig. 22.7   Response characteristics of second-order filters, normalized to $f_c = 1$ Hz and unity gain. (a) Low-pass. (b) High-pass.

respond with filter Q's of 0.5574, 0.7070, and 1.128, respectively. Filters with Q between these values would have intermediary response curves. These are "universal" curves, normalized to the cutoff frequency and passband amplitude, so that they are easily applied to any particular filter values. They apply to all the low- and high-pass circuits described below.

Circuits   There are a number of possible circuits for second-order active filters. The *single-feedback* circuits, also called the *voltage-controlled voltage source* (VCVS) of Fig. 22.8 and the *infinite gain multiple feedback* (IGMF) circuits of Fig. 22.9, are representative, widely used types. (Another very important type is the *variable-state* filter, which, because of its versatility and unique features, is described separately in Sec. 22.4.) These circuits may be analyzed by determining the natural frequency f_n, passband gain A_l or A_h, and quality

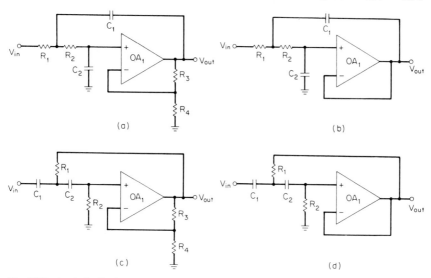

Fig. 22.8 Single-feedback second-order filters. (a) Low-pass, variable-gain filter. (b) Low-pass, unity-gain filter. (c) High-pass, variable-gain filter. (d) High-pass, unity-gain filter.

factor Q. Then, from Table 22.1 and Eq. (22.9) or (22.10), the cutoff frequency f_c is calculated or estimated.

Formulas for the circuits are as follows:

VCVS low-pass variable gain (Fig. 22.8a):

$$\omega_n = \frac{1}{\sqrt{R_1 R_2 C_1 C_2}} \qquad (22.11)$$

$$Q = \frac{1}{\omega_n[R_2 C_2 + R_1 C_2 - (A_l - 1)R_1 C_1]} \qquad (22.12)$$

Fig. 22.9 IGMF second-order filters. (a) Low-pass. (b) High-pass.

$$A_l = \frac{R_3 + R_4}{R_4} \tag{22.13}$$

VCVS low-pass unity gain (Fig. 22.8b): $A_l = 1$, Eq. (22.11) applies and

$$Q = \frac{1}{\omega_n C_2(R_1 + R_2)} \tag{22.14}$$

VCVS high-pass variable gain (Fig. 22.8c): Eqs. (22.11) and (22.13) apply, and

$$Q = \frac{1}{\omega_n[R_1C_1 + R_1C_2 - (A_h - 1)R_2C_2]} \tag{22.15}$$

VCVS high-pass unity gain (Fig. 22.8d): $A_h = 1$, Eq. (22.11) applies, and

$$Q = \frac{1}{\omega_n R_1(C_1 + C_2)} \tag{22.16}$$

IGMF low-pass (Fig. 22.9a):

$$\omega_n = \frac{1}{\sqrt{R_2R_3C_1C_2}} \tag{22.17}$$

$$A_l = -\frac{R_2}{R_1} \tag{22.18}$$

$$Q = \frac{\omega_n C_1}{1/R_1 + 1/R_2 + 1/R_3} \tag{22.19}$$

IGMF high-pass (Fig. 22.9b):

$$\omega_n = \frac{1}{\sqrt{R_1R_2C_2C_3}} \tag{22.20}$$

$$A_h = -\frac{C_1}{C_2} \tag{22.21}$$

$$Q = \frac{1}{\omega_n R_1(C_1 + C_2 + C_3)} \tag{22.22}$$

example 22.3   Find the characteristics f_c, A_h, and Q of the filter shown in Fig. 22.10.

solution   Comparing this circuit with Fig. 22.8 indicates that it is a high-pass, single-feedback, variable-gain filter. Equations (22.11), (22.13), and (22.15) apply:

$$\omega_n = \frac{1}{\sqrt{140\ \text{k}\Omega \cdot 140\ \text{k}\Omega \cdot 1.0\ \text{nF} \cdot 1.0\ \text{nF}}} = 7143\ \text{rads/s}$$

$$A_h = \frac{200\ \text{k}\Omega + 168\ \text{k}\Omega}{200\ \text{k}\Omega} = 1.840$$

Fig. 22.10   Circuit for Example 22.3.

$$Q = \frac{1}{(7143 \text{ rads/s})[140 \text{ k}\Omega \cdot 1.0 \text{ nF} + 140 \text{ k}\Omega \cdot 1.0 \text{ nF} - (1.840 - 1)140 \text{ k}\Omega \cdot 1.0 \text{ nF}]}$$
$$= 0.8621$$

Comparison of this value of Q with the listing in Table 22.1 indicates a Chebyshev characteristic with about 0.5-dB peak and passband extension factor $k_c \approx 1.128$. From Eq. (22.10), the cutoff frequency is $\omega_c \approx (7.143 \text{ krad/s})/1.128 \approx 6.332 \text{ krad/s}$, corresponding to $f_c \approx (6.332 \text{ krad/s})/2\pi \approx 1.0 \text{ kHz}$.

Design data Filters may be designed by using the above equations for the various circuit types. The process is difficult, however, and may involve complex mathematical manipulations. Design is simplified with design aids, such as the data in Tables 22.2 through 22.5, applied to the circuits of Figs. 22.11, 22.12, and 22.13.

The design steps, by using the tables, are as follows:

1. Assume reasonable values for the components, as indicated on the referenced figure showing the schematic diagram.

2. Calculate other component values by using the table. Primed symbols indicate *normalized* values (i.e., the R' value is in ohms when $f_c = 1$ Hz and $C = 1$ F; to find R, divide R' by the actual values of f_c and C).

3. If any of the resulting component values are too large or too small, they may be rescaled by multiplying resistance and dividing capacitance by the same factor.

example 22.4 Using the circuit of Fig. 22.13a, design an IGMF low-pass filter with a Butterworth characteristic, cutoff frequency $f_c = 2.5$ kHz, and gain $A_l = -2$.

solution Select a value for C, say 1.0 μF. Then from Table 22.4

$$R_1 = \frac{R_1'}{f_c C} = \frac{2.633}{2.5 \text{ kHz} \cdot 1.0 \text{ }\mu\text{F}} = 1.05 \text{ k}\Omega$$

(a)

(b)

Fig. 22.11 Single-feedback variable-gain circuits for design data (Table 22.2). (*a*) Low-pass. (*b*) High-pass. °Assume values for C and R_4.

TABLE 22.2 Design Data for Single-Feedback Variable-Gain Filters

| Characteristic | Either type (low-/high-pass) | | Low-pass, Fig. 22.11a | High-pass, Fig. 22.11b |
	A_l/A_h	R_3	R'	R'
Bessel	1.268	$0.2681R_4$	0.1250	0.2027
Butterworth	1.586	$0.5856R_4$	0.1592	0.1592
Chebyshev 0.5-dB peak	1.842	$0.8422R_4$	0.1795	0.1411
Chebyshev 1-dB peak	1.954	$0.9544R_4$	0.1845	0.1373
Chebyshev 2-dB	2.113	$1.113R_4$	0.1884	.1344

Note:$R = R'/f_cC$.

TABLE 22.3 Design Data for Single-Feedback Unity-Gain Filters

| Characteristic | Low-pass, Fig. 22.12a | | High-pass, Fig. 22.12b | |
	C'_1	C'_2	R'_1	R'_2
Bessel	0.1443	0.1082	0.1755	0.2341
Butterworth	0.2250	0.1126	0.1126	0.2250
Chebyshev 0.5-dB peak	0.3102	0.1040	0.08168	0.2437
Chebyshev 1-dB peak	0.3530	0.09646	0.07179	0.2627
Chebyshev 2-dB peak	0.4253	0.08349	0.05958	0.3033

Note:$C = C'/f_cR$; $R = R'/f_cC$.

TABLE 22.4 Design Data for IGMF Low-Pass Filters, Fig. 22.13a

| Characteristic | Unity gain, $A_l = -1$ | | $A_l = -2$ | |
| | $R_2 = R_1$ | | $R_2 = 2R_1$ | |
	R'_1	R'_3	R'_1	R'_3
Bessel	5.263	0.07471	2.593	0.07528
Butterworth	5.393	0.1174	2.633	0.1203
Chebyshev 0.5-dB peak	4.865	0.1656	2.340	0.1722
Chebyshev 1-dB peak	4.438	0.1917	2.108	0.2017
Chebyshev 2-dB peak	3.695	0.2403	1.696	0.2618

Note:$R = R'/f_cC$.

TABLE 22.5 Design Data for IGMF High-Pass Filters, Fig. 22.13b

| Characteristic | Unity gain, $A_h = -1$ | | $A_h = -2$ | |
| | $C_1 = C$ | | $C_1 = 2C$ | |
	R'_1	R'_2	R'_1	R'_2
Bessel	0.1170	0.3512	0.08777	0.4682
Butterworth	0.07514	0.3376	0.05628	0.4501
Chebyshev 0.5-dB peak	0.05445	0.3656	0.04084	0.4875
Chebyshev 1-dB peak	0.04786	0.3940	0.03590	0.5253
Chebyshev 2-dB peak	0.03972	0.4549	0.02979	0.6065

Note: $R = R'/f_cC$.

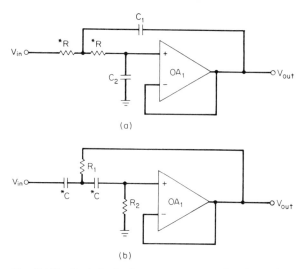

Fig. 22.12 Single-feedback unity-gain circuits for design data (Table 22.3). (*a*) Low-pass (°assume a value for *R*). (*b*) High-pass (°assume a value for *C*).

$$R_3 = \frac{R_3'}{f_c C} = \frac{0.1203}{2.5 \text{ kHz} \cdot 1.0 \ \mu\text{F}} = 48.1 \ \Omega$$
$$R_2 = 2R_1 = 2 \cdot 1.05 \text{ k}\Omega = 2.10 \text{ k}\Omega$$

From Fig. 22.13*a*, $C_1 = 0.04C = 40$ nF. These resistor values may be too low and cause excessive loading of the signal source. To rescale, multiply the resistance and divide the capacitance by 100. This yields $C = 10$ nF, $C_1 = 400$ pF, $R_1 = 105$ kΩ, $R_2 = 210$ kΩ, and $R_3 = 4.81$ kΩ.

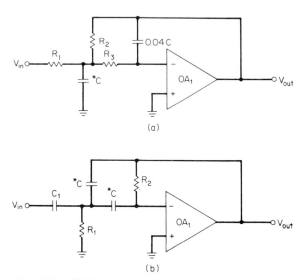

Fig. 22.13 IGMF circuits for design data. (*a*) Low-pass (Table 22.4). (*b*) High-pass (Table 22.5). °Assume a value for *C*.

example 22.5   A noninverting unity-gain low-pass filter with a Chebyshev 1.0-dB-peak characteristic and a cutoff frequency of 50 Hz is needed. How much attenuation will it have at 125 Hz?

solution   Use the single-feedback unity-gain low-pass filter circuit of Fig. 22.12a. Let $R = 100 \text{ k}\Omega$. Then, from Table 22.3,

$$C_1 = \frac{C_1'}{f_c R} = \frac{0.3530}{50 \text{ Hz} \cdot 100 \text{ k}\Omega} = 70.6 \text{ nF}$$

$$C_2 = \frac{C_2'}{f_c R} = \frac{0.09646}{50 \text{ Hz} \cdot 100 \text{ k}\Omega} = 19.3 \text{ nF}$$

At 125 Hz, $f/f_c = 125 \text{ Hz}/50 \text{ Hz} = 2.5$. Then, from Fig. 22.7a, attenuation is between -16 dB (for Butterworth) and -18.3 dB (for Chebyshev, 2-dB peak). A good estimate would be -17 dB.

Circuit features   In designing filters, the choice of circuits may be influenced by the following features:

Single-feedback (VCVS) unity-gain filter:
  Noninverting
  Fewest components
Single-feedback (VCVS) variable-gain filter:
  Noninverting
  Quality factor Q may be adjusted without changing the cutoff frequency
IGMF
  Inverting
  Fewest components for nonunity gain

THIRD- AND HIGHER-ORDER FILTERS   Higher-order filters are more difficult to design and more expensive to construct, but they have better performance. As the order of the filter is increased, the sharpness of the passband edge increases, yielding more attenuation of the frequencies beyond.

Single op amp third-order filter   Unity-gain low- and high-pass filters with three poles, or RC time constants, are shown in Fig. 22.14. Design data are shown in Tables 22.6 and 22.7. Frequency responses are plotted in Fig. 22.15.

Fig. 22.14   Third-order filters. (a) Low-pass (for design data, see Table 22.6). (b) High-pass (for design data, see Table 22.7).

TABLE 22.6   Design Data for Low-Pass Filter, Fig. 22.14a

Characteristic	C_1'	C_2'	C_3'
Bessel	0.1572	0.1958	0.04039
Butterworth	0.2215	0.5644	0.03221
Chebyshev 0.5-dB peak	0.3581	1.787	0.01424
Chebyshev 1-dB peak	0.4086	2.575	0.01023
Chebyshev 2-dB peak	0.4954	4.428	0.006194

Note:$C = C'/f_c R$.

example 22.6   For the high-pass third-order filter circuit shown in Fig. 22.14b, determine the component values needed for cutoff frequency $f_c = 200$ Hz with a Chebyshev characteristic having a 1.0-dB peak.

solution   Select a reasonable value for the capacitors, say $C = 1.0$ nF. Then, from Table 22.7,

$$R_1 = \frac{0.009836}{1.0 \text{ nF} \cdot 200 \text{ Hz}} = 49.2 \text{ k}\Omega$$

$$R_2 = \frac{0.06201}{1.0 \text{ nF} \cdot 200 \text{ Hz}} = 310 \text{ k}\Omega$$

$$R_3 = \frac{2.476}{1.0 \text{ nF} \cdot 200 \text{ Hz}} = 12.4 \text{ M}\Omega$$

Resistor R_3 is probably too large for a general-purpose op amp. Components may be rescaled by dividing resistances and multiplying capacitances by 20. Then $C = 20$ nF, $R_1 = 2.46$ kΩ, $R_2 = 15.5$ kΩ, and $R_3 = 620$ kΩ.

Analysis of this type of circuit is difficult and is not considered here.

Fourth-order filters using a single op amp also may be designed, but the procedure is difficult and there are more possible problems with component values. The cascaded filter approach, described below, is usually preferred.

Cascaded filters   A filter of any order may be devised by cascading first- and second-order filter circuits. For even-order filters, only second-order circuits are used, while one first-order circuit must be included to make an odd-order filter. Examples are shown in Fig. 22.16. Transfer functions, from Eqs. (22.1), (22.7), and (22.8), are indicated for each section. The overall transfer function is the product of the individual responses. Thus that for the third-order low-pass filter of Fig. 22.16a,

$$\frac{V_{out}(s)}{V_{in}(s)} = \frac{A_{l1}A_{l2}\omega_1^2\omega_2}{(s^2 + \omega_1 s/Q_1 + \omega_1^2)(s + \omega_2)} \tag{22.23}$$

and for the sixth-order high-pass filter of Fig. 22.16b,

$$\frac{V_{out}(s)}{V_{in}(s)} = \frac{A_{h1}A_{h2}s^6}{(s^2 + \omega_1 s/Q_1 + \omega_1^2)(s^2 + \omega_2 s/Q_2 + \omega_2)(s^2 + \omega_3 s/Q_3 + \omega_3^2)} \tag{22.24}$$

For filters of other orders, filter sections and transfer functions are similarly combined.

The Q values and pole frequencies, for various characteristics and orders of filters, are given in Table 22.8. These, together with the formulas for first- and second-order filters,

TABLE 22.7   Design Data for High-Pass Filter, Fig. 22.14b

Characteristic	R_1'	R_2'	R_3'
Bessel	0.1118	0.1611	0.6271
Butterworth	0.04488	0.1143	0.7864
Chebyshev 0.5-dB peak	0.01417	0.07073	1.778
Chebyshev 1-dB peak	0.009836	0.06201	2.476
Chebyshev 2-dB peak	0.005722	0.05112	4.089

Note:$R = R'/f_c C$.

Fig. 22.15 Response characteristics of third-order filters, normalized to $f_c = 1$ Hz and unity gain. (*a*) Low-pass. (*b*) High-pass.

may be used for analysis and design. However, design with only the formulas is very difficult. Extensive tables and other design aids are available in the literature. An easier and perhaps better design technique is to use state-variable filters. These, with design examples, are described in Sec. 22.4.

Attenuation rate   As the order is increased, the attenuation rate of the response beyond cutoff also increases, as shown in Fig. 22.17. For the Butterworth characteristic, attenuation may be estimated with this formula:

$$\text{dB attenuation} = -20N \log \frac{f_1}{f_2} \tag{22.25}$$

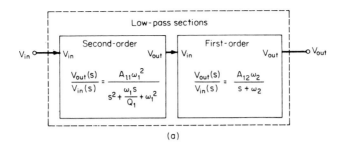

$$\frac{V_{out}(s)}{V_{in}(s)} = \frac{A_{11}\omega_1^2}{s^2 + \frac{\omega_1 s}{Q_1} + \omega_1^2}$$ (Second-order)

$$\frac{V_{out}(s)}{V_{in}(s)} = \frac{A_{12}\omega_2}{s + \omega_2}$$ (First-order)

(a)

$$\frac{V_{out}(s)}{V_{in}(s)} = \frac{A_{h1}s^2}{s^2 + \frac{\omega_1 s}{Q_1} + \omega_1^2}$$

$$\frac{V_{out}(s)}{V_{in}(s)} = \frac{A_{h2}s^2}{s^2 + \frac{\omega_2 s}{Q_2} + \omega_2^2}$$

$$\frac{V_{out}(s)}{V_{in}(s)} = \frac{A_{h3}s^2}{s^2 + \frac{\omega_3 s}{Q_3} + \omega_3^2}$$

(b)

Fig. 22.16 Higher-order cascaded filters. (*a*) Third-order, low-pass. (*b*) Sixth-order, high-pass.

TABLE 22.8 Design Data for Low-Pass Cascaded Filter Sections, Normalized to $f_c = 1$ Hz

Order	Characteristic	ω_1'	Q_1	ω_2'	Q_2	ω_3'	Q_3	ω_4'	Q_4
2	Bessel	6.283	0.5774						
	Butterworth	6.283	0.7071						
	Chebyshev 0.5-dB peak	7.737	0.8637						
	Chebyshev 2-dB peak	5.700	1.129						
3	Bessel	6.778	0.6910	6.192					
	Butterworth	6.283	1.000	6.283					
	Chebyshev 0.5-dB peak	6.716	1.706	3.936					
	Chebyshev 2-dB peak	5.915	2.552	2.318					
4	Bessel	6.779	0.8055	6.046	0.9286				
	Butterworth	6.283	1.307	6.283	0.5412				
	Chebyshev 0.5-dB peak	6.480	2.941	3.751	0.7051				
	Chebyshev 2-dB peak	6.055	4.594	2.958	0.9294				
5	Bessel	6.818	0.9165	6.044	0.5635	5.835			
	Butterworth	6.283	1.618	6.283	0.6180	6.283			
	Chebyshev 0.5-dB peak	6.395	4.545	4.338	1.178	2.277			
	Chebyshev 2-dB peak	5.700	7.232	3.940	1.775	1.372			
6	Bessel	6.866	1.023	6.088	0.6112	5.781	0.5103		
	Butterworth	6.283	1.932	6.283	0.7071	6.283	0.5176		
	Chebyshev 0.5-dB peak	6.355	6.513	4.826	1.810	2.490	0.6836		
	Chebyshev 2-dB peak	6.175	10.46	4.587	2.844	1.986	0.9016		

continued

TABLE 22.8   Design Data for Low-Pass Cascaded Filter Sections, Normalized to $f_c = 1$ Hz (*Cont.*)

Order	Characteristic	ω_1'	Q_1	ω_2'	Q_2	ω_3'	Q_3	ω_4'	Q_4
7	Bessel	6.914	1.126	6.148	0.6608	5.790	0.5324	5.682	
	Butterworth	6.283	2.247	6.283	0.8019	6.283	0.5550	6.283	
	Chebyshev 0.5-dB peak	6.334	8.842	5.169	2.576	3.166	1.092	1.610	
	Chebyshev 2-dB peak	6.203	14.28	5.008	4.115	2.896	1.646	0.9760	
8	Bessel	6.914	1.226	6.170	0.7019	5.788	0.5596	5.618	0.5060
	Butterworth	6.283	2.563	6.283	0.9000	6.283	0.6013	6.283	0.5098
	Chebyshev 0.5-dB peak	6.321	11.53	5.410	3.466	3.763	1.611	1.864	0.6766
	Chebyshev 2-dB peak	6.221	18.69	5.293	5.584	3.594	2.532	1.494	0.8924

Note: $\omega_{n(\text{low-pass})} = \omega' f_c$; for high-pass filter sections, $\omega_{n(\text{high-pass})} = 4\pi^2 f_c/\omega'$.

where $f_c = f_2$ (low-pass) or f_1 (high-pass) and N is the order. There is an error of 3 dB maximum at f_c, which becomes negligible an octave or more beyond cutoff. The attenuation is less for Bessel and greater for Chebyshev characteristics (see Figs. 22.7 and 22.15) and increases similarly at higher orders.

> **example 22.7**   A Butterworth high-pass filter, with cutoff frequency $f_c = 4.0$ kHz, must attenuate by at least -20 dB at 2.2 kHz. Find the order necessary.
>
> **solution**   From Eq. (22.25), -20 dB $= -20N \log (4.0 \text{ kHz}/2.2 \text{ kHz}) = -20N(0.260)$. Thus $N = 3.85$, and a fourth-order filter will suffice. (Although this is an estimate, any error would result in greater attenuation.)

22.3   BANDPASS AND BANDSTOP FILTERS

Band filters are characterized as *broadband* or *narrowband*, according to the width of the frequency band that is rejected or passed. Each requires different circuitry. The bandwidth is from f_{c1} to f_{c2}, the 3-dB cutoff frequencies. If the ratio $f_{c2}/f_{c1} < 1.5$, narrowband filters usually are best; if greater, broadband types may be used.

BROADBAND FILTERS   For wide-frequency bands ($f_{c2}/f_{c1} > 1.5$), filters are formed by combining the high- and low-pass filters described in Sec. 22.2. For bandpass, shown in Fig. 22.18, the low- and high-pass filters are cascaded. The low-pass cutoff frequency is f_{c2}, and the high-pass cutoff is f_{c1}. For bandstop or band-reject filters, the filters are connected in shunt as shown in Fig. 22.19. Op amp OA_1 is used to combine the two outputs in a summing circuit. Here, the cutoff of the low-pass filter is the lower cutoff f_{c1} of the bandstop function.

In each case, the *skirt* of the response (the region of attenuation near cutoff) is determined by the response of its associated low- or high-pass filter. The other filter has very little effect,

Fig. 22.17   Attenuation rates of passband skirt, low-pass Butterworth characteristic, for various orders.

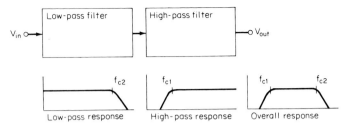

Fig. 22.18 Broadband bandpass filter.

except near the center frequency of the band-reject response. At this point, the filter outputs are about equal and are added, which can increase response (or reduce attenuation) by up to 6 dB. This effect can be reversed, so as to increase attenuation at the center frequency, with circuitry that causes phase reversal of the signals. Using odd-order filters will do this, or if an even-order filter is used, one filter can be of an inverting type while the other is noninverting.

NARROWBAND FILTERS When the cutoff frequencies are close together ($f_{c2}/f_{c1} < 1.5$), a specially designed filter, which has a response similar to that of a tuned LC circuit, is usually used.

Bandpass filter The usual transfer function for a narrow-bandpass filter is

$$\frac{V_{out}(s)}{V_{in}(s)} = \frac{A_p s \omega_n}{s^2 + s\omega_n/Q + \omega_n^2} \tag{22.26}$$

Normalized frequency response is shown in Fig. 22.20 for various values of Q. These equations are useful:
Bandwidth:

$$f_{c2} - f_{c1} = \frac{f_n}{Q} \tag{22.27}$$

Natural resonant frequency:

$$f_n = \sqrt{f_{c1}f_{c2}} \tag{22.28}$$

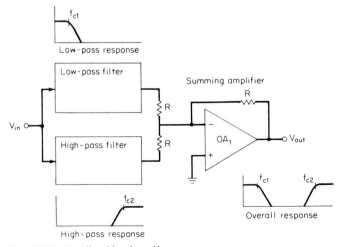

Fig. 22.19 Broadband bandstop filter.

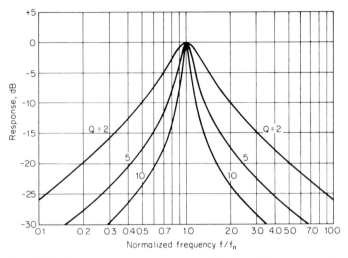

Fig. 22.20 Response characteristics, narrowband-bandpass filter, for various Q values.

Cutoff frequencies:

$$f_{c1} = f_n \frac{\sqrt{1 + 4Q^2} - 1}{2Q} \tag{22.29}$$

$$f_{c2} = f_n \frac{\sqrt{1 + 4Q^2} + 1}{2Q} \tag{22.30}$$

The attenuation in dB at any frequency f is

$$\text{dB attenuation} = -10 \log \left[1 + \frac{Q^2(f^2 - f_n^2)^2}{f^2 f_n^2} \right] \tag{22.31}$$

Two versions of a widely used circuit are given in Fig. 22.21. For the three-resistor version, Fig. 22.21a, the equations for analysis are

$$\omega_n = \left(\frac{R_1 + R_3}{C_1 C_2 R_1 R_2 R_3} \right)^{1/2} \tag{22.32}$$

$$A_p = \frac{C_2 R_2}{R_1(C_1 + C_2)} \tag{22.33}$$

$$Q = \left[\frac{R_1 + R_3}{R_1 R_2} \frac{R_2 C_1 C_2}{(C_1 + C_2)^2} \right]^{1/2} \tag{22.34}$$

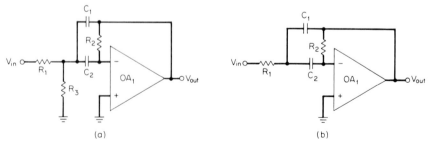

(a)   (b)

Fig. 22.21 Narrowband-bandpass filters. (a) Three-resistor circuit. (b) Two-resistor circuit.

For the two-resistor circuit of Fig. 22.21b, Eq. (22.33) applies. The others become

$$\omega_n = \frac{1}{\sqrt{R_1 R_2 C_1 C_2}} \tag{22.35}$$

$$Q = \left[\frac{R_2 C_1 C_2}{R_1 (C_1 + C_2)^2} \right]^{1/2} \tag{22.36}$$

example 22.8 A bandpass filter with the circuit configuration of Fig. 22.21b has $R_1 = 20$ kΩ, $R_2 = 1.0$ MΩ, and $C_1 = C_2 = 3.5$ nF. Find the bandpass gain A_p, natural resonant frequency f_n, quality factor Q, bandwidth, cutoff frequencies f_{c1} and f_{c2}, and the attenuation at 1.0 kHz (relative to the maximum output at f_n).

solution Equations (22.33), (22.35) and (22.36) apply. Substituting values, we find

$$A_p = -\frac{1.0 \text{ M}\Omega \cdot 3.5 \text{ nF}}{20 \text{ k}\Omega(3.5 \text{ nF} + 3.5 \text{ nF})} = -25$$

$$\omega_n = \frac{1}{\sqrt{20 \text{ k}\Omega \cdot 1.0 \text{ M}\Omega \cdot 3.5 \text{ nF} \cdot 3.5 \text{ nF}}} = 2.02 \text{ krad/s}$$

$$f_n = \frac{2.02 \text{ krad/s}}{2\pi} = 322 \text{ Hz}$$

$$Q = \left[\frac{1.0 \text{ M}\Omega \cdot 3.5 \text{ nF} \cdot 3.5 \text{ nF}}{20 \text{ k}\Omega(3.5 \text{ nF} + 3.5 \text{ nF})^2} \right]^{1/2} = 3.54$$

From Eqs. (22.27), (22.29), (22.30), and (22.31), we know that bandwidth = 322 Hz/3.54 = 90.8 Hz and the cutoff frequencies are

$$f_{c1} = 322 \text{ Hz} \left(\frac{\sqrt{1 + 4 \cdot 3.54^2} - 1}{2 \cdot 3.54} \right) = 280 \text{ Hz}$$

$$f_{c2} = 322 \text{ Hz} \left(\frac{\sqrt{1 + 4 \cdot 3.54^2} + 1}{2 \cdot 3.54} \right) = 371 \text{ Hz}$$

$$\text{Attenuation at 1.0 kHz} = -10 \log \left\{ 1 + 3.54^2 \frac{[(1.0 \text{ kHz})^2 - (322 \text{ Hz})^2]^2}{(322 \text{ Hz})^2 \cdot (1.0 \text{ kHz})^2} \right\} \text{ dB}$$

$$= -19.9 \text{ dB}$$

For design convenience, let $C_1 = C_2 = C$. The circuit of Fig. 22.21a is more versatile, allowing independent setting of gain A_p and quality factor Q. Design steps and formulas are as follows:

1. Assume a reasonable value for R_2, usually 100 kΩ to 1 MΩ.
2. Calculate R_1 for the desired gain A_p:

$$R_1 = \frac{R_2}{2A_p} \tag{22.37}$$

3. Calculate R_3 for the desired quality factor Q:

$$R_3 = \frac{R_1 R_2}{4R_1 Q^2 - R_2} \tag{22.38}$$

4. Calculate capacitance C for the desired natural resonant frequency f_n:

$$C = \frac{1}{2\pi f_n} \left(\frac{R_1 + R_3}{R_1 R_2 R_3} \right)^{1/2} \tag{22.39}$$

The circuit of Fig. 22.21b has fewer components, but Q and A_p are interdependent. Design steps and formulas are as follows:

1. Select a value for R_2, as above.
2. Calculate R_1 for the desired quality factor Q:

$$R_1 = \frac{R_2}{4Q^2} \tag{22.40}$$

Alternatively, R_1 may be calculated for a desired gain A_p with Eq. (22.37). In either case, A_p and Q are related by

$$A_p = 2Q^2 \tag{22.41}$$

3. Calculate C for the desired natural resonant frequency:

$$C = \frac{1}{2\pi f_n \sqrt{R_1 R_2}} \tag{22.42}$$

example 22.9   Design a narrow-bandpass filter with unity gain (at f_n) and with cutoff frequencies of 900 Hz and 1.11 kHz.

solution   From Eq. (22.28), $f_n = \sqrt{900 \text{ Hz} \cdot 1.11 \text{ kHz}} = 1.0 \text{ kHz}$. From Eq. (22.27), $Q = f_n/(f_{c2} - f_{c1}) = 1.0 \text{ kHz}/(1.11 \text{ kHz} - 900 \text{ Hz}) = 4.76$. Equation (22.42) is not satisfied; therefore, the circuit of Fig. 22.21a must be used. We apply the design steps:

1. Assume $R_2 = 1.0 \text{ M}\Omega$.
2. From Eq. (22.37), $R_1 = \frac{1}{2} \cdot 1.0 \text{ M}\Omega/1 = 500 \text{ k}\Omega$.
3. From Eq. (22.38),

$$R_3 = \frac{1.0 \text{ M}\Omega \cdot 500 \text{ k}\Omega}{4 \cdot 500 \text{ k}\Omega \cdot 4.76^2 - 1.0 \text{ M}\Omega}$$
$$= 11.3 \text{ k}\Omega$$

4. From Eq. (22.39),

$$C = \frac{1}{2\pi \cdot 1.0 \text{ kHz}} \left(\frac{500 \text{ k}\Omega + 11.3 \text{ k}\Omega}{500 \text{ k}\Omega \cdot 1.0 \text{ M}\Omega \cdot 11.3 \text{ k}\Omega} \right)^{1/2}$$
$$= 1.51 \text{ nF}$$

These circuits work well for Q values up to 10, with 1 percent resistors and 2 percent capacitors. For higher Q's the state-variable circuits are better (see Sec. 22.4).

Bandstop filter   These are also called *band-elimination*, or *notch*, filters. The usual transfer function is

$$\frac{V_{\text{out}}(s)}{V_{\text{in}}(s)} = \frac{A_s(s^2 + \omega_n^2)}{s^2 + s\omega_n/Q + \omega_n^2} \tag{22.43}$$

Bandwidth and cutoff frequency considerations are the same as for the bandpass filter, so that Eqs. (22.27), (22.28), (22.29), and (22.30) apply. The attenuation is different. The normalized attenuation at any frequency f is

$$\text{dB attenuation} = -10 \log \left[1 + \frac{f^2 f_n^2}{Q^2(f^2 - f_n^2)^2} \right] \tag{22.44}$$

The normalized frequency response is shown in Fig. 22.22 for various values of Q. Equation (22.44) indicates infinite attenuation at f_n. However, the practical maximum is -30 to -40 dB, depending on component precision.

The bandpass filters of Fig. 22.21 may be used, with an op amp summing amplifier, to form a bandstop filter as shown in Fig. 22.23. The bandpass response must subtract from the flat response, with $R_2 = |A_p| R_1$ (A_p is the passband gain of the bandpass filter). The bandstop characteristic has the same natural resonant frequency f_n and quality factor Q as the bandpass filter. The passband gain, of the overall bandstop is $A_s = R_2/R_1$ and may be adjusted independently to any desired value.

Another circuit offering high performance is the *twin-T* configuration of Fig. 22.24. The single op amp version, Fig. 22.24a has a very low quality factor Q which is difficult to calculate, usually about 0.3 or somewhat less. In the more complex circuit of Fig. 22.24b, Q is increased by the bootstrapping function of op amp OA_2. Quality factor Q may be estimated by

$$Q \approx \frac{0.3(R_4 + R_5)}{R_4} \tag{22.45}$$

and has a practical upper limit of about 50, with precision components.

Both filter versions have unity passband gain, that is, $A_s = 1$. The natural resonant frequency, or notch frequency, f_n is

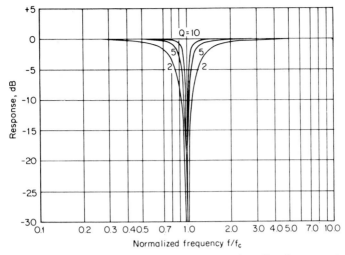

Fig. 22.22 Response characteristics, narrowband bandstop filter, for various Q values.

$$f_n = \frac{1}{2\pi} \left(\frac{C_1 + C_2}{R_1 R_2 C_1 C_2 C_3} \right)^{1/2} \qquad (22.46)$$

For design convenience with either circuit, use $C_1 = C_2 = C_3 = C$, $R_1 = R_2 = R$, and $R_3 = R/4$. The design steps and formulas are as follows:

1. Select a value for R. Calculate C:

$$C = \frac{1}{\sqrt{2}\,\pi R f_n} \qquad (22.47)$$

(The R and C values may be rescaled by maintaining the same RC product.)

2. The quality factor Q of the simpler circuit, Fig. 22.24a, is 0.238. This should be used if it is suitable.

3. For higher Q, use the circuit of Fig. 22.24b. Select a value for R_5. Calculate R_4 for the desired Q:

$$R_4 = \frac{\sqrt{2}\,R_5}{6Q - \sqrt{2}} \qquad (22.48)$$

example 22.10 Design a twin-T notch filter to have a notch frequency $f_n = 15$ Hz and bandwidth = 1.3 Hz. Find its attenuation at 13 Hz.

Fig. 22.23 Narrowband bandstop filter using a bandpass filter.

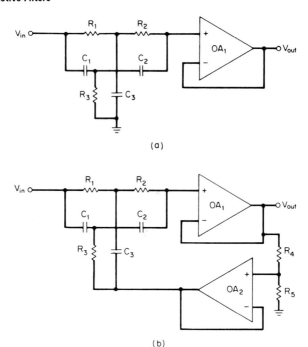

Fig. 22.24 Twin-T bandstop filter. (*a*) Low-*Q* circuit. (*b*) Bootstrapped circuit.

solution From Eq. (22.27), $Q = f_n/\text{bandwidth} = 15 \text{ Hz}/1.3 \text{ Hz} = 11.5$. Select $R_1 = R_2 = R = 400 \text{ k}\Omega$ and $R_3 = R/4 = 100 \text{ k}\Omega$. From Eq. (22.47), $C_1 = C_2 = C_3 = C = 1/(\sqrt{2}\pi \cdot 400 \text{ k}\Omega \cdot 15 \text{ Hz}) = 37.5 \text{ nF}$.

Choose $R_5 = 10 \text{ k}\Omega$. Then, from Eq. (22.48), $R_4 = (\sqrt{2} \cdot 10 \text{ k}\Omega)/(6 \cdot 11.5 - \sqrt{2}) = 209 \text{ }\Omega$. The attenuation at 13 Hz, from Eq. (22.44), is

$$\text{dB} = -10 \log \left\{ 1 + \frac{(13 \text{ Hz})^2 (15 \text{ Hz})^2}{(11.5)^2 [(13 \text{ Hz})^2 - (15 \text{ Hz})^2)]^2} \right\} = -0.38 \text{ dB}$$

Narrowband filters may also be implemented with state-variable circuits.

22.4 STATE-VARIABLE FILTERS

The basic state-variable filter unit is second-order and has three outputs which provide low-pass, high-pass, and bandpass functions. With minor circuit changes or additions, almost any other second-order function can be performed. For this reason, it is sometimes termed a *universal active filter*, and it is also referred to as a *biquad filter*. Other advantages are that the frequency usually can be changed without affecting the gain or *Q*; high values of quality factor *Q*, up to several hundred, are possible; and for some filter types, the design process is much simpler than for alternate circuits. The circuitry is somewhat complex, requiring at least three op amps. However, integrated-circuit (IC) versions are available that reduce both cost and space requirements.

BASIC CIRCUIT The simplest state-variable circuit uses op amps and discrete components, as shown in Fig. 22.25. There are three outputs for the low-pass, high-pass, and band-pass functions. Parameters for the transfer functions, Eqs. (22.7), (22.8), and (22.26), are calculated with these equations:

Natural resonant frequency:

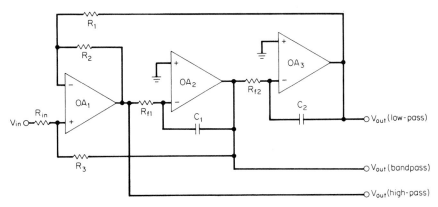

Fig. 22.25 State-variable filter.

$$\omega_n = \left(\frac{R_1}{R_2 R_{f1} R_{f2} C_1 C_2} \right)^{1/2} \tag{22.49}$$

Quality factor:

$$Q = \frac{R_1}{R_2} \frac{R_{in} + R_3}{R_1 + R_2} \left(\frac{R_2 R_{f1} C_1}{R_1 R_{f2} C_2} \right)^{1/2} \tag{22.50}$$

Passband gain, low-pass function:

$$A_1 = \frac{R_3(R_1 + R_2)}{R_2(R_{in} + R_3)} \tag{22.51}$$

Bandpass gain, high-pass function:

$$A_h = \frac{R_3(R_1 + R_2)}{R_1(R_{in} + R_3)} \tag{22.52}$$

Passband gain, bandpass function:

$$A_b = -\frac{R_3}{R_{in}} \tag{22.53}$$

After these parameters have been determined, the particular filter characteristics are found by using the data in Secs. 22.2 and 22.3.

example 22.11   A state-variable filter with the circuit configuration of Fig. 22.25 has the following component values: $R_1 = 45$ kΩ, $R_2 = 30$ kΩ, $R_3 = 40$ kΩ, $R_{in} = 25$ kΩ, $R_{f1} = 56$ kΩ, $R_{f2} = 100$ kΩ, $C_1 = 1.0$ nF, and $C_2 = 500$ pF. Find the filter characteristics at all three outputs.

solution   Quality factor, from Eq. (22.50), is

$$Q = \frac{45 \text{ k}\Omega}{30 \text{ k}\Omega} \frac{25 \text{ k}\Omega + 40 \text{ k}\Omega}{45 \text{ k}\Omega + 30 \text{ k}\Omega} \left(\frac{30 \text{ k}\Omega \cdot 56 \text{ k}\Omega \cdot 1.0 \text{ nF}}{45 \text{ k}\Omega \cdot 100 \text{ k}\Omega \cdot 500 \text{ pF}} \right)^{1/2}$$
$$= 1.123$$

Natural resonant frequency, from Eq. (22.49), is

$$\omega_n = \left(\frac{45 \text{ k}\Omega}{30 \text{ k}\Omega \cdot 56 \text{ k}\Omega \cdot 100 \text{ k}\Omega \cdot 1.0 \text{ nF} \cdot 500 \text{ pF}} \right)^{1/2}$$
$$= 23.1 \text{ krad/s}$$

Referring to the second-order filter characteristics data in Table 22.8, we find the Q value corresponds to a Chebyshev characteristic with 2-dB peak response. (See also Fig. 22.7.) From the ω_1'

value, cutoff frequencies are calculated: $f_{c(\text{low-pass})}$ = (23.1 krad/s)/(5.700 rads/Hz) = 4.05 kHz, and $f_{c(\text{high-pass})}$ = (23.1 krad/s)(5.700)/($4\pi^2$ rads/Hz) = 3.34 kHz. For the bandpass output, the natural frequency is f_n = $\omega_n/(2\pi)$ = (23.1 krad/s)/2π = 3.68 kHz, and the bandwidth, from Eq. (22.27), is (3.68 kHz)/1.123 = 3.27 kHz. Passband gain values, from Eqs. (22.51), (22.52), and (22.53), are

$$A_1 = \frac{40 \text{ k}\Omega \ (45 \text{ k}\Omega + 30 \text{ k}\Omega)}{30 \text{ k}\Omega \ (25 \text{ k}\Omega + 40 \text{ k}\Omega)} = = 1.54 \qquad \text{(low-pass)}$$

$$A_h = \frac{40 \text{ k}\Omega \ (45 \text{ k}\Omega + 30 \text{ k}\Omega)}{45 \text{ k}\Omega \ (25 \text{ k}\Omega + 40 \text{ k}\Omega)} = 1.03 \qquad \text{(high-pass)}$$

$$A_b = -\frac{40 \text{ k}\Omega}{25 \text{ k}\Omega} = -1.6 \qquad \text{(bandpass)}$$

The negative value of A_b indicates phase reversal of the signal, in the passband, at this output.

IC FILTERS   Integrated-circuit state-variable filters are designed for maximum versatility with a minimum number of discrete components. A widely used circuit type is shown in Fig. 22.26 with certain external component connections. These illustrate only a few of many possible ways of using the ICs.

Low- and high-pass functions   The configuration of Fig. 22.26a is convenient for designing most second-order low- and high-pass filters or for second-order modules of higher-order filters. Transfer functions are expressed by Eqs. (22.7) and (22.8). Parameter values are determined with these equations:
Natural resonant frequency:

$$\omega_n = \left(\frac{R_2}{R_1 R_{f1} R_{f2} C_1 C_2} \right)^{1/2} \qquad (22.54)$$

Quality factor:

$$Q = \frac{1 + R_3/R_q}{(R_1 + R_2)/R_1 + R_2/R_{\text{in}}} \left(\frac{R_2 R_{f1} C_1}{R_1 R_{f2} C_2} \right)^{1/2} \qquad (22.55)$$

Passband gain:

$$A_1 = -\frac{R_1}{R_{\text{in}}} \qquad \text{(low-pass)} \qquad (22.56)$$

$$A_h = -\frac{R_2}{R_{\text{in}}} \qquad \text{(high-pass)} \qquad (22.57)$$

Values for the bandpass function, having the transfer function of Eq. (22.26), may be needed also. Equations (22.54) and (22.55) apply for ω_n and Q. The passband gain is

$$A_p = \frac{(R_2/R_1)(1 + R_3/R_q)}{(R_1 + R_2)/R_1 + R_2/R_{\text{in}}} \qquad (22.58)$$

These design procedure steps are suggested:
1. From Table 22.8, or similar data, find values needed for ω_n and Q.
2. In the ICs, C_1 and C_2 are usually equal. Use $R_{f1} = R_{f2}$. Then, from Eq. (22.54),

$$R_{f1} = R_{f2} = \frac{1}{\omega_n C_1} \left(\frac{R_2}{R_1} \right)^{1/2} \qquad (22.59)$$

3. Determine R_{in} for the desired passband gain, with Eq. (22.56) and (22.57).
4. Calculate R_q for the desired quality factor Q. From Eq. (22.55), with $R_{f1} = R_{f2}$ and C_1 and C_2,

$$R_q = \frac{R_3}{Q[(R_1 + R_2)/R_1 + R_2/R_{\text{in}}](R_1/R_2)^{1/2} - 1} \qquad (22.60)$$

5. Check the signal amplitudes at all outputs for overdrive or limiting. Reduce the signal level or redesign for lower gain, if necessary.

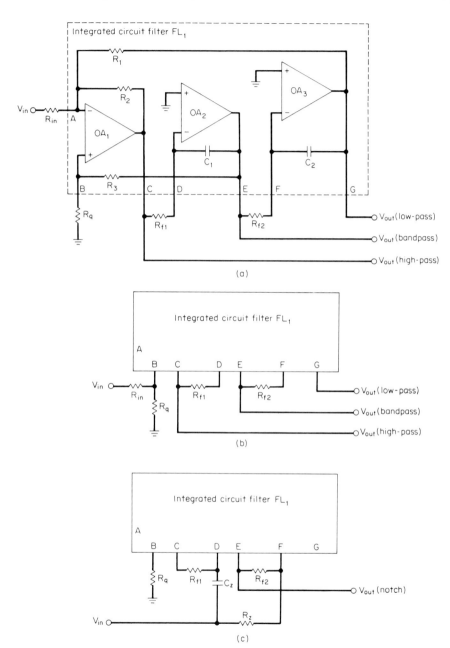

Fig. 22.26 Integrated-circuit state-variable filters. (*a*) Recommended circuit for low- and high-pass. (*b*) Recommended circuit for bandpass. (*c*) Recommended circuit for bandstop (notch).

example 22.12   Design a unity-gain high-pass second-order filter, with a 2-dB Chebyshev characteristic and cutoff frequency f_c = 400 Hz, using a National Semiconductor AF 100 active filter IC. The AF 100 has the following values: R_1 = R_3 = 100 kΩ, R_2 = 10 kΩ, and C_1 = C_2 = 1.0 nF.

solution
1. From Table 22.8, Q = 1.129, $\omega'_{\text{low-pass}}$ = 5.700 (r/s)/Hz, $\omega'_{\text{high-pass}}$ = $4\pi^2/5.700$ = 6.926 (rad/s)/Hz, and ω_n = 400 Hz · 6.926 (rad/s)/Hz = 2.77 krad/s.
2. From Eq. (22.59), R_{f1} = R_{f2} = (10 kΩ/100 kΩ)$^{1/2}$/(1.0 nF · 2.77 krad/s) = 114 kΩ.
3. From Eq. (22.57), R_{in} = 100 kΩ/1 = 100 kΩ.
4. From Eq. (22.60),

$$R_q = \frac{100\ \text{k}\Omega}{1.129[(100\ \text{k}\Omega + 10\ \text{k}\Omega)/100\ \text{k}\Omega + 10\ \text{k}\Omega/100\ \text{k}\Omega](100\ \text{k}\Omega/10\ \text{k}\Omega)^{1/2} - 1}$$
$$= 30.4\ \text{k}\Omega$$

Bandpass function   The circuit of Fig. 22.26a may be used for the bandpass function also. However, when unity gain is needed (or when gain elsewhere can be adjusted easily to accommodate unity gain in the filter), the alternate circuit of Fig. 22.27b is easier to design. This circuit also may be preferred for low- or high-pass if noninversion of the passband signal is needed.

Natural resonant frequency ω_n is given by Eq. (22.54). Other values for the bandpass transfer function, Eq. (22.26), from the circuit of Fig. 22.26b are as follows:
Quality factor:

$$Q = \frac{R_1}{R_1 + R_2}\left(1 + \frac{R_3}{R_{\text{in}}} + \frac{R_3}{R_q}\right)\left(\frac{R_2 R_{f1} C_1}{R_1 R_{f2} C_2}\right)^{1/2} \tag{22.61}$$

Bandpass gain:

$$A_p = -\frac{1 + R_3/R_q + R_3/R_{\text{in}}}{1 + R_{\text{in}}/R_3 + R_{\text{in}}/R_q} \tag{22.62}$$

For the low- and high-pass functions, Eqs. (22.54) and (22.61) apply. Passband gains are

$$A_l = \frac{R_1 + R_2}{R_2}\left(1 + \frac{R_3 + R_q}{R_{\text{in}}}\right) \qquad \text{(low-pass)} \tag{22.63}$$

$$A_h = \frac{R_1 + R_2}{R_1}\left(1 + \frac{R_3 + R_q}{R_{\text{in}}}\right) \qquad \text{(high-pass)} \tag{22.64}$$

These design procedure steps are suggested:
1. Determine the desired frequency ω_n and quality factor Q by using Eqs. (22.27), (22.28), (22.29), and (22.30) as needed.
2. Find R_{f1} = R_{f2} with Eq. (22.59).
3. For unity gain A_p = 1, set R_{in} = R_3.
4. Calculate R_q for the desired quality factor Q. From Eq. (22.61), with $R_{f1}C_1 = R_{f2}C_2$ and R_{in} = R_3,

$$R_q = \frac{R_3}{Q[(R_1 + R_2)/R_1](R_1/R_2)^{1/2} - 1 - R_3/R_1} \tag{22.65}$$

5. Check the signal levels at all outputs for limiting, and reduce the input signal level, if necessary.

example 22.13   Design a bandpass filter equivalent to that of Example 22.9 (A_p = −1, ω_n = 6.28 krad/s, Q = 4.76) using the AF 100 filter (R_1 = R_3 = 100 kΩ, R_2 = 10 kΩ, C_1 = C_2 = 1.0 nF).

solution
1. From Eq. (22.59), R_{f1} = R_{f2} = (10 kΩ/100 kΩ)$^{1/2}$/(1.0 nF · 6.28 krad/s) = 50.4 kΩ.
2. For unity gain, R_{in} = R_3 = 100 kΩ.
3. From Eq. (22.65),

$$R_q = \frac{100\ \text{k}\Omega}{4.76[(100\ \text{k}\Omega + 10\ \text{k}\Omega)/100\ \text{k}\Omega](100\ \text{k}\Omega/10\ \text{k}\Omega)^{1/2} - 1 - 100\ \text{k}\Omega/100\ \text{k}\Omega}$$
$$= 6.87\ \text{k}\Omega$$

Bandstop (notch) function   The circuit of Fig. 22.26c has the bandstop transfer function of Eq. (22.43) when $R_z = R_{f2}/A_s$ and $C_z = A_s C_2$, where A_s is the passband gain. The natural resonant frequency ω_n is given by Eq. (22.54). Values of other transfer function parameters are as follows:
Quality factor:

$$Q = \frac{R_1(R_3 + R_q)}{R_q(R_1 + R_2)} \left(\frac{R_2 R_{f1} C_1}{R_1 R_{f2} C_2} \right)^{1/2} \tag{22.66}$$

Passband gain:

$$A_s = \frac{R_{f2}}{R_z} = \frac{C_z}{C_2} \tag{22.67}$$

These design procedure steps are recommended:
1. Determine the desired frequency ω_n and quality factor Q, using Eqs. (22.27), (22.28), (22.29), and (22.30) as needed.
2. Find $R_{f1} = R_{f2}$ with Eq. (22.59).
3. Calculate R_z and C_z for the desired passband gain with Eq. (22.67).
4. Calculate R_q for the desired quality. From Eq. (22.66),

$$R_q = \frac{R_1 R_3}{Q(R_1 + R_2)(R_1/R_2)^{1/2} - R_1} \tag{22.68}$$

example 22.14   Design a bandstop filter with notch frequency $f_n = 250$ Hz, quality factor $Q = 15$, and unity passband gain $A_s = 1$, using the AF 100 filter ($R_1 = R_3 = 100$ kΩ, $R_2 = 10$ kΩ, and $C_1 = C_2 = 1.0$ nF).

solution
1. Here $\omega_n = 2\pi f_n = 2\pi \cdot 250$ Hz $= 1.57$ krad/s. From Eq. (22.59), $R_{f1} = R_{f2} = (10$ kΩ/ 100 kΩ)$^{1/2}$/(1.0 nF $\cdot$ 1.57 krad/s) $= 201$ kΩ.
2. For unity gain ($A_s = 1$), using Eq. (22.67), we have $1 = 201$ kΩ/$R_z = C_z$/1.0 nF. Thus $R_z = 201$ kΩ and $C_z = 1.0$ nF.
3. From Eq. (22.68),

$$R_q = \frac{100 \text{ k}\Omega \cdot 100 \text{ k}\Omega}{15\,(100 \text{ k}\Omega + 10 \text{ k}\Omega)(100 \text{ k}\Omega/10 \text{ k}\Omega)^{1/2} - 100 \text{ k}\Omega}$$
$$= 1.95 \text{ k}\Omega$$

Higher-order filters   State-variable filter sections may be cascaded, together with a first-order filter if needed, to form a filter of any desired order.

example 22.15   Design a fifth-order low-pass filter with a cutoff frequency $f_c = 650$ Hz and a 0.5; dB Chebyshev characteristic.

solution   Two second-order modules and one first-order module in cascade are needed, as indicated in Fig. 22.27a. The values of ω_n and Q for each module are found in Table 22.8, with $\omega_n = \omega' f_c$.
Implementation is shown in Fig. 22.27b, with AF 100 ICs for the second-order modules and the circuit of Fig. 22.3 for the first-order module. Components are calculated as in examples 22.12 and 22.1 for unity passband gain.

Low-frequency operation   Examination of Eq. (22.59) shows that as ω_n decreases, R_f increases. However, a maximum value of R_f, specified by the IC manufacturer, may not be exceeded. For low-frequency operation, R_f may be reduced by effectively increasing the capacitance values, as shown in Fig. 22.28, with added external capacitors in parallel with the internal capacitors. In the various formulas, the values for C_1 and C_2 are replaced by the sum of the capacitances in shunt with each other.
An alternate technique for low frequencies is the use of special resistive networks. Data for these usually are given by the IC manufacturer in the specifications and application recommendations.

Voltage-tuned filters   The development of IC multipliers, together with state-variable filters, makes compact, relatively low-cost, voltage-tuned filters possible. Circuit connections are shown in Fig. 22.29. Changing the signal amplitudes at R_{f1} and R_{f2} has the same effect as changing the resistor values, which determine the frequency. These amplitudes are acted on by the multiplier, so that they are proportional to the multiplier control input V_m.

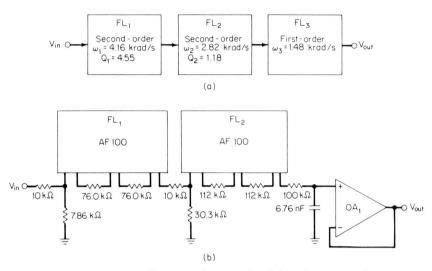

(a)

(b)

Fig. 22.27   Fifth-order low-pass filter (Example 22.15). (*a*) Modules and parameters. (*b*) Circuit using AF 100 ICs.

For circuit configurations such as Fig. 22.26*a* and *b*, where R_{f1} and R_{f2} are the only other connections to terminals *D* and *F*, this arrangement works very well, and voltage tuning will not affect passband gain or the quality factor *Q*. It is not, however, recommended for circuits, such as Fig. 22.26*c*, where additional components connect to these terminals.

22.5   OTHER FILTERS

There are many filter types that have very specialized applications or that are not used widely enough to be included here. There are also a few, described below, that need some mention.

ELLIPTIC FILTERS   Elliptic function filters may be devised for any of the usual low- and high-pass, bandpass, and band-reject responses. Their transfer functions are different from those described here, and the circuits are much more complex. The advantage of the elliptic characteristic is that, for a given order of filter, the passband skirt (rate of attenuation with frequency change) is steeper than that of other types. Disadvantages are that the design is more difficult, requiring many tables or other design aids; more expensive precision components are needed; and "tuning" or other modifications are more difficult and more likely to result in complete redesign.

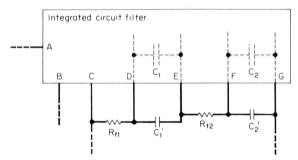

Fig. 22.28   Added capacitors for low-frequency operation (partial circuit).

Fig. 22.29   Added multipliers M1 and M2 for voltage control of frequency (frequency is proportional to V_m) (partial schematic).

When high performance is needed, usually it is easier and more economical to use higher-order Butterworth or Chebyshev characteristics. If an elliptic function is considered, the best approach is probably with state-variable ICs such as the AF 100. These usually require only discrete precision resistors, and not the more costly precision capacitors. Also, computer programs are usually available from the IC manufacturer, which greatly facilitates the design.

ALL-PASS NETWORKS   These are sometimes referred to as filters, but in the usual sense of the word they are not, for they pass all frequencies unattenuated. They are used to phase-shift the signal as a function of frequency. Frequency-dependent delay, which is related to phase shift, also results. An important application is to compensate, with an equalizing delay at various frequencies, for the frequency-dependent delays of conventional filters and other circuits. Equalizing the delay improves transient response, so that the signal shape is preserved better.

A first-order all-pass network is shown in Fig. 22.30. Its transfer function is

$$\frac{V_{\text{out}}(s)}{V_{\text{in}}(s)} = \frac{A_a(\omega_n + s)}{\omega_n - s} \quad (22.69)$$

Fig. 22.30   First-order all-pass network.

For this circuit, $A_a = 1$ and $\omega_n = 1/R_2C$. Phase shift at any frequency may be calculated with Eq. (22.69). The frequency-normalized phase shift is plotted in Fig. 22.31.

Fig. 22.31   Phase shift of first-order all-pass network.

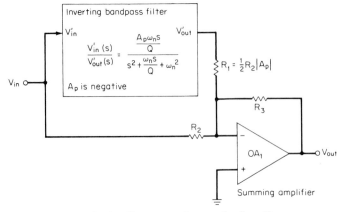

Fig. 22.32   Second-order all-pass network using a bandpass filter.

A bandpass filter and summing amplifier may be combined to form a second-order network, as shown in Fig. 22.32. The transfer function is

$$\frac{V_{out}(s)}{V_{in}(s)} = \frac{A_a(s^2 - \omega_n s/Q + \omega_n^2)}{s^2 + \omega_n s/Q + \omega_n^2}$$

(22.70)

Parameters ω_n and Q are determined by, and the same as, those of the bandpass amplifier FL_1. Gain $A_a = R_3/R_2$. Phase shift at any frequency is calculated from Eq. (22.70). The normalized phase shift is plotted in Fig. 22.33.

DIGITAL FILTERS   Filtering functions performed by active or passive networks often may be accomplished with digital techniques. The system is shown in Fig. 22.34. The sampling rate of the A/D converter must be greater than the highest input signal of interest, and the digital processing must be consistent with the sampling rate. Advantages are that there are no precision analog components, ideal filter characteristics usually can be more

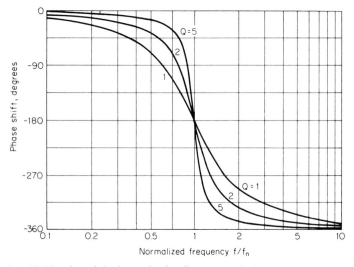

Fig. 22.33   Phase shift of second-order all-pass network (for the circuit of Fig. 22.32, the negative gain A_a causes an additional 180° shift at all frequencies).

Fig. 22.34 Digital filter system.

closely approached, characteristics are insensitive to temperature or aging (which causes drift in conventional filter parameters), and programmability may permit easy changes or adjustments. Disadvantages are that the useful frequency range is, at this time, limited to the low portion of the audio spectrum and below; circuits are exceedingly complex; and cost and space requirements usually have been excessive.

At present, technological progress is very rapid in digital electronics, especially in very large-scale integration (VLSI) techniques. With the decreasing costs of increasingly complex chips, digital filters could become important devices.

REFERENCES AND BIBLIOGRAPHY

Berlin, H. M.: *Design of Active Filters, with Experiments*, Sams, Indianapolis, 1978.
Daniels, R. W.: *Approximation Methods for Electronic Filter Design*, Artech House, Dedham, Mass., 1975.
Daryanani, G.: *Principles of Active Network Synthesis and Design*, Wiley, New York, 1976.
Ghausi, M., and K. Laker: *Modern Filter Design: Active RC and Switched Capacitor*, Prentice-Hall, Englewood Cliffs, N.J., 1982.
Johnson, D. E.: *Introduction to Filter Theory*, Prentice-Hall, Englewood Cliffs, N.J., 1976.
Johnson, D. E.: *A Handbook of Active Filters*, Prentice-Hall, Englewood Cliffs, N.J., 1980.
Johnson, D. E., and J. L. Hilburn: *Manual of Active Filter Design*, McGraw-Hill, New York, 1973.
Johnson, D. E., and J. L. Hilburn: *Rapid Practical Design of Active Filters*, Wiley, New York, 1975.
Schaumann, R.: *Modern Active Filter Design*, Wiley, New York, 1981.
Schaumann, R.: *Modern Active Filter Design*, Institute of Electrical and Electronics Engineers, New York, 1981.
Williams, A. B.: *Active Filter Design*, Artech House, Dedham, Mass., 1975.

Chapter 23

Digital Multimeters

Kenneth Jessen

Material Engineer

Hewlett-Packard Co.

Loveland, Colorado

23.1 BASIC BUILDING BLOCKS

The job of a digital multimeter (DMM) is to convert an analog signal to its digital equivalent. The analog signal might be a dc voltage, ac voltage, or ac or dc current or resistance. The process of conversion can be divided into four functional blocks, as shown in Fig. 23.1.

The input signal first passes through some type of signal conditioner or converter. For dc voltages, the signal conditioner is composed of an attenuator, operational amplifier (op amp), and precision feedback resistors. The attenuator is used to divide higher voltages above the range of the op amp while lower voltages are amplified. Ranges in a DMM are usually in decade increments. If the input is an ac signal, an ac-to-dc (A/D) converter is used to convert the ac signal to its equivalent dc value. To measure resistance, an ohm converter supplies a constant current to the unknown resistor. Then the voltage drop across the resistor is measured as a dc voltage. In all cases, it is the job of the signal conditioner or converter to normalize the input to the DMM into a dc voltage within the range of the A/D converter.

The A/D converter takes the prescaled dc voltage and converts it to digits. These single-range devices operate at a full-scale voltage such as 1, 3, or 10 V depending on the type of DMM.

At this point it is wise to take a brief look at the mechanics of the conversion process. Suppose a 250-V ac signal is applied to the DMM input which has an A/D converter with a 1-V full-scale input. First the ac signal must be attenuated on the 1000-V ac range to 0.25 V ac (100:1 division ratio). Next the ac signal is converted to its dc equivalent. The op amp with its gain set to unity passes the dc signal to the A/D converter which digitizes the voltage to 2500. On the DMM display, range and function information is added to produce a display of 250.0 V ac.

If the signal happened to be a dc voltage of say 0.05 V, the op amp would be set to a gain of X10 such that 0.5 V dc would be passed on to the A/D converter. The digitized value would be 500, and again range and function information would be added by the display to produce 0.05 V dc, or 50 mV dc.

The first two building blocks govern most of the characteristics of a DMM such as accuracy, number of digits, ranges, sensitivity, and so on. The logic manages the flow of information to and from the instrument as well as internal functions. The job of the display is to visually communicate the result of a given measurement.

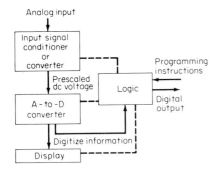

Analog input

Input signal conditioner or converter

Prescaled dc voltage

A-to-D converter

Logic

Programming instructions

Digital output

Digitize information

Display

Fig. 23.1 A digital multimeter may be divided into four basic blocks as shown here. The input signal conditioner or converter normalizes the input to a dc voltage within the range of the A/D converter. The digitized output from the A/D converter is sent to the display as a visual output, and the logic manages the internal timing and flow of information into and out of the instrument.

example 23.1 Show how a 500-V ac signal would be handled by a DMM with an A/D converter which had a 10-V full-scale input.

solution The 500 V ac would be attenuated on the 1000-V ac range to 5 V and converted to its dc equivalent. The op amp with its gain set to unity would pass this signal to the A/D converter. The signal would be digitized to 5000 and displayed as 500.0 V ac, using range and function information.

23.2 DC INPUT AMPLIFIER/ATTENUATOR

The signal conditioning for dc voltages includes both amplification and attenuation. If the full-scale input to the A/D converter is 10 V, then amplification is used on the lower ranges such as 100 mV and 1 V. Unity gain is used on the 10-V range; for the higher ranges, such as 100 and 1000 V, attenuation would be used. To make the DMM's "front end" useful, it must also act as an impedance-matching device. An ideal DMM has infinite input impedance and will not load the circuit being measured. As a practical matter, an op amp with a FET input stage will produce impedances of 10^{10} Ω. For the higher ranges where an attenuator is necessary, an input impedance of 10 MΩ is common. Figure 23.2 illustrates a typical dc input amplifier/attenuator.

Gains are selected by precision resistors in the feedback loop of the op amp. By changing the value of the resistors, the gain may be changed according to the following expression: $(R_1 + R_2)/R_2$. The precision resistors used are selected for excellent temperature tracking characteristics, and many are produced by using thin-film technology.

One of the sources of error in a DMM is the offset of the op amp. This directly affects accuracy and ability to measure small voltages. An automatic-zero circuit is commonly employed to overcome this problem, as illustrated in Fig. 22.3a and b. The dc offset is measured, stored, and then subtracted from the reading. The technique relies on the fact that

Input

Operational amplifier

To A-to-D converter

R_1

Attenuator

These resistors represent a set of resistors used for the lower ranges

R_2

Fig. 23.2 The dc input section to a DMM consists of an op amp with the gain controlled by precision resistors R1 and R2. An attenuator is switched in for voltage ranges higher than the capability of the op amp.

Fig. 23.3 The automatic-zero circuit cancels the effect of dc offsets by storing the offset on a capacitor during intervals when the input amplifier is not in use for a measurement. The stored offset is subtracted from the reading by placing the capacitor in series with the feedback resistors.

the input amplifier is used only during part of the measurement sequence. During auto-zero, the input is short-circuited to ground by switch S2. Switch S1 opens so that the input to the instrument from the user is not short-circuited to ground.

Any offset is stored on capacitor C relative to ground. When the amplifier is used for a measurement, switches S2, S3, and S5 are opened. The capacitor with the stored offset is introduced in series with the feedback loop but in opposite polarity. The capacitor acts as a battery, nulling the offset voltage.

The entire automatic-zero function can be handled digitally also. This has become quite practical through the use of a microprocessor. Many modern DMMs accomplish automatic-zeroing by short-circuiting the input between normal measurement cycles, measuring the value of the offset, and storing the value in random-access memory (RAM). By using the computational power of the microprocessor, the stored offset is subtracted from the subsequent measurement mathematically.

23.3 INTERNAL REFERENCE VOLTAGE

Many DMMs employ a zener reference diode as an internal voltage standard. Measurements are calibrated against this internal reference. A resistive divider allows the reference voltage from the zener diode to be calibrated periodically. The diode used is carefully selected for low temperature coefficient, and some instruments have the reference diode in a temperature-controlled chamber.

Some of the newer DMMs using microprocessors have eliminated all internal adjustments and employ electronic calibration, as shown in Fig. 23.4. During calibration the voltage standard is applied to the input, and the DMM is placed in a calibration routine. The calibration constants needed to produce a correct reading relative to the internal zener refer-

$$y \text{ (calibrated reading)} = mx \text{ (uncalibrated reading)} + b$$

$m = $ slope
$b = $ constant } both stored in RAM

Fig. 23.4 Electronic calibration makes use of a microprocessor controller to read and store the calibration constants necessary to produce correct readings. The reference voltage within the DMM need only be stable. This technique eliminates the need for internal adjustments.

ence diode are stored in a nonvolatile memory. During normal use, the calibration constants for that particular range and function are recalled from memory and applied mathematically to compute the correct reading. The nonvolatile memory typically consists of a CMOS RAM powered by a lithium battery with a life expectancy of up to 10 years.

23.4 AC CONVERSION

There are vast differences among ac converters used in DMMs. The ideal converter measures the rms value of the incoming signal and converts it to its equivalent dc voltage for digitization by the A/D converter. By definition, the rms value is the square root of the sum of the squares of the components of the ac signal, or the equivalent dc voltage required to produce the same heating effect as the ac signal.

Some ac converters use a simple, inexpensive diode bridge, as shown in Fig. 23.5. The output from the bridge is filtered and scaled to produce a dc voltage proportional to the average value of the ac input. The scale factor selected applies to a pure sine wave (without harmonic content):

$$\frac{V_{rms}}{V_{avg}} = \frac{0.707}{0.637} = 1.11$$

Should the odd harmonic distortion enter the signal, this scale factor changes (odd harmonics add an extra half-cycle of energy during the period of the fundamental). Even harmonics, however, cause little or no change in the scale factor to produce an rms reading. Since the scale factor is fixed, significant errors are produced by odd harmonic content. For an idea of just how severe the problem is, refer to Fig. 23.6 which shows percentage error versus phase for 1 percent odd harmonic content (relative to the fundamental) for average responding converters. The same type of information is illustrated in Fig. 23.7 for distortion as a function of amplitude.

Since even harmonics add as much as they subtract to the rectified waveform over the period of the fundamental, they contribute only small errors. For example, even harmonic content up to 10 percent produces an error below ±0.5 percent.

An ac converter that overcomes these limitations is a true rms converter. It responds to the dc heating value of the input signal independent of harmonic content up to the full bandwidth of the converter. One approach uses a combination thermocouple and heater and is illustrated in Fig. 23.8. The input signal is presented to the heater element, and the thermocouple measures the heating effect. The two components are in close thermal proximity. Since the dc output from the thermocouple is equivalent to the heating value of the incoming signal, by definition, this approach measures true rms.

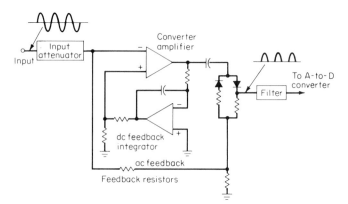

Fig. 23.5 Average responding ac converters use a diode bridge to rectify the signal. Filtering smoothes the signal into a dc value equivalent to the rms value for a pure sine wave. Errors in this type of converter become significant when the input signal contains small amounts of odd harmonic distortion.

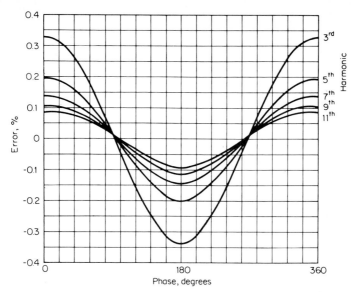

Fig. 23.6 Percentage of error versus phase for 1 percent odd harmonic distortion is shown for an average responding ac converter.

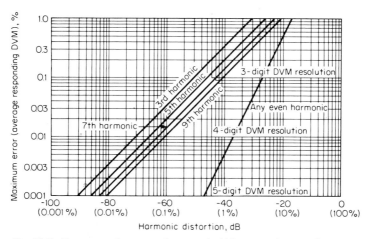

Fig. 23.7 Percentage of error as a function of odd harmonic distortion for a worst-case phase relationship with the fundamental.

Fig. 23.8 True rms ac converters use a combination thermocouple and heater. The ac signal is presented to the heater element, and the thermocouple converts the heating effect to its dc equivalent voltage, which is digitized. A reference thermocouple is necessary to linearize the output and compensate for changes in ambient temperature.

Thermocouples are nonlinear as a function of input signal amplitude. They also respond to any change in ambient temperature. To make the thermocouple-heater approach useful, a second matched thermocouple operated in opposition is necessary to linearize the output and compensate for changes in ambient temperature. The sensitivity of the single thermocouple to small changes in the input signal can be improved by adding more junctions in series, but all in proximity to the same heater element. This is called a *thermopile*, and such devices are manufactured by using thin-film technology to reduce thermal inertia, thereby improving response time.

As a side note, thermocouples and thermopiles are square-law devices. Their output decreases at a rate much greater than proportional changes in the level of the input signal. For this reason, typically true rms converters are not specified for operation below one-tenth of full scale.

There is a third type of converter used in DMMs which has become quite popular. It is called a *quasi-rms*, or *computing rms*, converter. It simulates true rms action by using a series of op amps to square, take the average of the square, and then take the square root. In other words, it mathematically models true rms action. Since the synthesis is not perfect, such converters are limited to symmetric waveshapes. Square waves, triangular waves, and sine waves with up to 3 percent odd harmonic distortion can be measured effectively.

A figure of merit called the *crest factor* tells the user how much of a pulse can be measured by a given ac converter. It is defined as the peak value divided by the true rms value, or in terms of duty cycle, as follows:

$$\text{Crest factor} = \frac{V_{\text{peak}}}{V_{\text{rms}}} = \sqrt{\frac{1 - D}{D}}$$

where D = duty cycle = $T_1/(T_1 + T_2)$.

Crest factors of 5:1 to 7:1 can be handled by quasi-rms converters while 10:1 crest factors are commonly specified for true rms converters.

example 23.2   For third-harmonic distortion with a 140° phase shift with respect to the fundamental, what is the reading error on an average responding ac converter?

solution   Using Fig. 23.6, we find the error is −2.5 percent.

example 23.3   If a true rms voltmeter could accurately measure a pulse with a 10 percent duty cycle, what would its crest factor be rated at?

solution   The square root of 0.9/0.1 is a crest factor of 3:1.

23.5   MEASURING RESISTANCE

The basic approach used to measure the value of an unknown resistor is to supply a constant dc current through the unknown and then, by using the DMM dc input, to measure the voltage drop across the unknown. The value of the resistor can be computed given the amount of current supplied and the voltage reading. The simplest approach, called a two-wire measurement, is shown in Fig. 23.9a.

When a DMM is used in a system where wire runs from the instrument are often long, the lead wire resistance introduces an error, especially for low-resistance readings. For example, a 6-ft wire can have 0.060 Ω of lead resistance. If a measurement is being made on a 100-Ω resistor with a five-digit DMM, the lead resistance will produce a 60-count error. A four-wire ohm converter solves this problem. As shown in Fig. 23.9b, sense wires which carry virtually no current to the DMM are used for the measurement. A separate pair of wires is used to supply current to the unknown resistor.

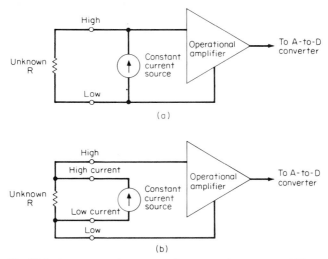

Fig. 23.9 In systems applications, simple two-wire ohm converters (*a*) have yielded to four-wire converters (*b*) to eliminate errors caused by lead resistance.

23.6 A/D CONVERSION TECHNIQUES

The A/D converter in a DMM governs some of the instrument's basic characteristics, including accuracy, speed, and noise rejection. A/D converters may be divided into two major classes: integrating and nonintegrating. The integrating techniques mathematically reduce the effect of power-line-related noise. Nonintegrating techniques offer advantages in speed and broadband noise rejection when they are combined with filtering.

One of the most important and popular A/D conversion techniques is dual-slope integration. Owing to its simplicity, it is widely used; a block diagram is shown in Fig. 23.10.

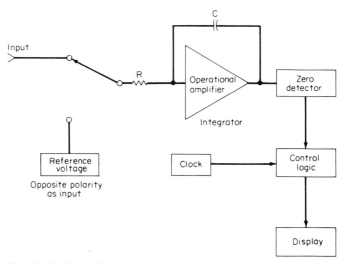

Fig. 23.10 The dual-slope A/D conversion technique uses an integrator as its key element. This integrating technique offers high noise rejection for line-related frequencies.

The input voltage is connected to an integrator via switch S for the duration of the integration period. Integration periods are selected to be related to the line frequency period; i.e., for 60-Hz line frequency, $\frac{1}{10}$ and $\frac{1}{60}$-s integration periods are common. During this time interval, C is charged at a rate governed by R. At the end of the integration period, the charge on C is proportional to the input voltage. The integrator action of a capacitor in the feedback loop of an op amp causes the voltage to charge in a linear fashion. The charging rate is governed by the current through R.

In the second phase of dual-slope integration, S is connected to a reference voltage of the opposite polarity as the input voltage. Then C is discharged linearly by the reference voltage until zero crossover is reached. At the beginning of the discharge cycle, a clock is gated on and counts are accumulated until zero crossover. The total count at the end of the measurement period is proportional to the input voltage and is scaled and displayed as the reading.

If the input voltage is doubled, as illustrated in Fig. 23.11, then C charges twice as fast to twice the value. When the reference voltage is applied, it takes twice as long to discharge C, and the accumulated counts are, therefore, doubled.

Considerable time can be saved during the measurement cycle if the reading is developed during discharge. An obvious solution is to increase the discharge slope to speed up the reading rate. As the slope increases, however, being able to accurately detect zero crossover becomes more difficult. A shallow slope produces more accurate readings. A modification to dual slope can be made by using a steep slope initially; then just prior to zero crossover, the slope is changed to a shallow one.

As microprocessors dropped in price and became more practical for instrumentation, a variation on the dual-slope one called a multislope DMM came into being. The multislope employs a steep rundown initially, as illustrated in Fig. 23.12. A finite amount of overshoot is allowed beyond the zero crossover. A slope exactly a decade less in steepness is applied next to the remaining voltage until a second zero crossover occurs, also with a certain amount of overshoot. At this point, a third slope yet another decade less in steepness is applied. Finally, a fourth slope one-thousandth as steep is applied for the final zero crossover.

At the beginning of each slope, a clock is started and then stopped at the zero crossover. The accumulated clock pulses for each slope represent the four least significant digits in the reading (10 000 counts). The digits are developed as the discharge cycle takes place, improving the instrument's reading speed.

Note that the steepest slope (with the least definition in its zero crossover) is the least critical to the reading. The shallowest slope with the greatest definition in zero crossover is used to determine the most sensitive digit.

Multislope DMMs can offer a tradeoff in resolution versus reading speed by reducing the integration period and terminating the reading during the rundown period for the desired

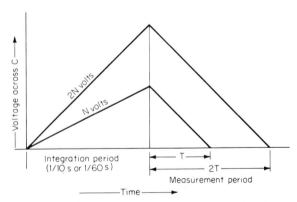

Fig. 23.11 The voltage on the integrator capacitor C is directly proportional to the magnitude of the input voltage. This example shows two different input voltages, one twice the magnitude of the other. During the measurement period, the number of counts accumulated at zero crossover is exactly double.

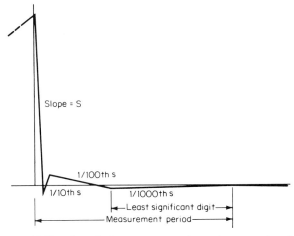

Fig. 23.12   The multislope integrating technique is an expansion of
the dual-slope technique. The least significant digits in the reading
are developed during the rundown period. Slopes exactly a decade
less in steepness are successively applied to the voltage on the integrating capacitor. A certain amount of overshoot is allowed each time
until the final zero crossover occurs.

number of digits. Fewer digits require less time to develop. Table 23.1 shows an example
of the specifications for a specific commercially available unit. The tradeoff is simple: the
greatest resolution and sensitivity are achieved at the greatest sacrifice in reading speed.

During the integration period, the most significant digits can be developed. With the
simple dual-slope technique, the input voltage is allowed to ramp during the integration
period in a linear fashion. For very small input voltages, the charge remaining on the integrator capacitor is proportionally less and, therefore, susceptible to noise. A sawtooth
approach can be used during the integration period to fully utilize the dynamic range of
the integrator. The sawtooth is created by allowing the input voltage to ramp up; then by
switching in the reference voltage (of opposite polarity), the voltage level is forced to ramp
down. At a predetermined level, the input voltage is applied again, the voltage ramps up,
and so on until the end of the integration period. The number of transitions necessary during
the integration period required to keep the voltage within range of the integrator is used to
develop the two most significant digits. When they are combined with the digits developed
during the discharge period, an accurate six-digit reading can be developed.

Successive approximation represents one of the most important nonintegrating A/D techniques. A succession of level changes is generated by a D/A converter and compared with
the input voltage, as shown in Fig. 23.13. If the ultimate reading is to have binary-coded
decimal (BCD) weighting, then the D/A converter produces a sequence of voltages which
have the same weighting (8:4:2:1). Other weighting schemes may be employed also.

For each digit, four "tries," or approximations, are made until the number of desired
digits is developed. The comparator is used to decide whether the try is to be retained as
part of the final reading or rejected. Accuracy is limited by the comparator, and the tech-

TABLE 23.1   Reading Speed of Multislope DMM

	Reading rate per second		
	Three digits	Four digits	Five digits
Automatic-zero off	71	33	4.4
Automatic-zero on	53	20	2.3

nique is quite vulnerable to noise. It is simple and extremely fast, making it popular for systems applications in which reading speed is important.

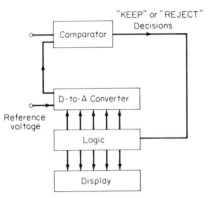

Fig. 23.13 Successive approximation is a technique used in DMMs where reading speed is important. A succession of level changes are generated by a D/A converter and compared with the input voltage to derive the final reading.

As an example of how successive approximation works, assume that the full-scale input to the A/D converter is 15 V and that an 11.250-V signal is applied. The first digit is established through the generation of 8, 4, 2, and 1 V. Refer to Fig. 23.14. The comparator keeps the 8 V, rejects the combination of 8 V and 4 V, keeps the combination of 8 and 2 V, and keeps the combination of 8, 2, and 1 V. Then the magnitude of the voltage levels is dropped by a decade to produce 0.8, 0.4, 0.2, and 0.1 V. The 11 V from the first approximation is combined sequentially with the second approximation. The combination of 11 and 0.8 V is rejected, the combination of 11 and 0.4 V is also rejected, but the combination of 11 and 0.2 V is kept. The combination of 11.2 and 0.1 V is rejected. After application of a sequence a decade lower (0.08, 0.04, 0.02, and 0.01 V), the final reading is derived as the sum of 8, 2, 1, 0.2, 0.04, and 0.01 V, or 11.250 V.

example 23.4   What is the greatest advantage of an integrating DMM?

solution   Rejection of power-line-related noise.

23.7   NOISE REJECTION

Noise is a big factor in the disruption of accurate measurements. It comes in the form of noise common to both input lines (common-mode) and noise between the inputs (normal-mode). Noise is frequently line-related and induced in the input wires from electromagnetic fields or from the device being measured. Other sources of noise include "white noise" covering a wide-frequency spectrum. There is no such thing as typical noise since it is so dependent on what is being measured and how long the wire runs are. Figure 23.15 is used to represent a ripple superimposed on a dc signal. The ripple could contain nonperiodic noise spikes and a 60-Hz line-related "hum." There are two techniques for reducing noise, and both would have to be employed for the type of noise shown in Fig. 23.15.

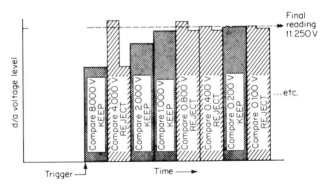

Fig. 23.14   This is an example of how successive approximation works for digitizing an input voltage equal to 11.250 V. The comparator either keeps or rejects the voltage generated for each "try," or approximation. Through a series of successive tries the final reading is developed.

Fig. 23.15 This is an example of line-related noise and noise spikes superimposed on a dc signal. To accurately measure the dc signal, the DMM must have some way of rejecting the noise.

Fig. 23.16 For integrating DMMs, this graph shows the relationship between noise rejection and deviation from a power-line frequency of 60 Hz. A maximum deviation of 0.15 percent is expected in North America.

Integration is an A/D measurement technique that will integrate power-line-related noise. The integration period is selected as an integral number of periodic noise cycles. For 60 Hz, integration periods of 1/10 and 1/60 s are common. An integrating DMM with a 1/60-s integration period will average out one complete 60-Hz cycle, two complete 120-Hz cycles, and so on. In North America the actual line frequency has a short-term fluctuation of less then ±0.15 percent. As shown in Fig. 23.16, a DMM with a 1/60-s integration period achieves 56 dB of rejection for 0.15 percent deviation from 60 Hz. This magnitude of common-mode rejection is more than adequate for most measurements. Higher degrees of rejection can be achieved by increasing the integration period, but at the sacrifice of reading speed.

Nonintegrating DMMs use filtering for common-mode noise rejection, but the reading rate suffers as the amount of filtering is increased. Quite often different degrees of filtering are selectable to allow the user to trade reading speed for noise rejection. Table 23.2 is an example of a successive-approximation DMM with a three-position active filter.

Integration is better at rejecting line-related noise while filtering does a better job on broadband noise. The differences are shown in Fig. 23.17 for two DMMs with almost identical characteristics except for the A/D conversion technique. The integrating DMM has a 1/10-s integration period with cusps in the noise rejection at 10, 20, 30 Hz, and so on.

Common-mode noise is common to both input terminals of a DMM. Such noise can come from ground currents flowing from the devices being measured, and its magnitude can range from a few millivolts to hundreds of volts. If the high and low input terminals of a DMM had identical impedances to ground (a balanced input), then common-mode noise rejection would not be a problem. Balanced inputs are not the rule for most DMMs because manufacturers want to achieve extremely high input impedances from high to low, and the impedances with respect to ground usually differ.

A floating measurement is illustrated in Fig. 23.18 where the measurement is made on a device far removed from the instrument (such as a transducer mounted to a mechanical structure apart from the instrumentation). There are two potential sources of common-mode voltages: E_{cm1} from the floating measurement and E_{cm2} from differences in grounding between the instrument and the point of measurement. Good design practices will keep Z_1 and Z_3 higher than Z_2. Because of the floating input, impedance Z_2 will be high and the effect of the circulating common-mode current will be reduced. However, the majority of the current will flow through the low input terminal and be dropped across the low lead resistance R_b. This will superimpose the unwanted common-mode voltage across the input terminals where it will produce an error in the readings.

TABLE 23.2   Reading Speed of Successive Approximation DMM

Filtering	Rejection at 60 Hz	Reading rate per second
None	None	1000
Medium	30 dB	5
Maximum	80 dB	1

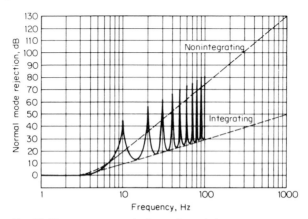

Fig. 23.17 Two DMMs with almost identical characteristics, except that one was integrating and the other nonintegrating with a filter, were selected for this graph of noise rejection. Note that integration is best for line-related noise while filtering works best for broadband noise.

Usually DMMs designed for systems include an inner shield called a *guard*. An external connection is made from the guard to the instrument's front panel. The guard increases the low-to-ground impedance (Z_2 and Z_3), as illustrated in Fig. 23.19. The real strength in guarding comes when it is connected out at the point of the measurement. The guard shunts away common-mode currents by providing a path through Z_3 to ground. By connecting the guard to the same point as the low input, both inputs are kept at about the same potential.

Guarded DMMs are more costly to construct because of the need for the floating inner shield and the electrical isolation between the in-guard and out-guard circuitry. All circuitry associated with analog signals is placed within the guard. Digital information from the A/D converter is passed out of the guard by optical isolators, pulse transformers, or relays. Instructions needed to program the DMM's analog functions flow inward by using similar devices.

Noise rejection is specified in decibels:

$$dB = 20 \log \frac{\text{noise voltage}}{\text{voltage error}}$$

Fig. 23.18 A floating measurement with common-mode voltages originating at the point of measurement (E_{cm1}) and between ground points (E_{cm2}) is illustrated. The majority of the common-mode voltage is dropped across the low-input lead resistance (R_b) and superimposed on the input of the DMM.

Fig. 23.19 Guarding is a technique used to shunt unwanted common-mode currents away from the input terminals of a DMM. It involves the use of an inner floating shield with a connection from this shield out to the point of the measurement.

Common-mode rejection (CMR) improves as lead resistance in the low input decreases. If the size of the lead resistance is decreased by a decade, the CMR appears to improve by 20 dB. For this reason most manufacturers specify CMR with a 1-kΩ resistor in the low lead in an attempt to present worst-case conditions.

There are two types of common-mode rejection: "pure" CMR from guarding alone and "effective" CMR (or ECMR) from the combination of pure CMR and the normal-mode rejection added by the instrument from integration and filtering (see Fig. 23.20).

example 23.5   If a 100-mV noise voltage causes a 1-mV error in a DMM reading, what is its noise rejection?

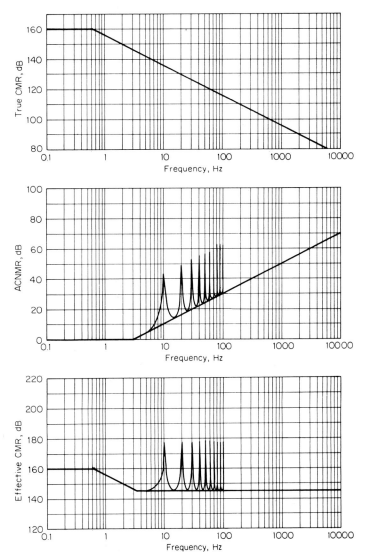

Fig. 23.20   Effective common-mode rejection (ECMR) is the summation of "pure" CMR and the instrument's normal-mode rejection, as shown graphically for an integrating DMM.

solution

$$20 \log \frac{100 \text{ mV}}{1 \text{ mV}} = 20 \log 100 = 40 \text{ dB of rejection}$$

23.8 HOW TO READ DMM SPECIFICATIONS

Often there is a great deal of confusion about the most fundamental of all DMM character-istics, the number of digits. The number of digits equals the number of 9s that the instrument is capable of displaying. The confusion comes when the instrument is able to over-range, adding a digit in the most significant place in the reading. The overrange digit is not a "whole" digit, however.

If we look at the example of a DMM able to display a maximum reading of 1999, it has three full digits and the 1 is used for overranging. The overrange digit allows the user to read beyond the normal full-scale reading. For example, a signal that changes from 9.99 to 10.01 V can be read without loss of sensitivity or upranging. If the instrument lacked over-ranging (maximum display of 999), as the voltage changed from 9.99 to 10.01 V, the user would be forced to uprange and read 10.0 V. The 0.01-V change in the least significant digit could not be observed.

Overranging is expressed in percentage. So for an instrument with a maximum reading of 1999 it would be a 3-digit DMM with 100 percent overranging. Many manufacturers would call such an instrument a 3½-digit DMM, which is an inexact way of saying that it has 3 full digits plus some overranging capability.

The number of digits also may be defined as resolution, which is unitless and equal to the maximum display. A four-digit DMM with 200 percent overranging would have a resolution of 1 in 29 999. For five- and six-digit instruments, resolution may be expressed in parts per million (ppm).

Sensitivity is the ability to respond to small changes in the input. For example, a five-digit DMM with a lowest range of 100 mV has as its least significant digit 1 μV (100.000 mV). Some manufacturers would specify sensitivity as 1 μV, but in reality the instrument may or may not be able to detect changes that small. The internal noise and digitizing error typi-cally cause at least a one count error in the reading.

Accuracy is the exactness to which a voltage can be determined relative to the legal volt maintained by the U.S. Bureau of Standards. To specify accuracy, DMM manufacturers must maintain calibration standards traceable to the legal volt. Any error associated with traceability is added to the accuracy specification. An equal burden is placed on the user. Purchasing a DMM and then failing to keep it in calibration defeats the instrument's accu-racy specification.

To specify accuracy, all the errors associated with the measurement must be tallied. This list could include the following elements:

Drift of the internal reference
Linearity of the integrator
Attenuator errors
Digital ambiguity
Internal noise
Zero drift
Detection of zero crossover
Traceability to the legal volt

To be meaningful, accuracy must be stated with the condition under which it applies. These conditions include time, temperature, power line variation, and relative humidity. Many manufacturers have standardized on time and temperature of 25°C ± 5°C for 90 days. Others add a 1-year specification.

Accuracy is generally expressed as a percentage of reading plus a percentage of range (or full scale). Two identical accuracy specifications for a four-digit DMM are

$$\pm(0.01\% \text{ of reading} + 0.01\% \text{ of range})$$

or

$$\pm 0.01\% \text{ of reading} \pm 1 \text{ digit}$$

Fig. 23.21 The difference between accuracy and short-term stability is not only in terms of time and temperature; accuracy is determined relative to the legal volt, maintained by the U.S. Bureau of Standards. Short-term stability is a measure of how well a DMM can make relative measurements.

Fig. 23.22 The time for a DMM to take a reading is divided into settling time and digitizing time. Total measurement time is typically specified for a full-scale step input and takes into account all the elements necessary to make a correct reading.

The absolute accuracy of a DMM may be considerably less than its resolution. For example, a four-digit DMM has a resolution of 1 count in 10 000 (excluding overranging); yet accuracy will typically run 0.05 percent to 0.08 percent at full scale. If the instrument has good linearity, stability, and low noise, it can make repeatable measurements by using nearly all its resolution. Absolute accuracy, then, may not be as important to some users as the ability to make relative, repeatable measurements.

Short-term stability is not a specification referenced to the legal volt but rather an indication of how well the instrument can make relative measurements. Time is usually expressed in hours over a restricted temperature range. The combination of 24 h and $\pm 1°C$ is quite common among manufacturers. The difference between accuracy and short-term stability is shown in Fig. 23.21.

The temperature coefficient (TC) is the amount of change in accuracy per degree change in temperature *outside* the temperature band given in the accuracy specification. Errors calculated by using the TC are additive to the basic accuracy specification. To make this calculation, users must know over what temperature range the product is designed to operate and to derate the accuracy specification to fit their needs.

Often DMMs are purchased for their speed, especially for applications in a data acquisition system or production test system. Two elements contribute to the time to take a reading: settling or response time of the signal conditioning or converter and digitizing time to convert the analog input to digits. The settling time and digitizing time are usually combined for a full-scale input voltage to give the user an overall reading rate. This point is illustrated in Fig. 23.22. As filtering is increased or integration time lengthened, the reading rate will drop. Some DMMs have variable resolution to allow high reading rates for three-digit operation and progressively slower rates for four- and five-digit operation. Automatic-range and automatic-zero functions add time to the reading; and on some instruments there are ways of inhibiting these functions when high reading rates are desired.

As covered previously, modern DMMs use a FET input stage which yields input impedances of 10^{10} Ω or greater on the lower ranges. For higher ranges (typically 100 and 1000 V) a resistive divider or attenuator is required, and input impedances will drop to 10 MΩ. Associated with a FET input is some leakage current. It is usually less than 10 pA at 25°C but can produce an error in the reading if the source resistance of the device being measured is high.

example 23.6 If the full-scale reading of a DMM is 3999, how many digits does it have and what percentage of overranging is it capable of?

solution Since it can display three 9s, it has three full digits plus 300 percent overranging.

Chapter **24**

Oscilloscope Measurements

Chet Heyberger

Portable Instruments Div.
Tektronix, Inc.
Beaverton, Oregon

.

24.1 INTRODUCTION*

Almost anything that can be measured, can be measured with the two-dimensional graph drawn by an oscilloscope. In most applications the oscilloscope shows a graph of voltage on the vertical axis versus time on the horizontal axis. This general-purpose display presents far more information than is available from other test and measurement instruments such as frequency counters or multimeters. For example, with an oscilloscope you can determine how much of a signal is direct current, how much is alternating current, how much is noise, whether the noise is changing with time, and what the frequency of the signal is. Using an oscilloscope lets you see everything at once rather than through many separate tests.

The usefulness of an oscilloscope is broad because most electric signals can be connected easily to the oscilloscope with either probes or cables. And to measure nonelectrical phenomena, transducers are available. Transducers, such as speakers and microphones, change one kind of energy into another. Other typical transducers can transform mechanical stress, pressure, light, or heat to electric signals.

The oscilloscope is a true multipurpose instrument. The front panel controls allow you to pick from a wide range of vertical sensitivities, time references, display modes, and triggering possibilities.

In other words, given the proper transducer and a modern oscilloscope, your test and measurement capabilities are almost limitless. To tap that measurement potential, you need to know how to use an oscilloscope. This chapter provides basic information on oscilloscope measurement techniques. It is divided into two main parts describing first the controls of an oscilloscope and then some common measurement techniques. In concluding sections we describe the effects of the scope's performance on your measurements; storage oscilloscopes, specialized instruments that offer measurement solutions not found anywhere else; and how to select an oscilloscope.

Wherever it is necessary for the sake of clarity to refer to an actual instrument instead of oscilloscopes in general, either the Tektronix 2213 or the 2215 portable oscilloscope is used as an example.

° This chapter has been adapted from the Tektronix publication *The XYZs of Using a Scope*, copyright 1981, with the permission of Tektronix, Inc., Beaverton, Oregon.

Fig. 24.1 Block diagram of a basic oscilloscope.

24.2 OSCILLOSCOPE CONTROLS

Making oscilloscope measurements will be easier if you understand the basics of how an oscilloscope works and which controls enable you to perform the operations you want. With that goal in mind, this section describes the controls of a triggered-sweep, dual-channel, general-purpose oscilloscope.

Think of the oscilloscope in terms of the functional blocks illustrated in Fig. 24.1; these include vertical, trigger, horizontal, and display systems. The vertical system controls the vertical axis of the display. Any time the electron beam that draws the graph moves up or down, it does so under control of the vertical system. The horizontal system controls the left-to-right movement of the beam. The trigger system determines the precise moment when the oscilloscope draws a display by "triggering" the beginning of the horizontal sweep across the screen. The display system contains the cathode-ray tube (CRT) on which the graph is drawn.

DISPLAY SYSTEM The oscilloscope draws a graph by moving an electron beam across a phosphor coating on the inside of the CRT. The result is a glow for a short time afterward, tracing the path of the beam. A grid of lines etched or silkscreened on the inside of the faceplate serves as the reference for your measurements; this is the graticule shown in Fig. 24.2.

The graticule is a grid of lines typically etched or silkscreened on the inside of the CRT faceplate. Putting the graticule inside—on the same plane as the trace drawn by the electron beam, and not on the outside of the glass—eliminates measurement inaccuracies called *parallax* errors. Parallax errors occur when the trace and the graticule are on different planes and the observer is shifted slightly from the direct line of sight. Although different-sized CRTs may be used, graticules are usually laid out in an 8 × 10 pattern. Each of the 8 vertical and 10 horizontal lines blocks off major divisions (or just divisions) of the screen. The labeling on the oscilloscope controls always refers to major divisions. The tick marks on the center graticule lines represent minor divisions, or subdivisions. Since risetime measurements are very common, most oscilloscope graticules include risetime measurement mark-

Fig. 24.2 Graticule of an oscilloscope.

Fig. 24.3   Front panel of a dual-channel 60-MHz oscilloscope.

ings: dashed lines for 0 and 100 percent points and labeled graticule lines for 10 and 90 percent points.

Common controls for display systems include intensity and focus; less often, you will find beam finder and trace rotation controls. The functions of these controls are described below, and their positions on the Tektronix 2213 portable oscilloscope are illustrated in Fig. 24.3. A functional diagram of the display section is shown in Fig. 24.4.

The display system of an oscilloscope consists of the CRT and its controls. To draw a graph of the measurements, the vertical system of the scope supplies the Y, or vertical, coordinates, and the horizontal system supplies the X, or horizontal, coordinates. There is also a Z axis in an oscilloscope; it determines whether the electron beam is turned on and how bright it is.

Fig. 24.4   Display system and controls of an oscilloscope.

Beam finder The beam finder is a convenience control that allows you to locate the electron beam any time it is off screen. When you push the BEAM FIND button, you reduce the vertical and horizontal deflection voltages (more about deflection voltages later) and override the intensity control, so that the beam always appears within the 8 × 10 cm screen. When you see which quadrant of the screen the beam appears in, you know which direction to turn the horizontal and vertical position controls to reposition the trace for normal operations.

Intensity An intensity control adjusts the brightness of the trace. It is necessary because you use an oscilloscope in different ambient light conditions and with many kinds of signals. For instance, with square waves you might want to change the trace brightness because the slower horizontal components always appear brighter than the faster vertical components.

Intensity controls are also useful because the intensity of a trace is a function of two things: how bright the beam is, and how long it is on screen. As you select different sweep speeds (a sweep is one movement of the electron beam across the oscilloscope screen) with the SEC/DIV switch, the beam on and off times change, and the beam has more or less time to excite the CRT's phosphor.

Focus The oscilloscope's electron beam is focused on the CRT faceplate by an electrical grid within the tube. The focus control adjusts that grid for optimum trace focus.

Trace rotation Another display adjustment on the front panel of some oscilloscopes is trace rotation. This control allows you to electrically align the horizontal deflection, or baseline, of the trace with the fixed graticule lines. To avoid accidental misalignments when the oscilloscope is in use, this control is often recessed and must be adjusted with a small screwdriver.

Normally trace rotation can be set once in most oscilloscope applications, but the earth's magnetic field affects the trace alignment. And when an oscilloscope is used in many different positions, as a service oscilloscope will be, it is very handy to have a front panel trace rotation adjustment.

VERTICAL SYSTEM The vertical system of the oscilloscope supplies the display system with the y, or vertical axis of information for the graph on the CRT screen. To do this, the vertical system uses the input signals to develop deflection voltages. Then the display system uses these deflection voltages to control or deflect the electron beam.

The vertical system also gives you a choice of how you connect the input signals (called *coupling* and described later). And the vertical system provides internal signals for the trigger circuit (described later). Figure 24.5 illustrates the vertical system schematically.

The vertical system of a dual-channel oscilloscope consists of two identical channels (although only one is shown in Fig. 24.5). Each channel has circuits to couple an input signal to that channel, attenuate (reduce) the input signal when necessary, preamplify it, delay it, and finally amplify the signal for use by the display system. The delay line lets you see the beginning of a waveform even when the oscilloscope is triggering on it.

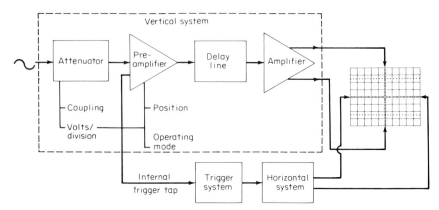

Fig. 24.5 Vertical system of a dual-channel (only one is shown here) oscilloscope.

Vertical position   The oscilloscope's position controls let you place the trace exactly where you want it on the screen. The two vertical position controls (there's one for each channel in a dual-trace oscilloscope) change the vertical placement of the traces from each vertical channel; the horizontal position control changes the horizontal position of both channels at once.

Input coupling   The input coupling switch for each vertical channel lets you control how the input signal is coupled to the vertical channel. The direct-current (dc) input coupling lets you see an entire input signal. Alternating-current (ac) coupling blocks the constant or dc signal components and permits only the ac components of the input signal to reach the channel.

The middle position of the coupling switches is marked GND, for ground. Choosing this position disconnects the input signal from the vertical system and shows a trace representing the oscilloscope's chassis ground if the scope is triggered. The position of the trace on the screen in this mode is the ground reference level. Switching from AC or DC to GND and back is a handy way to measure signal voltage levels with respect to chassis ground. Note that using the GND position does not ground the circuit you are measuring.

VOLTS/DIV switch   The rotary VOLTS/DIV switch controls the sensitivity of each vertical channel. Having different sensitivities extends the range of the oscilloscope's applications. With a VOLTS/DIV switch, a multipurpose oscilloscope is capable of accurately displaying input signal levels from millivolts to many volts.

Using the VOLTS/DIV switch to change sensitivity also changes the scale factor, the value of each major vertical division of the graticule. Each setting of the VOLTS/DIV control is marked with a number that represents the scale factor for that channel. For example, with a setting of 10 V each of the eight vertical major divisions represents 10 V, and the entire screen can show 80 V from bottom to top. With a VOLTS/DIV setting of 2 mV, the eight vertical divisions can display 16 mV.

The probe you use influences the scale factor. Note that there are two marked areas under the skirts of the VOLTS/DIV switches on most Tektronix scopes. The right-hand area shows the scale factor when you use a standard 10X probe. The left-hand area shows the factor for a 1X probe.

Variable volts per division   The red CAL (for calibration) control in the center of the VOLTS/DIV switch provides a continuously variable change in the scale factor to a maximum greater than 2.5 times the VOLTS/DIV setting. The control is labeled CAL to remind you that only one position of this control is calibrated; that switch position usually has a detent.

A variable sensitivity control is useful when you want to make quick amplitude comparisons on a series of signals. You could, for example, take a known signal of almost any amplitude and use the CAL control to make sure the waveform fits exactly on major division graticule lines. Then as you use the same vertical channel to look at other signals, you can quickly see whether the later signals have the same amplitude.

Channel 2 inversion   To make differential measurements (described later), you have to invert the polarity of one of your input channels. The INVERT control on the vertical amplifier for channel 2 provides this capability. When you push it in, the signal on channel 2 is inverted. When the switch is out, both channels have the same polarity.

Vertical operating modes   Oscilloscopes are more useful if they have more than one vertical display mode. With dual-channel oscilloscopes, the choices are usually channel 1 alone, channel 2 alone, both channels in either the alternate or chopped mode, and both channels algebraically summed.

Dual-channel oscilloscopes are more useful than single-channel instruments because they allow you to compare a known good signal on one channel with an unknown signal on the other. Two channels also allow you to compare two signals that are either related to or dependent on each other. Of course, you can still use a dual-channel oscilloscope one channel at a time.

Switches in the vertical system area of the front panel allow you to choose the vertical operating mode.

To make the oscilloscope display only channel 1, use the CH1 vertical-mode switch setting. To display only channel 2, use the CH2 setting. To display two separate signals, use either the ALT (alternate) or CHOP (chopped) mode.

Both CHOP and ALT modes are provided so that you can look at two signals at any sweep speed. The alternate mode draws first one trace and then the other, completing one before starting the other. This works well at faster sweep speeds where your eyes cannot see the alternation. To see two signals at slower sweeps, use the chopped mode.

When you want to see two input signals combined into one waveform on the screen, use the ADD mode. This gives you an algebraically combined signal: either channels 1 and 2 added, CH 1 + CH 2; or channel 1 minus channel 2 when channel 2 is inverted, +CH 1 + (−CH 2).

HORIZONTAL SYSTEM   To draw a graph, the oscilloscope needs horizontal as well as vertical data. The horizontal system supplies the second dimension by providing the deflection voltages to move the electron beam horizontally. The horizontal system contains a sweep generator that produces a sawtooth waveform, or ramp (see Fig. 24.6), that is used to control the oscilloscope's sweep rates.

The sweep generator makes the unique functions of the modern oscilloscope possible. The circuit that made the rate of rise in the ramp linear (a refinement pioneered by Tektronix) was one of the most important advances in oscillography. It meant that the horizontal beam movement could be calibrated directly in units of time. That advance makes it possible for you to measure time between events much more accurately on the oscilloscope screen.

Because it is calibrated in units of time, the sweep generator is often called the *time base*. It lets you pick the time units, observing the signal for either very short times (measured in nanoseconds or microseconds) or relatively long times of several seconds.

The sawtooth waveform is a voltage ramp produced by the sweep generator. The rising portion of the waveform is called the *ramp*, the falling edge is the *retrace*, and the time between ramps is the *holdoff* time. The sweep of the electron beam across the screen of an oscilloscope is controlled by the ramp, and the return of the beam to the left side of the screen takes place during the retrace.

The two functional blocks of the horizontal system, the horizontal amplifier and sweep generator, are diagramed in Fig. 24.7. The sweep generator produces a sawtooth waveform that is processed by the amplifier and applied to the horizontal deflection plates of the CRT. The horizontal system also provides the z axis of the oscilloscope; the z axis determines whether the electron beam is turned on and how bright it is.

Horizontal position   Like the vertical position controls, you use the horizontal position control to change the location of the waveforms on the screen.

Seconds per division   The SEC/DIV switch lets you select the rate at which the beam sweeps across the screen; changing the SEC/DIV settings allows you to look at longer or shorter time intervals of the input signal. Like the vertical system VOLTS/DIV switch, the SEC/DIV markings refer to the screen's scale factors. If the SEC/DIV setting is 1 ms, each horizontal major division represents 1 ms, and the total screen shows you 10 ms.

On dual-time-base oscilloscopes like the Tektronix 2215, there are two SEC/DIV controls. The A sweep offers all the possible settings while the SEC/DIV switch for the delayed B sweep usually lacks one or two of the slower settings. (On the 2215, there are B sweep settings for 0.05 μs to 50 ms per division.)

Oscilloscopes also have an XY setting on the SEC/DIV switch for making the XY measurements described later.

Fig. 24.6   Sawtooth waveform (or ramp) that controls the horizontal sweep of an oscilloscope.

Fig. 24.7   Horizontal system of a dual-channel oscilloscope.

Variable seconds per division   Besides providing calibrated speeds, most oscilloscopes allow you to change any sweep speed by turning the red CAL (calibrated) control in the center of the SEC/DIV switch counterclockwise. This control slows the sweep speed by at least 2.5:1, making the slowest sweep you have $0.5 \times 2.5 = 1.25$ s per division. Remember that the detent (usually in the extreme clockwise direction) is the calibrated position.

Horizontal magnification   Most oscilloscopes offer some means of horizontally magnifying the waveforms on the screen. The effect of magnification is to multiply the sweep speed by the amount of magnification. On 2200 series oscilloscopes, there is a 10X horizontal magnification that you engage by pulling out the red CAL switch. The 10X horizontal magnification gives you a sweep speed 10 times faster than the SEC/DIV switch setting; for example, 0.05 μs per division magnified is 5 ns per division.

The 10X magnification is useful when you want to look at signals and see details that occur very closely together in time. The magnification control of an oscilloscope always gives you the fastest possible sweep speeds.

Horizontal operating mode   Most general-purpose oscilloscopes do not have a horizontal operating mode control; they have a single time base and only one operating mode. But when an oscilloscope is capable of delayed-sweep measurements, there are horizontal operating mode options.

The front panels of such oscilloscopes are marked A, or NORM, for the normal undelayed sweep and B or DELAYED, for the delayed B sweep. Combinations of the two are possible also.

For more detail, see the description of delayed-sweep measurements under "Measurement Techniques."

TRIGGER SYSTEM   So far you have found that the display system draws the waveforms on the screen, the vertical system supplies the vertical information for the drawing, and the horizontal system provides the time axis. At this point the only thing that is missing is the "when"—when should the other circuits of the oscilloscope start drawing the signal and when should they not?

The "when" is the trigger, and it is important for a number of reasons. First, getting time-related information is one reason you use an oscilloscope. Second, and equally important, each drawing starts with the same "when."

Each sweep will start at the right time if you make the right trigger system control settings. You tell the trigger circuit which trigger signal to select with the source switches. With an external signal, you connect the trigger signal to the trigger system circuit with the external coupling controls. Next you set the trigger circuit to recognize a particular voltage level on the trigger signal with the slope and level controls. Then every time that level occurs, the sweep generator is turned on. The process is diagramed in Fig. 24.8.

Triggering gives you a stable display because the same trigger point starts the sweep each time. SLOPE and LEVEL controls define the trigger points on the trigger signal. When you look at a waveform on the screen, you see all those sweeps overlaid onto what appears to be one display.

Besides the trigger controls already mentioned, you have controls that determine how the trigger system operates (trigger operating mode) and how long the oscilloscope waits between triggers (holdoff). On some dual-time-base scopes, there is a separate trigger for the B sweep.

The trigger circuit and its controls are shown schematically in Fig. 24.9. The trigger source control describes whether the trigger signal is internal or external to the oscilloscope. The coupling controls determine the connection of an external trigger to the trigger circuit. The level and slope controls determine where the trigger point will be on the trigger signal. And the trigger operating-mode control determines how the trigger circuit operates.

Trigger level and slope   These controls define the trigger point. The slope control determines whether the trigger point is found on the rising or falling edge of a signal. It specifies either a positive (also called the rising, or positive-going) edge or a negative (falling, or negative-going) edge. The level control determines at what voltage level on that edge the trigger point occurs. See Fig. 24.10.

Variable trigger holdoff   Not every trigger event can be accepted as a trigger. The trigger system will not recognize a trigger during the sweep or the retrace or for a short time afterward, called the holdoff period. The retrace is the time it takes the electron beam to return to the left side of the screen. The holdoff period provides additional time beyond the retrace to ensure that the oscilloscope is ready to start another sweep.

Sometimes the normal holdoff period is not long enough to ensure a stable display; this possibility exists when the trigger signal is a complex waveform with many possible trigger points on it. Although the waveform is repetitive, a simple trigger might yield a series of patterns on the screen instead of the same pattern each time. Digital pulse trains are a good example; each pulse is very much like any other, so there are many possible trigger points, not all of which result in the same display.

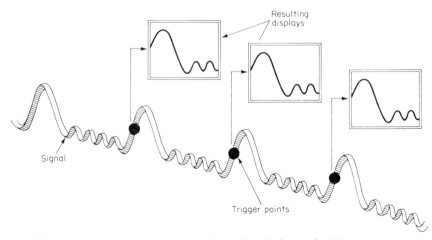

Fig. 24.8   How specific trigger points produce the resultant displays on the CRT.

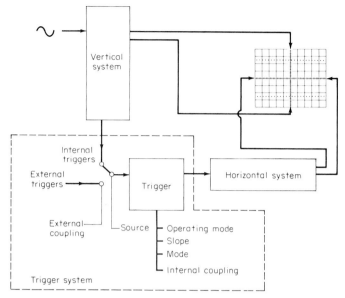

Fig. 24.9   Trigger system for a dual-channel oscilloscope.

What you need now is some way to control when a trigger point is accepted. The variable trigger holdoff control provides the capability. (The control is actually part of the horizontal system—because it adjusts the holdoff time of the sweep generator—but its function interacts with the trigger controls.) Figure 24.11 diagrams a situation in which the variable holdoff is useful. The variable holdoff control lets you make the oscilloscope ignore some potential trigger points. In the figure, all the possible trigger points in the input signal would result in an unstable display. Changing the holdoff time to make sure that the trigger point appears on the same pulse in each repetition of the input signal is the only way to ensure a stable waveform.

Trigger sources   Trigger sources are grouped into two categories that depend on whether the trigger signal is provided internally or externally. The source makes no difference in how the trigger circuit operates, but internal triggering usually means the oscilloscope is triggering on the same signal that it is displaying. That has the obvious advantage of letting you see where you are triggering.

Triggering on the displayed signal is not always what you need, so external triggering is also available. It often gives you more control over the display. To use an external trigger, you set the SOURCE switch on its EXT (external) position and connect the triggering signal to the input connector marked EXT INPUT on the front panel. Occasions when external triggering is useful often occur in digital design and repair; you might want to look at a long

Fig. 24.10   Effect of the slope and level controls on the trigger signal.

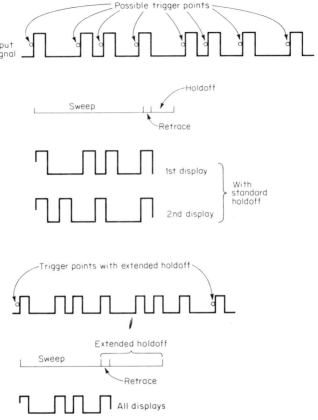

Fig. 24.11   Effect of the variable trigger holdoff control.

train of very similar pulses while triggering with an external clock or with a signal from another part of the circuit.

The line position on the source switch gives you another triggering possibility: the power line. Line triggering is useful any time you are looking at circuits that depend on the power-line frequency. Examples include devices such as light dimmers and power supplies.

Trigger operating modes   Trigger circuits of oscilloscopes like the Tektronix 2213 and 2215 can operate in four modes: normal, automatic, television, and vertical.

One of the most useful is the normal trigger mode (usually marked NORM on the front panel) because it can handle a wider range of trigger signals than any other triggering mode. The normal mode does not permit a trace to be drawn on the screen if there is no trigger. The normal mode usually gives you the widest range of triggering signals.

In the automatic, or "bright baseline," mode (labeled AUTO on front panels), a trigger starts a sweep, the sweep ends, and the holdoff period expires; at that point a timer begins to run. If another trigger is not found before the timer runs out, a trigger is generated anyway, causing the bright baseline to appear even when there is no waveform on the channel. The automatic mode lets you trigger on signals with changing voltage amplitudes or waveshapes without adjusting the LEVEL control.

Another useful operating mode is television triggering. Most oscilloscopes with this mode let you trigger on TV fields at sweeps of 100 μs per division and slower and TV lines at 50 μs per division or faster. Others can trigger on either fields or lines at any sweep speed.

You will probably use the normal and automatic modes the most, the AUTO mode because it is essentially totally automatic and the normal mode because it is the most ver-

TABLE 24.1   Types of Trigger Couplings

Coupling	Description
Direct current	Direct current couples all elements of the triggering signal (both ac and dc) to the trigger circuit.
Direct current with attenuation	If you want dc coupling and the external trigger is too large for the trigger system, move the trigger coupling switch to its DC ÷ 10 setting.
Alternating current	This coupling blocks dc components of the trigger signal and couples only the ac components.
High-/low-frequency rejection	These couplings prevent the passage of either high or low frequencies in the triggering signal. You use them to prevent triggering on parts of the waveform that would result in unstable displays.

satile. For example, it is possible to have a low-frequency signal with a repetition rate that is mismatched to the runout of an automatic-mode timer. When that happens, the signal will not be steady in the automatic mode. Moreover, automatic modes are often signal-seeking and cannot trigger on very low-frequency trigger signals. The normal mode, however, gives you a steady signal at any repetition rate.

A less common trigger circuit operating mode is the alternate, or vertical, mode (vertical mode on the Tektronix 2213 and 2215). This operating mode has unique advantages; the oscilloscope triggers alternately on the two vertical channels. Thus you can look at two completely unrelated signals. Most oscilloscopes trigger on only one channel or the other when the two signals are not synchronous.

Trigger coupling   Just as you may select either ac or dc coupling when you connect an input signal to your oscilloscope's vertical system, you can select the kind of coupling you need when you connect a signal to the trigger system. Some frequently offered choices are shown in Table 24.1.

24.3   OSCILLOSCOPE MEASUREMENTS

In previous sections we described the oscilloscope controls used to select the exact oscilloscope functions you need to make the measurements you want. Now we discuss the procedures and techniques you will need when actually making measurements.

SAFETY   Before you make any oscilloscope measurement, remember that you must be careful when you work with electric equipment. Always observe all safety precautions described in the operator's or service manual for the equipment you are using. Some general rules about servicing electric equipment are worth repeating here. Do not service electric devices alone. Know the symbols for dangerous circuits, and observe the safety instructions for your equipment. Do not operate an electric device in an explosive atmosphere. Always ground the oscilloscope to the circuit, and ground both the oscilloscope and the circuit under test. (Remember that if you lose the ground, all accessible conductive parts, including knobs that appear to be insulated, can give you a shock.) To avoid personal injury, do not touch exposed connections and components in the circuit under test when the power is on. And remember to consult the service manual for the equipment you are working on.

There are also rules about the oscilloscope itself. To avoid a shock, plug the power cord of the oscilloscope into a properly wired receptacle before connecting your probes; use only the power cord for your oscilloscope, and do not use one that is not in good condition (if it is cracked or broken, is missing a ground pin, etc.). Use the right fuse to avoid fire hazards. Do not remove covers and panels on your oscilloscope.

GETTING SET UP   Accurate oscilloscope measurements require that you make sure your system is properly set up each time you begin to use your oscilloscope. A very common and frustrating source of inaccurate oscilloscope measurements is the result of forgetting to check the controls to make sure they are where you think they are. Check the oscilloscope's controls before you begin making a measurement. Use the functional sections of the instrument as the outline for your checklist.

1. Check the settings on all the vertical system controls: variable controls (CH 1 and CH 2, CAL, VOLTS/DIV) should be in their calibrated detent positions; make sure CH 2

is not inverted (unless you want it to be). Check the vertical-mode switches to make sure the signal from the proper channel(s) will be displayed. Check the two vertical system VOLTS/DIV switches to see whether their settings are right (and do not forget to use the VOLTS/DIV readout that matches the probe, either 1X or 10X. Check the input coupling levers, too.

2. Check the horizontal system control settings: magnification is off (push in the red CAL switch in the middle of the SEC/DIV switch), and variable SEC/DIV is in its calibrated detent position.

3. Check your trigger system controls to make sure the oscilloscope will pick the right slope on the trigger signal, that the right coupling is selected, and that the correct operating mode will be used. Also make sure that the trigger variable holdoff control is at its minimum position.

4. Adjust the display section controls (focus, intensity, trace rotation, beam finder), if necessary.

PROBES Connecting measurement test points to the inputs of your oscilloscope is best done with a probe like the one illustrated in Fig. 24.12. Probes are the usual connection between an oscilloscope and the circuit under test. Tektronix probes consist of a patented resistive cable and a grounded shield. Probes like the Tektronix P6120 are usually supplied when you purchase an oscilloscope. The P6120 probe is a high-impedance, minimum-loading 10X passive probe. The accessories pictured with it include (from left to right) a grabber tip for integrated circuits (ICs) and small-diameter leads, a retractable hook tip, an IC tester tip cover, an insulating ground cover, marker bands, and the ground lead in the center. All are worthwhile additions to your test and measurement system.

Fig. 24.12 An oscilloscope probe and its accessories.

Probe compensation Most measurements you make with an oscilloscope require an attenuator probe, which is any probe that reduces signal amplitude. The most common are 10X passive probes which reduce the amplitude of the signal by a factor of 10:1.

Before you make any measurement with an attenuator probe, you should make sure it is compensated. Compensation matches that particular probe to the inputs of the oscilloscope you are using. To make compensation easy, oscilloscopes have a compensation signal output jack on their front panels, usually marked CAL, CAL OUTPUT, or PROBE ADJUST.

Figure 24.13 illustrates what can happen to the waveforms you measure when the probe is not properly compensated. Improperly compensated probes can distort the waveforms you see on the screen. In the figure, a square wave, a pulse, and a sine wave are shown as they will appear with proper and improper compensations.

Note that you should compensate your probe as it will be used when you make the measurement: compensate it with the accessory tip you will be using. Do not compensate the probe in one vertical channel and then use it on another.

Probe handling Before you probe a circuit, make sure you have the right probe tips and adaptors for the circuits you will be working on. Then make sure that the ground in the circuit under test is the same as the scope ground—do not just assume it is. The oscil-

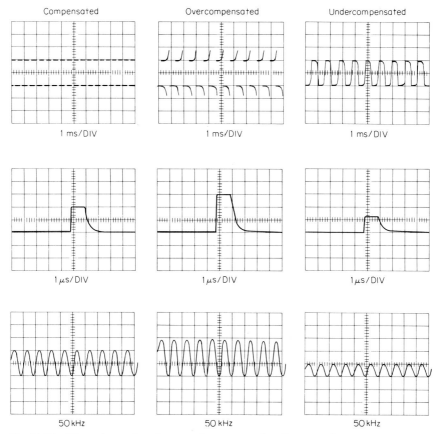

Fig. 24.13 Effects of properly and improperly compensated oscilloscope probes.

loscope ground will always be earth ground as long as you are using the proper power cord and plug. Check the circuit ground by touching the probe tip to the point you think is ground before you make a "hard" ground by attaching the ground strap of your probe.

If you are going to be probing a lot of different points in the same circuit to measure low frequencies, you can ground that circuit to the oscilloscope once instead of each time you move the probe. Connect the circuit ground to the jack marked GND on the front panel (if this jack was provided by the oscilloscope manufacturer).

BASIC WAVEFORMS What you will actually see on the screen of an oscilloscope is called a *waveform*, a graphic representation of an electric signal. The changes in the waveform with time form the waveshape, the most readily identifiable characteristic of a waveform. Figure 24.14 illustrates some common waveshapes.

Basic waveshapes include sine waves and various nonsinusoidal waves such as triangular waves, rectangular waves like the square wave, and sawtooth waves. A square wave has equal amounts of time for its two states. Triangular and sawtooth waves are usually the result of circuits designed to control voltage with respect to time, such as the sweep of an oscilloscope and some television circuits. In these waveforms, one or both transitions from state to state are made with a steady variation at a constant rate, a ramp. (Changes from one state to another on all waveforms except sine waves are called transitions.) The last two drawings represent aperiodic, single-shot waveforms. The first is a pulse; all pulses are marked by a rise, a finite duration, and a decay. The second one is a step, which is actually a single transition.

Waveshapes tell you a great deal about the signal. As a matter of fact, before oscilloscopes

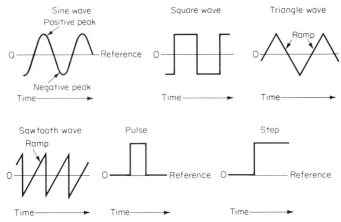

Fig. 24.14   Some common waveforms that may be seen on an oscilloscope.

had a linear time base for making time measurements, they were still acceptable design and service instruments because so much information is contained in just the shape of a signal.

In examining waveshapes, remember that a change in the amplitude, or vertical dimension, of a signal represents a change in signal voltage. Any time a flat horizontal line is displayed, there was no change for that length of time. Straight diagonal lines indicate a linear change, equal rise or fall of voltage for equal amounts of time. Sharp angles on a waveform indicate a sudden change.

But waveshapes alone are not the whole story. To completely describe an electric signal, you will want to find the parameters of that particular waveform. Depending on the signal, these parameters might be amplitude, period, frequency, width, risetime, or phase. You can review these signal parameters and the necessary measurements in the following paragraphs.

MEASUREMENT TECHNIQUES   Rather than attempt to describe how to make all the possible measurements, in this section we describe common measurement techniques you will use in almost every application.

Direct and derived measurements   The two most basic measurements you can make are the two direct oscilloscope measurements: amplitude and time. Almost every other measurement is based on one or both of these direct oscilloscope measurements.

Since the oscilloscope is a voltage-measuring device, voltage is shown as amplitude on the screen. Of course, voltage, current, resistance, and power are related:

$$\text{Current} = \frac{\text{voltage}}{\text{resistance}}$$

$$\text{Resistance} = \frac{\text{voltage}}{\text{current}}$$

$$\text{Power} = \text{current} \times \text{voltage}$$

Figure 24.15 illustrates the amplitude parameter. Amplitude is a characteristic of all waveforms. It is the amount of displacement from equilibrium at a particular point in time.

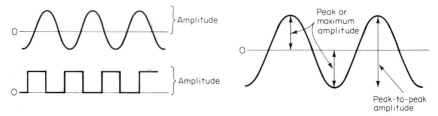

Fig. 24.15   Amplitude parameter of two waveforms.

Make
amplitude
measurements
on the center
vertical
graticule line

Make time measurements on the
center horizontal graticule line

Fig. 24.16 Method of making an amplitude measurement by using the graticule.

Fig. 24.17 How to make a time measurement by using the graticule.

Note that without a modifier, the word means the maximum change from a reference without regard to the direction of the change. In the first two drawings (sine wave and square wave), the amplitude is the same even though the sine wave is larger from peak to peak. In the third drawing, an ac waveform is shown with peak (or maximum) amplitude and peak-to-peak amplitude parameters annotated. In oscilloscope measurements, amplitude usually means peak-to-peak amplitude.

Amplitude measurements are best made with a signal that covers most of the screen vertically such as in Fig. 24.16. Time measurements are also more accurate when the signal covers a large area of the screen; see Fig. 24.17. Use the oscilloscope's vertical position control to place the waveform so that its topmost horizontal component is on a major division. Then use the horizontal position control to make the lowest horizontal component of the waveform intersect the center vertical graticule line. With a display like that in the figure, you can count major and minor divisions down the center vertical graticule line and multiply by the VOLTS/DIV setting to make the measurement. For example, 6.0 divisions times 0.1 V equals 0.6 V.

Time measurements are also more accurate when the signal covers a large area of the screen (see Fig. 24.17). Use the horizontal position control to place a vertical portion of the waveform on a major division of the graticule. Then count major and minor divisions across the center horizontal graticule line to make your time measurement. Multiply by the SEC/DIV setting; for example, 5.3 divisions times 0.2 ms equals 1.06 ms.

Derived measurements are any measurements that are calculated from direct measurements; see Fig. 24.18. For example, ac measurements require an amplitude measurement first. The easiest place to start is with a peak-to-peak amplitude measurement of the voltage—in this case, 330 V because peak-to-peak measurements ignore plus and minus signs. The peak voltage is one-half that value (when there is no dc offset) and is also called a maximum value; it is 165 V in this case. The average value is the total area under the voltage curves divided by the period in radians; for a sine wave, the average value is zero because the positive and negative values are equal. The root mean square (rms) voltage for this sine wave, which represents the line voltage in the United States, is equal to the maximum value divided by $\sqrt{2}$: 165/1.414 = 117 V.

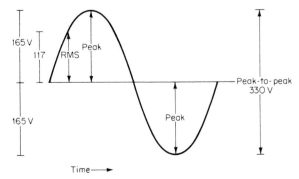

Fig. 24.18 A derived measurement.

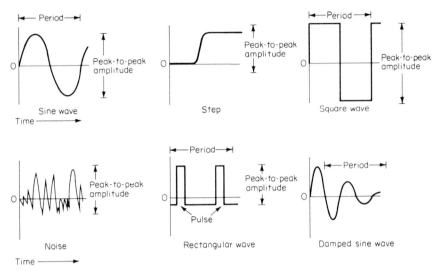

Fig. 24.19   Period measurements.

example 24.1   Use the derived measurements described in Fig. 24.18, and compute the peak-to-peak, peak, and rms voltage of the waveform.

solution   Since peak-to-peak (p-p) voltage is measured without regard to positive or negative, the waveform is 1 V p-p. The peak voltage is −1 V because the waveform is negative-going. The rms voltage is 0.707 V.

Period measurements   One of the most frequently used parameters of a signal is the period; see Fig. 24.19. The technique used to measure period is a direct time measurement. Period is the time required for one cycle of a signal if the signal repeats itself. Period is a parameter whether the signal is symmetrically shaped like the sine and square waves in the figure or whether it has a more complex and asymmetric shape like the rectangular wave and damped sine wave. Period is always expressed in units of time. One-time signals such as the step or signals without a time relation such as noise have no period.

example 24.2   If the oscilloscope showing the waveform in Fig. 24.19 has a SEC/DIV switch setting of 0.5 ms, what is the period of the waveform?

solution   The waveform takes five major divisions on the screen to repeat itself, so its period is 5 × 0.5 ms = 2.5 ms.

Frequency measurements   Frequency is an example of a derived measurement. One technique requires first making a direct period measurement. While period is the time required to complete one cycle of a periodic waveform, frequency is the number of cycles that take place in 1 s. The measurement unit is 1 Hz (1 cycle per second), and it is the reciprocal of the period. So a period of 0.00114 s (or 1.14 ms) means a frequency of 877 Hz. Figure 24.20 contains an illustration.

Period = 0.143 sec   Frequency = 7 Hz

Fig. 24.20   A frequency measurement.

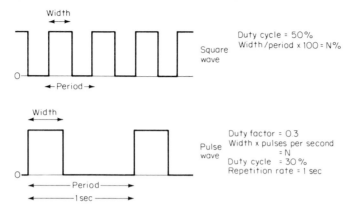

Fig. 24.21 Rectangular-wave measurements.

example 24.3   Calculate the frequency of the waveform pictured in Example 24.2.

solution   Period and frequency are reciprocal, so a period of 2.5 ms means a frequency of 400 Hz (1/0.0025).

Another method of making a frequency measurement is described under "XY Measurements."

Rectangular-wave measurements   Rectangular waves—square waves are the most common example—have many parameters, as illustrated in Fig. 24.21. The parameters described are all derived from direct time measurements.

Duty cycle, duty factor, and repetition rate are parameters of all rectangular waves. They are particularly important in digital circuitry. Duty cycle is the ratio of pulse width to signal period, expressed as a percentage. For square waves, the ratio is always 50 percent, as you can see; for the pulse wave in the second drawing, it is 30 percent. Duty factor is the same thing as duty cycle except it is expressed as a decimal, not a percentage. A repetition rate describes how often a pulse train occurs and is used instead of frequency to describe waveforms like that in the second drawing.

example 24.4   Use the measurements described and the waveform below to calculate duty cycle, duty factor, and repetition rate. The instrument's SEC/DIV setting is 1 ms.

solution   Duty cycle and duty factor depend on the ratio of pulse width to period. Without referring to the instrument settings, you can simply count major divisions:

Duty cycle: ¾ × 100 = 75%
Duty factor: ¾ = 0.75

To find the repetition rate, you need the SEC/DIV setting. Note that the waveform repeats itself in four major divisions: 4 × 1 ms per division = 4 ms.

Pulse measurements   Pulse measurements are important when you work with digital equipment and data communications devices. Some of the signal parameters of a pulse are shown in Fig. 24.22, an illustration of an ideal pulse. The most important parameters of a real pulse are shown in Fig. 24.23.

Pulse parameters can be important in a number of different applications. (See Fig. 24.22.) Digital circuitry, x-ray equipment, and data communications are examples. Pulse specifications include transition times measured on the leading edge of a positive-going transition; this is the risetime. The falltime is the transition time on a negative-going trailing edge. Pulse width is measured at the 50 percent points and amplitude from 0 to 100 percent. Any displacement from 0 V for the base of the pulse is the baseline offset.

The measurement of real pulses includes a few more parameters than those for an ideal pulse. In Fig. 24.23, several are shown. Preshooting is a usually small change of amplitude preceding and in the opposite direction to the pulse rise. Overshooting and rounding are changes that occur after the initial transition. Ringing is a series of amplitude changes—usually a damped sinusoid—that follow overshooting. All are expressed as percentages of

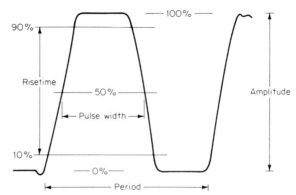

Fig. 24.22 Some signal parameters of an ideal pulse.

amplitude. Droop is a decrease in the maximum amplitude with time. And nonlinearity is any variation from a straight line drawn through the 10 and 90 percent points of a transition.

example 24.5   What are the risetime and pulse width of the waveform pictured if the oscilloscope's SEC/DIV switch setting is 0.2 ms?

solution   The risetime is measured from the 10 to the 90 percent point of the rising edge of the waveform. It takes two major divisions for the waveform to cross those points, so the risetime is 2 × 0.2 ms = 0.4 ms. A pulse width is always measured at the 50 percent point of a pulse; here the waveform takes six major and four minor divisions: 6.8 × 0.2 ms = 1.36 ms. (Note that each minor division is 0.2 of a major division.)

Phase measurements   One of the characteristics of an ac waveform is phase. Phase is the amount of time that has passed since the cycle began, measured in degrees, where one complete cycle is 360°. This parameter is illustrated by Fig. 24.24.

There is also a phase relationship between two or more waveforms, as illustrated by Fig. 24.24. There are two ways to measure the phase shift between two waveforms. One way is to put one waveform on each channel of a dual-channel oscilloscope and view them directly in the chopped or alternate vertical mode. Trigger on either channel; then adjust the trigger LEVEL control for a stable display and measure the period of the waveforms. Next increase the sweep speed so that you have a display something like the second drawing in Fig. 24.24. Measure the horizontal distance between the same points on the two waveforms. The phase shift is the difference in time divided by the period and multiplied by 360°:

$$\theta = \frac{\Delta T}{P} \times 360°$$

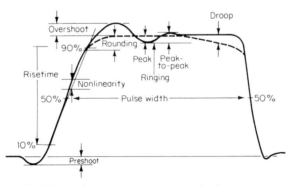

Fig. 24.23 Signal parameters present in a real pulse.

Fig. 24.24 Phase measurement as seen on a dual-channel oscilloscope.

Phase is best explained with a sine wave. Remember that this waveform is based on the sine of all the angles from 0° through 360°. The result is a plot that changes from 0 at 0°, i at 90°, 0 again at 180°, −1 at 270°, and to 0 again at 360°. Consequently, it is useful to refer to the phase angle (or simply phase, when there is no ambiguity) of a sine wave when you want to describe how much of the period has elapsed.

Another use of phase is found when you want to describe a relationship between two signals. Picture two clocks with their second hands sweeping the dial every 60 s. If the second hands touch the 12 at the same time, the clocks are said to be *in phase;* if the hands do not touch, the clocks are *out of phase.* How far out of phase they are can be expressed as degrees of phase shift. To illustrate, the waveform labeled CURRENT in Fig. 24.24 is said to be 90° out of phase with the voltage waveform. Another way of reporting the same information is to say that the current waveform has a 90° phase angle with respect to the voltage waveform or that the current waveform lags the voltage waveform by 90°. Note that there is always a reference to another waveform; in this case, between the voltage and current waveforms of an inductor.

example 24.6 If two sine waves have the same period of 12 ms but one starts 0.5 ms later, what is the phase shift of the second sine wave?

solution Phase shift is the difference in time divided by the period and multiplied by 360°:

$$\frac{0.5}{12} \times 360° = 15°$$

XY measurements Displaying two periodic ac waveforms and measuring when one starts with respect to another are possible with any dual-trace oscilloscope, but that is not the only way to make a phase measurement. Look at the front panel, and you will see that the vertical channel input connectors are labeled X and Y. The last position on the SEC/DIV switch is XY; and when you use it, the oscilloscope's time base is bypassed. The channel 1 input signal becomes the horizontal axis of the oscilloscope's display, and the signal on channel 2 becomes the vertical axis. In the XY mode, you can input one sinusoidal waveform on each channel, and your screen will display what is called a *Lissajous pattern.* The shape of the pattern will indicate the phase difference between the two signals. Some examples of Lissajous patterns are shown in Fig. 24.25.

Note that general-purpose oscilloscope Lissajous pattern phase measurements are usually limited by the frequency response of the horizontal amplifier (typically designed with far less bandwidth than vertical channels). Specialized XY oscilloscopes or monitors will have almost identical vertical and horizontal systems.

Finding the phase shift of two sinusoidal signals with a Lissajous pattern is one example of an XY measurement. The XY capability can be used for other measurements as well. The Lissajous patterns also can be used to determine the frequency of an unknown signal when you have a known signal on the other channel. This is a very accurate frequency measurement as long as your known signal is accurate and both signals are sine waves. The patterns you can see are illustrated in Fig. 24.25, where the effects of both frequency and phase differences are shown.

Frequency measurements with Lissajous patterns require a known sine wave on one channel. If there is no phase shift, the ratio between the known and unknown signals will correspond to the ratio of horizontal and vertical lobes of the pattern. When the frequencies

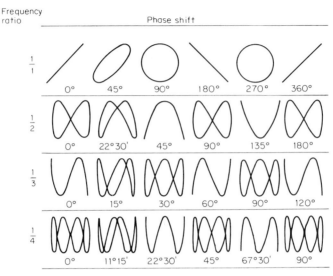

Fig. 24.25 Examples of Lissajous patterns.

are the same, only the shifts in phase will affect the pattern. In Fig. 24.25, both phase and frequency differences are shown.

There are many other applications for *XY* measurements in TV servicing, component checking, engine analysis, and two-way radio servicing, for example. In fact, any time you have physical phenomena that are interdependent and not time-dependent, *XY* measurements are the answer. With the proper transducer, you can use an oscilloscope's *XY* capability to make measurements of aerodynamic lift and drag, motor speed and torque, or pressure and volume of liquids and gases.

Differential measurements   The ADD vertical mode and the channel 2 inversion switch of your oscilloscope allow you to make differential measurements. Often differential measurements let you eliminate undesirable components from a signal that you are trying to measure. If you have a signal that is very similar to the unnecessary noise, the setup is simple. Put the signal with the spurious information on channel 1. Connect the signal that is like the unwanted components to channel 2. Set both input coupling switches to DC (use AC if the dc components of the signal are too large), and select the alternate vertical mode.

Set your VOLTS/DIV switches so that the two signals are about equal in amplitude. Then you can change the vertical operating-mode switch to ADD and press the inversion switch, so that the common-mode signals have opposite polarities.

If you use the channel 2 VOLTS/DIV switch and CAL control for maximum cancellation of the common-mode signal, the signal that remains on screen will contain only the desired

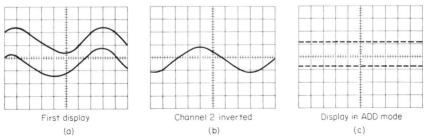

First display   Channel 2 inverted   Display in ADD mode
    (a)   (b)   (c)

Fig. 24.26   Differential measurement.

part of the channel 1 input signal. The two common-mode signals have canceled, leaving only the difference between the two.

Differential measurements allow you to remove unwanted information from a signal any time you have another signal that closely resembles the unwanted components. (See Fig. 24.26.) For example, the first drawing shows a square wave contaminated by a sine wave. Once the common-mode component (the sine wave) is inputted to channel 2 and that channel is inverted, the signals can be added with the ADD vertical mode. The result is shown in the second drawing.

Z axis measurements Remember that the CRT in the oscilloscope has three axes of information: X is the horizontal component of the display, Y is the vertical, and Z is the brightness of the electron beam. When an oscilloscope has an external Z axis input connector on the back of the instrument, you can change the brightness (modulate the intensity) of the signal on the screen with an external signal. The Z axis input usually will accept a signal of up to 30 V where positive voltages decrease the brightness and negative voltages increase it; usually 5 V will cause a noticeable change.

The Z axis input is an advantage to users that have their instruments set up for a long series of tests. One example is the testing of high-fidelity equipment illustrated by Fig. 24.27.

Using the Z axis can provide additional information on the oscilloscope screen. In the setup shown in Fig. 24.27, a function generator sweeps through the frequencies of interest during the product testing, 20 to 20 000 Hz in this case. Then an adjustable notch filter is used to generate a marker, at 15 kHz, for instance, and this marker signal is applied to the Z axis input to brighten the trace. This setup allows the tester to evaluate the product's performance at a glance.

Delayed-sweep measurements Delayed sweep is a technique that adds a precise amount of time between the trigger point and the beginning of a horizontal sweep. Delayed sweep is often used as a convenient way to make a measurement, but sometimes it is the only way to make a measurement. Suppose, for example, that the part of the waveform you want to measure is so far from the only available trigger point that it will not show on the

Fig. 24.27 Use of the Z axis of a CRT.

screen. The problem can be solved with delayed sweep. Trigger where you have to, and add delay out to where you want the sweep to start. The delayed-sweep feature that makes this easy is the intensified sweep mode; it lets you use the delayed sweep as a positionable magnifier.

You trigger normally and then switch to the oscilloscope's intensified sweep mode. Now the signal on the screen will show a brighter zone after the delay time. Run the delay time (and the intensified zone) out to the part of the signal that interests you. Then switch to the delayed mode and increase the sweep speed to magnify the selected waveform portion so that you can examine it in detail. Figure 24.28 illustrates the process. The drawings above show a common measurement problem solved with the use of delayed sweep. The trigger event in this case is too far from the area of interest on the signal. Adding a delay between the trigger event and the start of the oscilloscope sweep is the answer. The intensified sweep feature of the oscilloscope makes setting the correct delay time very simple.

Dual-Time-Base Oscilloscopes.   Delayed sweep is normally found on dual-time-base oscilloscopes like the Tektronix 2215; these instruments have two totally separate horizontal sweep generators. In dual-time-base instruments, one sweep is triggered in the normal fashion, and the start of the second sweep is delayed. To distinguish between these two sweeps, the delaying sweep is called the *A sweep* and the delayed sweep is called the *B sweep*. The time between the start of the A sweep and the start of the B sweep is called the *delay time*.

Dual-time-base oscilloscopes offer you all the measurement capabilities of single-time-base instruments plus the following:
- Convenient comparisons of signals at two different sweep speeds
- Jitter free triggering of delayed sweeps
- Timing measurement accuracy of 1.5 percent

Most of this increase in measurement performance is available because you can separately control the two sweep speeds and use them to examine two time slices—one large and one small—of the same signal.

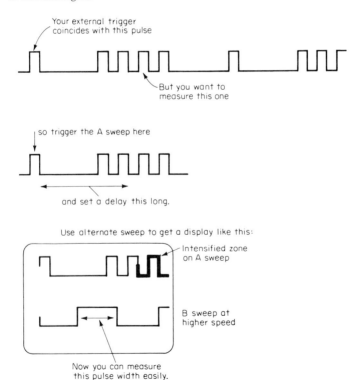

Fig. 24.28   A measurement problem that is solved by the use of delayed sweep.

Another great measurement advantage of dual-time-base oscilloscopes is the separate B trigger. It eliminates jitter, which can prevent an accurate measurement any time you want to look at a signal that is not perfectly periodic. With two time bases and delayed sweep, you trigger the A sweep normally and move the intensified zone out to the portion of the waveform you want to measure. Then you set the oscilloscope up for a triggered B sweep, rather than letting the B sweep simply run after the delay time.

Besides examining signals at two different sweep speeds and seeing a jitterfree B sweep, you get increased timing measurement accuracy with a dual-time-base oscilloscope. On the front panel, the delay time control (labeled B DELAY TIME POSITION on the Tektronix 2215) is a measuring indicator as well as a B-sweep positioning device. Numbers in a window at the top of the dial are calibrated to the major divisions of the oscilloscope screen. The numbers around the circumference divide the major division into hundreds.

To make very accurate timing measurements (to ±1.5 percent) with this control (see Fig. 24.29), do the following:

• Use the B-runs-after-delay horizontal operating mode.

• Place the intensified zone (or the B sweep waveform) where the timing measurement begins, and note the B DELAY TIME POSITION dial setting.

• Dial back to where the measurement ends, and note the reading there.

• Subtract the first reading from the second, and multiply by the A sweep SEC/DIV setting.

example 24.7   You are using a delayed-sweep oscilloscope to measure the delay time between two gates in a digital circuit. With a SEC/DIV setting of 0.1 ms, you position the intensified zone

Use B sweep fast enough to make
your intensified zone a spot

Then move the spot to the beginning of the measurement and note the delay dial setting.

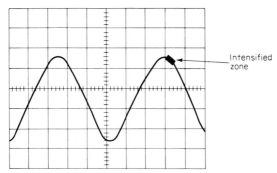

Move the spot to the end of the measurement and
read the dial. Subtract first reading from second
and multiply by A sweep SEC/DIV switch setting.

Fig. 24.29   Time measurement using the calibrated delay dial.

at the rising edge of the first waveform and read the calibrated DELAY TIME MULTIPLIER dial as 3.14. Next you position the intensified zone at the rising edge of the second waveform. This time the dial reads 6.34. How much time has elapsed between the two events?

solution DELAY TIME dials on dual-time-base oscilloscopes are always calibrated in terms of the SEC/DIV settings. You find the difference between the two readings and multiply by the setting: $(6.34 - 3.14) \times 0.1$ ms = 0.32 ms.

Single-Time-Base Oscilloscopes. A few single-time-base oscilloscopes offer delayed-sweep measurements. Those that do may have measurement capabilities similar to those of the Tektronix 2213, which has three possible horizontal operating modes annotated on the front panel as NO DLY, INTENS, and DLY'D.

When you set the horizontal mode switch to NO DLY (no delay), only the normal sweep functions.

When you choose INTENS (intensified sweep), the oscilloscope will display the normal sweep, and the trace also will be intensified after a delay time. The amount of delay is determined by both the delay time switch (you can use 0.5 μs, 10 μs, or 0.2 ms) and the delay time multiplier control. The multiplier lets you pick from 1 to 20 times the switch setting.

You can use the intensified-after-delay mode to make sure the oscilloscope triggers where you want and then use the DLY'D (delayed) sweep mode to make the sweep start after the delay time chosen. After selecting this position, you can move the SEC/DIV control to a faster sweep speed and examine the waveform in greater detail.

24.4 EFFECTS OF INSTRUMENT PERFORMANCE

The nature of the world is such that any measurement affects the thing being measured. Most often the effect is negligible; in the case of oscilloscope measurements, the effects are also predictable.

There are two aspects to oscilloscope performance and its effects on your measurements: the design parameters of the instrument and its conformance to those parameters at the time of measurement. Making an instrument conform to its design parameters simply means proper calibration, including making sure the probe is properly compensated. But even with proper calibration, there will be some effect of the designed performance on your measurements.

INSTRUMENT RISETIME AND MEASURED RISETIMES The risetime of an oscilloscope is a very important specification because the measuring instrument's risetime affects the accuracy of your measured risetimes, as expressed by the following approximation:

$$T_{r(\text{meas})} = T_{r(\text{signal})^2} + T_{r(\text{meas system})^2}$$

In practical terms, this means that the accuracy of a measured signal will be predictable and will depend on how much faster the oscilloscope is than the risetime you are measuring. If the measuring scope is 5 times faster than the observed signal, the measurement error can be as low as 2 percent. For measurement accuracies of 1 percent, it takes an oscilloscope 7 times faster, as you can see from Fig. 24.30. If the risetimes are equal, the error is a 41 percent increase.

Fig. 24.30 This chart enables you to calculate the percentage of risetime error.

example 24.8 Measured risetimes depend on the risetime being measured *and* on the risetime of the measurement system. If the oscilloscope you are using has a risetime of 10 ns and you have measured a pulse risetime of 25 ns, what is the actual risetime?

solution   Measured risetimes follow this formula:

$$\sqrt{T_{r(meas)}} = T_{r(signal)2} + T_{r(meas\ system)2}$$

Applying the values, you have

$$25 = \sqrt{T_{r(signal)2} + 10^2}$$
$$25 = \sqrt{T_r^2 + 100}$$
$$625 = T_r^2 + 100$$
$$T_r^2 = 525$$
$$T_r = 22.9 \text{ ns}$$

example 24.9   If the measuring oscilloscope has a 50-MHz bandwidth (BW), what will be the measured risetimes and percentage of error with signals having risetimes of 50, 35, 21, and 7 ns?

solution   The scope risetime is 7 ns because

$$T_{r(nanoseconds)} = \frac{350}{BW\ (megahertz)}$$

Then

$$T_{r(meas)} = \sqrt{T_{r(signal)2} + T_{r(meas\ system)2}}$$

and

$$50.5 = \sqrt{50^2 + 7^2}$$
$$35.7 = \sqrt{35^2 + 7^2}$$
$$22.1 = \sqrt{21^2 + 7^2}$$
$$9.9 = \sqrt{7^2 + 7^2}$$

The percentage of error, respectively, can be found by

$$\text{Measured risetime} - \text{signal risetime/Signal risetime} \times 100$$

or

$$1\% = \frac{50.5 - 50}{50} \times 100$$
$$2\% = \frac{35.7 - 35}{35} \times 100$$
$$5.4\% = \frac{22.1 - 21}{21} \times 100$$
$$41.4\% = \frac{9.9 - 7}{7} \times 100$$

RELATING BANDWIDTH AND RISETIME   The vertical channels of an oscilloscope are designed for a broad bandpass, generally from some low frequency or direct current to a much higher frequency. This is the oscilloscope's bandwidth, which is specified by listing the frequency at which a sinusoidal input signal's amplitude has been attenuated to 0.707 of the amplitude of signals in the middle frequencies; this is called the -3-dB point.

A bandwidth specification gives you an idea of the instrument's ability to handle high-frequency signals with a specified attenuation. But bandwidth specifications are derived from the instrument's ability to display sine waves. A 35-MHz oscilloscope will show a 35-MHz sine wave with only -3-dB attenuation, but the effects on a square wave at or near the oscilloscope's upper bandwidth limit will be much more severe because high-frequency information in the square wave will not be reproduced accurately. See Fig. 24.31 for an example.

Bandwidth specifications are based on the oscilloscope's ability to reproduce sine waves. The upper bandwidth is the frequency at which a sine wave is reduced to 0.707 of the amplitude shown at middle frequencies. Although this specification tells you how well the instrument reproduces sine waves, not every signal you examine is sinusoidal. Square waves, for example, have a great deal of high-frequency information in their rising and falling edges that will be lost as you approach the bandwidth limits of the instrument. To illustrate, the two CRT drawings in the figure show a 15-MHz square wave as reproduced by 60- and 35-MHz oscilloscopes.

The frequency response of most oscilloscopes is designed so that there is a constant that allows you to relate the bandwidth and risetime of the instrument. This constant is 0.35, and the risetime T_r and bandwidth BW are related by this approximation:

(a)

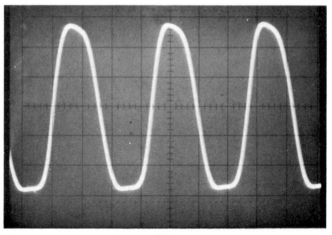

(b)

Fig. 24.31   How a 15-MHz square wave is reproduced by (*a*) a 60-MHz oscilloscope and (*b*) a 35-MHz oscilloscope.

$$T_r = \frac{0.35}{\text{BW}}$$

This is a simple way to apply the formula:

$$T_{r(\text{nanoseconds})} = \frac{350}{\text{BW (megahertz)}}$$

example 24.10   What are the risetimes of 15-, 25-, 50-, and 100-MHz oscilloscopes?

solution   Bandwidth and risetimes are related by

$$T_{r(\text{nanoseconds})} = \frac{350}{\text{BW (megahertz)}}$$

Therefore,

Bandwidth, MHz	Risetime, ns
15	23.3
25	14
50	7
100	3.5

MEASUREMENT SYSTEM BANDWIDTH   Like oscilloscopes, probes have a bandwidth; each has a specified range within which it does not attenuate the signal's amplitude more than -3 dB (0.707 of the original value). But do not assume that a 35-MHz probe and a 35-MHz oscilloscope give you a 35-MHz measurement capability. The combination will approximately equal the square root of the sum of the squares of the risetimes. In the 35-MHz examples, both the probe and the oscilloscope have risetimes of 10 ns:

$$T_{r(system)} = T_{r(oscilloscope)2} + T_{r(probe)2}$$
$$T_r = 100 + 100 = 200 \text{ ns}$$

That works out to 14.14 ns, equivalent to a bandwidth of 24.75 MHz.

To get the full bandwidth from your oscilloscope, you need more bandwidth from the probe. Or you should use the particular probe designed for that instrument.

example 24.11   If you have a 35-MHz oscilloscope and a 50-MHz probe, what are the bandwidth and risetime of your measurement system?

solution   The oscilloscope will have a risetime of 10 ns, and the probe will have one of 7 ns because

$$T_{r(nanoseconds)} = \frac{350}{\text{BW (megahertz)}}$$

Scope: $10 \text{ ns} = \dfrac{350}{35} \text{ MHz}$

Probe: $7 \text{ ns} = \dfrac{350}{50} \text{ MHz}$

The risetime of the combination is then computed by

$$T_{r(meas\ system)} = \sqrt{T_{r(oscilloscope)2} + T_{r(probe)2}}$$
$$T_r = \sqrt{10^2 + 7^2} = 12.2 \text{ ns}$$

Translating that to bandwidth yields 28.7 MHz.

CIRCUIT LOADING   The probe designed for use with your oscilloscope is always the best choice, but even so there is an effect resulting from putting a probe in a circuit. This is called *circuit loading*. Circuit loading modifies the environment of the signals in the circuit you want to measure and changes the signals in the circuit under test. You minimize circuit loading when you use the correct probe and oscilloscope combination and when you have compensated the probe. But occasionally you must use another probe. In these cases, remember that circuit loading is resistive, capacitive, and inductive. For signals with low frequencies, the most important component of loading is resistance. To avoid significant circuit loading here, all you need is a probe with a resistance at least two orders of magnitude greater than the circuit impedance (100-MΩ probes for 1-MΩ sources, 1-MΩ probes for 10-kΩ sources, and so on).

When you are making measurements on a circuit that contains high-frequency signals, inductance and capicitance become important. You cannot avoid adding capacitance when you make connections, but you can avoid adding more capacitance than is necessary. One way to do that is to use an attenuator probe; its design greatly reduces loading. Instead of loading the circuit with capacitance from the probe tip plus the cable plus the oscilloscope's own input, the 10X attenuator probe introduces about 10 times less capacitance, as little as 10 to 14 pF. The only penalty is the reduction in signal amplitude from the 10:1 attenuation.

If you are not going to use the probes that came with your oscilloscope, also remember that proper termination is important to avoid unwanted reflections within the cable of the signal you want to measure. Probe-cable combinations designed to drive 1-MΩ inputs are engineered to suppress these reflections. But for 50-Ω oscilloscopes, 50-Ω probes should be used. The proper termination is also necessary when you use a coaxial cable instead of a probe. If you use a 50-Ω cable and a 1-MΩ oscilloscope, be sure you also use a 50-Ω terminator at the oscilloscope input.

For all the reasons mentioned (probe bandwidth, loading, termination), the best way to ensure that the oscilloscope and probe measurement system has the least effect on your measurements is to use the probe recommended for your oscilloscope. And always make sure it is compensated.

24.5 STORAGE OSCILLOSCOPES

As useful as a conventional oscilloscope is, sometimes something more is needed. Maybe the signal you want to see happens so slowly that it is only a moving dot on the oscilloscope screen and you cannot see the waveshape. Maybe the signal is fast but happens only once in a while, too rarely for you to make a decent measurement. A storage oscilloscope could be the answer.

Conventional oscilloscopes allow you to evaluate events that are happening now; storage oscilloscopes let you save events and study them later. Because they store waveforms for later study, storage oscilloscopes also can be used to compare a waveform you saved earlier with signals occurring now. Additionally, storage oscilloscopes allow easy and accurate measurements of slowly changing phenomena on rapidly changing nonrepetitive waveforms whose images would otherwise flash across the CRT too quickly to be evaluated. Storage can also reduce the time it takes you to photograph waveforms because you can compose your pictures before taking out the oscilloscope camera.

Your measurement applications will determine which kind of storage oscilloscope you should use; read about CRT and digital storage oscilloscopes below to find the storage technology that provides the best instrument for your needs.

CRT STORAGE OSCILLOSCOPES There are two broad classes of storage oscilloscopes: CRT storage and digital storage. CRT storage oscilloscopes store the captured waveform within the CRT; digital oscilloscopes within a digital memory. In each category there are different technologies, and each has its own set of features and benefits.

Bistable CRT storage Bistable storage is the easiest CRT storage type to use. It is also the least expensive storage technology at present and will continue to be so until advances in digital storage change the picture. It features bright, long-lasting displays, but bistable storage displays have less contrast in comparison with other storage technologies.

This type of storage is named for the phosphor it uses. The phosphor coating the faceplate of a conventional oscilloscope has a single stable state, one in which the phosphor does not glow. When the electron beam strikes this phosphor, the phosphor glows for a short time afterward and then returns to its original state. The phosphor used in a bistable CRT storage oscilloscope, however, has two stable states: written and unwritten. Once this phosphor is struck by the electron beam, the phosphor will continue to glow; this allows waveforms to be displayed for as long as several hours or until they are erased by the operator.

The advantages of bistable storage make it particularly useful in applications involving signal comparisons, mechanical equipment measurements, and data recording.

In addition, most bistable storage oscilloscopes have split screens; you can control the top and bottom halves of the screen separately. This allows a reference waveform to be stored on one half the screen while you use the other half to see the effects of changes you are making on the circuit. The half of the screen that you are not using as a reference can be in either the storage or nonstorage mode.

Variable-persistence CRT storage If you do not need hours of stored waveforms and can use a storage oscilloscope that will hold a trace for a maximum of 30 to 60 min, variable-persistence CRT storage has advantages. In a variable-persistence CRT, the electron beam writes on a storage mesh first; then flood guns in the CRT paint waveforms on the screen where the storage mesh permits. Controls on the oscilloscope vary the charge on the mesh, allowing you to change the contrast between the trace and the background and to fine-tune how long the trace is stored.

The first capability provides easy viewing with high contrast between the dark background and bright waveforms, so variable-persistence storage provides the best displays for viewing traces with varying intensities (such as delaying and delayed sweeps or traces with Z axis intensity modulation).

Varying the length of time the stored waveform stays on the screen permits you to set up the oscilloscope so that the entire waveform can be viewed, yet the stored trace will fade from view just as a new waveform is being drawn. You could also view several new traces before the first one fades. Then you could see signal response variations as you make changes in the circuit.

You can also set up a variable-persistence storage oscilloscope to act as a display integrator; it can show a series of waveforms so that only the coincident portions are displayed. Any aberration or jitter not common to all traces will not be stored or displayed. A similar use is showing low-repetition-rate, fast-risetime signals that are not discernible on conventional CRTs; you can accomplish this by allowing each repetition to build up the trace brightness.

All these applications make variable-persistence CRT storage a good choice any time you do not need to keep waveforms on the screen for hours at a time.

Fast-transfer CRT storage Fast-transfer storage oscilloscopes use a CRT with a special intermediate mesh target designed to capture fast-moving sweeps of the electron beam. This target captures the waveform and then transfers it to another mesh, one optimized for long-term storage. As the name implies, the fast-transfer storage mode provides increased writing speed, letting you capture and store waveforms too fast for either bistable or variable-persistence oscilloscopes.

Since the second target can also be designed to offer either bistable or variable-persistence modes in combination with the transfer mesh or by itself, these oscilloscopes can be multi-mode storage instruments. With just the front panel controls, you can select the exact storage operating mode suited to your specific measurement situation.

Stored writing speed For CRT storage oscilloscopes, the most important specification (besides those common to all oscilloscopes) is the stored writing speed. This figure is expressed as either centimeters per unit of time or graticule divisions per unit of time. Often divisions per microsecond are more meaningful in terms of your measurements because some oscilloscopes have nonstandard sized graticules (i.e., other than 1-cm^2 major divisions). However, centimeters per microsecond is an absolute measurement which allows you to compare different families of storage technologies.

In either case, the specification is dependent on the speed of the input signal and on the size of the trace that the oscilloscope must draw. If you know the pulse risetime or sine wave frequency of the input signal and the size of the waveform you want displayed, you can use the chart of Fig. 24.32 to determine which writing speed you need.

Fig. 24.32 Chart to tell whether a CRT storage oscilloscope can display the desired signals.

The stored writing speed is the specification to use to determine whether a CRT storage oscilloscope can display the signals you must see. This specification depends on the phosphor and the electronics of the CRT as well as the input signal. To use the chart, start with either frequency (for sine waves) or risetime (for step signals) and follow the vertical axis to the point where it intersects the size of the display you want. Use the diagonal writing speed line to the left of the intersection to find the specification you need for your measurements.

DIGITAL STORAGE OSCILLOSCOPES   Digital storage oscilloscopes are easy to use and give you crisp, clear displays. Because the data are stored in a digital memory, no fading or blooming of the trace on the CRT phosphor will occur, and storage time is essentially unlimited. Because the waveforms you capture are stored as 1s and 0s in a digital memory and not in a CRT that will eventually have to be replaced, the life-cycle cost of a digital storage oscilloscope is likely to be lower.

Also you can do some things with digital storage that you cannot do with any other kind of storage, such as

• *Pretrigger viewing*. Because the trigger of a digital oscilloscope can be used to stop (as well as start) data acquisition, you can actually capture events that happened before the trigger.

• *Babysitting*. Because of its triggering and data acquisition techniques, you can use a digital storage oscilloscope in an untended mode to trap and save events that happen during the night or any other time when there is no operator.

• *Record keeping*. You can attach a digital storage oscilloscope to a controller and dump the waveforms it captures into a computer memory for permanent storage.

• *Signal processing*. Once the data are stored digitally, you can manipulate them, performing calculations such as rms voltage, averaging a number of samples of the input signal to reduce the effects of noise, etc.

• *Glitch catching*. Some digital storage oscilloscopes offer the capability of recording signals at what amounts to more than one sweep speed. You can look at signals over long periods yet catch quick transient events.

The Tektronix 468 portable scope shown in Fig. 24.39 (at the end of this chapter) is one example of a digital storage oscilloscope with all the features mentioned above.

How digital storage works   Because digital storage is a relatively new concept for oscilloscopes, we take a brief look at how it works. Start with the block diagram in Fig. 24.33. The front end of a digital storage oscilloscope is like one on a conventional oscilloscope or a CRT storage oscilloscope in that input signals are conditioned and coupled by a vertical system. Then the differences start. The analog input signal is digitized by an analog-to-digital converter (ADC) and stored in a memory. The rate at which the signal is digitized and stored is timed by the oscilloscope's digital clock and controlled with your time-base control selections. Then the stored data are converted back to analog form by a waveform reconstruction technique and displayed on the oscilloscope's CRT.

Within digital storage oscilloscopes there are two main techniques of digitizing signals: real-time sampling and equivalent-time sampling. Which technique is used has a direct effect on the applications of the instruments. Only real-time sampling can capture single-shot signals, while both can capture repetitive signals.

In real-time sampling, the oscilloscope digitizes the input waveform sequentially at equal intervals of time during a single sweep.

In equivalent-time sampling, two methods are used. One, random (or asynchronous) sampling, takes a sample of a different part of the input waveforms each time. The other

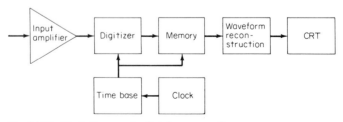

Fig. 24.33   Block diagram of a digital storage oscilloscope.

method, sequential equivalent-time sampling, samples the input waveforms a little later in the waveform after each trigger. In both cases, eventually a complete picture of the input signal is built during the repetitions.

The advantage of real-time sampling is that it can capture single-shot signals. The advantage of equivalent-time sampling is that it can capture signals that are 10 times or more faster than those that real-time sampling can handle.

Comparing digital and CRT storage   Along with the many advantages of digital storage oscilloscopes over CRT storage oscilloscopes, there are some drawbacks. One is the nature of the errors you can get with digital storage. For example, if you are trying to look at a signal that is faster than the bandwidth specification of a conventional or CRT storage oscilloscope, the amplitude of the displayed signal is reduced. However, digital storage oscilloscopes have what is called a *Nyquist frequency*, which is one-half the digitizing rate of the oscilloscope. If you try to measure a signal at or above the Nyquist limit, you get *aliasing*. An aliased signal does not look wrong (the amplitude is not reduced as it would be by an analog oscilloscope), and if you do not know it happened, your measurements will be unreliable. See Fig. 24.34 for an example.

It is possible to misinterpret a digital storage oscilloscope display when "perceptual aliasing" occurs; it is a sort of optical illusion that occurs with dot displays. For example, the original sine wave input signal is not apparent in the first drawing; but when vectors are added to connect the data points, the sine wave is easier to see. In a metaphorical sense, the aliasing of an input signal greater than the Nyquist limit of the oscilloscope is similar. Then the samples taken beyond the storage bandwidth limit result in a display that looks as if the input signal were lower than the bandwidth limit. Unfortunately in this case, the data in the memory are wrong, the frequency information is lost, and adding vectors cannot correct the mistake.

Another difference between CRT and digital storage involves bandwidth. In analog oscilloscopes, the bandwidth of the instrument never changes. But the storage bandwidth of a digital oscilloscope is not always the same, because when you change time-base settings, you also change the digitizing rate (unless the oscilloscope has a feature like the envelope mode of the Tektronix 468 digital storage oscilloscope).

Last, there is resolution. Resolution is the ability to discriminate between items. On

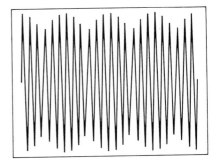

Fig. 24.34   An example of "perpetual aliasing" on a digital storage oscilloscope.

analog oscilloscope screens, the resolution is continuous. It depends on how sharp a trace the electron beam can draw on the screen and how carefully you make the measurement; but you can achieve great resolution. For digital oscilloscopes, however, the resolution is dependent on the hardware. The vertical resolution you need to discriminate between small voltage differences depends on the resolution of the analog-to-digital converter. The horizontal (time) resolution is dependent on how many data words are available in the oscilloscope's memory. See Table 24.2 to translate bits of digital storage oscilloscope resolution to percentages.

Use the columns in Table 24.2 to relate digital storage oscilloscope resolution specifications to percentages. For bits of vertical resolution, use the first column. To determine horizontal resolution in percentages, find the memory size used to store one waveform in the middle column: if 128 data words are used to store a waveform, you will be able to resolve 1 in 128 parts, the equivalent of 0.8 percent. Remember, though, that the resolution may not be the

TABLE 24.2 Relation of Digital Storage Oscilloscope Resolution to Percentages

Bits	Parts	Percentage
1	2	50.0
2	4	25.0
3	8	12.5
4	16	6.25
5	32	3.125
6	64	1.563
7	128	0.781
8	256	0.391
9	512	0.195
10	1,024	0.098
11	2,048	0.049
12	4,096	0.024
13	8,192	0.012
14	16,384	0.006

limiting specification; for example, a digital storage oscilloscope could have an accuracy of 3 percent in the analog amplifiers of its vertical system. In that case, accuracy will be 3 percent, even if the instrument has 15 bits of vertical resolution.

Digital storage oscilloscope specifications For real-time sampling digital storage oscilloscopes, there is a useful storage bandwidth specification that expresses the highest frequency sine wave that can be captured in a single sweep and displayed so that you can make measurements. Both the digitizing rate (how often the oscilloscope takes samples) and the display reconstruction technique (how the oscilloscope displays what is in its memory) must be taken into account in the useful storage bandwidth (see Fig. 24.35).

Figure 24.35 illustrates the effects of reconstruction techniques on the usefulness of a digital storage oscilloscope display. A 10-MHz sine wave was digitized at 25 MHz, and each sample is displayed on the screen. In the first case, only dots are used to indicate the data points, and the sine wave is not readily apparent. In the second drawing, vectors have been generated to join each data point, and the waveshape is more easily seen. And in the last drawing, the vectors have been smoothed by a sine interpolator, and the sine wave is obvious. Had the input signal been a step function—as opposed to a sine wave—a vector display would be more accurate than sine interpolation.

For digital oscilloscopes using equivalent-time sampling, the specification is equivalent-time bandwidth; this is the highest frequency signal that can be stored and displayed with less than 3-dB signal amplitude loss.

Other specifications of interest to digital oscilloscope users—besides the specifications, such as accuracy and linearity, that are common to all oscilloscopes—are as follows:

Maximum digitizing rate, or how often the instrument takes samples of the input signal.

Vertical resolution (usually expressed in bits of resolution) describes how finely the instrument can discriminate between signals very much alike in voltage.

Input signal: 10 MHz

  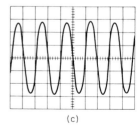

(a)   (b)   (c)

Fig. 24.35 Effects of reconstruction techniques with a digital storage oscilloscope. (*a*) Dot display. (*b*) Pulse interpolator. (*c*) Sine interpolator. Digitizing rate is 25 MHz.

Data words per waveform. Horizontal resolution depends on how many words of digital memory are used to store the captured waveforms.

24.6 SELECTING AN OSCILLOSCOPE

Picking a test and measurement instrument involves three parameters: suitability, service, and support. Suitability pertains to the instrument itself, while service and support are attributes of the manufacturer. The last two parameters are easier to evaluate. For service, simply find out how many service centers there are, where they are located, what kinds of service contracts are available, and whether the centers service the brand you want to buy exclusively (a plus). To determine how much support you can expect, find out whether the instrument manuals are usable, whether there are application notes and collateral literature available, and whether customer operation and service training are possible. Even if you do not take advantage of all the customer support, you can have confidence in the manufacturer when it is there.

Suitability is another matter. You know the applications best, but the manufacturer's sales engineers know the instruments best. Investigate what is available, read the literature, and then work with your local sales engineer to make the best decision. Figures 24.36 through 24.39 illustrate the broad range of oscilloscopes available today.

The Tektronix 7104 general-purpose oscilloscope is an example of the plug-in oscilloscope concept (Fig. 24.36). By adding time bases and vertical amplifiers to a mainframe, you can build the exact oscilloscope necessary for your applications. The configuration pictured will give you state-of-the-art capabilities: 1-GHz bandwidth, 200-ps sweep speeds, and a horizontal bandwidth of 350 MHz for *XY* measurements.

The Tektronix 2215 is an example of a portable scope (Fig. 24.37). It features dual 60-MHz channels, two time bases (for delayed-sweep timing measurements accurate to ± 1.5 percent), light weight (13.5 lb), and a fastest sweep of 5 ns.

The Tektronix 2336 is one of three 100-MHz oscilloscopes designed for field service in harsh environments (Fig. 24.38). These instruments are rugged and compact and have an

Fig. 24.36 Example of the plug-in oscilloscope concept. *(Courtesy Tektronix.)*

Fig. 24.37 Example of a portable oscilloscope. *(Courtesy Tektronix.)*

outstanding resistance to electromagnetic interference. A liquid crystal display on the 2336 gives you a digital readout of delta time measurements; the 2337 adds an LCD readout of digital multimeter measurements.

The Tektronix 468 digital storage oscilloscope offers the best of two worlds (Fig. 24.39). Used as a normal real-time oscilloscope, it is a 100-MHz portable instrument. But for applications where signals must be captured now and viewed later, the 468 is a 10-MHz storage

Fig. 24.38 A 100-MHz oscilloscope designed for field service in harsh environments. *(Courtesy Tektronix.)*

Fig. 24.39 Example of a digital storage oscilloscope. *(Courtesy Tektronix.)*

oscilloscope. Best of all, it is a digital storage oscilloscope, which means that the waveforms are recorded as digital data in a memory, giving users more measurement opportunities than any analog storage oscilloscope.

24.7 BIBLIOGRAPHY

Douglas-Young, John: *Practical Oscilloscope Handbook*, Parker Publishing Co., West Nyack, N.Y., 1979.

Lenk, John, D.: *Handbook of Oscilloscopes: Theory and Application*, Prentice-Hall, Englewood Cliffs, N.J., 1982.

Roth, Charles H., Jr.: *Use of the Dual Trace Oscilloscope: A Programmed Text*, Prentice-Hall, Englewood Cliffs, N.J., 1982.

Appendix

A. The Greek Alphabet

Name	Large	Small
Alpha	A	α
Beta	B	β
Gamma	Γ	γ
Delta	Δ	δ
Epsilon	E	ϵ
Zeta	Z	ζ
Eta	H	η
Theta	Θ	θ
Iota	I	ι
Kappa	K	κ
Lambda	Λ	λ
Mu	M	μ
Nu	N	ν
Xi	Ξ	ξ
Omicron	O	o
Pi	Π	π
Rho	P	ρ
Sigma	Σ	σ
Tau	T	τ
Upsilon	Y	υ
Phi	Φ	ϕ
Chi	X	χ
Psi	Ψ	ψ
Omega	Ω	ω

B. Probable Values of General Physical Constants*

Constant	Symbol	Value
Electronic charge	q	1.602×10^{-19} C
Electronic mass	m	9.109×10^{-31} kg
Ratio of charge to mass of an electron	q/m	1.759×10^{11} C/kg
Mass of atom of unit atomic weight (hypothetical)	.....	1.660×10^{-27} kg
Mass of proton	m_p	1.673×10^{-27} kg
Ratio of proton to electron mass	m_p/m	1.837×10^3
Planck's constant	h	6.626×10^{-34} J-s
Boltzmann constant	$\bar{k}$	1.381×10^{-23} J/K
	k	8.620×10^{-5} eV/K
Stefan-Boltzmann constant	σ	5.670×10^{-8} W/(m²)(K⁴)
Avogadro's number	N_A	6.023×10^{23} molecules/mol
Gas constant	R	8.314 J/(deg)(mol)
Velocity of light	c	2.998×10^8 m/s
Faraday's constant	F	9.649×10^3 C/mol
Volume per mole	V_o	2.241×10^{-2} m³
Acceleration of gravity	g	9.807 m/s²
Permeability of free space	μ_o	1.257×10^{-6} H/m
Permittivity of free space	ϵ_o	8.849×10^{-12} F/m

° E. A. Mechtly, "The International System of Units: Physical Constants and Conversion Factors," National Aeronautics and Space Administration, NASA SP-7012, Washington, D.C., 1964.
From J. Millman and C. Halkias, "Integrated Electronics: Analog Digital Circuits and Systems," McGraw-Hill Book Company, 1972.

C. Conversion Factors and Prefixes*

1 ampere (A)	$= 1$ C/s	1 lumen per square foot	$= 1$ ft-candle (fc)
1 angstrom unit (Å)	$= 10^{-10}$ m	mega (M)	$= \times 10^{6}$
	$= 10^{-4}$ μm	1 meter (m)	$= 39.37$ in
1 atmosphere pressure	$= 760$ mm Hg	micro (μ)	$= \times 10^{-6}$
1 coulomb (C)	$= 1$ A-s	1 micron	$= 10^{-6}$ m
1 electronvolt (eV)	$= 1.60 \times 10^{-19}$ J		$= 1$ μm
1 farad (F)	$= 1$ C/V	1 mil	$= 10^{-3}$ in
1 foot (ft)	$= 0.305$ m		$= 25$ μm
1 gram-calorie	$= 4.185$ J	1 mile	$= 5,280$ ft
giga (G)	$= \times 10^{9}$		$= 1.609$ km
1 henry (H)	$= 1$ V-s/A	milli (m)	$= \times 10^{-3}$
1 hertz (Hz)	$= 1$ cycle/s	nano (n)	$= \times 10^{-9}$
1 inch (in)	$= 2.54$ cm	1 newton (N)	$= 1$ kg-m/s^2
1 joule (J)	$= 10^{7}$ ergs	pico (p)	$= \times 10^{-12}$
	$= 1$ W-s	1 pound (lb)	$= 453.6$ g
	$= 6.25 \times 10^{18}$ eV	1 tesla (T)	$= 1$ Wb/m^2
	$= 1$ N-m	1 ton	$= 2,000$ lb
	$= 1$ C-V	1 volt (V)	$= 1$ W/A
kilo (k)	$= \times 10^{3}$	1 watt (W)	$= 1$ J/s
1 kilogram (kg)	$= 2.205$ lb	1 weber (Wb)	$= 1$ V-s
1 kilometer (km)	$= 0.622$ mi	1 weber per square	
1 lumen	$= 0.0016$ W	meter (Wb/m^2)	$= 10^{4}$ gauss
	(at 0.55 μm)		

From J. Millman and C. Halkias, "Integrated Electronics: Analog Digital Circuits and Systems," McGraw-Hill Book Company, 1972.

D. Decibel Table

Db	Current and voltage ratio		Power ratio		Db	Current and voltage ratio		Power ratio	
	Gain	Loss	Gain	Loss		Gain	Loss	Gain	Loss
0.1	1.01	0.989	1.02	0.977	8.0	2.51	0.398	6.31	0.158
0.2	1.02	0.977	1.05	0.955	8.5	2.66	0.376	7.08	0.141
0.3	1.03	0.966	1.07	0.933	9.0	2.82	0.355	7.94	0.126
0.4	1.05	0.955	1.10	0.912	9.5	2.98	0.335	8.91	0.112
0.5	1.06	0.944	1.12	0.891	10.0	3.16	0.316	10.00	0.100
0.6	1.07	0.933	1.15	0.871	11.0	3.55	0.282	12.6	0.079
0.7	1.08	0.923	1.17	0.851	12.0	3.98	0.251	15.8	0.063
0.8	1.10	0.912	1.20	0.832	13.0	4.47	0.224	19.9	0.050
0.9	1.11	0.902	1.23	0.813	14.0	5.01	0.199	25.1	0.040
1.0	1.12	0.891	1.26	0.794	15.0	5.62	0.178	31.6	0.032
1.1	1.13	0.881	1.29	0.776	16.0	6.31	0.158	39.8	0.025
1.2	1.15	0.871	1.32	0.759	17.0	7.08	0.141	50.1	0.020
1.3	1.16	0.861	1.35	0.741	18.0	7.94	0.126	63.1	0.016
1.4	1.17	0.851	1.38	0.724	19.0	8.91	0.112	79.4	0.013
1.5	1.19	0.841	1.41	0.708	20.0	10.00	0.100	100.0	0.010
1.6	1.20	0.832	1.44	0.692	25.0	17.8	0.056	3.16×10^2	3.16×10^{-3}
1.7	1.22	0.822	1.48	0.676	30.0	31.6	0.032	10^3	10^{-3}
1.8	1.23	0.813	1.51	0.661	35.0	56.2	0.018	3.16×10^3	3.16×10^{-4}
1.9	1.24	0.803	1.55	0.646	40.0	100.0	0.010	10^4	10^{-4}
2.0	1.26	0.794	1.58	0.631	45.0	177.8	0.006	3.16×10^4	3.16×10^{-5}
2.2	1.29	0.776	1.66	0.603	50.0	316	0.003	10^5	10^{-5}
2.4	1.32	0.759	1.74	0.575	55.0	562	0.002	3.16×10^5	3.16×10^{-6}
2.6	1.35	0.741	1.82	0.550	60.0	1,000	0.001	10^6	10^{-6}
2.8	1.38	0.724	1.90	0.525	65.0	1,770	0.0006	3.16×10^6	3.16×10^{-7}
3.0	1.41	0.708	1.99	0.501	70.0	3,160	0.0003	10^7	10^{-7}
3.2	1.44	0.692	2.09	0.479	75.0	5,620	0.0002	3.16×10^7	3.16×10^{-8}
3.4	1.48	0.676	2.19	0.457	80.0	10,000	0.0001	10^8	10^{-8}
3.6	1.51	0.661	2.29	0.436	85.0	17,800	0.00006	3.16×10^8	3.16×10^{-9}
3.8	1.55	0.646	2.40	0.417	90.0	31,600	0.00003	10^9	10^{-9}
4.0	1.58	0.631	2.51	0.398	95.0	56,200	0.00002	3.16×10^9	3.16×10^{-10}
4.2	1.62	0.617	2.63	0.380	100.0	100,000	0.00001	10^{10}	10^{-10}
4.4	1.66	0.603	2.75	0.363	105.0	178,000	0.000006	3.16×10^{10}	3.16×10^{-11}
4.6	1.70	0.589	2.88	0.347	110.0	316,000	0.000003	10^{11}	10^{-11}
4.8	1.74	0.575	3.02	0.331	115.0	562,000	0.000002	3.16×10^{11}	3.16×10^{-12}
5.0	1.78	0.562	3.16	0.316	120.0	1,000,000	0.000001	10^{12}	10^{-12}
5.5	1.88	0.531	3.55	0.282	130.0	3.16×10^6	3.16×10^{-7}	10^{13}	10^{-13}
6.0	1.99	0.501	3.98	0.251	140.0	10^7	10^{-7}	10^{14}	10^{-14}
6.5	2.11	0.473	4.47	0.224	150.0	3.16×10^7	3.16×10^{-8}	10^{15}	10^{-15}
7.0	2.24	0.447	5.01	0.199	160.0	10^8	10^{-8}	10^{16}	10^{-16}
7.5	2.37	0.422	5.62	0.178	170.0	3.16×10^8	3.16×10^{-9}	10^{17}	10^{-17}

From F. E. Terman, "Radio Engineers' Handbook," McGraw-Hill Book Company, 1943.

E. Natural Sines and Cosines

NOTE.—For cosines use right-hand column of degrees and lower line of tenths.

Deg	°0.0	°0.1	°0.2	°0.3	°0.4	°0.5	°0.6	°0.7	°0.8	°0.9	
0°	0.0000	0.0017	0.0035	0.0052	0.0070	0.0087	0.0105	0.0122	0.0140	0.0157	89
1	0.0175	0.0192	0.0209	0.0227	0.0244	0.0262	0.0279	0.0297	0.0314	0.0332	88
2	0.0349	0.0366	0.0384	0.0401	0.0419	0.0436	0.0454	0.0471	0.0488	0.0506	87
3	0.0523	0.0541	0.0558	0.0576	0.0593	0.0610	0.0628	0.0645	0.0663	0.0680	86
4	0.0698	0.0715	0.0732	0.0750	0.0767	0.0785	0.0802	0.0819	0.0837	0.0854	85
5	0.0872	0.0889	0.0906	0.0924	0.0941	0.0958	0.0976	0.0993	0.1011	0.1028	84
6	0.1045	0.1063	0.1080	0.1097	0.1115	0.1132	0.1149	0.1167	0.1184	0.1201	83
7	0.1219	0.1236	0.1253	0.1271	0.1288	0.1305	0.1323	0.1340	0.1357	0.1374	82
8	0.1392	0.1409	0.1426	0.1444	0.1461	0.1478	0.1495	0.1513	0.1530	0.1547	81
9	0.1564	0.1582	0.1599	0.1616	0.1633	0.1650	0.1668	0.1685	0.1702	0.1719	80°
10°	0.1736	0.1754	0.1771	0.1788	0.1805	0.1822	0.1840	0.1857	0.1874	0.1891	79
11	0.1908	0.1925	0.1942	0.1959	0.1977	0.1994	0.2011	0.2028	0.2045	0.2062	78
12	0.2079	0.2096	0.2113	0.2130	0.2147	0.2164	0.2181	0.2198	0.2215	0.2232	77
13	0.2250	0.2267	0.2284	0.2300	0.2317	0.2334	0.2351	0.2368	0.2385	0.2402	76
14	0.2419	0.2436	0.2453	0.2470	0.2487	0.2504	0.2521	0.2538	0.2554	0.2571	75
15	0.2588	0.2605	0.2622	0.2639	0.2656	0.2672	0.2689	0.2706	0.2723	0.2740	74
16	0.2756	0.2773	0.2790	0.2807	0.2823	0.2840	0.2857	0.2874	0.2890	0.2907	73
17	0.2924	0.2940	0.2957	0.2974	0.2990	0.3007	0.3024	0.3040	0.3057	0.3074	72
18	0.3090	0.3107	0.3123	0.3140	0.3156	0.3173	0.3190	0.3206	0.3223	0.3239	71
19	0.3256	0.3272	0.3289	0.3305	0.3322	0.3338	0.3355	0.3371	0.3387	0.3404	70°
20°	0.3420	0.3437	0.3453	0.3469	0.3486	0.3502	0.3518	0.3535	0.3551	0.3567	69
21	0.3584	0.3600	0.3616	0.3633	0.3649	0.3665	0.3681	0.3697	0.3714	0.3730	68
22	0.3746	0.3762	0.3778	0.3795	0.3811	0.3827	0.3843	0.3859	0.3875	0.3891	67
23	0.3907	0.3923	0.3939	0.3955	0.3971	0.3987	0.4003	0.4019	0.4035	0.4051	66
24	0.4067	0.4083	0.4099	0.4115	0.4131	0.4147	0.4163	0.4179	0.4195	0.4210	65
25	0.4226	0.4242	0.4258	0.4274	0.4289	0.4305	0.4321	0.4337	0.4352	0.4368	64
26	0.4384	0.4399	0.4415	0.4431	0.4446	0.4462	0.4478	0.4493	0.4509	0.4524	63
27	0.4540	0.4555	0.4571	0.4586	0.4602	0.4617	0.4633	0.4648	0.4664	0.4679	62
28	0.4695	0.4710	0.4726	0.4741	0.4756	0.4772	0.4787	0.4802	0.4818	0.4833	61
29	0.4848	0.4863	0.4879	0.4894	0.4909	0.4924	0.4939	0.4955	0.4970	0.4985	60°
30°	0.5000	0.5015	0.5030	0.5045	0.5060	0.5075	0.5900	0.5105	0.5120	0.5135	59
31	0.5150	0.5165	0.5180	0.5195	0.5210	0.5225	0.5240	0.5255	0.5270	0.5284	58
32	0.5299	0.5314	0.5329	0.5344	0.5358	0.5373	0.5388	0.5402	0.5417	0.5432	57
33	0.5446	0.5461	0.5476	0.5490	0.5505	0.5519	0.5534	0.5548	0.5563	0.5577	56
34	0.5592	0.5606	0.5621	0.5635	0.5650	0.5664	0.5678	0.5693	0.5707	0.5721	55
35	0.5736	0.5750	0.5764	0.5779	0.5793	0.5807	0.5821	0.5835	0.5850	0.5864	54
36	0.5878	0.5892	0.5906	0.5920	0.5934	0.5948	0.5962	0.5976	0.5990	0.6004	53
37	0.6018	0.6032	0.6046	0.6060	0.6074	0.6088	0.6101	0.6115	0.6129	0.6143	52
38	0.6157	0.6170	0.6184	0.6198	0.6211	0.6225	0.6239	0.6252	0.6266	0.6280	51
39	0.6293	0.6307	0.6320	0.6334	0.6347	0.6361	0.6374	0.6388	0.6401	0.6414	50°
40°	0.6428	0.6441	0.6455	0.6468	0.6481	0.6494	0.6508	0.6521	0.6534	0.6547	49
41	0.6561	0.6574	0.6587	0.6600	0.6613	0.6626	0.6639	0.6652	0.6665	0.6678	48
42	0.6691	0.6704	0.6717	0.6730	0.6743	0.6756	0.6769	0.6782	0.6794	0.6807	47
43	0.6820	0.6833	0.6845	0.6858	0.6871	0.6884	0.6896	0.6909	0.6921	0.6934	46
44	0.6947	0.6959	0.6972	0.6984	0.6997	0.7009	0.7022	0.7034	0.7046	0.7059	45
	°1.0	°0.9	°0.8	°0.7	°0.6	°0.5	°0.4	°0.3	°0.2	°0.1	Deg

From "Standard Handbook for Electrical Engineers," 7th ed.

E. Natural Sines and Cosines (*Concluded*)

Deg	°0.0	°0.1	°0.2	°0.3	°0.4	°0.5	°0.6	°0.7	°0.8	°0.9	
45	0.7071	0.7083	0.7096	0.7108	0.7120	0.7133	0.7145	0.7157	0.7169	0.7181	44
46	0.7193	0.7206	0.7218	0.7230	0.7242	0.7254	0.7266	0.7278	0.7290	0.7302	43
47	0.7314	0.7325	0.7337	0.7349	0.7361	0.7373	0.7385	0.7396	0.7408	0.7420	42
48	0.7431	0.7443	0.7455	0.7466	0.7478	0.7490	0.7501	0.7513	0.7524	0.7536	41
49	0.7547	0.7559	0.7570	0.7581	0.7593	0.7604	0.7615	0.7627	0.7638	0.7649	40°
50°	0.7660	0.7672	0.7683	0.7694	0.7705	0.7716	0.7727	0.7738	0.7749	0.7760	39
51	0.7771	0.7782	0.7793	0.7804	0.7815	0.7826	0.7837	0.7848	0.7859	0.7869	38
52	0.7880	0.7891	0.7902	0.7912	0.7923	0.7934	0.7944	0.7955	0.7965	0.7976	37
53	0.7986	0.7997	0.8007	0.8018	0.8028	0.8039	0.8049	0.8059	0.8070	0.8080	36
54	0.8090	0.8100	0.8111	0.8121	0.8131	0.8141	0.8151	0.8161	0.8171	0.8181	35
55	0.8192	0.8202	0.8211	0.8221	0.8231	0.8241	0.8251	0.8261	0.8271	0.8281	34
56	0.8290	0.8300	0.8310	0.8320	0.8329	0.8339	0.8348	0.8358	0.8368	0.8377	33
57	0.8387	0.8396	0.8406	0.8415	0.8425	0.8434	0.8443	0.8453	0.8462	0.8471	32
58	0.8480	0.8490	0.8499	0.8508	0.8517	0.8526	0.8536	0.8545	0.8554	0.8563	31
59	0.8572	0.8581	0.8590	0.8599	0.8607	0.8616	0.8625	0.8634	0.8643	0.8652	30°
60°	0.8660	0.8669	0.8678	0.8686	0.8695	0.8704	0.8712	0.8721	0.8729	0.8738	29
61	0.8746	0.8755	0.8763	0.8771	0.8780	0.8788	0.8796	0.8805	0.8813	0.8821	28
62	0.8829	0.8838	0.8846	0.8854	0.8862	0.8870	0.8878	0.8886	0.8894	0.8902	27
63	0.8910	0.8918	0.8926	0.8934	0.8942	0.8949	0.8957	0.8965	0.8973	0.8980	26
64	0.8988	0.8996	0.9003	0.9011	0.9018	0.9026	0.9033	0.9041	0.9048	0.9056	25
65	0.9063	0.9070	0.9078	0.9085	0.9092	0.9100	0.9107	0.9114	0.9121	0.9128	24
66	0.9135	0.9143	0.9150	0.9157	0.9164	0.9171	0.9178	0.9184	0.9191	0.9198	23
67	0.9205	0.9212	0.9219	0.9225	0.9232	0.9239	0.9245	0.9252	0.9259	0.9265	22
68	0.9272	0.9278	0.9285	0.9291	0.9298	0.9304	0.9311	0.9317	0.9323	0.9330	21
69	0.9336	0.9342	0.9348	0.9354	0.9361	0.9367	0.9373	0.9379	0.9385	0.9391	20°
70°	0.9397	0.9403	0.9409	0.9415	0.9421	0.9426	0.9432	0.9438	0.9444	0.9449	19
71	0.9455	0.9461	0.9466	0.9472	0.9478	0.9483	0.9489	0.9494	0.9500	0.9505	18
72	0.9511	0.9516	0.9521	0.9527	0.9532	0.9537	0.9542	0.9548	0.9553	0.9558	17
73	0.9563	0.9568	0.9573	0.9578	0.9583	0.9588	0.9593	0.9598	0.9603	0.9608	16
74	0.9613	0.9617	0.9622	0.9627	0.9632	0.9636	0.9641	0.9646	0.9650	0.9655	15
75	0.9659	0.9664	0.9668	0.9673	0.9677	0.9681	0.9686	0.9690	0.9694	0.9699	14
76	0.9703	0.9707	0.9711	0.9715	0.9720	0.9724	0.9728	0.9732	0.9736	0.9740	13
77	0.9744	0.9748	0.9751	0.9755	0.9759	0.9763	0.9767	0.9770	0.9774	0.9778	12
78	0.9781	0.9785	0.9789	0.9792	0.9796	0.9799	0.9803	0.9806	0.9810	0.9813	11
79	0.9816	0.9820	0.9823	0.9826	0.9829	0.9833	0.9836	0.9839	0.9842	0.9845	10°
80°	0.9848	0.9851	0.9854	0.9857	0.9860	0.9863	0.9866	0.9869	0.9871	0.9874	9
81	0.9877	0.9880	0.9882	0.9885	0.9888	0.9890	0.9893	0.9895	0.9898	0.9900	8
82	0.9903	0.9905	0.9907	0.9910	0.9912	0.9914	0.9917	0.9919	0.9921	0.9923	7
83	0.9925	0.9928	0.9930	0.9932	0.9934	0.9936	0.9938	0.9940	0.9942	0.9943	6
84	0.9945	0.9947	0.9949	0.9951	0.9952	0.9954	0.9956	0.9957	0.9959	0.9960	5
85	0.9962	0.9963	0.9965	0.9966	0.9968	0.9969	0.9971	0.9972	0.9973	0.9974	4
86	0.9976	0.9977	0.9978	0.9979	0.9980	0.9981	0.9982	0.9983	0.9984	0.9985	3
87	0.9986	0.9987	0.9988	0.9989	0.9990	0.9990	0.9991	0.9992	0.9993	0.9993	2
88	0.9994	0.9995	0.9995	0.9996	0.9996	0.9997	0.9997	0.9997	0.9998	0.9998	1
89	0.9998	0.9999	0.9999	0.9999	0.9999	1.000	1.000	1.000	1.000	1.000	0°
	°1.0	°0.9	°0.8	°0.7	°0.6	°0.5	°0.4	°0.3	°0.2	°0.1	Deg

From "Standard Handbook for Electrical Engineers," 7th ed.

F. Natural Tangents and Cotangents

NOTE.—For cotangents use right-hand column of degrees and lower line of tenths.

Deg	°0.0	°0.1	°0.2	°0.3	°0.4	°0.5	°0.6	°0.7	°0.8	°0.9	
0°	0.0000	0.0017	0.0035	0.0052	0.0070	0.0087	0.0105	0.0122	0.0140	0.0157	89
1	0.0175	0.0192	0.0209	0.0227	0.0244	0.0262	0.0279	0.0297	0.0314	0.0332	88
2	0.0349	0.0367	0.0384	0.0402	0.0419	0.0437	0.0454	0.0472	0.0489	0.0507	87
3	0.0524	0.0542	0.0559	0.0577	0.0594	0.0612	0.0629	0.0647	0.0664	0.0682	86
4	0.0699	0.0717	0.0734	0.0752	0.0769	0.0787	0.0805	0.0822	0.0840	0.0857	85
5	0.0875	0.0892	0.0910	0.0928	0.0945	0.0963	0.0981	0.0998	0.1016	0.1033	84
6	0.1051	0.1069	0.1086	0.1104	0.1122	0.1139	0.1157	0.1175	0.1192	0.1210	83
7	0.1228	0.1246	0.1263	0.1281	0.1299	0.1317	0.1334	0.1352	0.1370	0.1388	82
8	0.1405	0.1423	0.1441	0.1459	0.1477	0.1495	0.1512	0.1530	0.1548	0.1566	81
9	0.1584	0.1602	0.1620	0.1638	0.1655	0.1673	0.1691	0.1709	0.1727	0.1745	80°
10°	0.1763	0.1781	0.1799	0.1817	0.1835	0.1853	0.1871	0.1890	0.1908	0.1926	79
11	0.1944	0.1962	0.1980	0.1998	0.2016	0.2035	0.2053	0.2071	0.2089	0.2107	78
12	0.2126	0.2144	0.2162	0.2180	0.2199	0.2217	0.2235	0.2254	0.2272	0.2290	77
13	0.2309	0.2327	0.2345	0.2364	0.2382	0.2401	0.2419	0.2438	0.2456	0.2475	76
14	0.2493	0.2512	0.2530	0.2549	0.2568	0.2586	0.2605	0.2623	0.2643	0.2661	75
15	0.2679	0.2698	0.2717	0.2736	0.2754	0.2773	0.2792	0.2811	0.2830	0.2849	74
16	0.2867	0.2886	0.2905	0.2924	0.2943	0.2962	0.2981	0.3000	0.3019	0.3038	73
17	0.3057	0.3076	0.3096	0.3115	0.3134	0.3153	0.3172	0.3191	0.3211	0.3230	72
18	0.3249	0.3269	0.3288	0.3307	0.3327	0.3346	0.3365	0.3385	0.3404	0.3424	71
19	0.3443	0.3463	0.3482	0.3502	0.3522	0.3541	0.3561	0.3581	0.3600	0.3620	70°
20°	0.3640	0.3659	0.3679	0.3699	0.3719	0.3739	0.3759	0.3779	0.3799	0.3819	69
21	0.3339	0.3859	0.3879	0.3899	0.3919	0.3939	0.3959	0.3979	0.4000	0.4020	68
22	0.4040	0.4061	0.4081	0.4101	0.4122	0.4142	0.4163	0.4183	0.4204	0.4224	67
23	0.4245	0.4265	0.4286	0.4307	0.4327	0.4348	0.4369	0.4390	0.4411	0.4431	66
24	0.4452	0.4473	0.4494	0.4515	0.4536	0.4557	0.4578	0.4599	0.4621	0.4642	65
25	0.4663	0.4684	0.4706	0.4727	0.4748	0.4770	0.4791	0.4813	0.4834	0.4856	64
26	0.4877	0.4899	0.4921	0.4942	0.4964	0.4986	0.5008	0.5029	0.5051	0.5073	63
27	0.5095	0.5117	0.5139	0.5161	0.5184	0.5206	0.5228	0.5250	0.5272	0.5295	62
28	0.5317	0.5340	0.5362	0.5384	0.5407	0.5430	0.5452	0.5475	0.5498	0.5520	61
29	0.5543	0.5566	0.5589	0.5612	0.5635	0.5658	0.5681	0.5704	0.5727	0.5750	60°
30°	0.5774	0.5797	0.5820	0.5844	0.5867	0.5890	0.5914	0.5938	0.5961	0.5985	59
31	0.6009	0.6032	0.6056	0.6080	0.6104	0.6128	0.6152	0.6176	0.6200	0.6224	58
32	0.6249	0.6273	0.6297	0.6322	0.6346	0.6371	0.6395	0.6420	0.6445	0.6469	57
33	0.6494	0.6519	0.6544	0.6569	0.6594	0.6619	0.6644	0.6669	0.6694	0.6720	56
34	0.6745	0.6771	0.6796	0.6822	0.6847	0.6873	0.6899	0.6924	0.6950	0.6976	55
35	0.7002	0.7028	0.7054	0.7080	0.7107	0.7133	0.7159	0.7186	0.7212	0.7239	54
36	0.7265	0.7292	0.7319	0.7346	0.7373	0.7400	0.7427	0.7454	0.7481	0.7508	53
37	0.7536	0.7563	0.7590	0.7618	0.7646	0.7673	0.7701	0.7729	0.7757	0.7785	52
38	0.7813	0.7841	0.7869	0.7898	0.7926	0.7954	0.7983	0.8012	0.8040	0.8069	51
39	0.8098	0.8127	0.8156	0.8185	0.8214	0.8243	0.8273	0.8302	0.8332	0.8361	50°
40°	0.8391	0.8421	0.8451	0.8481	0.8511	0.8541	0.8571	0.8601	0.8632	0.8662	49
41	0.8693	0.8724	0.8754	0.8785	0.8816	0.8847	0.8878	0.8910	0.8941	0.8972	48
42	0.9004	0.9036	0.9067	0.9099	0.9131	0.9163	0.9195	0.9228	0.9260	0.9293	47
43	0.9325	0.9358	0.9391	0.9424	0.9457	0.9490	0.9523	0.9556	0.9590	0.9623	46
44	0.9657	0.9691	0.9725	0.9759	0.9793	0.9827	0.9861	0.9896	0.9930	0.9965	45
	°1.0	°0.9	°0.8	°0.7	°0.6	°0.5	°0.4	°0.3	°0.2	°0.1	Deg

From "Standard Handbook for Electrical Engineers," 7th ed.

F. Natural Tangents and Cotangents (*Concluded*)

Deg	°0.0	°0.1	°0.2	°0.3	°0.4	°0.5	°0.6	°0.7	°0.8	°0.9	
45	1.0000	1.0035	1.0070	1.0105	1.0141	1.0176	1.0212	1.0247	1.0283	1.0319	44
46	1.0355	1.0392	1.0428	1.0464	1.0501	1.0538	1.0575	1.0612	1.0649	1.0686	43
47	1.0724	1.0761	1.0799	1.0837	1.0875	1.0913	1.0951	1.0990	1.1028	1.1067	42
48	1.1106	1.1145	1.1184	1.1224	1.1263	1.1303	1.1343	1.1383	1.1423	1.1463	41
49	1.1504	1.1544	1.1585	1.1626	1.1667	1.1708	1.1750	1.1792	1.1833	1.1875	40°
50°	1.1918	1.1960	1.2002	1.2045	1.2088	1.2131	1.2174	1.2218	1.2261	1.2305	39
51	1.2349	1.2393	1.2437	1.2482	1.2527	1.2572	1.2617	1.2662	1.2708	1.2753	38
52	1.2799	1.2846	1.2892	1.2938	1.2985	1.3032	1.3079	1.3127	1.3175	1.3222	37
53	1.3270	1.3319	1.3367	1.3416	1.3465	1.3514	1.3564	1.3613	1.3663	1.3713	36
54	1.3764	1.3814	1.3865	1.3916	1.3968	1.4019	1.4071	1.4124	1.4176	1.4229	35
55	1.4281	1.4335	1.4388	1.4442	1.4496	1.4550	1.4605	1.4659	1.4715	1.4770	34
56	1.4826	1.4882	1.4938	1.4994	1.5051	1.5108	1.5166	1.5224	1.5282	1.5340	33
57	1 5399	1.5458	1.5517	1.5577	1.5637	1.5697	1.5757	1.5818	1.5880	1.5941	32
58	1.6003	1.6066	1.6128	1.6191	1.6255	1.6319	1.6383	1.6447	1.6512	1.6577	31
59	1.6643	1.6709	1.6775	1.6842	1.6909	1.6977	1.7045	1.7113	1.7182	1.7251	30°
60°	1.7321	1.7391	1.7461	1.7532	1.7603	1.7675	1.7747	1.7820	1.7893	1.7966	29
61	1.8040	1.8115	1.8190	1.8265	1.8341	1.8418	1.8495	1.8572	1.8650	1.8728	28
62	1.8807	1.8887	1.8967	1.9047	1.9128	1.9210	1.9292	1.9375	1.9458	1.9542	27
63	1.9626	1.9711	1.9797	1.9883	1.9970	2.0057	2.0145	2.0233	2.0323	2.0413	26
64	2.0503	2.0594	2.0686	2.0778	2.0872	2.0965	2.1060	2.1155	2.1251	2.1348	25
65	2.1445	2.1543	2.1642	2.1742	2.1842	2.1943	2.2045	2.2148	2.2251	2.2355	24
66	2.2460	2.2566	2.2673	2.2781	2.2889	2.2998	2.3109	2.3220	2.3332	2.3445	23
67	2.3599	2.3673	2.3789	2.3906	2.4023	2.4142	2.4262	2.4383	2.4504	2.4627	22
68	2.4751	2.4876	2.5002	2.5129	2.5257	2.5386	2.5517	2.5649	2.5782	2.5916	21
69	2.6051	2.6187	2.6325	2.6464	2.6605	2.6746	2.6889	2.7034	2.7179	2.7326	20°
70°	2.7475	2.7625	2.7776	2.7929	2.8083	2.8239	2.8397	2.8556	2.8716	2.8878	19
71	2.9042	2.9208	2.9375	2.9544	2.9714	2.9887	3.0061	3.0237	3.0415	3.0595	18
72	3.0777	3.0961	3.1146	3.1334	3.1524	3.1716	3.1910	3.2106	3.2305	3.2506	17
73	3.2709	3.2914	3.3122	3.3332	3.3544	3.3759	3.3977	3.4197	3.4420	3.4646	16
74	3.4874	3.5105	3.5339	3.5576	3.5816	3.6059	3.6305	3.6554	3.6806	3.7062	15
75	3.7321	3.7583	3.7848	3.8118	3.8391	3.8667	3.8947	3.9232	3.9520	3.9812	14
76	4.0108	4.0408	4.0713	4.1022	4.1335	4.1653	4.1976	4.2303	4.2635	4.2972	13
77	4.3315	4.3662	4.4015	4.4374	4.4737	4.5107	4.5483	4.5864	4.6252	4.6646	12
78	4.7046	4.7453	4.7867	4.8288	4.8716	4.9152	4.9594	5.0045	5.0504	5.0970	11
79	5.1446	5.1929	5.2422	5.2924	5.3435	5.3955	5.4486	5.5026	5.5578	5.6140	10°
80°	5.6713	5.7297	5.7894	5.8502	5.9124	5.9758	6.0405	6.1066	6.1742	6.2433	9
81	6.3138	6.3859	6.4596	6.5350	6.6122	6.6912	6.7720	6.8548	6.9395	7.0264	8
82	7.1154	7.2066	7.3002	7.3962	7.4947	7.5958	7.6996	7.8062	7.9158	8.0285	7
83	8.1443	8.2636	8.3863	8.5126	8.6427	8.7769	8.9152	9.0579	9.2052	9.3572	6
84	9.5144	9.677	9.845	10.02	10.20	10.39	10.58	10.78	10.99	11.20	5
85	11.43	11.66	11.91	12.16	12.43	12.71	13.00	13.30	13.62	13.95	4
86	14.30	14.67	15.06	15.46	15.89	16.35	16.83	17.34	17.89	18.46	3
87	19.08	19.74	20.45	21.20	22.02	22.90	23.86	24.90	26.03	27.27	2
88	28.64	30.14	31.82	33.69	35.80	38.19	40.92	44.07	47.74	52.08	1
89	57.29	63.66	71.62	81.85	94.49	114.6	143.2	191.0	286.5	573.0	0°
	°1.0	°0.9	°0.8	°0.7	°0.6	°0.5	°0.4	°0.3	°0.2	°0.1	Deg

From "Standard Handbook for Electrical Engineers," 7th ed.

G. Common Logarithms of Numbers

N	0	1	2	3	4	5	6	7	8	9
10	0000	0043	0086	0128	0170	0212	0253	0294	0334	0374
11	0414	0453	0492	0531	0569	0607	0645	0682	0719	0755
12	0792	0828	0864	0899	0934	0969	1004	1038	1072	1106
13	1139	1173	1206	1239	1271	1303	1335	1367	1399	1430
14	1461	1492	1523	1553	1584	1614	1644	1673	1703	1732
15	1761	1790	1818	1847	1875	1903	1931	1959	1987	2014
16	2041	2068	2095	2122	2148	2175	2201	2227	2253	2279
17	2304	2330	2355	2380	2405	2430	2455	2480	2504	2529
18	2553	2577	2601	2625	2648	2672	2695	2718	2742	2765
19	2788	2810	2833	2856	2878	2900	2923	2945	2967	2989
20	3010	3032	3054	3075	3096	3118	3139	3160	3181	3201
21	3222	3243	3263	3284	3304	3324	3345	3365	3385	3404
22	3424	3444	3464	3483	3502	3522	3541	3560	3579	3598
23	3617	3636	3655	3674	3692	3711	3729	3747	3766	3784
24	3802	3820	3838	3856	3874	3892	3909	3927	3945	3962
25	3979	3997	4014	4031	4048	4065	4082	4099	4116	4133
26	4150	4166	4183	4200	4216	4232	4249	4265	4281	4298
27	4314	4330	4346	4362	4378	4393	4409	4425	4440	4456
28	4472	4487	4502	4518	4533	4548	4564	4579	4594	4609
29	4624	4639	4654	4669	4683	4698	4713	4728	4742	4757
30	4771	4786	4800	4814	4829	4843	4857	4871	4886	4900
31	4914	4928	4942	4955	4969	4983	4997	5011	5024	5038
32	5051	5065	5079	5092	5105	5119	5132	5145	5159	5172
33	5185	5198	5211	5224	5237	5250	5263	5276	5289	5302
34	5315	5328	5340	5353	5366	5378	5391	5403	5416	5428
35	5441	5453	5465	5478	5490	5502	5514	5527	5539	5551
36	5563	5575	5587	5599	5611	5623	5635	5647	5658	5670
37	5682	5694	5705	5717	5729	5740	5752	5763	5775	5786
38	5798	5809	5821	5832	5843	5855	5866	5877	5888	5899
39	5911	5922	5933	5944	5955	5966	5977	5988	5999	6010
40	6021	6031	6042	6053	6064	6075	6085	6096	6107	6117
41	6128	6138	6149	6160	6170	6180	6191	6201	6212	6222
42	6232	6243	6253	6263	6274	6284	6294	6304	6314	6325
43	6335	6345	6355	6365	6375	6385	6395	6405	6415	6425
44	6435	6444	6454	6464	6474	6484	6493	6503	6513	6522
45	6532	6542	6551	6561	6571	6580	6590	6599	6609	6618
46	6628	6637	6646	6656	6665	6675	6684	6693	6702	6712
47	6721	6730	6739	6749	6758	6767	6776	6785	6794	6803
48	6812	6821	6830	6839	6848	6857	6866	6875	6884	6893
49	6902	6911	6920	6928	6937	6946	6955	6964	6972	6981
50	6990	6998	7007	7016	7024	7033	7042	7050	7059	7067
51	7076	7084	7093	7101	7110	7118	7126	7135	7143	7152
52	7160	7168	7177	7185	7193	7202	7210	7218	7226	7235
53	7243	7251	7259	7267	7275	7284	7292	7300	7308	7316
54	7324	7332	7340	7348	7356	7364	7372	7380	7388	7396

From "Standard Handbook for Electrical Engineers," 7th ed.

G. Common Logarithms of Numbers (*Concluded*)

N	0	1	2	3	4	5	6	7	8	9
55	7404	7412	7419	7427	7435	7443	7451	7459	7466	7474
56	7482	7490	7497	7505	7513	7520	7528	7536	7543	7551
57	7559	7566	7574	7582	7589	7597	7604	7612	7619	7627
58	7634	7642	7649	7657	7664	7672	7679	7686	7694	7701
59	7709	7716	7723	7731	7738	7745	7752	7760	7767	7774
60	7782	7789	7796	7803	7810	7818	7825	7832	7839	7846
61	7853	7860	7868	7875	7882	7889	7896	7903	7910	7917
62	7924	7931	7938	7945	7952	7959	7966	7973	7980	7987
63	7993	8000	8007	8014	8021	8028	8035	8041	8048	8055
64	8062	8069	8075	8082	8089	8096	8102	8109	8116	8122
65	8129	8136	8142	8149	8156	8162	8169	8176	8182	8189
66	8195	8202	8209	8215	8222	8228	8235	8241	8248	8254
67	8261	8267	8274	8280	8287	8293	8299	8306	8312	8319
68	8325	8331	8338	8344	8351	8357	8363	8370	8376	8382
69	8388	8395	8401	8407	8414	8420	8426	8432	8439	8445
70	8451	8457	8463	8470	8476	8482	8488	8494	8500	8506
71	8513	8519	8525	8531	8537	8543	8549	8555	8561	8567
72	8573	8579	8585	8591	8597	8603	8609	8615	8621	8627
73	8633	8639	8645	8651	8657	8663	8669	8675	8681	8686
74	8692	8698	8704	8710	8716	8722	8727	8733	8739	8745
75	8751	8756	8762	8768	8774	8779	8785	8791	8797	8802
76	8808	8814	8820	8825	8831	8837	8842	8848	8854	8859
77	8865	8871	8876	8882	8887	8893	8899	8904	8910	8915
78	8921	8927	8932	8938	8943	8949	8954	8960	8965	8971
79	8976	8982	8987	8993	8998	9004	9009	9015	9020	9025
80	9031	9036	9042	9047	9053	9058	9063	9069	9074	9079
81	9085	9090	9096	9101	9106	9112	9117	9122	9128	9133
82	9138	9143	9149	9154	9159	9165	9170	9175	9180	9186
83	9191	9196	9201	9206	9212	9217	9222	9227	9232	9238
84	9243	9248	9253	9258	9263	9269	9274	9279	9284	9289
85	9294	9299	9304	9309	9315	9320	9325	9330	9335	9340
86	9345	9350	9355	9360	9365	9370	9375	9380	9385	9390
87	9395	9400	9405	9410	9415	9420	9425	9430	9435	9440
88	9445	9450	9455	9460	9465	9469	9474	9479	9484	9489
89	9494	9499	9504	9509	9513	9518	9523	9528	9533	9538
90	9542	9547	9552	9557	9562	9566	9571	9576	9581	9586
91	9590	9595	9600	9605	9609	9614	9619	9624	9628	9633
92	9638	9643	9647	9652	9657	9661	9666	9671	9675	9680
93	9685	9689	9694	9699	9703	9708	9713	9717	9722	9727
94	9731	9736	9741	9745	9750	9754	9759	9763	9768	9773
95	9777	9782	9786	9791	9795	9800	9805	9809	9814	9818
96	9823	9827	9832	9836	9841	9845	9850	9854	9859	9863
97	9868	9872	9877	9881	9886	9890	9894	9899	9903	9908
98	9912	9917	9921	9926	9930	9934	9939	9943	9948	9952
99	9956	9961	9965	9969	9974	9978	9983	9987	9991	9996

From "Standard Handbook for Electrical Engineers," 7th ed.

H. Natural, Napierian, or Hyperbolic Logarithms

N	0	1	2	3	4	5	6	7	8	9
0	— ∞	0.0000	0.6931	1.0986	1.3863	1.6094	1.7918	1.9459	2.0794	2.1972
10	2.3026	2.3979	2.4849	2.5649	2.6391	2.7081	2.7726	2.8332	2.8904	2.9444
20	2.9957	3.0445	3.0910	3.1355	3.1781	3.2189	3.2581	3.2958	3.3322	3.3673
30	3.4012	3.4340	3.4657	3.4965	3.5264	3.5553	3.5835	3.6109	3.6376	3.6636
40	3.6889	3.7136	3.7377	3.7612	3.7842	3.8067	3.8286	3.8501	3.8712	3.8918
50	3.9120	3.9318	3.9512	3.9703	3.9890	4.0073	4.0254	4.0431	4.0604	4.0775
60	4.0943	4.1109	4.1271	4.1431	4.1589	4.1744	4.1897	4.2047	4.2195	4.2341
70	4.2485	4.2627	4.2767	4.2905	4.3041	4.3175	4.3307	4.3438	4.3567	4.3694
80	4.3820	4.3944	4.4067	4.4188	4.4308	4.4427	4.4543	4.4659	4.4773	4.4886
90	4.4998	4.5109	4.5218	4.5326	4.5433	4.5539	4.5643	4.5747	4.5850	4.5951
100	4.6052	4.6151	4.6250	4.6347	4.6444	4.6540	4.6634	4.6728	4.6821	4.6913
110	4.7005	4.7095	4.7185	4.7274	4.7362	4.7449	4.7536	4.7622	4.7707	4.7791
120	4.7875	4.7958	4.8040	4.8122	4.8203	4.8283	4.8363	4.8442	4.8520	4.8598
130	4.8675	4.8752	4.8828	4.8903	4.8978	4.9053	4.9127	4.9200	4.9273	4.9345
140	4.9416	4.9488	4.9558	4.9628	4.9698	4.9767	4.9836	4.9904	4.9972	5.0039
150	5.0106	5.0173	5.0239	5.0304	5.0370	5.0434	5.0499	5.0562	5.0626	5.0689
160	5.0752	5.0814	5.0876	5.0938	5.0999	5.1059	5.1120	5.1180	5.1240	5.1299
170	5.1358	5.1417	5.1475	5.1533	5.1591	5.1648	5.1705	5.1761	5.1818	5.1874
180	5.1930	5.1985	5.2040	5.2095	5.2149	5.2204	5.2257	5.2311	5.2364	5.2417
190	5.2470	5.2523	5.2575	5.2627	5.2679	5.2730	5.2781	5.2832	5.2883	5.2933
200	5.2983	5.3033	5.3083	5.3132	5.3181	5.3230	5.3279	5.3327	5.3375	5.3423
210	5.3471	5.3519	5.3566	5.3613	5.3660	5.3706	5.3753	5.3799	5.3845	5.3891
220	5.3936	5.3982	5.4027	5.4072	5.4116	5.4161	5.4205	5.4250	5.4293	5.4337
230	5.4381	5.4424	5.4467	5.4510	5.4553	5.4596	5.4638	5.4681	5.4723	5.4765
240	5.4806	5.4848	5.4889	5.4931	5.4972	5.5013	5.5053	5.5094	5.5134	5.5175
250	5.5215	5.5255	5.5294	5.5334	5.5373	5.5413	5.5452	5.5491	5.5530	5.5568
260	5.5607	5.5645	5.5683	5.5722	5.5759	5.5797	5.5835	5.5872	5.5910	5.5947
270	5.5984	5.6021	5.6058	5.6095	5.6131	5.6168	5.6204	5.6240	5.6276	5.6312
280	5.6348	5.6384	5.6419	5.6454	5.6490	5.6525	5.6560	5.6595	5.6630	5.6664
290	5.6699	5.6733	5.6768	5.6802	5.6836	5.6870	5.6904	5.6937	5.6971	5.7004
300	5.7038	5.7071	5.7104	5.7137	5.7170	5.7203	5.7236	5.7268	5.7301	5.7333
310	5.7366	5.7398	5.7430	5.7462	5.7494	5.7526	5.7557	5.7589	5.7621	5.7652
320	5.7683	5.7714	5.7746	5.7777	5.7807	5.7838	5.7869	5.7900	5.7930	5.7961
330	5.7991	5.8021	5.8051	5.8081	5.8111	5.8141	5.8171	5.8201	5.8230	5.8260
340	5.8289	5.8319	5.8348	5.8377	5.8406	5.8435	5.8464	5.8493	5.8522	5.8551
350	5.8579	5.8608	5.8636	5.8665	5.8693	5.8721	5.8749	5.8777	5.8805	5.8833
360	5.8861	5.8889	5.8916	5.8944	5.8972	5.8999	5.9026	5.9054	5.9081	5.9108
370	5.9135	5.9162	5.9189	5.9216	5.9243	5.9269	5.9296	5.9322	5.9349	5.9375
380	5.9402	5.9428	5.9454	5.9480	5.9506	5.9532	5.9558	5.9584	5.9610	5.9636
390	5.9661	5.9687	5.9713	5.9738	5.9764	5.9789	5.9814	5.9839	5.9865	5.9890
400	5.9915	5.9940	5.9965	5.9989	6.0014	6.0039	6.0064	6.0088	6.0113	6.0137
410	6.0162	6.0186	6.0210	6.0234	6.0259	6.0283	6.0307	6.0331	6.0355	6.0379
420	6.0403	6.0426	6.0450	6.0474	6.0497	6.0521	6.0544	6.0568	6.0591	6.0615
430	6.0638	6.0661	6.0684	6.0707	6.0730	6.0753	6.0776	6.0799	6.0822	6.0845
440	6.0868	6.0890	6.0913	6.0936	6.0958	0.0981	6.1003	6.1026	6.1048	6.1070
450	6.1092	6.1115	6.1137	6.1159	6.1181	6.1203	6.1225	6.1247	6.1269	6.1291
460	6.1312	6.1334	6.1356	6.1377	6.1399	6.1420	6.1442	6.1463	6.1485	6.1506
470	6.1527	6.1549	6.1570	6.1591	6.1612	6.1633	6.1654	6.1675	6.1696	6.1717
480	6.1738	6.1759	6.1779	6.1800	6.1821	6.1841	6.1862	6.1883	6.1903	6.1924
490	6.1944	6.1964	6.1985	6.2005	6.2025	6.2046	6.2066	6.2086	6.2106	6.2126

From "Standard Handbook for Electrical Engineers," 7th ed.

NOTE 1: Moving the decimal point n places to the right (or left) in the number is equivalent to adding (or subtracting) n times 2.3026.	n	$n \times 2.3026$
	1	$2.3026 = 0.6974{-}3$
	2	$4.6052 = 0.3948{-}5$
	3	$6.9078 = 0.0922{-}7$
NOTE 2:	4	$9.2103 = 0.7897{-}10$
$\log_e x = 2.3026 \log_{10} x$	5	$11.5129 = 0.4871{-}12$
$\log_{10} x = 0.4343 \log_e x$	6	$13.8155 = 0.1845{-}14$
$\log_e 10 = 2.3026$	7	$16.1181 = 0.8819{-}17$
$\log_{10} e = 0.4343$	8	$18.4207 = 0.5793{-}19$
	9	$20.7233 = 0.2767{-}21$

H. Natural, Napierian, or Hyperbolic Logarithms (*Concluded*)

N	0	1	2	3	4	5	6	7	8	9
500	6.2146	6.2166	6.2186	6.2206	6.2226	6.2246	6.2265	6.2285	6.2305	6.2324
510	6.2344	6.2364	6.2383	6.2403	6.2422	6.2442	6.2461	6.2480	6.2500	6.2519
520	6.2538	6.2558	6.2577	6.2596	6.2615	6.2634	6.2653	6.2672	6.2691	6.2710
530	6.2729	6.2748	6.2766	6.2785	6.2804	6.2823	6.2841	6.2860	6.2879	6.2897
540	6.2916	6.2934	6.2953	6.2971	6.2989	6.3008	6.3026	6.3044	6.3063	6.3081
550	6.3099	6.3117	6.3135	6.3154	6.3172	6.3190	6.3208	6.3226	6.3244	6.3261
560	6.3279	6.3297	6.3315	6.3333	6.3351	6.3368	6.3386	6.3404	6.3421	6.3439
570	6.3456	6.3474	6.3491	6.3509	6.3256	6.3544	6.3561	6.3578	6.3596	6.3613
580	6.3630	6.3648	6.3665	6.3682	6.3699	6.3716	6.3733	6.3750	6.3767	6.3784
590	6.3801	6.3818	6.3835	6.3852	6.3869	6.3886	6.3902	6.3919	6.3936	6.3953
600	6.3969	6.3986	6.4003	6.4019	6.4036	6.4052	6.4069	6.4085	6.4102	6.4118
610	6.4135	6.4151	6.4167	6.4184	6.4200	6.4216	6.4232	6.4249	6.4265	6.4281
620	6.4297	6.4313	6.4329	6.4345	6.4362	6.4378	6.4394	6.4409	6.4425	6.4441
630	6.4457	6.4473	6.4489	6.4505	6.4520	6.4536	6.4552	6.4568	6.4583	6.4599
640	6.4615	6.4630	6.4646	6.4661	6.4677	6.4693	6.4708	6.4723	6.4739	6.4754
650	6.4770	6.4785	6.4800	6.4816	6.4831	6.4846	6.4862	6.4877	6.4892	6.4907
660	6.4922	6.4938	6.4953	6.4968	6.4983	6.4998	6.5013	6.5028	6.5043	6.5058
670	6.5073	6.5088	6.5103	6.5117	6.5132	6.5147	6.5162	6.5177	6.5191	6.5206
680	6.5221	6.5236	6.5250	6.5265	6.5280	6.5294	6.5309	6.5323	6.5338	6.5352
690	6.5367	6.5381	6.5396	6.5410	6.5425	6.5439	6.5453	6.5468	6.5482	6.5497
700	6.5511	6.5525	6.5539	6.5554	6.5568	6.5582	6.5596	6.5610	6.5624	6.5639
710	6.5653	6.5667	6.5681	6.5695	6.5709	6.5723	6.5737	6.5751	6.5765	6.5779
720	6.5793	6.5806	6.5820	6.5834	6.5848	6.5862	6.5876	6.5889	6.5903	6.5917
730	6.5930	6.5944	6.5958	6.5971	6.5985	6.5999	6.6012	6.6026	6.6039	6.6053
740	6.6067	6.6080	6.6093	6.6107	6.6120	6.6134	6.6147	6.6161	6.6174	6.6187
750	6.6201	6.6214	6.6227	6.6241	6.6254	6.6267	6.6280	6.6294	6.6307	6.6320
760	6.6333	6.6346	6.6359	6.6373	6.6386	6.6399	6.6412	6.6425	6.6438	6.6451
770	6.6464	6.6477	6.6490	6.6503	6.6516	6.6529	6.6542	6.6554	6.6567	6.6580
780	6.6593	6.6606	6.6619	6.6631	6.6644	6.6657	6.6670	6.6682	6.6695	6.6708
790	6.6720	6.6733	6.6746	6.6758	6.6771	6.6783	6.6796	6.6809	6.6821	6.6834
800	6.6846	6.6859	6.6871	6.6884	6.6896	6.6908	6.6921	6.6933	6.6946	6.6958
810	6.6970	6.6983	6.6995	6.7007	6.7020	6.7032	6.7044	6.7056	6.7069	6.7081
820	6.7093	6.7105	6.7117	6.7130	6.7142	6.7154	6.7166	6.7178	6.7190	6.7202
830	6.7214	6.7226	6.7238	6.7250	6.7262	6.7274	6.7286	6.7298	6.7310	6.7322
840	6.7334	6.7346	6.7358	6.7370	6.7382	6.7393	6.7405	6.7417	6.7429	6.7441
850	6.7452	6.7464	6.7476	6.7488	6.7499	6.7511	6.7523	6.7534	6.7546	6.7558
860	6.7569	6.7581	6.7593	6.7604	6.7616	6.7627	6.7639	6.7650	6.7662	6.7673
870	6.7685	6.7696	6.7708	6.7719	6.7731	6.7742	6.7754	6.7765	6.7776	6.7788
880	6.7799	6.7811	6.7822	6.7833	6.7845	6.7856	6.7867	6.7878	6.7890	6.7901
890	6.7912	6.7923	6.7935	6.7946	6.7957	6.7968	6.7979	6.7991	6.8002	6.8013
900	6.8024	6.8035	6.8046	6.8057	6.8068	6.8079	6.8090	6.8101	6.8112	6.8123
910	6.8134	6.8145	6.8156	6.8167	6.8178	6.8189	6.8200	6.8211	6.8222	6.8233
920	6.8244	6.8255	6.8265	6.8276	6.8287	6.8298	6.8309	6.8320	6.8330	6.8341
930	6.8352	6.8363	6.8373	6.8384	6.8395	6.8405	6.8416	6.8427	6.8437	6.8448
940	6.8459	5.8469	6.8480	6.8491	6.8501	6.8512	6.8522	6.8533	6.8544	8.8554
950	6.8565	6.8575	6.8586	6.8596	6.8607	6.8617	6.8628	6.8638	6.8648	6.8659
960	6.8669	6.8680	6.8690	6.8701	6.8711	6.8721	6.8732	6.8742	6.8752	6.8763
970	6.8773	6.8783	6.8794	6.8804	6.8814	6.8824	6.8835	6.8845	6.8855	6.8865
980	6.8876	6.8886	6.8896	6.8906	6.8916	6.8926	6.8937	6.8947	6.8957	6.8967
990	6.8977	6.8987	6.8997	6.9007	6.9017	6.9027	6.9037	6.9047	6.9057	6.9068

From "Standard Handbook for Electrical Engineers," 7th ed.

I. Degrees and Minutes Expressed in Radians

Degrees						Hundredths				Minutes	
1°	0.0175	61°	1.0647	121°	2.1118	0°.01	0.0002	0°.51	0.0089	1'	0.0003
2	0.0349	62	1.0821	122	2.1293	0.02	0.0003	0.52	0.0091	2'	0.0006
3	0.0524	63	1.0996	123	2.1468	0.03	0.0005	0.53	0.0093	3'	0.0009
4	0.0698	64	1.1170	124	2.1642	0.04	0.0007	0.54	0.0094	4'	0.0012
5°	0.0873	65°	1.1345	125°	2.1817	0.05	0.0009	0.55	0.0096	5'	0.0015
6	0.1047	66	1.1519	126	2.1991	0.06	0.0010	0.56	0.0098	6'	0.0017
7	0.1222	67	1.1694	127	2.2166	0.07	0.0012	0.57	0.0099	7'	0.0020
8	0.1396	68	1.1868	128	2.2340	0.08	0.0014	0.58	0.0101	8'	0.0023
9	0.1571	69	1.2043	129	2.2515	0.09	0.0016	0.59	0.0103	9'	0.0026
10°	0.1745	70°	1.2217	130°	2.2689	0°.10	0.0017	0°.60	0.0105	10'	0.0029
11	0.1920	71	1.2392	131	2.2864	0.11	0.0019	0.61	0.0106	11'	0.0032
12	0.2094	72	1.2566	132	2.3038	0.12	0.0021	0.62	0.0108	12'	0.0035
13	0.2269	73	1.2741	133	2.3213	0.13	0.0023	0.63	0.0110	13'	0.0038
14	0.2443	74	1.2915	134	2.3387	0.14	0.0024	0.64	0.0112	14'	0.0041
15°	0.2618	75°	1.3090	135°	2.3562	0.15	0.0026	0.65	0.0113	15'	0.0044
16	0.2793	76	1.3265	136	2.3736	0.16	0.0028	0.66	0.0115	16'	0.0047
17	0.2967	77	1.3439	137	2.3911	0.17	0.0030	0.67	0.0117	17'	0.0049
18	0.3142	78	1.3614	138	2.4086	0.18	0.0031	0.68	0.0119	18'	0.0052
19	0.3316	79	1.3788	139	2.4260	0.19	0.0033	0.69	0.0120	19'	0.0055
20°	0.3491	80°	1.3963	140°	2.4435	0°.20	0.0035	0°.70	0.0122	20'	0.0058
21	0.3665	81	1.4137	141	2.4609	0.21	0.0037	0.71	0.0124	21'	0.0061
22	0.3840	82	1.4312	142	2.4784	0.22	0.0038	0.72	0.0126	22'	0.0064
23	0.4014	83	1.4486	143	2.4958	0.23	0.0040	0.73	0.0127	23'	0.0067
24	0.4189	84	1.4661	144	2.5133	0.24	0.0042	0.74	0.0129	24'	0.0070
25°	0.4363	85°	1.4835	145°	2.5307	0.25	0.0044	0.75	0.0131	25'	0.0073
26	0.4538	86	1.5010	146	2.5482	0.26	0.0045	0.76	0.0133	26'	0.0076
27	0.4712	87	1.5184	147	2.5656	0.27	0.0047	0.77	0.0134	27'	0.0079
28	0.4887	88	1.5359	148	2.5831	0.28	0.0049	0.78	0.0136	28'	0.0081
29	0.5061	89	1.5533	149	2.6005	0.29	0.0051	0.79	0.0138	29'	0.0084
30°	0.5236	90°	1.5708	150°	2.6180	0°.30	0.0052	0°.80	0.0140	30'	0.0087
31	0.5411	91	1.5882	151	2.6354	0.31	0.0054	0.81	0.0141	31'	0.0090
32	0.5585	92	1.6057	152	2.6529	0.32	0.0056	0.82	0.0143	32'	0.0093
33	0.5760	93	1.6232	153	2.6704	0.33	0.0058	0.83	0.0145	33'	0.0096
34	0.5934	94	1.6406	154	2.6878	0.34	0.0059	0.84	0.0147	34'	0.0099
35°	0.6109	95°	1.6581	155°	2.7053	0.35	0.0061	0.85	0.0148	35'	0.0102
36	0.6283	96	1.6755	156	2.7227	0.36	0.0063	0.86	0.0150	36'	0.0105
37	0.6458	97	1.6930	157	2.7402	0.37	0.0065	0.87	0.0152	37'	0.0108
38	0.6632	98	1.7104	158	2.7576	0.38	0.0066	0.88	0.0154	38'	0.0111
39	0.6807	99	1.7279	159	2.7751	0.39	0.0068	0.89	0.0155	39'	0.0113
40°	0.6981	100°	1.7453	160°	2.7925	0°.40	0.0070	0°.90	0.0157	40'	0.0116
41	0.7156	101	1.7628	161	2.8100	0.41	0.0072	0.91	0.0159	41'	0.0119
42	0.7330	102	1.7802	162	2.8274	0.42	0.0073	0.92	0.0161	42'	0.0122
43	0.7505	103	1.7977	163	2.8449	0.43	0.0075	0.93	0.0162	43'	0.0125
44	0.7679	104	1.8151	164	2.8623	0.44	0.0077	0.94	0.0164	44'	0.0128
45°	0.7854	105°	1.8326	165°	2.8798	0.45	0.0079	0.95	0.0166	45'	0.0131
46	0.8029	106	1.8500	166	2.8972	0.46	0.0080	0.96	0.0168	46'	0.0134
47	0.8203	107	1.8675	167	2.9147	0.47	0.0082	0.97	0.0169	47'	0.0137
48	0.8378	108	1.8850	168	2.9322	0.48	0.0084	0.98	0.0171	48'	0.0140
49	0.8552	109	1.9024	169	2.9496	0.49	0.0086	0.99	0.0173	49'	0.0143
50°	0.8727	110°	1.9199	170°	2.9671	0°.50	0.0087	1°.00	0.0175	50'	0.0145
51	0.8901	111	1.9373	171	2.9845	....	......	....	......	51'	0.0148
52	0.9076	112	1.9548	172	3.0020	....	......	....	......	52'	0.0151
53	0.9250	113	1.9722	173	3.0194	....	......	....	......	53'	0.0154
54	0.9425	114	1.9897	174	3.0369	....	......	....	......	54'	0.0157
55°	0.9599	115°	2.0071	175°	3.0543	....	......	....	......	55'	0.0160
56	0.9774	116	2.0246	176	3.0718	....	......	....	......	56'	0.0163
57	0.9948	117	2.0420	177	3.0892	....	......	....	......	57'	0.0166
58	1.0123	118	2.0595	178	3.1067	....	......	....	......	58'	0.0169
59	1.0297	119	2.0769	179	3.1241	....	......	....	......	59'	0.0172
60°	1.0472	120°	2.0944	180°	3.1416	....	......	....	......	60'	0.0175

Arc 1° = 0.01745. Arc 1' = 0.0002909. Arc 1" = 0.000004848.
1 radian = 57°.296 = 57° 17'.75 = 57° 17' 44".81.
From Lionel S. Marks, "Mechanical Engineers' Handbook."

J. Radians Expressed in Degrees

rad	deg	rad	deg	rad	deg	rad	deg	rad	deg
0.01	0°.57	0.64	36°.67	1.27	72°.77	1.90	108°.86	2.53	144°.96
0.02	1°.15	0.65	37°.24	1.28	73°.34	1.91	109°.43	2.54	145°.53
0.03	1°.72	0.66	37°.82	1.29	73°.91	1.92	110°.01	2.55	146°.10
0.04	2°.29	0.67	38°.39	1.30	74°.48	1.93	110°.58	2.56	146°.68
0.05	2°.86	0.68	38°.96	1.31	75°.06	1.94	111°.15	2.57	147°.25
0.06	3°.44	0.69	39°.53	1.32	75°.63	1.95	111°.73	2.58	147°.82
0.07	4°.01	0.70	40°.11	1.33	76°.20	1.96	112°.30	2.59	148°.40
0.08	4°.58	0.71	40°.68	1.34	76°.78	1.97	112°.87	2.60	148°.97
0.09	5°.16	0.72	41°.25	1.35	77°.35	1.98	113°.45	2.61	149°.54
0.10	5°.73	0.73	41°.83	1.36	77°.92	1.99	114°.02	2.62	150°.11
0.11	6°.30	0.74	42°.40	1.37	78°.50	2.00	114°.59	2.63	150°.69
0.12	6°.88	0.75	42°.97	1.38	79°.07	2.01	115°.16	2.64	151°.26
0.13	7°.45	0.76	43°.54	1.39	79°.64	2.02	115°.74	2.65	151°.83
0.14	8°.02	0.77	44°.12	1.40	80°.21	2.03	116°.31	2.66	152°.41
0.15	8°.59	0.78	44°.69	1.41	80°.79	2.04	116°.88	2.67	152°.98
0.16	9°.17	0.79	45°.26	1.42	81°.36	2.05	117°.46	2.68	153°.55
0.17	9°.74	0.80	45°.84	1.43	81°.93	2.06	118°.03	2.69	154°.13
0.18	10°.31	0.81	46°.41	1.44	82°.51	2.07	118°.60	2.70	154°.70
0.19	10°.89	0.82	46°.98	1.45	83°.08	2.08	119°.18	2.71	155°.27
0.20	11°.46	0.83	47°.56	1.46	83°.65	2.09	119°.75	2.72	155°.84
0.21	12°.03	0.84	48°.13	1.47	84°.22	2.10	120°.32	2.73	156°.42
0.22	12°.61	0.85	48°.70	1.48	84°.80	2.11	120°.89	2.74	156°.99
0.23	13°.18	0.86	49°.27	1.49	85°.37	2.12	121°.47	2.75	157°.56
0.24	13°.75	0.87	49°.85	1.50	85°.94	2.13	122°.04	2.76	158°.14
0.25	14°.32	0.88	50°.42	1.51	86°.52	2.14	122°.61	2.77	158°.71
0.26	14°.90	0.89	50°.99	1.52	87°.09	2.15	123°.19	2.78	159°.28
0.27	15°.47	0.90	51°.57	1.53	87°.66	2.16	123°.76	2.79	159°.86
0.28	16°.04	0.91	52°.14	1.54	88°.24	2.17	124°.33	2.80	160°.43
0.29	16°.62	0.92	52°.71	1.55	88°.81	2.18	124°.90	2.81	161°.00
0.30	17°.19	0.93	53°.29	1.56	89°.38	2.19	125°.48	2.82	161°.57
0.31	17°.76	0.94	53°.86	1.57	89°.95	2.20	126°.05	2.83	162°.15
0.32	18°.33	0.95	54°.43	1.58	90°.53	2.21	126°.62	2.84	162°.72
0.33	18°.91	0.96	55°.00	1.59	91°.10	2.22	127°.20	2.85	163°.29
0.34	19°.48	0.97	55°.58	1.60	91°.67	2.23	127°.77	2.86	163°.87
0.35	20°.05	0.98	56°.15	1.61	92°.25	2.24	128°.34	2.87	164°.44
0.36	20°.63	0.99	56°.72	1.62	92°.82	2.25	128°.92	2.88	165°.01
0.37	21°.20	1.00	57°.30	1.63	93°.39	2.26	129°.49	2.89	165°.58
0.38	21°.77	1.01	57°.87	1.64	93°.97	2.27	130°.06	2.90	166°.16
0.39	22°.35	1.02	58°.44	1.65	94°.54	2.28	130°.63	2.91	166°.73
0.40	22°.92	1.03	59°.01	1.66	95°.11	2.29	131°.21	2.92	167°.30
0.41	23°.49	1.04	59°.59	1.67	95°.68	2.30	131°.78	2.93	167°.88
0.42	24°.06	1.05	60°.16	1.68	96°.26	2.31	132°.35	2.94	168°.45
0.43	24°.64	1.06	60°.73	1.69	96°.83	2.32	132°.93	2.95	169°.02
0.44	25°.21	1.07	61°.31	1.70	97°.40	2.33	133°.50	2.96	169°.60
0.45	25°.78	1.08	61°.88	1.71	97°.98	2.34	134°.07	2.97	170°.17
0.46	26°.36	1.09	62°.45	1.72	98°.55	2.35	134°.65	2.98	170°.74
0.47	26°.93	1.10	63°.03	1.73	99°.12	2.36	135°.22	2.99	171°.31
0.48	27°.50	1.11	63°.60	1.74	99°.69	2.37	135°.79	3.00	171°.89
0.49	28°.07	1.12	64°.17	1.75	100°.27	2.38	136°.36	3.01	172°.46
0.50	28°.65	1.13	64°.74	1.76	100°.84	2.39	136°.94	3.02	173°.03
0.51	29°.22	1.14	65°.32	1.77	101°.41	2.40	137°.51	3.03	173°.61
0.52	29°.79	1.15	65°.89	1.78	101°.99	2.41	138°.08	3.04	174°.18
0.53	30°.37	1.16	66°.46	1.79	102°.56	2.42	138°.66	3.05	174°.75
0.54	30°.94	1.17	67°.04	1.80	103°.13	2.43	139°.23	3.06	175°.33
0.55	31°.51	1.18	67°.61	1.81	103°.71	2.44	139°.80	3.07	175°.90
0.56	32°.09	1.19	68°.18	1.82	104°.28	2.45	140°.37	3.08	176°.47
0.57	32°.66	1.20	68°.75	1.83	104°.85	2.46	140°.95	3.09	177°.04
0.58	33°.23	1.21	69°.33	1.84	105°.42	2.47	141°.52	3.10	177°.62
0.59	33°.80	1.22	69°.90	1.85	106°.00	2.48	142°.09	3.11	178°.19
0.60	34°.38	1.23	70°.47	1.86	106°.57	2.49	142°.67	3.12	178°.76
0.61	34°.95	1.24	71°.05	1.87	107°.14	2.50	143°.24	3.13	179°.34
0.62	35°.52	1.25	71°.62	1.88	107°.72	2.51	143°.81	3.14	179°.91
0.63	36°.10	1.26	72°.19	1.89	108°.29	2.52	144°.39	3.15	180°.48

Interpolation

0.0002	0°.01
0.0004	0°.02
0.0006	0°.03
0.0008	0°.05
0.0010	0°.06
0.0012	0°.07
0.0014	0°.08
0.0016	0°.09
0.0018	0°.10
0.0020	0°.11
0.0022	0°.13
0.0024	0°.14
0.0026	0°.15
0.0028	0°.16
0.0030	0°.17
0.0032	0°.18
0.0034	0°.19
0.0036	0°.21
0.0038	0°.22
0.0040	0°.23
0.0042	0°.24
0.0044	0°.25
0.0046	0°.26
0.0048	0°.28
0.0050	0°.29
0.0052	0°.30
0.0054	0°.31
0.0056	0°.32
0.0058	0°.33
0.0060	0°.34
0.0062	0°.36
0.0064	0°.37
0.0066	0°.38
0.0068	0°.39
0.0070	0°.40
0.0072	0°.41
0.0074	0°.42
0.0076	0°.44
0.0078	0°.45
0.0080	0°.46
0.0082	0°.47
0.0084	0°.48
0.0086	0°.49
0.0088	0°.50
0.0090	0°.52
0.0092	0°.53
0.0094	0°.54
0.0096	0°.55
0.0098	0°.56

Multiples of π

1	3.1416	180°
2	6.2832	360°
3	9.4248	540°
4	12.5664	720°
5	15.7080	900°
6	18.8496	1080°
7	21.9911	1260°
8	25.1327	1440°
9	28.2743	1620°
10	31.4159	1800°

From Lionel S. Marks, "Mechanical Engineers' Handbook."

K. Exponentials

$$(e^n \text{ and } e^{-n})$$

n	e^n	Diff.	n	e^n	Diff.	n	e^n	n	e^{-n}	Diff.	n	e^{-n}	n	e^{-n}
0.00	1.000	10	0.50	1.649	16	1.0	2.718†	0.00	1.000	−10	0.50	0.607	1.0	0.368
0.01	1.010	10	0.51	1.665	17	1.1	3.004	0.01	0.990	−10	0.51	0.600	1.1	0.333
0.02	1.020	10	0.52	1.682	17	1.2	3.320	0.02	0.980	−10	0.52	0.595	1.2	0.301
0.03	1.030	11	0.53	1.699	17	1.3	3.669	0.03	0.970	−9	0.53	0.589	1.3	0.273
0.04	1.041	10	0.54	1.716	17	1.4	4.055	0.04	0.961	−10	0.54	0.583	1.4	0.247
0.05	1.051	11	0.55	1.733	18	1.5	4.482	0.05	0.951	−9	0.55	0.577	1.5	0.223
0.06	1.062	11	0.56	1.751	17	1.6	4.953	0.06	0.942	−10	0.56	0.571	1.6	0.202
0.07	1.073	10	0.57	1.768	18	1.7	5.474	0.07	0.932	−9	0.57	0.566	1.7	0.183
0.08	1.083	11	0.58	1.786	18	1.8	6.050	0.08	0.923	−9	0.58	0.560	1.8	0.165
0.09	1.094	11	0.59	1.804	18	1.9	6.686	0.09	0.914	−9	0.59	0.554	1.9	0.150
0.10	1.105	11	0.60	1.822	18	2.0	7.389	0.10	0.905	−9	0.60	0.549	2.0	0.135
0.11	1.116	11	0.61	1.840	19	2.1	8.166	0.11	0.896	−9	0.61	0.543	2.1	0.122
0.12	1.127	12	0.62	1.859	19	2.2	9.025	0.12	0.887	−9	0.62	0.538	2.2	0.111
0.13	1.139	11	0.63	1.878	18	2.3	9.974	0.13	0.878	−9	0.63	0.533	2.3	0.100
0.14	1.150	12	0.64	1.896	20	2.4	11.02	0.14	0.869	−8	0.64	0.527	2.4	0.0907
0.15	1.162	12	0.65	1.916	19	2.5	12.18	0.15	0.861	−9	0.65	0.522	2.5	0.0821
0.16	1.174	11	0.66	1.935	19	2.6	13.46	0.16	0.852	−8	0.66	0.517	2.6	0.0743
0.17	1.185	12	0.67	1.954	20	2.7	14.88	0.17	0.844	−9	0.67	0.512	2.7	0.0672
0.18	1.197	12	0.68	1.974	20	2.8	16.44	0.18	0.835	−8	0.68	0.507	2.8	0.0608
0.19	1.209	12	0.69	1.994	20	2.9	18.17	0.19	0.827	−8	0.69	0.502	2.9	0.0550
0.20	1.221	13	0.70	2.014	20	3.0	20.09	0.20	0.819	−8	0.70	0.497	3.0	0.0498
0.21	1.234	12	0.71	2.034	20	3.1	22.20	0.21	0.811	−8	0.71	0.492	3.1	0.0450
0·22	1.246	13	0.72	2.054	21	3.2	24.53	0.22	0.803	−8	0.72	0.487	3.2	0.0408
0.23	1.259	12	0.73	2.075	21	3.3	27.11	0.23	0.795	−8	0.73	0.482	3.3	0.0369
0.24	1.271	13	0.74	2.096	21	3.4	29.96	0.24	0.787	−8	0.74	0.477	3.4	0.0334
0.25	1.284	13	0.75	2.117	21	3.5	33.12	0.25	0.779	−8	0.75	0.472	3.5	0.0302
0.26	1.297	13	0.76	2.138	22	3.6	36.60	0.26	0.771	−8	0.76	0.468	3.6	0.0273
0.27	1.310	13	0.77	2.160	21	3.7	40.45	0.27	0.763	−7	0.77	0.463	3.7	0.0247
0.28	1.323	13	0.78	2.181	22	3.8	44.70	0.28	0.756	−8	0.78	0.458	3.8	0.0224
0.29	1.336	14	0.79	2.203	23	3.9	49.40	0.29	0.748	−7	0.79	0.454	3.9	0.0202
0.30	1.350	13	0.80	2.226	22	4.0	54.60	0.30	0.741	−8	0.80	0.449	4.0	0.0183
0.31	1.363	14	0.81	2.248	22	4.1	60.34	0.31	0.733	−7	0.81	0.445	4.1	0.0166
0.32	1.377	14	0.82	2.270	23	4.2	66.69	0.32	0.726	−7	0.82	0.440	4.2	0.0150
0.33	1.391	14	0.83	2.293	23	4.3	73.70	0.33	0.719	−7	0.83	0.436	4.3	0.0136
0.34	1.405	14	0.84	2.316	24	4.4	81.45	0.34	0.712	−7	0.84	0.432	4.4	0.0123
0.35	1.419	14	0.85	2.340	23	4.5	90.02	0.35	0.705	−7	0.85	0.427	4.5	0.0111
0.36	1.433	15	0.86	2.363	24			0.36	0.698	−7	0.86	0.423		
0.37	1.448	14	0.87	2.387	24	5.0	148.4	0.37	0.691	−7	0.87	0.419	5.0	0.00674
0.38	1.462	15	0.88	2.411	24	6.0	403.4	0.38	0.684	−7	0.88	0.415	6.0	0.00248
0.39	1.477	15	0.89	2.435	25	7.0	1097.	0.39	0.677	−7	0.89	0.411	7.0	0.000912
0.40	1.492	15	0.90	2.460	24	8.0	2981.	0.40	0.670	−6	0.90	0.407	8.0	0.000335
0.41	1.507	15	0.91	2.484	25	9.0	8103.	0.41	0.664	−7	0.91	0.403	9.0	0.000123
0.42	1.522	15	0.92	2.509	26	10.0	22026.	0.42	0.657	−6	0.92	0.399	10.0	0.000045
0.43	1.537	16	0.93	2.535	25	$\pi/2$	4.810	0.43	0.651	−7	0.93	0.395		
0.44	1.553	15	0.94	2.560	26	$2\pi/2$	23.14	0.44	0.644	−6	0.94	0.391		
0.45	1.568	16	0.95	2.586	26	$3\pi/2$	111.3	0.45	0.638	−7	0.95	0.387	$\pi/2$	0.208
0.46	1.584	16	0.96	2.612	26	$4\pi/2$	535.5	0.46	0.631	−6	0.96	0.383	$2\pi/2$	0.0432
0.47	1.600	16	0.97	2.638	26	$5\pi/2$	2576	0.47	0.625	−6	0.97	0.379	$3\pi/2$	0.00898
0.48	1.616	16	0.98	2.664	27	$6\pi/2$	12392	0.48	0.619	−6	0.98	0.375	$4\pi/2$	0.00187
0.49	1.632	17	0.99	2.691	27	$7\pi/2$	59610	0.49	0.613	−6	0.99	0.372	$5\pi/2$	0.000388
0.50	1.649		1.00	2.718		$8\pi/2$	286751	0.50	0.607		1.00	0.368	$6\pi/2$	0.000081

$e = 2.71828.$ $1/e = 0.367879.$ $\log_{10}e = 0.4343.$ $1/(0.4343) = 2.3026.$
From Lionel S. Marks, "Mechanical Engineers' Handbook."
† NOTE: Do not interpolate in this column.

L. Trigonometric Relations

$$\sin x = \frac{A}{B}$$

$$\cos x = \frac{C}{B}$$

$$\tan x = \frac{A}{C} = \frac{\sin x}{\cos x}$$

$$\sec x = \frac{B}{C} = \frac{1}{\cos x}$$

$$\operatorname{cosec} x = \frac{B}{A} = \frac{1}{\sin x}$$

$$\cot x = \frac{C}{A} = \frac{1}{\tan x} = \frac{\cos x}{\sin x}$$

Fig. 1.

$$\sin (-x) = - \sin x$$
$$\cos (-x) = \cos x$$
$$\tan (-x) = - \tan x$$

$$\sin \left(\frac{\pi}{2} - x\right) = \cos x \qquad\qquad \sin \left(\frac{\pi}{2} + x\right) = \cos x$$

$$\cos \left(\frac{\pi}{2} - x\right) = \sin x \qquad\qquad \cos \left(\frac{\pi}{2} + x\right) = - \sin x$$

$$\tan \left(\frac{\pi}{2} - x\right) = \cot x \qquad\qquad \tan \left(\frac{\pi}{2} + x\right) = - \cot x$$

$$\sin (\pi - x) = \sin x \qquad\qquad \sin (\pi + x) = - \sin x$$
$$\cos (\pi - x) = - \cos x \qquad\qquad \cos (\pi + x) = - \cos x$$
$$\tan (\pi - x) = - \tan x \qquad\qquad \tan (\pi + x) = \tan x$$

$$\left.\begin{array}{l} \sin (x + 2\pi n) = \sin x \\ \cos (x + 2\pi n) = \cos x \\ \tan (x + 2\pi n) = \tan x \end{array}\right\} \text{(n a positive or negative integer)}$$

$$\sin (x + y) = \sin x \cos y + \cos x \sin y$$
$$\sin (x - y) = \sin x \cos y - \cos x \sin y$$
$$\cos (x + y) = \cos x \cos y - \sin x \sin y$$
$$\cos (x - y) = \cos x \cos y + \sin x \sin y$$

$$\tan (x + y) = \frac{\tan x + \tan y}{1 - \tan x \tan y}$$

$$\tan (x - y) = \frac{\tan x - \tan y}{1 + \tan x \tan y}$$

$$\sin 2x = 2 \sin x \cos x \qquad\qquad \cos 2x = 2 \cos^2 x - 1 = 1 - 2 \sin^2 x$$

$$\tan 2x = \frac{2 \tan x}{1 - \tan^2 x} \qquad\qquad \cot 2x = \frac{(\cot^2 x - 1)}{2 \cot x}$$

$$\sin 3x = 3 \sin x - 4 \sin^3 x \qquad\qquad \cos 3x = 4 \cos^3 x - 3 \cos x$$

$$\sin \frac{1}{2} x = \left[\frac{1}{2} (1 - \cos x)\right]^{\frac{1}{2}} = \frac{1}{2} (1 + \sin x)^{\frac{1}{2}} - \frac{1}{2} (1 - \sin x)^{\frac{1}{2}}$$

$$\cos \frac{1}{2} x = \left[\frac{1}{2} (1 + \cos x)\right]^{\frac{1}{2}} = \frac{1}{2} (1 + \sin x)^{\frac{1}{2}} + \frac{1}{2} (1 - \sin x)^{\frac{1}{2}}$$

$$\tan \frac{1}{2} x = \frac{(1 - \cos x)^{\frac{1}{2}}}{(1 + \cos x)^{\frac{1}{2}}} = \frac{1 - \cos x}{\sin x} = \frac{\sin x}{1 + \cos x}$$

$$\sin^2 x = \frac{1}{2} (1 - \cos 2x) \qquad\qquad \cos^2 x = \frac{1}{2} (1 + \cos 2x)$$

$$\sin^3 x = \frac{1}{4} (3 \sin x - \sin 3x) \qquad\qquad \cos^3 x = \frac{1}{4} (\cos 3x + 3 \cos x)$$

L. Trigonometric Relations (*Concluded*)

$$\sin x \sin y = \frac{1}{2} \cos (x - y) - \frac{1}{2} \cos (x + y)$$

$$\cos x \cos y = \frac{1}{2} \cos (x - y) + \frac{1}{2} \cos (x + y)$$

$$\sin x \cos y = \frac{1}{2} \sin (x - y) + \frac{1}{2} \sin (x + y)$$

$$\sin x + \sin y = 2 \sin \frac{1}{2} (x + y) \cos \frac{1}{2} (x - y)$$

$$\sin x - \sin y = 2 \cos \frac{1}{2} (x + y) \sin \frac{1}{2} (x - y)$$

$$\cos x + \cos y = 2 \cos \frac{1}{2} (x + y) \cos \frac{1}{2} (x - y)$$

$$\cos x - \cos y = -2 \sin \frac{1}{2} (x + y) \sin \frac{1}{2} (x - y)$$

$$\tan x + \tan y = \frac{\sin (x + y)}{\cos x \cos y} \qquad\qquad \tan x - \tan y = \frac{\sin (x - y)}{\cos x \cos y}$$

$$\sin^2 x - \sin^2 y = \sin (x + y) \sin (x - y)$$
$$\cos^2 x - \cos^2 y = -\sin (x + y) \sin (x - y)$$
$$\cos^2 x - \sin^2 y = \cos (x + y) \cos (x - y)$$
$$\sin^2 x + \cos^2 x = 1 \qquad\qquad \sec^2 x - \tan^2 x = 1$$

FIG. 2.

In any triangle (Fig. 2):

$$\frac{A}{\sin a} = \frac{B}{\sin b} = \frac{C}{\sin c}$$ (law of sines)

$$A^2 = B^2 + C^2 - 2BC \cos a$$ (law of cosines)

$$\frac{A + B}{A - B} = \frac{\tan \frac{1}{2} (a + b)}{\tan \frac{1}{2} (a - b)} = \frac{\sin a + \sin b}{\sin a - \sin b}$$ (law of tangents)

$$a + b + c = 180° \qquad\qquad A = B \cos c + C \cos b$$
$$B = C \cos a + A \cos c \qquad\qquad C = A \cos b + B \cos a$$

M. Properties of Hyperbolic Functions

$$\sinh x = \frac{\epsilon^x - \epsilon^{-x}}{2} = x + \frac{x^3}{3!} + \frac{x^5}{5!} + \cdots$$

$$\cosh x = \frac{\epsilon^x + \epsilon^{-x}}{2} = 1 + \frac{x^2}{2!} + \frac{x^4}{4!} + \cdots$$

$$\tanh x = \frac{\sinh x}{\cosh x}$$

$$\sinh (-x) = -\sinh x$$

$$\cosh (-x) = \cosh x$$

$$\cosh^2 x = 1 + \sinh^2 x$$

$$\frac{d (\sinh x)}{dx} = \cosh x$$

$$\frac{d (\cosh x)}{dx} = \sinh x$$

$$\int \sinh x \cdot dx = \cosh x$$

$$\int \cosh x \cdot dx = \sinh x$$

$$\sinh (x \pm j\pi) = -\sinh x$$

$$\cosh (x \pm j\pi) = -\cosh x$$

$$\sinh (x \pm j2\pi) = \sinh x$$

$$\cosh (x \pm j2\pi) = \cosh x$$

FIG. 1.—REACTANCE CHART°

Always obtain approximate value from Fig. 2 before using Fig. 1

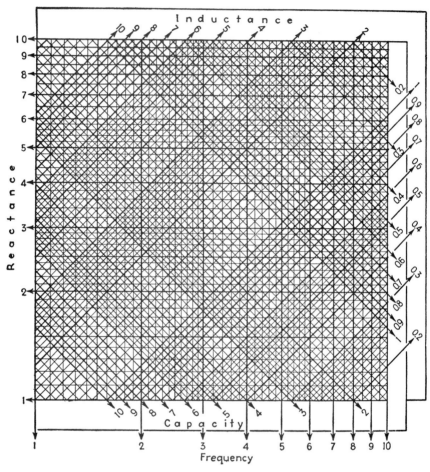

USE OF FIG. 1

Figure 1 is used to obtain additional precision of reading but does not place the decimal point which must be located from a preliminary entry on Fig. 2. Since the chart necessarily requires two logarithmic decades for inductance and capacitance for every single decade of frequency and reactance, unless the correct decade for L and C is chosen, erroneous results are obtained.

TYPICAL RESULTS

1. Find reactance of inductance of 0.00012 henry at 960 kc(kHz).

Answer: 720 ohms.

2. What capacity will have 265 ohms reactance at 7000 kc?

Answer: 86 $\mu\mu$f (pF).

3. What is the resonant frequency with $L = 21$ μh and $C = 45\mu\mu$f?

Answer: 5.18 mc(MHz).

* Figures 1 and 2 are from General Radio Company Reactance Chart.

FIG. 2.—REACTANCE CHART°

Always use corresponding scales

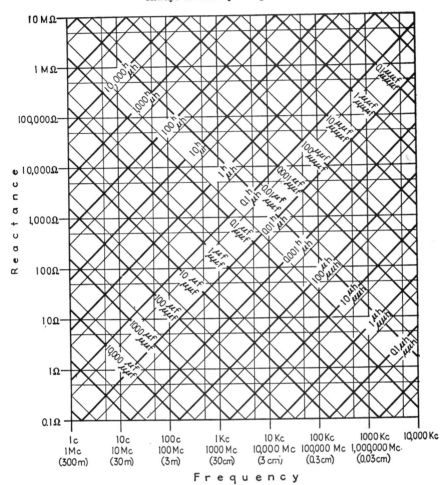

The accompanying chart may be used to find

(1) The reactance of a given inductance at a given frequency.

(2) The reactance of a given capacity at a given frequency.

(3) The resonant frequency of a given inductance and capacitance.

In order to facilitate the determination of magnitude of the quantities involved to two or three significant figures the chart is divided into two parts. Figure 2 is the complete chart to be used for rough calculations. Figure 1, which is a single decade of Fig. 2 enlarged approximately 7 times, is to be used where the significant two or three figures are to be determined.

To Find Reactance

Enter the charts vertically from the bottom (frequency) and along the lines slanting upward to the left (inductance) or to the right (capacity). Corresponding scales (upper or lower) must be used throughout. Project horizontally to the left from the intersection and read reactance.

To Find Resonant Frequency

Enter the slanting lines for the given inductance and capacity. Project downward from their intersection and read resonant frequency from the bottom scale. Corresponding scales (upper or lower) must be used throughout.

* Figures 1 and 2 are from General Radio Company Reactance Chart.

Index

1

ABOUT THE EDITORS

MILTON KAUFMAN, president of Electronics Writers and Editors, Inc., has extensive experience both as a working engineer and as a writer. He has been a senior engineer with CBS Columbia Color TV Lab, Sperry Gyroscope Co., and Grumman Aircraft (LM-Apollo Projects). He is editor of McGraw-Hill's *Handbook for Chemical Technicians, Handbook of Operational Amplifier Circuit Design, Handbook of Electronics Calculations for Engineers and Technicians,* and *Handbook of Microcircuit Design and Application.* He has also authored, coauthored, or edited numerous other books, articles and papers on electronics. ARTHUR H. SEIDMAN, Professor of Electrical Engineering and Acting Dean, School of Engineering, Pratt Institute, has over ten years' experience in industry, including a position as a senior engineer at Sperry Gyroscope Co. He has authored, coauthored, or edited numerous books, articles, and papers on electronics, including McGraw-Hill's *Handbook of Electronics Calculations for Engineers and Technicians.*